Mathematics

8

MYP 3

third edition

Michael Haese

Mark Humphries

Ngoc Vo

for use with
IB Middle Years
Programme

MATHEMATICS 8 MYP 3 third edition

Michael Haese   B.Sc.(Hons.), Ph.D.
Mark Humphries   B.Sc.(Hons.)
Ngoc Vo   B.Ma.Sc.

Published by Haese Mathematics
152 Richmond Road, Marleston, SA 5033, AUSTRALIA
Telephone: +61 8 8210 4666
Email: info@haesemathematics.com
Web: www.haesemathematics.com

National Library of Australia Card Number & ISBN 978-1-922416-32-2

© Haese & Harris Publications 2021

First Edition	2008
Reprinted	2009 (twice), 2010, 2011, 2012
Second Edition	2014
Reprinted	2015, 2016 (twice), 2017, 2018, 2019 (twice)
Third Edition	2021
Reprinted	2022, 2023, 2024

Cartoon artwork by Yi-Tung Huang, John Martin, and Ngoc Vo.

Artwork by Hannah Coleman, Brian Houston, and Bronson Mathews.

Computer software by Patrick French, Ben Hensley, Huda Kharrufa, Rachel Lee, Bronson Mathews, Linden May, and Han Zong Ng.

Audio recorded by Eloise Quinn-Valentine.

Production work by Hannah Coleman and Michael Mampusti.

Typeset in Australia by Charlotte Frost and Deanne Gallasch. Typeset in Times Roman $10\frac{1}{2}$.

Printed in China by Prolong Press Limited.

FOREWORD

Mathematics 8 MYP 3 third edition has been designed and written for the International Baccalaureate Middle Years Programme (IB MYP) Mathematics framework, providing complete coverage of the content and expectations outlined.

Discussions, Activities, Investigations, and Research exercises are used throughout the chapters to develop conceptual understanding. Material is presented in a clear, easy-to-follow style to aid comprehension and retention, especially for English Language Learners. Each chapter ends with extensive review sets and an online multiple choice quiz.

The associated digital Snowflake subscription supports the textbook content with interactive and engaging resources for students and educators.

The Global Context projects highlight the use of mathematics in understanding history, culture, science, society, and the environment. We have aimed to provide a diverse range of topics and styles to create interest for all students and illustrate the real-world application of mathematics.

We have developed this book in consultation with experienced teachers of IB Mathematics internationally but independent of the International Baccalaureate Organisation (IBO). It is not endorsed by the IBO.

We have endeavoured to publish a stimulating and thorough textbook and digital resource to develop and encourage student understanding and nurture an appreciation of mathematics.

Many thanks to Rob Colaiacovo and our other contributors for their recommendations and advice.

We welcome your feedback.   Email: info@haesemathematics.com

Web: www.haesemathematics.com

ONLINE FEATURES

Each textbook comes with a 12 month subscription to the online edition and its range of interactive features. This can be accessed through the **SNOWFLAKE** online learning platform via a web browser or our offline viewer.

To activate your electronic textbook, please contact Haese Mathematics by emailing info@haesemathematics.com with your proof of purchase such as a copy of your receipt.

For general queries regarding **SNOWFLAKE** and online subscriptions:
- Visit our help page: snowflake.haesemathematics.com.au/help
- Contact Haese Mathematics: info@haesemathematics.com

SELF TUTOR

Self Tutor is an engaging feature that supports students learning independently and in classrooms.

Click on any example box to access a step-by-step animation with teacher voice support providing explanation and understanding.

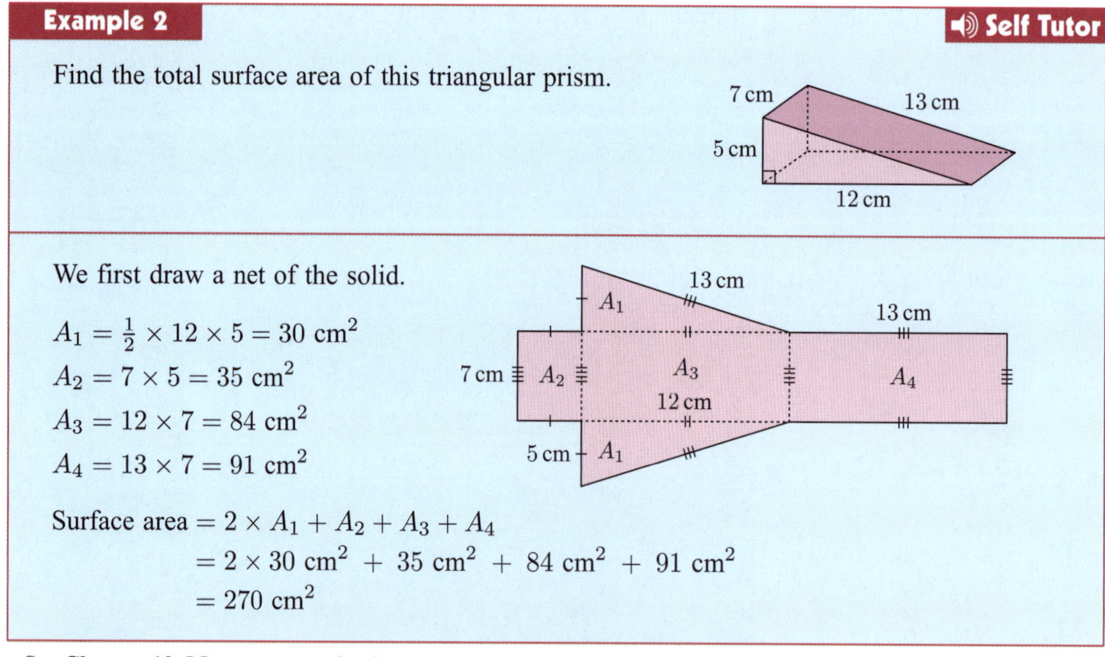

See **Chapter 12, Measurement: Surface area, volume, and capacity**, p. 243

INTERACTIVE LINKS

The **SNOWFLAKE** icons direct you to interactive tools to enhance learning and teaching.

These features include:
- demonstrations to illustrate and animate concepts
- multiple choice quizzes to test understanding
- games to practise and build skills
- tools for graphing and statistics
- printable pages for use in class.

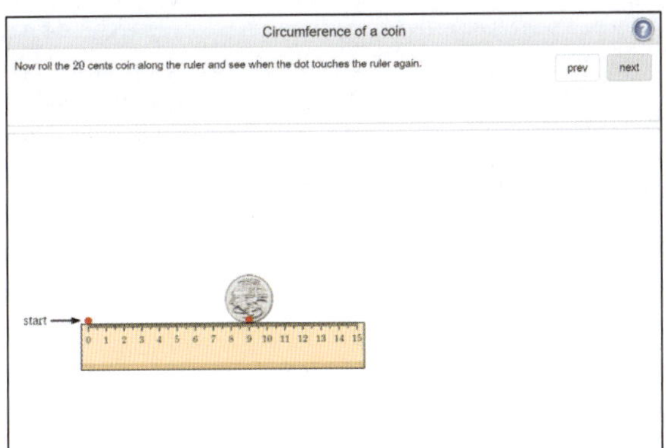

DEMO

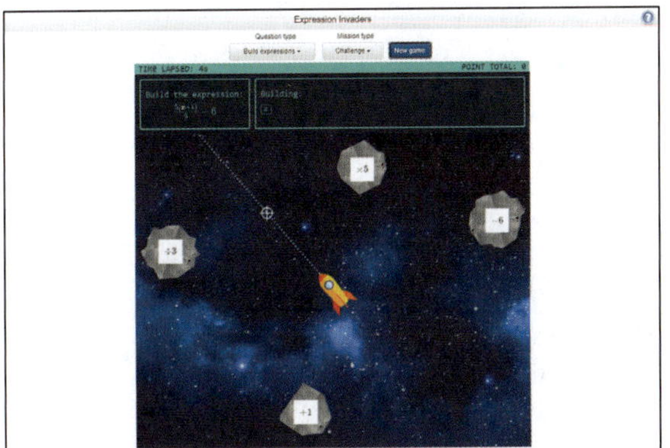

GAME

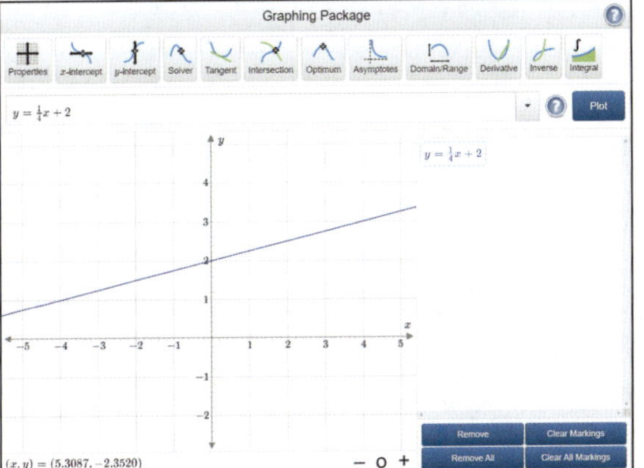

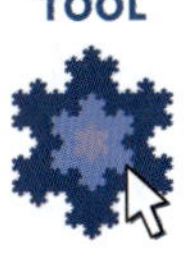

TOOL

TABLE OF CONTENTS

GLOBAL CONTEXTS

The International Baccalaureate Middle Years Programme focuses teaching and learning through six Global Contexts:

- Identities and relationships
- Orientation in space and time
- Personal and cultural expression
- Scientific and technical innovation
- Globalisation and sustainability
- Fairness and development

The Global Contexts help students to develop connections between different subject areas in the curriculum.

GLOBAL CONTEXT	MISLEADING GRAPHS

Global context: Fairness and development

Statement of inquiry: Knowing the ways in which statistical graphs can be misleading can help us to better interpret data.

Criterion: Communicating

Each project contains a series of questions, divided into:

- Factual questions (in green)
- Debatable questions (in red).
- Conceptual questions (in blue)

The projects are also accompanied by the general descriptor and a task-specific descriptor for one of the four assessment criteria.

GRAPHICS CALCULATOR INSTRUCTIONS

Graphics calculator instruction booklets are available for the **Casio fx-CG50**, **TI-84 Plus CE**, and the **HP Prime**. Click on the relevant icon below.

When additional calculator help may be needed, specific instructions are available from icons within the text.

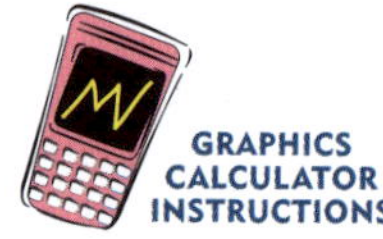

Number

Contents:

OPENING PROBLEM

Jasmine is preparing for her wedding reception. She has counted out 196 white roses, 140 apricot roses, and 84 pink roses which she will use to make identical table decorations, one for each table. The white roses cost $4 each, and the coloured roses cost $5 each.

Things to think about:

a What is the largest possible number of tables at the reception?

b Can you write *two* expressions involving numbers and symbols, which both give the total cost of the flowers?

c Can you evaluate each expression correctly, and show that they give the same answer?

People have used numbers since prehistoric times. We know this from ancient writings and drawings.

Today, we live in numbered houses and apartments, have telephone numbers, registration numbers, bank account numbers, credit card numbers, and tax file numbers.

We use numbers to count and measure things, and to understand the world around us.

Numbers are thus an essential part of our lives.

INTEGERS

In this Chapter we study the **integers** or **whole numbers**:

$$...., \ -5, \ -4, \ -3, \ -2, \ -1, \ 0, \ 1, \ 2, \ 3, \ 4, \ 5, \ ....$$

negative integers   zero   positive integers

The integers continue forever in either direction. We say they are **infinite**.

- In the list above, we show this using dots at either end.
- On a number line, we show it using arrows at either end.

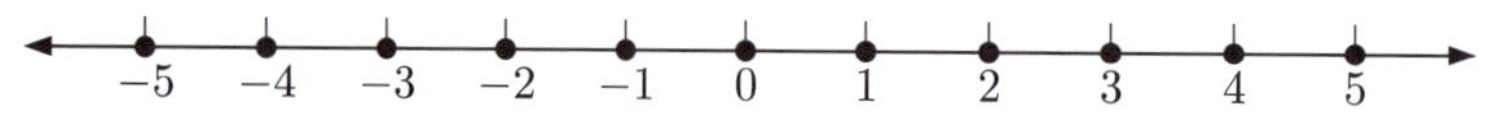

We also define:

- the **counting numbers** as the positive integers 1, 2, 3, 4, 5, ...

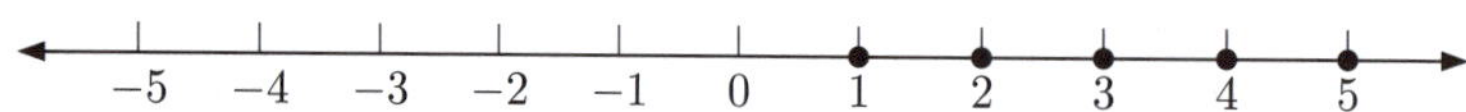

- the **natural numbers** as the non-negative integers 0, 1, 2, 3, 4, 5, ...

A   OPERATIONS WITH NEGATIVE NUMBERS

In this Section we revise rules you should have seen in previous years.

ADDITION AND SUBTRACTION

- **Adding** a **negative** number is equivalent to **subtracting** its opposite.
- **Subtracting** a **negative** number is equivalent to **adding** its opposite.

Example 1   ◀ๆ **Self Tutor**

Find the value of:

a $4 + -9$ **b** $4 - -9$ **c** $-3 + -5$ **d** $-3 - -5$

a $4 + -9$
   $= 4 - 9$
   $= -5$

b $4 - -9$
   $= 4 + 9$
   $= 13$

c $-3 + -5$
   $= -3 - 5$
   $= -8$

d $-3 - -5$
   $= -3 + 5$
   $= 2$

MULTIPLICATION

- $(\text{positive}) \times (\text{positive}) = (\text{positive})$
- $(\text{positive}) \times (\text{negative}) = (\text{negative})$
- $(\text{negative}) \times (\text{positive}) = (\text{negative})$
- $(\text{negative}) \times (\text{negative}) = (\text{positive})$

Example 2   ◀ๆ **Self Tutor**

Find the value of:

a 3×4 **b** 3×-4 **c** -3×4 **d** -3×-4

a $3 \times 4 = 12$ **b** $3 \times -4 = -12$ **c** $-3 \times 4 = -12$ **d** $-3 \times -4 = 12$

DIVISION

- $(\text{positive}) \div (\text{positive}) = (\text{positive})$
- $(\text{positive}) \div (\text{negative}) = (\text{negative})$
- $(\text{negative}) \div (\text{positive}) = (\text{negative})$
- $(\text{negative}) \div (\text{negative}) = (\text{positive})$

Example 3   ◀ **Self Tutor**

Find the value of:

  a $14 \div 2$ **b** $14 \div -2$ **c** $-14 \div 2$ **d** $-14 \div -2$

  a $14 \div 2 = 7$ **b** $14 \div -2 = -7$ **c** $-14 \div 2 = -7$ **d** $-14 \div -2 = 7$

EXERCISE 1A

1 Find the value of:

  a $14 - 5$ **b** $14 + -5$ **c** $14 + 5$ **d** $14 - -5$

  e $5 - 14$ **f** $-14 + 5$ **g** $-14 - 5$ **h** $-14 - -5$

2 Find the value of:

  a $13 + 27$ **b** $13 - 27$ **c** $13 + -27$ **d** $13 - -27$

  e $-13 - 27$ **f** $-13 + 27$ **g** $-13 + -27$ **h** $-13 - -27$

3 Find the value of:

  a $-3 + -8 - -2$ **b** $3 - 8 + -2$ **c** $-3 - -8 - -2$

  d $-12 + 7 - 4$ **e** $12 - -7 - 4$ **f** $-12 - -7 + 4$

4 Find the difference between:

  a 6 and 15 **b** -6 and 15 **c** 6 and -15 **d** -6 and -15

5 Find the value of:

  a 4×9 **b** 4×-9 **c** -4×9 **d** -4×-9

  e 3×11 **f** 3×-11 **g** -3×11 **h** -3×-11

  i $4 \times 2 \times 7$ **j** $4 \times -2 \times 7$ **k** $4 \times -2 \times -7$ **l** $-4 \times -2 \times -7$

6 Determine the missing number:

  a $\square \times -7 = -42$ **b** $-5 \times \square = 40$ **c** $\square \times 9 = 63$

  d $-12 \times \square = -36$ **e** $\square \times -12 = -132$ **f** $-15 \times \square = 60$

7 Find the value of:

  a $32 \div 8$ **b** $32 \div -8$ **c** $-32 \div 8$ **d** $-32 \div -8$

  e $35 \div 7$ **f** $35 \div -7$ **g** $-35 \div 7$ **h** $-35 \div -7$

  i $54 \div 6$ **j** $54 \div -6$ **k** $-54 \div 6$ **l** $-54 \div -6$

8 Determine the missing number:

  a $24 \div \square = -3$ **b** $\square \div 5 = -4$ **c** $-18 \div \square = 6$

  d $\square \div 11 = 5$ **e** $-108 \div \square = 12$ **f** $\square \div 6 = -8$

9 Determine the missing number:

  a $7 \times \square = -70$ **b** $\square - 3 = -4$ **c** $15 + \square = -1$

  d $25 \div \square = -5$ **e** $\square \times -8 = 40$ **f** $-8 + \square = 2$

  g $\square \div 2 = -20$ **h** $4 - \square = 6$ **i** $-18 + \square = 0$

  j $\square \div -6 = -12$ **k** $\square \times 10 = 100$ **l** $9 - \square = 18$

B   EXPONENT NOTATION

We often deal with numbers that are repeatedly multiplied together. Mathematicians use **exponents** or **indices** to represent such products more easily.

Rather than writing $2 \times 2 \times 2$, we can write 2^3. We call this **exponent notation**.

We say that 2 is the **base** and that 3 is the **exponent**, **index**, or **power**.

2^3 reads   *"two cubed"* or
  "two to the power three".

$2^3 \leftarrow$ exponent, index, or power
$2 \leftarrow$ base

> If n is a positive integer, then a^n is the product of n factors of a.
>
> $$a^n = \underbrace{a \times a \times a \times a \times \ldots \times a}_{n \text{ factors}}$$

Example 4   ◀) **Self Tutor**

Evaluate:

a 2^5 **b** $2^3 \times 5^2 \times 7$

a $\quad 2^5$
$= 2 \times 2 \times 2 \times 2 \times 2$
$= 32$

b $\quad 2^3 \times 5^2 \times 7$
$= 2 \times 2 \times 2 \times 5 \times 5 \times 7$
$= 8 \times 25 \times 7$
$= 1400$

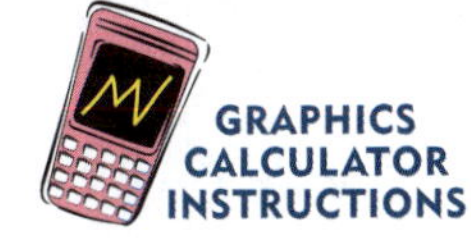
GRAPHICS CALCULATOR INSTRUCTIONS

NEGATIVE BASES

Consider the following patterns:

$(-1)^1 = -1$
$(-1)^2 = -1 \times -1 = 1$
$(-1)^3 = -1 \times -1 \times -1 = -1$
$(-1)^4 = -1 \times -1 \times -1 \times -1 = 1$

$(-2)^1 = -2$
$(-2)^2 = -2 \times -2 = 4$
$(-2)^3 = -2 \times -2 \times -2 = -8$
$(-2)^4 = -2 \times -2 \times -2 \times -2 = 16$

These patterns lead us to conclude that:

- A **negative** base raised to an **odd** power is **negative**.
- A **negative** base raised to an **even** power is **positive**.

Example 5   🔊 **Self Tutor**

Evaluate:

a -5^2 **b** $(-5)^2$ **c** -3^3 **d** $(-3)^3$

---

a -5^2
$= -5 \times 5$
$= -25$

b $(-5)^2$
$= -5 \times -5$
$= 25$

c -3^3
$= -3 \times 3 \times 3$
$= -27$

d $(-3)^3$
$= -3 \times -3 \times -3$
$= -27$

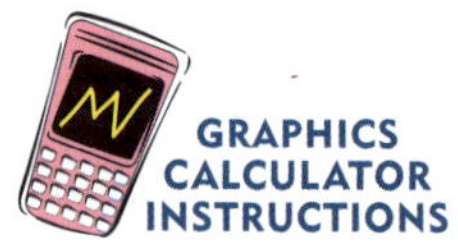

EXERCISE 1B

1 Copy and complete the values of these common powers. Try to remember them.

a $2^1 = \ldots\ldots,$ $2^2 = \ldots\ldots,$ $2^3 = \ldots\ldots,$ $2^4 = \ldots\ldots,$ $2^5 = \ldots\ldots,$ $2^6 = \ldots\ldots$

b $3^1 = \ldots\ldots,$ $3^2 = \ldots\ldots,$ $3^3 = \ldots\ldots,$ $3^4 = \ldots\ldots$

c $5^1 = \ldots\ldots,$ $5^2 = \ldots\ldots,$ $5^3 = \ldots\ldots,$ $5^4 = \ldots\ldots$

d $7^1 = \ldots\ldots,$ $7^2 = \ldots\ldots,$ $7^3 = \ldots\ldots$

2 Evaluate:

a 2×3^2 **b** $2^2 \times 3 \times 5$ **c** $2^2 \times 3^3$ **d** $3^2 \times 5^2$

e $2 \times 3^2 \times 5^3$ **f** $2^3 \times 3 \times 7^2$ **g** $3 \times 5^2 \times 11$ **h** $2^5 \times 5^3 \times 13$

3 Write using exponent notation:

a $2 \times 2 \times 3$ **b** $2 \times 3 \times 3 \times 5$ **c** $3 \times 3 \times 5 \times 5$

d $2 \times 2 \times 2 \times 5 \times 5$ **e** $3 \times 3 \times 5 \times 7 \times 7$ **f** $2 \times 2 \times 2 \times 2 \times 3 \times 3 \times 3 \times 5$

4 Evaluate:

a $(-1)^4$ **b** $(-1)^5$ **c** -1^5 **d** $-(-1)^5$

e -2^4 **f** $(-2)^4$ **g** $-(-2)^4$ **h** $-(-5)^2$

i $-(-5)^3$ **j** -5^3 **k** $(-3)^4$ **l** -7^2

5 Evaluate:

a $(-1)^3 \times 2^2 \times 3$ **b** $-2 \times 3^3 \times -5$ **c** $3^2 \times (-5)^3 \times -2$

6 Write as a power of 2:

a 4   **b** 8   **c** 32   **d** 128

7 Write as a power of 5:

a 25   **b** 125   **c** 625

8 By considering $3^1, 3^2, 3^3, 3^4, 3^5, \ldots.$ and looking for a pattern, find the last digit of 3^{20}.

9 Find the last digit of:

a 2^{87} **b** 9^{26} **c** 7^{41}

C   FACTORS

The **factors** of a positive integer are the positive integers which divide exactly into it.

For example:

- $65 \div 5 = 13$, so 5 is a factor of 65.
- $65 \div 13 = 5$, so 13 is a factor of 65.
- $65 \div 8 = 8$ remainder 1, so 8 is *not* a factor of 65.

We can write 65 as 5×13 where 5 and 13 are both factors of 65. We say that 5 and 13 are a **factor pair**.

When a number is written as a product of factors, we say it is **factorised**.

EXERCISE 1C

1 **a** Is 5 a factor of 40?   **b** Is 4 a factor of 50?   **c** Is 7 a factor of 26?

   d Is 8 a factor of 56?   **e** Is 6 a factor of 82?   **f** Is 3 a factor of 87?

2 List all the factors of:

   a 6   **b** 15   **c** 24   **d** 63   **e** 23   **f** 25

3 Write all the factor pairs of:

   a 45   **b** 48   **c** 72   **d** 100

4 Write in factorised form in as many ways as you can *without* using 1 as a factor:

   a 20   **b** 36   **c** 64

DISCUSSION

Is it sensible to talk about the factors of zero? If so, what are they?

D   PRIME AND COMPOSITE NUMBERS

A **prime number** is a positive integer which has exactly two distinct factors, 1 and itself.

A **composite number** is a positive integer which has more than two different factors.

For example:

- 17 is a prime number since it has only 2 factors, 1 and 17.
- 26 is a composite number because it has four factors: 1, 2, 13, and 26.
- 1 is neither prime nor composite.

The **Fundamental Theorem of Arithmetic** states that:

Apart from order, every composite number can be written as a **product of prime factors** in **one and only one way**.

For example, $60 = 2 \times 2 \times 3 \times 5$ is the only way of writing 60 as the product of prime factors.

Using exponent notation, we write $60 = 2^2 \times 3 \times 5$.

We call this the **prime factorisation** of 60, and say that 60 is written in **prime factored form**.

There are two methods we can use for writing a composite number in prime factored form:

- In **repeated division**, we systematically divide the number by prime numbers which are its factors, starting with the smallest.

- In a **factor tree**, we find a factor pair for the number and use these factors as branches of the tree. We continue finding factor pairs for each branch until we are left only with prime numbers.

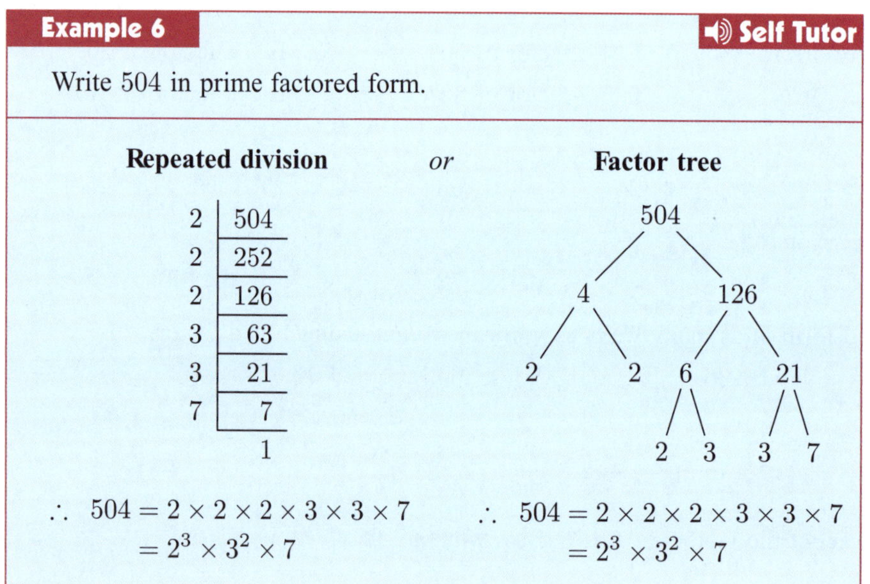

EXERCISE 1D

1 List the set of all prime numbers less than 50.

2 Find two consecutive odd numbers between 60 and 80 which are both prime.

3 A number between 20 and 30 has 2 prime factors and 3 composite factors. Find the number.

4 Write as a power of a prime:

    a 16   **b** 81   **c** 64   **d** 243   **e** 169   **f** 256

5 Write in prime factored form:

    a 24   **b** 40   **c** 36   **d** 54   **e** 66   **f** 84

    g 100   **h** 132   **i** 320   **j** 324   **k** 588   **l** 945

6 **a** Write 1960 in prime factored form.

  b Hence explain why $2^2 \times 7 = 28$ is a factor of 1960.

7 The **most abundant** number in a set of numbers is the number which has the highest power of 2 as a factor.

Find the most abundant number in each set:

  a 1, 2, 3, 4, 5, 6, 7, 8, 9, 10

  b 41, 42, 43, 44, 45, 46, 47, 48, 49, 50

  c 151, 152, 153, 154, 155, 156, 157, 158, 159, 160

8 Find integers a and b greater than 1 such that $a^b \times b^a = 800$.

GAME

Click on the icon to play a game which involves writing numbers as the product of prime factors. See if you can get the highest score in your class.

ACTIVITY 1   HIGHLY COMPOSITE NUMBERS

A **highly composite** number is a number which has more factors than any other number before it.

For example, the factors of 6 are 1, 2, 3, and 6. The numbers from 1 to 5 all have less than 4 factors, so 6 is a highly composite number.

What to do:

1 Find the only highly composite number that is also prime.

2 Find all highly composite numbers less than or equal to 100.

E HIGHEST COMMON FACTOR

The **highest common factor** or **HCF** of two or more numbers is the largest factor which is common to all of them.

For small numbers, we can find the HCF mentally or by listing the factors.

For example: The factors of 18 are 1, 2, 3, 6, 9, and 18.

The factors of 45 are 1, 3, 5, 9, 15, and 45.

So, the HCF of 18 and 45 is 9.

For larger numbers, we can find the HCF by first writing each number as a product of primes.

Example 7

$\blacktriangleleft$) Self Tutor

Find the HCF of 180 and 324.

2	180		2	324
2	90		2	162
3	45		3	81
3	15		3	27
5	5		3	9
	1		3	3
				1

$180 = 2 \times 2 \times 3 \times 3 \times 5$

$324 = 2 \times 2 \times 3 \times 3 \times 3 \times 3$

$2 \times 2 \times 3 \times 3$ is common to the factorisations.

So, the HCF of 180 and 324 is $2 \times 2 \times 3 \times 3 = 36$.

EXERCISE 1E

1 Find the HCF of:

   a 12 and 16
   b 9 and 15
   c 14 and 56

   d 16 and 40
   e 24 and 60
   f 35 and 50

   g 55 and 121
   h 24 and 42
   i 80 and 96

2 Farmer Giles has a field 45 m $\times$ 60 m. He wishes to divide it into square yards of equal size. What is the biggest size the yards could be?

3 Find the HCF of:

   a 12, 16, and 24
   b 18, 30, and 36
   c 32, 48, and 60

4 Find the HCF of:

   a 160 and 172
   b 124 and 156
   c 132 and 168

   d 169 and 208
   e 252 and 490
   f 280 and 308

5 In the **Opening Problem** on page **12**, what is the largest number of tables that could be at Jasmine's wedding reception?

6 Suppose a, b, and c are different prime numbers. Find the HCF of:

   a $a^2 b$ and ab^3
   b $a^3 b^2 c$ and $ab^3 c^2$

   c $a^3 b^3 c$, $a^2 b^5 c^3$, and $a^4 b^2 c^3$

F   MULTIPLES

The **multiples** of any natural number have that number as a factor. They are obtained by multiplying it by 1, then 2, then 3, then 4, and so on.

The **lowest common multiple** or **LCM** of two or more natural numbers is the smallest multiple which is common to all of them.

Example 8   ◀)) Self Tutor

Find the lowest common multiple of 9 and 12.

The multiples of 9 are: 9, 18, 27, **36**, 45, 54, 63, **72**, 81, ...

The multiples of 12 are: 12, 24, **36**, 48, 60, **72**, 84, ...

∴ the LCM of 9 and 12 is 36.

Another method for finding lowest common multiples is to first write each number as the product of its prime factors. We write matching factors on top of each other, then include each factor only once in the LCM.

Example 9   ◀)) Self Tutor

Find the LCM of:

a 9 and 12

b 15, 20, and 24.

a
$$9 = 3 \times 3$$
$$12 = 2 \times 2 \times 3$$
$$\therefore \ \text{LCM} = 2 \times 2 \times 3 \times 3$$
$$\therefore \ \text{LCM} = 36$$

b
$$15 = 3 \times 5$$
$$20 = 2 \times 2 \times 5$$
$$24 = 2 \times 2 \times 2 \times 3$$
$$\therefore \ \text{LCM} = 2 \times 2 \times 2 \times 3 \times 5$$
$$\therefore \ \text{LCM} = 120$$

EXERCISE 1F

1 List the first five multiples of:

   a 7   **b** 11   **c** 18   **d** 21   **e** 30

2 Find the:

   a smallest multiple of 9 which is greater than 400

   b largest multiple of 6 which is less than 800.

3 List the common multiples of 3 and 5 between 50 and 100.

4 Find the LCM of:

   a 5 and 8   **b** 4 and 6   **c** 8 and 10   **d** 15 and 18

   e 12 and 15   **f** 14 and 20   **g** 12 and 27   **h** 42 and 45

5 Find the LCM of:

   a 2, 3, and 4   **b** 5, 7, and 10   **c** 4, 6, and 8   **d** 5, 8, and 9

   e 8, 10, and 12   **f** 9, 15, and 20   **g** 14, 18, and 21   **h** 20, 25, and 30

6 Chris has a piece of rope. It can be cut exactly into either 10 metre or 18 metre lengths. Find the shortest length that Chris' rope could be.

7 A wildlife sanctuary has three shows:

- "Brilliant Birds" is 60 minutes long.
- "Meet a Monkey" is 90 minutes long.
- "Cuddle a Koala" is 45 minutes long.

The shows run continuously throughout the day.
If all three shows start together, how long will it be
before all three shows start together again?

8 Gloria is an avid coffee drinker. Every day, she drinks a coffee at each of these cafés:

Keen Beans:	"Buy 5 coffees, get 1 free!"
The Caffeine Club:	"Buy 7 coffees, get 1 free!"
Espresso Yourself:	"Buy 9 coffees, get 1 free!"

Today, Gloria had a free coffee at each of the cafés. How long will it be before she next has:

a 2 free coffees in one day
b 3 free coffees in one day?

9 Suppose a, b, and c are different prime numbers.
Find the LCM of:

a a^2b and abc
b ab^2c and a^2bc^3

10 Explain why the product of any two whole numbers is equal to the product of their HCF and
their LCM.

G   ORDER OF OPERATIONS

Some expressions contain more than one operation. To evaluate these expressions correctly, we use
the following rules:

- Perform the operations within **B**rackets first.
- Calculate any part involving **E**xponents.
- Starting from the left, perform all **D**ivisions
  and **M**ultiplications as you come to them.
- Restart from the left, performing all **A**dditions
  and **S**ubtractions as you come to them.

Brackets are **grouping symbols** which indicate a part of an expression which should be evaluated first.

- If an expression contains *one set* of brackets, evaluate that part first.
- If an expression contains *two or more sets* of brackets, one inside the other, evaluate the *innermost set* first.

Example 10   ◀) Self Tutor

Evaluate:

   a $3^2 - 15 \div 5 + 5$ **b** $5 + -8 \times 3$

a $3^2 - 15 \div 5 + 5$ {exponent first}

$= 9 - 15 \div 5 + 5$ {$\div$ before $-$ and $+$}

$= 9 - 3 + 5$ {$+$ and $-$ from left to right}

$= 11$

b $5 + -8 \times 3$ {$\times$ before $+$}

$= 5 + -24$

$= 5 - 24$

$= -19$

EXERCISE 1G

1 Evaluate:

  a $3 + 7 - 5$ **b** $6 + 9 \div 3$ **c** $8 - 3 + 2$

  d $5 - 4^2$ **e** $4 \times 3 - 11$ **f** $6 \div 2 \times 3$

  g $5 \times 8 \div 2$ **h** $2 \times 4 - 7^2$ **i** $12 \div 4 + 2 \times 5$

  j $13 - 2 \times 6 + 4$ **k** $3 \times 5 + 4 \times 6$ **l** $5 + 6 \times 3 \div 9$

  m $18 - 5 \times 2 + 7$ **n** $5 \times 4 - 24 \div 6$ **o** $3^2 - 8 \times 5 + 2^3$

2 Evaluate:

  a $3 + 4 \div -2$ **b** $-1 + -3 \times 2$ **c** $8 \div -2 + 5$

  d $-3 \times -2 - 4$ **e** $2 - 6 \div -3$ **f** $-2 \times 4 + -7$

  g $7 - 3 \times -3$ **h** $-4 \times -5 - 12$ **i** $3 - 6 \div -6$

Example 11   ◀) Self Tutor

Evaluate: $2 \times (3 \times 6 - 4) + 7$

$2 \times (3 \times 6 - 4) + 7$ {brackets first, $\times$ before $-$}

$= 2 \times (18 - 4) + 7$ {brackets first}

$= 2 \times 14 + 7$ {$\times$ before $+$}

$= 28 + 7$

$= 35$

3 Evaluate:

    a $(5+4) \div 3$ **b** $3 \times (4-2)$ **c** $(5+4)^2$

    d $(4+7) \times 8$ **e** $2 \times (4-7)^2$ **f** $12 + (3+7) \div 5$

    g $9 + 6 \times (8-5)$ **h** $18 - (7+4)$ **i** $(6-3) \times 11 - 12$

    j $16 + (17-11)$ **k** $(3+8) \times (6-2)$ **l** $(3-8) - (5+2)^2$

4 Evaluate:

    a $(12+3) \div 5 + 2 \times 4$ **b** $(13-5) \div (1+3) + 2$

    c $23 - (6 \div 2 + 7) + 4$ **d** $(5 \times 2 - 6) \times (3 - 6 \div 2)$

    e $7 - (4 \times 3 - 8) + 18 \div 3$ **f** $(3+4) \times 5 + 6 \times 7 - 8$

5 Evaluate using your calculator:

    a $87 + 27 \times 13$ **b** $(29+17) \times 19$ **c** $136 \div 8 + 16$

    d $136 \div (8+9)$ **e** $-67 + 64 \div -4$ **f** $3^3 + (17 - 2^4)$

Example 12　　　　　　　　　　🔊 **Self Tutor**

Evaluate:　$5 + [13 - (8 \div 4)]$

$$5 + [13 - (8 \div 4)] \quad \text{\{innermost brackets first\}}$$
$$= 5 + [13 - 2] \quad \text{\{remaining brackets\}}$$
$$= 5 + 11$$
$$= 16$$

---

6 Evaluate:

    a $[(3+5) \times 8] - 4$ **b** $3 + [(5 \times 8) - 4]$ **c** $3 + [5 \times (8-4)]$

    d $[(15-12) \div 3] + 3$ **e** $15 - [(12 \div 3) + 3]$ **f** $15 - [12 \div (3+3)]$

7 Replace each □ with either $+$, $-$, $\times$, or $\div$ to make each statement true:

    a $7 \,\square\, 3 \,\square\, 4 = 6$ **b** $4 \,\square\, 6 \,\square\, 3 = 21$ **c** $12 \,\square\, 4 \,\square\, 3 = 9$

    d $3 \,\square\, 5 \,\square\, 6 \,\square\, 4 = -2$ **e** $3 \,\square\, 8 \,\square\, 2 \,\square\, 5 = 2$ **f** $9 \,\square\, 3 \,\square\, 2 \,\square\, 4 = 11$

8 Insert brackets, if necessary, to make each statement true:

    a $9 - 7 \times 4 = 8$ **b** $80 \div 8 \times 2 = 5$ **c** $80 \div 8 \times 2 = 20$

    d $4 \times 8 - 7 - 1 = 26$ **e** $4 \times 8 - 7 - 1 = 3$ **f** $4 \times 8 - 7 - 1 = 0$

    g $5 + 2 \times 6 - 3 = 39$ **h** $5 + 2 \times 6 - 3 = 11$ **i** $5 + 2 \times 6 - 3 = 21$

ACTIVITY 2

Click on the icon to run the BEDMAS Challenge.

How fast can you go?

BEDMAS CHALLENGE

PUZZLE

Consider the statement $2 \,\square\, 2$.

If $\square$ is replaced by an operation, we get $2 + 2 = 4$, $2 - 2 = 0$, $2 \times 2 = 4$, or $2 \div 2 = 1$.

So, there are *three* possible results: 0, 1, or 4.

How many possible results are there if each $\square$ is replaced by an operation in:

a $2 \,\square\, 2 \,\square\, 2$ **b** $2 \,\square\, 2 \,\square\, 2 \,\square\, 2$?

H   PROBLEM SOLVING

In this Section we consider some real-world problems where more than one operation is involved.

We need to decide:

- what operations we need to perform
- which order we need to perform them in.

This allows us to write a **mathematical expression** involving numbers and operations, for what we want to calculate.

If necessary, we can use brackets to make sure the operations are performed in the correct order.

Example 13   ◀ Self Tutor

Gemma bought 2 bags of 50 marbles, and Jerome bought 3 bags of 35 marbles. Gemma and Jerome shared the marbles equally between themselves and 3 other friends.

a Write an *expression* for the number of marbles each child received.

b Calculate the number of marbles each child received.

---

a The total number of marbles was $2 \times 50 + 3 \times 35$.
The total number of children was $2 + 3$.
So, the number of marbles each child received was $(2 \times 50 + 3 \times 35) \div (2 + 3)$.

b $(2 \times 50 + 3 \times 35) \div (2 + 3)$
$= (100 + 105) \div 5$
$= 205 \div 5$
$= 41$

Each child received 41 marbles.

EXERCISE 1H

1 For Lunar New Year, Phuong received five $10 notes, three $20 notes, and a $50 note.

  a Write an *expression* for the total amount of money Phuong received.

  b Calculate the total amount of money Phuong received.

2 Ian aims to drink at least 4000 mL of water every day.

So far today, Ian has drunk two 900 mL bottles of water and five 250 mL glasses of water.

a Write an *expression* for the amount of water Ian still needs to drink to reach his goal.

b Calculate the amount of water Ian needs to drink.

3 Priya and Van are sending a 5 kg parcel and a 3 kg parcel to London. Postage costs $12 per kilogram. The girls will split the cost of postage equally between themselves.

a Write an *expression* for the amount of postage each girl will pay.

b Calculate the amount of postage each girl will pay.

4 On Saturday, Lucas cycled 45 km to visit his friends. On his way home, he cycled at 20 km per hour for 2 hours before he had a flat tyre.

a Write an *expression* for the distance Lucas was from home when he had the flat tyre.

b Calculate the distance Lucas was from home when he had the flat tyre.

5 Penelope baked a cake using the recipe alongside.
The cake is shared equally between 8 people.

a Write an *expression* for the amount each person receives.

b Calculate the amount each person receives.

flour	225 g
butter	200 g
sugar	175 g
4 eggs	60 g each

6 A cat had eight kittens. Each kitten grew up and had eight kittens of its own.

a Write an *expression* for the total number of kittens born.

b Calculate the total number of kittens born.

7 In the **Opening Problem**, Jasmine has 196 white roses, 140 apricot roses, and 84 pink roses. The white roses cost $4 each and the coloured roses cost $5 each.

a Write *two expressions* for the total cost of the flowers. One of your expressions should contain brackets.

b Use each expression to calculate the total cost of the flowers.

8 Mr D'Arcy has brought these two chocolate bars for his students to share at their volleyball practice.
Mr D'Arcy normally has 7 girls and 8 boys in his squad, but today 2 students are away sick.

a Write an *expression* for the number of chocolate blocks each student will receive. Use exponents in your expression.

b Calculate the number of chocolate blocks each student will receive.

GLOBAL CONTEXT   RUSSIAN PEASANT MULTIPLICATION

Global context: Scientific and technical innovation

Statement of inquiry: Awareness of alternative methods of multiplication can help people perform calculations.

Criterion: Communicating

QUICK QUIZ

MULTIPLE CHOICE QUIZ

REVIEW SET 1A

1 Find the value of:

  a $4 + -7$
  b -8×9
  c -3×-12
  d $54 \div -6$

2 Find the integer equal to:

  a $2^3 \times 5^2$
  b $2^1 \times 5^2 \times 7^1$
  c $(-5)^3$

3 List the factors of:

  a 21
  b 32
  c 37

4 List the prime numbers between 40 and 50.

5 Write in prime factored form:

  a 450
  b 212

6 Find the lowest common multiple of:

  a 6 and 15
  b 4 and 11
  c 5, 8, and 10

7 Find the HCF of:

  a 12 and 14
  b 24 and 56
  c 18, 27, and 45

8 Evaluate:

  a $24 \div 3 \times 2$
  b $13 - 5 \times 2 - 4^2$
  c $5 \times (8 - 2)$
  d $8 \div (5 - 1) \times 3$
  e $15 - [3 \times (8 - 6) - 2]$

9 A passenger train goes through a 2-track level crossing every 8 minutes. A freight train goes through the same level crossing every 52 minutes.
A passenger train and a freight train go through the level crossing at the same time. How long is it before a passenger train and freight train next pass through the level crossing together?

10 A greengrocer sells nectarines by the bag, and each bag contains the same number of nectarines. The greengrocer sold 126 nectarines yesterday, and 198 nectarines today. What is the greatest number of nectarines that could be in each bag?

11 Replace each $\square$ with either $+$, $-$, $\times$, or $\div$ to make each statement true:

  a $3 \square 5 \square 4 = 11$
  b $8 \square 6 \square 3 = 6$

12 An outdoor cinema has received bookings for 200 adults at \$20 each, and for 150 children at \$8 each.

Each customer must also pay a \$4 booking fee.

   a Write an *expression* for the total amount paid by the customers.

   b Calculate the total amount paid by the customers.

REVIEW SET 1B

1 Find the value of:

   a $15 - -9$ **b** $-6 + 7 - -3$ **c** $49 \div -7$

2 Determine the missing number:

   a $\square + 2 = -7$ **b** $\square \div -3 = -4$ **c** $5 \times \square = -55$

3 Write as a power of 3:

   a 9   **b** 27   **c** 81

4 List the first five multiples of 8.

5 Write 54 in prime factored form.

6 Find three consecutive odd numbers less than 100, which are all composite.

7 For the numbers 18 and 30, find:

   a the highest common factor   **b** the lowest common multiple.

8 List the square numbers between 700 and 800.

9 Evaluate:

   a $3 + 15 \div -3$ **b** $4^2 \div (7 - 5)$ **c** $[29 - (3 - -2)] \div 6$

10 Insert brackets to make each statement true:

   a $12 \div 6 - 2 = 3$ **b** $6 + 4 \div 2 + 3 = 2$ **c** $18 \div 1 + 2 \times 4 = 2$

11 Amy has some lollies in a bag, which she will share with her friends at a party. She knows she can share them equally whether there are four, five, or six children present. What is the smallest number of lollies Amy could have?

12 16 000 votes need to be counted for an election. There are 12 officials available to count the votes. The eight junior officials are each given 1200 votes to count, and the rest of the votes will be divided equally between the senior officials.

   a Write an *expression* for the number of votes each senior official will count.

   b Calculate the number of votes each senior official will count.

   c Given that the senior officials can count 8 votes each minute, how long will each senior official take to count their votes?

Sets and Venn diagrams

Contents:

OPENING PROBLEM

In a survey of 50 tea-drinkers, 32 people have their tea with milk, 19 have their tea with sugar, and 10 have their tea with both milk and sugar.

Things to think about:

a How can we represent this information using a diagram?

b How many of the people surveyed have their tea with:

   i milk but not sugar   **ii** milk or sugar   **iii** neither milk nor sugar?

The tea-drinkers in the **Opening Problem** can be placed into different groups according to what they have with their tea. We call these groups **sets**.

In this Chapter we will look at some properties of sets, and how we can describe relationships between them. We will also study **Venn diagrams** which we use to display sets and to solve problems.

A   SETS

A **set** is a collection of objects or things.

Each object is called an **element** or **member** of the set.

When we record a set, we write its members within curly brackets, separated by commas.

We often use a capital letter to represent a set so that we can refer to it easily.

For example:

- The colours of the French flag can be written as the set $C = \{$blue, white, red$\}$.
  We say "C is the set of colours of the French flag."
- The multiples of 4 which are less than 30 can be written as the set $M = \{4,\ 8,\ 12,\ 16,\ 20,\ 24,\ 28\}$.

We do not repeat an element in a set.

For example, the set of letters in the word BABY is $\{$B, A, Y$\}$ rather than $\{$B, A, B, Y$\}$.

SET NOTATION

$\in$ means "is an element of" or "is in".

$\notin$ means "is not an element of" or "is not in".

$n(A)$ means "the number of elements in set A".

For example, if $M = \{4,\ 8,\ 12,\ 16,\ 20,\ 24,\ 28\}$ then $12 \in M$, $19 \notin M$, and $n(M) = 7$.

EQUAL SETS

Two sets are **equal** if they contain exactly the same elements.

For example, $\{2, 3, 5, 7\} = \{5, 3, 7, 2\}$.

SUBSETS

Set A is a **subset** of set B if every element of A is also an element of B. We write $A \subseteq B$.

For example, if $A = \{1, 3, 6\}$ and $B = \{1, 2, 3, 5, 6, 7\}$, then every element of A is also an element of B. A is a subset of B, and we write $A \subseteq B$.

EMPTY SET

The **empty set** $\varnothing$ or $\{\ \}$ is a set which contains no elements.

For example, the set of multiples of 5 between 1 and 4 is the empty set.

The empty set is a subset of all other sets.

Example 1 ◄) **Self Tutor**

Let P be the set of all multiples of 6 less than 20, and Q be the set of all positive even numbers less than 20.

- **a** List the elements of P and Q.
- **b** True or false?   **i** $10 \in P$ **ii** $10 \notin Q$ **iii** $12 \in P$.
- **c** Find: **i** $n(P)$ **ii** $n(Q)$.
- **d** Is $P \subseteq Q$?

- **a** $P = \{6, 12, 18\}$, $Q = \{2, 4, 6, 8, 10, 12, 14, 16, 18\}$
- **b**
  - **i** 10 is not an element of P, so $10 \in P$ is false.
  - **ii** 10 is an element of Q, so $10 \notin Q$ is false.
  - **iii** 12 is an element of P, so $12 \in P$ is true.
- **c**
  - **i** $n(P) = 3$ $\{P$ has 3 elements$\}$
  - **ii** $n(Q) = 9$ $\{Q$ has 9 elements$\}$
- **d** Every element of P is also an element of Q, so $P \subseteq Q$.

EXERCISE 2A

1 List the elements of each set:

- **a** {days of the week}
- **b** {letters in the word FOOTBALL}
- **c** {prime numbers less than 20}
- **d** {vowels}
- **e** {factors of 21}
- **f** {colours of keys on a piano}
- **g** {whole numbers between 42 and 48}
- **h** {continents}

2 Let $S = \{2, 3, 5, 8, 11, 12\}$ and $T = \{2, 4, 5, 12\}$.

  a Find: **i** $n(S)$ **ii** $n(T)$.

  b True or false?

    i $5 \in S$ **ii** $5 \in T$ **iii** $12 \notin T$ **iv** $4 \notin S$

  c Is $T \subseteq S$?

3 Suppose $A = \{\text{positive square numbers less than } 10\}$,

  $B = \{\text{composite numbers less than } 20\}$, and $C = \{\text{factors of } 36\}$.

  a List the elements of: **i** A **ii** B **iii** C.

  b Find: **i** $n(A)$ **ii** $n(B)$ **iii** $n(C)$.

  c True or false?   **i** $A \subseteq B$ **ii** $A \subseteq C$

4 For each set S, list the elements of S and hence state $n(S)$:

  a $S = \{\text{factors of } 6\}$ **b** $S = \{\text{multiples of } 6 \text{ less than } 40\}$

  c $S = \{\text{factors of } 17\}$ **d** $S = \{\text{positive square numbers less than } 50\}$

  e $S = \{\text{prime numbers less than } 30\}$

  f $S = \{\text{composite numbers between } 10 \text{ and } 30\}$

5 Find $n(A)$ for:

  a $A = \{\text{months of the year}\}$ **b** $A = \{\text{consonants}\}$

6 List all the subsets of:

  a $\{x\}$ **b** $\{8, 9\}$ **c** $\{p, q, r\}$

7 Let A be the set of colours on spinner 1, and B be the set of colours on spinner 2.

  a List the elements of A and B.

  b Is green $\in B$?

  c Find: **i** $n(A)$ **ii** $n(B)$.

  d Is $A \subseteq B$?

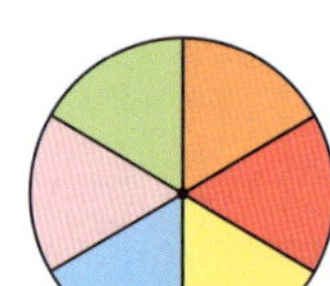

8 Suppose $F = \{2, 3, 5, 6, 7, 10, 11, 12\}$, $G = \{10, 5, x, 2, 12\}$, and $G \subseteq F$.

  What possible values could x have?

B   COMPLEMENT OF A SET

The **universal set** U is the set of all elements we are considering.

For example, if we are considering the letters of the English alphabet, the universal set is

$U = \{\text{a, b, c, d, e, f, g, h, i, j, k, l, m, n, o, p, q, r, s, t, u, v, w, x, y, z}\}$.

From this universal set we can define subsets of U, such as $V = \{\text{vowels}\} = \{\text{a, e, i, o, u}\}$ and

$C = \{\text{consonants}\} = \{\text{b, c, d, f, g, h, j, k, l, m, n, p, q, r, s, t, v, w, x, y, z}\}$.

The **complement** of a set A is the set of all elements of U that are *not* elements of A.

The complement of A is written A'. We say that A and A' are **complementary**.

For example:

- If $U = \{1, 2, 3, 4, 5, 6, 7, 8, 9\}$ and $A = \{2, 3, 5, 7\}$, then $A' = \{1, 4, 6, 8, 9\}$.
- If $U = \{$letters of the English alphabet$\}$, $V = \{$vowels$\}$, and $C = \{$consonants$\}$, then $V' = C$ and $C' = V$.

EXERCISE 2B

1 Let $U = \{1, 2, 3, 4, 5, 6, 7, 8, 9\}$.   Find the complement of:

  a $A = \{2, 4, 6\}$ **b** $B = \{1, 3, 5, 7, 9\}$ **c** $C = \{8, 4, 7, 3\}$

  d $D = \{1, 2, 3, 7, 8, 9\}$ **e** $E = \{1, 2, 3, 4, 5, 6, 7, 8, 9\}$

2 Suppose $U = \{$whole numbers between 0 and 20$\}$, $P = \{$factors of 12$\}$, and $Q = \{$prime numbers between 10 and 20$\}$.   List the elements of:

  a P **b** Q **c** P' **d** Q'.

3 Let $U = \{$letters of the English alphabet$\}$, $X = \{$letters in the word INHABITANT$\}$, and $Y = \{$letters in the word MILLION$\}$.
List the elements of:

  a X **b** Y **c** X' **d** Y'.

4 Alongside is a list of sports played at a school.
Let K be the set of sports that involve hitting a ball with a bat or racquet.

  a List the elements of:

  i U **ii** K **iii** K'.

  b What do the elements of K' represent?

5 **a** Suppose $U = \{2, 3, 4, 5, 6, 7, 8\}$, $A = \{2, 3, 4, 7\}$, and $B = \{2, 5\}$.   Find:

  i $n(U)$ **ii** $n(A)$ **iii** $n(A')$ **iv** $n(B)$ **v** $n(B')$.

  b Suppose $U = \{\triangle, \square, \bigcirc, \star, \diamondsuit, \heartsuit, \blacktriangle, \blacksquare, \bullet, \bigstar, \blacklozenge, \blacktriangledown\}$,
  $A = \{$shapes which are shaded$\}$, and $B = \{$polygons$\}$.   Find:

  i $n(U)$ **ii** $n(A)$ **iii** $n(A')$ **iv** $n(B)$ **v** $n(B')$.

  c Copy and complete:

  For any set S within a universal set U, $n(S) + n(S') = \ldots\ldots$

6 Let $U = \{$positive whole numbers$\}$, $E = \{$even numbers$\}$, and $O = \{$odd numbers$\}$.
Are E and O complementary? Explain your answer.

7 Let $U = \{$positive whole numbers$\}$, $P = \{$prime numbers$\}$, and $C = \{$composite numbers$\}$.
Are P and C complementary? Explain your answer.

C   INTERSECTION AND UNION

Sam and Tess are planning a holiday around the world.

The set of languages Sam can speak is
$S = \{$English, French, German, Spanish$\}$.

The set of languages Tess can speak is
$T = \{$English, German, Japanese$\}$.

INTERSECTION

Sam and Tess can communicate with each other in any language spoken by *both* of them.

By inspecting the sets S and T, we can see that Sam and Tess can both speak English and German.

The set {English, German} is called the **intersection** of sets S and T.

> The **intersection** of two sets A and B is the set of elements that are in **both** set A **and** set B.
>
> The intersection of sets A and B is written $A \cap B$.

For example, if $A = \{2, 5, 7, 9\}$ and $B = \{3, 5, 9, 10\}$, then $A \cap B = \{5, 9\}$.

> Two sets A and B are **disjoint** if they have no elements in common. We write $A \cap B = \varnothing$.

UNION

When deciding which countries to visit on their trip, Sam and Tess want to make sure that *at least* one of them can speak the native language.

By inspecting the sets, we can see that between them, Sam and Tess can speak English, French, German, Spanish, and Japanese.

The set {English, French, German, Spanish, Japanese} is called the **union** of sets S and T.

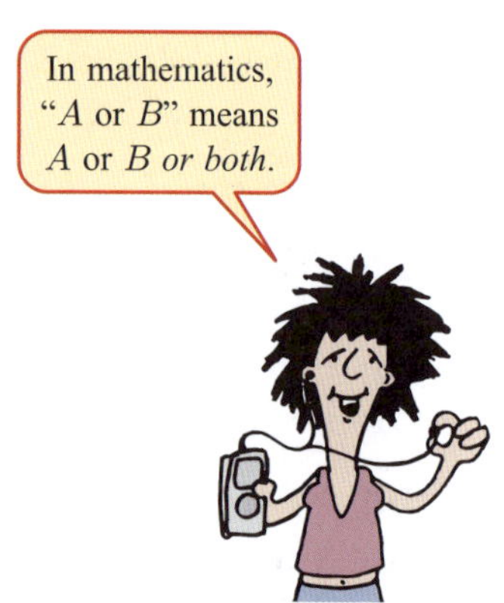

> The **union** of two sets A and B is the set of elements that are in **either** set A **or** set B.
>
> The union of sets A and B is written $A \cup B$.

For example, if $A = \{2, 5, 7, 9\}$ and $B = \{3, 5, 9, 10\}$, then
$A \cup B = \{2, 3, 5, 7, 9, 10\}$.

Notice that the elements 5 and 9 are in both A and B. They are included in the union $A \cup B$.

Example 2   ◄)) **Self Tutor**

Let $P = \{b, c, f, g, h\}$ and $Q = \{c, d, g, i\}$.   Find:

a $P \cap Q$ **b** $P \cup Q$.

a $P \cap Q = \{c, g\}$ {c and g are elements of both sets}

b $P \cup Q = \{b, c, d, f, g, h, i\}$ {elements of either P or Q}

EXERCISE 2C

1 Suppose $A = \{1, 3, 5, 7\}$ and $B = \{3, 5, 6, 9\}$.   Find:

a $A \cap B$ **b** $A \cup B$.

2 Find $P \cap Q$ and $P \cup Q$ for:

a $P = \{$Dragons, Tigers, Roosters, Raiders$\}$, $Q = \{$Tigers, Storm, Dragons, Knights$\}$

b $P = \{1, 3, 6, 10, 15\}$, $Q = \{1, 4, 9, 16\}$

c $P = \{d, e, g, k, m\}$, $Q = \{g, h, l, m, p\}$

3 Let $A = \{$blue, green, yellow$\}$, $B = \{$green, red, pink$\}$, and $C = \{$orange, blue, black$\}$.
Which pair of sets is disjoint?

4 Suppose $X = \{$prime numbers less than 20$\}$ and $Y = \{$factors of 20$\}$.

a List the elements of X and Y.

b Find:

  i $X \cap Y$ **ii** $n(X \cap Y)$ **iii** $X \cup Y$ **iv** $n(X \cup Y)$.

5 Suppose A is a set in a universal set U. Find:

a $A \cap \varnothing$ **b** $A \cup \varnothing$.

6 Suppose A and B are sets in a universal set U.
Explain why $n(A \cap B) \leqslant n(A) \leqslant n(A \cup B)$.

7 Sarah has gardenias and roses in her garden in New Zealand. Her gardenias flower every year from September to December, and her roses flower from October to March.

Let G be the set of months when the gardenias flower, and R be the set of months when the roses flower.

a State the universal set U in this case.

b List the elements of G and R.

c Find $G \cap R$. What does this set represent?

d Find $G \cup R$. What does this set represent?

e Find $(G \cup R)'$. What does this set represent?

8 Holly is baking scones and cake for a morning tea. The ingredients needed for her recipes are shown below:

Let S be the set of ingredients needed to make the scones, and C be the set of ingredients needed to make the cake.

  a List the elements of S and C.

  b Find $S \cap C$. What does this set represent?

  c Find $S \cup C$.

  d How many different ingredients will Holly use in her baking?

9 Suppose $A = \{$multiples of 4 which are less than 10$\}$ and $B = \{$factors of 16$\}$.

  a List the elements of A and B.

  b Is $A \subseteq B$?

  c Find: **i** $A \cap B$ **ii** $A \cup B$.

  d Copy and complete: "If $A \subseteq B$ then $A \cap B = \ldots\ldots$ and $A \cup B = \ldots\ldots$"

D   VENN DIAGRAMS

A **Venn diagram** consists of a universal set U represented by a rectangle, and sets within it that are usually represented by circles.

Venn diagrams are used to represent sets visually.

For example, this Venn diagram shows the set $A = \{1, 4, 8\}$ within the universal set $U = \{1, 2, 3, 4, 5, 6, 7, 8, 9\}$.

The **complement** of a set A is represented by the region outside the circle which represents A.

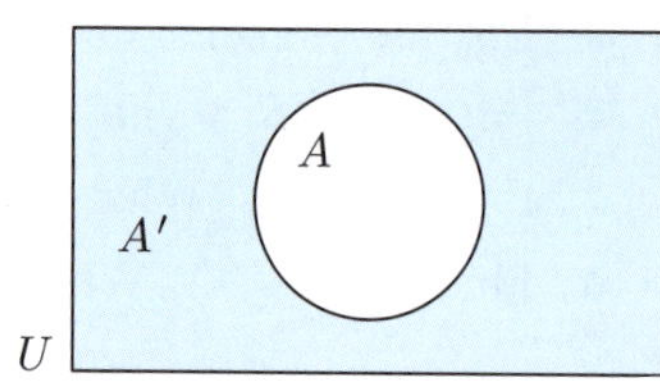

VENN DIAGRAMS FOR TWO SETS

If we are representing two sets A and B on the same Venn diagram, there are several layouts we can choose from.

- If A and B have elements in common, but neither is a subset of the other, the circles for the sets overlap.

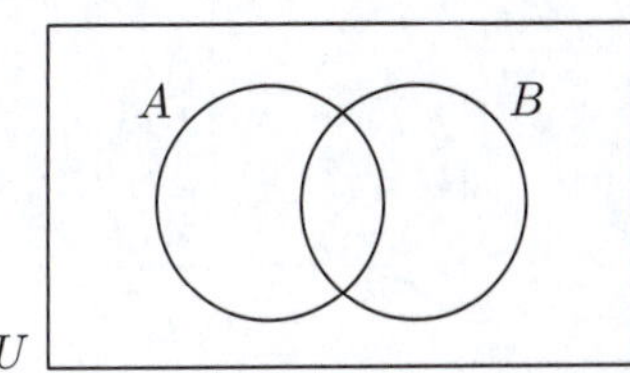

Notice that:

▶ The **intersection** of A and B is the region which is in *both* circles.

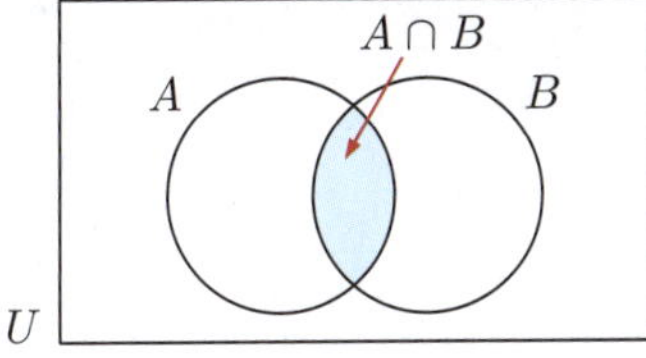

▶ The **union** of A and B is the region which is in *either* circle.

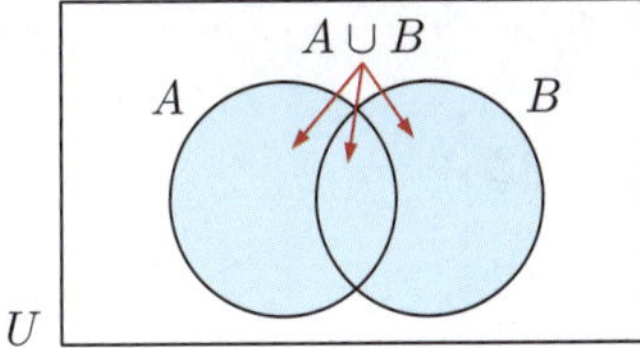

- If A is a **subset** of B, then every element of A is also an element of B.

  We place a circle representing A completely within the circle representing B.

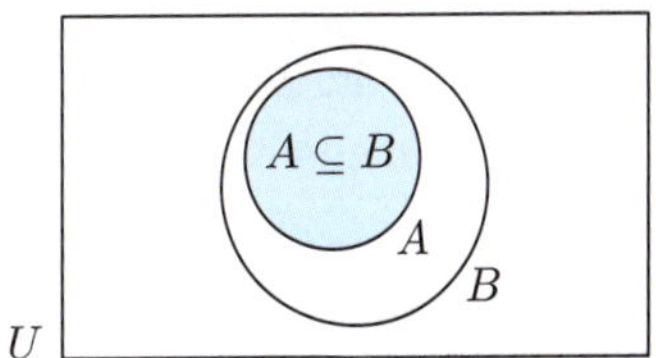

- If A and B are **disjoint** then there are no elements in both A and B.

  We represent A and B using circles which do not overlap.

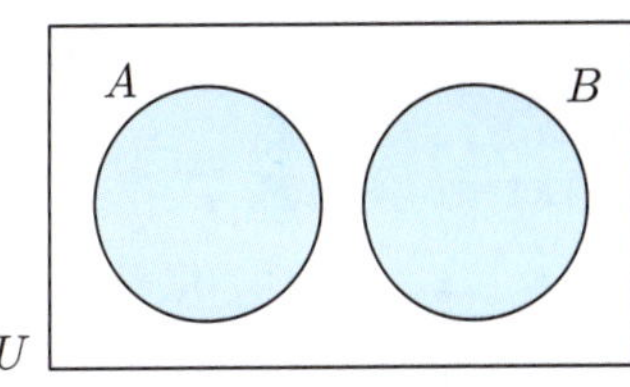

Example 3　　　　　　　　　　　◀ϡ **Self Tutor**

Let $U = \{1, 2, 3, 4, 5, 6, 7, 8\}$. Draw a Venn diagram to represent:

a　$C = \{2, 3, 7\}$　　　**b**　$A = \{1, 3, 4, 6\}$,　$B = \{2, 3, 6, 8\}$

a　$C = \{2, 3, 7\}$
　　$C' = \{1, 4, 5, 6, 8\}$

b　$A \cap B = \{3, 6\}$
　　$A \cup B = \{1, 2, 3, 4, 6, 8\}$

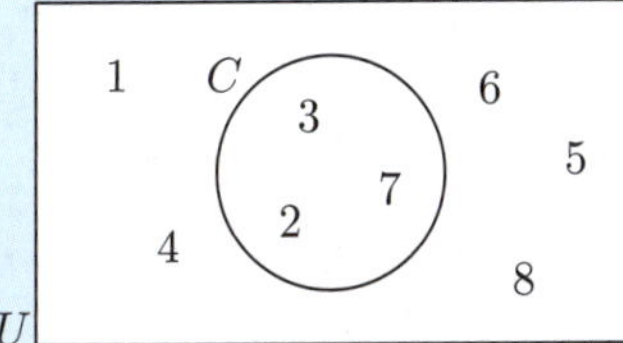

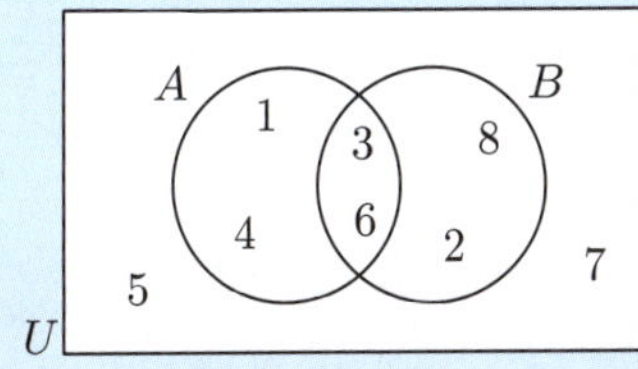

Example 4

Let $U = \{1, 2, 3, 4, 5, 6\}$. Draw a Venn diagram to represent:

a $A = \{2, 5\}$, $B = \{1, 2, 4, 5, 6\}$ **b** $A = \{1, 3, 6\}$, $B = \{2, 4\}$

a Every element in A is also in B, so $A \subseteq B$.

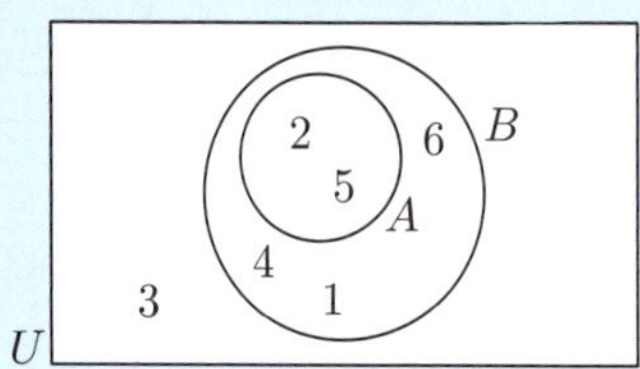

b $A \cap B = \varnothing$, so A and B are disjoint sets.

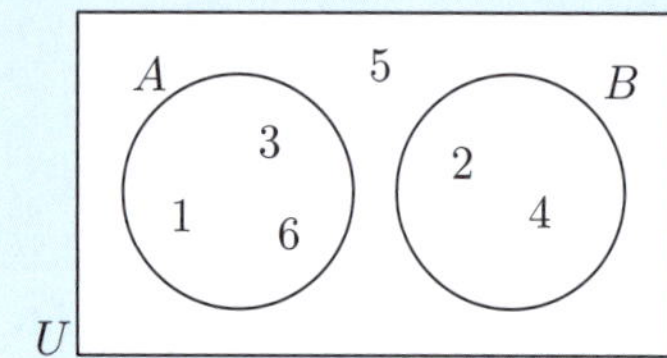

EXERCISE 2D

1 Let $U = \{1, 2, 3, 4, 5, 6, 7, 8\}$. Draw a Venn diagram to represent:

a $A = \{1, 2, 3\}$ **b** $A = \{2, 5\}$ **c** $A = \{3, 4, 7, 8\}$

2 Consider the Venn diagram alongside.
List the elements of:

a A **b** B

c $A \cap B$ **d** $A \cup B$

e A' **f** B'

g U

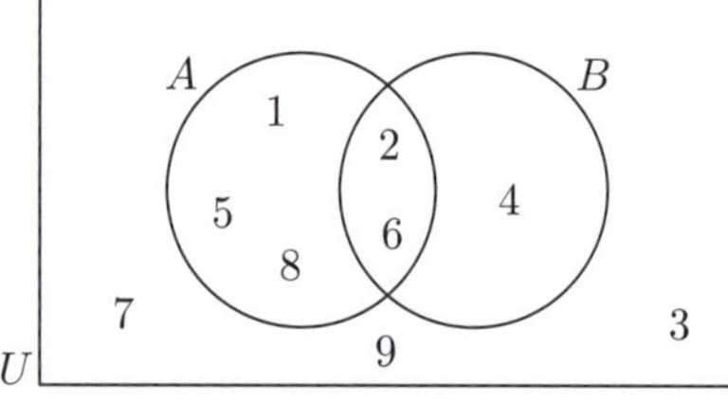

3 Let $U = \{1, 2, 3, 4, 5, 6, 7, 8, 9\}$. Draw a Venn diagram to represent:

a $A = \{3, 5, 6, 8\}$ **b** $A = \{1, 3, 5, 7\}$, $B = \{2, 3, 5, 8\}$

c $A = \{2, 3, 4, 5\}$, $B = \{4, 5, 6, 7, 8\}$ **d** $A = \{1, 4, 9\}$, $B = \{2, 6, 8\}$

e $A = \{3, 4, 8\}$, $B = \{1, 2, 3, 4, 7, 8\}$ **f** $A = \{2, 3, 6, 7\}$, $B = \{1, 3, 5, 6, 7, 9\}$

4 Let $U = \{a, b, c, d, e, f, g, h, i, j\}$. Draw a Venn diagram to represent:

a $P = \{c, e, f, g, i, j\}$, $Q = \{j, i, g\}$ **b** $P = \{b, e, a, d\}$, $Q = \{c, a, g, e\}$

5 George's cricket team has training on Mondays and Thursdays, and matches on Saturdays. Hugh's cricket team has training on Tuesdays and Thursdays, and matches on Sundays.

Let G be the set of days that George plays cricket, and H be the set of days that Hugh plays cricket.

a List the sets G and H.

b Draw a Venn diagram to illustrate G and H.

c List the elements of G'. Explain what this set represents.

d List the elements of $G \cap H$. Explain what this set represents.

6 Let $U = \{$whole numbers from 1 to 10$\}$, $A = \{$prime numbers less than 10$\}$, and $B = \{$factors of 10$\}$.

    a List the elements of A and B.   **b** Illustrate the sets on a Venn diagram.

    c List the elements of: **i** $A \cup B$ **ii** $(A \cup B)'$.

Example 5   🔊 **Self Tutor**

Shade the region of a Venn diagram representing:

    a in A and B **b** in A but not in B.

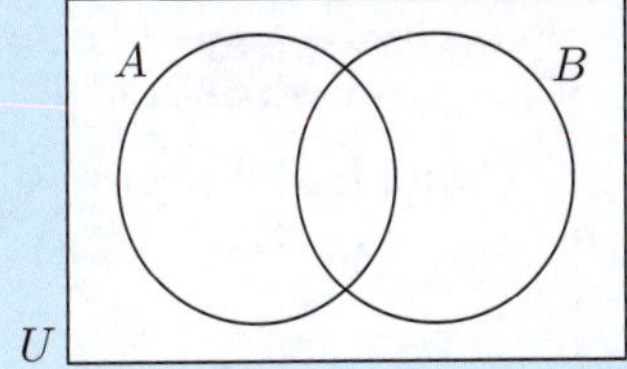

a

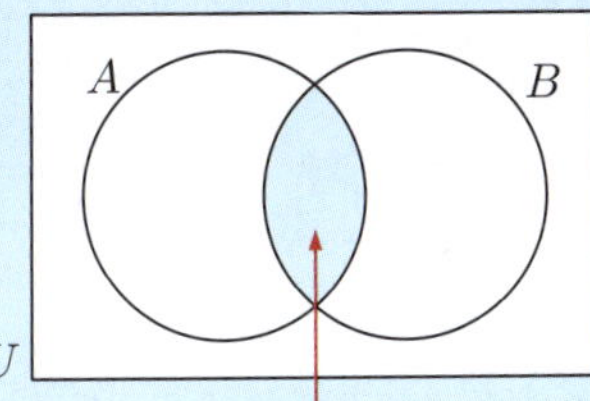

Elements in this region are in both A and B.

b

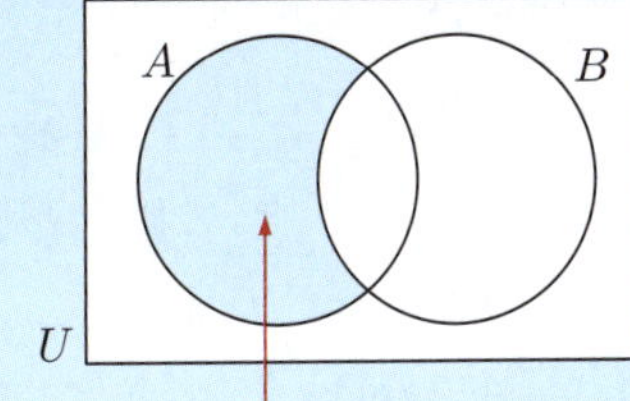

Elements in this region are in A but not in B.

---

7 On separate Venn diagrams like the one illustrated, shade the region representing:

    a not in A **b** in B

    c in B but not in A **d** in either A or B.

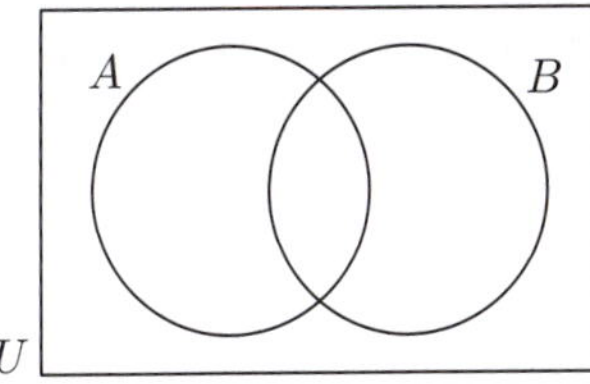

8 Describe, in words, the shaded region:

a 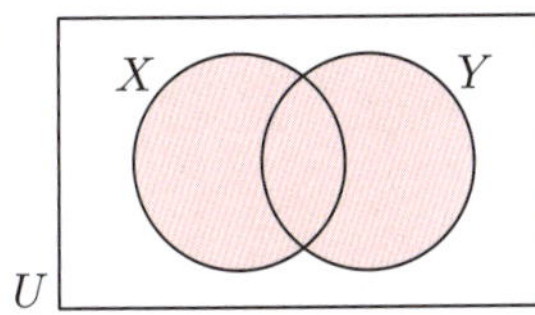   **b** 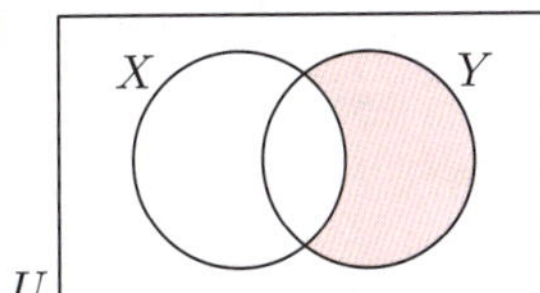   **c** 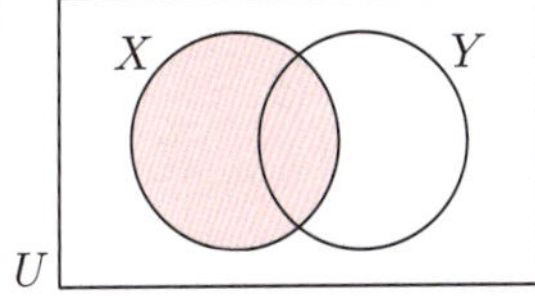

9 Suppose $U = \{$whole numbers from 2 to 12$\}$, $X = \{$multiples of 2 up to 12$\}$, and $Y = \{$prime numbers less than 13$\}$.

    a List the elements of X and Y.

    b Find: **i** $X \cap Y$ **ii** $X \cup Y$.

    c Draw a Venn diagram to illustrate the sets.

    d How many elements are in:

        i Y but not in X **ii** Y **iii** X or Y?

E   NUMBERS IN REGIONS

In many situations there are too many elements to list on a Venn diagram, and we are more interested in the **number of elements** in each region.

To indicate the number of elements in each region of a Venn diagram, we write the number in brackets.

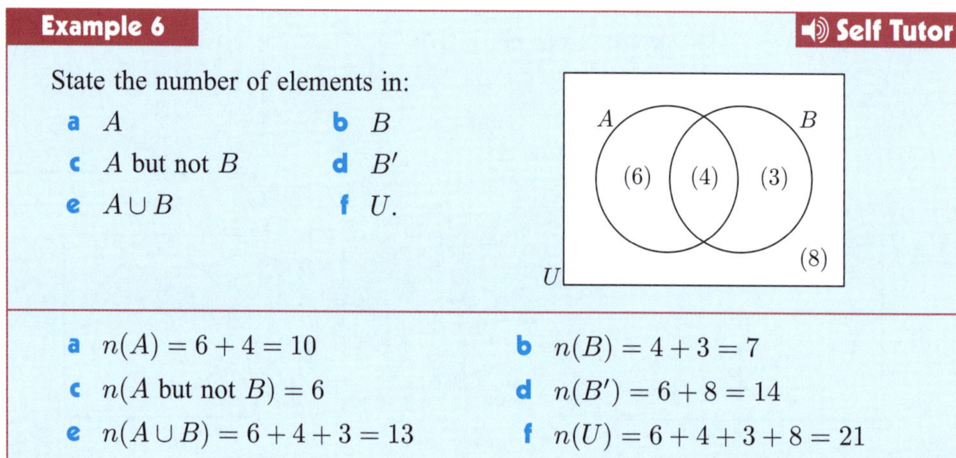

Example 6   ◀⑴ **Self Tutor**

State the number of elements in:

a A	**b** B
c A but not B	**d** B'
e $A \cup B$	**f** U.

a $n(A) = 6 + 4 = 10$	**b** $n(B) = 4 + 3 = 7$
c $n(A$ but not $B) = 6$	**d** $n(B') = 6 + 8 = 14$
e $n(A \cup B) = 6 + 4 + 3 = 13$	**f** $n(U) = 6 + 4 + 3 + 8 = 21$

EXERCISE 2E

1 State the number of elements in:

a A	**b** B
c $A \cap B$	**d** $A \cup B$
e A but not B	**f** U.

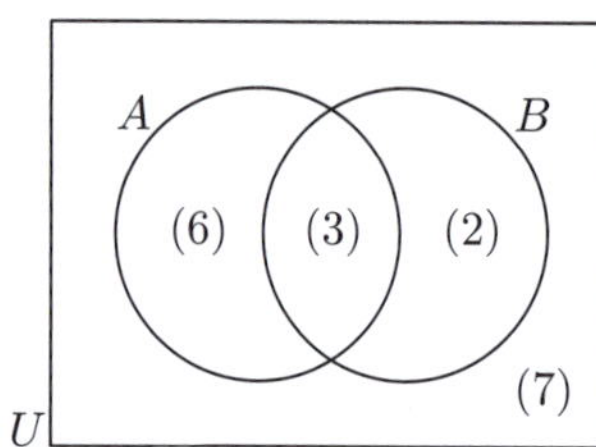

2 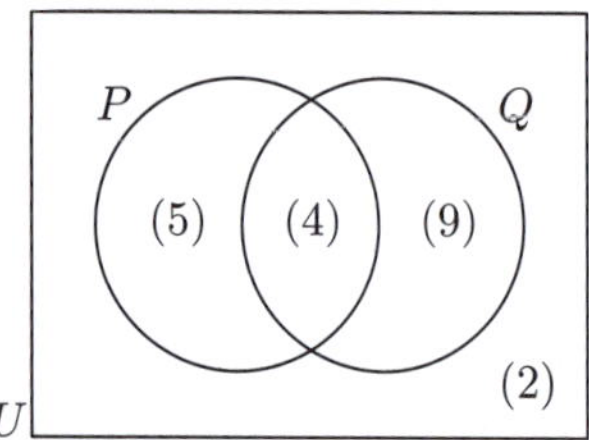 State the number of elements in:

a P	**b** Q'
c Q but not P	**d** $P \cup Q$
e P or Q but not both.	

3 Suppose $n(A) = 9$, $n(B) = 12$, $n(A \cap B) = 5$, and $n(U) = 20$.

Copy and complete the Venn diagram.

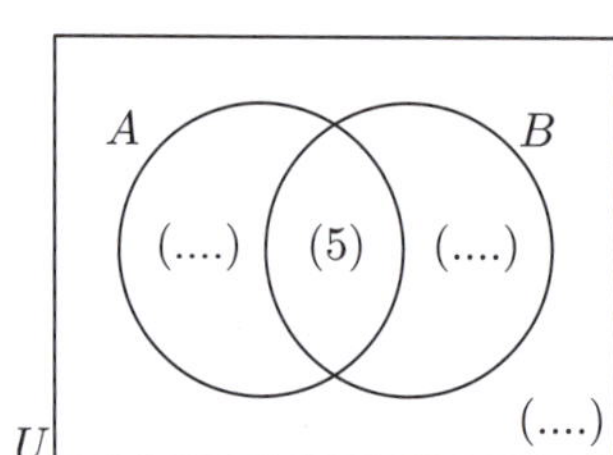

Example 7   ◀꒐ **Self Tutor**

Suppose $n(A) = 11$, $n(B) = 15$, $n(A \cap B) = 6$, and $n(U) = 24$.   Copy and complete the Venn diagram.

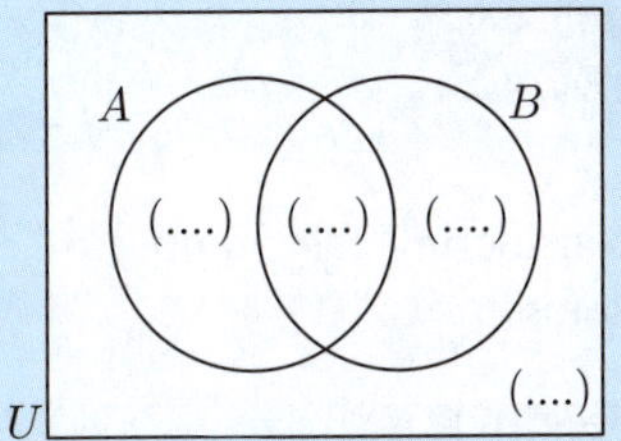

① $n(A \cap B) = 6$

② $n(A) = 11$, so there are $11 - 6 = 5$ elements in A but not B.

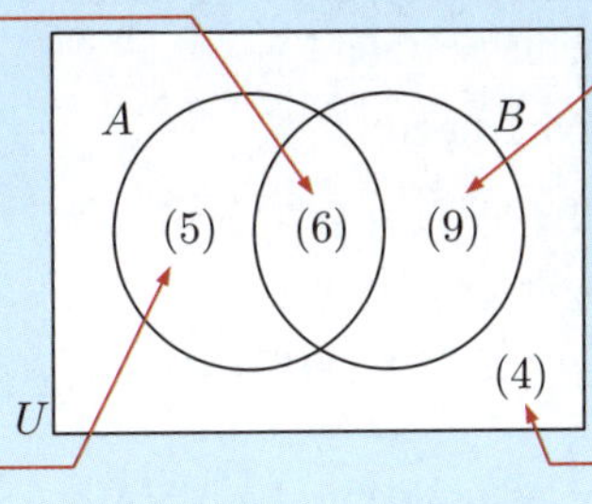

③ $n(B) = 15$, so there are $15 - 6 = 9$ elements in B but not A.

④ $n(U) = 24$, so there are $24 - 5 - 6 - 9 = 4$ elements in neither A nor B.

4

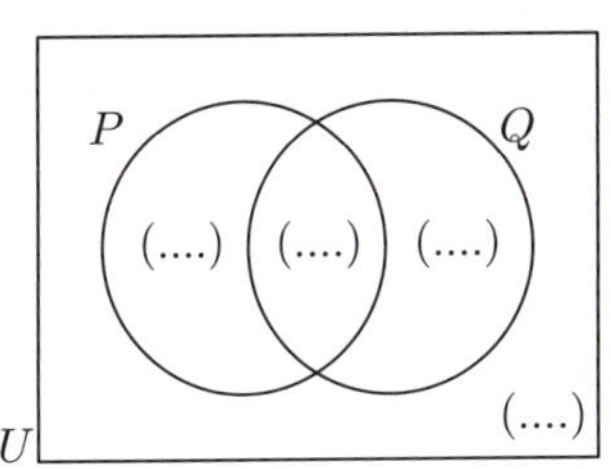

Suppose $n(P) = 14$, $n(Q) = 8$, $n(P \cap Q) = 3$, and $n(U) = 30$. Copy and complete the Venn diagram.

5 Suppose $n(A) = 10$, $n(B) = 13$, $n(A \cap B) = 6$, and $n(U) = 20$.

  a Represent this information on a Venn diagram.

  b Hence find the number of elements in:

  i B but not A **ii** $A \cup B$.

6 Suppose $n(A) = 7$, $n(A \cup B) = n(A \cap B) + 2$, and $n(U) = 10$.

Draw possible Venn diagrams to show the number of elements in each region.

7 Suppose $U = \{$positive integers from 1 to 100$\}$, $P = \{$prime numbers less than $x\}$, and $Q = \{$factors of $x\}$, where x is a positive integer.

It is known that $n(P) = 17$, $n(Q) = 12$, and $n(P \cap Q) = 3$.

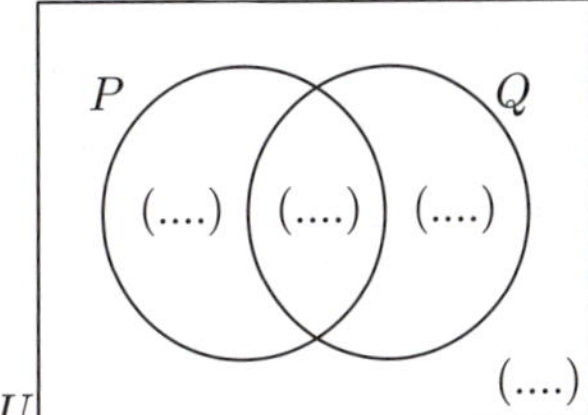

  a Copy and complete the Venn diagram.

  b Determine the value of x.

  c List the elements of:

  i $P \cap Q$ **ii** $P' \cap Q$.

F   PROBLEM SOLVING WITH VENN DIAGRAMS

By considering the number of elements in different regions, we can use Venn diagrams to solve problems.

Example 8

There are 30 houses on a street. 16 of the houses have a burglar alarm, 22 houses have a security door, and 10 houses have both a burglar alarm and a security door.

a Place this information on a Venn diagram.

b How many houses on the street have:

   i a burglar alarm but not a security door

   ii either a burglar alarm or a security door?

a Let B represent the set of houses with burglar alarms, and S represent the set of houses with security doors.

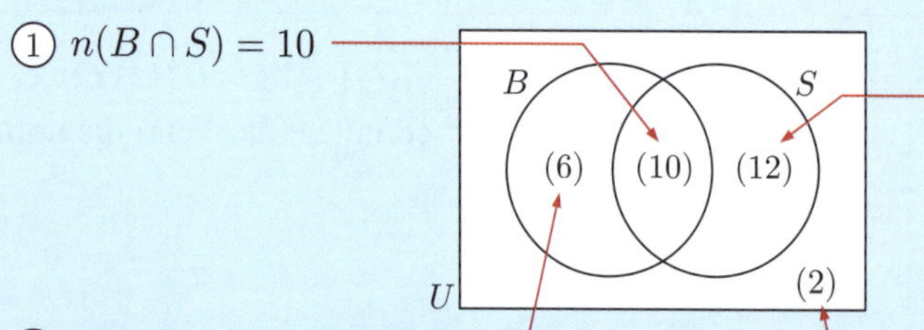

① $n(B \cap S) = 10$

③ $n(S) = 22$, so there are $22 - 10 = 12$ houses in S but not B.

② $n(B) = 16$, so there are $16 - 10 = 6$ houses in B but not S.

④ $n(U) = 30$, so there are $30 - 6 - 10 - 12 = 2$ houses in neither B nor S.

b i 6 houses have a burglar alarm but not a security door.

   ii $6 + 10 + 12 = 28$ houses have either a burglar alarm or a security door.

EXERCISE 2F

1 A group of 26 people are discussing which drinks they like. 18 of them like iced tea, 20 of them like water, and 13 of them like both of these drinks.

   a Place this information on a Venn diagram.

   b How many people in the group like:

   i iced tea or water

   ii water but not iced tea

   iii iced tea but not water

   iv neither iced tea nor water?

2 Nicola is inspecting her 45 apricot trees after a pest outbreak. She finds that 17 trees have ants and 26 have aphids. 9 of the trees have both ants and aphids.

   a Place this information on a Venn diagram.

   b How many trees have:

   i ants but not aphids   ii neither ants nor aphids   iii ants or aphids but not both?

3   Answer the **Opening Problem** on page **30**.

4   A pizza shop sells 15 types of pizza. 6 of them are vegetarian, and 5 are gluten free. 3 of them are both vegetarian and gluten free.

   a   Place this information on a Venn diagram.

   b   How many pizza types are:

      i   vegetarian but not gluten free   **ii**   neither vegetarian nor gluten free

      iii   either vegetarian or gluten free, but not both?

5   14 of the 32 nations that competed in the 2018 FIFA World Cup were European. 10 European nations advanced to the final 16 knockout stage of the World Cup.

   a   Display this information on a Venn diagram.

   b   How many non-European nations competed in the World Cup?

   c   How many European nations were eliminated before the knockout stage?

   d   How many non-European nations advanced to the knockout stage?

6   In one night, a hospital admitted 43 emergency patients. 24 patients required stitches, 21 required an X-ray, and 6 required neither stitches nor an X-ray.

   a   Display this information on a Venn diagram.

   b   How many of the patients required:

      i   both stitches and an X-ray   **ii**   stitches but not an X-ray

      iii   either stitches or an X-ray, but not both?

ACTIVITY   VENN DIAGRAMS FOR THREE SETS

Click on the icon to obtain this Activity.

VENN DIAGRAMS

QUICK QUIZ

MULTIPLE CHOICE QUIZ

REVIEW SET 2A

1   Suppose $A = \{$factors of 12$\}$ and $B = \{$multiples of 2 less than 10$\}$.

   a   List the elements of A and B.

   b   Find: **i** $n(A)$ **ii** $n(B)$.

   c   True or false?

      i $8 \in B$ **ii** $8 \in A$ **iii** $10 \notin A$ **iv** $B \subseteq A$

2 Let $U = \{$months of the year$\}$, and $A = \{$months which contain the letter R$\}$.

   a Describe the set A'.   **b** List the elements of A'.   **c** Find $n(A')$.

3 For the following sets, determine whether $X \subseteq Y$:

   a $X = \{2, 4, 7\}$, $Y = \{1, 2, 3, 4, 6, 8\}$

   b $X = \varnothing$, $Y = \{a, b, c, d, e\}$

   c $X = \{$cats, dogs, ferrets, birds$\}$, $Y = \{$birds, cats, dogs, ferrets$\}$

   d $X = \{$prime numbers between 10 and 20$\}$, $Y = \{$odd numbers between 10 and 20$\}$

4 List all subsets of $\{2, 4, 6\}$.

5 Suppose $S = \{$positive square numbers less than 50$\}$ and
$E = \{$positive even numbers less than 20$\}$.
Find:

   a $S \cap E$ **b** $S \cup E$.

6 Consider the set of whole numbers from 1 to 15.
Let $A = \{$factors of 15$\}$ and $B = \{$multiples of 3 less than 16$\}$.

   a List the elements of:

       i A **ii** B **iii** $A \cap B$ **iv** $A \cup B$ **v** $(A \cup B)'$.

   b Illustrate the sets A, B, and U on a Venn diagram.

7 Let $U = \{1, 2, 3, 4, 5, 6, 7, 8, 9\}$.   Draw a Venn diagram to represent:

   a $P = \{2, 5, 8, 9\}$ **b** $P = \{1, 4, 5, 6, 8\}$, $Q = \{2, 4, 6, 7\}$

   c $P = \{3, 7\}$, $Q = \{1, 3, 4, 7, 8\}$ **d** $P = \{2, 6, 7\}$, $Q = \{1, 4, 9\}$

8 On separate Venn diagrams, shade the region representing:

   a in Y but not in X

   b in neither X nor Y.

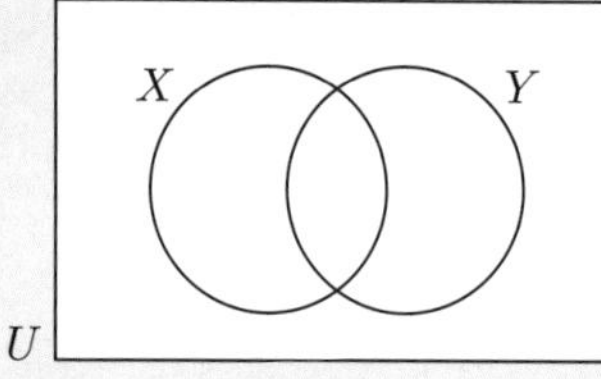

9

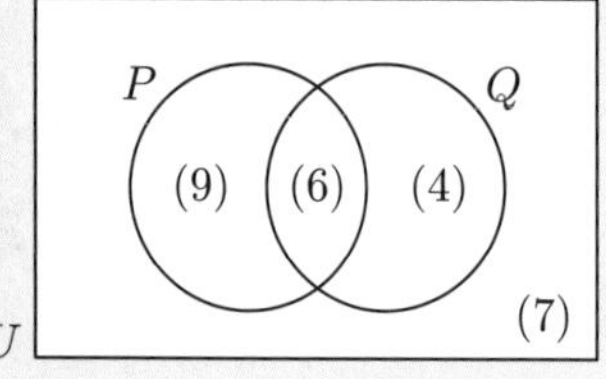

State the number of elements in:

   a P **b** Q

   c $P \cap Q$ **d** neither P nor Q

   e $P \cup Q$ **f** U

   g P but not Q **h** Q but not P

10 Judy has tickets numbered 1 to 20 in a bag. Let $P = \{$prime numbers from 1 to 20$\}$, and
$S = \{$square numbers from 1 to 20$\}$.

   a List the elements of: **i** P **ii** S.

   b Find $P \cap S$. Comment on your result.

   c Draw an appropriate Venn diagram to display the sets.

   d Find: **i** $n(P)$ **ii** $n(S)$ **iii** $n(P \cup S)$.

   e What do you notice about your answers in **d**?

11 Twins Mark and Stephen are organising a joint birthday party. Each makes a list of people they want to invite to the party.

Let M be the set of people Mark wants to invite, and S be the set of people Stephen wants to invite.

  a Find $M \cap S$. What does this set represent?

  b Find $M \cup S$.

  c How many guests will be invited to the party?

MARK	STEPHEN
Michael	William
Bradley	Nigel
Craig	Kylie
Sally	David
Alistair	Sam
Kylie	Craig
Emma	Luke
Nigel	

REVIEW SET 2B

1 List the elements of each set:

  a {multiples of 6 between 0 and 50}   **b** {factors of 40}

2 Suppose $C = $ {composite numbers less than or equal to 20}. Find $n(C)$.

3 Let $U = \{1, 2, 3, 4, 5, 6, 7, 8, 9\}$, $A = \{3, 5, 7\}$, and $B = \{2, 3, 4, 5, 6, 7\}$.

  a Find the complement of: **i** A **ii** B.

  b True or false?   **i** $A \subseteq B$ **ii** $B' \subseteq A'$

4 Suppose $A = \{2, 3, 4, 5, 7, 9, 12, 15\}$, $B = \{2, 12, x, 9\}$, and $B \subseteq A$. What possible values could x have?

5 Let $A = \{c, d, e, h\}$, $B = \{a, c, e, i\}$, $C = \{b, d, f, h\}$, and $D = \{a, b, f, i\}$. Which pairs of sets are disjoint?

6 Suppose $S = $ {square numbers less than 40}, and $P = $ {factors of 32}. Find:

  a $S \cup P$ **b** $S \cap P$.

7 Consider the set of uppercase letters in the English alphabet.

A B C D E F G H I J K L M N O P Q R S T U V W X Y Z

Let $A = $ {letters with straight edges only}, and $V = $ {vowels}.

  a List the elements of:

    i A **ii** V **iii** $A \cap V$.

  b Illustrate the sets A, V, and U on a Venn diagram.

8 Describe in words the shaded region:

  a

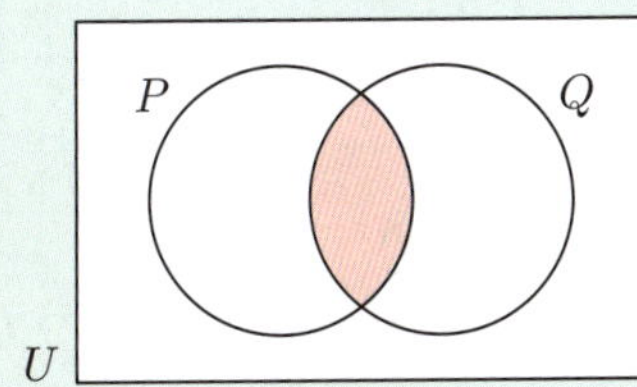

  b

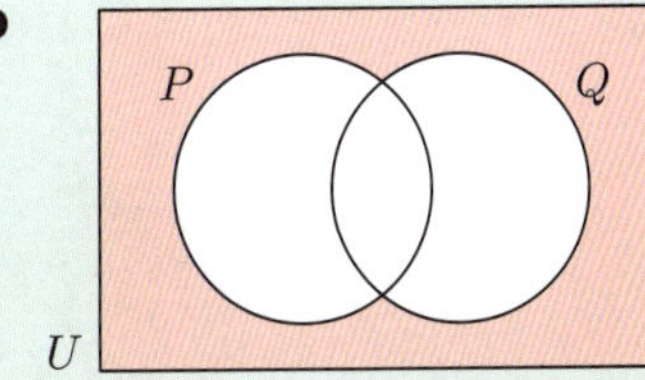

9

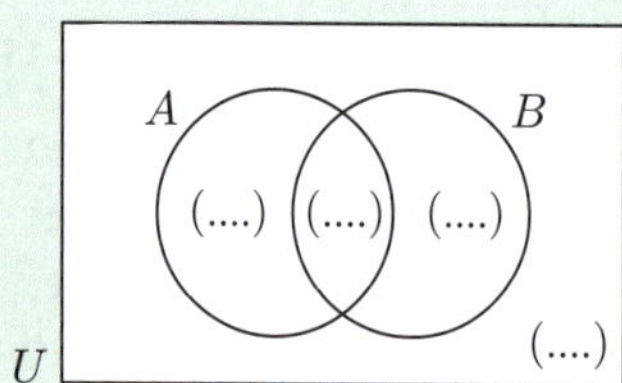

Suppose $n(A) = 12$, $n(B) = 10$, $n(A \cap B) = 3$, and $n(U) = 26$.

Copy and complete the Venn diagram.

10 A class of 40 students attended their school's swimming and athletics carnival.

28 students competed in swimming events.

37 competed in athletics events.

25 students competed in both swimming and athletics events.

  a Place this information on a Venn diagram.

  b How many students competed in:

  i athletics events only

  ii either swimming events or athletics events, but not both

  iii none of the events?

11 Tom and Margot have the choice of these cable television channels:

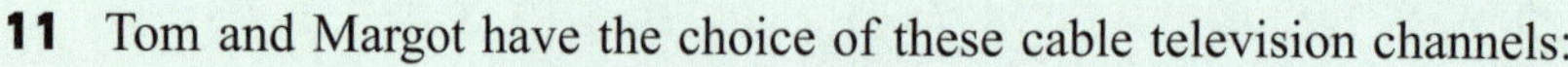

Drama	Movies Plus	Classics
Foreign Movies	Sport	News
History	Food	Music
Sci-Fi	Soapfest	Cartoons

The channels they wish to see are listed in the following sets:

Tom $T = \{$History, News, Movies Plus, Classics, Sci-Fi, Sport$\}$

Margot $M = \{$Movies Plus, Classics, Soapfest, Sport, Food$\}$

  a Display these sets on a Venn diagram.

  b Are the sets M and T disjoint? Explain your answer.

  c List the elements of $(M \cup T)'$. What does this set represent?

  d Which channels are on Tom's list but not on Margot's list?

Real numbers

Contents:

OPENING PROBLEM

Qi-Zheng has a circular pond with radius 1 m. He wants to know its surface area, so he asks his family for help. Their responses are given in the table below:

	Area (m^2)
Dad	3.14
Mum	$\dfrac{22}{7}$
Brother	$\approx 3.141\,592\,653\,....$

Things to think about:

a Can all of the answers be illustrated on a number line?

b **i** Use your calculator to write a decimal approximation for $\dfrac{22}{7}$.

    ii Can you write $\dfrac{22}{7}$ *exactly* as a decimal?

c Consider the three answers in decimal form. In which answer does the decimal:

    i keep repeating in a pattern   **ii** stop or terminate

    iii go on forever without repeating?

The set of **real numbers** includes all numbers which can be placed on the number line. It includes the set of whole numbers or **integers**, as well as the **fractions** and **decimals** between them.

In this Chapter we first revise fractions and decimals. We will see how the set of real numbers can also be divided into **rational numbers** and **irrational numbers**.

A   FRACTIONS

The division $a \div b$ can be written as the **fraction** or **common fraction** $\dfrac{a}{b}$.

$\dfrac{a}{b}$ The **numerator** is the number of parts we are looking at.

        The **bar** indicates division.

        The **denominator** is the number of equal parts in a whole.

A fraction is a **proper fraction** if its numerator is *less* than its denominator.

A fraction is an **improper fraction** if its numerator is *greater* than its denominator.

An improper fraction can also be written as a **mixed number**, which is the sum of a whole number and a proper fraction.

For example:

- $\frac{3}{5}$ is a proper fraction. A whole has been divided into 5 equal parts, and we are looking at 3.

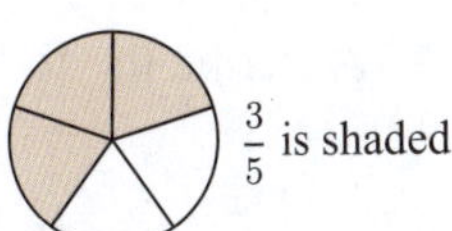

- $\frac{5}{2}$ is an improper fraction. It can also be written as the mixed number $2\frac{1}{2}$, which represents $2 + \frac{1}{2}$.

Example 1

Write:

a $3\frac{4}{5}$ as an improper fraction

b $-\frac{17}{3}$ as a mixed number.

a $3\frac{4}{5} = 3 + \frac{4}{5}$

$\phantom{3\frac{4}{5}} = \frac{15}{5} + \frac{4}{5}$

$\phantom{3\frac{4}{5}} = \frac{19}{5}$

b $\frac{17}{3} = \frac{15}{3} + \frac{2}{3}$ {the largest multiple of 3 less than 17 is 15}

$\phantom{\frac{17}{3}} = 5 + \frac{2}{3}$

$\phantom{\frac{17}{3}} = 5\frac{2}{3}$

$\therefore \quad -\frac{17}{3} = -5\frac{2}{3}$

NEGATIVE FRACTIONS

Since the bar of a fraction indicates division:

- the fraction $\frac{-1}{2}$ means $(-1) \div 2 = -\frac{1}{2}$

- the fraction $\frac{1}{-2}$ means $1 \div (-2) = -\frac{1}{2}$

So, $\frac{-1}{2} = \frac{1}{-2} = -\frac{1}{2}$.

In general: $\dfrac{-a}{b} = \dfrac{a}{-b} = -\dfrac{a}{b}$

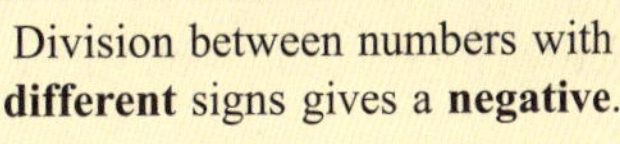

PLACING FRACTIONS ON A NUMBER LINE

We can represent fractions on a **number line** by dividing each whole into the number of parts in the denominator. By extending the number line either side of zero, we can represent positive and negative fractions.

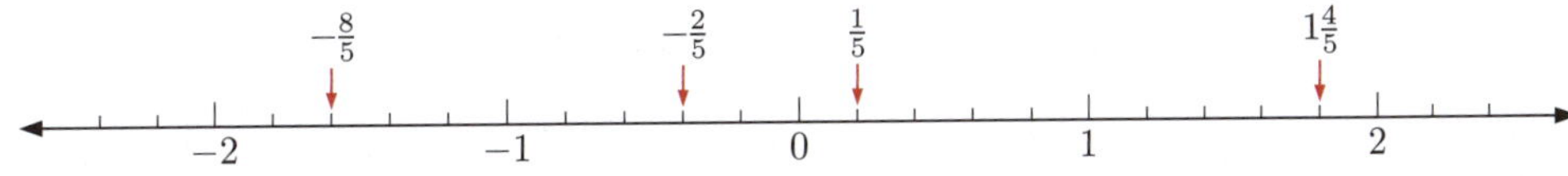

EXERCISE 3A

1 Describe the following as proper fractions, improper fractions, or mixed numbers:

   a $\dfrac{2}{7}$ **b** $\dfrac{8}{5}$ **c** $\dfrac{7}{8}$ **d** $5\dfrac{1}{3}$ **e** $\dfrac{10}{9}$ **f** $1\dfrac{1}{2}$

2 Convert these mixed numbers to improper fractions:

   a $1\dfrac{3}{4}$ **b** $3\dfrac{1}{2}$ **c** $4\dfrac{1}{5}$ **d** $2\dfrac{3}{8}$ **e** $-1\dfrac{1}{3}$ **f** $-3\dfrac{1}{5}$

3 Write as a division and hence evaluate:

   a $\dfrac{12}{4}$ **b** $\dfrac{42}{6}$ **c** $\dfrac{45}{5}$ **d** $\dfrac{108}{9}$ **e** $\dfrac{132}{11}$ **f** $\dfrac{85}{5}$

4 Write as a division and hence evaluate:

   a $\dfrac{56}{7}$ **b** $\dfrac{-56}{7}$ **c** $\dfrac{56}{-7}$ **d** $\dfrac{-56}{-7}$

   e $\dfrac{99}{11}$ **f** $\dfrac{-99}{11}$ **g** $\dfrac{99}{-11}$ **h** $\dfrac{-99}{-11}$

5 Write as a mixed number:

   a $\dfrac{11}{7}$ **b** $\dfrac{16}{5}$ **c** $\dfrac{19}{8}$ **d** $-\dfrac{10}{9}$ **e** $\dfrac{26}{5}$ **f** $-\dfrac{13}{4}$

6 Place on a number line:

   a $\dfrac{1}{3}, \; -\dfrac{2}{3}, \; \dfrac{5}{3}, \; -1\dfrac{1}{3}$ **b** $\dfrac{3}{7}, \; \dfrac{6}{7}, \; -\dfrac{2}{7}, \; 1\dfrac{4}{7}, \; -\dfrac{10}{7}$

7 **a** Place $\dfrac{3}{5}, \; -1\dfrac{4}{5}, \; 1\dfrac{1}{5}, \; -\dfrac{2}{5},$ and $-\dfrac{7}{5}$ on a number line.

   b Hence write the fractions in ascending order.

Example 2 🔊 **Self Tutor**

Evaluate: $\dfrac{12 + (5 - 7)}{18 \div (6 + 3)}$

$$\dfrac{12 + (5 - 7)}{18 \div (6 + 3)} = \dfrac{12 + (-2)}{18 \div 9} \quad \text{\{brackets first\}}$$

$$= \dfrac{10}{2} \quad \text{\{evaluate numerator and denominator\}}$$

$$= 5$$

8 Evaluate using the rule of BEDMAS:

   a $\dfrac{72}{4 \times 2}$ **b** $\dfrac{33}{15 - 4}$ **c** $\dfrac{32 \div 4}{6 - 4}$ **d** $\dfrac{13 + 7}{6 - 11}$

   e $\dfrac{16 - 64}{17 - 5}$ **f** $\dfrac{5 \times 2 + 6}{8}$ **g** $\dfrac{30}{(12 - 2) \div 2}$ **h** $\dfrac{(5 - 3) + 4}{5 - (3 + 4)}$

9 Evaluate using your calculator:

a $\dfrac{39}{3} - 18$

b $\dfrac{139 - 7}{4 \times 11}$

c $\dfrac{-15 \times 2}{7 - (16 \div 8)}$

d $\dfrac{118 + 8}{3 \times 7}$

e $\dfrac{-240 - 120}{3 \times 4 \times 5}$

f $\dfrac{(10 - 2) \times 3}{32 - (4 \times 5)}$

B   EQUAL FRACTIONS

Two fractions are **equal** if they have the same value.

Multiplying or dividing both the numerator and denominator by the same non-zero number produces an equal fraction.

Example 3 ◀ᵩ **Self Tutor**

Write: **a** $\dfrac{3}{4}$ with denominator 32   **b** $\dfrac{24}{33}$ with denominator 11

a $\dfrac{3}{4} = \dfrac{3 \times 8}{4 \times 8} = \dfrac{24}{32}$ **b** $\dfrac{24}{33} = \dfrac{24 \div 3}{33 \div 3} = \dfrac{8}{11}$

LOWEST TERMS

A fraction is written in **lowest terms** or **simplest form** if it is written with the smallest possible integer numerator and denominator.

To write a fraction in lowest terms, we divide both the numerator and denominator by their **highest common factor** (HCF). This is achieved either by writing the divisions or by cancelling common factors in the numerator and denominator.

Example 4 ◀ᵩ **Self Tutor**

Write $\dfrac{10}{25}$ in lowest terms.

$$\dfrac{10}{25} = \dfrac{10 \div 5}{25 \div 5} \quad \{\text{HCF of 10 and 25 is 5}\}$$
$$= \dfrac{2}{5}$$

or

$$\dfrac{10}{25} = \dfrac{2 \times \cancel{5}}{5 \times \cancel{5}}$$
$$= \dfrac{2}{5}$$

or

$$\dfrac{\overset{2}{\cancel{10}}}{\underset{5}{\cancel{25}}} = \dfrac{2}{5}$$

EXERCISE 3B

1 Write with denominator 20:

a $\dfrac{7}{10}$ **b** $\dfrac{3}{4}$ **c** $\dfrac{26}{40}$ **d** $\dfrac{15}{100}$

2 **a** Write the fractions $\frac{7}{10}$, $\frac{3}{5}$, and $\frac{2}{3}$ with denominator 30.

  b Hence write the fractions in descending order.

3 Write in lowest terms:

  a $\frac{8}{12}$ **b** $\frac{15}{18}$ **c** $\frac{25}{15}$ **d** $\frac{21}{35}$ **e** $-\frac{2}{12}$

  f $\frac{-5}{15}$ **g** $\frac{36}{48}$ **h** $\frac{8}{-20}$ **i** $\frac{22}{32}$ **j** $\frac{-42}{28}$

  k $\frac{12}{52}$ **l** $-\frac{15}{55}$ **m** $\frac{80}{24}$ **n** $\frac{-75}{30}$ **o** $\frac{120}{-45}$

4 Evaluate, giving your answer in lowest terms:

  a $\frac{6}{5+10}$ **b** $\frac{3\times 4}{20-2}$ **c** $\frac{50\div 2}{3+7}$ **d** $\frac{10-14}{32\div 2}$

  e $\frac{4\times 3-4}{8-3\times 6}$ **f** $\frac{18-6\div 3}{6\times(9-7)}$ **g** $\frac{6\times(5-12)}{6\times 5-12}$ **h** $\frac{14-7\times 12}{(3-6)\times 5}$

5 27 beginners, 25 amateurs, and 8 professionals competed in a golf tournament.

  a How many players competed in the tournament?

  b Find, in lowest terms, the fraction of competitors who were:

    i amateur   **ii** professional   **iii** either amateur or professional.

C   ADDING AND SUBTRACTING FRACTIONS

To **add** or **subtract** fractions:

- If necessary, write the fractions with the **lowest common denominator** (LCD). This is the lowest common multiple of the original denominators.
- Add or subtract the new numerators. The denominator stays the same.

Example 5   ◀⧽ **Self Tutor**

Find: **a** $\frac{3}{8}+\frac{1}{2}$ **b** $\frac{3}{4}-\frac{2}{3}+\frac{1}{2}$

a $\dfrac{3}{8}+\dfrac{1}{2}$

$=\dfrac{3}{8}+\dfrac{1\times 4}{2\times 4}$ $\{\text{LCD}=8\}$

$=\dfrac{3}{8}+\dfrac{4}{8}$

$=\dfrac{7}{8}$

b $\dfrac{3}{4}-\dfrac{2}{3}+\dfrac{1}{2}$

$=\dfrac{3\times 3}{4\times 3}-\dfrac{2\times 4}{3\times 4}+\dfrac{1\times 6}{2\times 6}$ $\{\text{LCD}=12\}$

$=\dfrac{9}{12}-\dfrac{8}{12}+\dfrac{6}{12}$

$=\dfrac{9-8+6}{12}$

$=\dfrac{7}{12}$

EXERCISE 3C

1 Find:

   a $\dfrac{2}{9} + \dfrac{5}{9}$ **b** $\dfrac{8}{11} - \dfrac{3}{11}$ **c** $\dfrac{3}{5} + \dfrac{1}{10}$ **d** $\dfrac{1}{3} - \dfrac{1}{4}$

   e $\dfrac{5}{8} + \dfrac{3}{4}$ **f** $\dfrac{5}{6} - \dfrac{3}{4}$ **g** $\dfrac{4}{7} + \dfrac{4}{5}$ **h** $\dfrac{1}{2} - \dfrac{5}{9}$

   i $-\dfrac{1}{2} + \dfrac{3}{4}$ **j** $-\dfrac{5}{7} + \dfrac{1}{3}$ **k** $-\dfrac{3}{8} - \dfrac{1}{2}$ **l** $-\dfrac{2}{3} + \dfrac{4}{5}$

2 Find:

   a $\dfrac{7}{9} + \dfrac{2}{3} - \dfrac{2}{9}$ **b** $\dfrac{7}{4} - \dfrac{7}{8} + \dfrac{1}{2}$ **c** $\dfrac{19}{15} - \dfrac{4}{5} + \dfrac{3}{10}$ **d** $\dfrac{1}{5} + \dfrac{1}{4} - \dfrac{1}{3}$

Example 6   ◀) **Self Tutor**

Find: $2\dfrac{1}{2} - 1\dfrac{2}{5}$

$2\dfrac{1}{2} - 1\dfrac{2}{5}$

$= \dfrac{5}{2} - \dfrac{7}{5}$ {write as improper fractions}

$= \dfrac{5 \times 5}{2 \times 5} - \dfrac{7 \times 2}{5 \times 2}$ {LCD = 10}

$= \dfrac{25}{10} - \dfrac{14}{10}$

$= \dfrac{11}{10}$ or $1\dfrac{1}{10}$

---

3 Find:

   a $1 - \dfrac{3}{7}$ **b** $1\dfrac{1}{4} + \dfrac{5}{8}$ **c** $3 - 1\dfrac{3}{5}$

   d $1 - \dfrac{1}{6} - \dfrac{3}{8}$ **e** $2\dfrac{2}{5} - 1\dfrac{4}{9}$ **f** $2 - \dfrac{3}{4} - \dfrac{4}{5}$

4 Carter read $\dfrac{1}{5}$ of a book during a flight from London to Glasgow, and another $\dfrac{3}{8}$ of the book during the flight back. What fraction of the book:

   a has Carter read   **b** remains for Carter to read?

5 A recipe requires $\dfrac{2}{3}$ cup of self-raising flour and $1\dfrac{1}{2}$ cups of plain flour.

   a In total, how much flour is used in the recipe?

   b How much more plain flour is used than self-raising flour?

6 Use your calculator to evaluate:

   a $\dfrac{1}{5} + \dfrac{1}{6} + \dfrac{1}{7}$ **b** $\dfrac{1}{8} + \dfrac{7}{10} - \dfrac{29}{30}$ **c** $3\dfrac{5}{11} - 2\dfrac{3}{4}$

D   MULTIPLYING FRACTIONS

To multiply a fraction by a whole number, the numerator is multiplied by the whole number and the denominator remains the same.

We usually look to cancel common factors so the result is written in lowest terms.

Example 7   ◆ Self Tutor

During a season, Joe hit $\frac{2}{5}$ of his team's 40 home runs.

How many home runs did Joe hit?

Joe hit $\frac{2}{5}$ of $40 = \frac{2}{5} \times 40$

$\phantom{Joe hit \frac{2}{5} of 40} = \frac{2 \times \cancel{40}^{\,8}}{\cancel{5}_{\,1}}$

$\phantom{Joe hit \frac{2}{5} of 40} = 16$ home runs.

To multiply two fractions, we multiply the numerators together and multiply the denominators together.

$$\frac{a}{b} \times \frac{c}{d} = \frac{a \times c}{b \times d}$$

To make multiplication easier, we can cancel any common factors in the numerator and denominator *before* we multiply.

Example 8   ◆ Self Tutor

Find: **a** $\frac{3}{7} \times \frac{2}{5}$ **b** $1\frac{1}{3} \times \frac{3}{5}$

a $\quad \frac{3}{7} \times \frac{2}{5}$

$\quad = \frac{3 \times 2}{7 \times 5}$

$\quad = \frac{6}{35}$

b $\quad 1\frac{1}{3} \times \frac{3}{5}$

$\quad = \frac{4}{\cancel{3}_{\,1}} \times \frac{\cancel{3}^{\,1}}{5}$

$\quad = \frac{4}{5}$

EXERCISE 3D

1 Find:

a $\frac{1}{3} \times 60$ **b** $\frac{3}{4} \times 16$ **c** $4 \times \frac{3}{5}$ **d** $\frac{1}{4} \times 5$

e $\frac{5}{6} \times 14$ **f** $21 \times \frac{2}{7}$ **g** $\frac{3}{8} \times 12$ **h** $18 \times \frac{7}{9}$

2 Find:

a $\frac{2}{5}$ of 50

b $\frac{3}{8}$ of 24

c $\frac{5}{7}$ of 35

d $\frac{1}{4}$ of \$60

e $\frac{5}{6}$ of 30 m

f $\frac{3}{10}$ of 80 kg

g $\frac{4}{9}$ of 54 kg

h $\frac{3}{4}$ of 28 ha

i $\frac{10}{11}$ of \$132

3 Lee is travelling 700 km from Barcelona to Monaco. He stops at Marseille, having travelled $\frac{7}{10}$ of the way. How far has Lee travelled?

4 Find:

a $\frac{2}{3} \times \frac{4}{5}$

b $\frac{3}{4} \times \frac{5}{7}$

c $\frac{3}{5} \times \frac{5}{6}$

d $\frac{11}{8} \times \frac{4}{11}$

e $\frac{4}{9} \times \frac{6}{7}$

f $1\frac{1}{2} \times \frac{2}{7}$

g $1\frac{5}{8} \times \frac{3}{10}$

h $\frac{1}{2} \times \frac{3}{5} \times \frac{2}{3}$

i $\left(\frac{1}{2}\right)^2$

j $\left(\frac{2}{3}\right)^3$

k $\left(2\frac{1}{5}\right)^2$

l $\left(\frac{1}{3}\right)^2 \times 6$

m $8 \times \left(\frac{1}{2}\right)^3$

n $\frac{3}{4} \times \left(\frac{2}{3}\right)^2$

o $1\frac{1}{2} \times 2\frac{2}{3}$

p $2\frac{1}{4} \times 1\frac{1}{3}$

5 Find:

a $-\frac{4}{5} \times \frac{2}{3}$

b $2\frac{1}{4} \times \left(-\frac{1}{3}\right)$

c $\left(-\frac{2}{3}\right) \times \left(-\frac{6}{5}\right)$

d $\left(-\frac{1}{2}\right)^4$

e $\left(-\frac{2}{5}\right)^3$

f $\left(\frac{3}{4}\right)^2 \times \left(-\frac{8}{27}\right)$

6 $\frac{2}{5}$ of the money raised at a charity event is given to the local hospital. The hospital spends $\frac{3}{4}$ of the money on a new X-ray machine. What fraction of the total money raised by the charity was spent on the X-ray machine?

7 Trevor ate $\frac{1}{9}$ of a lasagne. Eleanor ate $\frac{1}{6}$ of what remained. What fraction of the lasagne:

a did Eleanor eat

b is left over?

8 Emma buys two identical bottles of shampoo. She uses $\frac{3}{8}$ of one of them at home, and uses $\frac{7}{10}$ of all of the remaining shampoo while she is on holidays. In total, how much shampoo remains?

9 Use your calculator to evaluate:

a $\left(\frac{2}{3}\right)^4$

b $\frac{24}{35} \times \frac{77}{60}$

c $\frac{6}{11} \times \frac{143}{12} \times \frac{2}{13}$

E   DIVIDING FRACTIONS

For any fraction $\dfrac{c}{d}$ where $c \neq 0$, notice that $\dfrac{c}{d} \times \dfrac{d}{c} = 1$.

$\dfrac{c}{d}$ and $\dfrac{d}{c}$ are called **reciprocals** because their product is one.

> To **divide** by a fraction, we multiply by the **reciprocal** of that fraction.
>
> $$\frac{a}{b} \div \frac{c}{d} = \frac{a}{b} \times \frac{d}{c}$$

Example 9   ◀◗ Self Tutor

Find: **a** $4 \div \dfrac{1}{3}$ **b** $1\dfrac{1}{3} \div \dfrac{2}{5}$

a $\begin{aligned} 4 \div \frac{1}{3} &= 4 \times \frac{3}{1} \\ &= 4 \times 3 \\ &= 12 \end{aligned}$

b $\begin{aligned} 1\frac{1}{3} \div \frac{2}{5} &= \frac{4}{3} \div \frac{2}{5} \\ &= \frac{\overset{2}{4}}{3} \times \frac{5}{\underset{1}{2}} \\ &= \frac{10}{3} \\ &= 3\frac{1}{3} \end{aligned}$

EXERCISE 3E

1 Evaluate:

a $\dfrac{3}{4} \div \dfrac{5}{7}$ **b** $\dfrac{1}{3} \div \dfrac{1}{8}$ **c** $\dfrac{11}{12} \div \dfrac{2}{3}$ **d** $\dfrac{5}{6} \div 3$

e $8 \div \dfrac{4}{5}$ **f** $1\dfrac{1}{3} \div \dfrac{3}{4}$ **g** $\dfrac{3}{5} \div 1\dfrac{1}{2}$ **h** $1\dfrac{2}{3} \div 2\dfrac{3}{4}$

2 Evaluate:

a $-\dfrac{3}{7} \div \dfrac{5}{6}$ **b** $\dfrac{2}{3} \div \left(-\dfrac{4}{5}\right)$ **c** $\dfrac{3}{4} \div \left(-\dfrac{6}{5}\right)$ **d** $-2\dfrac{1}{3} \div \dfrac{4}{5}$

3 Tina uses $\dfrac{2}{3}$ tablespoon of butter to make an apricot loaf.
How many loaves can she make with 4 tablespoons of butter?

4 Use your calculator to evaluate:

a $5\dfrac{1}{9} \div \dfrac{2}{3}$ **b** $\dfrac{22}{7} \div \dfrac{132}{21}$ **c** $\dfrac{\frac{1}{3} + \frac{2}{5}}{6 - \frac{4}{7}}$

 **F** # DECIMAL NUMBERS

A **decimal number** is a number which contains a decimal point. The decimal point separates the whole number part from the fraction part.

The place value of each digit after the decimal point corresponds to a fraction of a whole, with a denominator that is a power of ten.

We can therefore write a decimal number in several equivalent forms:

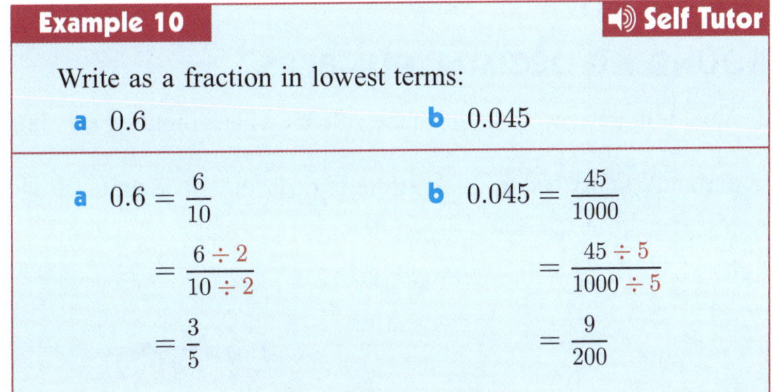

Decimal form	14.062
Expanded form	$14 + \dfrac{6}{100} + \dfrac{2}{1000}$
Mixed number	$14\dfrac{62}{1000}$

Place value table

tens	units	.	tenths	hundredths	thousandths
1	4	.	0	6	2

When we convert a decimal number to a fraction, we usually give the answer in lowest terms.

Example 10 ◄» **Self Tutor**

Write as a fraction in lowest terms:

a 0.6

b 0.045

a $0.6 = \dfrac{6}{10}$

$= \dfrac{6 \div 2}{10 \div 2}$

$= \dfrac{3}{5}$

b $0.045 = \dfrac{45}{1000}$

$= \dfrac{45 \div 5}{1000 \div 5}$

$= \dfrac{9}{200}$

DECIMAL NUMBERS ON A NUMBER LINE

To represent decimal numbers on a number line, we divide each interval according to the place value of the smallest decimal place.

For example, when representing 0.2, 0.9, 1.3, and 1.8, the place value of the smallest decimal place is tenths. We therefore divide each whole into tenths.

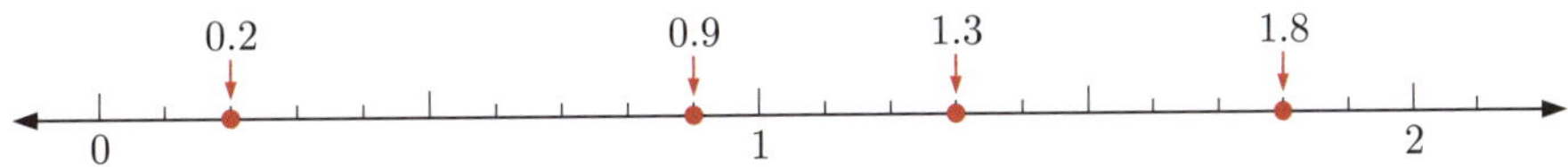

EXERCISE 3F

1 Write in expanded form:

a 1.2   **b** 1.02   **c** 1.0234   **d** 9.0909   **e** 0.0382

2 Write in decimal form:

    a $4 + \dfrac{5}{10}$ **b** $2 + \dfrac{6}{10} + \dfrac{9}{100}$ **c** $\dfrac{5}{10} + \dfrac{2}{1000}$

    d $\dfrac{3}{100} + \dfrac{8}{1000}$ **e** $2 + \dfrac{5}{1000}$ **f** $4 + \dfrac{1}{100} + \dfrac{4}{10\,000}$

3 State the value of the digit 7 in:

    a 7295   **b** 571   **c** 0.724   **d** 0.078   **e** 0.000 237

4 Write as a fraction in lowest terms:

    a 0.2   **b** 0.17   **c** 0.74   **d** 0.04

    e 0.025   **f** 0.008   **g** 0.625   **h** -0.8

5 Place on a number line:

    a 0.4, 0.7, 1.1, 1.5   **b** 0.5, -0.3, -0.6, 1.2, -1.4 **c** 0.1, 0.35, 0.6, 0.95

G   ROUNDING DECIMAL NUMBERS

We are often given measurements as decimal numbers. We can **approximate** decimal numbers by **rounding** to a certain number of **decimal places** or to a certain number of **significant figures**.

RULES FOR ROUNDING DECIMAL NUMBERS

When rounding decimal numbers, we use the same rule as when rounding whole numbers:

> To round to a particular place value, look at the digit in the place value to the right of it.
> - If this digit is 0, 1, 2, 3, or 4, we round down.
> - If this digit is 5, 6, 7, 8, or 9, we round up.

Example 11   ◀ෲ **Self Tutor**

Round 39.748 to:

    a 2 decimal places   **b** 3 significant figures.

    a $39.748 \approx 39.75$ {the third decimal place is 8, so we round up}

    b $39.748 \approx 39.7$ {the fourth significant figure is 4, so we round down}

EXERCISE 3G

1 Round to the nearest whole number:

    a 6.326   **b** 7.543   **c** 18.72   **d** 199.6

2 Round to 1 decimal place:

    a 4.21   **b** 6.39   **c** 11.54   **d** 6.87

3 Round to 2 decimal places:

    a 7.542   **b** 3.967   **c** 2.1649   **d** 8.995

4 Round to 3 significant figures:

    a 5.2374   **b** 37.1356   **c** 0.2409   **d** 0.019 25

5 Round to 4 significant figures:

    a 27.0394   **b** 0.257 24   **c** 4.001 87   **d** 0.010 200 38

6 **a** Use your calculator to find the value of $\frac{2}{43}$.

    b Round this value to:

      i 4 decimal places   **ii** 2 significant figures   **iii** 4 significant figures.

DISCUSSION

As a class:

- Round 0.945 to 1 decimal place.
- Round 0.945 to 2 decimal places, and then the *result* to 1 decimal place.

Discuss your results.

H ADDING AND SUBTRACTING DECIMAL NUMBERS

To **add** or **subtract** decimal numbers, we write the numbers under one another so the decimal points and place values line up. We then add or subtract as we do with whole numbers.

Example 12 — Self Tutor

Find $15.3 + 9.26$.

```
  1 5 . 3 0
+  ₁9 . 2 6
  2 4 . 5 6
```

EXERCISE 3H

1 Find:

    a $2.31 + 4.57$ **b** $4.9 + 6.3$ **c** $14.2 + 8.64$

    d $19.876 + 5.1$ **e** $3.2 + 5.1 + 4.8$ **f** $21.3 + 4.78 + 3.9$

    g $3.92 + 4.076$ **h** $22.019 + 9.3042$ **i** $37.9 + 1.576 + 0.49$

Example 13   ◀)) **Self Tutor**

Find:

   a $4.632 - 1.507$ **b** $8 - 0.706$

  a
$$\begin{array}{r} 2\ 12 \\ 4.6\cancel{3}\cancel{2} \\ -\ 1.507 \\ \hline 3.125 \end{array}$$

  b
$$\begin{array}{r} 7\ \ 9\ 9\ 10 \\ \cancel{8}.\cancel{0}\cancel{0}\cancel{0} \\ -\ 0.706 \\ \hline 7.294 \end{array}$$

---

2   Find:

  a $7.84 - 3.22$ **b** $9.8 - 5.36$ **c** $11.58 - 3.7$

  d $17 - 4.85$ **e** $14 - 9.227$ **f** $26.702 - 18.8$

  g $47.08 - 3.942$ **h** $1.9 - 0.137$ **i** $0.032 - 0.005\,79$

3   For an interstate holiday, Tony has packed a suitcase weighing 11.6 kg and a backpack weighing 6.75 kg. Find the total weight of Tony's luggage.

4   In a long jump competition, Samantha jumped 5.23 m. This was 1.37 m further than Theresa's jump. How far did Theresa jump?

5   At the supermarket, Margaret bought a tub of margarine for \$3.95, a carton of milk for \$1.85, and a bottle of sunscreen for \$12.90.

  a How much did Margaret spend?

  b How much change will she receive from a \$20 note?

  c How much more did the sunscreen cost than the other two items together?

I   MULTIPLYING AND DIVIDING BY POWERS OF 10

MULTIPLYING BY A POWER OF 10

When we multiply a number by 10, each digit moves one place to the left.

This has the effect of moving the decimal point one place to the right.

When multiplying by a power of 10, the power tells us how many places to move the decimal point to the right.

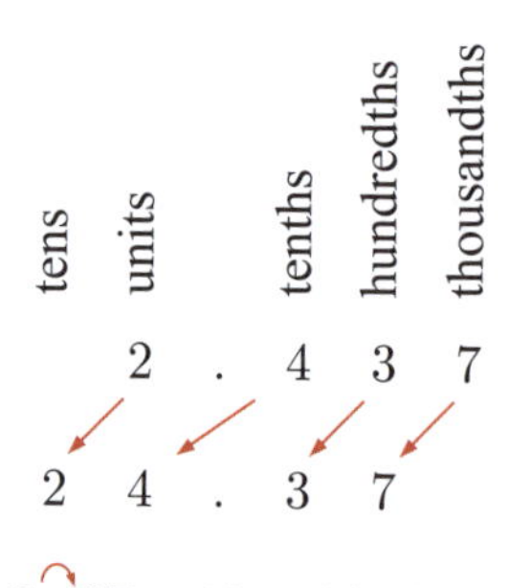

$2.437 \times 10 = 24.37$

DIVIDING BY A POWER OF 10

When we divide a number by 10, each digit moves one place to the right.

This has the effect of moving the decimal point one place to the left.

When dividing by a power of 10, the power tells us how many places to move the decimal point to the left.

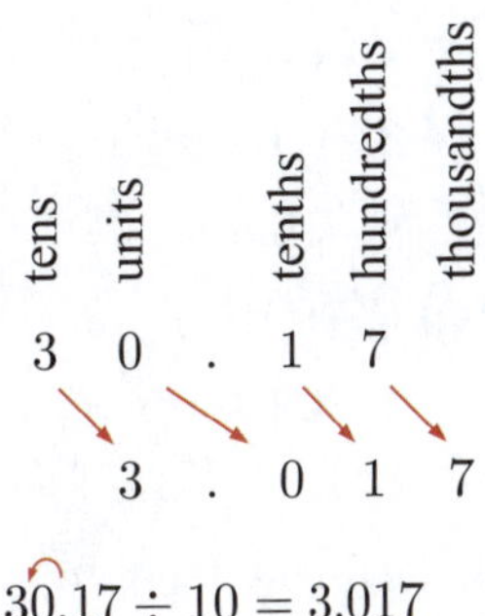

$$30.17 \div 10 = 3.017$$

EXERCISE 3I

1 Multiply 15.328 by:

  a 10   **b** 100   **c** 1000   **d** 10^4 **e** 10^5

2 Divide 936.7 by:

  a 10   **b** 100   **c** 10^3 **d** 10^4 **e** 10^5

3 Find:

  a 5.7×10 **b** 0.046×100 **c** $0.06 \times 10\,000$

  d $0.000\,01 \times 1000$ **e** $7 \div 100$ **f** $3.2 \div 1000$

  g $0.022 \div 10$ **h** $0.091 \div 10\,000$ **i** $2.005 \div 10\,000$

J   MULTIPLYING DECIMAL NUMBERS

To perform multiplication with decimal numbers, we first convert the decimal numbers to fractions with denominators that are powers of 10. We multiply the fractions *without cancelling*, then convert the result back to a decimal number.

Example 14   ◀) Self Tutor

Find: 2.57×4.3

$$2.57 \times 4.3$$
$$= \frac{257}{100} \times \frac{43}{10}$$
$$= \frac{257 \times 43}{100 \times 10}$$
$$= \frac{11\,051}{1000}$$
$$= 11.051$$

$$
\begin{array}{r}
2\;5\;7 \\
\times \quad 4\;3 \\
\hline
7\;7\;1 \\
+\; 1\;0\;2\;8\;0 \\
\hline
1\;1\;0\;5\;1 \\
\end{array}
$$

EXERCISE 3J

1 Find:

    a 6×0.8 **b** 0.7×9 **c** 0.9×0.4

    d 1.2×0.7 **e** 0.6×0.15 **f** 0.08×0.11

    g -0.5×12 **h** $0.3 \times (-0.4)$ **i** $(-0.9) \times (-0.8)$

2 Find:

    a 3.2×1.6 **b** -1.7×4.2 **c** 6.4×9.5

    d 4.1×0.58 **e** 10.4×6.9 **f** $1.84 \times (-3.8)$

3 Dominic drank 2.2 litres of water each day for 8 days. How much water did he drink in this time?

4 Each kilogram of beef contains 0.8 grams of cholesterol. How many grams of cholesterol are in a 0.250 kg serving of beef?

5 Bernard buys 4 bottles of water costing $2.35 each, and 3 fruit bars costing $1.15 each. How much money has Bernard spent altogether?

K   DIVIDING DECIMAL NUMBERS

To divide a decimal number by a whole number:

- We divide each place value in turn, starting with the largest place value.
- If a digit is too small to be divided on its own, we exchange it in the next column.
- We place the decimal point in the answer directly above the decimal point in the question.
- If necessary, we write extra zeros at the end of the decimal number we are dividing into.

For example: $1.36 \div 5 = 0.272$

$$\begin{array}{r} 0.272 \\ 5\overline{\smash{)}1.{}^{1}3{}^{3}6{}^{1}0} \end{array}$$

Rather than dividing two decimal numbers, we instead write the division as a fraction, then multiply the numerator and denominator by the same power of 10 to make the denominator a whole number. We then perform the division using the method above.

Example 15   ◀)) **Self Tutor**

Find:

    a $2.7 \div 0.3$ **b** $0.002 \div 0.08$

    a $\quad 2.7 \div 0.3$

$$= \frac{2.7 \times 10}{0.3 \times 10}$$

$$= \frac{27}{3}$$

$$= 9$$

    b $\quad 0.002 \div 0.08$

$$= \frac{0.002 \times 100}{0.08 \times 100}$$

$$= \frac{0.2}{8}$$

$$= 0.025$$

$$\begin{array}{r} 0.025 \\ 8\overline{\smash{)}0.2{}^{2}0{}^{4}0} \end{array}$$

EXERCISE 3K

1 Find:

   a $2.36 \div 4$ **b** $10.47 \div 3$ **c** $3.187 \div 5$

2 Find:

   a $1.6 \div 0.2$ **b** $4 \div 0.8$ **c** $0.56 \div 0.07$

   d $2 \div 0.05$ **e** $0.03 \div 0.2$ **f** $0.006 \div 0.03$

   g $0.0008 \div 0.05$ **h** $-1.7 \div 0.1$ **i** $(-4.2) \div (-0.7)$

3 Find:

   a $3.54 \div 0.3$ **b** $5.36 \div 0.8$ **c** $0.448 \div 0.07$

4 A mother duck weighs 2.8 kg. Her baby duckling weighs 0.7 kg. How many times heavier is the mother than her baby?

5 A drinks machine contains 90 litres of water. How many 0.3 litre cups of water can be filled from the machine?

INVESTIGATION 1   DIVISION BY ZERO

In this Investigation we think about whether it is meaningful to divide a number by zero.

What to do:

1 There are many pairs of numbers whose product is 60. Each of them corresponds to a division.

For example, $3 \times 20 = 60$ so $60 \div 3 = 20$.

   a Copy and complete:

$$15 \times \ldots\ldots = 60 \quad \text{so} \quad 60 \div 15 \;= \ldots\ldots$$

$$10 \times \ldots\ldots = 60 \quad \text{so} \quad 60 \div 10 \;= \ldots\ldots$$

$$5 \times \ldots\ldots = 60 \quad \text{so} \quad 60 \div 5 \;= \ldots\ldots$$

$$2 \times \ldots\ldots = 60 \quad \text{so} \quad 60 \div 2 \;= \ldots\ldots$$

$$\tfrac{1}{2} \times \ldots\ldots = 60 \quad \text{so} \quad 60 \div \tfrac{1}{2} \;= \ldots\ldots$$

$$0.1 \times \ldots\ldots = 60 \quad \text{so} \quad 60 \div 0.1 = \ldots\ldots$$

   b   **i** What is the result when you multiply any number by zero?

        ii *Is it possible* to complete the following statement?

$$0 \times \ldots\ldots = 60 \quad \text{so} \quad 60 \div 0 = \ldots\ldots$$

   c We say that $60 \div 0$ is *undefined*.

What can you say about the division of *any* non-zero number by zero?

2 Consider the statement: $\ldots\ldots \times 0 = 0$.

   a What numbers can you think of that will correctly complete the statement?

   b The division corresponding to the statement is $0 \div 0 = \ldots\ldots$

Can we define a *unique* value for $0 \div 0$?

L   SQUARE ROOTS

We have seen previously that we can multiply a number by itself to produce its **square**.

We now consider the reverse process:

What number do we need to multiply by itself to give a particular number?

The **square root** of the number a is the non-negative number which, when squared, gives a.

We write the square root of a as $\sqrt{a}$.

$$\sqrt{a} \times \sqrt{a} = a$$

The square root of a is a real number only if $a \geqslant 0$.

For example:

- Since $3^2 = 9$, $\sqrt{9} = 3$.
- Since $0^2 = 0$, $\sqrt{0} = 0$.
- There is no real number whose square is -1, so $\sqrt{-1}$ is not a real number.

DISCUSSION

Some people say that since $2^2 = 4$ and $(-2)^2 = 4$, both $2 = \sqrt{4}$ and $-2 = -\sqrt{4}$ should be regarded as "square roots" of 4. There are reasonable arguments for this.

However, in this course we choose to define "the square root of 4" as the *positive* value $\sqrt{4} = 2$ only. We do this to avoid confusion in what we *say*.

How could we be clear in distinguishing $\sqrt{4}$ from $-\sqrt{4}$ if they were both said as "the square root of 4"?

If a number is not a perfect square, then its square root will not be a whole number. In this case, we can estimate the square root of the number by considering the square numbers either side of it.

Example 16   ◄)) Self Tutor

Between which two consecutive integers does $\sqrt{11}$ lie?

Check your answer using a calculator.

The square numbers either side of 11 are 9 and 16

$\therefore$ $\sqrt{11}$ is between $\sqrt{9}$ and $\sqrt{16}$

$\therefore$ $\sqrt{11}$ is between 3 and 4.

Using a calculator, $\sqrt{11} \approx 3.3166$. ✓

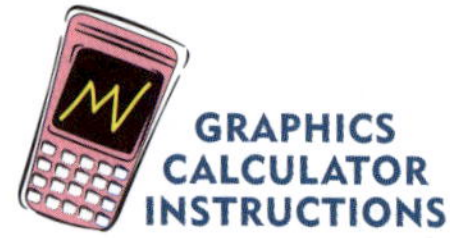
GRAPHICS
CALCULATOR
INSTRUCTIONS

EXERCISE 3L

1   Without using a calculator, find the value of:

   a $\sqrt{1}$ **b** $\sqrt{16}$ **c** $\sqrt{25}$ **d** $\sqrt{100}$

   e $\sqrt{49}$ **f** $\sqrt{64}$ **g** $\sqrt{121}$ **h** $\sqrt{81}$

2   Use your calculator to find:

   a $\sqrt{289}$ **b** $\sqrt{484}$ **c** $\sqrt{576}$ **d** $\sqrt{1521}$ **e** $\sqrt{2304}$

3   Between which two consecutive integers do the following values lie?

   a $\sqrt{6}$ **b** $\sqrt{12}$ **c** $\sqrt{28}$ **d** $\sqrt{73}$ **e** $\sqrt{103}$

   Check your answers using a calculator.

4   Use your calculator to find, correct to 4 decimal places:

   a $\sqrt{7}$ **b** $\sqrt{23}$ **c** $\sqrt{105}$ **d** $\sqrt{186}$ **e** $\sqrt{310}$

5   Use your calculator to find, correct to 4 decimal places:

   a $\sqrt{3}$ **b** $\sqrt{300}$ **c** $\sqrt{30\,000}$ **d** $\sqrt{3\,000\,000}$

M   CUBE ROOTS

In a similar way to how we defined *square* roots, we define:

> The **cube root** of the number a is the number which, when cubed, gives a.
>
> We write the cube root of a as $\sqrt[3]{a}$.
>
> $$\sqrt[3]{a} \times \sqrt[3]{a} \times \sqrt[3]{a} = a$$

For example:

- Since $4^3 = 64$, $\sqrt[3]{64} = 4$.
- Since $(-4)^3 = -64$, $\sqrt[3]{-64} = -4$.

Notice that, in contrast to square roots, the cube root of a negative number is itself a real negative number.

EXERCISE 3M

1   Find, without using a calculator:

   a $\sqrt[3]{1}$ **b** $\sqrt[3]{-1}$ **c** $\sqrt[3]{8}$ **d** $\sqrt[3]{-8}$

   e $\sqrt[3]{27}$ **f** $\sqrt[3]{-27}$ **g** $\sqrt[3]{125}$ **h** $\sqrt[3]{-125}$

   i $\sqrt[3]{0}$ **j** $\sqrt[3]{1000}$ **k** $\sqrt[3]{-1000}$

2   Use your calculator to find, correct to 3 decimal places if necessary:

   a $\sqrt[3]{216}$ **b** $\sqrt[3]{343}$ **c** $\sqrt[3]{729}$

   d $\sqrt[3]{2197}$ **e** $\sqrt[3]{2}$ **f** $\sqrt[3]{9}$

   g $\sqrt[3]{44}$ **h** $\sqrt[3]{250}$

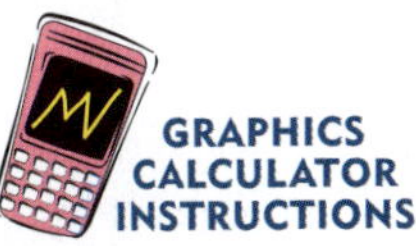

HISTORICAL NOTE   BRAHMAGUPTA

Brahmagupta (598 - 668) was an Indian mathematician and astronomer.

His most famous work, the *Brahmasphutasiddhanta* ("Correctly Established Doctrine of Brahma"), contained methods for computing square roots and cube roots, rules for summing the square numbers and cubic numbers, and techniques for solving linear and quadratic equations.

It also provided one of the first treatments of zero. Brahmagupta treated zero as a number, rather than simply as a placeholder digit. He defined zero as *the result of subtracting a number from itself*, and noted that:

- When zero is *added* to or *subtracted* from a quantity, the quantity remains unchanged.
- When a number is *multiplied* by zero, the result is zero.

Brahmagupta also believed that zero divided by zero gives zero, although the modern view is that zero divided by zero is undefined.

Brahmagupta's work on the number zero allowed him to develop rules for operating with positive and negative numbers.

N   RATIONAL NUMBERS

A **rational number** is a number that can be written in the form $\dfrac{a}{b}$ where a and b are integers and $b \neq 0$.

For example:

- $\dfrac{2}{5}$ and $\dfrac{7}{4}$ are rational numbers.

- Mixed numbers such as $1\frac{1}{2}$ are rational, since $1\frac{1}{2} = \dfrac{3}{2}$.

- Whole numbers such as 5 and -3 are also rational, since $5 = \dfrac{5}{1}$ and $-3 = \dfrac{-3}{1}$.

We can express any rational number in decimal form by performing the division of these integers.

When we convert a rational number to decimal form, the result may either be a **terminating decimal** or a **recurring decimal**.

TERMINATING DECIMALS

A **terminating decimal** has only a finite number of non-zero digits after the decimal point.

For example, 4.256 is a terminating decimal, as it finishes or terminates after 3 decimal places.

When written as a fraction in lowest terms, a rational number will convert to a **terminating decimal** if its denominator has no prime factors other than 2 or 5.

For example, $\frac{7}{20}$ has denominator $20 = 2^2 \times 5$, which has only 2 and 5 as its prime factors.

So, $\frac{7}{20}$ converts to a terminating decimal. In fact, $\frac{7}{20} = 0.35$.

We can convert fractions like these to terminating decimals by first writing the fraction so its denominator is a power of 10.

Example 17　　　　　　　　　　　　　　　　◀) **Self Tutor**

Write as a terminating decimal:

a $\dfrac{3}{5}$　　　　　　**b** $\dfrac{7}{25}$　　　　　　**c** $\dfrac{5}{8}$

a $\dfrac{3}{5}$

$= \dfrac{3 \times 2}{5 \times 2}$

$= \dfrac{6}{10}$

$= 0.6$

b $\dfrac{7}{25}$

$= \dfrac{7 \times 4}{25 \times 4}$

$= \dfrac{28}{100}$

$= 0.28$

c $\dfrac{5}{8}$

$= \dfrac{5 \times 125}{8 \times 125}$

$= \dfrac{625}{1000}$

$= 0.625$

RECURRING DECIMALS

If you enter $1 \div 3$ into your calculator, it will give the answer as either $\frac{1}{3}$ or $0.333\,333\,333$, depending on the settings.

If the answer is given as a decimal, your calculator can only show a finite number of decimal places. However, the series of 3s after the decimal actually continues forever. We say this is a *recurring decimal*.

In a **recurring decimal**, the same sequence of digits is repeated without stopping.

When written as a fraction in lowest terms, a rational number will convert to a **recurring decimal** if its denominator has at least one prime factor other than 2 or 5.

For example, $\frac{8}{11} = 0.727\,272\,72\ldots$ is a recurring decimal since the denominator has the prime factor 11.

We indicate a recurring decimal with a line over the repeated digits.

For example, $\frac{1}{3} = 0.\overline{3}$ and $\frac{8}{11} = 0.\overline{72}$.

There may be some non-repeating digits in the decimal before the repeating digits start.

For example, $\frac{5}{6} = 0.833\,33\ldots = 0.8\overline{3}$.

Example 18
Self Tutor

Write as a recurring decimal: **a** $\dfrac{7}{9}$ **b** $\dfrac{5}{11}$

a $\dfrac{7}{9}$

$= 0.7777....$

$= 0.\overline{7}$

$$\begin{array}{r} 0\,.\,7\ \ 7\ \ 7\ \ 7\ \ .... \\ 9\,\big)\,\overline{7\,.\,{}^{7}0\,{}^{7}0\,{}^{7}0\,{}^{7}0\ ....} \end{array}$$

b $\dfrac{5}{11}$

$= 0.454\,545....$

$= 0.\overline{45}$

$$\begin{array}{r} 0\,.\,4\ \ 5\ \ 4\ \ 5\ \ 4\ \ .... \\ 11\,\big)\,\overline{5\,.\,{}^{5}0\,{}^{6}0\,{}^{5}0\,{}^{6}0\,{}^{5}0\ ....} \end{array}$$

Some decimals have long strings of digits which repeat. For example,

$$\frac{1}{17} = 0.\overline{058\,823\,529\,411\,764\,7}$$

EXERCISE 3N

1 Show that each number is rational by writing it in the rational form $\dfrac{a}{b}$ where a and b are integers and $b \neq 0$.

   a $1\dfrac{1}{3}$ **b** 7 **c** $\sqrt{36}$ **d** -4 **e** 0.3

   f $-2\dfrac{3}{4}$ **g** $\dfrac{\sqrt{49}}{\sqrt{9}}$ **h** 0.01 **i** 0 **j** $-\sqrt{441}$

2 Without using your calculator, state whether each rational number will convert to a terminating decimal or a recurring decimal.

   a $\dfrac{3}{10}$ **b** $\dfrac{2}{3}$ **c** $\dfrac{4}{25}$ **d** $\dfrac{13}{40}$

   e $\dfrac{2}{7}$ **f** $\dfrac{5}{16}$ **g** $\dfrac{4}{45}$ **h** $\dfrac{75}{81}$

   Use your calculator to check your answers.

3 Oliver notices that $\dfrac{9}{15}$ has a prime factor of 3 in the denominator, so he thinks the fraction will convert to a recurring decimal.

   However, when he enters it into his calculator, he finds that $\dfrac{9}{15} = 0.6$.

   Can you explain Oliver's mistake?

4 Write as a terminating decimal:

   a $\dfrac{9}{10}$ **b** $\dfrac{17}{20}$ **c** $\dfrac{2}{5}$ **d** $\dfrac{3}{25}$

   e $\dfrac{31}{50}$ **f** $\dfrac{3}{8}$ **g** $\dfrac{197}{250}$ **h** $\dfrac{68}{125}$

5 Write as a recurring decimal:

   a $\dfrac{1}{3}$ **b** $\dfrac{2}{9}$ **c** $\dfrac{4}{11}$ **d** $\dfrac{1}{6}$

   e $\dfrac{4}{7}$ **f** $\dfrac{5}{9}$ **g** $\dfrac{9}{11}$ **h** $\dfrac{6}{7}$

6 Use your calculator to write as a recurring decimal:

a $\dfrac{8}{13}$ **b** $\dfrac{6}{37}$ **c** $\dfrac{20}{27}$ **d** $\dfrac{13}{30}$ **e** $\dfrac{15}{22}$ **f** $\dfrac{11}{14}$

DISCUSSION

- We have seen that $0.\overline{3} = \dfrac{1}{3}$. What can we say about $0.\overline{9}$?

- If your calculator only displays 10 digits, how can you tell which digits in the decimal form of $\dfrac{1}{19}$ are repeated?

O   IRRATIONAL NUMBERS

An **irrational number** is a number that **cannot** be written in the form $\dfrac{a}{b}$ where a and b are integers.

For example:

- $\sqrt{2} \approx 1.414\,213\,56....$ is an irrational number.

  Note that fractions such as $\dfrac{99}{70} = 1.4\overline{14\,285\,7}$ are sometimes used to *approximate* $\sqrt{2}$, but they can never be exact.

- $\pi \approx 3.141\,59....$, read as "pi", is also irrational.

  In the **Opening Problem**, Qi-Zheng was calculating the surface area of a circular pond with radius 1 m. His family helped him with different approximations, but the exact answer he needed was π m^2.

  In general, the area of any circle is π times the square of its radius, and the perimeter or distance around the boundary is 2π times its radius.

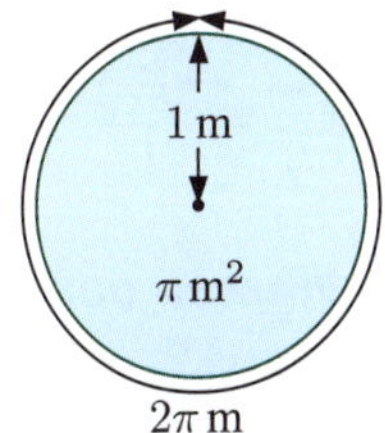

HISTORICAL NOTE

Proving that some numbers such as $\sqrt{2}$ are irrational is relatively easy, and can be done with school level mathematics. However, proving that π is irrational is particularly difficult!

One of the very greatest mathematicians **Leonhard Euler** believed π was irrational but was unable to prove it.

The first proof that π is irrational was achieved by the Swiss mathematician **Johann Heinrich Lambert** (1728 - 1777) in the 1760s.

Although π is irrational and its sequence of decimal digits will never recur, calculating the decimal places of π continues to be a topic of interest for mathematicians and computer scientists alike. The current world record is 50 trillion decimal places which took 303 days to compute!

Together, the rational numbers and irrational numbers make up the set of all **real numbers** which can be placed on a number line.

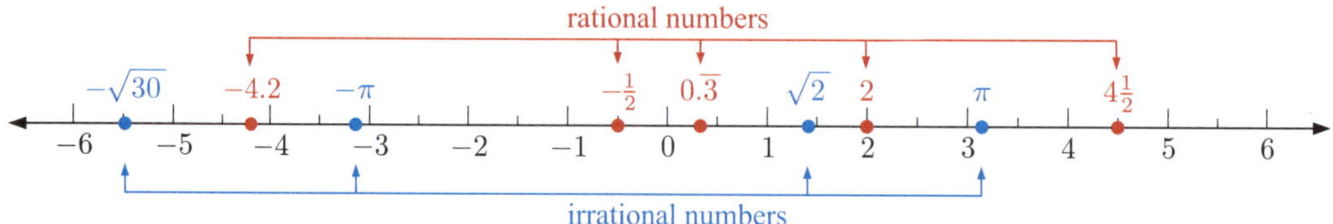

We have seen that every rational number can be written as either a terminating decimal or a recurring decimal.

By contrast:

> The decimal form of an irrational number neither terminates nor recurs.

For example, $\sqrt{2}$ to its first 30 decimal places is:

$$\sqrt{2} \approx 1.414\,213\,562\,373\,095\,048\,801\,688\,724\,209\ldots\ldots$$

We could keep writing these digits forever, and they would never terminate nor get into a loop of repeating digits.

INVESTIGATION 2   DECIMAL FORM OF $\sqrt{20}$

In this Investigation we will consider the decimal form of the irrational number $\sqrt{20}$.

$\sqrt{20}$ is the positive number such that $\left(\sqrt{20}\right)^2 = 20$.

Notice that:

- $4^2 = 16$ and $5^2 = 25$, so $\sqrt{20}$ is between 4 and 5.
- $4.4^2 = 19.36$ and $4.5^2 = 20.25$, so $\sqrt{20}$ is between 4.4 and 4.5.

What to do:

1 Show that $\sqrt{20}$ is between 4.47 and 4.48.

2 By trial and error, find the closest numbers either side of $\sqrt{20}$ which have:

    a 3 decimal places   **b** 4 decimal places

    c 5 decimal places   **d** 6 decimal places.

    Check your answers by evaluating $\sqrt{20}$ on your calculator.

EXERCISE 30

1   **a** State whether each number is rational or irrational. It may help to view the numbers in decimal form using your calculator.

      i $\dfrac{5}{6}$ **ii** $\sqrt{5}$ **iii** $0.\overline{2}$ **iv** π

      v $\sqrt{4}$ **vi** $\pi - 2$ **vii** -3 **viii** $\sqrt{2} - 2$

  b Place the numbers in **a** on a number line.

2 Decide whether each statement is true or false:

  a The negative of an irrational number must also be irrational.

  b 3 more than an irrational number must also be irrational.

  c An irrational number multiplied by 2 must also be irrational.

  d The square of an irrational number must also be irrational.

3 Give an example of an irrational number between 0 and 1.

4 Is the sum of two irrational numbers always irrational? Explain your answer.

5 Denes says that the reciprocal of any rational number is also rational.
There is one rational number that proves Denes wrong. What is it?

DISCUSSION

- Can you show that there are infinitely many rational numbers?
- Are there infinitely many irrational numbers?
- Are there more rational numbers than irrational numbers?

MULTIPLE CHOICE QUIZ

QUICK QUIZ

REVIEW SET 3A

1 Place the fractions $\dfrac{3}{4}$, $-\dfrac{1}{4}$, $-\dfrac{7}{4}$, $2\dfrac{1}{4}$ on a number line.

2 Write as a division and hence evaluate:

  a $\dfrac{54}{9}$

  b $\dfrac{88}{8}$

  c $\dfrac{120}{-12}$

3 Write in lowest terms:

  a $\dfrac{3}{18}$

  b $\dfrac{16}{40}$

  c $\dfrac{49}{28}$

  d $-\dfrac{42}{46}$

  e $\dfrac{-8}{36}$

4 Find:

  a $\dfrac{2}{3} + \dfrac{1}{9}$

  b $\dfrac{1}{6} - \dfrac{1}{4}$

  c $3\dfrac{1}{3} - 2\dfrac{2}{5}$

  d $1\dfrac{1}{4} \times \dfrac{5}{9}$

  e $\dfrac{2}{5} \div \left(-\dfrac{2}{3}\right)$

  f $9 \div 2\dfrac{1}{3}$

5 State the value of the digit 5 in:

  a 25.7

  b 1.252

  c 0.0005

  d 4.502

6 Round:

    a 2.5732 to 2 decimal places   **b** 10.0694 to 3 significant figures

    c 0.051549 to 4 decimal places.

7 Find:

    a $2.18 + 6.275$ **b** $13.46 - 5.8$ **c** $8 - 6.27$

8 Find:

    a 6.349×100 **b** $52.97 \div 1000$ **c** 0.7×8

    d $5.5 \times (-2.1)$ **e** $0.048 \div 0.6$ **f** $0.561 \div 0.03$

9 Carolyn bought 3 tumblers and a water jug from a homewares store. The tumblers cost $6.78 each and the water jug cost $18.90. How much did Carolyn spend?

10 In a hammer throw event, Brooke threw 34.5 m and Bronwyn threw 38.16 m.

    a What was the combined distance of the two throws?

    b How much farther was Bronwyn's throw than Brooke's throw?

    c Bronwyn's shotput distance was $\frac{1}{4}$ of her hammer throw distance. How far did she throw the shotput?

11 In my family there are four people. My father has an apricot pie which he divides between us. He takes a quarter of the pie for himself, and then gives my mother a third of what is left. My sister gets a half of what is left after that, and I get all of the rest.

    a How much pie is left after my father has taken his share?

    b How much of the pie does my mother get?

    c How much of the pie does my sister get?

    d Is my father's method of sharing fair? Explain your answer.

12 Between which two consecutive integers does $\sqrt{180}$ lie?

13 Find, without using a calculator:

    a $\sqrt[3]{1\,000\,000}$ **b** $\sqrt[3]{-1\,000\,000}$

14 Write as a terminating decimal:

    a $\frac{4}{5}$ **b** $\frac{19}{20}$ **c** $\frac{11}{25}$ **d** $\frac{27}{50}$ **e** $\frac{3}{40}$

15 **a** Without using your calculator, state whether each rational number will convert to a terminating decimal or a recurring decimal.

      i $\frac{16}{37}$ **ii** $\frac{7}{12}$ **iii** $\frac{5}{16}$ **iv** $\frac{119}{250}$ **v** $\frac{29}{88}$

    b Use your calculator to write each number in **a** as a terminating or recurring decimal.

REVIEW SET 3B

1 Write as a mixed number:

    a $\dfrac{11}{5}$ **b** $\dfrac{23}{6}$ **c** $-\dfrac{19}{8}$

2 Evaluate, giving your answer in lowest terms:

    a $\dfrac{12+8}{9-4}$ **b** $\dfrac{4\times(7-3)}{5+9\times3}$ **c** $\dfrac{36\div(1-4)}{3\times6+2}$

3 Find:

    a $3\dfrac{2}{5}+1\dfrac{2}{3}$ **b** $0.56+1.952$ **c** $\dfrac{4}{7}\div\dfrac{2}{21}$

    d $61.3\div4$ **e** $\dfrac{2}{3}\times2\dfrac{1}{4}$ **f** $21.3-18.52$

4 $\dfrac{7}{8}$ of Peter's garden beds are used to grow fruit and vegetables. Of these, $\dfrac{1}{4}$ of the growing space is taken up by eggplants. What fraction of Peter's garden beds is taken up by eggplants?

5 Place the decimals 0.2, 0.6, -0.4, 1.1, -1.3 on a number line.

6 Write as a fraction in lowest terms:

    a 0.22 **b** 0.65 **c** 0.092

7 Multiply 0.0359 by:

    a 10 **b** 100 **c** $10\,000$

8 Find:

    a 0.12×0.07 **b** $3\div0.6$ **c** $1.967\div0.07$

9 Mrs Austen is having trouble with her printer. The first quarter of her book is printed in red, the next sixth in blue, the next seventh in red again, and all of the rest in black.

    a What fraction of the book has black printing?

    b The book has 252 pages. What page numbers are in blue?

    c 100 of the pages were also creased. Find, in lowest terms, the fraction of pages that were creased.

10 Without using your calculator, write $\dfrac{5}{9}$ as a recurring decimal.

11 Between which two consecutive integers does $\sqrt{75}$ lie?

12 Use your calculator to find, correct to 3 decimal places if necessary:

    a $\sqrt[3]{1331}$ **b** $\sqrt[3]{1800}$

13 State whether each number is rational or irrational. If it is rational, write it in the form $\frac{a}{b}$, where a and b are integers.

    a   8   **b** $\sqrt{5}$ **c** $\sqrt{81}$

    d $6\frac{2}{3}$ **e** -3.5 **f** $\sqrt{3}+1$

14   **a**   Write $\frac{1}{3}$ as a recurring decimal.

     b   Hence write as a terminating or recurring decimal:

        i $0.2 \div 0.\overline{3}$ **ii** $0.2 \times 0.\overline{3}$ **iii** $0.\overline{3} \div 0.2$ **iv** $0.\overline{3} \times 0.2$

     c   Write $\frac{1}{6}$ and $\frac{1}{7}$ as recurring decimals.

     d   Use the numbers $\frac{1}{3}$, $\frac{1}{6}$, and $\frac{1}{7}$ to give an example where the quotient of two recurring decimals is:

        i   a recurring decimal   **ii**   a terminating decimal.

Algebraic expressions

Contents:

OPENING PROBLEM

Nadia's brother Tim is 2 years younger than she is. Her mother Melanie is three times Nadia's age. Nadia's father Peter is 4 years older than Melanie.

Things to think about:

a Suppose Nadia is 13 years old. How old is:

    i Tim   **ii** Melanie   **iii** Peter?

b Now suppose Nadia is n years old. Can you write an expression involving n, for the age of:

    i Tim   **ii** Melanie   **iii** Peter?

c Can you use your expressions in **b** to check your answers to **a**?

Algebra is a powerful tool used in mathematics.

In algebra we use letters or symbols to represent unknown values or **variables**.

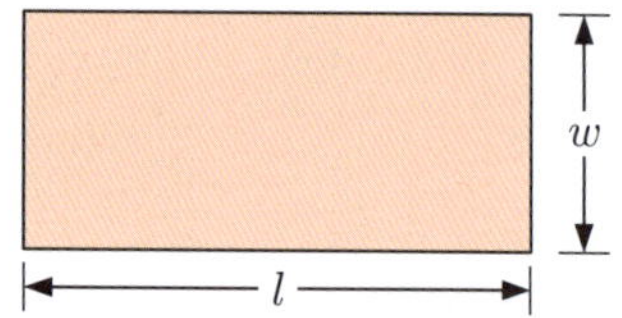

Together with numbers and operation signs, we use variables to construct mathematical **expressions**.

For example, the area of any rectangle is found by multiplying its length by its width.

If someone asked you to draw a rectangle, you would probably respond by asking how big the rectangle should be.

The length and width of the rectangle are unknown or variable.

If we let l represent the length of the rectangle, and w represent its width, the area of the rectangle can be written as $l \times w$, or lw.

A   PRODUCT NOTATION

In algebra the variables are included in expressions as though they are numbers.

However, we obey some rules which help to simplify the appearance of algebraic expressions.

Product notation is used to represent the *sum* of identical terms.

For example, just as $3 + 3 + 3 + 3 = 4 \times 3 = 12$ {4 lots of 3}

we can write $b + b + b + b = 4 \times b = 4b$ {4 lots of b}

Example 1   🔊 **Self Tutor**

Simplify: **a** $r + r + r + s + s$ **b** $d + d - (a + a + a + a)$

a $r + r + r + s + s$
    $= 3r + 2s$

b $d + d - (a + a + a + a)$
    $= 2d - 4a$

In algebra we agree:

- to **leave out** the $\times$ **signs** between multiplied quantities
- to write **numbers first** in any product
- where products contain two or more letters, we write them in **alphabetical order**.

For example:
- $3b$ is used rather than $3 \times b$ or $b3$
- $3bc$ is used rather than $3cb$.

Example 2 	◀)) **Self Tutor**

Write in product notation:

	a $t \times 6s$ 		**b** $4 \times k + m \times 3$ 		**c** $3 \times (r + s)$

	a 	$t \times 6s$ 		**b** 	$4 \times k + m \times 3$ 		**c** 	$3 \times (r + s)$

		$= 6st$ 			$= 4k + 3m$ 			$= 3(r + s)$

EXERCISE 4A

1 Write using product notation:

	a $b + b$ 		**b** $q + q + q$ 		**c** $x + x + y + y + y + y$

	d $c + c + c + e$ 		**e** $3 + z + y + y$ 		**f** $a + a + a + a + 7$

	g $g + g + 2 + g + g$ 		**h** $3 - (d + d + d)$ 		**i** $s - t + t$

	j $s - (t + t)$ 		**k** $4 + r + r + r + 1$ 		**l** $2 + a + a + b + b$

2 Write using product notation:

	a $5 \times x$ 		**b** $c \times 2$ 		**c** $q \times 7$ 		**d** $f \times 4g$

	e $6q \times p$ 		**f** $r \times 9s$ 		**g** $2a \times 3b$ 		**h** $m \times 4n$

	i $a \times 5 \times b$ 		**j** $q \times 2 \times p$ 		**k** $j \times k \times l$ 		**l** $p \times h \times d$

3 Write using product notation:

	a $p \times q + r$ 		**b** $4 \times x + 5 \times y$ 		**c** $2 \times a - b$

	d $b \times a - c$ 		**e** $b - a \times c$ 		**f** $f - g \times 7$

	g $c \times a + d \times a$ 		**h** $12 - r \times s \times 6$ 		**i** $x - 3 \times y \times x$

4 Write using product notation:

	a $3 \times (x + y)$ 		**b** $5 \times (d - 1)$ 		**c** $(w - x) \times 8$ 		**d** $p \times q \times (r - 2)$

## B 		EXPONENT NOTATION

Exponent notation is used to represent the *product* of identical factors.

For example, just as $3 \times 3 \times 3 \times 3 = 3^4$,

	we can write $b \times b \times b \times b = b^4$.

Example 3

🔊 **Self Tutor**

Simplify:

a $8 \times b \times b \times a \times a \times a$

b $k + k - 3 \times d \times d \times d$

a $\quad 8 \times b \times b \times a \times a \times a$
$\quad = 8a^3 b^2$

b $\quad k + k - 3 \times d \times d \times d$
$\quad = 2k - 3d^3$

EXERCISE 4B

1 Write using product and exponent notation:

a $x \times x \times x$

b $n \times n \times n \times n$

c $3 \times k \times k$

d $4 \times a \times a \times a$

e $2 \times d \times d \times d \times d$

f $4 \times p \times q \times q$

g $s \times t \times t \times t$

h $3 \times f \times g \times f \times g$

i $w \times w \times x \times y \times y \times y$

2 Write in expanded form:

a a^4

b $4p^3$

c $3t^5$

d $5x^2 y$

e $7f^2 g^3$

f $(5a)^2$

g $5a^2$

h $p^2 + 2q$

i $p^3 - 3q^2$

3 Write using product and exponent notation:

a $x \times x + 3$

b $c \times c + 2 \times d$

c $m + m \times m$

d $n \times n \times n + n$

e $y \times y - z \times z \times z \times z$

f $a \times a + 7 \times a$

g $8 \times b - b \times b \times b$

h $q \times p - 2 \times p$

i $2 \times p \times q \times q + 6 \times r \times r \times s$

j $h \times h \times 2 - h \times j$

k $3 \times x + 5 \times x \times x \times x$

l $a \times a + 2 \times b \times b \times b - a \times b \times b$

C   WRITING EXPRESSIONS

It is important to be able to convert algebraic expressions into words, and to concisely write worded expressions using algebraic symbols. This can often be done in several ways.

Examples for each operation are given in the following table:

Operation	Algebraic Expression	Written in Words
Addition	$x + 5$	x add 5, x plus 5, 5 more than x, the sum of x and 5, the total of x and 5
Subtraction	$3 - x$	3 minus x, x less than 3, 3 take x, 3 subtract x, subtract x from 3
Multiplication	$2x$	2 times x, 2 multiplied by x, twice x, the product of 2 and x
Division	$\dfrac{x}{5}$	x divided by 5, the quotient of x and 5, x over 5, x on 5

Operation	Algebraic Expression	Written in Words
Powers	x^2	x squared, the square of x, x to the power 2
	$(x-5)^2$	x minus 5 all squared, the square of all of x minus 5
Roots	$\sqrt{x}$	the square root of x

Example 4

◀)) Self Tutor

Write in words:

   a $3x - 5$ **b** $\dfrac{pr}{2}$ **c** $(p+q)^2$

   a $3x - 5$ is "5 less than the product of 3 and x".

   b $\dfrac{pr}{2}$ is "the product of p and r, divided by 2".

   c $(p+q)^2$ is "the sum of p and q, all squared".

EXERCISE 4C

1 Write in words:

a $y + 2$	**b** $x + y$	**c** $x - 2y$	**d** $3 - 2x$
e $\sqrt{m}$	**f** $4b^2$	**g** $3xy$	**h** $3y - 2x$
i xyz	**j** $\sqrt{x - 3}$	**k** $a + b + c$	**l** $\dfrac{b}{a}$
m $\dfrac{3}{\sqrt{x}}$	**n** $\dfrac{3a}{5b}$	**o** $\dfrac{5+a}{a}$	

2 Write in words:

a $x^2 + y$	**b** $(5x)^2$	**c** $3 - 2x^2$	**d** $y^2 - x^2$
e $\dfrac{x^2}{2}$	**f** $\left(\dfrac{x}{2}\right)^2$	**g** $2 + b^2$	**h** $(2 + b)^2$
i $a^2 - 7$	**j** $\sqrt{3a - 7}$	**k** $\sqrt{x^2 + y^2}$	**l** $x^3 - y^3$

Example 5

◀)) Self Tutor

Write as an algebraic expression:

   a the product of 3 and m, plus 5
   b the sum of m and the square of n
   c m minus n, all squared.

   a $3m + 5$ **b** $m + n^2$ **c** $(m - n)^2$

3 Write as an algebraic expression:

    a the sum of p and 7

    b x less than the product of y and z

    c the quotient of x and y

    d 3 times z, minus 15

4 Write as an algebraic expression:

    a the sum of a and b squared

    b the sum of a squared and b

    c the square of the sum of a and b

    d a plus the square of b

    e the sum of a and b, all squared

5 Write as an algebraic expression:

    a the product of 7 and x squared

    b the quotient of 7 and x, all squared

    c a squared minus b squared

    d a minus b, all squared

    e the square root of x, subtract p

6 Suppose we start with a number x. Write an algebraic expression for the final result, if we:

    a square it then add 4

    b subtract it from 3, then double the result

    c add 2 to it then take the square root of the result.

D   GENERALISING ARITHMETIC

To find algebraic expressions for many real world situations, we first think in terms of numbers or numerical cases. We then proceed to more general cases.

Example 6   🔊 **Self Tutor**

Find:

    a the cost of x bananas at y cents each

    b the change from \$50 when buying y books at \$6 each.

    a The cost of 7 bananas at 30 cents each is 7×30 cents.

        $\therefore$ the cost of x bananas at 30 cents each is $x \times 30 = 30x$ cents.

        $\therefore$ the cost of x bananas at y cents each is $x \times y = xy$ cents.

    b The change when buying 5 books costing \$6 each is $50 - (5 \times 6)$ dollars.

        $\therefore$ the change when buying y books costing \$6 each is

           $50 - (y \times 6) = (50 - 6y)$ dollars.

EXERCISE 4D

1 Find the total cost of buying:

    a 5 caps at \$20 each

    b a caps at \$20 each

    c a caps at \$$d$ each.

2 Rick is now 14 years old. How old was he:

    a 6 years ago

    b x years ago?

3 Find the change from \$100 when buying:

    a 3 hammers at \$15 each

    b h hammers at \$15 each

    c h hammers at \$$p$ each.

4 Patrick decided to go jogging each morning. As a result, he lost 6 kg. If he initially weighed w kg, how much does he weigh now?

5 There were 20 people at a party, then m more people arrived and n people left. How many people are now at the party?

6 Laura buys a apricots and p peaches. Find the total cost (in dollars) if each apricot costs 60 cents and each peach costs 90 cents.

7 Tia is walking to her friend's house, 600 m down the road. Each step she takes is 80 cm long. If Tia has walked x steps, find:

   a how far she has walked   **b** how far she is from her friend's house.

8 **a** A cyclist travels at an average speed of 15 km per hour for 3 hours. How far has the cyclist travelled?

   b How far would the cyclist travel at an average speed of s km per hour for t hours?

9 **a** Jan has 96 cupcakes to share amongst 8 tables. How many cupcakes does each table receive?

   b If Jan had c cupcakes to share amongst n tables, how many cupcakes would each table receive?

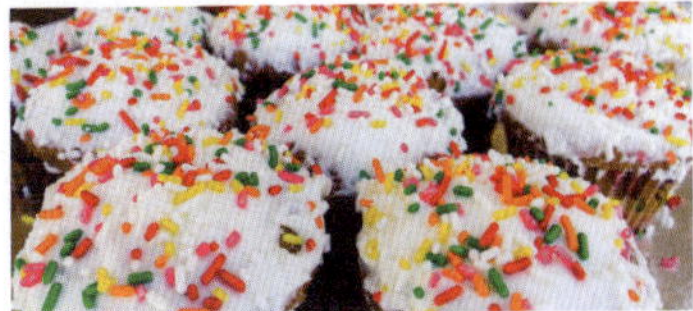

10 Suppose we have b bags which contain some apples. Find how many apples are present if each bag contains:

   a x apples, and there are 3 left over   **b** y apples, and there are 4 left over

   c t apples, and there are 7 left over   **d** m apples, and there are n left over.

11 Frank buys b balls at \$$m$ each and r racquets at \$$n$ each. Find the total cost of these items.

E   ALGEBRAIC SUBSTITUTION

If we know the values of the variables in an expression, we can **substitute** the values for the variables. This allows us to **evaluate** the expression.

> When we **evaluate** a mathematical expression, we calculate its value for particular numerical values of the variables or unknowns.

For example, the area of a triangle with base b units and height h units is given by the expression $\frac{1}{2}bh$ units^2.

If we are given the base and height of a triangle, we **substitute** the given values into the expression to find the area.

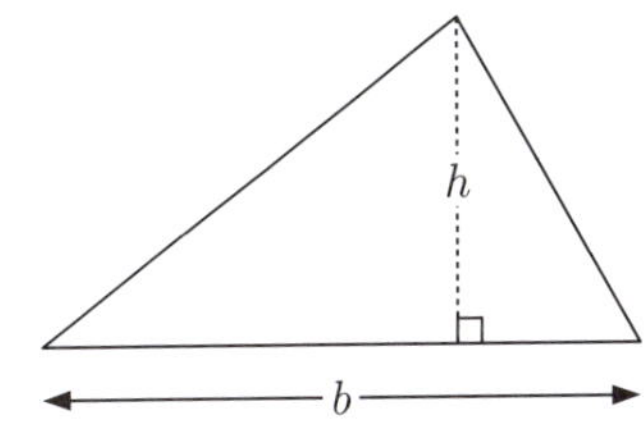

For example, if the base $b = 5$ cm and the height $h = 2$ cm, then the triangle has area $\frac{1}{2} \times 5 \times 2 = 5$ cm^2.

If we need to substitute a negative value, we write it in brackets. This helps us obtain the correct sign of each term, and ensures that we do not perform the wrong operation.

Example 7   ◀⟩ **Self Tutor**

If $a = 2$, $b = -1$, and $c = 3$, evaluate:

a $3a - 2b$ **b** $c^2 + b$

a $3a - 2b$
$= 3 \times 2 - 2 \times (-1)$
$= 6 + 2$
$= 8$

b $c^2 + b$
$= 3^2 + (-1)$
$= 9 - 1$
$= 8$

EXERCISE 4E

1 If $f = 2$, $g = 4$, and $h = 1$, evaluate:

a $f + h$ **b** $3g$ **c** $h - g$ **d** $7 + 4f$

e $2 - gh$ **f** $fg - h$ **g** $5(g - h)$ **h** $2g^2$

2 If $w = 3$, $x = -1$, $y = 4$, and $z = -2$, evaluate:

a $w + x$ **b** $5z$ **c** $2x + y$ **d** $4w - z$

e x^2 **f** $7 - w^2$ **g** $5x + 2y$ **h** $3z^2$

i $yz + 2w$ **j** $2(y - z)$ **k** $y^2 - wz$ **l** $z(y - 3x)$

Example 8   ◀⟩ **Self Tutor**

If $p = 3$, $q = -2$, and $r = -4$, evaluate:

a $\dfrac{p - q}{r}$ **b** $\dfrac{q^2 + pr}{p + r}$

a $\dfrac{p - q}{r} = \dfrac{3 - (-2)}{(-4)}$
$= \dfrac{3 + 2}{-4}$
$= \dfrac{5}{-4}$
$= -\dfrac{5}{4}$

b $\dfrac{q^2 + pr}{p + r} = \dfrac{(-2)^2 + 3 \times (-4)}{3 + (-4)}$
$= \dfrac{4 - 12}{-1}$
$= \dfrac{-8}{-1}$
$= 8$

3 If $j = 5$, $k = -3$, $l = 2$, and $m = 4$, evaluate:

a $\dfrac{m}{l}$ **b** $\dfrac{2j + l}{m}$ **c** $\dfrac{j}{l - k}$ **d** $\dfrac{k - j}{l}$

e $\dfrac{j - l}{2m}$ **f** $\dfrac{l + k}{j + m}$ **g** $\dfrac{3l}{j + 6}$ **h** $\dfrac{15 - m}{k - 1}$

i $\dfrac{k^2}{5l + m}$ **j** $\dfrac{j + k}{l^2 + 3}$ **k** $\dfrac{6 - k}{lm - 5}$ **l** $\dfrac{l^2 m}{j^2 - 21}$

4 Evaluate the expression $a(b^2 - 4)$ for:

a $a = 2$, $b = 3$ **b** $a = 4$, $b = 2$ **c** $a = -1$, $b = 5$ **d** $a = 7$, $b = -4$

5 If $a = 3$ and $b = 1$, evaluate:

   a $\sqrt{b}$ **b** $\sqrt{a+b}$ **c** $\sqrt{a^2 + 7b}$ **d** $\sqrt{4a - 3b}$

6 Answer the **Opening Problem** on page **76**.

7 The area of a triangle with base length b cm and height h cm is given by $\frac{1}{2}bh$ cm^2.

   Find the area of the triangle with base 6 cm and height 9 cm.

8 The *density* of an object is given by $\frac{m}{V}$ where m is the mass in grams, and V is the volume in cm^3. The units of density are g/cm^3.

   a Find the density of an object with mass 130 g and volume 26 cm^3.

   b A 3 cm by 5 cm by 2 cm block of iron has mass 234 g. Find the density of the block.

9 A stone is dropped off a cliff. Its speed after it has fallen s m is given by $\sqrt{19.6s}$ m/s.

   a Find the speed of the stone, correct to 2 decimal places, after it has fallen:

      i 1 m   **ii** 5 m.

   b The cliff is 12 m high. What is the stone's speed just before it hits the ground?

ACTIVITY 1   EQUAL EXPRESSIONS

Two expressions are **equal** if their numerical values are equal no matter what values of the variables are chosen.

For example, $2(x + 3)$ and $2x + 6$ are equal for *any* value of x.

We can get an idea whether two expressions are equal by choosing several values of the variables and trying each of them.

If *any* **counterexample** is found where a particular value of the variable results in the expressions taking *different* numerical values, then the expressions are *not* equal.

For example, $2(x + 3)$ and $2x + 3$ cannot be equal, because when $x = 1$,

$$
\begin{array}{lll}
2(x+3) & \text{whereas} & 2x + 3 \\
= 2(1 + 3) & & = 2 \times 1 + 3 \\
= 2 \times 4 & & = 2 + 3 \\
= 8 & & = 5
\end{array}
$$

What to do:

1 Consider the expressions $(x-1)^2$, $x^2 - 1$, and $x^2 - 2x + 1$.

   a Copy and complete this table:

x	$(x-1)^2$	$x^2 - 1$	$x^2 - 2x + 1$
1			
0			
4			
-1			

   b Which of the expressions do you think are equal?

2 Consider the expressions $a + \dfrac{b}{4}$, $\dfrac{a+b}{4}$, $\dfrac{a}{4} + b$, and $\dfrac{a}{4} + \dfrac{b}{4}$.

a Copy and complete this table:

a	b	$a + \dfrac{b}{4}$	$\dfrac{a+b}{4}$	$\dfrac{a}{4} + b$	$\dfrac{a}{4} + \dfrac{b}{4}$
8	4				
3	5				
6	-2				

b Hence decide which expressions you think are equal, and which are definitely not equal.

3 Use a table to test whether the following expressions are equal:

a $(2x)^2$ and $2x^2$ using $x = 0, 1, 2, 3$

b $180 - (a + b)$ and $180 - a - b$ using:

   i $a = 0$, $b = 10$ **ii** $a = 20$, $b = 30$ **iii** $a = 40$, $b = 75$

c $(x + y)^2$ and $x^2 + y^2$ using:

   i $x = 1$, $y = 2$ **ii** $x = -1$, $y = 3$ **iii** $x = 3$, $y = 5$

d $(x - y)^2$ and $x^2 - 2xy + y^2$ using:

   i $x = 3$, $y = 1$ **ii** $x = 2$, $y = 0$ **iii** $x = 4$, $y = -1$

F   THE LANGUAGE OF ALGEBRA

The table below defines some **key words** which we use in algebra:

Word	*Meaning*	*Example(s)*
variable	an unknown value that is represented by a letter or symbol	In $S = l^2 + 4hl$, the variables are S, h, and l.
expression	numbers and variables joined by operations	$2x + y - 7$, $\dfrac{2a + b}{c}$
equation	two expressions joined by an $=$ sign	$3x + 8 = -1$, $\dfrac{x - 1}{2} = -4$
terms	algebraic forms which are separated by $+$ or $-$ signs, the signs being included	$3x - 2y + xy - 7$ has four terms: $3x$, $-2y$, xy, and -7.
like terms	terms with the same variables to the same powers	In $4x + 3y + xy - 3x$: • $4x$ and $-3x$ are like terms • $4x$ and $3y$ are not like terms • xy and $3y$ are not like terms.
constant	a term which does not contain a variable	In $3x - y^2 + 7 + x^3$, the term 7 is a constant.
coefficient	the number factor of an algebraic term	In $4x + 2xy - y^3$: • 4 is the coefficient of x • 2 is the coefficient of xy • -1 is the coefficient of y^3.

Example 9

Consider $4y^2 - 6x + 2xy - 5 + x^2$.

a Is this an equation or an expression?

b How many terms does it contain?

c State the coefficient of: **i** x **ii** x^2.

d State the term which is a constant.

---

a This is an expression, as it does not contain an $=$ sign.

b There are five terms: $4y^2$, $-6x$, $2xy$, -5, and x^2.

c   **i** The coefficient of x is -6.   **ii** The coefficient of x^2 is 1.

d The constant is -5.

EXERCISE 4F

1 Decide whether each of the following is an expression or an equation:

a $xy + 2x + 1$ **b** $4x - y = 13$ **c** $4x - 5 = 1$

d $\dfrac{x}{2} = \dfrac{6}{x}$ **e** $\dfrac{x^2}{3} - 2x^2$ **f** $x^2 + 7x + 10$

2 State the number of terms in:

a $4x^2 + 4x + 1$ **b** $p^2 + q^2 - 5pq + 17$ **c** $x^3 - 2x^2 + 5x - \dfrac{1}{x} - 1$

3 State the coefficient of x in:

a $3x$ **b** $-8x$ **c** x **d** $-x$

e $3 + 4x$ **f** $xy - 5x$ **g** $3x - 4x^2$ **h** $2x^2 - 2x + 1$

4 State the coefficient of y in:

a $5y$ **b** $-5y$ **c** $14y$ **d** $-7y$

e $3x - y$ **f** $2x + 6y - 3$ **g** $y^2 + 2xy + 3y$ **h** $3y^2 - 2y + 5$

5 State the constant in:

a $3x + 2$ **b** $5x^2 - 4$ **c** $3 + x$ **d** $\dfrac{1}{x} + 2 - x$

6 Consider $2x^2 + 5x - 7xy + 5y^2 - 2y + 1$.

a Is this an equation or an expression?

b How many terms does it contain?

c State the coefficient of:

   i x^2 **ii** y^2 **iii** xy **iv** y.

d State the term which is a constant.

7 Identify the like terms in:

a $3x + 2 + 5x$ **b** $4x + 2y - 3 - 3y$

c $x^2 + 2x - 5x - 10$ **d** $xy + 3x^2y - 2xy + x^2y$

e $5x - 3x^2 + \dfrac{x}{2} - 1$ **f** $3x^2y^2 + 7 - xy^2 - 2 + x^2y^2$

G   COLLECTING LIKE TERMS

We have seen that like terms have the same variable form. They contain the same variables to the same powers.

Algebraic expressions can often be simplified by adding like terms. This process is called **collecting like terms**.

When we add like terms, we add the coefficients of the terms.

For example, since $2 + 4 = 6$,

$$2a + 4a = 6a.$$

We can see this by expanding the product notation:

$$2a + 4a = \underbrace{a + a}_{2a} + \underbrace{a + a + a + a}_{4a} = 6a$$

Example 10   ◄)) Self Tutor

Where possible, simplify by collecting like terms:

a $3x + 2x$ **b** $7a - 3a$ **c** $-2x + 3 - x$ **d** $3bc + bc$ **e** $2x - x^2$

---

a $3x + 2x = 5x$ {$3x$ and $2x$ are like terms}

b $7a - 3a = 4a$ {$7a$ and $-3a$ are like terms}

c $-2x + 3 - x = -3x + 3$ {$-2x$ and $-x$ are like terms}

d $3bc + bc = 4bc$ {$3bc$ and bc are like terms}

e $2x - x^2$ cannot be simplified   {$2x$ and $-x^2$ are not like terms}

EXERCISE 4G

1 Where possible, simplify by collecting like terms:

a $2 + x + 4$ **b** $q + 5 + 6$ **c** $b + 3 + b + b$

d $a + a + 7$ **e** $d + d$ **f** $q + 1 + q + 4$

g $5y - 3y$ **h** $4z - z$ **i** $g^2 + g^2$

j $5x + 5$ **k** $5w^2 - 4w^2$ **l** $3x - 3x^2$

m $3x - x$ **n** $3ab + 6ab$ **o** $m + m + m + m$

2 Simplify, where possible:

a $8p - 8p$ **b** $8p - p$ **c** $8p - 8$

d $7pq - pq$ **e** $ab + 3ab$ **f** $p^2 + 2p$

g $3w + 4w + 5w$ **h** $8xy + 5yx$ **i** $2z + 5z - 4z$

j $2m + 3m - 5m$ **k** $5d + 4d - 9$ **l** $n - 6n - 5n$

m $4j - 9j - 4$ **n** $2a + a^2 - a$ **o** $3g + 4g - 7g^2$

p $s + 3s + 4s^2$ **q** $2x^2 + 2x + 2$ **r** $2a^2 - b^2$

3 Simplify, where possible:

a $-x + 7x$

b $-7x - x$

c $-11m - 4m$

d $-y - -8y$

e $8y - -y$

f $y - -4y$

g $-2a + 3a - 4a$

h $3x - 4x - -2x$

Example 11 🔊 **Self Tutor**

Simplify by collecting like terms:

a $2 + 3a - 3 - 2a$

b $x^2 - 2x + 3x - 2x^2$

a
$$2 + 3a - 3 - 2a$$
$$= 3a - 2a + 2 - 3$$
$$= a - 1$$
$\{3a$ and $-2a$ are like terms,
2 and -3 are like terms$\}$

b
$$x^2 - 2x + 3x - 2x^2$$
$$= x^2 - 2x^2 - 2x + 3x$$
$$= -x^2 + x$$
$\{x^2$ and $-2x^2$ are like terms,
$-2x$ and $3x$ are like terms$\}$

---

4 Simplify by collecting like terms:

a $x + 5 - 3x - 6$

b $8t + 4 - 3t - 1$

c $x + 5y - 6y - 3x$

d $pq + 3 + 5pq - 7$

e $-cd + 2cd + 9cd$

f $2a - 6 + 6 - 3a$

g $12x^2 + 5 - 7x^2 - 7$

h $-5n + 3 + 2n - 6$

i $2v - 7v + w - 6w$

j $-3x^3 - 2x^2 + 3x^3 - x^2$

k $4a - 3b - -a - 4b$

l $-2z - 3 - 3z - 4$

m $2p + pq - 3pq - p$

n $3n - 5 - n - 6n$

o $-6mn + 3m - mn - 5m$

H ALGEBRAIC PRODUCTS

When two numbers are multiplied together, the result is called a **product**. If we are dealing with integers, we talk about writing a number as a product of **factors**.

We also use these words in algebra.

For example:

$$3x^2 \times -2x$$
$$= 3 \times x \times x \times -2 \times x \qquad \{\text{expanding each product}\}$$
$$= 3 \times -2 \times x \times x \times x \qquad \{\text{changing the order}\}$$
$$= -6x^3 \qquad \{\text{multiplying the numbers and using exponent notation}\}$$

We can therefore say that:

- the *product* of $3x^2$ and $-2x$ is $-6x^3$

- $3x^2$ and $-2x$ are *factors* of $-6x^3$.

To find an algebraic product, we:
- multiply the numbers, including any signs
- write the product of the variables using exponent notation where appropriate.

Example 12   ◆) Self Tutor

Simplify:

a $2a \times 3a$ **b** $-3x^2 \times 4xy$

a $\quad 2a \times 3a$	**b** $\quad -3x^2 \times 4xy$
$= 2 \times a \times 3 \times a$	$= -3 \times x \times x \times 4 \times x \times y$
$= 6 \times a^2$	$= -12 \times x^3 \times y$
$= 6a^2$	$= -12x^3y$

EXERCISE 4H

1 Simplify:

a $5 \times 4k$ **b** $x^2 \times x^3$ **c** $4x \times 7x^2$

d $5z \times 6z$ **e** $m \times 8m^3$ **f** $2g^2 \times 5g^2$

g $(5y)^2$ **h** $3x \times x \times 4$ **i** $4x \times 3x^2 \times 2x$

2 Simplify:

a $-2x \times 4x$ **b** $5a \times ab$ **c** $6c \times -c$

d $m^2 \times 3mn$ **e** $-4a^2b \times 3b$ **f** $-4p \times -5p$

g $7x^2 \times -5x$ **h** $(4g)^2 \times g$ **i** $8x^2y^2 \times 3xy$

j $7 \times -x^3$ **k** $(-3x)^2$ **l** $s^2 \times -2st$

m $x^3 \times -x^2$ **n** $5r \times (-r)^2$ **o** $-3x^2 \times -2x$

p $9xy^3 \times 3x^2y$ **q** $(2y)^3$ **r** $(-2z)^3$

DISCUSSION

Is there a way of simplifying algebraic products such as $x^4 \times x^6$ and $x^5 \times x^9$ without writing out all the factors?

I   ALGEBRAIC FRACTIONS

$\dfrac{8x}{5x^2}$ and $\dfrac{x^3}{4x}$ are examples of **algebraic fractions**, also known as **algebraic quotients**. They can be simplified by cancelling common factors in exactly the same way as for numerical fractions.

For example, just as $\quad \dfrac{8}{12} = \dfrac{2 \times \cancel{4}}{3 \times \cancel{4}} = \dfrac{2}{3},$

$$\dfrac{2x^2}{x^5} = \dfrac{2 \times \cancel{x} \times \cancel{x}}{x \times x \times x \times \cancel{x} \times \cancel{x}} = \dfrac{2}{x^3}.$$

Example 13

Simplify:

a $\dfrac{6x^3}{3x}$ **b** $\dfrac{4xy^2}{12x^2y}$

a $\dfrac{6x^3}{3x} = \dfrac{\overset{2}{\cancel{6}} \times x \times x \times \cancel{x}^{1}}{\underset{1}{\cancel{3}} \times \cancel{x}_{1}}$

$\phantom{\dfrac{6x^3}{3x}} = \dfrac{2x^2}{1}$

$\phantom{\dfrac{6x^3}{3x}} = 2x^2$

b $\dfrac{4xy^2}{12x^2y} = \dfrac{\overset{1}{\cancel{4}} \times \cancel{x}^{1} \times y \times \cancel{y}^{1}}{\underset{3}{\cancel{12}} \times \cancel{x}_{1} \times x \times \cancel{y}_{1}}$

$\phantom{\dfrac{4xy^2}{12x^2y}} = \dfrac{y}{3x}$

EXERCISE 4I

1 Simplify:

a $\dfrac{10x}{5}$ **b** $\dfrac{12}{4x}$ **c** $\dfrac{5a}{a}$ **d** $\dfrac{9k}{15k}$

e $\dfrac{x^2}{x}$ **f** $\dfrac{x}{x^2}$ **g** $\dfrac{2c^2}{c}$ **h** $\dfrac{3m}{m^3}$

2 Simplify:

a $\dfrac{x^5}{x^2}$ **b** $\dfrac{5x^3}{x}$ **c** $\dfrac{x^5}{3x^3}$ **d** $\dfrac{6x}{x^2}$

e $\dfrac{xy}{x}$ **f** $\dfrac{a}{ab}$ **g** $\dfrac{2cd}{c^2}$ **h** $\dfrac{8x^2}{10xy}$

i $\dfrac{a^3}{a^3}$ **j** $\dfrac{2mn^2}{14m^2n^2}$ **k** $\dfrac{15p^3q}{3p}$ **l** $\dfrac{3x^2y}{9y^3}$

m $\dfrac{12s^2t^2}{8s^2t}$ **n** $\dfrac{k^5}{k^5}$ **o** $\dfrac{x^2y^3z}{xyz^2}$ **p** $\dfrac{10a^2bc^3}{12ab^2}$

DISCUSSION

Is there a way of simplifying algebraic fractions such as $\dfrac{x^5}{x^2}$ and $\dfrac{x^7}{x^5}$ without writing out all of the factors?

J   MULTIPLYING ALGEBRAIC FRACTIONS

The rules used for multiplying algebraic fractions are identical to those for numerical fractions.

To multiply two fractions, we multiply the numerators together and multiply the denominators together.

$$\frac{a}{b} \times \frac{c}{d} = \frac{a \times c}{b \times d} = \frac{ac}{bd}$$

To write the result in lowest terms, we cancel any common factors in the numerator and denominator.

Example 14   ◀» **Self Tutor**

Simplify:

a $\dfrac{x}{5} \times \dfrac{y}{2}$ **b** $\dfrac{a^2}{6} \times \dfrac{3}{a}$

a $\dfrac{x}{5} \times \dfrac{y}{2} = \dfrac{x \times y}{5 \times 2}$

$\qquad\qquad = \dfrac{xy}{10}$

b $\dfrac{a^2}{6} \times \dfrac{3}{a} = \dfrac{a^2 \times 3}{6 \times a}$

$\qquad\qquad = \dfrac{\overset{1}{\cancel{a}} \times a \times \overset{1}{\cancel{3}}}{\underset{2}{\cancel{6}} \times \underset{1}{\cancel{a}}}$

$\qquad\qquad = \dfrac{a}{2}$

EXERCISE 4J

1 Simplify:

a $\dfrac{a}{3} \times \dfrac{b}{4}$ **b** $\dfrac{x}{y} \times \dfrac{x}{7}$ **c** $\dfrac{10}{m} \times \dfrac{m}{2}$ **d** $\dfrac{5}{x} \times \dfrac{6}{y}$

e $\dfrac{p^2}{6} \times \dfrac{2}{p}$ **f** $\dfrac{a}{b^2} \times \dfrac{b}{4}$ **g** $\dfrac{x}{y^2} \times \dfrac{y}{x^2}$ **h** $\dfrac{t^3}{r} \times \dfrac{r}{t}$

2 Simplify:

a $\dfrac{2x}{y} \times \dfrac{x}{8}$ **b** $\dfrac{x^2}{6} \times \dfrac{3y}{x}$ **c** $\dfrac{4x}{y} \times \dfrac{x}{6y^2}$ **d** $\dfrac{x^2}{3y} \times \dfrac{6y^3}{5}$

K   DIVIDING ALGEBRAIC FRACTIONS

To **divide** by a fraction, we multiply by the **reciprocal** of that fraction.

$$\frac{a}{b} \div \frac{c}{d} = \frac{a}{b} \times \frac{d}{c}$$

Example 15   ◀» **Self Tutor**

Simplify:

a $\dfrac{x}{5} \div \dfrac{y}{3}$ **b** $\dfrac{8}{m} \div \dfrac{2}{m}$

a $\dfrac{x}{5} \div \dfrac{y}{3}$

$\quad = \dfrac{x}{5} \times \dfrac{3}{y}$

$\quad = \dfrac{x \times 3}{5 \times y}$

$\quad = \dfrac{3x}{5y}$

b $\dfrac{8}{m} \div \dfrac{2}{m}$

$\quad = \dfrac{8}{m} \times \dfrac{m}{2}$

$\quad = \dfrac{\overset{4}{\cancel{8}} \times \overset{1}{\cancel{m}}}{\underset{1}{\cancel{m}} \times \underset{1}{\cancel{2}}}$

$\quad = 4$

EXERCISE 4K

1 Simplify:

a $\dfrac{a}{7} \div \dfrac{b}{4}$ **b** $\dfrac{5}{x} \div \dfrac{y}{8}$ **c** $\dfrac{c}{2} \div \dfrac{4}{c}$ **d** $\dfrac{p}{q} \div \dfrac{q}{6}$

e $\dfrac{x}{2} \div \dfrac{x}{3}$ **f** $\dfrac{k^2}{4} \div \dfrac{k}{j}$ **g** $\dfrac{3}{t} \div \dfrac{t^2}{s}$ **h** $\dfrac{x^2}{y} \div \dfrac{x^2}{y}$

2 Simplify:

a $\dfrac{a^2}{4} \div \dfrac{5}{3a}$ **b** $\dfrac{11}{k^2} \div \dfrac{5k}{m}$ **c** $\dfrac{12a}{b^2} \div \dfrac{9b}{a^2}$ **d** $\dfrac{3ab}{5} \div \dfrac{2b}{a}$

L   ALGEBRAIC COMMON FACTORS

When multiplying algebraic fractions, we cancelled common factors one at a time in the numerator and denominator.

Just as a numerical fraction is simplified to lowest terms by dividing the numerator and denominator by their highest common factor, the same is true for algebraic fractions.

In this Section we study algebraic common factors in more detail.

Example 16   ◀) Self Tutor

Find the highest common factor of:

a $8a$ and $12b$ **b** $4x^2$ and $6xy$

a $8a = 2 \times 2 \times 2 \times a$ **b** $4x^2 = 2 \times 2 \times x \times x$
  $12b = 2 \times 2 \times 3 \times b$ $6xy = 2 \times 3 \times x \times y$
  $\therefore$ HCF $= 2 \times 2$ $\therefore$ HCF $= 2 \times x$
  $= 4$ $= 2x$

EXERCISE 4L

1 Find the missing factor:

a $3 \times \square = 6a$ **b** $3 \times \square = 15b$ **c** $2 \times \square = 8xy$

d $2x \times \square = 8x^2$ **e** $\square \times 2x = 2x^2$ **f** $\square \times 5x = -10x^2$

g $-a \times \square = ab$ **h** $\square \times a^2 = 4a^3$ **i** $3x \times \square = -9x^2y$

2 Find the highest common factor of:

a $2a$ and 6 **b** $5c$ and $8c$ **c** $8r$ and 27

d $12k$ and $7k$ **e** 3 and $12a$ **f** $5x$ and $15x$

g $25x$ and $10x$ **h** $24y$ and $32y$ **i** $4x^2$ and $3x$

j $15x$ and $10x^2$ **k** $9r$ and r^3 **l** $6x^3$ and 21

3 Find the HCF of:

a $36b$ and $54d$

b $3q$ and qr

c $6x^2$ and xy

d $36ab$ and $12a$

e $23ab$ and $7ab$

f abc and $6abc$

g $3pq$ and $6pq^2$

h $2a^2b$ and $6ab$

i $6xy$ and $18x^2y^2$

j $15a$, $20ab$, and $30b$

k $12wxz$, $12wz$, and $24wxyz$

l $24p^2qr$ and $36pqr^2$

ACTIVITY 2   ALGEBRAIC COMMON FACTOR MAZE

Instructions:

You may move horizontally or vertically to an adjacent cell if it has a common factor (other than 1) with the cell you are currently on.

Find a path from the start to the exit.

PRINTABLE MAZE

$6m$	$2a$	3	$9c^2$	$3c$	c^2	np	$2p^2$
$4m$	mn	$6n$	$5c$	25	$5m$	n	$4p$
$8y$	xy	n	$6x$	$2x^2$	mn	$6n^2$	4
$7y$	21	$3z$	$5x$	$3y$	z^2	$3p$	p
ab	$5b$	yz	xy	$15x$	yz	p^2	$3b$
$4a$	$9q$	$3q$	q^2	63	$7y$	b^2	b
12	9	10	$10b$	12	$8b$	$9b$	3
$6a$	a^2	$5a$	$3a$	$4x$	xy	$2x$	x^2

start → (row 7, column 1: 12)

p → exit (row 4, column 8)

MULTIPLE CHOICE QUIZ

QUICK QUIZ

REVIEW SET 4A

1 Write using product and exponent notation:

a $7 \times a \times 3 \times a \times b$

b $b \times b \times b + b + b$

c $2 \times d \times d \times d - 3 \times d \times d$

d $a \times 2 \times a - 3 \times (b + b - a)$

2 Write in expanded form:

a $7t^4$

b $(3b)^3$

c $2b^2 - 3c^3$

3 Write in words:

a $\dfrac{m}{3}$

b $(3x)^2$

c $a + x^2$

4 Write as an algebraic expression:

a the total of a, b, and c

b 5 more than the square root of x

c the sum of k and the square of h

5 Gina bought a bag of 18 apples. How many apples will be left in the bag if she eats:

    a 2 apples per day for the next 3 days   **b** a apples per day for the next 3 days

    c a apples per day for the next d days?

6 If $p = -2$, $q = 6$, and $r = -3$, find:

    a r^2 **b** p^2qr **c** $\dfrac{p-r}{2q}$

7 An object with mass m kg and acceleration a m/s^2 has a force of ma newtons acting upon it. Find the force acting on:

    a a bicycle with mass 8 kg and acceleration 9 m/s^2

    b an arrow with mass 54 g and acceleration 5 m/s^2.

8 Decide whether each of the following is an equation or an expression:

    a $2x^2 - 4x + 7$ **b** $(x - 2)^2$ **c** $xy + 2y = -7$

9 State the coefficient of p in:

    a $3 + p^2 - p$ **b** $2p^2 + 7p + \dfrac{1}{p}$

10 Simplify by collecting like terms:

    a $7k + 6k - 6$ **b** $-x + 3 + 2x - 1$ **c** $9f + 3g - (-8f) - 8g$

11 Consider the expression $5g + 12gh - 2h + 3gh + 4$.

    a How many terms does it contain?   **b** State any like terms.

    c State the coefficient of h.   **d** State the term which is constant.

    e If possible, simplify the expression by collecting like terms.

12 Simplify:

    a $5x^2 \times 4x$ **b** $-a^3 \times 2a$ **c** $4y \times -2x^2y$

13 Simplify:

    a $\dfrac{2t^3}{t^3}$ **b** $\dfrac{5x^2y}{10xy^2}$ **c** $\dfrac{14a^2b}{8b^3}$

14 Simplify:

    a $\dfrac{m}{4} \times \dfrac{10}{m^2}$ **b** $\dfrac{2}{x} \div \dfrac{x}{3}$ **c** $\dfrac{5n}{p^2} \times \dfrac{2n^2}{7}$

15 Find the HCF of:

    a $30x$ and 24   **b** $3a^2b$ and $6ab$

REVIEW SET 4B

1 Write using product and exponent notation:

    a $a + a + a \times a$ **b** $4 \times p \times p - 6 \times p \times p \times q$

2 Write as an algebraic expression: q minus the square of p.

3 Write $a^3 - 7a^2b^3$ in expanded form.

4 Find the change from \$20 when buying:

  a 3 ice creams at \$2 each   **b** 3 ice creams at \$p each   **c** c ice creams at \$p each.

5 If $a = 3$, $b = -1$, and $c = -4$, evaluate:

  a $a^2 b$ **b** $(a - c)^2$

6 A cake stall had 20 muffins. n customers bought 2 muffins each, then a new batch of m muffins arrived. How many muffins does the stall now have?

7 State the:

  a coefficient of x in $3x^2 + 2x - 5$ **b** constant in $y^3 - 3 + y$.

8 Simplify by collecting like terms:

  a $3z - 8z + 2$ **b** $t + 5u - 6t + 7u$ **c** $3x^2 + 4x - 7x + 5$

9 Consider the expression $3s + 7t - 10st - 2 - 9t$.

  a How many terms does it contain?

  b State the coefficient of st.

  c State the term which is constant.

  d If possible, simplify the expression by collecting like terms.

  e Evaluate the expression if:

   i $s = 0$ and $t = 3$ **ii** $s = -1$ and $t = -2$

10 When Sinbad went fishing, he cast his net out n times and caught 6 fish each time. In addition, 3 flying fish jumped into the boat.

  a Write an expression for the total number of fish in Sinbad's boat.

  b How many fish did Sinbad return to shore with, if he cast his net out:

   i twice   **ii** seven times?

  c Suppose Sinbad cast his net out seven times. When he returned to shore, he shared his haul with his four brothers. How many fish did each brother get?

11 Simplify:

  a $(3d)^2$ **b** $-x^4 \times 5x$ **c** $x^2 y \times -xy$

12 Simplify:

  a $\dfrac{15x^3}{3x}$ **b** $\dfrac{14a^2 b}{4ab^2}$

13 Simplify:

  a $-\dfrac{4}{x} \div \dfrac{2x}{3}$ **b** $\dfrac{a^2}{b} \times \dfrac{b}{3a}$ **c** $\dfrac{(2c)^2}{d} \div \dfrac{c}{14d}$

14 Find the HCF of:

  a $6y^2$ and $8y$ **b** $10xy$ and $4x^2$

Chapter 5

Percentage

Contents:

OPENING PROBLEM

Theo carves wooden figurines and sells them online. The wood for each figurine costs \$4.50. He sells each figurine for \$12.

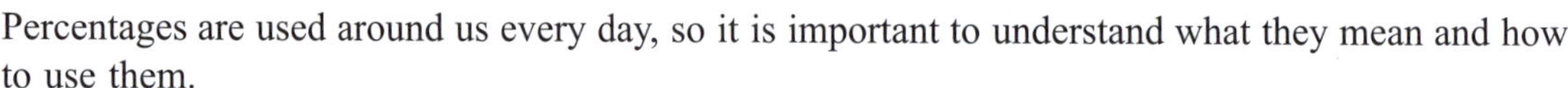

Things to think about:

a How much *profit* does Theo make on each figurine?

b How can we express this profit as a percentage?

c Why might Theo be interested in knowing the percentage profit rather than the profit in dollars?

d Suppose the price of the wood increases by 5%, and Theo sells his figurines at the same price. Does Theo's profit decrease by 5%? Explain your answer.

Percentages are used around us every day, so it is important to understand what they mean and how to use them.

A **percentage** is used to compare a portion with a whole amount.

The whole amount is represented by 100%, which has the value 1.

% reads **percent**, which means "in every hundred".

$$x\% = \frac{x}{100}$$

RESEARCH

The word *percent* is derived from the Latin phrase *per centum*. In Latin, *centum* means "hundred".

Explain how each of the following words is related to one hundred:

- centimetre
- cents
- century
- centigrade
- centenary
- centenarian
- centurion

A   CONVERTING PERCENTAGES INTO DECIMALS AND FRACTIONS

To convert a percentage into a decimal or a fraction, we divide by 100%.

$100\% = \frac{100}{100} = 1$, so dividing by 100% does not change the value of the number.

Example 1

🔊 **Self Tutor**

Write as a decimal:

| **a** | 88% | **b** | 116% |

a	88%	**b**	116%
	$= 88.\%$		$= 116.\%$
	$= 0.88$		$= 1.16$

EXERCISE 5A

1 Write as a decimal:

a	48%	**b**	110%	**c**	101%	**d**	62.5%
e	4.9%	**f**	2%	**g**	552%	**h**	0.01%
i	1000%	**j**	0.0025%	**k**	235.2%	**l**	27.3%

2 Write each percentage as a decimal:

  a A bank offers 6.5% interest on its savings account.

  b Football attendances are down 13.8% this year.

Example 2

🔊 **Self Tutor**

Write as a fraction in lowest terms:

| **a** | 115% | **b** | $12\frac{1}{2}\%$ |

a 115%
$= \dfrac{115}{100}$
$= \dfrac{115 \div 5}{100 \div 5}$
$= \dfrac{23}{20}$

b $12\frac{1}{2}\%$
$= \dfrac{25}{2} \div 100$
$= \dfrac{\overset{1}{25}}{2} \times \dfrac{1}{\underset{4}{100}}$
$= \dfrac{1}{8}$

3 Write as a fraction in lowest terms:

a	55%	**b**	32%	**c**	210%	**d**	16%
e	72%	**f**	$37\frac{1}{2}\%$	**g**	$3\frac{1}{4}\%$	**h**	144%
i	$33\frac{1}{3}\%$	**j**	120%	**k**	0.01%	**l**	0.75%

4 On an average day, 5% of students are absent and 24% of students are late.
Write down the fraction of students who are:

| **a** | absent | **b** | not absent | **c** | late | **d** | on time. |

B — CONVERTING DECIMALS AND FRACTIONS INTO PERCENTAGES

To convert a decimal or a fraction into a percentage, we multiply by 100%.

For some fractions, we may end up with a percentage with many decimal places. In this case we often round the percentage to 1 or 2 decimal places.

Example 3   ◄⑴ **Self Tutor**

Write as a percentage:

a 0.042
b $\dfrac{3}{5}$
c $\dfrac{1}{6}$

a $0.042 = 0.042 \times 100\%$
$= 4.2\%$

b $\dfrac{3}{5} = \dfrac{3}{5} \times 100\%$
$= 60\%$

c $\dfrac{1}{6} = \dfrac{1}{6} \times 100\%$
$= \dfrac{100}{6}\%$
$= 16\dfrac{2}{3}\%$
$\approx 16.7\%$

EXERCISE 5B

1 Write as a percentage:

a 0.02   **b** 0.4   **c** 0.37
d 0.65   **e** 0.416   **f** 3.2
g 1.25   **h** 0.015   **i** 0.333
j 0.0062   **k** 0.0304   **l** 1.057

2 Write as a percentage:

a $\dfrac{3}{10}$ **b** $\dfrac{9}{50}$ **c** $\dfrac{21}{25}$ **d** $\dfrac{1}{4}$

e 2   **f** $\dfrac{13}{20}$ **g** $\dfrac{4}{5}$ **h** $\dfrac{17}{40}$

i $\dfrac{7}{8}$ **j** $1\dfrac{3}{4}$ **k** $\dfrac{29}{20}$ **l** $1\dfrac{1}{8}$

3 Convert into a percentage, giving your answer exactly:

a $\dfrac{2}{3}$ **b** $\dfrac{2}{9}$ **c** $0.8\overline{3}$ **d** $\dfrac{8}{125}$ **e** $\dfrac{1}{16}$

4 Use your calculator to write these fractions as percentages, rounding your answers to 1 decimal place:

a $\dfrac{3}{7}$ **b** $\dfrac{8}{13}$ **c** $\dfrac{6}{89}$ **d** $\dfrac{18}{17}$

C  EXPRESSING ONE QUANTITY AS A PERCENTAGE OF ANOTHER

We can compare *like* quantities using percentages.

To express one quantity as a percentage of another, we write them as a fraction, then convert the fraction to a percentage.

We must make sure that the quantities are compared in the *same units*.

Example 4   **◄)) Self Tutor**

Express as a percentage:

  a Mike ran 10 km out of 50 km

  b Rani spent 5 months of the last two years overseas.

---

a $\dfrac{10 \text{ km}}{50 \text{ km}} = \dfrac{10 \times 2}{50 \times 2}$

$\qquad = \dfrac{20}{100}$

$\qquad = 20\%$

So, Mike ran 20% of 50 km.

b $\dfrac{5 \text{ months}}{2 \text{ years}} = \dfrac{5 \text{ months}}{24 \text{ months}}$

$\qquad = \dfrac{5}{24} \times 100\%$

$\qquad = \dfrac{500}{24}\%$

$\qquad \approx 20.8\%$

So, Rani spent about 20.8% of the last two years overseas.

EXERCISE 5C

1 Express as a percentage:

  a 17 marks out of 20 marks   **b** $10 out of $25

  c 8 km out of 60 km   **d** 9 months out of 3 years

  e 4 L out of 25 L   **f** 800 g out of 30 kg

  g 3 hours out of 2 days   **h** 20 cm out of 1.2 m

  i 28 mL out of 2 L   **j** 138 mm out of 3 m

2 Of the 40 passengers on a tour bus, 12 are from South Africa. What percentage of the passengers are South African?

3 Last week, Jemima was given $15 pocket money and Henry was given $18. Of this money, Jemima saved $6 while Henry saved $7. Who saved the greater percentage of their pocket money?

4 At the cinema, Michael spent €15 on his movie ticket, €8 on popcorn, and €3 on a drink. He paid for these items out of the €30 he had in his wallet. What percentage of his money did Michael spend on:

  a the ticket   **b** a drink   **c** popcorn   **d** his outing to the cinema?

5 Every year, firefighters take part in The Sky Tower Climb in Auckland, New Zealand. There are 1103 steps in the climb. The progress of four firefighters is shown in the table below. Copy and complete the table by finding the percentage of the race each person has run, and the percentage he or she has remaining.

Name	Steps climbed	% already run	% remaining
Marcel	813		
Ariel	672		
Shane	901		
Emma	866		

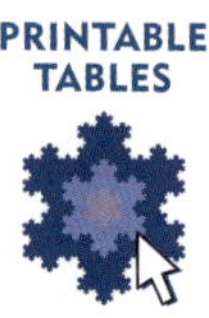

6 This table shows the number of votes and registered voters in the most recent Romanian presidential elections.

Year	Votes	Registered voters	Voter turnout %
2019	9 359 673	18 286 865	
2014	9 723 232	18 284 066	
2009	9 946 748	18 293 277	
2004	10 791 215	18 449 344	
2000	11 559 458	17 699 727	

a Copy and complete the table, rounding each percentage to 1 decimal place.

b In which year was the:

   i highest voter turnout   **ii** lowest voter turnout?

D   FINDING A PERCENTAGE OF A QUANTITY

To find a percentage of a quantity, we convert the percentage to a decimal, then **multiply** to find the required amount.

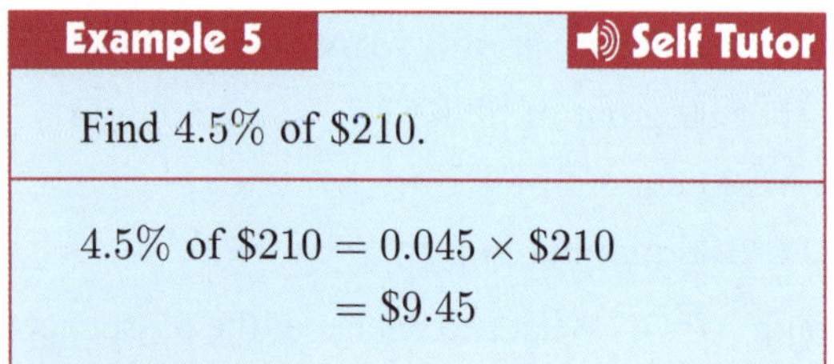

Example 5   ◄୬ **Self Tutor**

Find 4.5% of $210.

4.5% of $\$210 = 0.045 \times \$210$
$\qquad\qquad\qquad = \$9.45$

EXERCISE 5D

1 Find:

  a 80% of 30   **b** 15% of 60   **c** 25% of 110 kg

  d 70% of 400 m   **e** 35% of $250   **f** 6% of 30 L

  g 7.5% of 12 km   **h** 85% of 8 m   **i** 38.5% of £360

  j $4\frac{1}{2}\%$ of ¥64 000 000   **k** 0.5% of 3500 kg   **l** $8\frac{1}{4}\%$ of $900

2 Annette spends 32% of her wage on her mortgage. If Annette is paid \$52 000 in a year, how much does she spend on her mortgage?

3 A mixture of water and lime cordial concentrate contains 12% cordial concentrate. How much water is required to make 6 litres of this lime cordial mixture?

4 A car salesman receives 3.5% commission for every used car he sells. How much does he receive for a car which he sells for \$12 500?

5 Kerrie harvested 4900 pears from her orchard. Of these, she found that 12% were rotten. How many rotten pears did Kerrie need to throw away?

6 Tax on vehicles imported into a country is 33%. If the total value of vehicles imported is \$52 500 000, how much tax will be charged?

7 A whitegoods manufacturer receives 35% of the sale price of their goods sold by a department store. How much does the manufacturer receive if the store sells whitegoods worth:

    a \$45 000   **b** \$70 000?

8 Tom ate 40% of a 175 g bag of crisps. Teresa ate 85% of an 80 g bag of crisps.

    a Find the total amount of crisps Tom and Teresa ate.

    b Who ate more crisps?

E   THE UNITARY METHOD FOR PERCENTAGES

Sometimes we know a part or a percentage of a quantity, but we do not know the whole amount.

For example, suppose Michelle and Brigette own a business. Brigette receives 25% of the profits each month. Last month, Brigette received \$2080. How can she work out the total profit made by the business last month?

To answer this question, we can use a method called the *unitary method*.

In the **unitary method**, we first use division to find 1% of the quantity. We then multiply by 100 to find the whole amount, or by another number to find that percentage of the quantity.

Example 6　　　　　　　　　　　　　　　　🔊 **Self Tutor**

Brigette receives 25% of the company profits each month. Last month, she received \$2080. Find the total company profit last month.

$$25\% \text{ of the profit} = \$2080$$
$$\therefore \quad 1\% \text{ of the profit} = \$2080 \div 25 = \$83.20$$
$$\therefore \quad 100\% \text{ of the profit} = \$83.20 \times 100$$

So, the total company profit last month was \$8320.

EXERCISE 5E

1 Find 100% if:

    a 10% is 40 mL   **b** 45% is 225 g   **c** 6% is €72

    d 70% is 49 kg   **e** 53% is \$159   **f** 95% is 38 minutes

2 22% of students at a school ride their bikes to school. If 132 students ride their bikes to school, how many students attend the school?

3 The area of Wales is approximately $20\,800$ km^2. Wales is approximately 9% of Great Britain. Estimate the total area of Great Britain.

Example 7   ◀ϡ **Self Tutor**

18% of the students at a school are in Year 8, and 16% of the students are in Year 9. If the school has 126 Year 8 students, determine the number of Year 9 students.

$$18\% \text{ of the school} = 126 \text{ students}$$

$$\therefore \quad 1\% \text{ of the school} = \frac{126}{18} = 7 \text{ students}$$

$$\therefore \quad 16\% \text{ of the school} = 7 \times 16 = 112 \text{ students}$$

So, the school has 112 Year 9 students.

---

4 **a** Find 40% of an injection if 5% of the injection is 7 mL.

    b Find 84% of a packet of nuts if 14% of the packet is 21 g.

    c Find 72% of the contents of a bottle if 9% of the contents is 80 mL.

    d Find 18% of a wage if 54% of the wage is \$630.

5 An alloy contains 15% manganese and 85% iron. 37.5 kg of manganese is used to make the alloy.

    a How much alloy is produced?   **b** How much iron is used?

6 When Vivian bakes cookies, she always burns 20% of them. How many cookies does Vivian need to bake so that she finishes with 28 unburnt cookies?

7 The children at a camp each chose either canoeing, hiking, or swimming for their afternoon activity. 45% of the children chose canoeing, 18 children chose hiking, and 15 children chose swimming. How many children were at the camp?

8 When churning cream to make butter, 48% of the cream comes out as butter, and the remainder as buttermilk.

    a How much butter will you get from 3.5 kg of cream?

    b Amy needs 120 g of butter for a recipe. How much cream does she need to churn?

9 Maddie operates an ice cream van. The table alongside shows the sales for each flavour in the last week, expressed as a percentage of the total number of ice creams sold.

Flavour	% of sales
Chocolate	36%
Vanilla	31%
Strawberry	22%
Mint	7.5%
Banana	3.5%
Total	100%

a What percentage of ice creams sold were strawberry flavoured?

b Given that Maddie sold 124 vanilla ice creams, determine:

  i the number of chocolate ice creams sold

  ii the number of mint ice creams sold

  iii the total number of ice creams sold.

DISCUSSION

Looking back over the unitary method, is there a way to make the process quicker? Can you perform the two operations *together* by multiplying by a single fraction?

GLOBAL CONTEXT — NUTRITION INFORMATION

Global context:	Identities and relationships
Statement of inquiry:	Reading nutrition labels can help us identify the foods we should eat as part of a healthy diet.
Criterion:	Applying mathematics in real-life contexts

F   PERCENTAGE INCREASE OR DECREASE

There are many situations where quantities are either increased or decreased by a certain percentage.

For example:

- class sizes are increased by 10%
- the price of goods increases by 4%
- a man on a diet reduces his weight by 8%
- a store has a 25% discount sale.

We will now examine *two* methods for increasing or decreasing a quantity by a given percentage.

PERCENTAGE CHANGE USING TWO STEPS

To apply a percentage increase or decrease, we can:

- find the *size* of the increase or decrease, then
- *apply* this change to the original quantity by addition or subtraction.

Example 8   ◀ Self Tutor

Use two steps to:

  a increase $5000 by 20%   **b** decrease $80 000 by 12%.

---

a 20% of $5000 $= 0.2 \times \$5000$
$= \$1000$

$\therefore$ the new amount
$= \$5000 + \$1000$
$= \$6000$

b 12% of $80 000 $= 0.12 \times \$80\,000$
$= \$9600$

$\therefore$ the new amount
$= \$80\,000 - \$9600$
$= \$70\,400$

EXERCISE 5F.1

1 Perform these operations using two steps:

  a increase 160 kg by 10%   **b** decrease 12 km by 25%

  c increase $28 500 by 5%   **d** decrease 650 mL by 40%

  e increase 200 L by 2.5%   **f** decrease €450 by 18%

2 At the end of last year, Arthur's international student group had 65 members. It has increased this year by 20%. Find the number of:

  a new students who have joined the group   **b** students now in the group.

3 The annual profit of a telecommunications company fell by 65% this year. If the profit in the previous year was $324 million, find:

  a the decrease in profit this year   **b** the actual profit this year.

PERCENTAGE CHANGE USING A MULTIPLIER

Before we change a quantity, its original amount is regarded as 100%.

- If we **increase** a quantity by 20%, then in total we have $100\% + 20\% = 120\%$ of the original amount. So, to increase a quantity by 20%, we multiply the original amount by 120% or 1.2.
- If we **decrease** a quantity by 20%, then we are left with $100\% - 20\% = 80\%$ of the original amount. So, to decrease a quantity by 20%, we multiply the original amount by 80% or 0.8.

The value we multiply by is called a **multiplier**.

Example 9   ◀ Self Tutor

Use a multiplier to:

  a increase $5000 by 20%   **b** decrease $80 000 by 12%.

---

a To increase by 20%, we multiply by $100\% + 20\% = 120\%$.

$\therefore$ the new amount $= 120\%$ of $5000
$= 1.2 \times \$5000$
$= \$6000$

b To decrease by 12%, we multiply by $100\% - 12\% = 88\%$.

$\therefore$ the new amount $= 88\%$ of $80 000
$= 0.88 \times \$80\,000$
$= \$70\,400$

In summary, when calculating the change in a quantity, the

$$\textbf{original amount} \times \textbf{multiplier} = \textbf{new amount}$$

EXERCISE 5F.2

1 Find the multiplier for:

  a an increase of 5%
  b a decrease of 6%
  c an increase of 12%
  d a decrease of 25%
  e a decrease of 49%
  f an increase of 34%
  g a decrease of $7\frac{1}{2}\%$
  h an increase of 3.8%.

2 Perform these operations using a multiplier:

  a increase $750 by 12%
  b decrease 145 kg by 8%
  c decrease 800 m by 16%
  d increase 800 L by 65%
  e increase $850 000 by 32%
  f decrease 12.5 L by 85%
  g decrease 240 m by $9\frac{1}{2}\%$
  h increase $328 by 7.5%

3 A Pacific island has area 850 km^2. Due to rising sea levels, its area is predicted to decrease by 7% by 2040. Find the predicted area of the island in 2040.

4 Rita currently pays $250 rent per week, but has been told her rent will increase by 6%. How much rent will she pay after the increase?

5 Ed is 160 cm tall. His sister Peggy is 8.5% taller. How tall is Peggy?

6 In March, the price of a bus ticket was $5.

  a If the price rose by 10% in May, what was the new price?
  b If the price rose again by 10% in November, what was the new price?
  c Explain why the price increase was greater in November than in May.

7 When increasing a quantity by 10% and then decreasing the result by 10%, the overall result is to decrease the original quantity by 1%. Use multipliers to explain why this occurs.

8 In 2017, farmer Chris produced 1000 tonnes of wheat. The table below describes the percentage change in his wheat harvest each year in comparison with the previous year.

2018	2019	2020
+20%	−37.5%	+44%

  a How much wheat did Chris produce in 2020?
  b In which year did Chris produce:
    i the most wheat
    ii the least wheat?

G   FINDING A PERCENTAGE CHANGE

When calculating the change in a quantity, the

$$\text{original amount} \times \text{multiplier} = \text{new amount}.$$

Dividing both sides by the original amount, the multiplier for a change is given by:

$$\textbf{multiplier} = \frac{\textbf{new amount}}{\textbf{original amount}}$$

We can then use the multiplier to determine the percentage change.

Example 10   ◄ゥ **Self Tutor**

Determine the percentage change when:

a 50 kg is increased to 70 kg   **b** $160 is decreased to $120.

a $\text{multiplier} = \dfrac{\text{new amount}}{\text{original amount}}$ **b** $\text{multiplier} = \dfrac{\text{new amount}}{\text{original amount}}$

$\qquad\qquad = \dfrac{70 \text{ kg}}{50 \text{ kg}}$ $\qquad\qquad = \dfrac{\$120}{\$160}$

$\qquad\qquad = 1.4$ $\qquad\qquad = 0.75$

This corresponds to a 40% increase.   This corresponds to a 25% decrease.

EXERCISE 5G

1 Write down the percentage change corresponding to the multiplier:

    a 1.2   **b** 0.8   **c** 1.35   **d** 0.65

    e 1.63   **f** 0.37   **g** 2.4   **h** 0.995

2 Find the percentage change when:

    a $20 is increased to $22

    b 80 mL is decreased to 68 mL

    c 45 g is decreased to 27 g

    d 90 cm is increased to 1.35 m

    e 2 minutes is decreased to 102 seconds

    f 840 g is increased to 1.22 kg.

3 Describe the percentage change when:

    a the price of a haircut rises from $30 to $34.50

    b 150 people attended a community picnic last year, but it rained this year so only 108 people attended

    c Jorge bought a house for 3 200 000 pesos, and it is now worth 3 600 000 pesos

    d Casey threw the javelin 56.33 m, breaking the previous school record of 52.40 m

    e John completed a half-marathon in 1 hour and 52 minutes, improving on his previous best time of 2 hours and 8 minutes.

4   The population of swallows migrating this year was 29 400, compared with 34 300 last year. Find the percentage change.

5   This year at school there are 186 Year 8 students and 169 Year 9 students.
Last year there were 147 Year 8 students and 138 Year 9 students.
Which year group increased by a greater percentage?

6   Harriot really loves snakes. For her birthday in 2017, she was given a pet carpet python by her uncle. Since then, she has measured its length on her birthday each year.

Year	2017	2018	2019	2020
Length	58 cm	79 cm	94 cm	107 cm

  a   Calculate the percentage by which the snake's length increased from:

     i   2017 to 2018   **ii**   2018 to 2019   **iii**   2019 to 2020.

  b   Calculate the overall percentage increase for the 3-year period.

7   The table alongside shows the estimated number of internet users in different years.
Calculate the percentage by which the number of internet users increased from:

  a   1995 to 2000   **b**   2010 to 2016

  c   1990 to 1995   **d**   2000 to 2005.

Year	Internet users
1990	2.6 million
1995	44.4 million
2000	412.8 million
2005	1.026 billion
2010	1.992 billion
2016	3.408 billion

Source: ourworldindata.org

8   The table alongside summarises the top host destinations for higher education students studying abroad in 2000 and 2020.

  a   Use the data to estimate the number of students going to the USA for higher education in each year.

  b   Hence estimate the percentage change in the number of students going to the USA for higher education between 2000 and 2020.

  c   Estimate the percentage change in the number of students going to each of these countries for higher education between 2000 and 2020:

     i   UK   **ii**   Australia   **iii**   France

2000		2020	
1.6 million students		5.6 million students	
USA	28%	USA	20%
UK	14%	UK	10%
Germany	12%	Canada	9%
France	8%	China	9%
Australia	7%	Australia	8%
Japan	4%	France	6%

Source: OECD, Project Atlas, UNESCO

9   The attendance at council meetings has dropped by 15% each month for the last 3 months. Find the overall percentage decrease in attendance over this period.

H   FINDING THE ORIGINAL AMOUNT

When calculating the change in a quantity, the

$$\text{original amount} \times \text{multiplier} = \text{new amount}.$$

Dividing both sides by the multiplier, the original amount is given by:

$$\text{original amount} = \frac{\text{new amount}}{\text{multiplier}}$$

Example 11
 Self Tutor

Between 2009-10 and 2019-20, the number of students in universities in Germany during winter semesters rose by 36.3% to 2.89 million.   (*Source*:   Statistica, 2021)

Estimate the number of students in Germany during the winter semester in 2009-10.

$\text{original amount} = \dfrac{\text{new amount}}{\text{multiplier}}$

$\qquad\qquad = \dfrac{2.89 \text{ million}}{1.363} \qquad \{100\% + 36.3\% = 136.3\%\}$

$\qquad\qquad \approx 2.12 \text{ million}$

During the winter semester 2009-10, there were about 2.12 million students in universities in Germany.

EXERCISE 5H

1 Find the original amount given that:

   a after an increase of 20%, the length was 24 cm

   b after a decrease of 15%, the mass was 51 kg

   c after an increase of 3.6%, the amount was $129.50

   d after an increase of 130%, the capacity was 9200 L

   e after a decrease of 0.8%, the attendance was 49 600 people.

2 In 2018, 92 nations participated in the Winter Olympics. This was a 15% increase from the 2006 Winter Olympics. How many nations participated in the 2006 Winter Olympics?

3 A clothing store is having a 15% off sale. The price of a coat has been reduced to £119. How much does the coat usually cost?

4 In 2020, a music festival included 747 songs. The festival director said this was an increase of 66% since 2016. How many songs were included at the 2016 festival?

5 Joan has just received an electricity bill of $283.50. This is 32.5% less than her previous bill. How much was Joan's previous bill?

6 From 1993 to 2020, the recorded number of black rhinoceroses has increased by 127.5% to 5630. Estimate the black rhinoceros population in 1993.

PROFIT AND LOSS

In order to run a successful business, a shopkeeper must sell their products at prices that are greater than what they paid for the products. For each individual product, we define:

> The **cost price** is the price at which the business buys a product.
>
> The **selling price** is the price at which the business sells a product.

If the selling price is *greater than* the cost price, the business has made a **profit** on that product.

If the selling price is *less than* the cost price, the business has made a **loss** on that product.

$$\text{profit or loss} = \text{selling price} - \text{cost price}$$

Example 12
◀)) **Self Tutor**

Find the profit or loss when:

a an electronics store buys a television for $180 and sells it for $295

b a car yard buys a car for $900 and sells it for $650.

a selling price − cost price	**b** selling price − cost price
$= \$295 - \$180$	$= \$650 - \$900$
$= \$115$	$= -\$250$
So, a profit of $115 is made.	So, a loss of $250 is made.

PERCENTAGE PROFIT OR LOSS

By itself, knowing that you made $2 profit on an item is not always very useful. $2 would be a very small profit on an expensive item like a washing machine, or a very large profit on a cheap item like a chocolate bar.

It is often more useful to express a profit or loss as a **percentage of the cost price**.

- $\text{percentage profit} = \dfrac{\text{profit}}{\text{cost price}} \times 100\%$

- $\text{percentage loss} = \dfrac{\text{loss}}{\text{cost price}} \times 100\%$

EXERCISE 5I

1 Copy and complete:

	Cost price	Selling price	Profit or loss
	$600	$450	$150 loss
a	$35	$65	
b	£225	£160	
c	$520	$670	
d	€26 500	€33 000	

2 Copy and complete:

	Cost price	Selling price	Profit or loss
a	$56		$20 profit
b	$420		$75 loss
c		$580	$195 profit
d		$200	$65 loss

Example 13 ◀)) **Self Tutor**

A microscope was purchased for $600. It sold two years later for $450.
Find the profit or loss as a percentage of the cost price.

selling price − cost price
$= \$450 - \$600$
$= -\$150$

So, a loss of $150 was made.

Percentage loss $= \dfrac{\text{loss}}{\text{cost price}} \times 100\%$

$= \dfrac{150}{600} \times 100\%$

$= 25\%$

---

3 For each of the following transactions, find:

    i the profit or loss   **ii** the percentage profit or loss, rounded to 1 decimal place.

  a Janet bought a toy car set for $45 and sold it for $65.

  b Dave bought a motorbike for €7000 and sold it for €5850.

  c I bought a kettle for $38 and sold it for $15.

  d Lana sold a stereo for $250 which cost her $190.

  e Robert sold a chainsaw for $330 which cost him $240.

4 A boat was purchased for $26 000 and three years later was sold for $19 000. Find the loss as
a percentage of the cost price. Round your answer to 1 decimal place.

5 A retailer buys a lounge suite for $680 and sells it for $1120. Find the profit as a percentage
of the cost price. Round your answer to 1 decimal place.

6 A used car salesman buys a car for RM13 500, then sells it three days later for RM18 000.

  a Find the profit made on the sale of the car.

  b Express this profit as a percentage of the cost price.

7 A department store purchased 15 handbags for a total of CNY1650. The store then sold them
for CNY450 per handbag.

  a Find the total profit made by the department store.

  b Express this profit as a percentage of the cost price.

8 A sweets shop buys 1600 fruit gums for 20 cents each. 910 of the fruit gums are sold for
35 cents each. The remainder are thrown out because they have reached their expiry date.

  a Determine whether the shop has made a profit or loss on the fruit gums.

  b Express the profit or loss as a percentage of the cost price.

Example 14

Self Tutor

Find the selling price of goods purchased for $150 and sold at a 20% profit.

For a 20% profit we must increase $150 by 20%.

To increase by 20%, we multiply by 120%.

So, the selling price $= 120\%$ of $150

$$= 1.2 \times \$150$$
$$= \$180$$

9 Find the selling price of goods bought for:

  a $600 and sold at a 30% profit

  b €450 and sold at a 20% loss

  c $3500 and sold at a 15% loss

  d $25 000 and sold at a 12% profit.

10 Deanna bought a bedroom suite for $2400 and sold it for a profit of 8%. At what price did she sell the bedroom suite?

11 Aaron bought a wardrobe for $200, but was forced to sell it at a 30% loss. At what price did he sell the wardrobe?

12 Answer the **Opening Problem** on page **96**.

13 Andy bought a broken washing machine for $50, then spent a further $26 on parts to fix it. He sold the washing machine for a 120% profit. At what price did he sell the washing machine?

J   DISCOUNT

In order to attract customers or to clear old stock, many businesses reduce the price of items from the **marked price** shown on the price tag.

The amount of money by which the marked price of the item is reduced is called **discount**.

selling price = marked price − discount

Example 15

Self Tutor

A dishwasher has a marked price of $495. During a sale, a $120 discount is offered. Find the sale price of the dishwasher.

selling price $=$ marked price $-$ discount

$$= \$495 - \$120$$
$$= \$375$$

So, the dishwasher is on sale at $375.

Discount is often stated as a percentage of the marked price. It is thus a **percentage decrease**.

Example 16   ◀ Self Tutor

The marked price of a video projector is $1060. If a 22% discount is offered, find the actual selling price.

For a 22% discount we must decrease $1060 by 22%.

To decrease by 22%, we multiply by $100\% - 22\% = 78\%$.

So, the selling price $= 78\%$ of $1060
$$= 0.78 \times \$1060$$
$$= \$826.80$$

EXERCISE 5J

1 Find the selling price of:

   a a lamp with a marked price of $49, if a $15 discount is offered

   b a cordless drill which has been discounted by $80 from its marked price of $329.

2 The marked price of a billiard table is $2450, and an 8% discount is offered. Find the actual selling price of the billiard table.

3 A television marketing company advertises exercise equipment for £180. It offers a 12% discount for the first 50 callers. Find the actual selling price if you are one of the first 50 callers.

4 A toaster with a marked price of $50 has been discounted to $42. Find the percentage discount.

5 A department store is offering the following discounts:

   • 5% off for customers who spend more than $100

   • 7.5% off for customers who spend more than $200

   • 10% off for customers who spend more than $500.

Find the final price paid by a customer who buys:

   a a television set marked at $180   **b** a washing machine marked at $320

   c a dishwasher marked at $890.

6 Copy and complete:

	Marked price	Discount	Selling price	Discount as a % of marked price
a	$160	$40		
b	$500			34%
c	$2.40			15%
d		$0.75	$3.40	
e	$252		$163	
f		€88.20		18%
g			$3379	38%

K   VAT AND GST

In most countries a tax is payable on the sale of items. This tax has various names depending on the country where the items are sold. The two most common names are:

- **value-added tax (VAT)**
- **goods and services tax (GST)**.

The standard rate of tax varies from one country to another. Examples are shown in the table.

The tax is applied as a **percentage increase** to an item's price.

Country	VAT or GST rate
Chile	19%
Sweden	25%
Australia	10%
Ghana	12.5%
Russia	20%
Singapore	7%

EXERCISE 5K

1 Find the amount of tax which needs to be paid on an item which costs:

  a 624 krona plus 25% VAT

  b 8550 rubles plus 20% VAT.

2 Find the total amount paid by the customer for an item which costs:

  a $26.40 plus 10% GST

  b 4800 cedi plus 12.5% VAT.

3 Find the percentage tax which has been added if the price of the item increases from:

  a $250 to $267.50

  b 82 000 pesos to 97 580 pesos.

4 Find the price of the item before tax was added, if the customer pays:

  a HRK76 250 and the tax was 25%

  b CHF2154 and the tax was 7.7%.

DISCUSSION

In some countries, shops are required to include the tax in the *marked price* of an item.

- Do you think this helps customers understand how much they are spending?
- Do you think this should be done in all countries?

GLOBAL CONTEXT   INFLATION

Global context:	Fairness and development
Statement of inquiry:	Using percentages to measure change allows us to make more meaningful comparisons.
Criterion:	Applying mathematics in real-life contexts

MULTIPLE CHOICE QUIZ   QUICK QUIZ

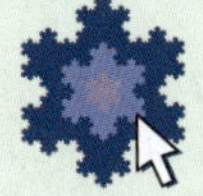

REVIEW SET 5A

1 Write as a percentage, rounded to 2 decimal places if necessary:

   a $\dfrac{3}{8}$ **b** 2.5   **c** $\dfrac{5}{12}$ **d** 0.015   **e** $\dfrac{1}{40}$ **f** $\dfrac{8}{9}$

2 Write as a decimal:

   a 83%   **b** 27.4%   **c** 152%   **d** 0.4%

3 Write as a fraction in lowest terms:

   a 48%   **b** 15%   **c** $5\frac{1}{2}\%$ **d** 0.1%

4 Vince brought 25 pumpkins to the market, and sold 18 of them. What percentage of his pumpkins did Vince sell?

5 An elephant eats 5% of its body weight in vegetation each day. Find the weight in kilograms of vegetation eaten each day by:

   a a 3 tonne elephant

   b a 4.8 tonne elephant.

6 A hospital has 36 doctors who make up 24% of the total hospital staff. How many people work at the hospital?

7 A group of hotel guests were asked to rate the hotel's performance in a number of categories. The results are presented in the table below:

	Excellent	Good	Fair	Poor
Cleanliness	20%	31%	35%	14%
Staff	32%	27%	15%	26%
Restaurant	9%	19%	37%	35%
Value for money	12%	16%	28%	44%
Location	33%	39%	18%	10%
Facilities	27%	34%	26%	13%

   a What percentage of guests rated the hotel's staff as either "Good" or "Excellent"?

   b What fraction of guests rated the hotel's cleanliness as "Fair"?

   c What was the most common response when rating the hotel's value for money?

   d 130 guests rated the hotel's facilities as "Fair". How many guests were surveyed?

8 Find the multiplier for:

   a an increase of 45%   **b** a decrease of 75%   **c** an increase of 227%.

9 Increase 512 kg by 25%.

10 Find the percentage change when:

   a 75 points is increased to 80 points   **b** 51 230 votes decreases to 38 975 votes.

11 Due to a cyclone, the price of bananas rose 85% to $9.99 per kg. How much did bananas cost before the cyclone?

12 A football stadium normally has maximum capacity 60 000 people. However, one stand is under renovation, so the stadium's capacity has been reduced by 8.5%. How many people can the stadium currently hold?

13 Julie bought a phone for $600, then sold it three years later for $120.

    a Find Julie's profit or loss.

    b Find the profit or loss as a percentage of the cost price.

14 Find the selling price of a book that is bought for $60 and sold at a 30% profit.

15 A lawnmower with marked price $595 has been discounted to $399. Find the percentage discount applied.

16 A bottle of perfume in Singapore sells for $64.20 including 7% GST. Find the price of the perfume before the tax was added.

REVIEW SET 5B

1 Copy and complete:

	Percentage	Decimal	Fraction (in lowest terms)
a	32%		
b			$\frac{7}{8}$

2 86% of trees in a forest were damaged in a forest fire. What fraction of the trees were damaged?

3 From a class of 30 French students, 6 had visited Germany. What percentage of the class had visited Germany?

4 Increase 230 g by 120%.

5 In a mathematics competition, the top 0.3% of participants are awarded a prize. Last year 600 000 students took part. How many students were awarded a prize?

6 A class of 25 students had a maths test out of 40 marks. 24% of students gained an A, and 8 students gained a B.

    a How many students gained an A?

    b What percentage of the class gained a B?

    c Carrie got 25 out of 40 for her test. Express this as a percentage.

7 Anwen scored 82% in her last test. Her friend Bree scored 76%, which was 38 marks.

    a How many marks did Anwen score?   **b** How many marks were in the test?

8 The income of a bicycle shop last month was $8614. This month, the income fell by 24%.

    a Find the decrease in income this month.

    b Find the actual income this month.

9 Find the percentage change when a bird bath is filled with 2.8 L of water, but only 2.5 L is left at the end of the day.

10 A guitar was bought for $250 and sold for $295. Find the profit as a percentage of the cost price.

11 As part of a road safety campaign, fines for all traffic offences will increase by 5%.

Copy and complete the table alongside, showing the changes to each fine.

Offence	Old fine	New fine
Speeding	$200	
Drink driving	$840	
Not wearing seatbelt		$273
Illegal parking		$52.50

12 In 2012, the price of a theatre ticket was $50. The price rose 5% in 2013, and then a further 8% in 2014. Find the cost of a theatre ticket in:

    a 2013   **b** 2014.

13 A pair of jeans has a marked price of $80. If a 25% discount is offered, find the actual selling price.

14 **a** At tennis practice one day, Marcos hit only 40% of his serves in. If Marcos served 70 times, how many went in?

    b Marcos decides to go to a specialist serving clinic. He has two options to choose from: the Tennis World clinic is offering 20% discount off $480, and SportsLife is offering 15% discount off $450. Which clinic is cheaper?

    c As a result of the clinic, Marcos' serving speed increased by 8%. If Marcos served at 189 km/h after attending the clinic, what was his serving speed before the clinic?

15 A tour company has organised a bus tour around Europe. So far, 22 Australians, 5 Americans, and 3 Canadians have booked.

    a What percentage of the passengers are Australians?

    b A new booking has confirmed 12 New Zealanders will be on the tour.

        i Find the percentage increase in passengers.

        ii What percentage of the passengers are Canadian?

    c The Australians booking through an Australian agency must pay $2600 plus 10% GST per person. What is the total amount each Australian must pay?

16 Find the percentage VAT which has been added to a souvenir from Rio de Janeiro if the tax increases its price from R$17.00 to R$19.89.

Chapter 6

Laws of algebra

Contents:

OPENING PROBLEM

Rhonda notices that when she multiplies the square numbers 4 and 9 together, the result 36 is also a square number.

Things to think about:

- Is this just a coincidence, or will this always happen when two square numbers are multiplied together?
- Can you explain *why* this occurs?

We have seen how algebraic symbols or **variables** can be used to represent unknown numbers.

We perform operations in algebra using the same principles as we use with numbers. In fact, the use of algebra is particularly valuable for *generalising* properties we observe with numbers. These general results are **algebraic laws**.

A   EXPONENT LAWS

We have seen how exponent notation can be used to represent repeated multiplication of the same number:

If n is a positive integer, then a^n is the product of n factors of a.

$$a^n = \underbrace{a \times a \times a \times \text{....} \times a}_{n \text{ factors}}$$

INVESTIGATION 1   EXPONENT LAWS

We can discover some laws for exponents by considering several examples and looking for patterns.

WORKSHEET

What to do:

1 Consider the multiplication $3^2 \times 3^4 = (3 \times 3) \times (3 \times 3 \times 3 \times 3) = 3^6$.
   Copy and complete:

   a $2^2 \times 2^3 = (2 \times 2) \times (2 \times 2 \times 2)\ =\ \boxed{}$

   b $3^3 \times 3^1 = \boxed{}\ =\ \boxed{}$

   c $x^2 \times x^4 = \boxed{}\ =\ \boxed{}$

   d $a^4 \times a^2 = \boxed{}\ =\ \boxed{}$

   In general, $a^m \times a^n = \boxed{}$.

2 Consider the division $\dfrac{2^5}{2^3} = \dfrac{2 \times 2 \times 2 \times 2 \times 2}{2 \times 2 \times 2} = 2^2$.

Copy and complete:

a $\dfrac{5^4}{5^1} = \boxed{} = \boxed{}$ **b** $\dfrac{3^6}{3^4} = \boxed{} = \boxed{}$

c $\dfrac{a^5}{a^2} = \boxed{} = \boxed{}$ **d** $\dfrac{x^7}{x^4} = \boxed{} = \boxed{}$

In general, $\dfrac{a^m}{a^n} = \boxed{}$.

3 Consider the power $(2^2)^3 = 2^2 \times 2^2 \times 2^2 = (2 \times 2) \times (2 \times 2) \times (2 \times 2) = 2^6$.

Copy and complete:

a $(7^3)^2 = 7^3 \times 7^3 \ = \boxed{} = \boxed{}$

b $(3^2)^4 = \boxed{} = \boxed{} = \boxed{}$

c $(a^4)^3 = \boxed{} = \boxed{} = \boxed{}$

d $(x^3)^3 = \boxed{} = \boxed{} = \boxed{}$

In general, $(a^m)^n = \boxed{}$.

From the **Investigation** you should have found these **exponent laws** for **positive exponents**:

If m and n are positive integers, then:

- $a^m \times a^n = a^{m+n}$ To **multiply** numbers with the **same base**, keep the base and **add** the exponents.

- $\dfrac{a^m}{a^n} = a^{m-n}, \quad a \neq 0$ To **divide** numbers with the **same base**, keep the base and **subtract** the exponents.

- $(a^m)^n = a^{m \times n}$ When **raising** a **power** to a **power**, keep the base and **multiply** the exponents.

Example 1   ◀) Self Tutor

Simplify using the exponent laws:

a $2^3 \times 2^2$ **b** $x^4 \times x^5$

a $\quad 2^3 \times 2^2$	**b** $\quad x^4 \times x^5$
$\quad = 2^{3+2}$	$\quad = x^{4+5}$
$\quad = 2^5$	$\quad = x^9$
$\quad = 32$	

EXERCISE 6A

1 Simplify using the exponent laws:

  a $3^2 \times 3^3$ **b** $2^2 \times 2^2$ **c** $5^2 \times 5^4$ **d** $3^4 \times 3^3$

  e $a \times a^4$ **f** $n^2 \times n$ **g** $x^3 \times x^6$ **h** $y^3 \times y^5$

Example 2   🔊 **Self Tutor**

Simplify using the exponent laws:

  a $\dfrac{3^5}{3^3}$ **b** $\dfrac{p^7}{p^3}$

  a $\quad \dfrac{3^5}{3^3}$

  $= 3^{5-3}$

  $= 3^2$

  $= 9$

  b $\quad \dfrac{p^7}{p^3}$

  $= p^{7-3}$

  $= p^4$

2 Simplify using the exponent laws:

  a $\dfrac{2^4}{2^2}$ **b** $\dfrac{3^3}{3}$ **c** $\dfrac{7^5}{7^2}$ **d** $\dfrac{10^4}{10^3}$

  e $\dfrac{x^6}{x^2}$ **f** $\dfrac{y^9}{y^5}$ **g** $c^6 \div c^4$ **h** $b^8 \div b^5$

Example 3   🔊 **Self Tutor**

Simplify using the exponent laws:

  a $(2^3)^2$ **b** $(x^4)^5$

  a $\quad (2^3)^2$

  $= 2^{3 \times 2}$

  $= 2^6$

  $= 64$

  b $\quad (x^4)^5$

  $= x^{4 \times 5}$

  $= x^{20}$

---

3 Simplify using the exponent laws:

  a $(2^2)^3$ **b** $(10^4)^2$ **c** $(3^3)^2$ **d** $(2^4)^3$

  e $(x^5)^2$ **f** $(p^3)^3$ **g** $(t^3)^4$ **h** $(z^2)^6$

4 Simplify using the exponent laws:

  a $c^2 \times c^4$ **b** $b^8 \div b^3$ **c** $(y^5)^3$ **d** $y^4 \times y^6$

  e $q \times q^6$ **f** $(z^6)^5$ **g** $t^{10} \div t^7$ **h** $a^2 \times a^n$

  i $g^4 \div g$ **j** $n^2 \times n^3 \times n^5$ **k** $(k^4)^2 \div k$ **l** $(p^2)^2 \times p^2$

5 Copy and complete, replacing each $\square$ and $\triangle$ with a number or operation:

   a $(7^2)^{\square} = 7^6$
   b $2^4 \square 2 = 2^3$
   c $2^2 \square 2^7 = 2^9$

   d $(x^{\square})^4 = x^{12}$
   e $a^5 \square a^5 = a^{10}$
   f $5^9 \square 5^3 = 5^6$

   g $b^4 \square b^3 \triangle b^2 = b^5$
   h $c^8 \square c^2 \triangle c^3 = c^3$
   i $(x^3)^{\square} = x^{10} \triangle x^4$

Example 4   ◀ΰ **Self Tutor**

Simplify using the exponent laws:

  a $2b^2 \times 3a^2b^3$
      b $\dfrac{6x^4y^3}{3x^2y}$

  a $2b^2 \times 3a^2b^3$
$= 2 \times 3 \times a^2 \times b^2 \times b^3$
$= 6 \times a^2 \times b^{2+3}$
$= 6a^2b^5$

  b $\dfrac{6x^4y^3}{3x^2y}$
$= \dfrac{6}{3} \times x^{4-2} \times y^{3-1}$
$= 2x^2y^2$

---

6 Simplify using the exponent laws:

   a $\dfrac{5a^3}{a}$
   b $3q^2 \times 5q$
   c $8x^2y \times 2xy^3$

   d $\dfrac{21t^3}{3t^2}$
   e $\dfrac{24w^2z^4}{6w^2z^3}$
   f $\dfrac{j^5k^3}{j^4k}$

   g $\dfrac{3x^5}{6x^3} \times 2x^2$
   h $\dfrac{m^{30}}{(m^5)^5}$
   i $\dfrac{h^{14} \times h^2}{(h^4)^2}$

7 Express as a power of 2:

   a $2^2 \times 2^5$
   b $2^3 \times (2^2)^4$

   c $\dfrac{2^3 \times 2^7}{2 \times 2^2}$
   d $\dfrac{(2^{11})^2 \times 2^3}{(2^6)^3 \times 2^4}$

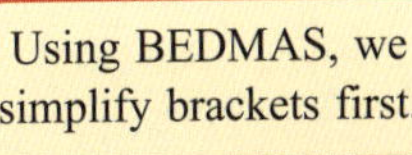

8 Express as a power of x:

   a $x^4 \times x^9$
   b $x^2 \times (x^3)^2$

   c $\dfrac{x^{11}}{x^2 \times x^3}$
   d $(x^6 \times x^7)^2$

   e $\dfrac{x^{10} \times (x^2)^5}{x}$
   f $\dfrac{(x^2)^4 \times x^5}{x^3 \times x^4}$

9 Describe the error in each statement, then write the statement correctly:

   a $2^5 \times 2^3 = 4^8$
   b $(3^2)^3 = 3^5$
   c $x^5 \times x^3 = x^{15}$

   d $\dfrac{x^{12}}{x^3} = x^4$
   e $\dfrac{3^5}{3^3} = 1^2$

B   EXPANSION LAWS

We next consider exponent laws for raising a product or quotient to a power.

For example, given $(3a)^4$ or $\left(\dfrac{2}{y}\right)^4$, we need laws which allow us to write the expressions without brackets. We call these **expansion laws**.

INVESTIGATION 2   EXPANSION LAWS

Look for any patterns as you complete the following Investigation.

WORKSHEET

What to do:

1 Consider the expansion

$$(3a)^4 = 3a \times 3a \times 3a \times 3a = 3 \times 3 \times 3 \times 3 \times a \times a \times a \times a = 81a^4.$$

Copy and complete:

a $(ab)^4 = ab \times ab \times ab \times ab = a \times a \times a \times a \times b \times b \times b \times b = \boxed{}$

b $(ab)^3 = \boxed{} = \boxed{} = \boxed{}$

c $(2a)^5 = \boxed{} = \boxed{} = \boxed{}$

d $(5x)^3 = \boxed{} = \boxed{} = \boxed{}$

In general, $(ab)^n = \boxed{}$.

2 Consider the expansion $\left(\dfrac{2}{y}\right)^4 = \dfrac{2}{y} \times \dfrac{2}{y} \times \dfrac{2}{y} \times \dfrac{2}{y} = \dfrac{2 \times 2 \times 2 \times 2}{y \times y \times y \times y} = \dfrac{16}{y^4}$.

Copy and complete:

a $\left(\dfrac{a}{b}\right)^3 = \dfrac{a}{b} \times \dfrac{a}{b} \times \dfrac{a}{b} = \dfrac{a \times a \times a}{b \times b \times b} = \boxed{}$

b $\left(\dfrac{a}{b}\right)^4 = \boxed{} = \boxed{} = \boxed{}$

c $\left(\dfrac{5}{a}\right)^2 = \boxed{} = \boxed{} = \boxed{}$

d $\left(\dfrac{3}{x}\right)^3 = \boxed{} = \boxed{} = \boxed{}$

In general, $\left(\dfrac{a}{b}\right)^n = \boxed{}$ provided $b \neq 0$.

From **Investigation 2** you should have found these **expansion laws** for **positive exponents**:

If n is a positive integer, then:

- $(ab)^n = a^n b^n$

- $\left(\dfrac{a}{b}\right)^n = \dfrac{a^n}{b^n}$ provided $b \neq 0$.

DISCUSSION

Why do we state "provided $b \neq 0$"?

Example 5

🔊 **Self Tutor**

Expand the brackets and simplify:

a $(ab)^5$ **b** $(2xy)^3$

a $(ab)^5 = a^5 b^5$ **b** $(2xy)^3$
$= 2^3 \times x^3 \times y^3$
$= 8x^3 y^3$

EXERCISE 6B

1 Expand the brackets and simplify:

a $(pq)^2$ **b** $(xy)^4$ **c** $(ab)^6$ **d** $(abc)^3$

e $(2a)^3$ **f** $(3d)^5$ **g** $(2k)^5$ **h** $(5gh)^2$

Example 6

🔊 **Self Tutor**

Expand the brackets and simplify:

a $\left(\dfrac{m}{n}\right)^4$ **b** $\left(\dfrac{2}{b}\right)^3$

a $\left(\dfrac{m}{n}\right)^4 = \dfrac{m^4}{n^4}$ **b** $\left(\dfrac{2}{b}\right)^3 = \dfrac{2^3}{b^3}$
$= \dfrac{8}{b^3}$

2 Expand the brackets and simplify:

a $\left(\dfrac{a}{b}\right)^2$ **b** $\left(\dfrac{b}{2}\right)^3$ **c** $\left(\dfrac{j}{k}\right)^4$ **d** $\left(\dfrac{2}{z}\right)^4$

e $\left(\dfrac{4}{x}\right)^2$ **f** $\left(\dfrac{2}{b}\right)^5$ **g** $\left(\dfrac{q}{2}\right)^4$ **h** $\left(\dfrac{3}{b}\right)^3$

3 Find, using the expansion laws:

a $\left(\dfrac{2}{5}\right)^2$ b $\left(\dfrac{3}{4}\right)^3$ c $\left(\dfrac{2}{3}\right)^4$ d $\left(\dfrac{1}{2}\right)^5$

Example 7   ◀)) **Self Tutor**

Write in simplest form, without brackets:

a $(3a^3b)^4$ b $\left(\dfrac{x^2}{2y}\right)^3$

---

a $(3a^3b)^4 = 3^4 \times (a^3)^4 \times b^4$ b $\left(\dfrac{x^2}{2y}\right)^3 = \dfrac{(x^2)^3}{2^3 \times y^3}$

$ = 81 \times a^{3\times 4} \times b^4$ $\phantom{b \left(\dfrac{x^2}{2y}\right)^3} = \dfrac{x^{2\times 3}}{8 \times y^3}$

$ = 81a^{12}b^4$ $\phantom{b \left(\dfrac{x^2}{2y}\right)^3} = \dfrac{x^6}{8y^3}$

---

4 Write in simplest form, without brackets:

a $(2a^2)^2$ b $(3b^3)^3$ c $(2c^2)^4$ d $(2d^2)^5$

e $(jk^3)^2$ f $(xy^2)^3$ g $(3g)^2 \times 2g$ h $(5r^2s)^2$

i $(8ab^3)^2$ j $(3h^2k)^3$ k $(2bc^2)^3$ l $(7e^3f)^2$

5 Write in simplest form, without brackets:

a $\left(\dfrac{jk}{2}\right)^2$ b $\left(\dfrac{2}{cd}\right)^2$ c $\left(\dfrac{3p}{q}\right)^3$ d $\left(\dfrac{z^2}{5}\right)^2$

e $\left(\dfrac{7d^2}{e}\right)^2$ f $\left(\dfrac{w^2}{3v}\right)^2$ g $\left(\dfrac{4r}{3s^2}\right)^2$ h $\left(\dfrac{5g}{2h^3}\right)^3$

6 Consider the **Opening Problem** on page **118**. Show that when two square numbers are multiplied together, the result is also a perfect square.

7 Suppose two cubic numbers are multiplied together. Is the result a cubic number? Explain your answer.

C   THE ZERO EXPONENT LAW

For all positive integers n, a^n is defined as the product of n factors of a:

$$a^n = \underbrace{a \times a \times a \times \ldots \times a}_{n \text{ factors}}$$

But what if $n = 0$? In the following **Investigation** we will discover how to find values such as 2^0 and 5^0.

INVESTIGATION 3   THE ZERO EXPONENT LAW

What to do:

1 Find the value of:

   a $\dfrac{2}{2}$ **b** $\dfrac{3}{3}$ **c** $\dfrac{-5}{-5}$ **d** $\dfrac{57}{57}$ **e** $\dfrac{-23}{-23}$

2 Copy and complete:

    When a non-zero value is divided by itself, the result is always $\boxed{}$.

    For any $a \neq 0$, $\dfrac{a^3}{a^3} = \boxed{}$.

3 Use an exponent law to show that $\dfrac{a^3}{a^3} = a^0$.

4 Hence complete: $a^0 = \boxed{}$ for all $a \neq 0$.

5 Check your answer by evaluating 2^0 and 5^0 on your calculator.

From the **Investigation** you should have discovered that $\boxed{a^0 = 1 \quad \text{for all} \quad a \neq 0.}$

DISCUSSION

Why do we state "for all $a \neq 0$"?

EXERCISE 6C

1 Simplify:

   a 7^0 **b** 41^0 **c** x^0 **d** 5×2^0

   e $8 + 10^0$ **f** $6 - 6^0$ **g** 11×11^0 **h** $p^6 \times p^0$

   i $(2^3)^0$ **j** $(2^0)^3$ **k** 7×3^0 **l** $(7 \times 3)^0$

2 Simplify:

   a $\dfrac{n^2}{n^2}$ **b** $\dfrac{k^6}{k^6}$ **c** $\dfrac{xy}{y}$ **d** $\dfrac{6x}{2x}$

   e $\dfrac{a^3 b^2}{b^2}$ **f** $\dfrac{m^3 n^4}{m^3 n^2}$ **g** $\dfrac{pq^3 r^2}{pq^2 r^2}$ **h** $\dfrac{2x^2 y^3}{8x^2 y}$

PUZZLE

Click on the icon to print a crossword puzzle for practising the exponent laws.   **CROSSWORD**

D   THE NEGATIVE EXPONENT LAW

INVESTIGATION 4   NEGATIVE EXPONENTS

What to do:

1 Consider the fraction $\dfrac{7^3}{7^5}$.

   a By expanding and then cancelling common factors, show that $\dfrac{7^3}{7^5} = \dfrac{1}{7^2}$.

   b Use an exponent law to show that $\dfrac{7^3}{7^5} = 7^{-2}$.

   c Hence copy and complete: $7^{-2} = \ldots\ldots$

2 Use the fact that $a^0 = 1$ to copy and complete: $a^{-n} = a^{0-n} = \dfrac{a^0}{\square} = \dfrac{\square}{\square}$.

You should have discovered the following law for **negative exponents**:

> If a is any non-zero number and n is an integer, then $a^{-n} = \dfrac{1}{a^n}$.
>
> This means that a^n and a^{-n} are **reciprocals** of one another.
>
> In particular, notice that $a^{-1} = \dfrac{1}{a}$.

Example 8   ◀)) **Self Tutor**

Simplify:

   a 3^{-1} **b** 5^{-2}

   a $3^{-1} = \dfrac{1}{3^1}$ **b** $5^{-2} = \dfrac{1}{5^2}$

         $= \dfrac{1}{3}$ $= \dfrac{1}{25}$

EXERCISE 6D

1 Simplify:

  a 5^{-1} **b** 4^{-1} **c** 8^{-1} **d** 3^{-2} **e** 2^{-2}

  f 11^{-2} **g** 7^{-2} **h** 3^{-3} **i** 2^{-5} **j** 2^{-7}

2 Write as a fraction and hence as a decimal:

  a 10^{-1} **b** 10^{-2} **c** 10^{-3} **d** 10^{-4} **e** 10^{-5}

3 Simplify:

  a $3^1 - 3^{-1}$ **b** $3^0 + 3^{-1}$ **c** $7^0 - 7^{-1}$ **d** $5^0 + 5^1 - 5^{-1}$

4 Write as a fraction:

  a x^{-1} **b** k^{-1} **c** a^{-2} **d** t^{-3} **e** r^{-5}

5 Write as a power with a negative exponent:

a $\dfrac{3}{3^3}$ **b** $\dfrac{5^2}{5^3}$ **c** $\dfrac{2^3}{2^5}$ **d** $\dfrac{3^2}{3^4}$ **e** $\dfrac{7^4}{7^5}$

Example 9   ◀)) **Self Tutor**

Write without brackets or negative exponents:

a $8ab^{-1}$ **b** $8(ab)^{-1}$

---

a $8ab^{-1} = 8a \times \dfrac{1}{b}$ **b** $8(ab)^{-1} = 8 \times \dfrac{1}{ab}$

$\qquad\quad = \dfrac{8a}{b}$ $\qquad\qquad = \dfrac{8}{ab}$

---

6 Write without brackets or negative exponents:

a $2x^{-1}$ **b** $(2x)^{-1}$ **c** st^{-1} **d** $(st)^{-1}$

e $7a^{-2}$ **f** $(5z)^{-2}$ **g** gh^{-3} **h** $(gh)^{-3}$

i $4cd^{-2}$ **j** $(4cd)^{-2}$ **k** $4(cd)^{-2}$

7 Simplify, writing your answer without negative exponents:

a $x \times x^{-1}$ **b** $x^2 \times x^{-1}$ **c** $x^{-2} \times x$ **d** $x^{-2} \times x^3$

INVESTIGATION 5   A FRACTION TO A NEGATIVE EXPONENT

We have seen that a^n and a^{-n} are reciprocals of one another, and in particular that $a^{-1} = \dfrac{1}{a}$ is the reciprocal of a.

What to do:

1 Since $a^{-1} = \dfrac{1}{a} = 1 \div a$, we can write $\left(\dfrac{2}{3}\right)^{-1} = 1 \div \dfrac{2}{3} = 1 \times \dfrac{3}{2} = \dfrac{3}{2}$.

Write as a fraction without exponents:

a $\left(\dfrac{1}{4}\right)^{-1}$ **b** $\left(\dfrac{3}{5}\right)^{-1}$ **c** $\left(\dfrac{2}{7}\right)^{-1}$ **d** $\left(\dfrac{a}{b}\right)^{-1}$

2 Using the exponent laws and **1**, $\left(\dfrac{2}{3}\right)^{-2} = \left(\left(\dfrac{2}{3}\right)^{-1}\right)^2 = \left(\dfrac{3}{2}\right)^2$.

Write as a fraction raised to a *positive* exponent:

a $\left(\dfrac{1}{4}\right)^{-2}$ **b** $\left(\dfrac{3}{5}\right)^{-2}$ **c** $\left(\dfrac{2}{7}\right)^{-2}$ **d** $\left(\dfrac{a}{b}\right)^{-2}$

e $\left(\dfrac{1}{4}\right)^{-3}$ **f** $\left(\dfrac{3}{5}\right)^{-3}$ **g** $\left(\dfrac{2}{7}\right)^{-3}$ **h** $\left(\dfrac{a}{b}\right)^{-3}$

3 If n is a positive integer, write $\left(\dfrac{a}{b}\right)^{-n}$ as a fraction raised to a *positive* exponent.

 E   THE DISTRIBUTIVE LAW

When we write real life problems in terms of algebra, we often obtain expressions containing brackets.

In order to work with these expressions, we often want to rewrite them without brackets. The process of writing expressions without brackets is called **expansion**.

To expand the brackets in an expression like $5(x + 3)$ or $x(x - 2)$, we use a rule called the **distributive law**.

INVESTIGATION 6   THE DISTRIBUTIVE LAW

What to do:

1 Copy and complete:

a $2 \times (5 + 3) = 2 \times 8$ and $2 \times 5 + 2 \times 3 = 10 + 6$

$= \boxed{}$ $= \boxed{}$

b $4 \times (5 + 3) = 4 \times \boxed{}$ and $4 \times 5 + 4 \times 3 = \boxed{} + \boxed{}$

$= \boxed{}$ $= \boxed{}$

c $7 \times (5 + 3) = \boxed{} \times \boxed{}$ and $7 \times 5 + 7 \times 3 = \boxed{} + \boxed{}$

$= \boxed{}$ $= \boxed{}$

d $a \times (5 + 3) = \boxed{} \times \boxed{}$ and $a \times 5 + a \times 3 = \boxed{} + \boxed{}$

$= \boxed{}$ $= \boxed{}$

2 Calculate:

a $3 \times (2 + 4)$ and $3 \times 2 + 3 \times 4$ **b** $3 \times (6 + 1)$ and $3 \times 6 + 3 \times 1$

c $3 \times (5 + 3)$ and $3 \times 5 + 3 \times 3$

Comment on your results.

3 Calculate:

a $5 \times (6 - 2)$ and $5 \times 6 + 5 \times (-2)$ **b** $5 \times (7 - 1)$ and $5 \times 7 + 5 \times (-1)$

c $5 \times (5 - 3)$ and $5 \times 5 + 5 \times (-3)$

Comment on your results.

4 Using your observations from **1**, **2**, and **3**, what do you suspect about $a(b+c)$ and $ab+ac$?

From the **Investigation** you should have discovered the **distributive law**:

> When a coefficient is multiplied by an expression in brackets, we remove the brackets by multiplying each term inside them by the coefficient, then adding the results.
>
> $$a(b + c) = ab + ac$$

We can explain why the distributive law works using geometry:

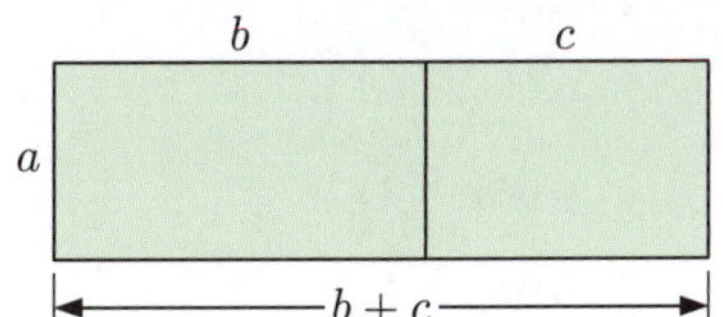

The overall area is $a(b + c)$.

However, this could also be found by adding the areas of the two small rectangles: $ab + ac$.

So, $a(b + c) = ab + ac$.   {equating areas}

Example 10   ◀》 **Self Tutor**

Expand and simplify:

a $5(x + 4)$

b $4(y - 3)$

a $5(x + 4)$
$= 5 \times x + 5 \times 4$
$= 5x + 20$

b $4(y - 3)$
$= 4(y + {-3})$
$= 4 \times y + 4 \times (-3)$
$= 4y - 12$

Multiply each term inside the brackets by the factor outside the brackets.

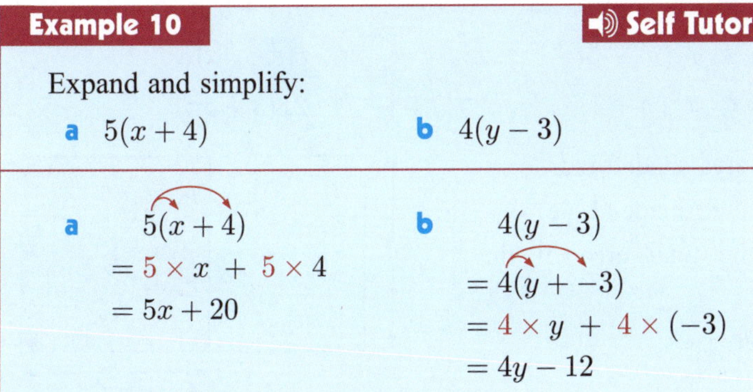

EXERCISE 6E

1 Expand and simplify:

a $2(x + 7)$	**b** $3(x - 2)$	**c** $4(a + 3)$	**d** $5(a + c)$
e $6(b - 3)$	**f** $7(m + 4)$	**g** $2(n - p)$	**h** $4(p - q)$
i $3(5 + x)$	**j** $5(y - x)$	**k** $8(t - 8)$	**l** $6(d + e)$
m $4(10 - j)$	**n** $7(y + n)$	**o** $2(n - 12)$	**p** $8(11 - d)$

Example 11   ◀》 **Self Tutor**

Expand and simplify:

a $3(2a + 7)$

b $2x(2 - 3x)$

a $3(2a + 7)$
$= 3 \times 2a + 3 \times 7$
$= 6a + 21$

b $2x(2 - 3x)$
$= 2x(2 + {-3x})$
$= 2x \times 2 + 2x \times (-3x)$
$= 4x - 6x^2$

2 Expand and simplify:

a	$9(2x+1)$	**b**	$3(1-3x)$	**c**	$5(2a+3)$
d	$11(1-2n)$	**e**	$6(3x+y)$	**f**	$5(x-2y)$
g	$4(3b+c)$	**h**	$2(a-2b)$	**i**	$7(a-5b)$
j	$12(2+3d)$	**k**	$8(3-4y)$	**l**	$6(5b+3a)$
m	$11(2x-y)$	**n**	$7(c-9d)$	**o**	$6(m+7n)$
p	$8(8a-c)$	**q**	$4(x^2-1)$	**r**	$5(2+3a^2)$

3 Expand and simplify:

a	$x(x+2)$	**b**	$x(5-x)$	**c**	$a(2a+4)$
d	$b(5-3b)$	**e**	$a(b+2c)$	**f**	$a(a^2+1)$
g	$x(3-4x)$	**h**	$3x(6-x)$	**i**	$5x(x-4)$
j	$4a(1-a)$	**k**	$7b(b+2)$	**l**	$a(a+b)$
m	$b(3-8b)$	**n**	$m(m-3n)$	**o**	$c(c-4a)$
p	$6p(4-7p)$	**q**	$x^2(x+4)$	**r**	$2x(3-x^2)$

4

  a Write an expression for the area shaded red.

  b Write an expression for the area shaded blue.

  c Write an expression for the total area of the rectangle.

  d Explain why $a(b-c)=ab-ac$.

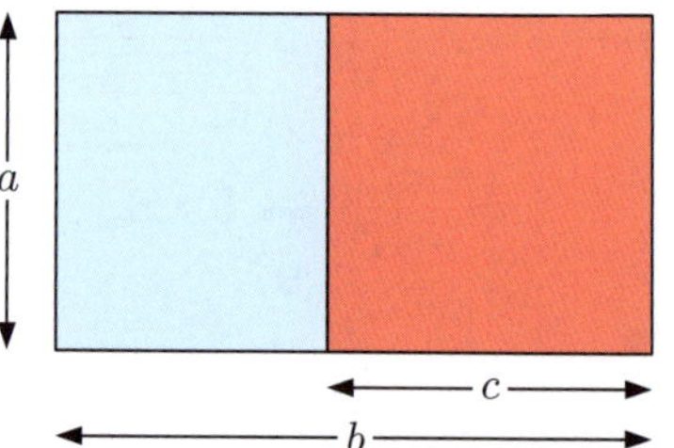

5 Expand and simplify:

a	$3(x+y+4)$	**b**	$7(2x+3z-y)$	**c**	$5(2l-4m+7n)$
d	$2a(a+b-8)$	**e**	$3x(x+2y-5)$	**f**	$4p(2p+3q+5)$

Example 12   ◄ঠ **Self Tutor**

Expand and simplify:

  a $-4(x+3)$ **b** $-(2x-4)$ **c** $-a(a+7)$

a $-4(x+3)$

$= -4 \times x + -4 \times 3$

$= -4x - 12$

b $-(2x-4)$

$= -1(2x-4)$

$= -1 \times 2x + -1 \times (-4)$

$= -2x + 4$

c $-a(a+7)$

$= -a \times a + -a \times 7$

$= -a^2 - 7a$

6 Expand and simplify:

    a $-2(x+2)$ **b** $-3(x+4)$ **c** $-4(x-2)$

    d $-5(5-x)$ **e** $-(a+2)$ **f** $-(x-3)$

    g $-(5-x)$ **h** $-(2x+1)$ **i** $-3(4-x)$

    j $-4(5x-2)$ **k** $-5(3-4c)$ **l** $-2(7-5x)$

7 Expand and simplify:

    a $-a(a+1)$ **b** $-b(b+4)$ **c** $-c(5-c)$

    d $-x(2x+4)$ **e** $-2x(1-x)$ **f** $-3y(y+2)$

    g $-4a(5-a)$ **h** $-6b(3-2b)$ **i** $-xy(2y-x)$

Example 13   ◄) **Self Tutor**

Expand and then simplify by collecting like terms:

    a $4+2(x+3)$ **b** $3x-2(2x-1)$

    a $\begin{aligned} 4+2(x+3) &= 4+2\times x+2\times 3 \\ &= 4+2x+6 \\ &= 2x+10 \end{aligned}$

    b $\begin{aligned} 3x-2(2x-1) &= 3x+-2(2x-1) \\ &= 3x+-2\times 2x+-2\times(-1) \\ &= 3x-4x+2 \\ &= -x+2 \end{aligned}$

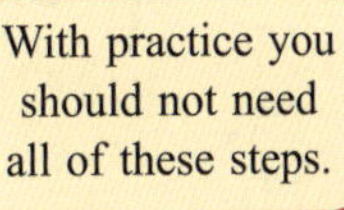

---

8 Expand and then simplify by collecting like terms:

    a $3(x+2)+5$ **b** $3x+2(2x+1)$ **c** $7-6(2x-3)$

    d $11x-(2+x)$ **e** $6+5(1-2x)$ **f** $11-(3-2x)$

    g $16-7(1-3x)$ **h** $x+6+3(4+x)$ **i** $8x+1+2(3-2x)$

    j $7-(1-2x)$ **k** $2x-(8+7x)+3$ **l** $8-5(11-3x)$

    m $5x+x(x+2)$ **n** $8x+x(x-1)$ **o** $7x-x(x+3)$

    p $x^2-x(2-x)$ **q** $4x-x(x-3)+2x^2$ **r** $3x^2-2x(x-5)-6x$

9 Expand and then simplify by collecting like terms:

    a $5(x+2)+3(6+x)$ **b** $2(4+x)+x(x+2)$

    c $2(x-3)+4(x+6)$ **d** $x(3-x)+2(x+5)$

    e $5(a+3)-2(4-a)$ **f** $-2(x+1)-x(2-x)$

    g $2a(a-2)+3(a+5)$ **h** $x(2x-3)+5x(x+2)$

    i $-2y(y+3)+5(y-1)$ **j** $x(2x+3)-2x(x+1)$

    k $-x(2-x)-2x(x-3)$ **l** $-(7-2n)-3(n+5)$

F   FACTORISATION

Factorisation is the process of writing an expression as a **product** of its **factors**.

Factorisation is the reverse process of expansion.

When we expand an expression, we remove its brackets.

When we factorise an expression, we insert brackets.

$3(x + 2)$ is the product of two factors, 3 and $(x + 2)$.

$$\text{expansion}$$
$$3(x + 2) = 3x + 6$$
$$\text{factorisation}$$

To factorise an algebraic expression involving a number of terms, we look for the HCF of the terms. We write it in front of a set of brackets, then use the reverse of the distributive law to complete the factorisation.

For example: $5x^2$ and $10xy$ have $HCF = 5x$.

$$\therefore \ 5x^2 + 10xy = 5x \times x + 5x \times 2y$$
$$= 5x(x + 2y)$$

An expression is **fully factorised** if none of its algebraic factors can be factorised further. This is why we must write the *highest* common factor in front of the brackets.

For example: $4a + 12 = 2(2a + 6)$ is not fully factorised since $(2a + 6)$ still has the common factor 2.

$4a + 12 = 4(a + 3)$ is fully factorised.

EXERCISE 6F

1   Copy and complete:

  a $\ 2x + 4 = 2(x + \text{......})$ **b** $\ 3a - 12 = 3(a - \text{......})$

  c $\ 15 - 5p = 5(\text{......} - p)$ **d** $\ 18x + 12 = 6(\text{......} + 2)$

  e $\ 4x^2 - 8x = 4x(x - \text{......})$ **f** $\ 2m + 8m^2 = 2m(\text{......} + 4m)$

  g $\ 4x + 16 = 4(\text{......} + \text{......})$ **h** $\ 10 + 5d = 5(\text{......} + \text{......})$

  i $\ 5c - 5 = 5(\text{......} - \text{......})$ **j** $\ cd - de = d(\text{......} - \text{......})$

  k $\ 6a + 8ab = \text{......}(3 + 4b)$ **l** $\ 6x - 2x^2 = \text{......}(3 - x)$

Example 14   ◀)) **Self Tutor**

Fully factorise:

  a $\ 3a + 6$ **b** $\ ab - 2bc$

  a $\quad 3a + 6$
      $= 3 \times a + 3 \times 2$
      $= 3(a + 2) \quad \{HCF = 3\}$

  b $\quad ab - 2bc$
      $= a \times b - 2 \times b \times c$
      $= b(a - 2c) \quad \{HCF = b\}$

2 Factorise:

  a $3x + 9$ **b** $4x + 12$ **c** $7x + 35$

  d $8x + 4$ **e** $10x + 15$ **f** $28b + 8$

  g $x + ax$ **h** $ab + ac$ **i** $xy + y$

3 Factorise:

  a $4x - 8$ **b** $5y - 10$ **c** $3b - 24$

  d $8a - 32$ **e** $10n - 15$ **f** $12a - 18$

  g $a - ab$ **h** $xy - xz$ **i** $6b - 2ab$

4 Factorise:

  a $3a + 3b$ **b** $8x - 16$ **c** $3p + 18$

  d $28 - 14x$ **e** $7x - 14$ **f** $12 + 6x$

  g $ac + bc$ **h** $12y - 6a$ **i** $5a + ab$

  j $bc - 6cd$ **k** $7x - xy$ **l** $a + ab$

  m $xy - yz$ **n** $3pq + pr$ **o** $cd - c$

Example 15   ◄ΰ **Self Tutor**

Factorise:

  a $3x^2 + 12x$ **b** $4y - 2y^2$

a $\quad 3x^2 + 12x$	**b** $\quad 4y - 2y^2$
$= 3 \times x \times x + 4 \times 3 \times x$	$= 2 \times 2 \times y - 2 \times y \times y$
$= 3x(x + 4) \quad \{HCF = 3x\}$	$= 2y(2 - y) \quad \{HCF = 2y\}$

---

5 Factorise:

  a $x^2 + 2x$ **b** $3x^2 + 12x$ **c** $2x^2 - 8x$

  d $9x - x^2$ **e** $18x - 9x^2$ **f** $14x - 6x^2$

  g $2x^3 + 6x^2$ **h** $3x^3 + 7x^2$ **i** $4ab^2 - 6a^2b$

6 Factorise:

  a $2x^3 + 4x^2 + 4x$ **b** $x^4 + 3x^3 + 6x^2$ **c** $ab^2 + ab + a^2b$

7 Explain why $x(3x - x^2)$ is not *fully* factorised.

MULTIPLE CHOICE QUIZ

QUICK QUIZ

INVESTIGATION 7 — FURTHER EXPANSION RULES

In this Investigation we see how the distributive law can be applied to products where we have two factors, each contained in brackets.

For example, consider this expansion:

$$(x + 5)(x - 3)$$
$$= (x + 5)(x + -3)$$
$$= (x + 5) \times x + (x + 5) \times (-3) \qquad \text{\{distributive law\}}$$
$$= x(x + 5) - 3(x + 5) \qquad \text{\{rearranging\}}$$
$$= x \times x + x \times 5 + -3 \times x + -3 \times 5 \qquad \text{\{distributive law\}}$$
$$= x^2 + 5x - 3x - 15$$
$$= x^2 + 2x - 15 \qquad \text{\{collecting like terms\}}$$

What to do:

1 Use the distributive law to expand each product, then simplify the result by collecting like terms:

   a $(x + 2)(x + 3)$ **b** $(x + 4)(x - 1)$ **c** $(x - 2)(x - 5)$

2 **a** Use the distributive law to expand $(a + b)(c + d)$ as a sum of four terms.

   b Use your expansion to check your answers to **1**.

3 **a** Use the distributive law to expand $(a + b)^2$, then simplify the result by collecting like terms.

   b Use your result to expand:

     i $(x + 2)^2$ **ii** $(x - 1)^2$ **iii** $(3 + a)^2$ **iv** $(2 - b)^2$

4 **a** Use the distributive law to expand $(a + b)(a - b)$, then simplify the result by collecting like terms.

   b Use your result to expand:

     i $(x + 2)(x - 2)$ **ii** $(1 + x)(1 - x)$ **iii** $(a - 3)(a + 3)$

5 Look back at the rules you have generated for expanding $(a + b)^2$ and $(a + b)(a - b)$. Use these rules to help you *factorise*:

   a $m^2 - 16$ **b** $m^2 + 6m + 9$ **c** $m^2 - 4m + 4$

REVIEW SET 6A

1 Simplify using the exponent laws:

   a $k^3 \times k^6$ **b** $(b^4)^3$ **c** $\dfrac{p^{13}}{p^5}$

2 Expand the brackets and simplify:

   a $(ab)^4$ **b** $(3z)^3$ **c** $(2xy)^3$

3 Simplify using the exponent laws:

   a $6g^5 \times 7g$ **b** $40a^2b^3 \div 5ab$ **c** $\dfrac{m^4 \times m^5}{(m^2)^3}$

4 Write as a power of 5:

   a $\dfrac{5^4 \times 5^3}{5^2}$ **b** $(5^4 \times 5^3)^2$ **c** 1

5 Expand the brackets and simplify:

   a $\left(\dfrac{c}{d}\right)^2$ **b** $\left(\dfrac{q}{4}\right)^3$ **c** $\left(\dfrac{ab}{8}\right)^2$

6 Evaluate:

   a 8^0 **b** 13×13^0 **c** $7 - 4^0$ **d** $3^0 - 3^{-1}$

7 Write as a fraction:

   a 9^{-1} **b** 6^{-2} **c** y^{-3} **d** $(xy)^{-2}$

8 Expand and simplify:

   a $7(x + 6)$ **b** $-5(x + 2)$ **c** $y(4y - 7)$

9 Expand and simplify:

   a $(q + 5)p$ **b** $2(a + 3b + 5)$ **c** $3x(x^2 + 7x - 5)$

10 Expand and simplify:

   a $6(x - 4) + 5x$ **b** $2(x - 4) + 3(x + 4)$ **c** $4(5 + a) + 5(a - 2)$

   d $-y(y - 3) - 5(y + 1)$ **e** $3x(x + 7) - 4(x - 9)$ **f** $p(p + 2) + 4p(p - 3)$

11 Factorise:

   a $4x + 20$ **b** $6x - 36$ **c** $7x - xy$

12 Factorise:

   a $a^2 + 7a$ **b** $12x^2 - 18x$ **c** $20x^3 + 15x^2$

   d $4x + 24y$ **e** $2x^2 - 8x$ **f** $3a + 6ab + 9a^2$

13 Consider the rug shown.

   a Write an expression for the area of:

      i one of the brown squares

      ii the white rectangle.

   b Write an expression for the total area of the rug.

   c Hence explain why $\;2a^2 + ab = a(2a + b)$.

REVIEW SET 6B

1 Simplify using the exponent laws:

   a $x^3 \times x^3$ **b** $\dfrac{c^{12}}{c^7}$ **c** $(d^{11})^3$

2 Copy and complete, replacing each $\square$ and $\triangle$ with a number or operation:

   a $3^2 \square 3^3 = 3^5$ **b** $(x^4)^{\square} = x^8$ **c** $a^5 \square a^2 \triangle a^3 = a^6$

3 Simplify using the exponent laws:

   a $2ab \times 5a^2$ **b** $\dfrac{15x^3y}{3x^2y}$ **c** $\dfrac{t^5 \times t^2}{(t^2)^3}$

4 Expand the brackets and simplify:

   a $(xy)^6$ **b** $\left(\dfrac{5a}{b}\right)^3$ **c** $(3x^2y)^3$

   d $(10xy^2)^2$ **e** $\left(\dfrac{m}{4n}\right)^3$ **f** $(2k^3)^2 \times k^{32}$

5 Simplify:

   a $\dfrac{m^3}{m^3}$ **b** $\dfrac{x^5y^3}{x^2y^3}$ **c** $\dfrac{a^2b^3c}{a^2bc}$

6 Write as a fraction:

   a 9^{-2} **b** 2^{-4} **c** 10^{-6} **d** x^{-3}

7 Write without brackets or negative exponents:

   a $(xy)^{-2}$ **b** $2(ab)^{-1}$ **c** $3xy^{-1}$ **d** $y \times y^{-2}$

8 **a** Find the result when the number 2 is:

     i tripled, then squared   **ii** squared, then tripled.

   b Find the result when the number -3 is:

     i tripled, then squared   **ii** squared, then tripled.

   c Hence copy and complete:

    When a number is tripled then squared, the result is ...... times larger than when the number is squared then tripled.

   d Prove algebraically that your statement in **c** is true for any number a.

9 Expand and simplify:

   a $3(x - 6)$ **b** $4(2x + 3y)$ **c** $-3x(x - 8)$

10 Expand and simplify:

   a $6x + 3(x + 9)$ **b** $10z - 3(z - 6)$ **c** $6x(x - 4) + 5x$

11 Expand and simplify:

   a $3(x - 2) + 3(x + 1)$ **b** $-y(y + 2) - 4(y - 1)$ **c** $5x(x + 6) - 7(x - 3)$

   d $3(x + 4) - 2(1 - x)$ **e** $-x(2x + 1) + 2(x - 3)$

12 Factorise:

   a $3x + 27$ **b** $8x - 48$ **c** $ab - ad$

13 Factorise:

   a $x^2 - 10x$ **b** $a^2 + 8a$ **c** $5y^2 + 30y$

   d $8x^2 - 24x$ **e** $10z^2 - 15z$ **f** $8x - 22x^2$

Equations

Contents:

OPENING PROBLEM

We are often faced with problems where we need to work out the value of an unknown quantity.

For example, consider the following questions:

   i   When 6 is added to a number, the result is 8. What is the number?

   ii   When 3 is subtracted from a number, the result is 11. What is the number?

   iii   When a number is multiplied by 4, the result is 28. What is the number?

   iv   When a number is divided by 5, the result is 3. What is the number?

Things to think about:

  a   Can you think of a real-world situation corresponding to each question?

  b   Can you summarise the information in each question as an *equation*?

  c   What techniques can you use to *solve* these problems?

Equations are a fundamental part of mathematics, and an important tool for problem solving.

The use of equations dates back to at least the 16th century BC, when the Egyptian scribe **Ahmes** or **Ahmose** recorded a series of problems which were solved using equations. These were discovered on the **Rhind papyrus** pictured, along with arithmetic, fractions, geometry, and other mathematics.

An **equation** is a mathematical statement which indicates that two expressions have the same value.

The expressions are connected by an **equal sign** $=$.

The **left hand side** (LHS) of an equation is on the left of the $=$ sign.

The **right hand side** (RHS) of an equation is on the right of the $=$ sign.

For example,
$$\underbrace{5x - 3}_{\text{LHS}} \underset{\text{equals}}{=} \underbrace{2x + 6}_{\text{RHS}}$$

A   SOLUTIONS OF AN EQUATION

A **solution** of an equation is a value of the unknown or variable which makes the equation true.

Consider the equation $5x - 3 = 2x + 6$.

When $x = 3$,   LHS $= 5(3) - 3$ and   RHS $= 2(3) + 6$

$$\begin{aligned}
\text{LHS} &= 15 - 3 & \text{RHS} &= 6 + 6 \\
&= 12 & &= 12
\end{aligned}$$

Since LHS $=$ RHS, $x = 3$ is a solution of the equation.

SOLVING BY INSPECTION

Some simple equations can be solved by **inspection**.

This means we look at the equation and compare it with known number facts, in order to find any solutions.

Example 1 ◀)) **Self Tutor**

Solve by inspection, if possible:

a $x - 7 = 4$ **b** $\dfrac{x}{3} = -2$ **c** $x^2 = 9$ **d** $x - 2 = x$

a $x - 7 = 4$

We know $11 - 7 = 4$

$\therefore\ x = 11$

b $\dfrac{x}{3} = -2$

We know $\dfrac{-6}{3} = -2$

$\therefore\ x = -6$

c $x^2 = 9$

We know $3^2 = 9$

and $(-3)^2 = 9$

$\therefore\ x = -3$ or 3

d $x - 2 = x$

The LHS is always 2 less than the RHS.

$\therefore$ there are no solutions.

NUMBER OF SOLUTIONS

From the Example above, we can see that an equation may have no solutions, one solution, or two solutions.

The equation $3x - x = 2x$ is true for *all* values of the variable x. We call it an *identity*.

> An **identity** is an equation which is true for all values of the variable.
>
> An identity has *infinitely many* solutions.

DISCUSSION

What do we mean by *infinitely many* solutions?

Can you write down some other equations which are identities?

EXERCISE 7A

1 Solve by inspection:

a $x + 2 = 3$ **b** $x - 7 = 2$ **c** $10 + x = 15$

d $3 - x = -2$ **e** $x - 4 = -6$ **f** $2x = 8$

g $-7x = 28$ **h** $4x = 52$ **i** $\dfrac{x}{8} = 5$

j $\dfrac{x}{5} = -35$ **k** $\dfrac{10}{x} = -2$ **l** $\dfrac{-12}{x} = 3$

2 One of the numbers in brackets is the solution of the equation. Find the solution.

a $3x + 8 = 14$ $\{0, 1, 2, 3\}$ **b** $5 - 3x = -4$ $\{0, 1, 2, 3\}$

c $7x + 3 = -11$ $\{-4, -3, -2, -1\}$ **d** $2x - 5 = x$ $\{-5, 0, 5, 10\}$

e $\dfrac{k}{4} = k + 3$ $\{-8, -4, 4, 8\}$ **f** $7 - 2m = m + 4$ $\{-2, -1, 0, 1, 2\}$

3 Match each equation to its solution(s):

a $x + 4 = 0$ **A** any value of x

b $4 + x = x - 4$ **B** $x = -4, 0,$ or 4

c $4 - x = 0$ **C** $x = -4$

d $4 + x = x + 4$ **D** $x = -4$ or 4

e $x^3 = 16x$ **E** $x = 4$

f $16 - x^2 = 0$ **F** no solutions

4 What real values of the variable, if any, make the equation true?

a $z \times z = z$ **b** $3 - k = 0$ **c** $h - h = 5$

d $x \times x = -1$ **e** $4 - q = q - 4$ **f** $d + d = d$

g $t \times 1 = t$ **h** $10 + g = g + 10$ **i** $w - w = w$

5 State whether each equation is true for:

  A *no* real values of x **B** *exactly one* real value of x

  C *exactly two* real values of x **D** *all* real values of x.

a $x + 2 = 6$ **b** $2x - 2x = 0$ **c** $\dfrac{x}{5} = 10$

d $x + 8 = 8 + x$ **e** $8 - x = 9$ **f** $x^2 = 4$

g $x^2 = 0$ **h** $x^2 = x$ **i** $9x - 7x = 2x$

j $x \times x = -9$ **k** $x \times -1 = x$ **l** $x + 2 = x - 2$

6 Determine whether each equation is an identity. If it is not, give a value of the variable for which the equation is false.

a $x + 6 = 6 + x$ **b** $k + k = k^2$ **c** $m - m = 0$

d $y + 4y = 5y$ **e** $8 - t = t - 8$ **f** $p \div p = 1$

g $6x - 2x = 4x$ **h** $x \times 1 = x$ **i** $5x - 4 = x$

B

MAINTAINING BALANCE

For any equation, the LHS must always equal the RHS. We can therefore think of an equation as a set of scales that must always be in **balance**.

The balance of an equation is maintained provided we perform the same operation on **both sides** of the equals sign.

For example, notice how the balance is maintained when we add 3 to both sides of the following scales:

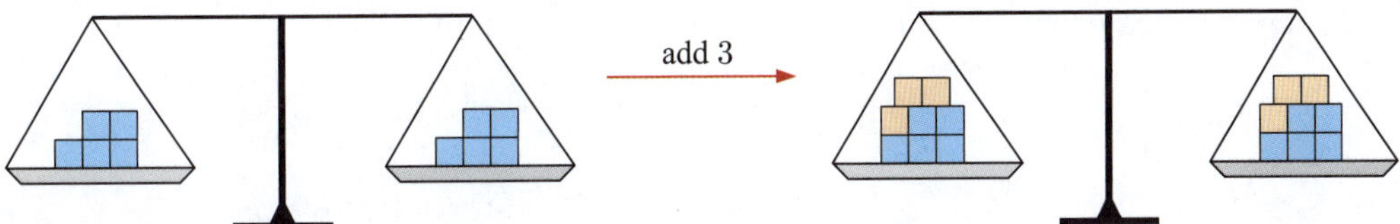

The balance of an equation will be maintained if we:

- add the same amount to both sides
- subtract the same amount from both sides
- multiply both sides by the same amount
- divide both sides by the same amount.

By maintaining the balance of an equation, we will not change its solutions.

Example 2   ◀ **Self Tutor**

Write down the equation which results when:

a 3 is added to both sides of $x - 3 = 8$

b 4 is subtracted from both sides of $2x + 4 = 18$

c both sides of $5x = 15$ are divided by 5

d both sides of $\dfrac{x}{4} = -7$ are multiplied by 4.

a
$$x - 3 = 8$$
$$\therefore \ x - 3 + 3 = 8 + 3$$
$$\therefore \ x = 11$$

b
$$2x + 4 = 18$$
$$\therefore \ 2x + 4 - 4 = 18 - 4$$
$$\therefore \ 2x = 14$$

c
$$5x = 15$$
$$\therefore \ \dfrac{5x}{5} = \dfrac{15}{5}$$
$$\therefore \ x = 3$$

d
$$\dfrac{x}{4} = -7$$
$$\therefore \ \dfrac{x}{4} \times 4 = -7 \times 4$$
$$\therefore \ x = -28$$

EXERCISE 7B

1 Write down the equation which results when we add:

a 3 to both sides of $x - 3 = 2$

b 9 to both sides of $x - 9 = 0$

c 4 to both sides of $5x - 4 = 11$

d 5 to both sides of $7x - 5 = x + 1$.

2 Write down the equation which results when we subtract:

a 1 from both sides of $x + 1 = 5$

b 6 from both sides of $2x + 6 = 10$

c 5 from both sides of $3x + 5 = 2$

d 9 from both sides of $4x + 9 = 3x + 11$.

3 Write down the equation which results when we multiply both sides of:

a $\dfrac{x}{2} = 8$ by 2

b $\dfrac{x-1}{5} = 1$ by 5

c $\dfrac{3x}{7} = 2$ by 7

d $\dfrac{3x-4}{4} = -10$ by 4.

4 Write down the equation which results when we divide both sides of:

a $4x = -40$ by 4

b $-2x = 18$ by -2

c $3(2-x) = 15$ by 3

d $-5(2x-1) = -55$ by -5.

5 Consider the equation $2x - 5 = 19$.

    a Show that $x = 12$ is a solution of the equation.

    b Write down the equation which results when 5 is added to both sides.

    c Use inspection to solve the equation in **b**.

    d Did adding 5 to both sides of the equation change its solution?

6 Consider the equation $\dfrac{x+1}{2} = 3$.

    a Show that $x = 5$ is a solution of the equation.

    b Write down the equation which results when both sides are multiplied by 2.

    c Use inspection to solve the equation in **b**.

    d Did multiplying both sides of the equation by 2 change its solution?

C — INVERSE OPERATIONS

When I woke up this morning there were 4 eggs in my refrigerator.

I found my chickens had already laid 2 eggs, so I then had $4 + 2 = 6$ eggs in total.

I fried 2 eggs to have with my breakfast, so I now have $6 - 2 = 4$ eggs left.

Adding 2 and subtracting 2 have the opposite effect. One *undoes* the other.

> Addition and subtraction are **inverse operations**.

Over the next week my chickens lay very well. The number of eggs in my fridge doubles, so I now have $4 \times 2 = 8$ eggs.

On the weekend I bake some cakes. Only half of the eggs now remain. This is $8 \div 2 = 4$ eggs.

Multiplying by 2 and dividing by 2 have the opposite effect. One *undoes* the other.

> Multiplication and division are **inverse operations**.

SOLVING EQUATIONS

We can use inverse operations to solve simple equations.

To keep the equation balanced, we must perform the same operation on both sides of the equation.

Example 3 ◢ **Self Tutor**

Solve for x using an inverse operation:

a $x + 6 = 13$ **b** $y - 4 = -1$ **c** $4g = 20$ **d** $\dfrac{h}{7} = -6$

a $x + 6 = 13$

$\therefore \;\; x + 6 - 6 = 13 - 6$ {the inverse of $+6$ is -6, so we subtract 6 from both sides}

$\therefore \;\; x = 7$

b $y - 4 = -1$

$\therefore \;\; y - 4 + 4 = -1 + 4$ {the inverse of -4 is $+4$, so we add 4 to both sides}

$\therefore \;\; y = 3$

c $4g = 20$

$\therefore \;\; \dfrac{4g}{4} = \dfrac{20}{4}$ {the inverse of $\times 4$ is $\div 4$, so we divide both sides by 4}

$\therefore \;\; g = 5$

d $\dfrac{h}{7} = -6$

$\therefore \;\; \dfrac{h}{7} \times 7 = -6 \times 7$ {the inverse of $\div 7$ is $\times 7$, so we multiply both sides by 7}

$\therefore \;\; h = -42$

EXERCISE 7C

1 State the inverse of:

a $+3$ **b** -8 **c** $\times 2$ **d** $\div 5$

e $+12$ **f** $\div 6$ **g** -5 **h** $\times 9$

i $+\dfrac{2}{3}$ **j** $\div 13$ **k** $\times 15$ **l** $-\dfrac{4}{5}$

2 Simplify:

a $x - 3 + 3$ **b** $x + 5 - 5$ **c** $x \div 12 \times 12$

d $x \times 9 \div 9$ **e** $7 + x - 7$ **f** $p + 4 - 4$

g $3q \div 3$ **h** $\dfrac{8r}{7} \times 7$ **i** $\dfrac{3}{4}s \div \dfrac{3}{4}$

j $-3x \div -3$ **k** $-2 + x + 2$ **l** $\dfrac{3y}{-4} \times -4$

3 Solve using an inverse operation:

a $\dfrac{a}{5} = -2$ **b** $3b = 4$ **c** $-c = 9$ **d** $4 + d = 1$

e $e - 7 = 3$ **f** $\dfrac{1}{4}f = 15$ **g** $z - 7 = -9$ **h** $\dfrac{w}{6} = -12$

4 Solve for x using an inverse operation:

a $x + 4 = 10$ **b** $\dfrac{x}{10} = 1$ **c** $x - 9 = -2$ **d** $2x = 12$

e $x - 5 = 1$ **f** $x + 2 = 0$ **g** $\dfrac{x}{7} = -5$ **h** $5x = -55$

i $\dfrac{x}{8} = -4$ **j** $-20x = 60$ **k** $x + 5 = -2$ **l** $x - 7 = -16$

m $\dfrac{x}{-3} = 3$ **n** $x - 4 = 0$ **o** $-9x = -81$ **p** $x + 9 = -9$

q $12x = -48$ **r** $x + 9 = -18$ **s** $x - 8 = -12$ **t** $\dfrac{x}{15} = -6$

u $x + 13 = 49$ **v** $13x = 0$ **w** $\dfrac{x}{-11} = -4$ **x** $x - 21 = -7$

5 Solve using an inverse operation: $1 = \dfrac{3}{x}$

ACTIVITY 1

Click on the icon to practise solving equations using a single inverse operation.

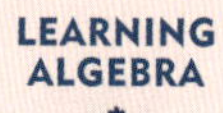

ACTIVITY 2

We are often faced with problems where we need to work out the value of an unknown quantity.

Consider these problems:

A Each of my cats has 4 kittens. There are 28 kittens in total. How many cats do I have?

B My training course runs for 6 hours. It finishes at 8 pm. What time does it start?

C This morning I bought some plants from the nursery. When I divided them equally into 5 large pots, I found I had 3 plants in each. How many plants did I buy this morning?

D Bernadette tells me that 3 years ago she was 11 years old. How old is she now?

What to do:

1 Match each of the problems above to the corresponding "find the number" question in the **Opening Problem**.

2 Summarise the information in each problem as an equation.

3 Solve each problem. Write your answer as a sentence.

D ALGEBRAIC FLOWCHARTS

To solve more complicated equations, we need to know how algebraic expressions are built up. We can study this using an **algebraic flowchart**.

For example, to "build up" the expression $3x + 2$, we start with x, multiply it by 3, then add on 2.

$$x \xrightarrow{\times 3} 3x \xrightarrow{+2} 3x + 2$$

To "undo" the expression $3x + 2$, we perform **inverse operations** in the **reverse order**.

$$3x + 2 \xrightarrow{-2} 3x \xrightarrow{\div 3} x$$

Example 4 ◀⑴ Self Tutor

Use flowcharts to show how to "build up" and "undo":

a $4x - 7$ **b** $4(x - 7)$

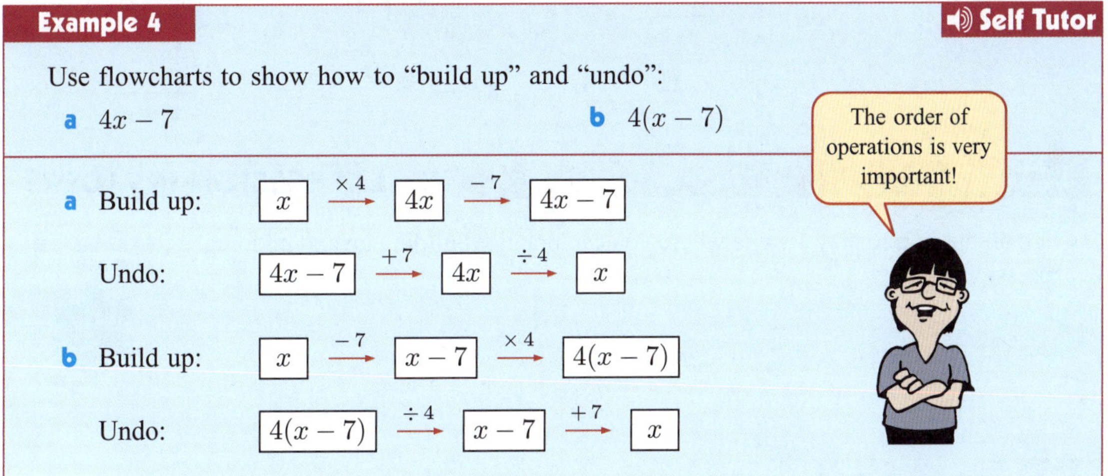

a Build up: $\quad x \xrightarrow{\times 4} 4x \xrightarrow{-7} 4x - 7$

Undo: $\quad 4x - 7 \xrightarrow{+7} 4x \xrightarrow{\div 4} x$

b Build up: $\quad x \xrightarrow{-7} x - 7 \xrightarrow{\times 4} 4(x - 7)$

Undo: $\quad 4(x - 7) \xrightarrow{\div 4} x - 7 \xrightarrow{+7} x$

Example 5 ◀⑴ Self Tutor

Use flowcharts to show how to "build up" and "undo":

a $\dfrac{3x + 4}{7}$ **b** $1 - \dfrac{x}{2}$

a Build up: $\quad x \xrightarrow{\times 3} 3x \xrightarrow{+4} 3x + 4 \xrightarrow{\div 7} \dfrac{3x + 4}{7}$

Undo: $\quad \dfrac{3x + 4}{7} \xrightarrow{\times 7} 3x + 4 \xrightarrow{-4} 3x \xrightarrow{\div 3} x$

b Build up: $\quad x \xrightarrow{\div -2} -\dfrac{x}{2} \xrightarrow{+1} 1 - \dfrac{x}{2}$

Undo: $\quad 1 - \dfrac{x}{2} \xrightarrow{-1} -\dfrac{x}{2} \xrightarrow{\times -2} x$

EXERCISE 7D

1 Use flowcharts to show how to "build up" and "undo":

   a $7x + 3$ **b** $7(x + 3)$ **c** $5(x - 2)$ **d** $5x - 2$

   e $\dfrac{x}{3} + 1$ **f** $\dfrac{x + 1}{3}$ **g** $\dfrac{x}{8} - 5$ **h** $\dfrac{x - 5}{8}$

2 Use flowcharts to show how to "build up" and "undo":

   a $2x - 6$ **b** $\dfrac{x}{-3} + 10$ **c** $8(x - 7)$ **d** $\dfrac{x - 3}{4}$

3 Use flowcharts to show how to "build up" and "undo":

   a $\dfrac{3x + 2}{5}$ **b** $\dfrac{3x}{5} + 2$ **c** $\dfrac{3(x + 2)}{5}$ **d** $\dfrac{7x - 1}{6}$

   e $\dfrac{7x}{6} - 1$ **f** $\dfrac{7(x - 1)}{6}$ **g** $\dfrac{5x}{6} - 3$ **h** $\dfrac{5(x - 3)}{6}$

   i $\dfrac{5x - 3}{6}$ **j** $1 - \dfrac{2x}{3}$ **k** $\dfrac{1 - 2x}{3}$ **l** $\dfrac{2(1 - x)}{3}$

ACTIVITY 3   EXPRESSION INVADERS

Click on the icon to play a game where you can practise building up and undoing expressions.

E   SOLVING EQUATIONS

When we are given an equation containing a **built up** expression, we use **inverse operations** in the **reverse order** to **isolate** the unknown.

Example 6   ◀) Self Tutor

Solve for x: $4x - 5 = 25$

$$4x - 5 = 25$$
$$\therefore\ 4x - 5 + 5 = 25 + 5 \qquad \text{\{adding 5 to both sides\}}$$
$$\therefore\ 4x = 30$$
$$\therefore\ \frac{4x}{4} = \frac{30}{4} \qquad \text{\{dividing both sides by 4\}}$$
$$\therefore\ x = 7\tfrac{1}{2}$$

$Check$: $\text{LHS} = 4\left(7\tfrac{1}{2}\right) - 5 = 30 - 5 = 25 = \text{RHS}$ ✓

EXERCISE 7E

1 Solve for x:

a $2x + 1 = 5$ **b** $4x + 7 = 27$ **c** $7 + 3x = 19$

d $3x + 1 = -23$ **e** $5x - 9 = 11$ **f** $-3 + 8x = 0$

g $2x - 7 = -4$ **h** $2x - 11 = 23$ **i** $7 + 8x = -9$

j $6 + 3x = 0$ **k** $8 + 13x = 34$ **l** $11 + 4x = -6$

Example 7 🔊 **Self Tutor**

Solve for x: $32 - 5x = 8$

$$32 - 5x = 8$$
$$\therefore \quad 32 - 5x - 32 = 8 - 32 \qquad \text{\{subtracting 32 from both sides\}}$$
$$\therefore \quad -5x = -24$$
$$\therefore \quad \frac{-5x}{-5} = \frac{-24}{-5} \qquad \text{\{dividing both sides by } -5\text{\}}$$
$$\therefore \quad x = 4\frac{4}{5}$$

Check: LHS $= 32 - 5\left(4\frac{4}{5}\right) = 32 - 5\left(\frac{24}{5}\right) = 32 - 24 = 8 =$ RHS   ✓

---

2 Solve for x:

a $6 - x = 9$ **b** $-x + 3 = -11$ **c** $-2x + 7 = 13$

d $15 - 2x = 7$ **e** $2 - 3x = 8$ **f** $1 - 4x = -15$

g $-5x + 4 = -21$ **h** $16 - 8x = 11$ **i** $22 - 3x = 1$

j $14 - x = -1$ **k** $19 - 4x = -10$ **l** $-5x + 12 = -9$

Example 8 🔊 **Self Tutor**

Solve for x: $\dfrac{x}{4} + 5 = -8$

$$\frac{x}{4} + 5 = -8$$
$$\therefore \quad \frac{x}{4} + 5 - 5 = -8 - 5 \qquad \text{\{subtracting 5 from both sides\}}$$
$$\therefore \quad \frac{x}{4} = -13$$
$$\therefore \quad \frac{x}{4} \times 4 = -13 \times 4 \qquad \text{\{multiplying both sides by 4\}}$$
$$\therefore \quad x = -52$$

Check: LHS $= \dfrac{-52}{4} + 5 = -13 + 5 = -8 =$ RHS   ✓

3 Solve for x:

   a $\dfrac{x}{2} + 1 = 3$ **b** $\dfrac{x}{2} - 5 = 6$ **c** $\dfrac{x}{8} + 3 = 5$

   d $-4 + \dfrac{x}{3} = -1$ **e** $\dfrac{x}{3} - 4 = -11$ **f** $2 + \dfrac{x}{6} = -2$

   g $\dfrac{x}{9} - 7 = 0$ **h** $5 + \dfrac{x}{7} = -3$ **i** $\dfrac{x}{11} + 31 = 33$

4 Solve for x:

   a $\dfrac{x}{-3} + 2 = -1$ **b** $\dfrac{x}{-2} - 3 = 0$ **c** $-\dfrac{x}{2} - 5 = 1$

   d $3 - \dfrac{x}{2} = 5$ **e** $5 - \dfrac{x}{5} = 1$ **f** $6 - \dfrac{x}{3} = -2$

Example 9   ◀)) **Self Tutor**

Solve for x: $\dfrac{2x - 3}{3} = -2$

$$\dfrac{2x - 3}{3} = -2$$

$$\therefore \ \dfrac{2x - 3}{3} \times 3 = -2 \times 3 \qquad \{\text{multiplying both sides by 3}\}$$

$$\therefore \ 2x - 3 = -6$$

$$\therefore \ 2x - 3 + 3 = -6 + 3 \qquad \{\text{adding 3 to both sides}\}$$

$$\therefore \ 2x = -3$$

$$\therefore \ \dfrac{2x}{2} = \dfrac{-3}{2} \qquad \{\text{dividing both sides by 2}\}$$

$$\therefore \ x = -\dfrac{3}{2}$$

Check: $\text{LHS} = \dfrac{2(-\frac{3}{2}) - 3}{3} = \dfrac{-6}{3} = -2 = \text{RHS}$ ✓

---

5 Solve for x:

   a $\dfrac{x + 1}{3} = 4$ **b** $\dfrac{4x - 1}{5} = 7$ **c** $\dfrac{2x - 5}{2} = 1$

   d $\dfrac{3x + 1}{4} = -5$ **e** $\dfrac{6 + 5x}{-2} = 7$ **f** $\dfrac{1 + 2x}{-5} = 11$

   g $\dfrac{11x - 1}{8} = -7$ **h** $\dfrac{-2 + 6x}{-5} = -2$ **i** $\dfrac{11 + 4x}{3} = -11$

6 Solve for x:

   a $\dfrac{1 - x}{3} = 5$ **b** $\dfrac{-2x + 1}{3} = -3$ **c** $\dfrac{3 - 4x}{-5} = 1$

   d $\dfrac{-3x + 2}{-2} = 5$ **e** $\dfrac{6 - 2x}{3} = 2$ **f** $\dfrac{-x + 7}{-2} = 13$

Example 10   ◀)) **Self Tutor**

Solve for x: $3(2x - 1) = -21$

$$3(2x - 1) = -21$$
$$\therefore \ \frac{3(2x - 1)}{3} = \frac{-21}{3} \qquad \{\text{dividing both sides by 3}\}$$
$$\therefore \ 2x - 1 = -7$$
$$\therefore \ 2x - 1 + 1 = -7 + 1 \quad \{\text{adding 1 to both sides}\}$$
$$\therefore \ 2x = -6$$
$$\therefore \ \frac{2x}{2} = \frac{-6}{2} \qquad \{\text{dividing both sides by 2}\}$$
$$\therefore \ x = -3$$

Check: LHS $= 3(2(-3) - 1) = 3(-7) = -21 =$ RHS   ✓

7 Solve for x:

  a $2(x - 1) = 18$ **b** $3(1 + 2x) = 15$ **c** $5(2x - 7) = 10$

  d $-4(5 - 3x) = -28$ **e** $4(2 - 3x) = 44$ **f** $7(-7 + 3x) = -49$

  g $6(3x - 2) = 12$ **h** $-5(4x + 1) = -15$ **i** $-6(3 + 8x) = -18$

  j $3(5 - 2x) = 21$ **k** $-2(6 - x) = 3$ **l** $3(2x + 1) = 4$

8 Solve:

  a $3a + 5 = 14$ **b** $\dfrac{x}{8} - 1 = 55$ **c** $\dfrac{3x - 1}{2} = 7$

  d $-2(3 - x) = 14$ **e** $4(x + 5) = 24$ **f** $6(n - 2) = 12$

  g $9 + 5a = -31$ **h** $\dfrac{1 - 3x}{2} = -4$ **i** $\dfrac{2x - 5}{4} = 0$

  j $\dfrac{x}{5} - 3 = 12$ **k** $\dfrac{x + 15}{3} = 6$ **l** $-5(1 - 2n) = -35$

  m $\dfrac{3k + 5}{2} = 13$ **n** $-8(5z + 1) = 24$ **o** $\dfrac{7 - 2x}{3} = -1$

9 Solve for x:

  a $\dfrac{7 + \frac{3 - x}{2}}{4} = 2$ **b** $\dfrac{1 + \frac{1 + \frac{1 + x}{2}}{2}}{2} = 0$

ACTIVITY 4   SOLVING EQUATIONS

Click on the icon to practise solving linear equations.

LEARNING ALGEBRA

F   EQUATIONS WITH A REPEATED UNKNOWN

If the unknown appears more than once in an equation, we may need to take extra steps in its solution. However:

- Our aim is still to isolate the unknown on one side of the equation.
- We will achieve this by performing a series of operations on *both* sides of the equation.

DISCUSSION

Consider the equation $3x + 1 = x + 7$.

Suppose each unit is represented by a block 

and that a bag   contains x blocks.

We can represent the equation on a set of scales: 

- What operation do we need to perform on both sides, to remove x from the RHS?
- When we perform this operation, what equation are we left with?
- Can we now solve this equation using inverse operations?

In general, we follow these steps to solve equations:

Step 1:   If necessary, **expand** any brackets and **collect** like terms.

Step 2:   If necessary, **remove** the unknown from one side of the equation. Remember to balance the other side.

Step 3:   Use **inverse operations** to **isolate** the unknown.

Example 11   ◀ӱ Self Tutor

Solve for d: $2d + 3(d - 1) = 7$

$2d + 3(d - 1) = 7$

$\therefore \quad 2d + 3d - 3 = 7$ {expanding the brackets}

$\therefore \quad 5d - 3 = 7$ {collecting like terms}

$\therefore \quad 5d - 3 + 3 = 7 + 3$ {adding 3 to both sides}

$\therefore \quad 5d = 10$

$\therefore \quad \dfrac{5d}{5} = \dfrac{10}{5}$ {dividing both sides by 5}

$\therefore \quad d = 2$

Check: LHS $= 2(2) + 3(2 - 1) = 4 + 3 = 7 =$ RHS   ✓

EXERCISE 7F

1 Solve:

a $2x + 3x = 10$

b $5x + 4 + 3x = -36$

c $3x - 6 - 2x + 7 = 16$

d $4y - 9 + 3y - 5 = -14$

e $5x + 3(x + 2) = 30$

f $2m + 4(m - 3) = 3$

g $n - (3 - n) = 5$

h $7x + 2(x + 1) = 28$

i $3 + 2(2 - b) = -1$

j $-3(p + 2) + 4(p - 2) = 1$

k $t + \dfrac{t}{3} = 8$

l $d - \dfrac{d}{5} = -8$

m $2x + \dfrac{x}{3} = -14$

n $\dfrac{x}{2} - \dfrac{x}{3} = 2$

Example 12 ◀)) **Self Tutor**

Solve for x: $3x + 2 = x + 14$

$$3x + 2 = x + 14$$
$$\therefore \ 3x + 2 - x = x + 14 - x \qquad \text{\{subtracting } x \text{ from both sides\}}$$
$$\therefore \ 2x + 2 = 14$$
$$\therefore \ 2x + 2 - 2 = 14 - 2 \qquad \text{\{subtracting 2 from both sides\}}$$
$$\therefore \ 2x = 12$$
$$\therefore \ \frac{2x}{2} = \frac{12}{2} \qquad \text{\{dividing both sides by 2\}}$$
$$\therefore \ x = 6$$

Check: LHS $= 3(6) + 2 = 18 + 2 = 20$
RHS $= 6 + 14 = 20$ ✓

---

2 Solve for x:

a $2x - 1 = 5 + x$

b $3x + 1 = x + 5$

c $2x - 6 = 7x + 14$

d $4x - 5 = 2x + 1$

e $7 + 7x = 3x + 23$

f $-6 + x = 5x + 4$

g $x + 2 = 4 - x$

h $2x + 5 = 10 - 3x$

i $2 + 7x = 3 - 3x$

j $3x - 11 = 9 - 2x$

k $8 - 9x = 3 - 4x$

l $9 - 3x = -6x + 5$

3 Solve:

a $2(x + 1) = x + 5$

b $3(t - 2) = t + 4$

c $6(2x - 3) = x - 7$

d $4(3y - 1) = 2y + 1$

e $2(a - 5) = 5 - 3a$

f $8(p + 2) = 1 + 3p$

4 Solve:

a $x = 3(x - 2)$

b $x - 2 = 2(x - 4)$

c $9 - x = -2(x - 7)$

d $3 + 2x = 7(3 - x)$

e $2x + 12 = 5(1 - x)$

f $3x + 8 = -2(x + 3)$

5 **a** Try to solve $7(a + 3) = 21 + 7a$.   What do you notice?

   b How many values of a satisfy the equation?

   c What name do we give to an equation like this?

6 **a** Try to solve $3(a + 1) = 4 + 3a$.   What do you notice?

   b How many values of a satisfy the equation?

7 Solve for x:

   a $4 - x = 2(x + 1) + 1$

   b $3x + 1 = 2(1 - 3x) + 19$

   c $10 - x = 5(x - 3) + 7$

   d $7 + 6x = 3 + 2(1 - x)$

   e $3(x - 5) = 5x + 1$

   f $2x + 7 + 4(2x + 3) = -1$

   g $x - 2 + 4(x - 1) = 2$

   h $5(x + 1) + 5(x - 1) = 30$

   i $4 - 3x - (x + 5) = 3$

   j $7(1 - x) - 6(2 - x) = 8$

   k $3x - 2(x + 3) - 4x = 0$

   l $8(2x - 1) - 5(2x + 3) = -5$

8 Solve for x:

   a $3(x - 6) = 2(2 - x) + 3$

   b $2(x - 8) = 3(x - 5)$

   c $4(2x + 3) - 10 = 5(3 - x)$

   d $5(1 - 2x) + 3(x + 5) = 2(1 + x) - 9$

   e $x - 7 - (3x + 2) = 8x + 11$

   f $7 - 6x = 8(1 + x) - (1 - x)$

   g $5x + 2(3 - x) = 4 - (2 - 5x)$

   h $7 - (2 - 3x) = 12 - (5 - 4x)$

   i $4x = 4(3x - 1) - 2(3 - x)$

   j $3(x - 1) - 4(5 + x) = 4(2x + 1)$

DISCUSSION

Leigh's teacher asks him to expand $5(2 + 3x)$ using the distributive law. His answer is that $5(2 + 3x) = 10x + 15$.

- Is there a value of x for which $5(2 + 3x) = 10x + 15$?

- If there is a solution of $5(2 + 3x) = 10x + 15$,   does that mean Leigh's expansion is correct?

- For Leigh's expansion to be correct, how many solutions would it need to have?

INVESTIGATION   RECURRING DECIMALS

In **Chapter 3** we encountered **recurring decimals** such as $0.\overline{5} = 0.5555\ldots$.

We can now use our skills with equations to convert recurring decimals such as $0.\overline{5}$ to fractions.

What to do:

1 Let $x = 0.\overline{5}$,   so $x = 0.5555\ldots$.

   a Multiply both sides of this equation by 10 and hence explain why $10x = 5 + x$.

   b Solve this equation to find the value of $0.\overline{5}$ as a fraction.

2 Write each recurring decimal as a fraction:

   a $0.\overline{7}$

   b $0.\overline{8}$

   c $0.\overline{3}$

3 Let $x = 0.\overline{26}$, so $x = 0.2626....$.

    a Multiply both sides of this equation by 100 and hence explain why $100x = 26 + x$.

    b Hence write $0.\overline{26}$ as a fraction.

4 Write each recurring decimal as a fraction:

    a $0.\overline{51}$ **b** $1.\overline{09}$ **c** $2.\overline{15}$

5 Write each recurring decimal as a fraction:

    a $0.\overline{318}$ **b** $0.0\overline{4}$ **c** $0.2\overline{46}$

G   POWER EQUATIONS

A **power equation** is an equation of the form $x^n = k$ where $n \neq 0$.

In this Section we consider the cases $n = 2$ and $n = 3$.

EQUATIONS OF THE FORM $x^2 = k$

Consider the equation $x^2 = 7$.

We know that $\sqrt{7} \times \sqrt{7} = 7$

and $-\sqrt{7} \times -\sqrt{7} = 7$.

So, $x = \sqrt{7}$ and $x = -\sqrt{7}$ are both solutions of the equation.

We write the solutions as $x = \pm\sqrt{7}$, which reads "plus or minus the square root of 7".

Now consider the equation $x^2 = -7$.

We know that when any non-zero real number is multiplied by itself, the result is positive.

So, there are no real numbers which satisfy $x^2 = -7$.

In general, we conclude that:

$$\text{If } x^2 = k \text{ then } \begin{cases} x = \pm\sqrt{k} & \text{if } k > 0 \\ x = 0 & \text{if } k = 0 \\ x \text{ has no real solutions if } k < 0. \end{cases}$$

EQUATIONS OF THE FORM $x^3 = k$

In contrast to $x^2 = k$, there is always exactly one real solution of $x^3 = k$.

We know that $\sqrt[3]{7} \times \sqrt[3]{7} \times \sqrt[3]{7} = 7$, so $x = \sqrt[3]{7}$ is the solution of $x^3 = 7$.

We know that $\sqrt[3]{-7} \times \sqrt[3]{-7} \times \sqrt[3]{-7} = -7$, so $x = \sqrt[3]{-7}$ is the solution of $x^3 = -7$.

In general,

$$\text{If } x^3 = k \text{ then } x = \sqrt[3]{k}.$$

Example 13 ◀)) **Self Tutor**

Solve for x:

  a $x^2 = 4$ **b** $x^2 = -5$ **c** $x^3 = 45$

a $x^2 = 4$

$\therefore \ x = \pm\sqrt{4}$

$\therefore \ x = \pm 2$

b $x^2 = -5$ has

no real solutions.

c $x^3 = 45$

$\therefore \ x = \sqrt[3]{45}$

$\therefore \ x \approx 3.56$

EXERCISE 7G

1 Solve for x:

  a $x^2 = 9$ **b** $x^2 = 49$ **c** $x^2 = 36$ **d** $x^2 = 0$

  e $x^2 = 1$ **f** $x^2 = 17$ **g** $x^2 = 23$ **h** $x^2 = 100$

  i $x^2 = -4$ **j** $x^2 = -7$ **k** $x^2 = 27$ **l** $x^2 = -27$

2 Solve for x:

  a $x^3 = 8$ **b** $x^3 = 27$ **c** $x^3 = 0$ **d** $x^3 = 64$

  e $x^3 = -1$ **f** $x^3 = -125$ **g** $x^3 = \frac{1}{8}$ **h** $x^3 = 36$

Example 14 ◀)) **Self Tutor**

Solve for x: **a** $x^2 + 4 = 9$ **b** $x^3 + 27 = 0$

a $x^2 + 4 = 9$

$\therefore \ x^2 + 4 - 4 = 9 - 4$ {subtracting 4 from both sides}

$\therefore \ x^2 = 5$

$\therefore \ x = \pm\sqrt{5}$

b $x^3 + 27 = 0$

$\therefore \ x^3 + 27 - 27 = 0 - 27$ {subtracting 27 from both sides}

$\therefore \ x^3 = -27$

$\therefore \ x = \sqrt[3]{-27}$

$\therefore \ x = -3$

3 Solve for x:

  a $x^2 + 5 = 9$ **b** $x^2 + 16 = 25$ **c** $x^2 + 2 = 27$

  d $x^2 + 7 = 23$ **e** $8 + x^2 = 44$ **f** $x^2 + 14 = 39$

  g $10 + x^2 = 60$ **h** $x^2 + 5 = 49$ **i** $6 + x^2 = 18$

4 Solve for x:

  a $x^3 - 1 = 0$ **b** $x^3 + 8 = 0$ **c** $x^3 - 216 = 0$

  d $x^3 - 1\,000\,000 = 0$ **e** $x^3 - 16 = 0$ **f** $x^3 + 800 = 0$

5   Solve for x:

    a $2x^2 = 18$ **b** $3x^2 = 48$ **c** $5x^2 = 20$

    d $x^2 + x^2 = 32$ **e** $x^2 + 2x^2 = 90$ **f** $x^2 + 5x^2 = 120$

    g $x^2 + (2x)^2 = 45$ **h** $x^2 + (3x)^2 = 70$ **i** $x^2 + (2x)^2 = 1$

ACTIVITY 5   ARITHMAGONS

The diagram opposite is an **arithmagon**. The number in each square is equal to the sum of the numbers in the two circles connected to it.

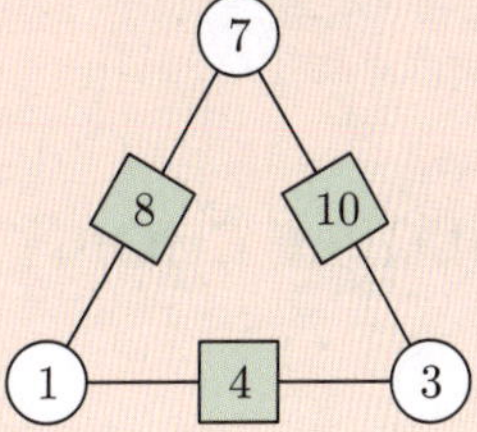

What to do:

1   Copy and complete these arithmagons:

  a 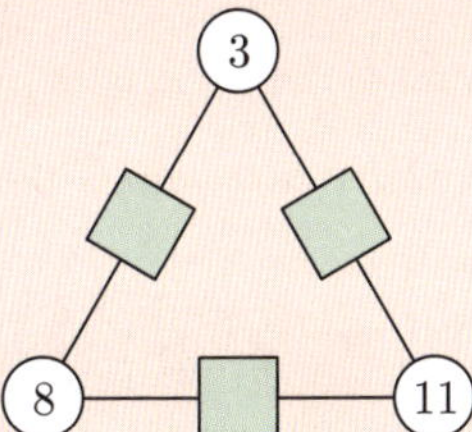   **b** 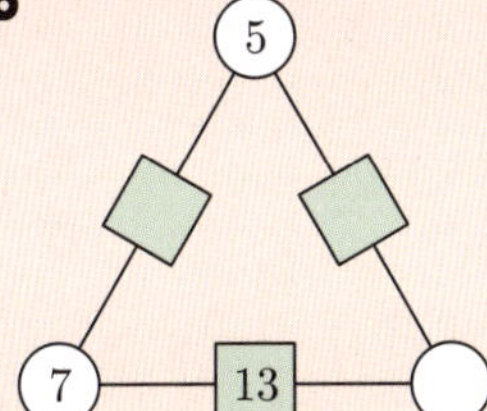   **c** 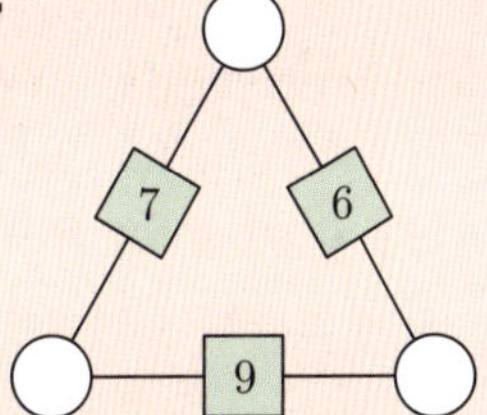

2   Which arithmagon in **1** did you find hardest to complete? Why was it the hardest?

3   Consider the following approach for **1 c**:

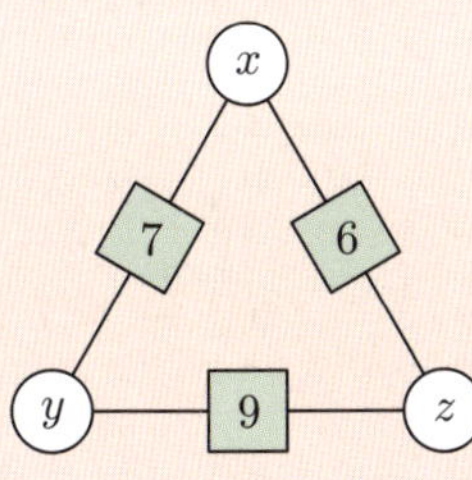

Let the numbers in the circles be x, y, and z as shown.

$$\therefore \quad x + y = 7$$
$$y + z = 9$$
$$x + z = 6$$

Adding these three equations gives

$$2x + 2y + 2z = 22$$
$$\therefore \quad x + y + z = 11 \quad \text{\{dividing both sides by 2\}}$$

We know that $\ y + z = 9$, so $\ x = 2$

$$\therefore \quad y = 5 \ \text{ and } \ z = 4.$$

Use this method to complete these arithmagons:

  a 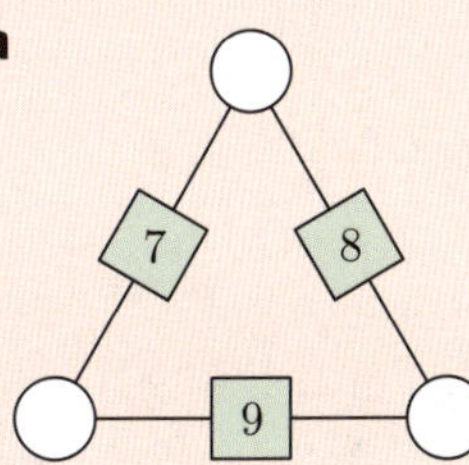   **b** 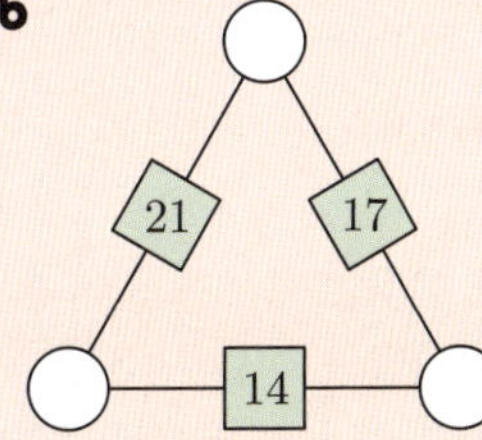   **c** 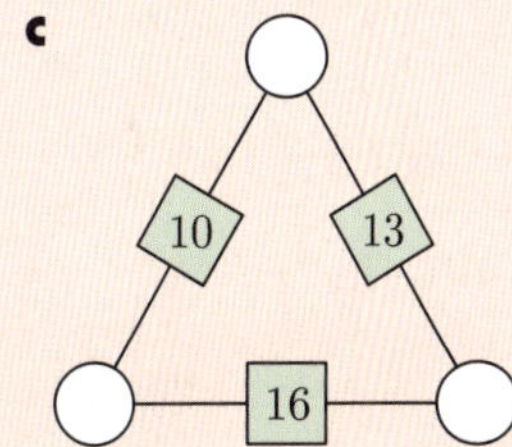

GLOBAL CONTEXT   REPRESENTING ALGEBRA

Global context: Orientation in space and time

Statement of inquiry: The way that mathematicians write algebraic
                           expressions has changed over time.

Criterion: Communicating

MULTIPLE CHOICE QUIZ

REVIEW SET 7A

1   State whether each equation is:

    A true for *exactly one value* of x **B** true for *exactly two values* of x

    C true for *all values* of x **D** *never* true.

  a $x + 3 = x$ **b** $x - 4 = 1$ **c** $x^2 = 1$ **d** $2x - x = x$

2   One of the numbers 6, 10, -6, or -4 is the solution of the equation $8x + 3 = 3(x - 9)$.
Find the solution.

3   Find the equation which results when:

  a 2 is added to both sides of $3x - 2 = -11$

  b 9 is subtracted from both sides of $4x + 9 = -1$.

4   Solve using an inverse operation:

  a $x - 11 = 4$ **b** $6x = 42$ **c** $a + 4 = -9$ **d** $\dfrac{t}{-3} = 7$

5   Use flowcharts to show how to "build up" and "undo":

  a $\dfrac{x}{6} + 1$ **b** $4(3x - 4)$ **c** $\dfrac{2 - 4x}{3}$

6   Solve for x:

  a $10x - 7 = 13$ **b** $5 + 4x = 29$ **c** $-3(3x - 4) = 30$

7   Solve for x:

  a $-4x + 9 = -19$ **b** $6 - \dfrac{x}{5} = 0$ **c** $\dfrac{1 - 5x}{3} = 4$

8   Solve for x:

  a $x + 3x - 5 = 31$ **b** $7x - 4(x - 3) = -2$ **c** $\dfrac{x}{2} - \dfrac{x}{5} = 3$

9 Solve:

    a $7a - 6 = 6a - 1$ **b** $3 - b = 3(b + 5) + 8$ **c** $4c - 2(3c - 1) = 5 - 7c$

10 Solve for x:

    a $3x + 2(3 - x) = -3$ **b** $3(x + 6) - 4(4 - 2x) = 7x + 6$

    c $9 - 5(x - 1) = 2(x + 4)$

11 Mark and Laura were given the equation $2(x - 3) = 14$.

    a Mark wanted to solve the equation by first dividing each side by 2. Solve the equation using Mark's method.

    b Laura wanted to solve the equation by first expanding the left hand side. Solve the equation using Laura's method.

    c What did you notice? Whose method was correct?

12 Solve for x:

    a $x^2 = 64$ **b** $x^3 = -10$ **c** $15 + x^2 = 50$

REVIEW SET 7B

1 Determine whether each equation is an identity. If it is not, give a value of the variable for which the equation is false.

    a $x - 5 = 5 - x$ **b** $8 \times m = m \times 8$ **c** $2t + 2t = 4t^2$

2 Solve $x + 5 = 2$ by inspection.

3 Find the equation which results when we divide both sides of $-2(2x + 3) = 4$ by -2.

4 State the inverse of:

    a multiplying by 5   **b** subtracting 7.

5 Solve using an inverse operation:

    a $a - 3 = 4$ **b** $-5b = 45$ **c** $c + 17 = 7$ **d** $\dfrac{d}{8} = -12$

6 Use flowcharts to show how to "build up" and "undo" $\dfrac{1 - 3x}{8}$.

7 Solve for x:

    a $3x + 5 = 17$ **b** $\dfrac{x}{4} + 1 = -11$ **c** $5 + 2x = 3$

8 Solve for x:

    a $\dfrac{1 - 2x}{3} = 1$ **b** $-2(3 - x) = -10$ **c** $-1 - \dfrac{x}{4} = -6$

9 Solve for x:

    a $2x + 1 = x + 8$ **b** $x - 4 = 5x - 1$ **c** $3(x - 3) = 8 - x$

10 Solve for x:

    a $2(4x - 3) + x = 3(2x - 1) + 2$ **b** $2(x - 3) - 3(4 - x) = 4(2x - 5)$

11 Ms Maxwell wrote the equation $\frac{x}{3} + 2 = 7$ on the board. She told her class there were *two* good ways to solve the equation.

    a **i** Find the equation that results when 2 is subtracted from both sides of $\frac{x}{3} + 2 = 7$.

        ii Hence solve the equation.

    b **i** Write down the equation that results when both sides of $\frac{x}{3} + 2 = 7$ are multiplied by 3. Your answer should include brackets.

        ii Apply the distributive law to expand the brackets.

       iii Hence solve the equation.

12 Solve for x:

    a $x^2 = -2$ **b** $x^3 = 50$ **c** $x^2 + (4x)^2 = 85$

Chapter 8

Lines and angles

Contents:

OPENING PROBLEM

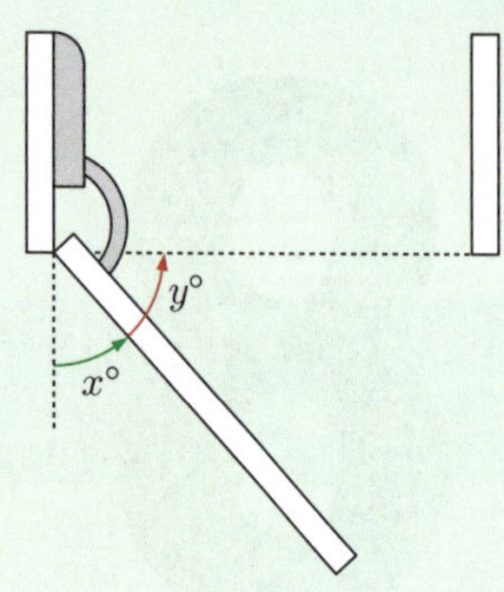

A cupboard door can open on its hinge to a maximum of a right angle.

Things to think about:

a What is the relationship between x and y?

b What is the value of x when the cupboard is:
    i closed   **ii** fully open?

c Find the value of x when $y = 76$.

d Find the value of y when $x = 55$.

In previous years, you should have seen that:

- A **point** is used to mark a position or location.
  We often label a point with a capital letter.

- A **line** is a continuous infinite collection of points in a particular direction. A line has no beginning and no end.

  ▶ (AB) is the **line** which passes through A and B.

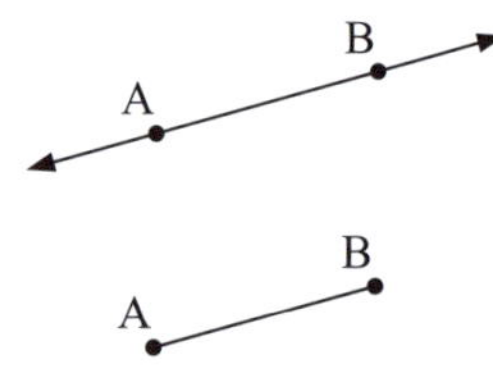

  ▶ [AB] is the **line segment** which joins the points A and B.
    It is only part of the line (AB).
    AB is the **length** of line segment [AB].

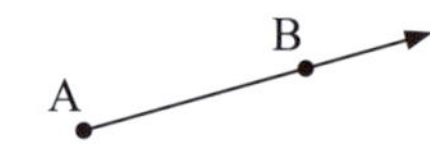

  ▶ [AB) is the **ray** which starts at A, passes through B, and continues on endlessly.

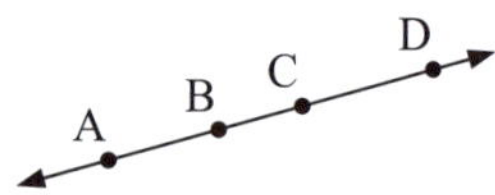

- A set of points is **collinear** if they lie on the same line.

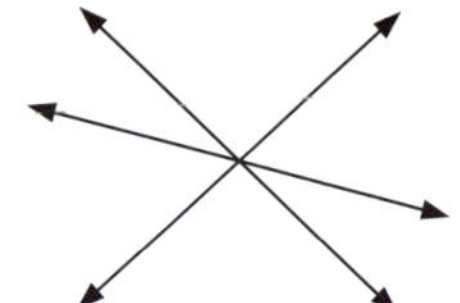

- A set of lines is **concurrent** if they pass through the same point.

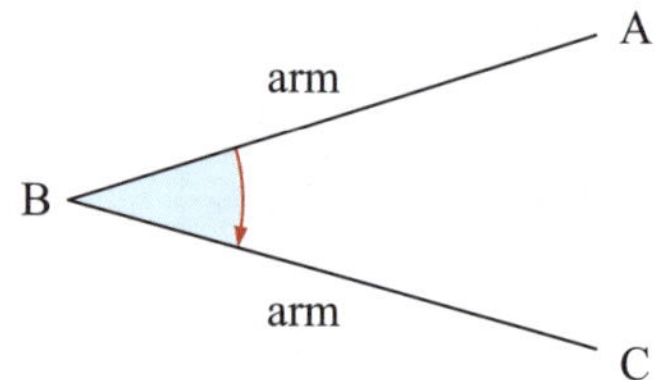

A   ANGLES

An **angle** is formed where two straight lines meet.

The point where the lines meet is called the **vertex** of the angle, and the lines are called the **arms**.

In this diagram, the shaded angle is $\widehat{ABC}$. Its vertex is B.

CLASSIFYING ANGLES

The **size** or **measure** of an angle is the amount of *turn* between its arms.

We measure the turn in **degrees**, and use the symbol °.

Revolution	Straight angle	Right angle
One complete turn = 360°	$\frac{1}{2}$ turn = 180°	$\frac{1}{4}$ turn = 90°

Acute angle	Obtuse angle	Reflex angle
Less than $\frac{1}{4}$ turn. Between 0° and 90°.	Between $\frac{1}{4}$ turn and $\frac{1}{2}$ turn. Between 90° and 180°.	Between $\frac{1}{2}$ turn and a complete turn. Between 180° and 360°.

COMPLEMENTARY AND SUPPLEMENTARY ANGLES

Angles which add to 90° are called **complementary angles**.

Angles which add to 180° are called **supplementary angles**.

EXERCISE 8A

1 True or false?

  a An angle measuring 88° is an acute angle.

  b An angle measuring 92° is an obtuse angle.

  c The size of an angle depends on the lengths of its arms.

  d When a vertical line crosses a horizontal line, the angle formed is a straight angle.

  e $\frac{2}{3}$ of a straight angle is an acute angle.

  f A right angle is neither an acute angle nor an obtuse angle.

2 Name and classify each angle:

  a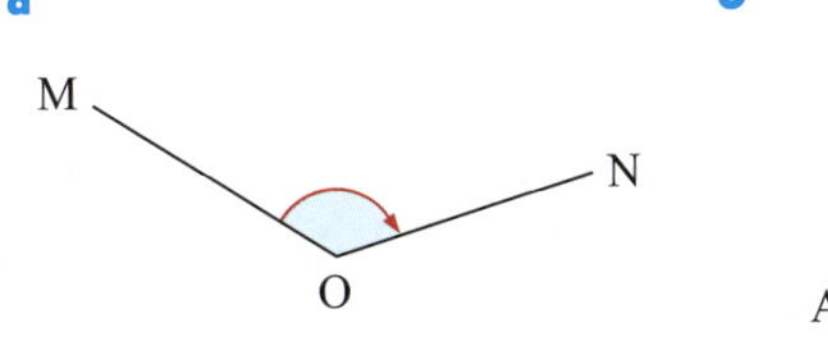

  b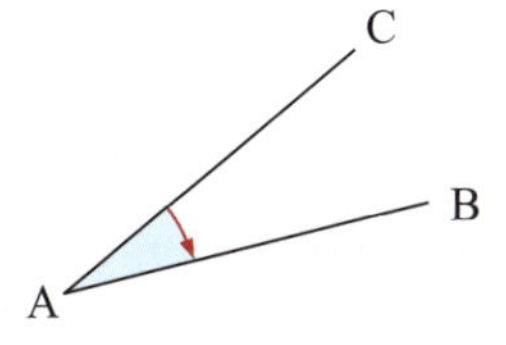

  c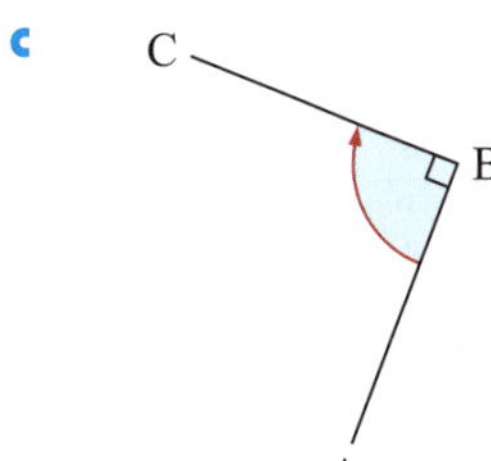

3 Find the angle which is complementary to:

    a $28°$ **b** $77°$ **c** $y°$ **d** $(90 - x)°$

4 Find the angle which is supplementary to:

    a $85°$ **b** $127°$ **c** $q°$ **d** $(180 - n)°$ **e** $(x + 90)°$

5 $A\widehat{O}C = 132°$ and $B\widehat{O}C = 48°$.

    a Describe the relationship between $A\widehat{O}C$ and $B\widehat{O}C$.

    b Draw *two* diagrams which could illustrate points O, A, B, and C. Classify $A\widehat{O}B$ in each case.

B   PARALLEL AND PERPENDICULAR LINES

- Two lines in a plane are **parallel** if they never meet. We use arrowheads to show lines are parallel.

  We use the symbol $\parallel$ to mean "is parallel to".

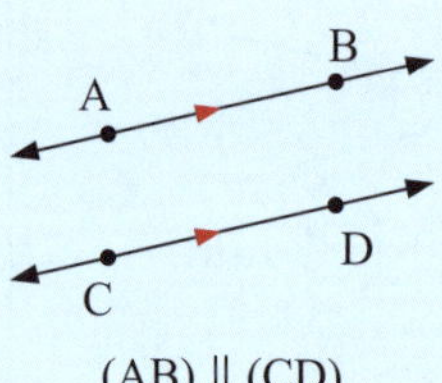

(AB) $\parallel$ (CD)

- Two lines in a plane are **perpendicular** if they intersect at right angles.

  We use the symbol $\perp$ to mean "is perpendicular to".

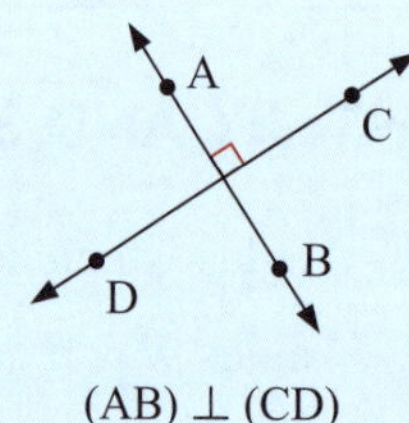

(AB) $\perp$ (CD)

EXERCISE 8B

1 Use $\parallel$ or $\perp$ to complete each statement:

    a (PQ) ...... (SR)   **b** (SQ) ...... (PR)

    c (PX) ...... (XR)   **d** (XR) ...... (PS)

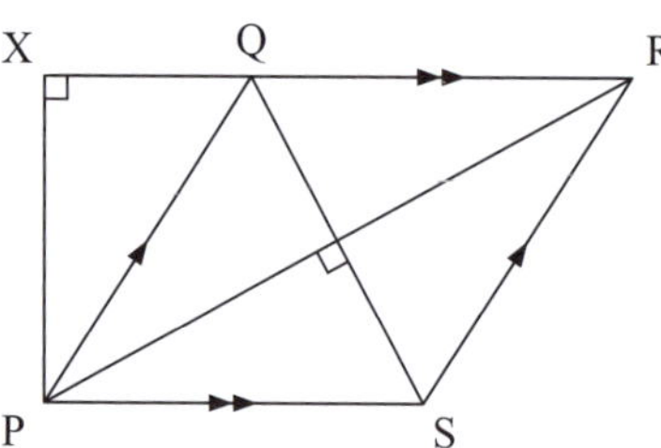

2 What can you say about P, Q, and R if (PR) $\parallel$ (RQ)?

DISCUSSION

In question **1** above, [QR] $\perp$ [PX] even though these line segments never actually meet.

How do we need to modify our definitions for parallel and perpendicular so they are written for *line segments*?

C   ANGLE PROPERTIES

Title	Theorem	Example
Angles at a point	Angles at a point add to $360°$.	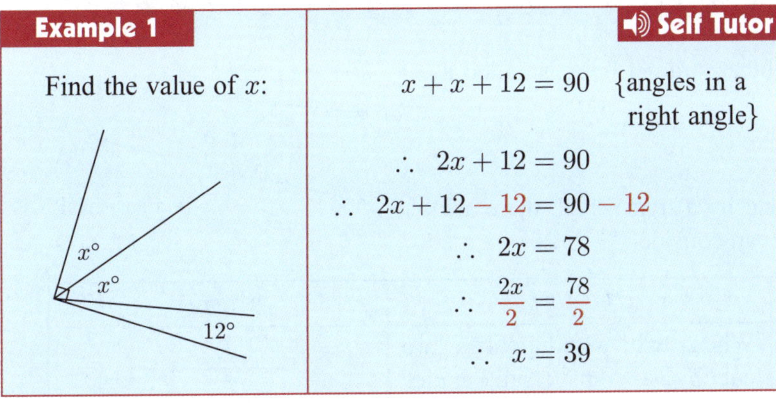 $a + b + c = 360$
Angles on a straight line	Angles on a line are supplementary.	$a + b = 180$
Angles in a right angle	Angles in a right angle are complementary.	$a + b = 90$
Vertically opposite angles	Vertically opposite angles are equal in size.	$a = b$

Example 1   ◀) Self Tutor

Find the value of x:

$x + x + 12 = 90$ {angles in a right angle}

$\therefore \ 2x + 12 = 90$

$\therefore \ 2x + 12 - 12 = 90 - 12$

$\therefore \quad 2x = 78$

$\therefore \quad \dfrac{2x}{2} = \dfrac{78}{2}$

$\therefore \quad x = 39$

EXERCISE 8C

1 Find the value of the unknown, giving brief reasons for your answers:

a
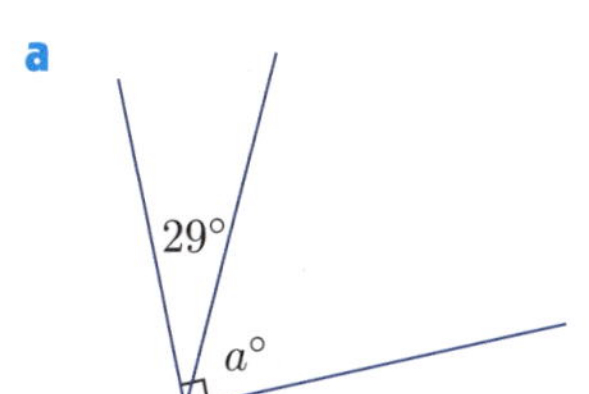

b
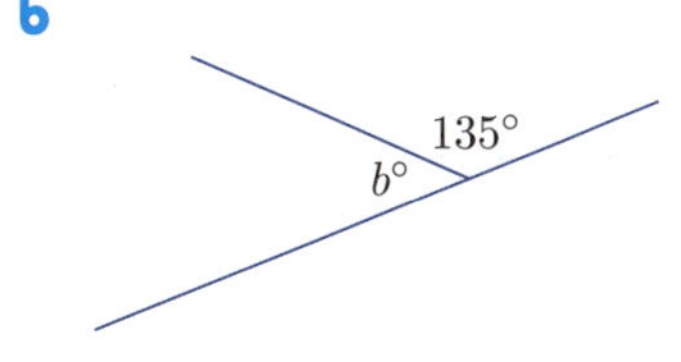

c
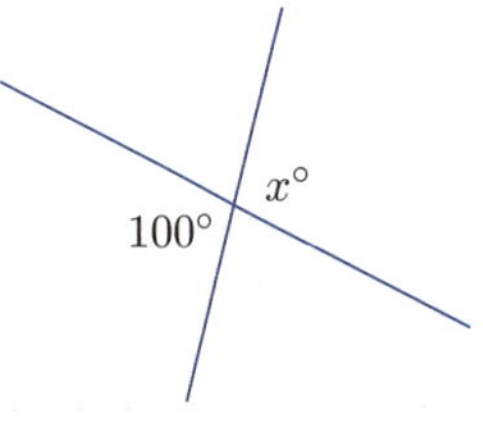

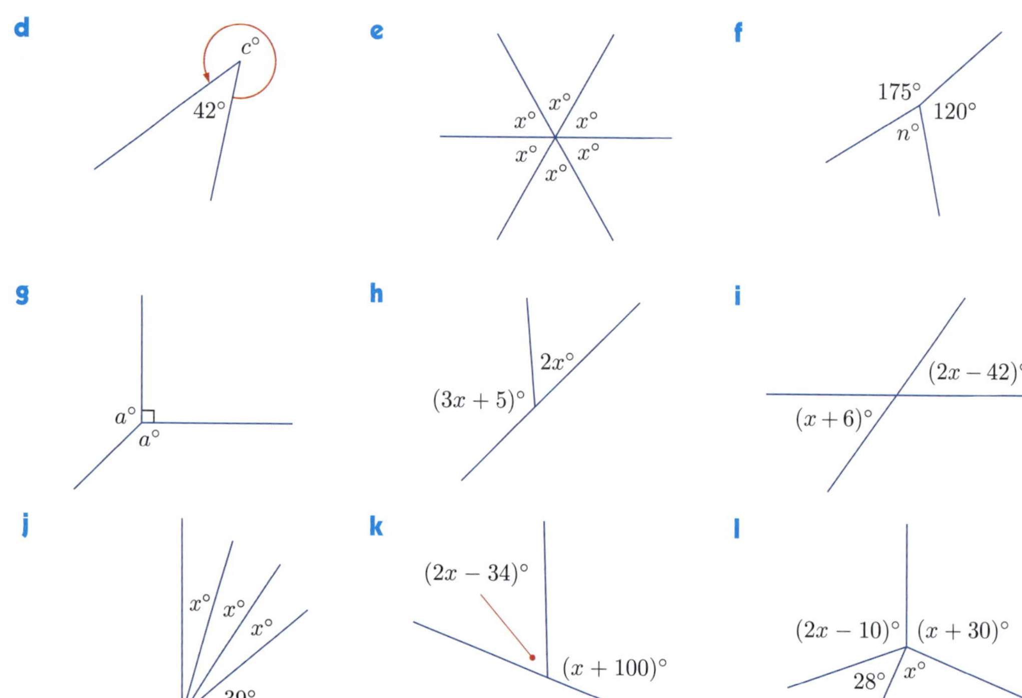

2   Answer the **Opening Problem** on page **160**.

D   LINES CUT BY A TRANSVERSAL

A third line that crosses two other straight lines is called a **transversal**.

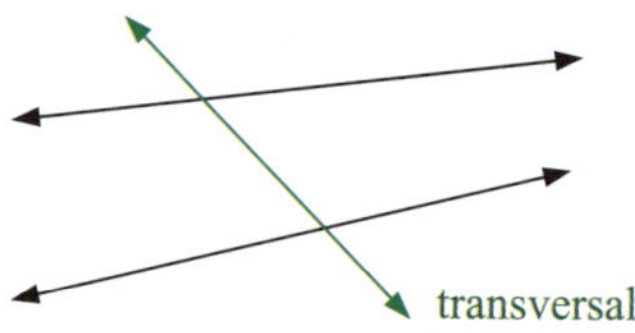

When two *parallel* lines are cut by a transversal, there are three different angle pairs we can consider:

Title	Theorem	Example
Corresponding angles	When two parallel lines are cut by a transversal, angles in corresponding positions are equal.	$a°$ $b°$ $a = b$
Alternate angles	When two parallel lines are cut by a transversal, angles in alternate positions are equal.	$a°$ $b°$ $a = b$

Title	Theorem	Example
Co-interior or allied angles	When two parallel lines are cut by a transversal, co-interior angles are supplementary.	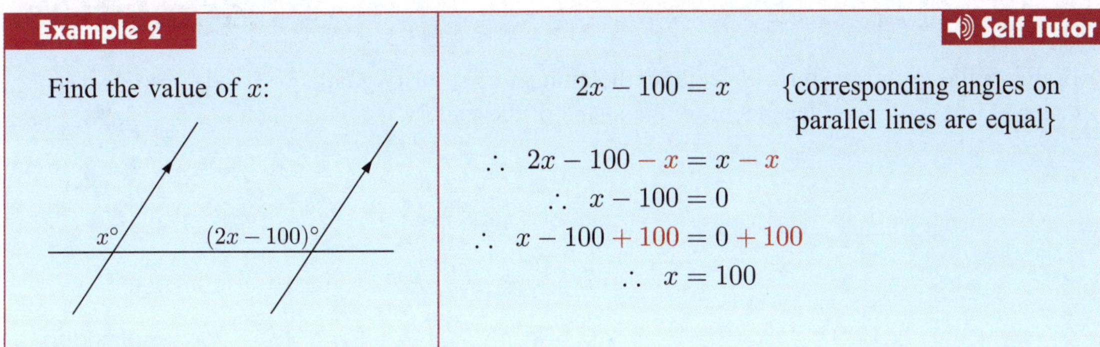 $a + b = 180$

The special properties are only true if the lines cut by the transversal are parallel. We can therefore use the sizes of corresponding, alternate, or co-interior angles to *test* for parallelism.

Example 2 ◀ঃ **Self Tutor**

Find the value of x:

$2x - 100 = x$ {corresponding angles on parallel lines are equal}

$\therefore\ 2x - 100 - x = x - x$

$\therefore\ x - 100 = 0$

$\therefore\ x - 100 + 100 = 0 + 100$

$\therefore\ x = 100$

EXERCISE 8D

1 Find the value of the unknown, giving brief reasons for your answers:

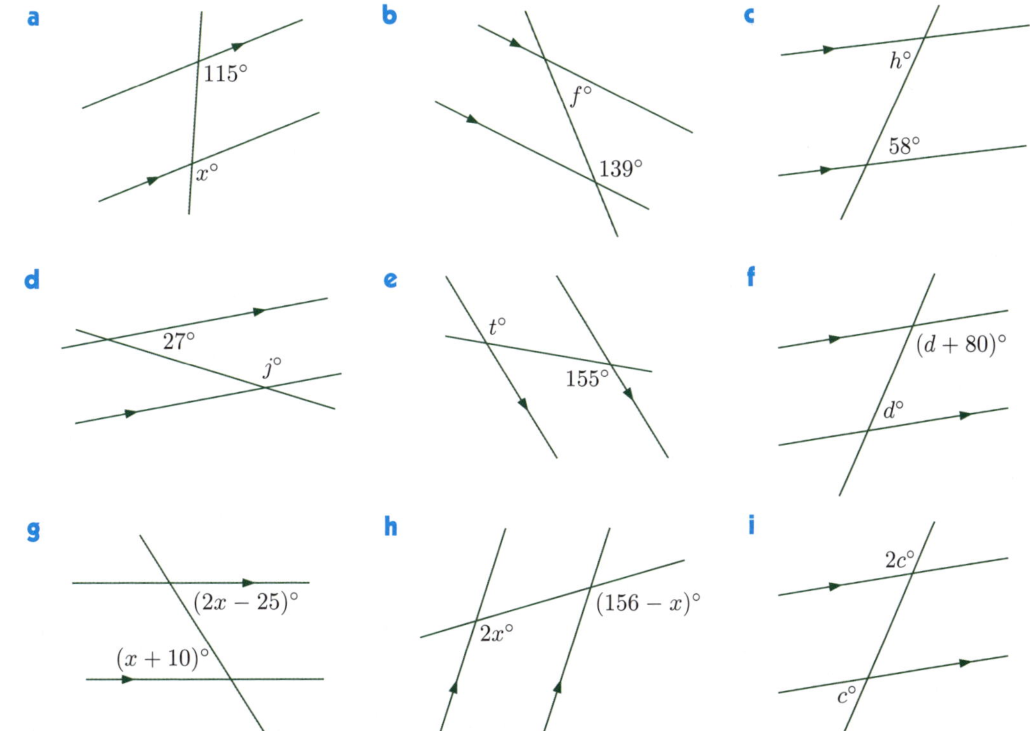

2 These diagrams are not drawn to scale, but the information on them is correct. Decide whether [AB] is parallel to [CD], giving a brief reason for your answer.

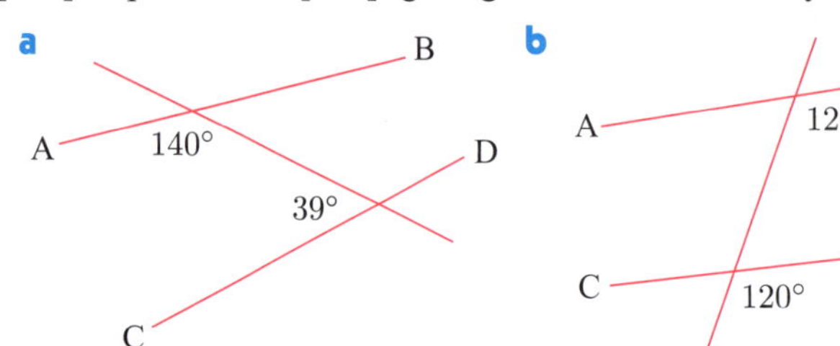

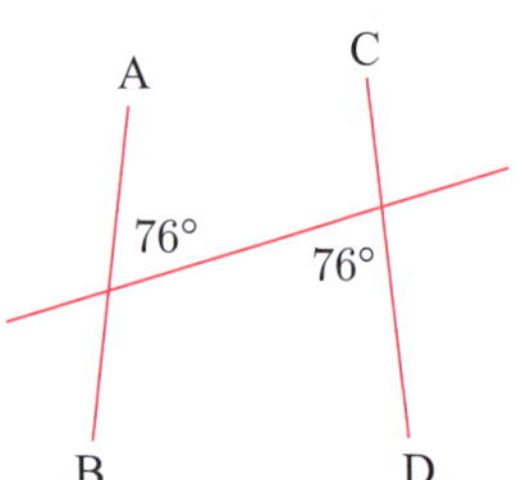

ACTIVITY   ART GALLERIES

In this online Activity you will see how the number of security guards or fixed cameras in a gallery is dependent on the shape of the gallery.

MULTIPLE CHOICE QUIZ

REVIEW SET 8A

1 Find the value of the unknown, giving brief reasons for your answers:

a

$a°$

$142°$

b

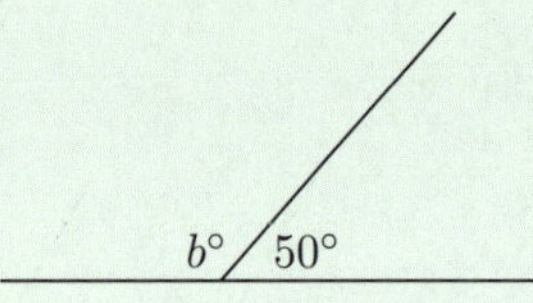

c

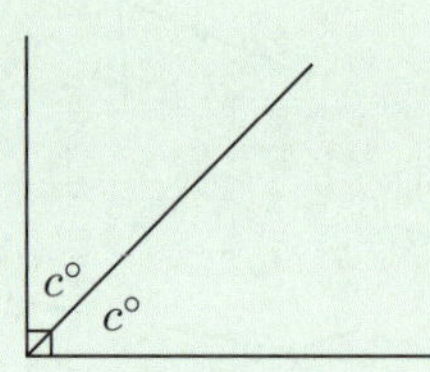

2 Find the angle which is:

  a complementary to $63°$ **b** supplementary to $26°$.

3 Use $\parallel$ or $\perp$ to complete each statement:

  a (AB) ...... (DC)   **b** (BE) ...... (AD)

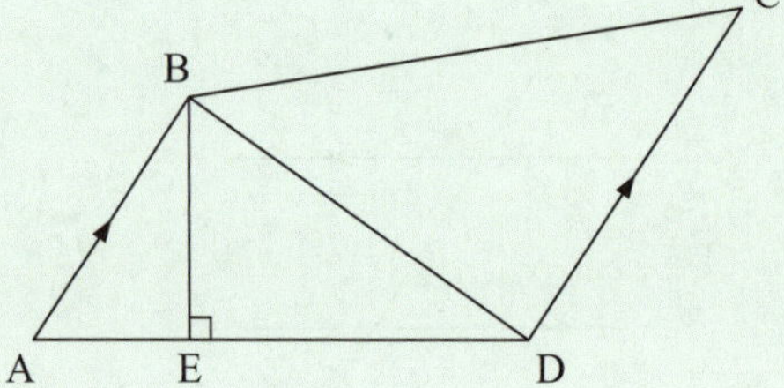

4 Find the value of the unknowns, giving brief reasons for your answers:

a

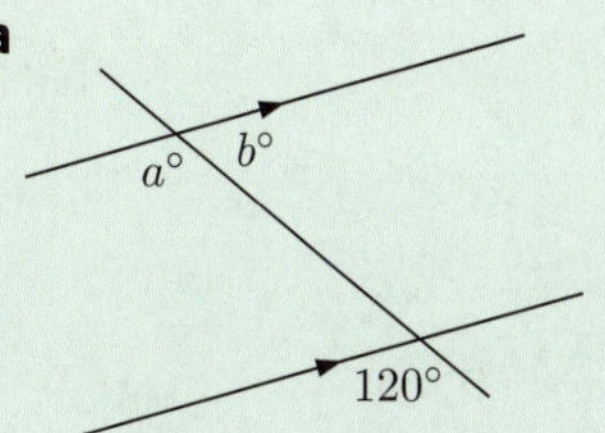

b

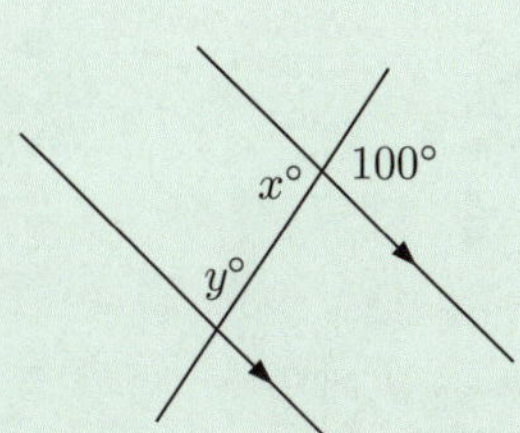

c

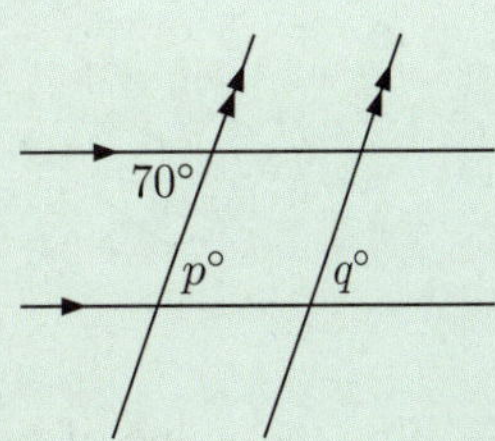

5 These diagrams are not drawn to scale, but the information on them is correct. Decide whether [AB] is parallel to [CD], giving a brief reason for your answer.

a

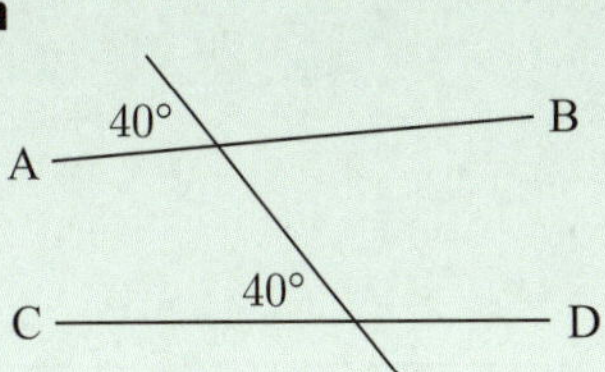

b

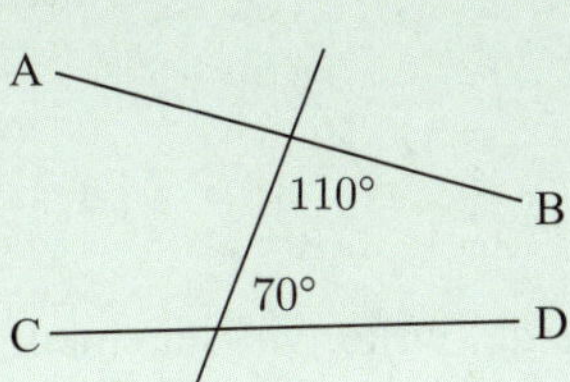

c

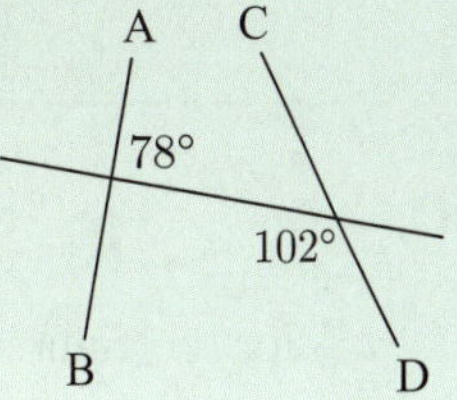

6 Four birds A, B, C, and D fly away from their nest O in different directions, such that $\hat{AOB} = 60°$, $\hat{BOC} = 100°$, and $\hat{COD} = 120°$.

a Describe the relationship between $\hat{AOB}$ and $\hat{COD}$.

b Draw *four* diagrams which could illustrate points O, A, B, C, and D.

c Given that $\hat{AOD}$ is obtuse, find the size of $\hat{BOD}$.

REVIEW SET 8B

1 True or false?

a Angles measuring 62° and 118° are complementary.

b An angle measuring 205° is a reflex angle.

c Vertically opposite angles are supplementary.

2 Name and classify each angle:

a

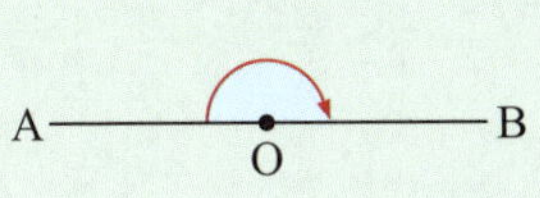

b

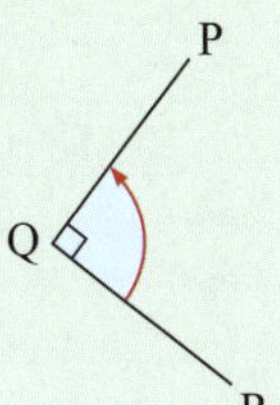

c

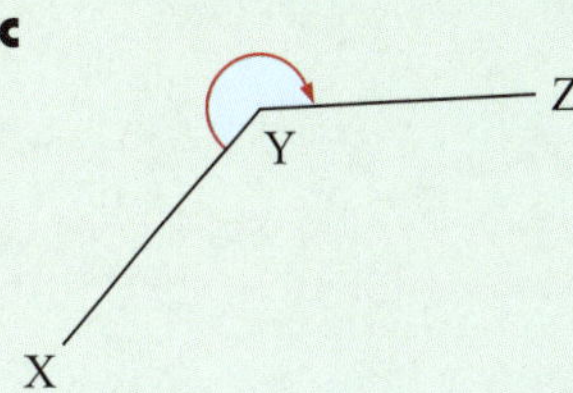

3

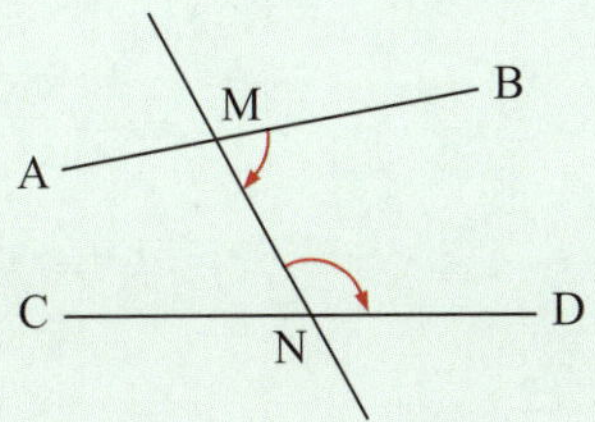

The given figure is not drawn to scale.

a What name describes the two marked angles?

b If the marked angles are supplementary, what can we say about [AB] and [CD]?

4 Find the value of x:

a

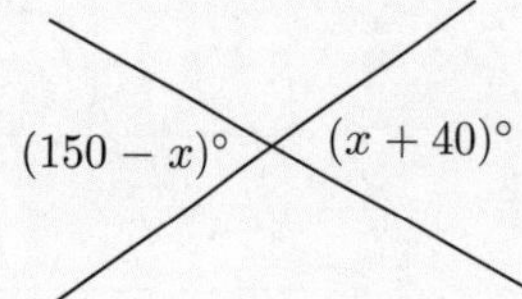

b

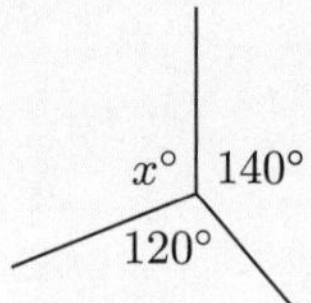

5 Find the value of the unknown:

a

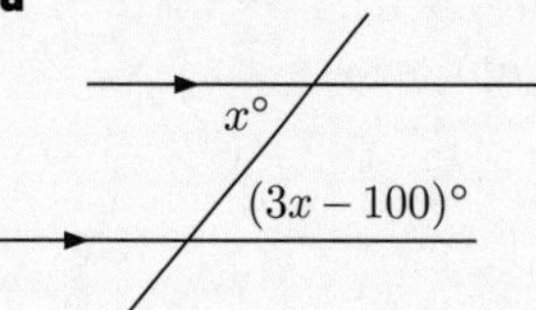

b

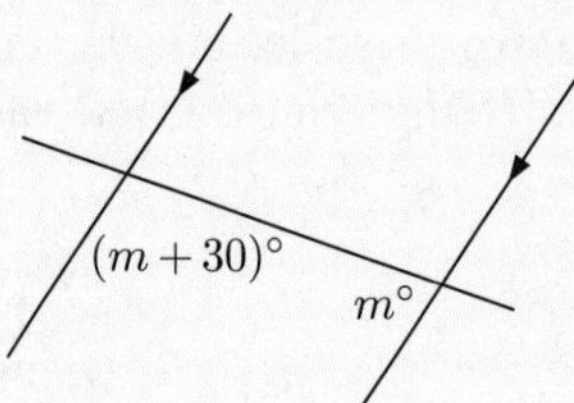

c

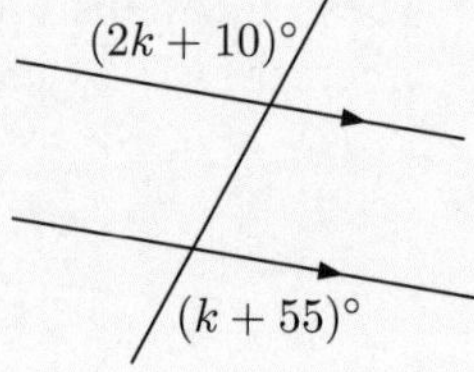

6 **a** For the figure alongside, write two statements involving $\perp$.

 b Is [AB] $\parallel$ [DC]? Explain your answer.

 c Explain why the sum of the angles of quadrilateral ABCD is $360°$.

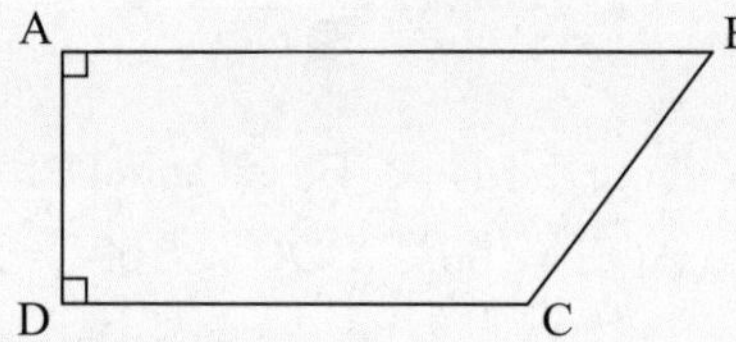

Plane geometry

Contents:

OPENING PROBLEM

The figure alongside contains two pairs of parallel lines.

Things to think about:

a What name is given to this shape?

b Which angles in the figure are equal?

c What is the sum of the sizes of:
   i the red angles   **ii** the green angles?

d What is the total sum of the angles of this figure?

In mathematics, a **plane** is a flat two-dimensional surface.

In this Chapter we will study **circles** and **polygons**, which are examples of **plane figures**.

A   CIRCLES

A **circle** is the set of points which are the same distance from a fixed point called its **centre**.

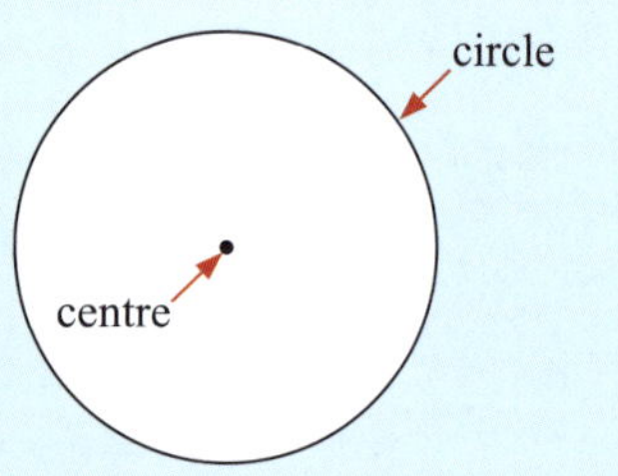

These terms are used to describe different parts of a circle:

- A **chord** is a line which joins any two points on the circle.
- A **diameter** is a chord which passes through the circle's centre.
- A **radius** is a straight line segment which joins the circle's centre to any point on the circle. **Radii** is the plural of radius.
- A **semi-circle** is a half of a circle.
- An **arc** is a part of a circle. It joins any two different points on the circle.
- A **segment** of a circle is the region between a chord and the circle.
- A **sector** of a circle is the region between two radii and the circle.
- A **tangent** to a circle is a line which *touches* the circle but does not enter it. A tangent is always perpendicular to the radius at that point.

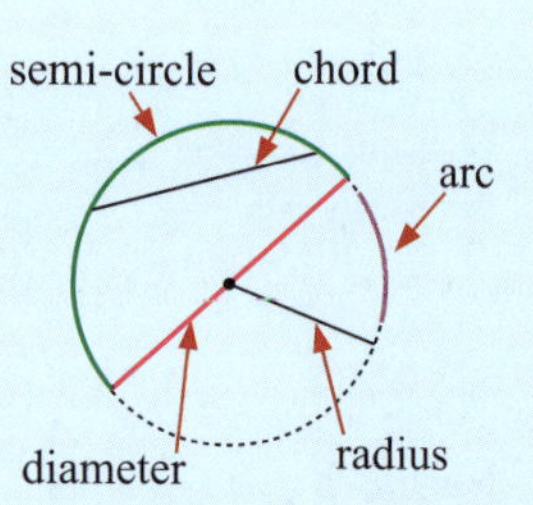

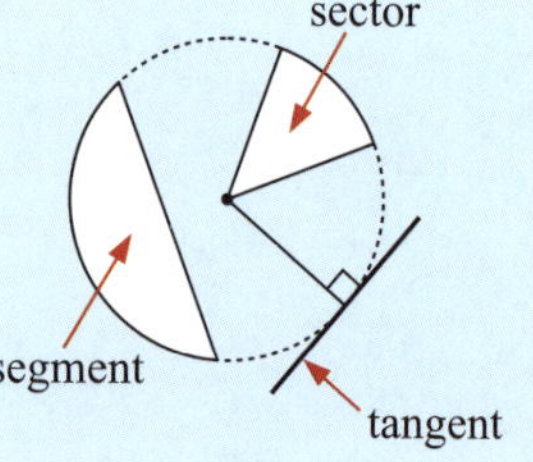

It is common to refer to the radius of a circle as the length of any of its radii, and the diameter of a circle as the length of any of its diameters. For example, we say that the radius of this circle is 3 cm.

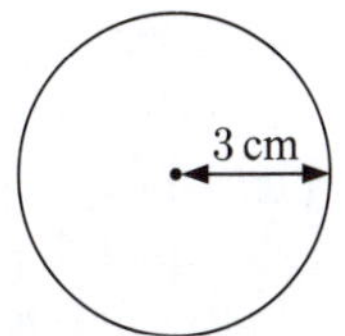

EXERCISE 9A

1 Match the part of the figure indicated to the phrase which best describes it:

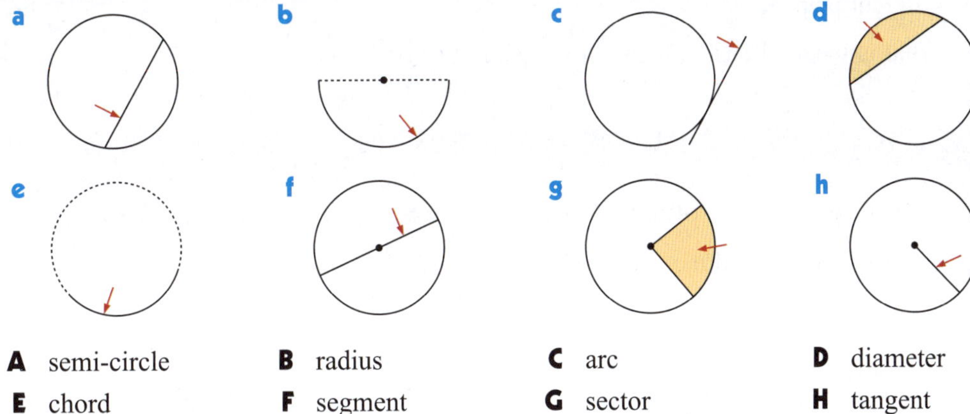

A semi-circle	**B** radius	**C** arc	**D** diameter
E chord	**F** segment	**G** sector	**H** tangent

2 What name can be given to the longest chord that you can draw in a circle?

3 **a** Explain why the diameter of a circle is always twice as long as its radius.

    b Find:

        i the diameter of a circle with radius 4 cm

        ii the radius of a circle with diameter 12 cm.

4 The circles shown both have centre C. The larger circle has radius 5 cm, and the smaller circle has radius 3 cm.
Points A, B, C, D, and E are collinear.
Find the distance between:

    a C and B   **b** C and A   **c** B and D

    d A and E   **e** A and B   **f** E and B.

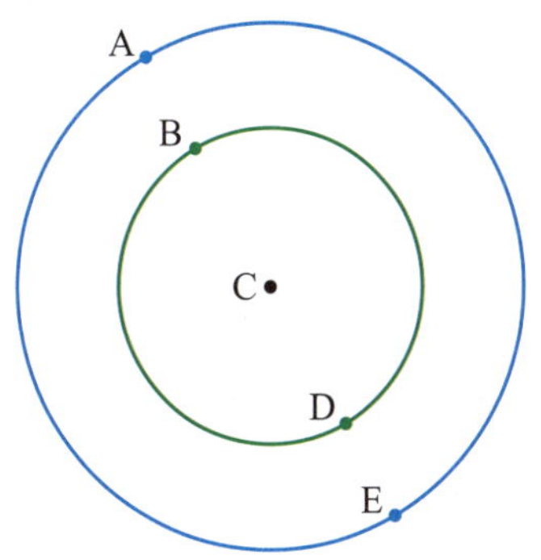

5 **a** Use a compass to draw a circle with radius 4 cm. Draw a diameter [AB] on the circle.

    b Mark a point X on the circle which is not A or B, and draw [AX] and [BX].

    c Use a protractor to measure $A\hat{X}B$. What do you notice?

    d Repeat **b** and **c** with a different point X.

    e Copy and complete:

        "The angle in a semi-circle is a ...... angle."

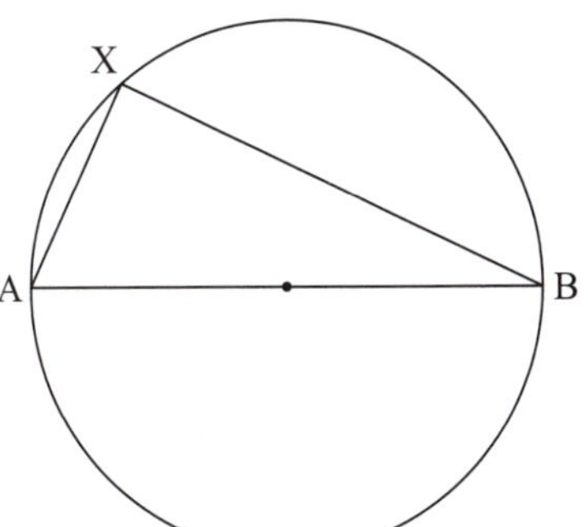

6   **a**   Use a compass to draw a circle with radius 5 cm and centre O.

   b   Draw a chord [PQ], and mark a point R on the circle which lies on the same side of the chord as O. Draw [PO], [QO], [PR], and [QR].

   c   Measure $P\hat{O}Q$ and $P\hat{R}Q$. What do you notice?

   d   Repeat **b** and **c** with a different chord [PQ] and a different point R.

   e   What is the relationship between $P\hat{O}Q$ and $P\hat{R}Q$?

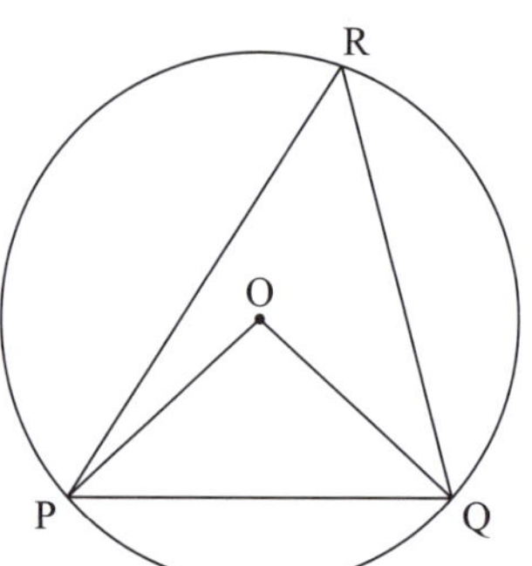

PUZZLE

You are given a circular disc of paper which does not have the circle's centre marked.

Explain how you could find the centre of the circle by folding the paper.

B   TRIANGLES

A **polygon** is a closed plane figure which has only straight line sides which do not cross.

Each corner of a polygon is called a **vertex**.

The plural of vertex is *vertices*.

In this Section we consider triangles, which are the first member of the polygon family.

A **triangle** is a polygon with three sides.

We can classify triangles according to the number of sides which are equal in length, or according to their angles.

CLASSIFICATION BY SIDE LENGTHS

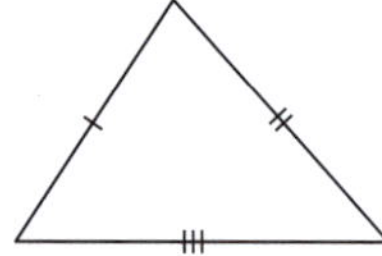
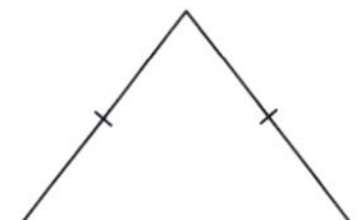
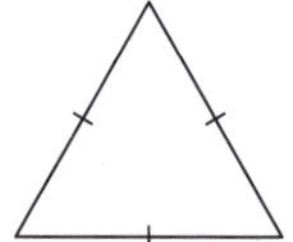
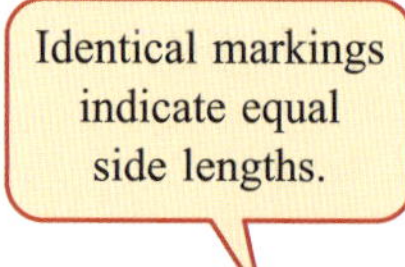

A **scalene triangle** has no equal sides.

An **isosceles triangle** has at least two equal sides.

An **equilateral triangle** has three equal sides.

CLASSIFICATION BY ANGLES

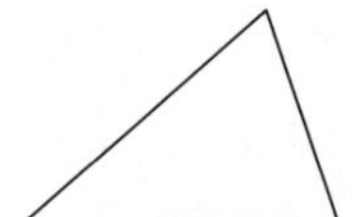

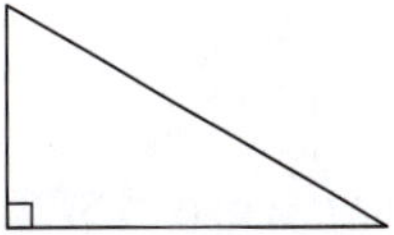

An **acute angled triangle** has all acute angles.

An **obtuse angled triangle** has one obtuse angle.

A **right angled triangle** has one right angle.

INVESTIGATION 1

In this Investigation you will discover some properties of triangles.

What to do:

1 Draw 6 large triangles on a sheet of paper.
   Label the vertices of each triangle A, B, and C.

2 Measure the side lengths and angles of each triangle. Record your results in a table like the one alongside.

Triangle	AB	AC	BC	$A\widehat{C}B$	$A\widehat{B}C$	$B\widehat{A}C$
1						
2						
3						
4						
5						
6						

3 For each triangle in your table, where is:

   a the longest side positioned relative to the largest angle

   b the shortest side positioned relative to the smallest angle?

From the **Investigation**, you should have concluded that:

- The longest side of any triangle is opposite the largest angle.
- The shortest side of any triangle is opposite the smallest angle.

EXERCISE 9B

1 Classify each triangle as scalene, isosceles, or equilateral:

a
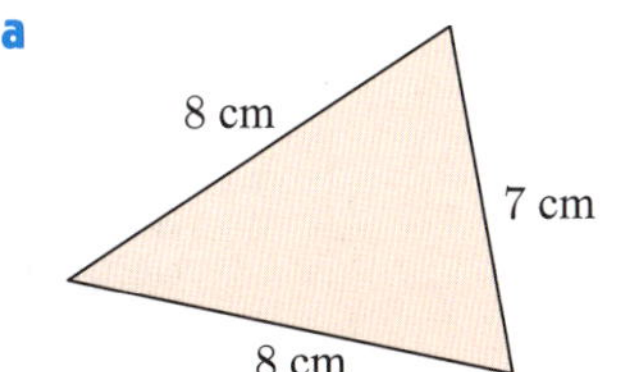

b
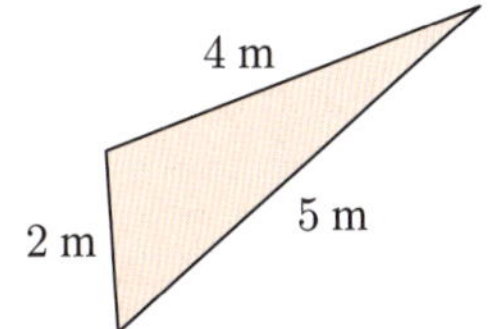

c
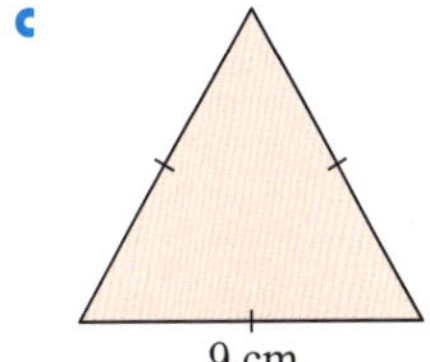

2 Classify each triangle as acute angled, obtuse angled, or right angled:

a

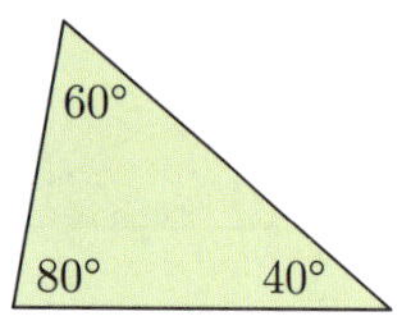

b

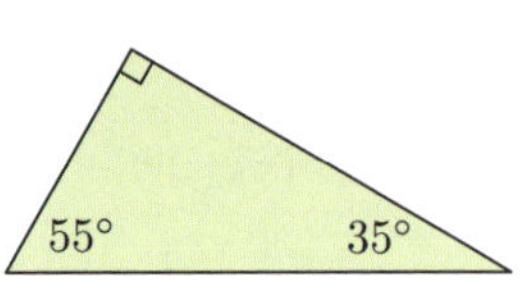

c

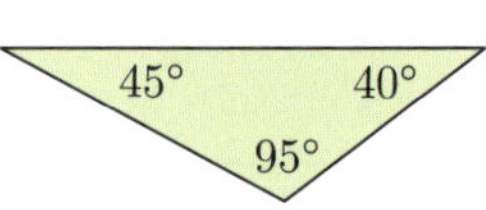

3 The triangle alongside is not drawn to scale. Identify its:

  a longest side

  b shortest side.

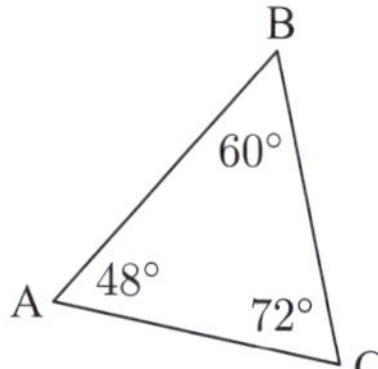

4 The triangle alongside is not drawn to scale. Identify its:

  a largest angle

  b smallest angle.

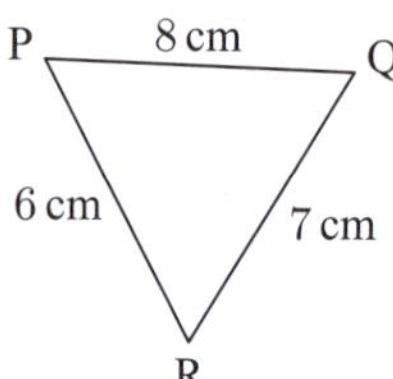

C   TRIANGLE THEOREMS

When we talk about the angles of a triangle, we are usually referring to the **interior angles** which are *inside* the triangle.

If we extend any side of a triangle, we can also create **exterior angles** which are *outside* the triangle.

For example, $\widehat{BAD}$ is an exterior angle of the triangle shown.

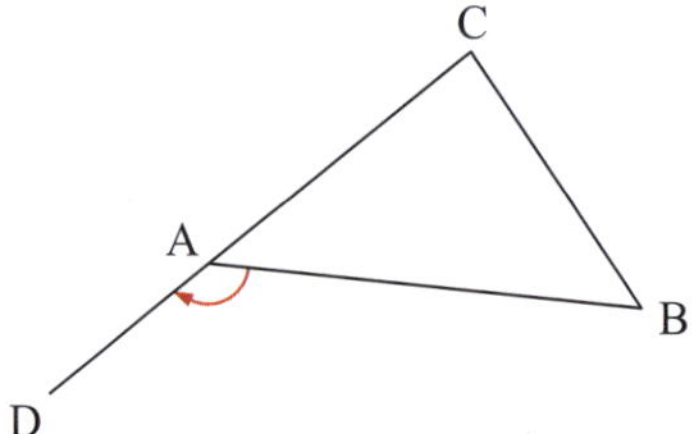

Name	Theorem	Diagram
Angle sum of a triangle	The sum of the interior angles of a triangle is 180°.	a° b° c° **GEOMETRY PACKAGE** $a + b + c = 180$
Exterior angle of a triangle	Any exterior angle of a triangle is equal to the sum of the opposite interior angles.	b° a° x° **GEOMETRY PACKAGE** $x = a + b$

Proof of the angle sum of a triangle theorem:

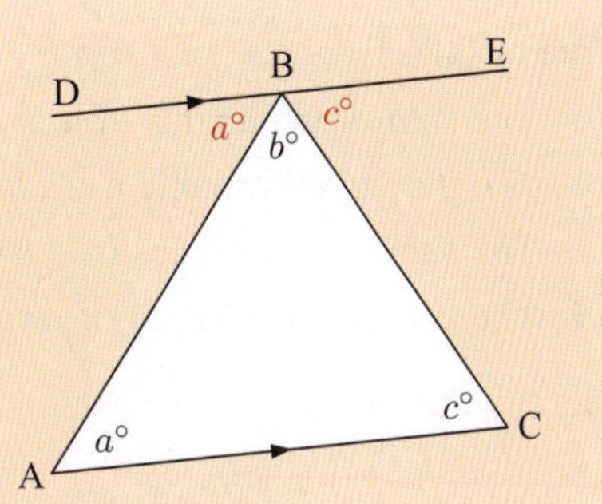

Consider a triangle ABC with interior angles $a°$, $b°$, and $c°$.

Draw a line segment [DE] which is parallel to [AC] and passes through B.

Since alternate angles on parallel lines are equal,

$\widehat{ABD} = a°$ and $\widehat{CBE} = c°$.

But $a + b + c = 180$ {angles on a line}

  $\therefore$ $a + b + c = 180$

Example 1   ◀) Self Tutor

Find the value of the unknown:

a
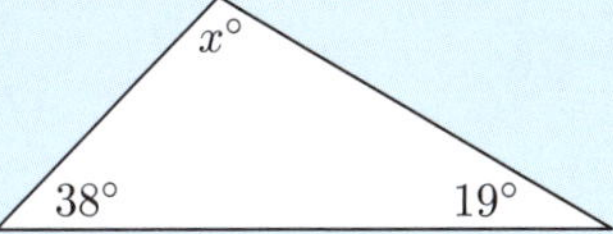

b
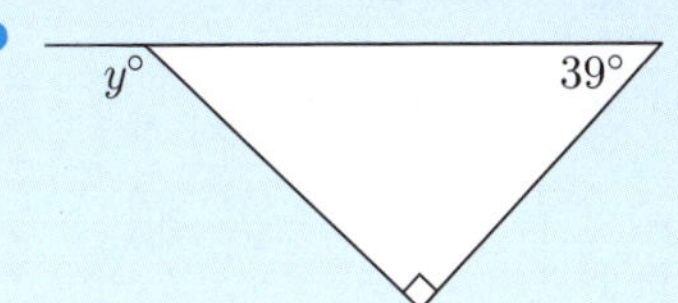

a $x + 38 + 19 = 180$ {angle sum of a triangle}

  $\therefore$ $x + 57 = 180$

  $\therefore$ $x + 57 - 57 = 180 - 57$

  $\therefore$ $x = 123$

b $y = 39 + 90$ {exterior angle of a triangle}

  $\therefore$ $y = 129$

EXERCISE 9C

1 Find the value of the unknown, giving a brief reason for your answer:

a
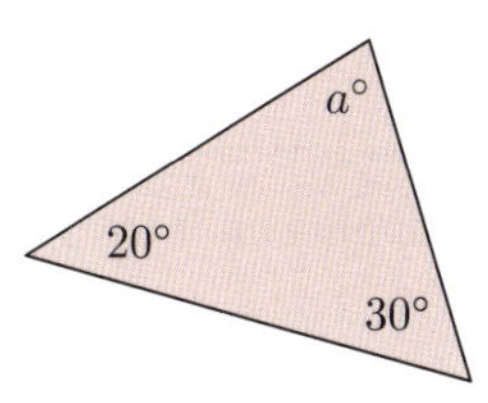

b
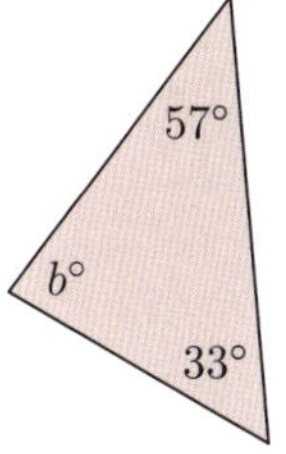

c
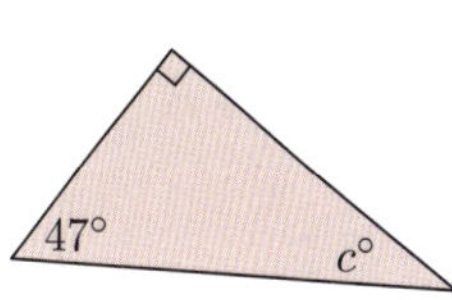

d
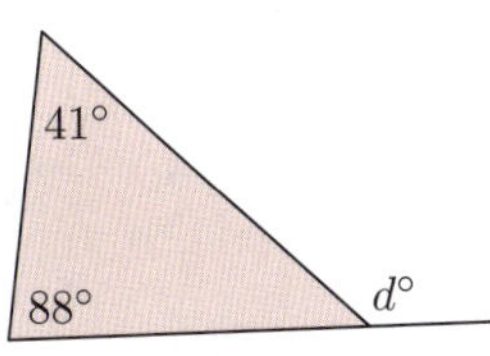

e
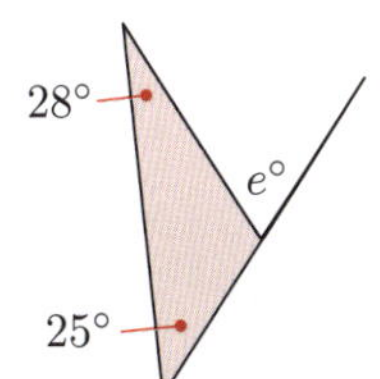

f
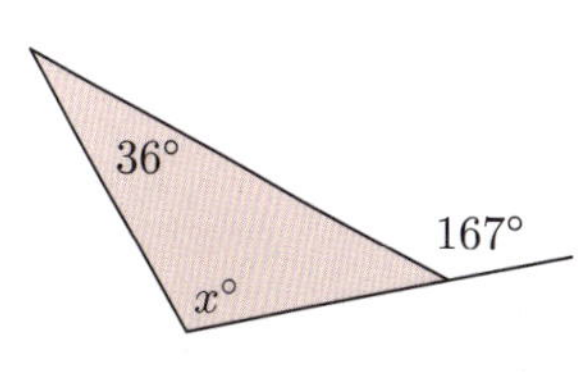

2 Two of the angles in Nancy's triangular pizza slice are $72°$ and $58°$. Find the measure of the third angle.

3 True or false?

a The sum of the angles of a triangle is equal to two right angles.

b A right angled triangle can contain an obtuse angle.

c The sum of two angles of a triangle is always greater than the third angle.

d The two smaller angles of a right angled triangle are supplementary.

4 Identify the longest side of each triangle:

a 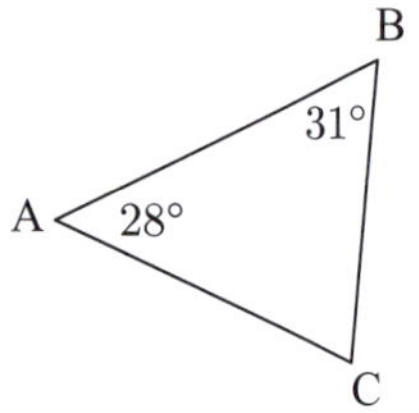

b 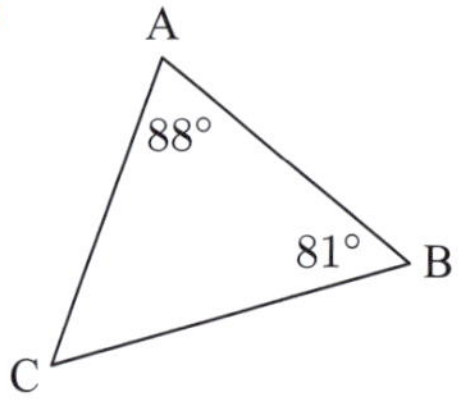

c 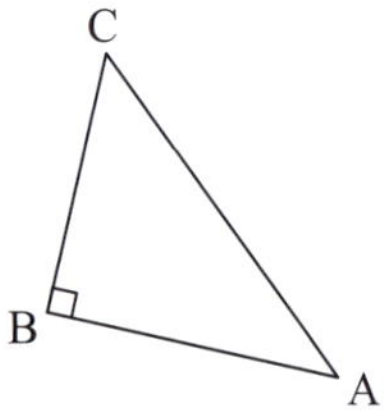

5 Identify the shortest side of each triangle:

a 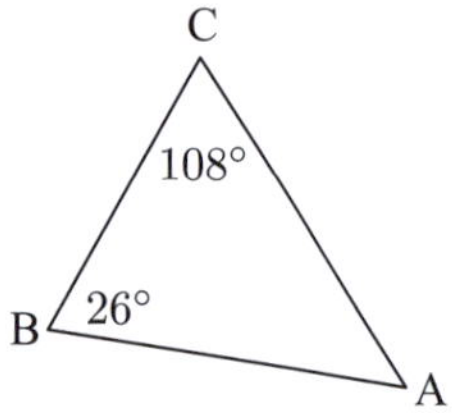

b 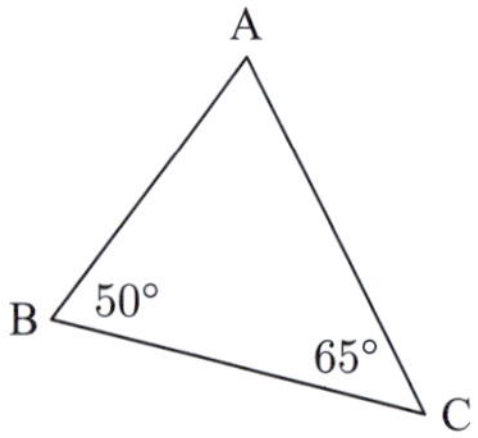

c 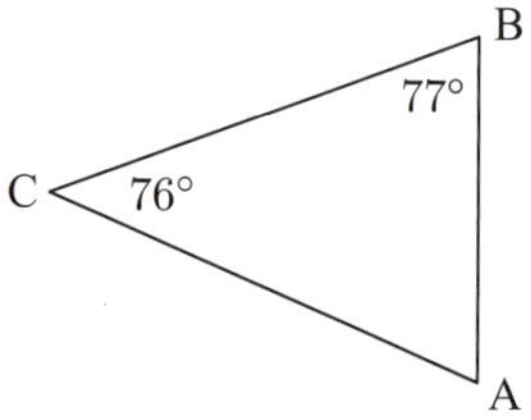

Example 2 — Self Tutor

Find the values of the unknowns:

a 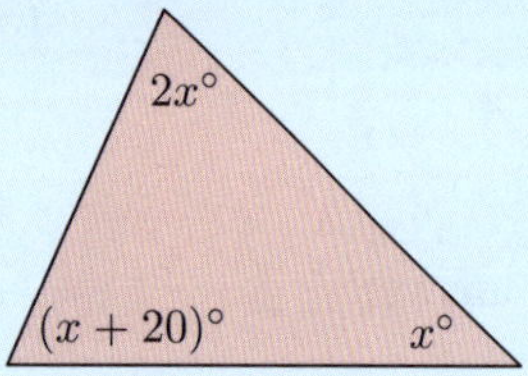

b 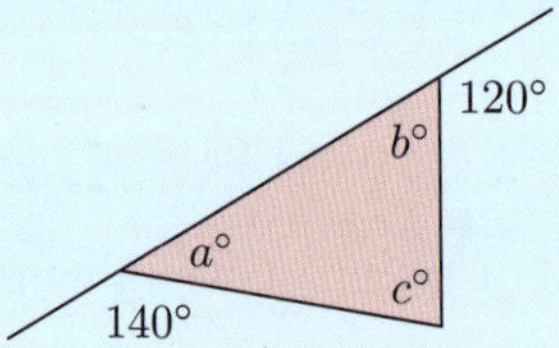

a $2x + x + (x + 20) = 180$ {angle sum of a triangle}

$\therefore\ 4x + 20 = 180$

$\therefore\ 4x + 20 - 20 = 180 - 20$

$\therefore\ 4x = 160$

$\therefore\ \frac{4x}{4} = \frac{160}{4}$

$\therefore\ x = 40$

b Now $a = 180 - 140 = 40$ {angles on a line}

and $b = 180 - 120 = 60$ {angles on a line}

But $a + b + c = 180$ {angle sum of a triangle}

$\therefore\ 40 + 60 + c = 180$

$\therefore\ 100 + c = 180$

$\therefore\ 100 + c - 100 = 180 - 100$

$\therefore\ c = 80$

6 Find the values of the unknowns:

a
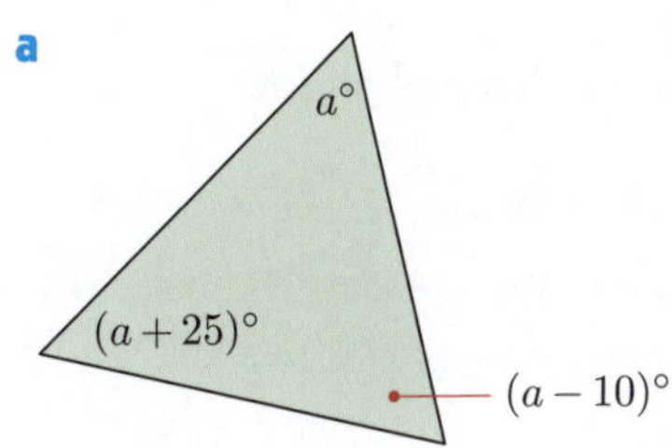

b
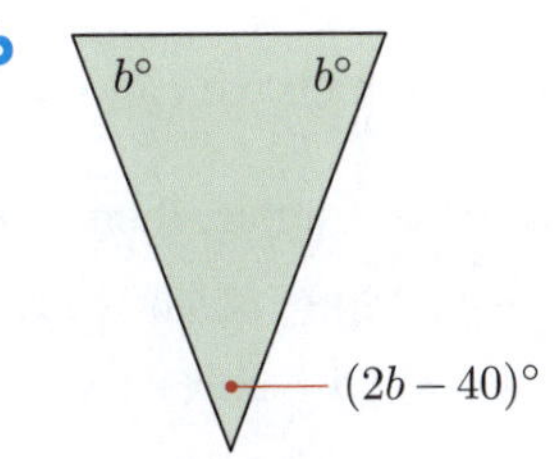

c
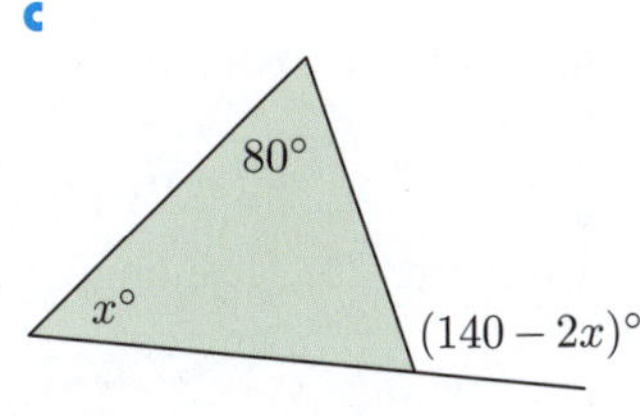

d
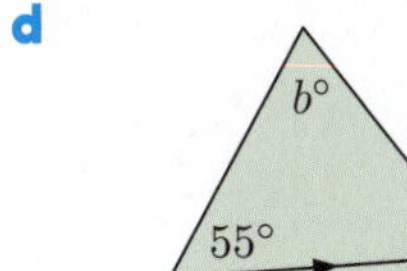

e
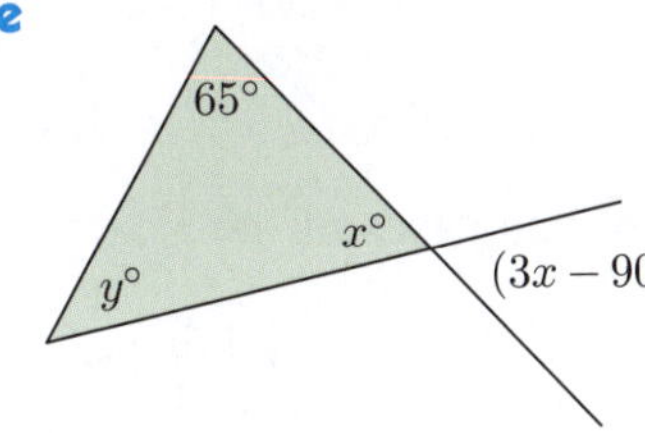

f
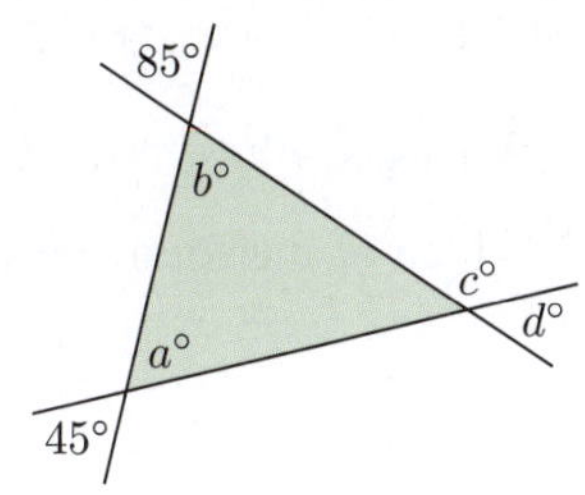

D ISOSCELES TRIANGLES

An **isosceles triangle** is a triangle which has at least two sides equal in length.

Having identified two equal sides of an isosceles triangle, we label the triangle as follows:

- The third side is called the **base**.
- The vertex between the equal sides is called the **apex**.
- The angle at the apex is called the **vertical angle**.
- The angles opposite the equal sides are called the **base angles**.

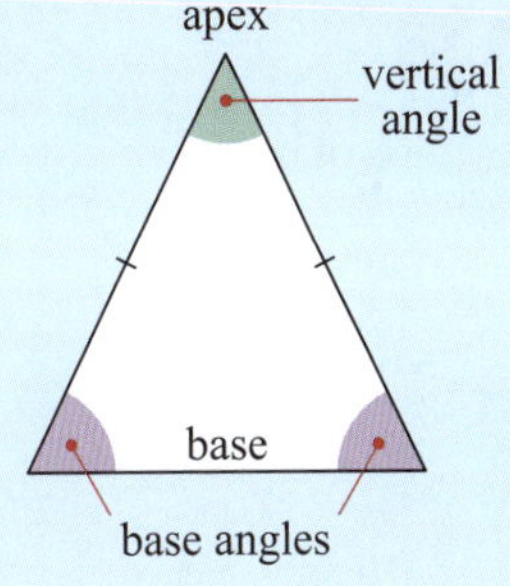

THE ISOSCELES TRIANGLE THEOREM

In any isosceles triangle:

- the base angles are equal
- the line joining the apex to the midpoint of the base bisects the vertical angle and meets the base at right angles.

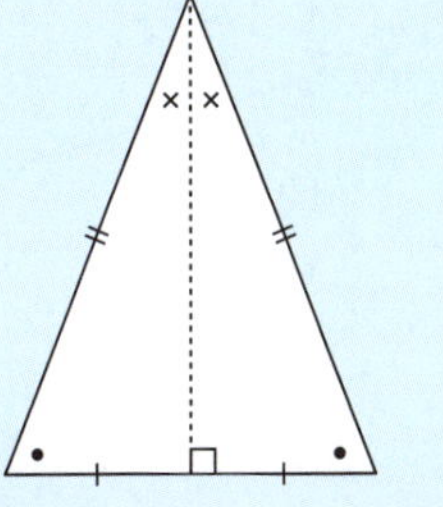

CONVERSES OF THE ISOSCELES TRIANGLE THEOREM

With many theorems there are *converses* which we can use in problem solving.

Converse 1:   If a triangle has two equal angles, then it is isosceles.

Converse 2:   The angle bisector of the apex of an isosceles triangle bisects the base at right angles.

DISCUSSION

What does the word *converse* mean?

Which of the following are also converses of the isosceles triangle theorem?

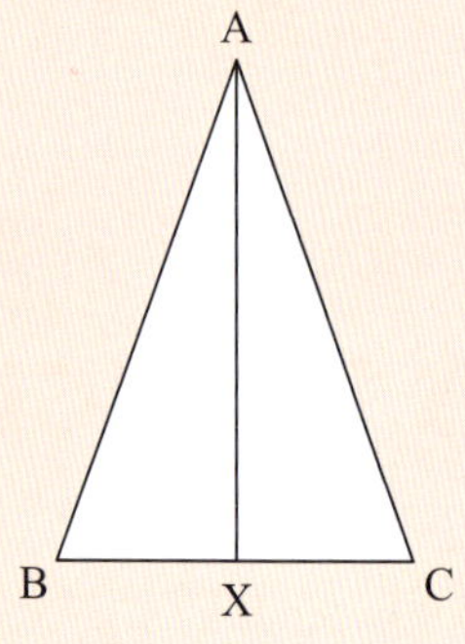

1 If the line joining one vertex to the midpoint of the opposite side is perpendicular to that side, then the triangle is isosceles.

2 If the line joining one vertex to the midpoint of the opposite side bisects the angle at the vertex, then the triangle is isosceles.

3 If a perpendicular to one side of the triangle passes through a vertex and bisects the angle at that vertex, then the triangle is isosceles.

Example 3   ◀) **Self Tutor**

Find the value of x:

a
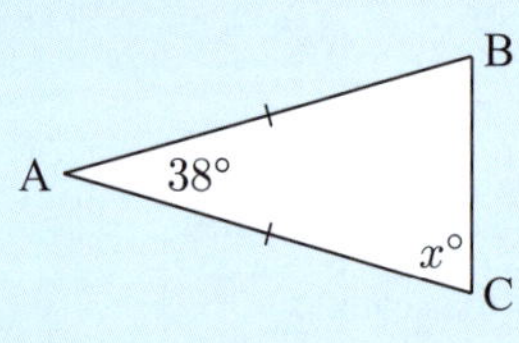

b
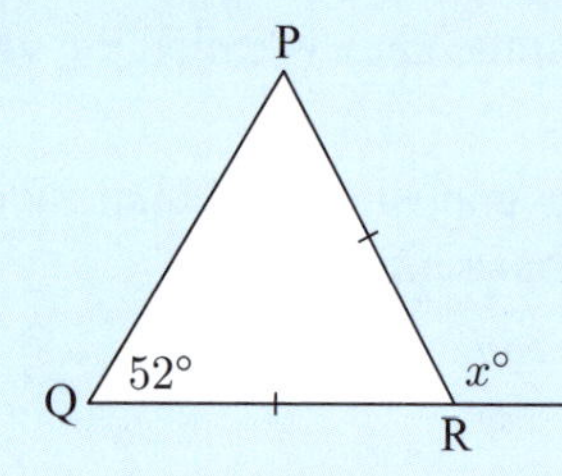

---

a
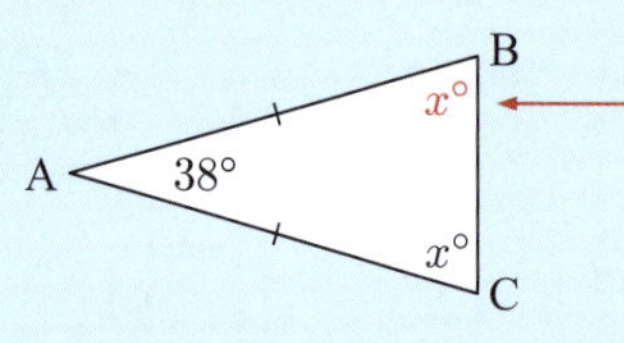

Since $AB = AC$, the triangle is isosceles.

$\therefore \ \hat{ABC} = x^\circ$ {isosceles triangle theorem}

Now $x + x + 38 = 180$ {angle sum of a triangle}

$\therefore \ 2x + 38 = 180$

$\therefore \ 2x + 38 - 38 = 180 - 38$

$\therefore \ 2x = 142$

$\therefore \ \dfrac{2x}{2} = \dfrac{142}{2}$

$\therefore \ x = 71$

b

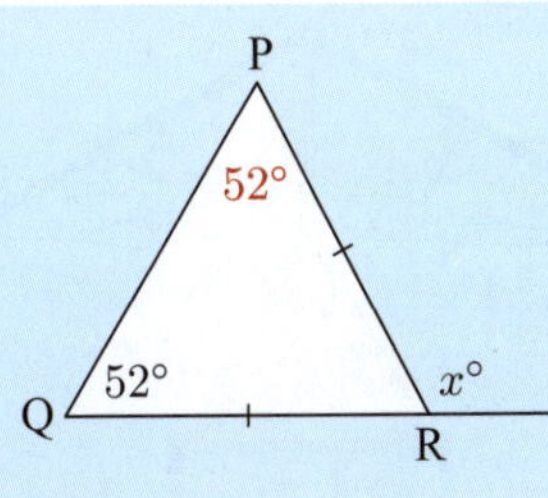

Since $PR = QR$, the triangle is isosceles.

$\therefore \ Q\widehat{P}R = 52°$ {isosceles triangle theorem}

$\therefore \ x = 52 + 52$ {exterior angle of a triangle}

$\therefore \ x = 104$

EXERCISE 9D

1 Find the value of x:

a

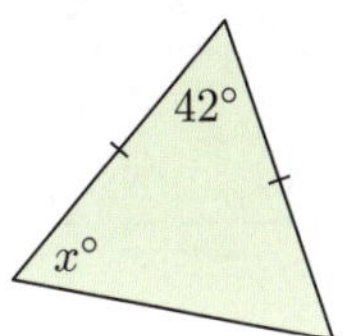

b

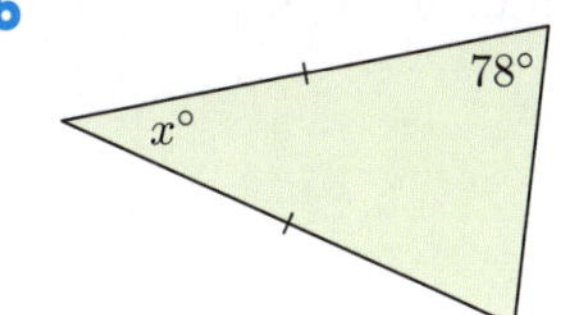

c

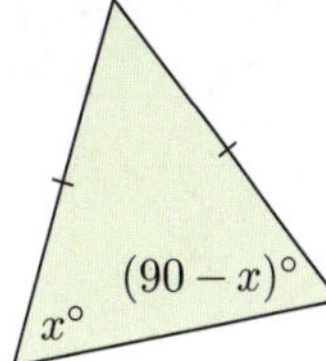

d

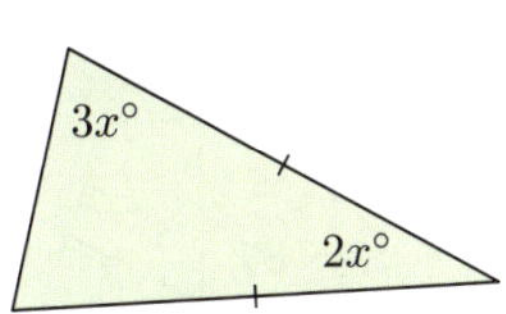

e

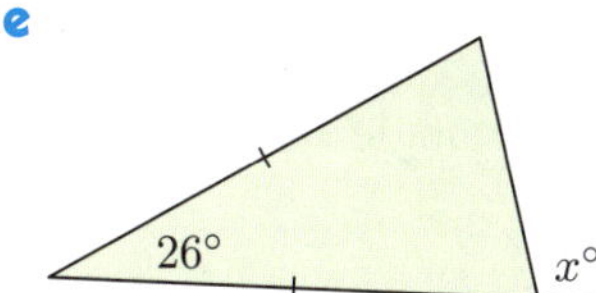

f

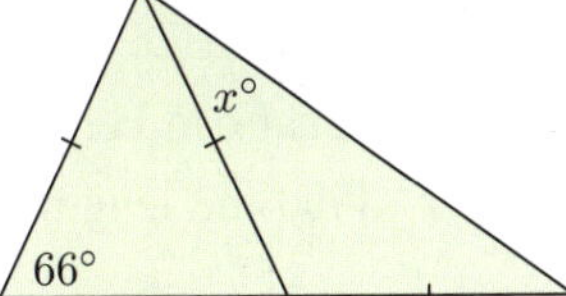

2 Find the value of x:

a

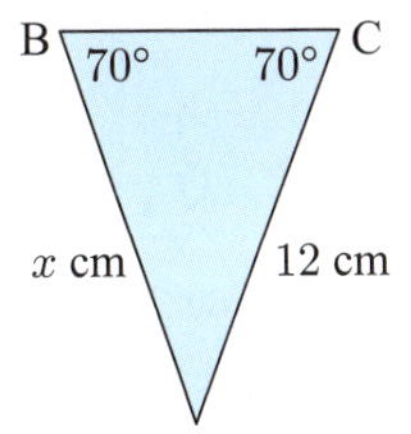

b

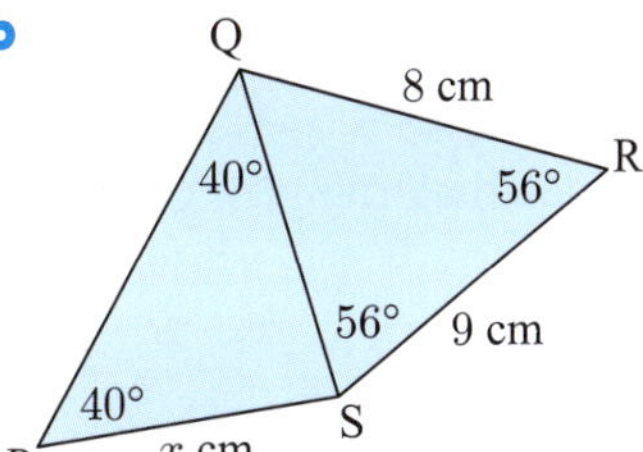

c

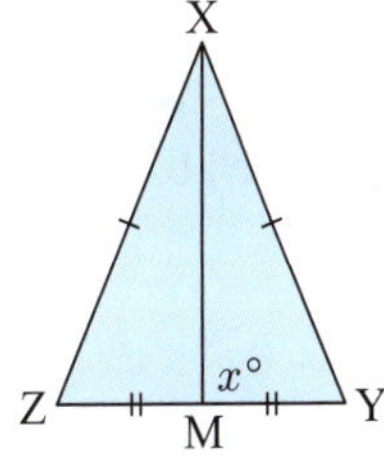

d

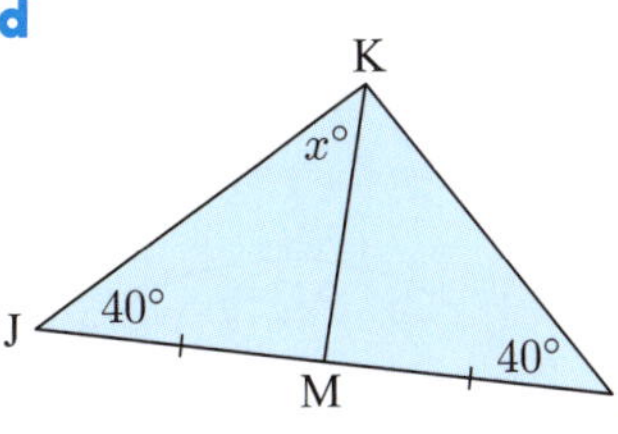

e

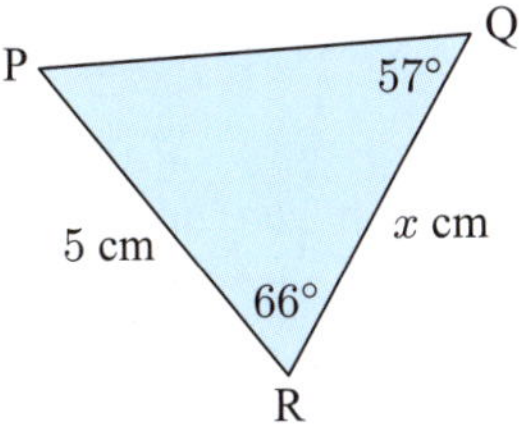

f

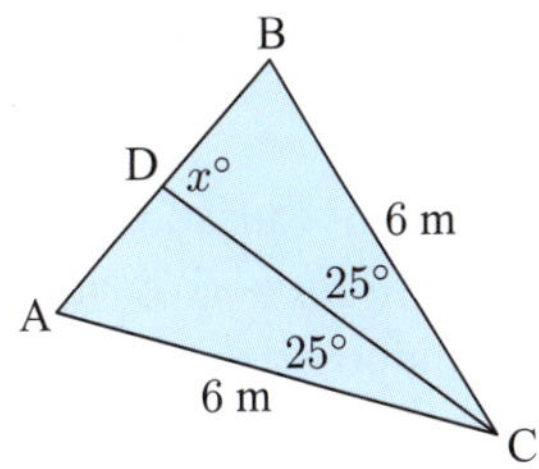

g

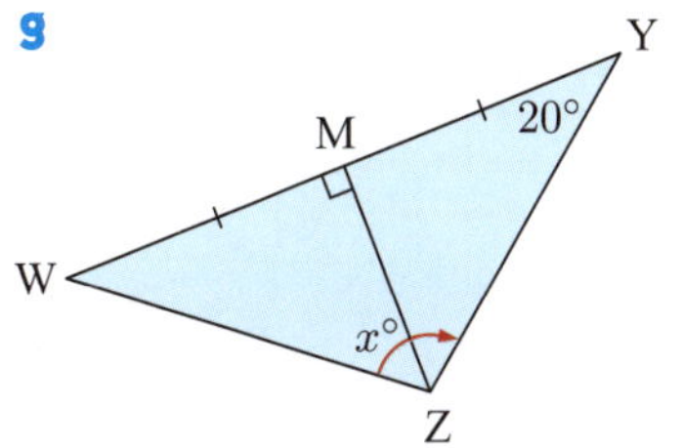

h

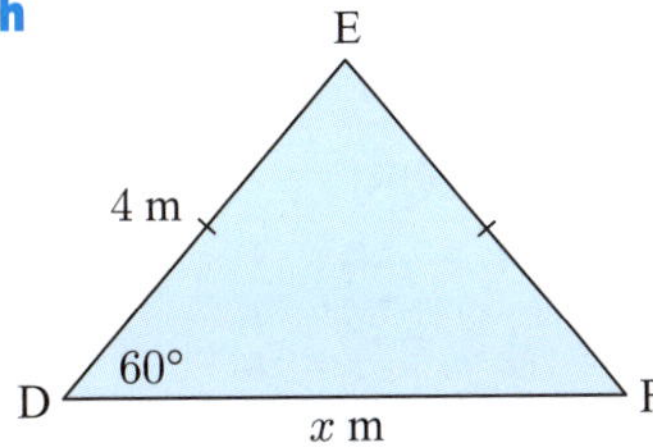

i

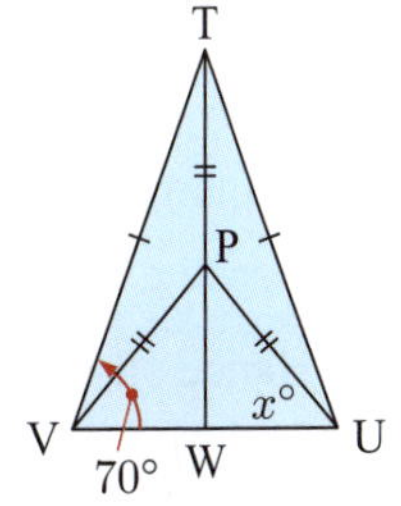

3 The triangular control frame of a hang glider has an angle of 46° between the two equal sides. Find the measure of the other two angles.

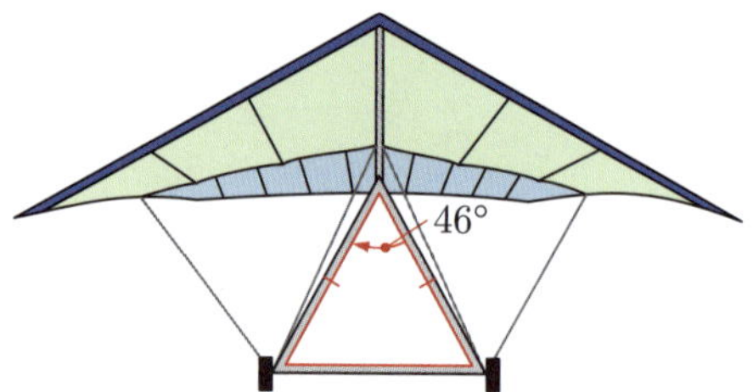

4 The figure alongside has not been drawn to scale, but the information given is correct.

  a Find the value of x.

  b What can be deduced about the triangle?

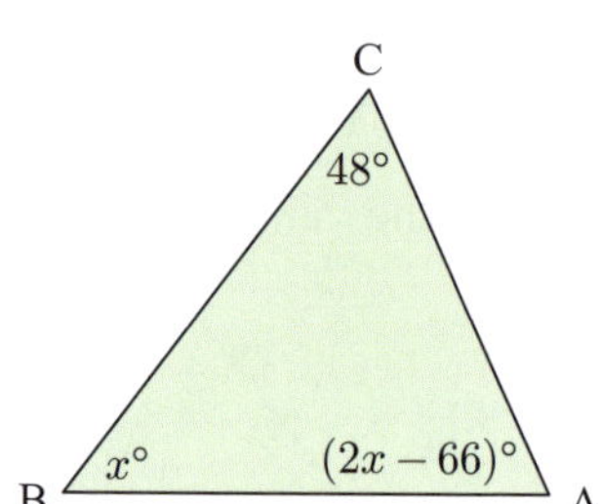

5 The figure alongside has not been drawn to scale, but the information given is correct.

  a Find $\widehat{ABD}$.

  b What can be deduced about triangle ABD?

  c Find the length BX.

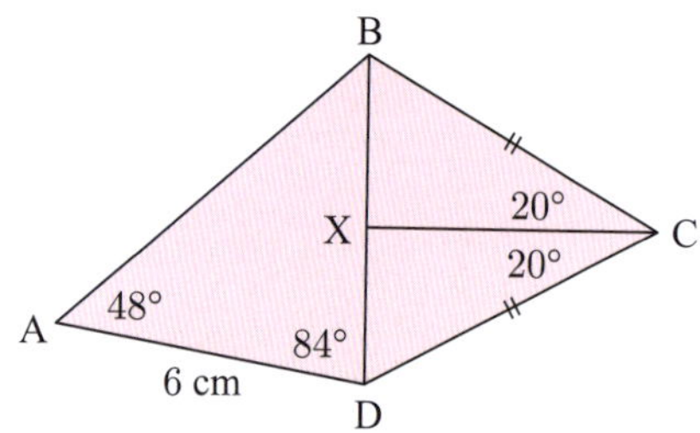

6 The base angles of an isosceles triangle each measure $(5x - 12)°$, and the vertical angle measures $(4x + 8)°$. What are the sizes of these angles?

7 Find the possible values of x such that triangle PQR is isosceles.

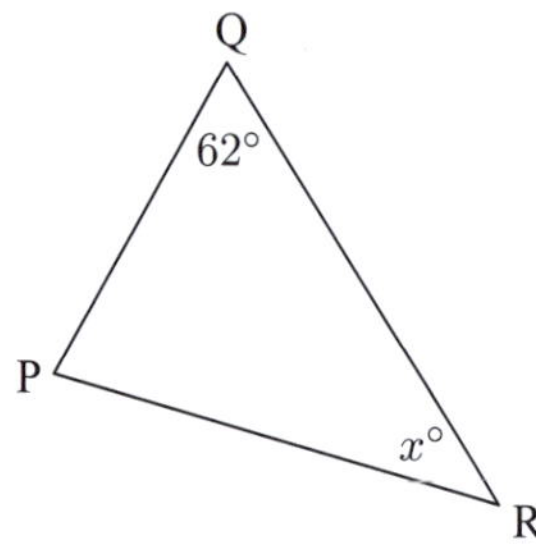

8 In this question we will prove that *the angle in a semi-circle is a right angle.*

Suppose [PQ] is a diameter of a circle with centre O, and R lies on the circle. We will prove that $\widehat{PRQ} = 90°$.

Let $\widehat{QPR} = \alpha°$ and $\widehat{PQR} = \beta°$ as shown.

  a Explain why $\widehat{PRO} = \alpha°$ and $\widehat{QRO} = \beta°$.

  b Hence explain why $2\alpha + 2\beta = 180$.

  c Hence show that $\widehat{PRQ} = (\alpha + \beta)° = 90°$.

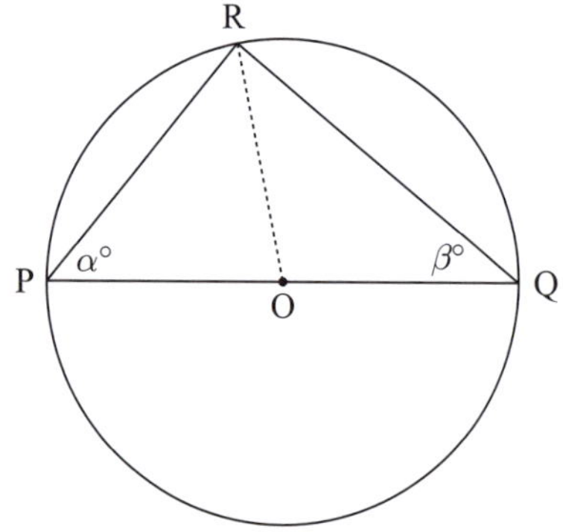

9 Copy and complete this proof that the longest side of a triangle is opposite its largest angle:

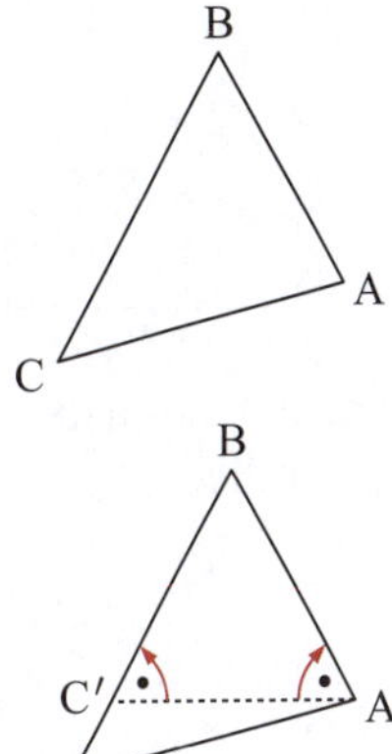

Label a triangle ABC so that $\widehat{BAC}$ is the largest angle.

Suppose we construct an isosceles triangle BAC′ with vertical angle $\widehat{ABC}$ and base [AC′].

The base angles are the average of $\widehat{BAC}$ and $\widehat{BCA}$. They are ...... than $\widehat{BAC}$ since $\widehat{BAC}$ is the ...... angle.

So, C′ must lie on [BC].

But   AB = BC′   {...... triangle}

   $\therefore$ AB < BC

Using the same procedure, we can construct an isosceles triangle B′AC with vertical angle $\widehat{ACB}$ and base [AB′]. Following the same argument, ... < BC.

So, BC is the longest side, and it is opposite the ....

E

QUADRILATERALS

A **quadrilateral** is a polygon which has four sides.

There are six special quadrilaterals we consider:

1 A **parallelogram** is a quadrilateral whose opposite sides are parallel.

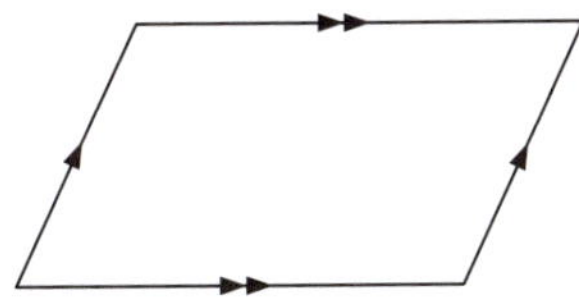

Properties:

- opposite sides are equal in length
- opposite angles are equal in size
- diagonals bisect each other.

2 A **rectangle** is a parallelogram with right angled corners.

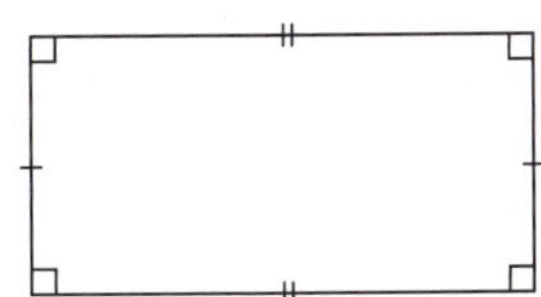

Properties:

- opposite sides are parallel and equal
- diagonals are equal in length
- diagonals bisect each other.

3 A **rhombus** is a quadrilateral with all sides equal in length.

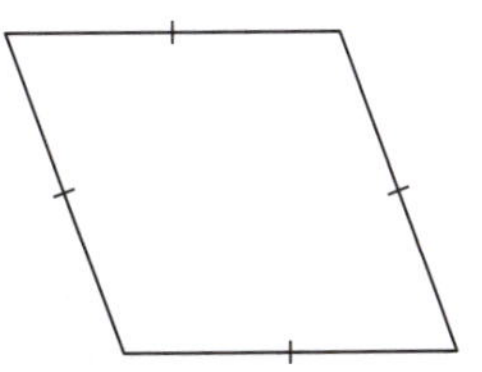

Properties:

- opposite sides are parallel
- opposite angles are equal in size
- diagonals bisect each other at right angles
- diagonals bisect the angles at each vertex.

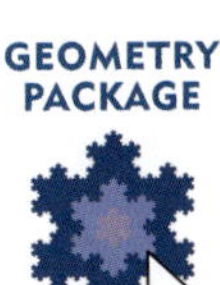

4   A **square** is a rectangle with all sides equal in length.

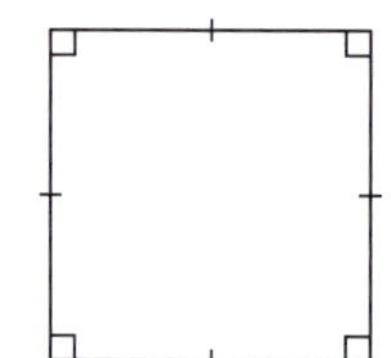

Properties:

- opposite sides are parallel
- diagonals bisect each other at right angles
- diagonals bisect the angles at each vertex
- diagonals are equal in length.

GEOMETRY PACKAGE

5   A **trapezium** is a quadrilateral with one pair of opposite sides parallel.

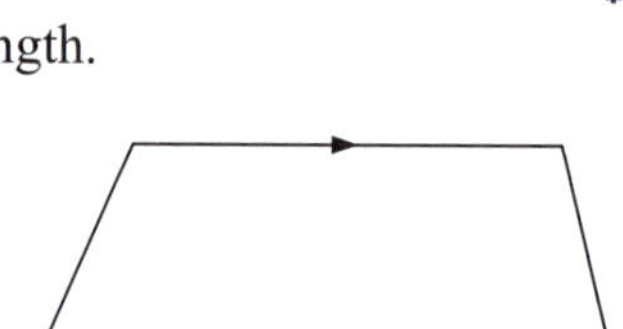

6   A **kite** is a quadrilateral with two pairs of adjacent sides equal in length.

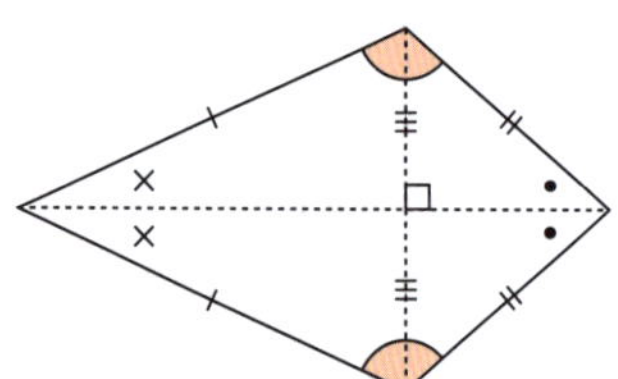

Properties:

- one diagonal is a line of symmetry
- one pair of opposite angles are equal
- diagonals cut each other at right angles
- *one* diagonal bisects *one* pair of angles at the vertices
- *one* of the diagonals bisects the other.

Example 4   ◀⑴ **Self Tutor**

Draw three diagrams to show the properties of a parallelogram.

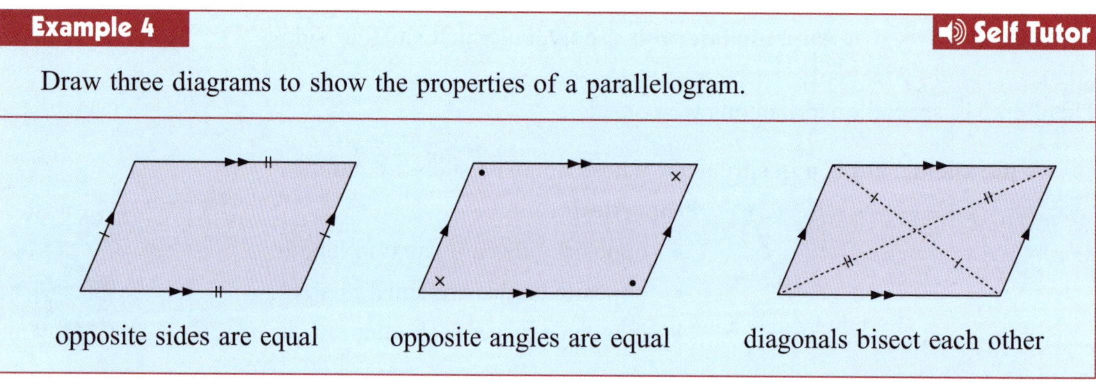

opposite sides are equal   opposite angles are equal   diagonals bisect each other

---

Example 5   ◀⑴ **Self Tutor**

Find the value of x, giving brief reasons for your answer:

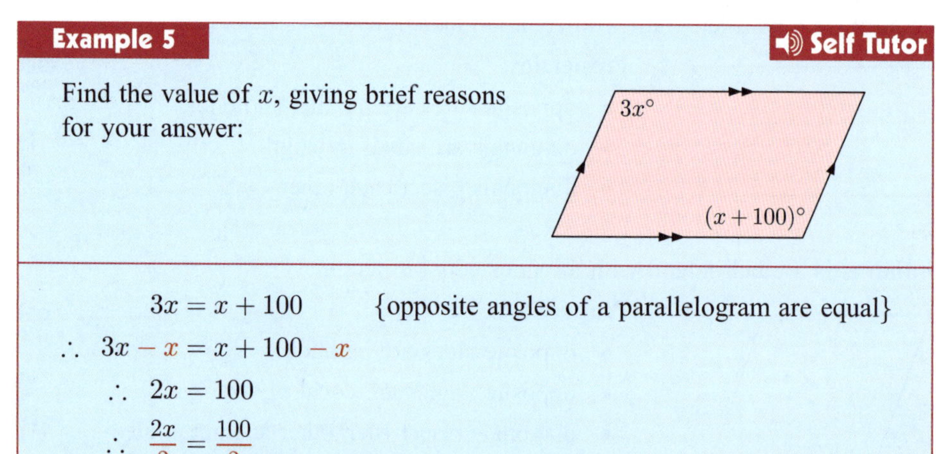

$$3x = x + 100 \quad \text{\{opposite angles of a parallelogram are equal\}}$$
$$\therefore \ 3x - x = x + 100 - x$$
$$\therefore \ 2x = 100$$
$$\therefore \ \frac{2x}{2} = \frac{100}{2}$$
$$\therefore \ x = 50$$

EXERCISE 9E

1   Draw a set of diagrams which show all the properties of:

   a a square
   b a kite
   c a rhombus.

2   Find the values of the unknowns:

a
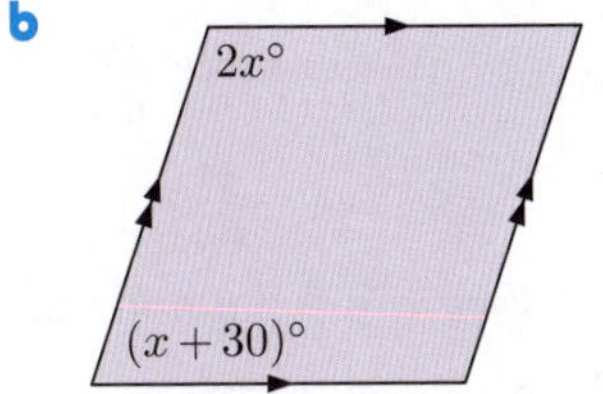

b
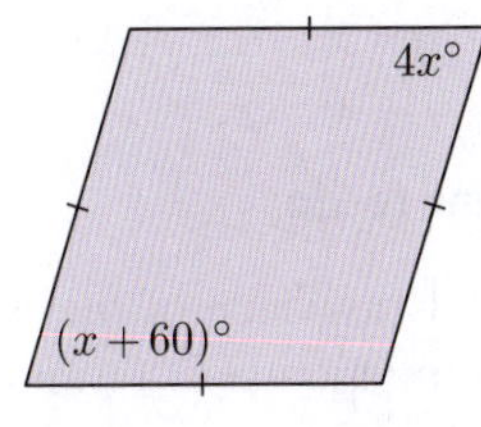

c
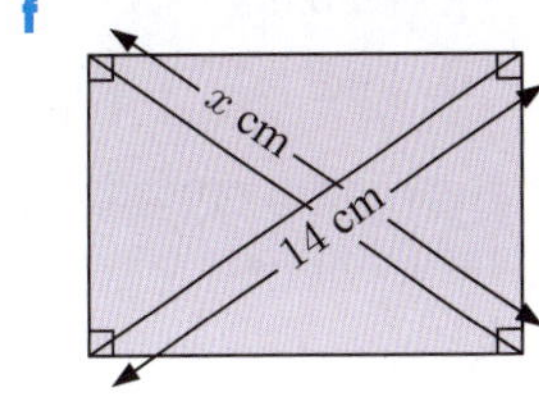

d
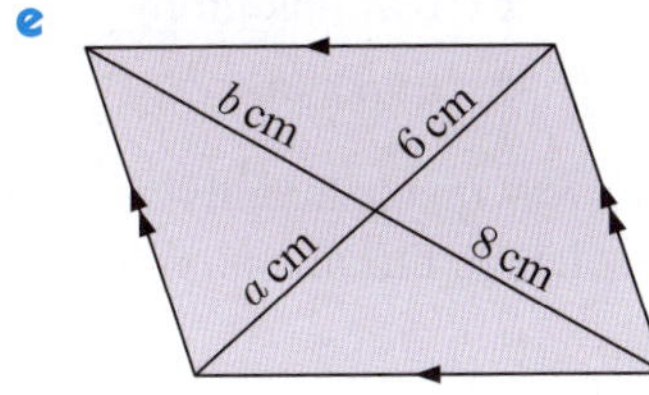

e
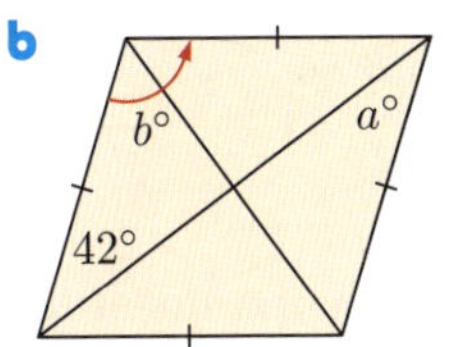

f
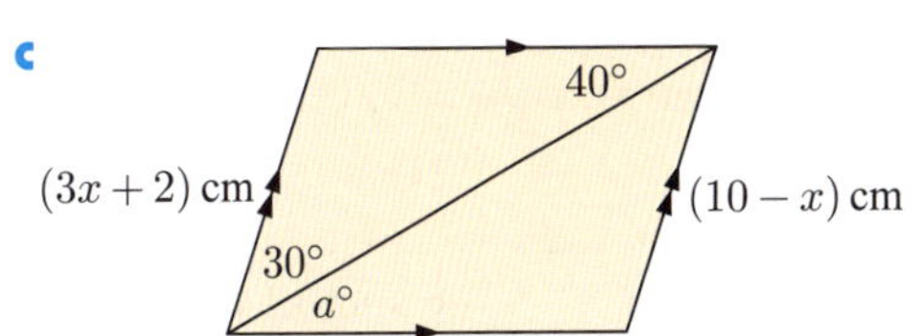

3   True or false?

   a A square is a quadrilateral in which all sides are equal in length.

   b A quadrilateral in which all sides are equal is a rhombus.

   c The diagonals of a parallelogram are equal in length.

   d The diagonals of a kite intersect at right angles.

4   Jarrod draws a quadrilateral ABCD and its diagonals [AC] and [BD]. He notices that the diagonals bisect each other at right angles. The shorter diagonal [AC] is 2 cm long, and the other diagonal is twice this length.

   a What type of quadrilateral is ABCD? Explain your answer.

   b Sketch and label quadrilateral ABCD.

5   Find the values of the unknowns:

a
b
c

6   The figures below are not drawn to scale, but the information on them is correct. What can you deduce about each quadrilateral?

a

b
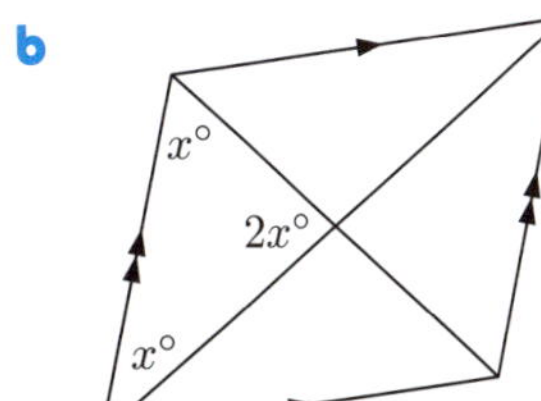

INVESTIGATION 2 — THE MIDPOINTS OF A QUADRILATERAL

In this Investigation we consider the quadrilateral formed when
the midpoints of the adjacent sides of a quadrilateral are joined.

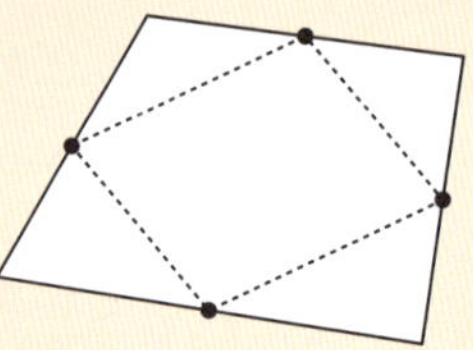

What to do:

1   Draw each of the following shapes. Find the midpoint of each side, and join the adjacent
    midpoints to form a quadrilateral.

    a   rectangle   **b**   parallelogram   **c**   rhombus

    d   kite   **e**   trapezium

2   Repeat the procedure with a few quadrilaterals of your own,
    including non-convex ones such as the one alongside.

3   Copy and complete: "When the midpoints of adjacent sides of a quadrilateral are joined,
    the resulting figure is always a ......".

F — ANGLE SUM OF A QUADRILATERAL

Suppose a quadrilateral is drawn
on a piece of paper.

If the four angles are torn off and
reassembled at a point, we notice that
the angle sum is always $360°$.

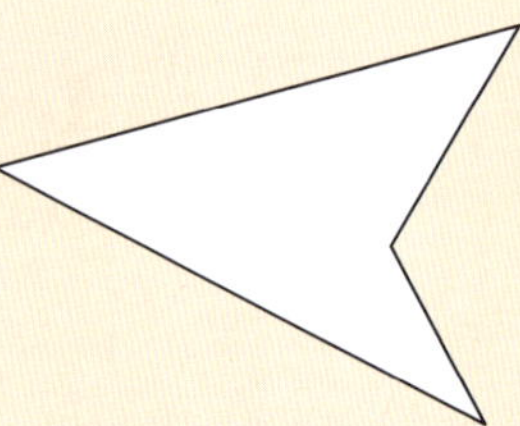

GEOMETRY PACKAGE

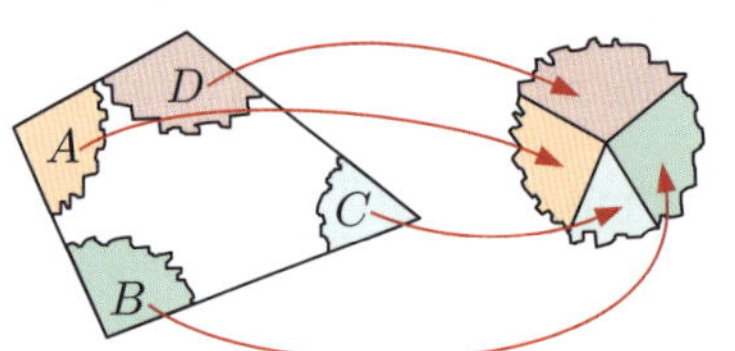

The sum of the interior angles of a quadrilateral is $360°$.

Proof:

Suppose we divide the quadrilateral into two triangles.

The sum of the interior angles of each triangle is $180°$,
so the angle sum of the quadrilateral $= 2 \times 180° = 360°$.

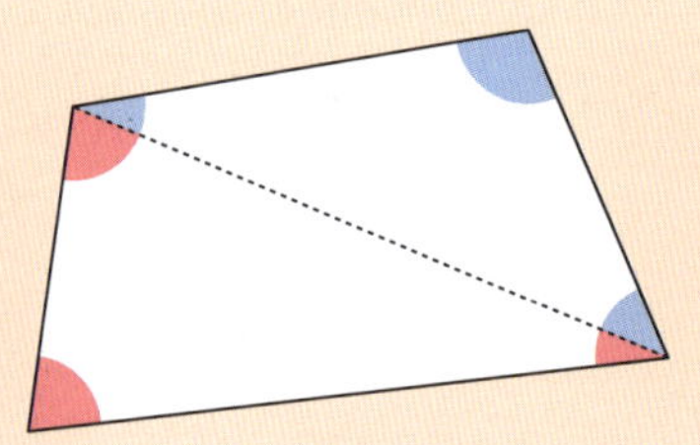

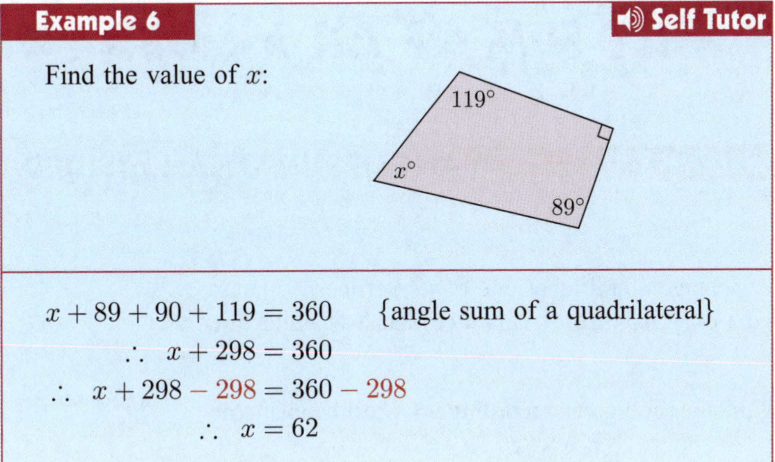

Example 6   ◄ᴺ) **Self Tutor**

Find the value of x:

$x + 89 + 90 + 119 = 360$ {angle sum of a quadrilateral}

$\therefore\ \ x + 298 = 360$

$\therefore\ \ x + 298 - 298 = 360 - 298$

$\therefore\ \ x = 62$

EXERCISE 9F

1 Find the value of x:

a
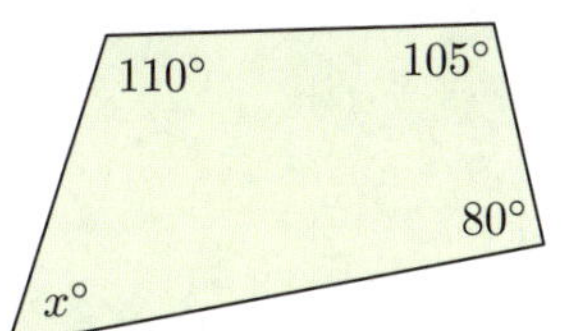

b
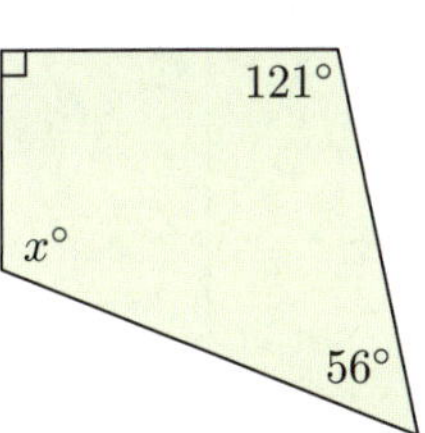

c
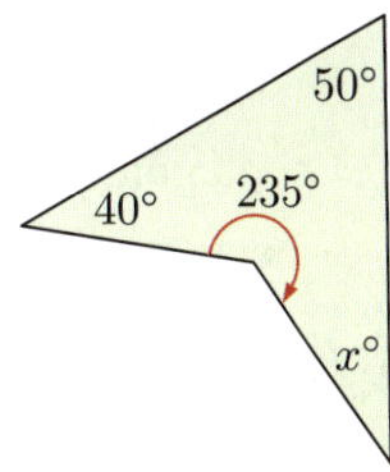

d
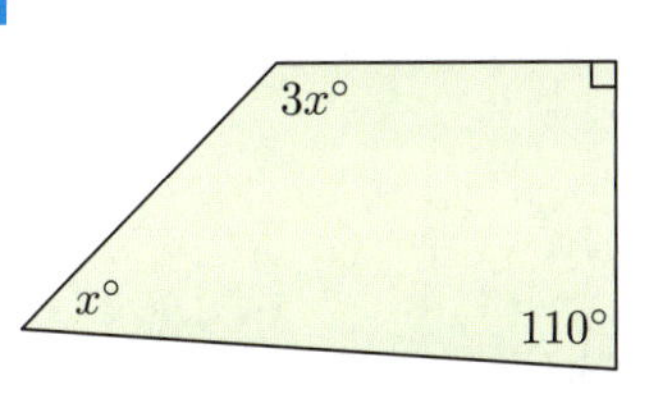

e
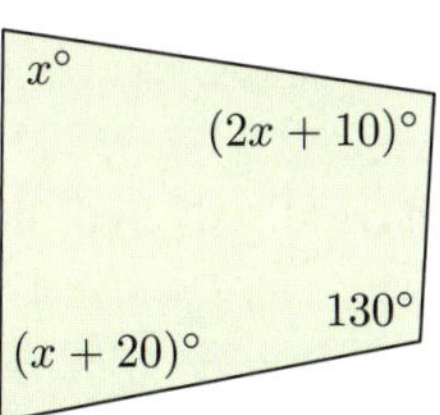

f
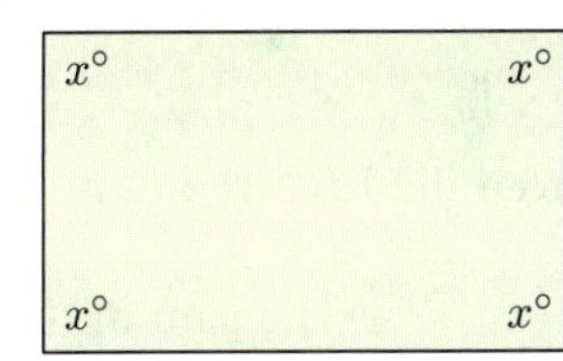

2 Find the values of the unknowns:

a
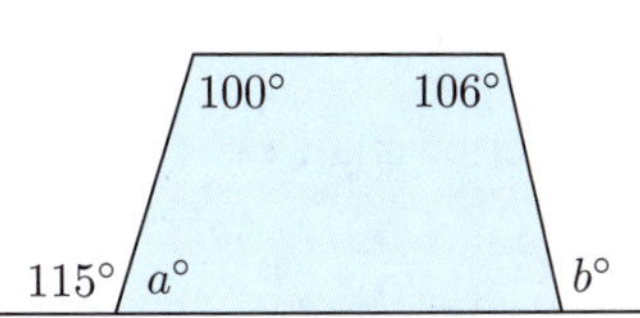

b
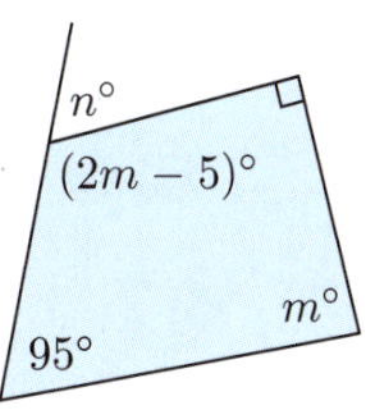

c
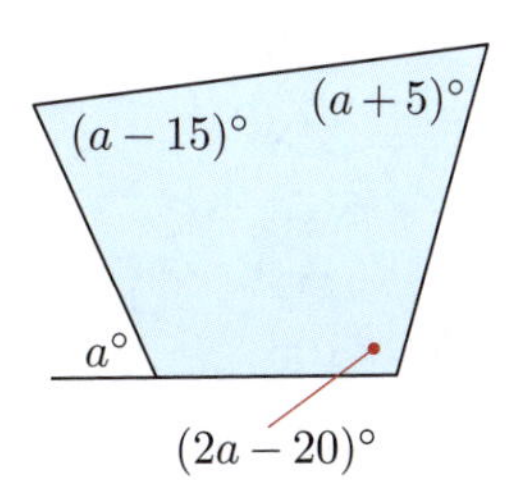

d
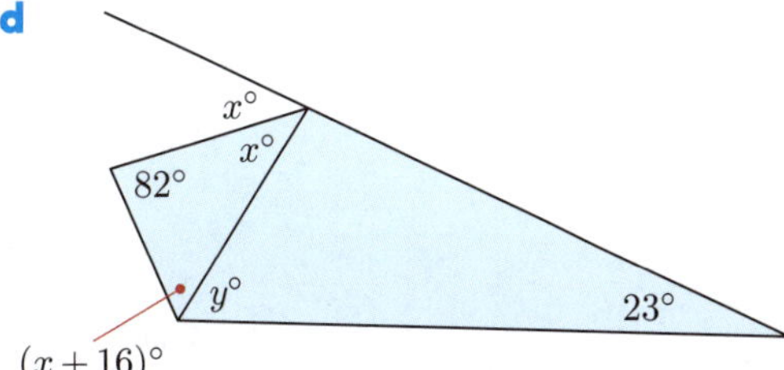

G ANGLE SUM OF AN n-SIDED POLYGON

INVESTIGATION 3 — ANGLE SUM OF AN n-SIDED POLYGON

What to do:

1 Draw any pentagon, and label one of its vertices A. Draw in all of the diagonals from A. Notice that 3 triangles are formed.

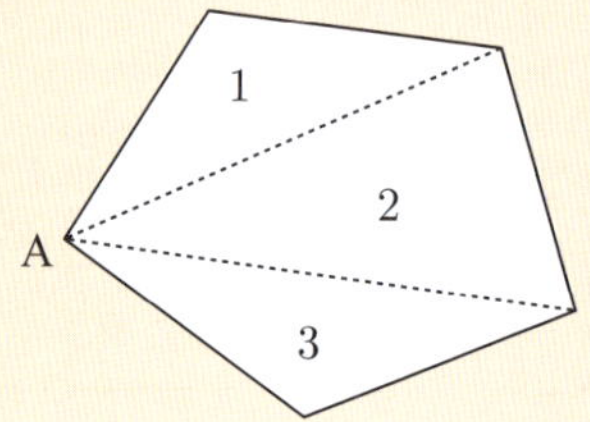

2 Repeat with other polygons, recording your results in a table like this:

Polygon	Number of sides	Number of triangles	Angle sum of polygon
quadrilateral	4	2	$2 \times 180° = 360°$
pentagon	5	3	
hexagon			
heptagon			
octagon			
20-gon			

3 Copy and complete:

"*The sum of the interior angles of any n-sided polygon is ...... $\times 180°$.*"

From the **Investigation** you should have discovered that:

> The sum of the interior angles of any n-sided polygon is $(n - 2) \times 180°$.

Example 7 | ◀ Self Tutor

Find the value of x:

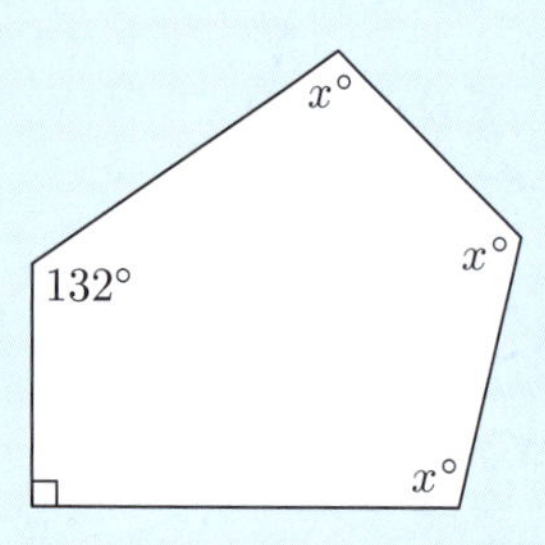

The figure has 5 sides.

$\therefore$ the sum of its interior angles is
$3 \times 180° = 540°$

$\therefore \quad x + x + x + 132 + 90 = 540$

$\therefore \quad 3x + 222 = 540$

$\therefore \quad 3x + 222 - 222 = 540 - 222$

$\therefore \quad 3x = 318$

$\therefore \quad \dfrac{3x}{3} = \dfrac{318}{3}$

$\therefore \quad x = 106$

EXERCISE 9G

1 Find the sum of the angles of:

a 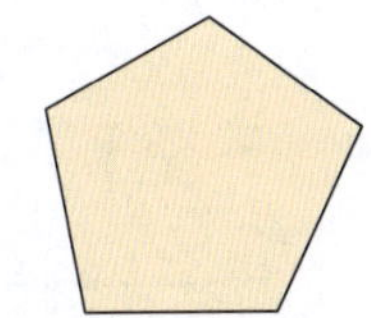   **b** 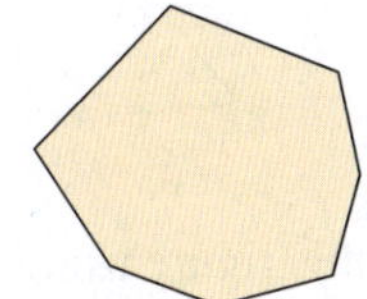   **c** 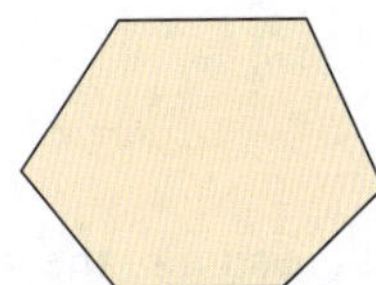

d a polygon with 12 sides   **e** a 15-gon.

2 Find the value of x:

a 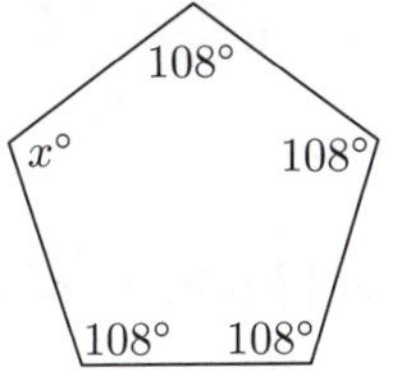   **b** 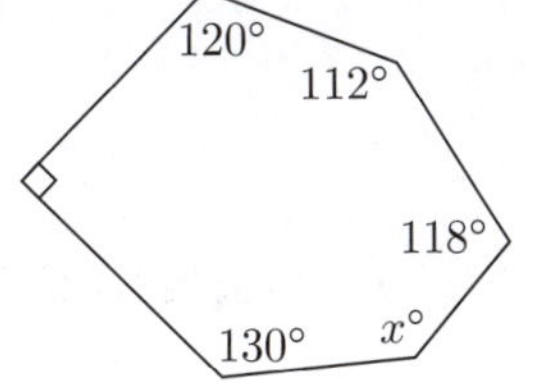   **c** 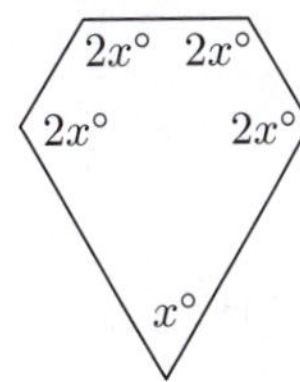

d 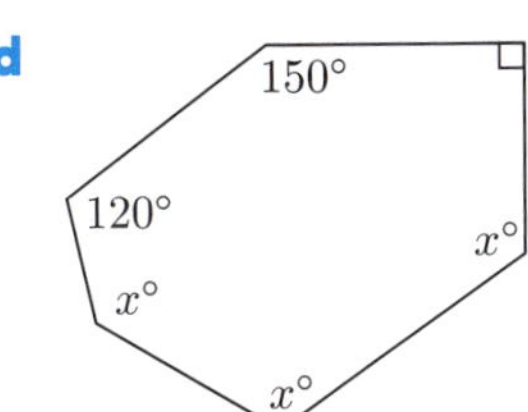   **e** 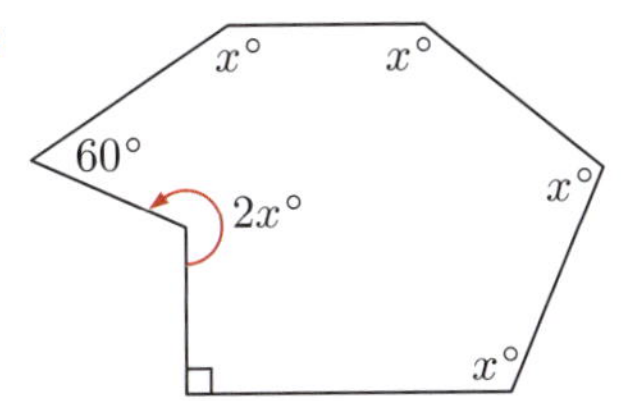   **f** 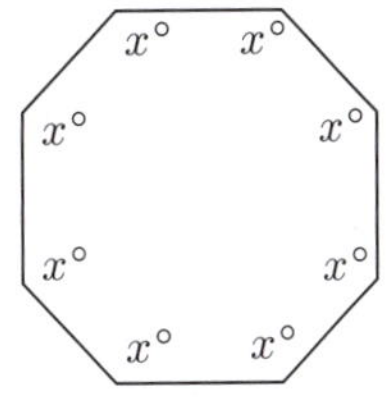

3 A pentagon has three right angles and two other equal angles. Find the size of each of the two equal angles.

4 The sum of the angles of a polygon is $1980°$. How many sides does the polygon have?

5 A **regular** polygon has all sides of equal length and all angles of equal size.

  a Copy and complete the following table:

Regular polygon	Number of sides	Sum of angles	Size of each angle
triangle			
quadrilateral			
pentagon			
hexagon			
octagon			
decagon			

  b Copy and complete:

    i The sum of the angles of an n-sided polygon is ......

    ii The size of each angle of a regular n-sided polygon is ......

  c Find the size of each angle of a regular 12-sided polygon.

6 **a** What is the maximum number of reflex angles that a hexagon can have?

  b Draw a hexagon with this number of reflex angles.

7   Find the sum of the shaded angles.

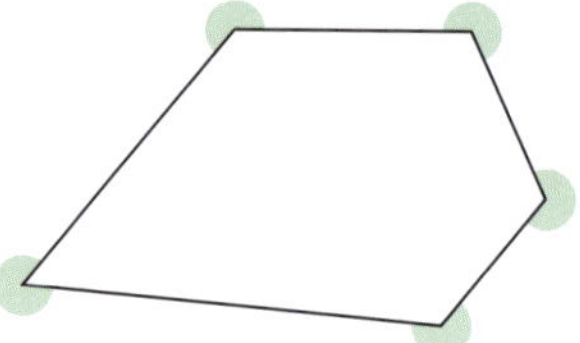

8   Find the sum of the interior angles of this figure using:

   a   the angle sum of a polygon formula

   b   properties of angles and parallel lines.

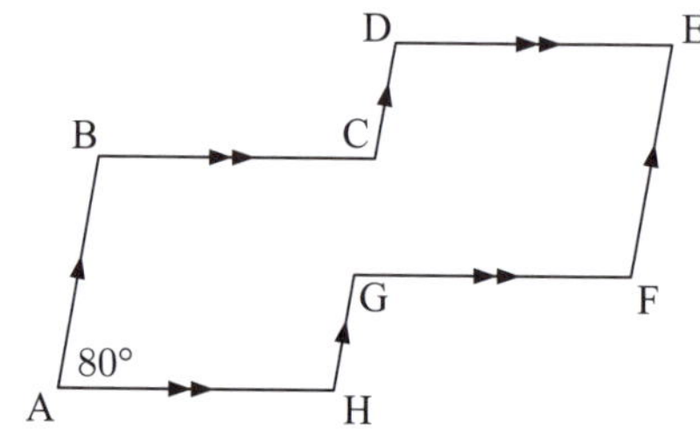

INVESTIGATION 4   **ANGLE SUM OF EXTERIOR ANGLES**

We have seen that the angle sum of the *interior* angles of an n-sided polygon is $(n-2) \times 180°$.

In this Investigation we consider the angle sum of the *exterior* angles of a convex n-sided polygon.

What to do:

1   **a**   Draw any triangle.
         Measure one exterior angle from each vertex.
         Find the sum of these angles.

   b   Repeat this procedure with two other triangles.
         Summarise your results.

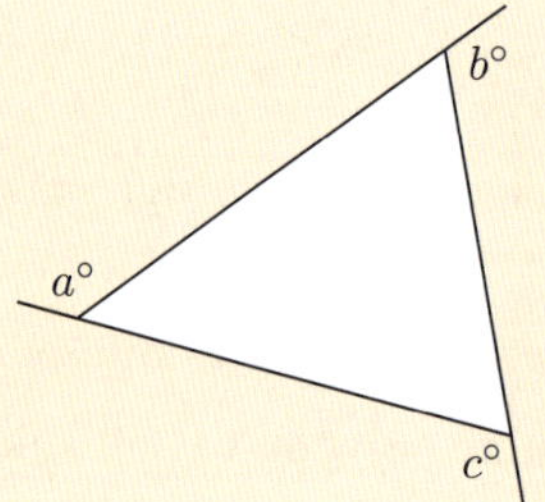

2   Find the sum of the exterior angles (one from each vertex) of *three* different convex quadrilaterals. Summarise your results.

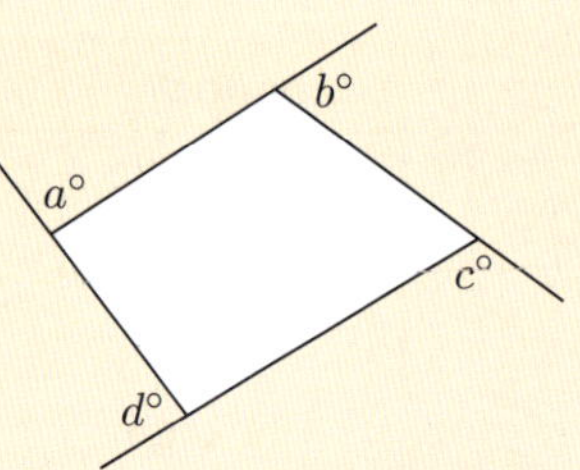

3   Predict the sum of the exterior angles of *any* convex polygon.
      Check that your answer is correct for a pentagon and a hexagon of your choosing.

4   *Explain* the sum of the exterior angles of a convex polygon by supposing you are an ant who walks in a clockwise direction around a polygon, starting from point A which is the midpoint of one side.

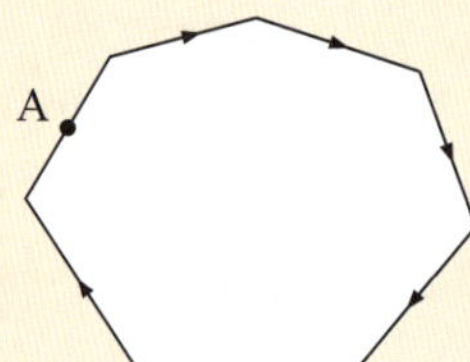

5 Copy and complete this proof:

At each vertex of an n-sided polygon,

the interior angle + the exterior angle = ...   {.....................}

∴ the sum of the n interior angles + the sum of the n exterior angles = ...

∴ ...... + the sum of the n exterior angles = ...

∴ the sum of the n exterior angles = ...

QUICK QUIZ

MULTIPLE CHOICE QUIZ

REVIEW SET 9A

1 **a** Use a compass to draw a circle with radius 22 mm.

  b State the diameter of the circle.

  c On the circle, draw a chord [AB] of length 3 cm, and draw a tangent to the circle at A.

2 Find the values of the unknowns, giving reasons for your answers:

a 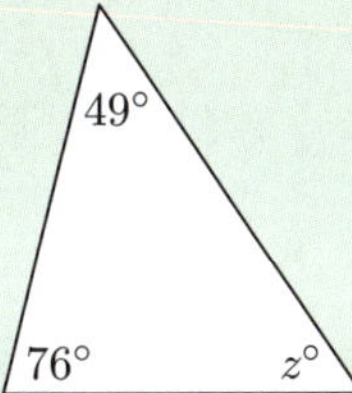

b 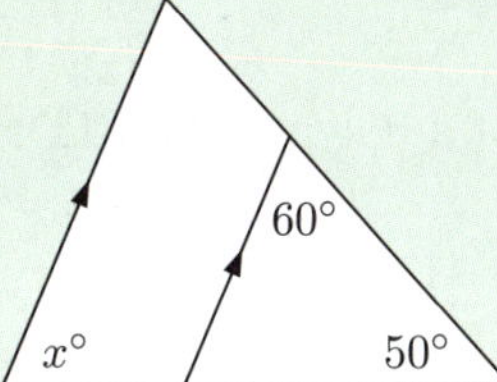

c 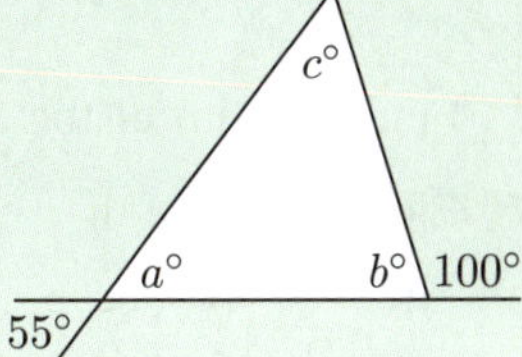

3 The triangle alongside is not drawn to scale. Identify its:

  a longest side

  b shortest side.

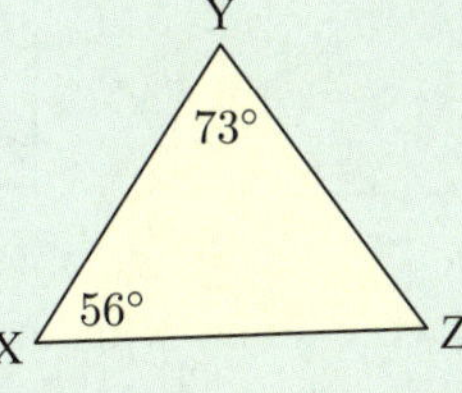

4 The quadrilaterals below are not drawn to scale, but the information on them is correct. Classify each quadrilateral.

a 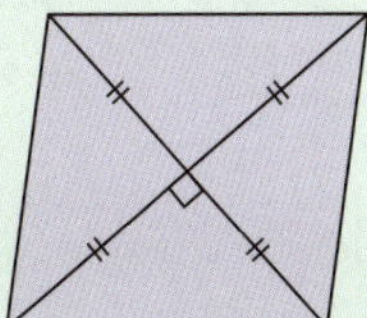

b 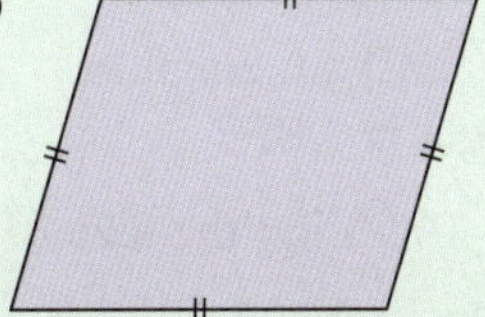

c 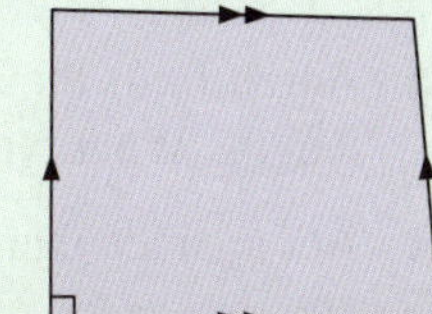

5 Classify each triangle by side lengths and by angles:

a
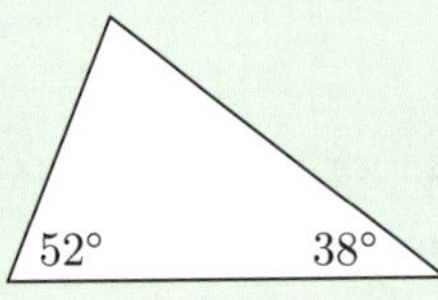

b
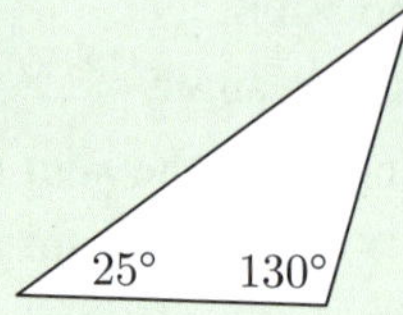

c
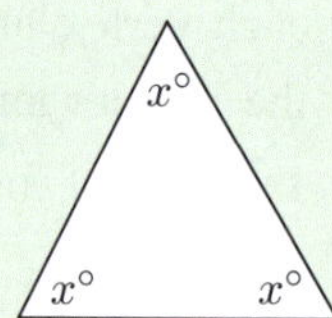

6 Find the values of the unknowns, giving brief reasons for your answers:

a
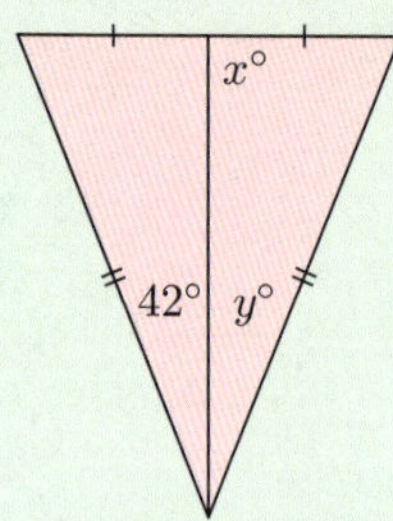

b
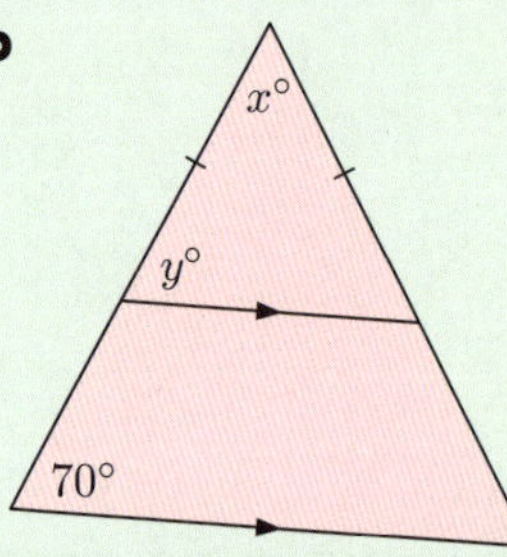

c
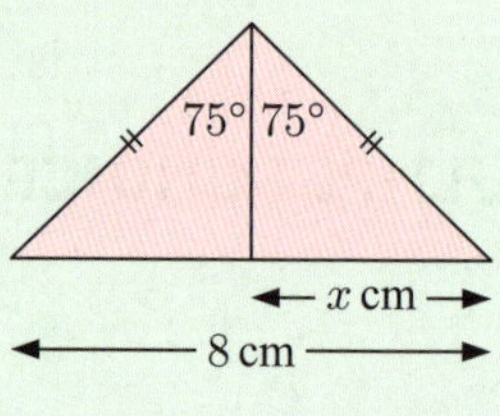

7 Find the values of the unknowns:

a
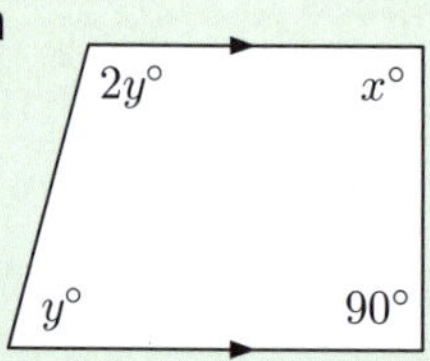

b
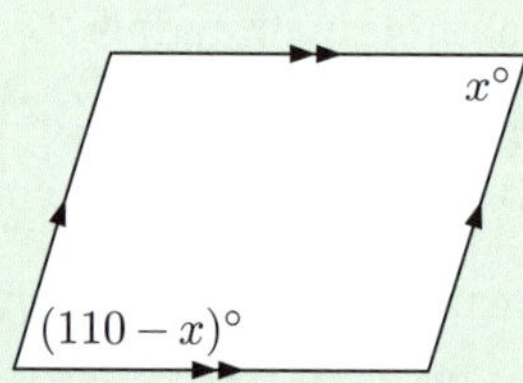

c
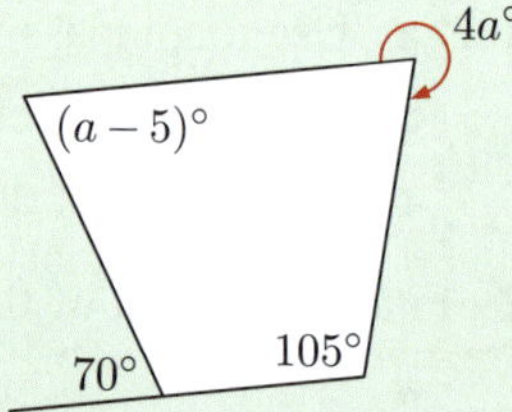

8 Find the sum of the angles of a polygon with 11 sides.

9 Find the value of x:

a
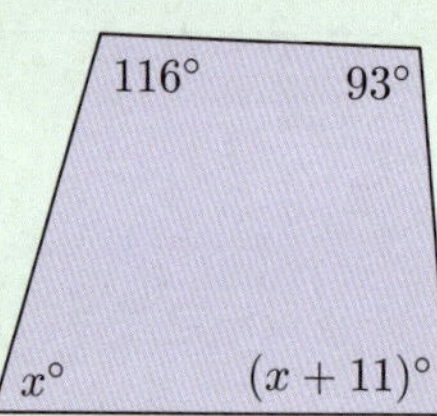

b
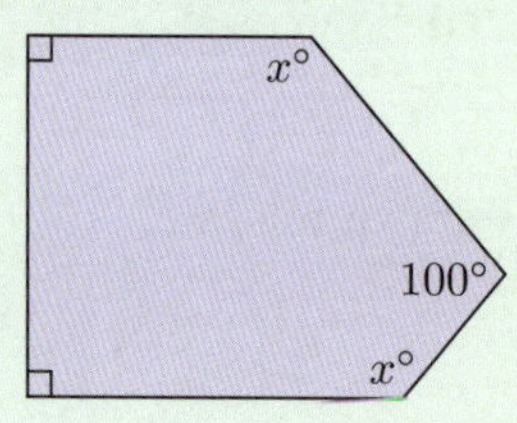

c
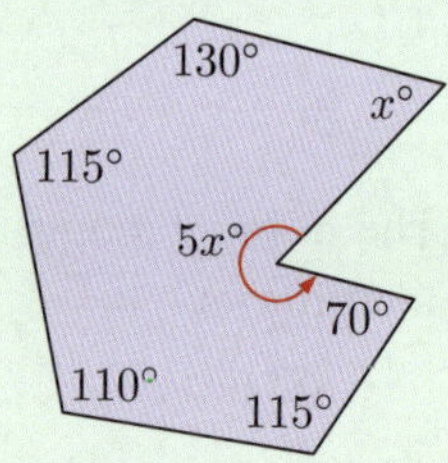

10 Aaron's bicycle has a frame with the side view shown. Strut [BE] is parallel to the front fork [CD].

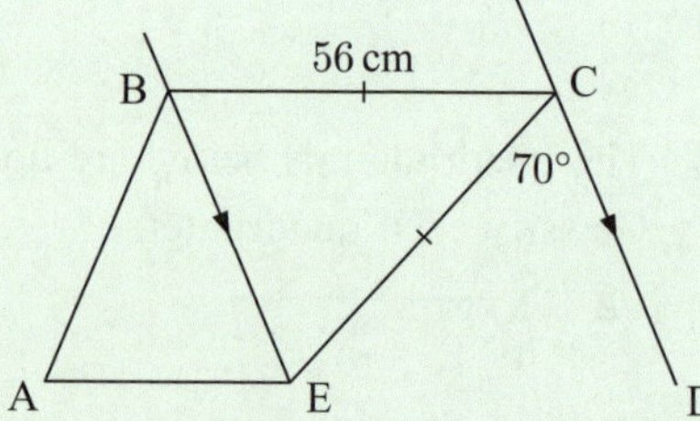

a Find the measure of:

   i $\widehat{CEB}$ **ii** $\widehat{CBE}$ **iii** $\widehat{BCE}$.

b Aaron finds that [AB] and [BE] are equal in length. What type of triangle is ABE?

c Aaron finds that $\widehat{ABE} = 40°$. Find the size of the base angles of triangle ABE.

d What can be said about line segments [BC] and [AE]?

e Classify quadrilateral ABCE.

11 The seven pieces of a tangram puzzle fit together to form a square as shown.

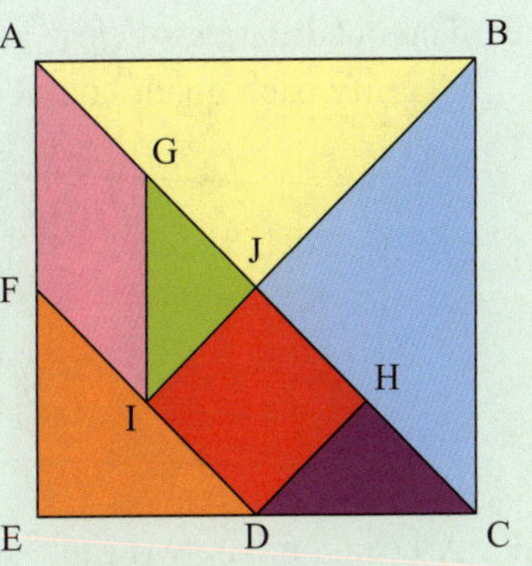

a Classify triangle ABC, and hence find the size of $\widehat{BAC}$.

b What is the measure of $\widehat{GAF}$? Give a reason for your answer.

c B, J, and E are collinear. Use this fact to explain why $\widehat{BJC}$ is a right angle.

d Triangle GJI is isosceles. Find, giving reasons, the measure of $\widehat{IGA}$.

REVIEW SET 9B

1 Find the radius of a circle with diameter 16 cm.

2 Classify this triangle as:

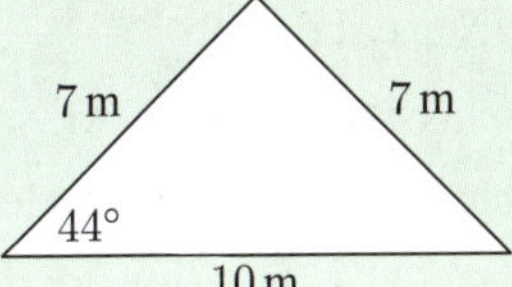

a scalene, isosceles, or equilateral

b acute angled, obtuse angled, or right angled.

3 Draw two diagrams to illustrate the properties of a rectangle.

4 Find the values of the unknowns:

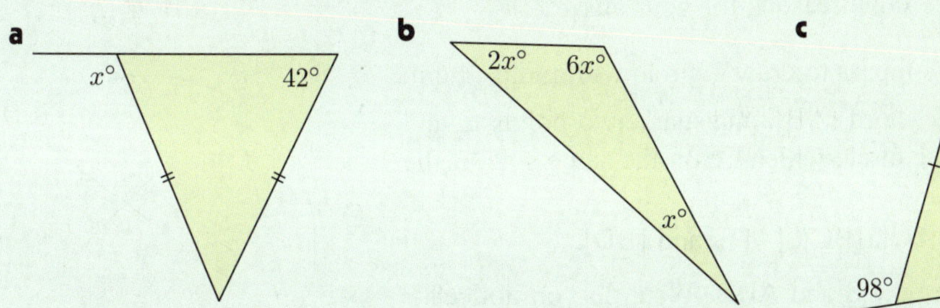

5 True or false?

a The diagonals of a square are equal in length.

b The diagonals of a rhombus intersect at right angles.

6 The vertical angle of an isosceles triangle is $82°$. Find the size of each base angle.

7 Find the values of the unknowns, giving brief reasons for your answers:

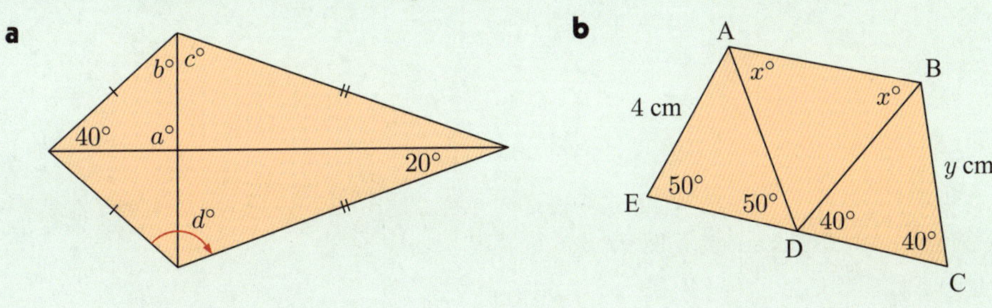

8 The quadrilaterals below are not drawn to scale, but the information on them is correct. Classify each quadrilateral.

a

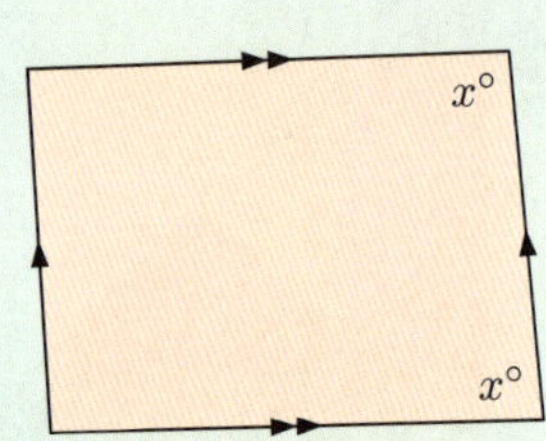

b

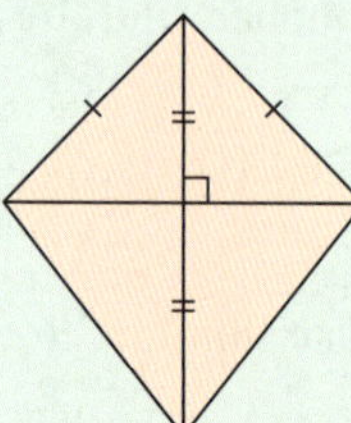

c

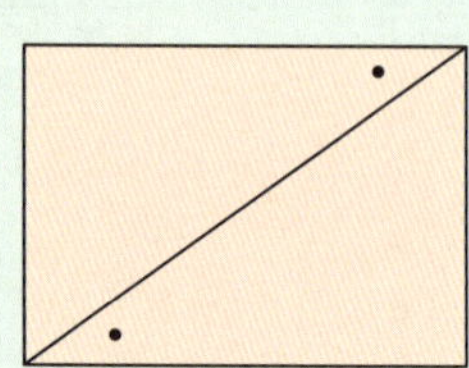

9 A hexagon has two right angles and four other equal angles. What is the size of each of the four equal angles?

10 **a** What is the maximum number of reflex angles a pentagon can have?

  b Draw a pentagon with this number of reflex angles.

11 Lisa is making a plane from a rectangular piece of paper. She starts by folding two corners to point X as shown.

  a Classify triangles AFX and EFX.

  b Hence, classify triangle AFE.

  c Is [AE] parallel to [BD]? Explain your answer.

  d Lisa suspects that ABDE is a square. Is she correct? Give reasons for your answer.

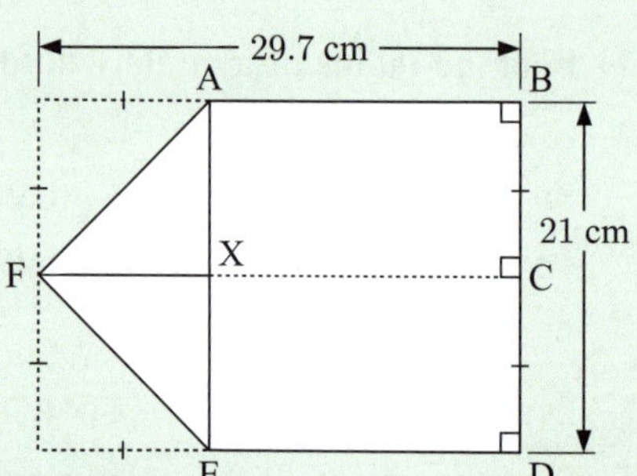

12 **a** Use a compass to draw a circle with radius 4 cm.

  b Draw a chord [AB], and mark two points C and D on the circle which lie on the same side of the chord.
  Draw [AC], [BC], [AD], and [BD].

  c Measure $A\hat{C}B$ and $A\hat{D}B$. What do you notice?

  d Repeat **b** and **c** with a different chord [AB] and different points C and D.

  e Find the relationship between $A\hat{C}B$ and $A\hat{D}B$.

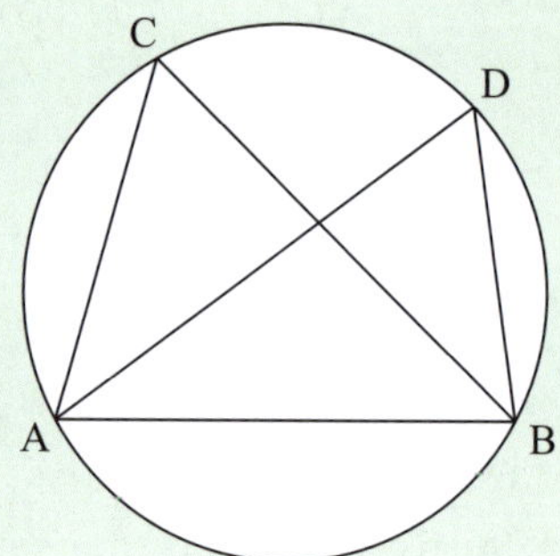

Algebra: Formulae

Contents:

OPENING PROBLEM

Leonard builds ladders using pieces of metal which are all the same length. The more pieces Leonard uses, the bigger the ladder he makes.

Things to think about:

a How many metal pieces does Leonard require to make a:

   i 4 rung ladder   **ii** 5 rung ladder

   iii 12 rung ladder?

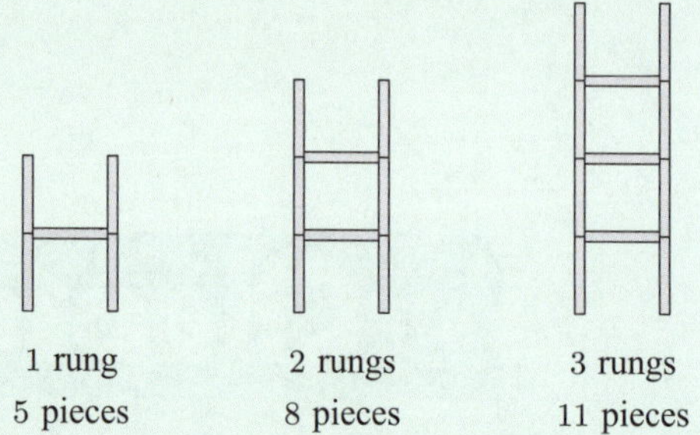

b How can we connect the *number of metal pieces* and the *number of rungs* in a ladder using a mathematical equation?

c What size ladder can Leonard build with 50 metal pieces?

In the **Opening Problem**, the *number of metal pieces* required is related to the *number of rungs* in a ladder. In this Chapter we will learn how to write a **formula** to describe the relationship between these variables.

> A **formula** is an equation which connects two or more variables.
>
> The plural of formula is **formulae**.

A   NUMBER CRUNCHING MACHINES

Consider a machine which operates on numbers.

For any **input number** fed into the machine, the machine calculates an **output number** according to a rule.

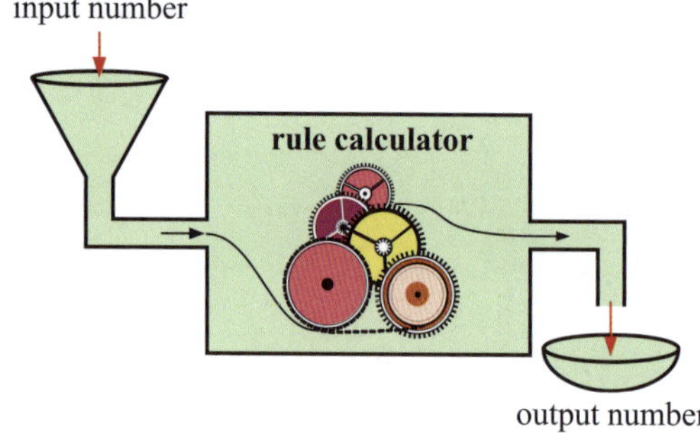

For example, consider the rule "the output number M is two times the input number n, plus seven".

We can use this rule to calculate the output number for the input numbers 1, 2, 3, ...

Input (n)	Calculation	Output (M)
1	$2 \times 1 + 7$	9
2	$2 \times 2 + 7$	11
3	$2 \times 3 + 7$	13
$\vdots$	$\vdots$	$\vdots$

We can also write a **formula** connecting the input number n and the output number M.

The formula is $M = 2 \times n + 7$ or $M = 2n + 7$.

We say that M is the *subject* of the formula, and that M is written *in terms of* n.

The **subject** of a formula is the variable which appears by itself on one side of the equation.

EXERCISE 10A

1   Copy and complete each table by applying the rule to each input number:

a   *The input number plus six*

Input	Calculation	Output
1		
2		
3		
4		

b   *Three times the input number*

Input	Calculation	Output
2		
3		
6		
9		

c   *Multiply by 4 then subtract 3*

Input	Calculation	Output
1		
4		
9		
12		

d   *Add 5 then double*

Input	Calculation	Output
0		
3		
7		
10		

e   *Multiply the input number by itself*

Input	Calculation	Output
2		
5		
8		
10		

f   *Halve the input number*

Input	Calculation	Output
4		
8		
12		
20		

g   *Subtract 4 then multiply by 10*

Input	Calculation	Output
5		
7		
9		
11		

h   *Add 8 then divide by 3*

Input	Calculation	Output
7		
16		
22		
28		

2   For each rule, write a formula connecting the input number n and the output number M.

   a   The output number is five times the input number.

   b   The output number is three times the input number, minus 4.

   c   The output number is 1 plus the input number, divided by 2.

   d   The output number is the input number multiplied by itself, plus 7.

B   FINDING THE FORMULA

Given a table of input and output numbers, we can often work out the formula that was used.

Consider these rules with input and output numbers:

- Rule: *Two times the input number.*

Input	Output
1	2
2	4
3	6
4	8

When the input number is increased by 1, the output number increases by 2.

- Rule: *Three times the input number, plus one.*

Input	Output
1	4
2	7
3	10
4	13

When the input number is increased by 1, the output number increases by 3.

- Rule: *Five times the input number, minus two.*

Input	Output
1	3
2	8
3	13
4	18

When the input number is increased by 1, the output number increases by 5.

- Rule: *Minus two times the input number, plus nine.*

Input	Output
1	7
2	5
3	3
4	1

When the input number is increased by 1, the output number decreases by 2.

> If the output number increases by k each time the input number increases by 1, the rule contains "k times the input number".

Example 1   ◀ Self Tutor

Find the formula which connects each input number n with its corresponding output number M.

Input (n)	Output (M)
1	6
2	10
3	14
4	18

When the input number is increased by 1, the output number increases by 4.

So, the rule contains "four times the input number".

We therefore compare $4n$ with M.

$4n$	4	8	12	16
M	6	10	14	18

M is always 2 more than $4n$, so the rule is *four times the input number, plus two.*

The formula is $M = 4n + 2$.

EXERCISE 10B

1 Find the formula which connects each input number n with its corresponding output number M:

a

Input (n)	Output (M)
1	4
2	5
3	6
4	7

b

Input (n)	Output (M)
1	5
2	10
3	15
4	20

c

Input (n)	Output (M)
1	5
2	7
3	9
4	11

d

Input (n)	Output (M)
1	6
2	10
3	14
4	18

e

Input (n)	Output (M)
1	12
2	11
3	10
4	9

f

Input (n)	Output (M)
1	2
2	5
3	8
4	11

g

Input (n)	Output (M)
1	15
2	10
3	5
4	0

h

Input (n)	Output (M)
1	5
2	11
3	17
4	23

2 For each table, write a formula connecting the variables.

a

n	1	2	3	4	5	6
M	6	8	10	12	14	16

b

x	1	2	3	4	5	6
y	11	22	33	44	55	66

c

a	1	2	3	4	5	6
P	3	10	17	24	31	38

d

m	1	2	3	4	5	6
N	23	19	15	11	7	3

e

d	1	2	3	4	5	6
C	1	4	7	10	13	16

f

h	1	2	3	4	5	6
L	8	17	26	35	44	53

g

t	1	3	5	7	9	11
D	5	13	21	29	37	45

h

x	1	3	5	7	9	11
y	31	25	19	13	7	1

i

x	2	4	6	8	10	12
y	14	11	8	5	2	-1

j

n	1	3	4	6	8	11
P	-13	-1	5	17	29	47

C   SUBSTITUTING INTO FORMULAE

If we know the values of all but one variable in a formula, we can **substitute** these known values
and hence find the value of the remaining variable.

Example 2   ◀)) **Self Tutor**

Consider the formula $D = 3x - 11$. Find the value of:

   a D when $x = 10$ **b** x when $D = 7$.

a When $x = 10$,
$$D = 3 \times 10 - 11$$
$$\therefore \ D = 30 - 11$$
$$\therefore \ D = 19$$

b When $D = 7$,
$$7 = 3x - 11$$
$$\therefore \ 7 + 11 = 3x - 11 + 11$$
$$\therefore \ 18 = 3x$$
$$\therefore \ \frac{18}{3} = \frac{3x}{3}$$
$$\therefore \ x = 6$$

EXERCISE 10C

1 Consider the formula $P = 2t + 7$. Find the value of:

   a P when $t = 6$ **b** t when $P = 25$.

2 Consider the formula $C = 18 - 5d$.

   a Find the value of C when:

      i $d = 2$ **ii** $d = 6$ **iii** $d = -4$.

   b Find the value of d when:

      i $C = 3$ **ii** $C = -10$.

3 Consider the formula $A = \frac{1}{2}b - 5$.

   a Find the value of A when:

      i $b = 12$ **ii** $b = 20$ **iii** $b = 38$.

   b Find the value of b when:

      i $A = 9$ **ii** $A = -3$ **iii** $A = -17$.

4 The cost of hiring a bus for p people is given by $C = 12p + 80$ dollars.
Find the cost of hiring the bus for:

   a 12 people   **b** 20 people   **c** 45 people.

5 The number of bread rolls in a bakery h hours after the bakery opens for the day is given by
$B = 400 - 50h$.

   a Find the value of B when $h = 0$. Explain what this value means.

   b Find the number of bread rolls in the bakery after:

      i 2 hours   **ii** 5 hours   **iii** 30 minutes.

   c How long will it take for the bakery to run out of bread rolls?

6 The cost C of an international phone call is given by $C = 60t + 120$ cents, where t is the length of the call in minutes.

    a Find the cost in *dollars* for a call that lasts:

       i 5 minutes   **ii** 20 minutes.

    b Find the length of a call which costs $15.

7 John has mass 62.5 kg and Jessica has mass x kg. Their average mass is $M = \dfrac{62.5 + x}{2}$ kg.

    a If Jessica has mass $x = 58.3$ kg, find the average mass M.

    b If the average mass $M = 59.1$ kg, find Jessica's mass x.

8 Consider the formula $M = 2n^2 - 1$.

    a Find the value of M when:

       i $n = 1$ **ii** $n = 2$ **iii** $n = 4$.

    b Find the possible values of n when:

       i $M = -1$ **ii** $M = 17$ **iii** $M = 71$.

9 Match each formula to the correct table of values:

    a $y = x^2 + 2$

A

x	0	2	4	6
y	4	6	8	10

    b $y = 2x + 4$

B

x	0	1	2	3
y	2	6	10	14

    c $y = 4x + 2$

C

x	0	1	2	3
y	4	9	16	25

    d $y = 4 + x$

D

x	0	2	4	6
y	2	6	18	38

    e $y = (x + 2)^2$

E

x	0	1	2	3
y	4	6	8	10

10 Consider the formula $P = 5x - 2y$.

    a Find the value of P when:

       i $x = 4$ and $y = 3$ **ii** $x = 3$ and $y = 7$.

    b Find the value of x when:

       i $P = 23$ and $y = 6$ **ii** $P = 44$ and $y = -2$.

    c Find the value of y when:

       i $P = 8$ and $x = 4$ **ii** $P = 37$ and $x = 11$.

11 A restaurant with a tables for 2 and b tables for 4 can seat $P = 2a + 4b$ people.

    a Find the number of people who can be seated if $a = 12$ and $b = 8$.

    b Find a if $P = 38$ and $b = 6$.

    c Find b if $P = 104$ and $a = 26$.

12 The size of a force when a pressure p Pa is applied over an area A m^2 is given by $F = pA$ newtons.

   a When a bulldozer is parked on a bridge, 8000 Pa of pressure is applied over 5 m^2. Find the size of the resulting force.

   b When a vase is placed on a table, 12 newtons of force is spread over 0.04 m^2. Find the amount of pressure applied.

13 The density of an object with mass m g and volume V cm^3 is given by $D = \dfrac{m}{V}$ g/cm^3.

   a Find the density of a block of chocolate with mass 300 g and volume 226 cm^3.

   b Find the mass of a window pane with density 2.89 g/cm^3 and volume 8640 cm^3.

14 If n dots are placed on each side of a triangle, there are $L = 3n^2$ lines which can be drawn within the triangle connecting the dots.

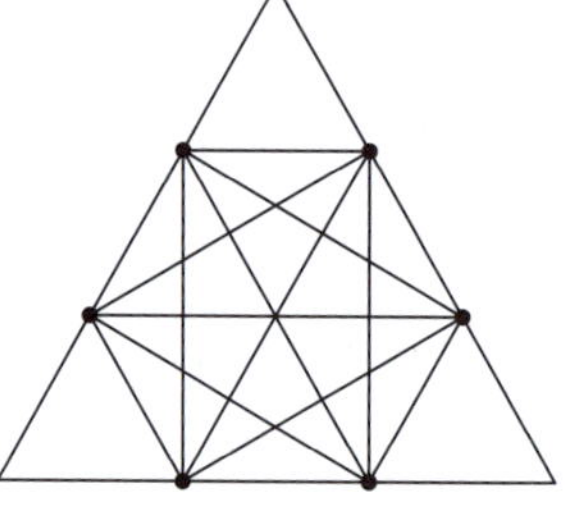

   a What possible values can n take?

   b Find the value of L when:

      i $n = 3$ **ii** $n = 5$ **iii** $n = 8$.

   c Find the value of n such that:

      i $L = 108$ **ii** $L = 147$ **iii** $L = 363$.

15 If n overlapping circles are drawn on a sheet of paper, the maximum number of regions which can be formed is given by $R = n^2 - n + 2$.

For example, 2 circles can create a maximum of $R = 2^2 - 2 + 2 = 4$ regions.

Notice that outside of the circles is considered a region.

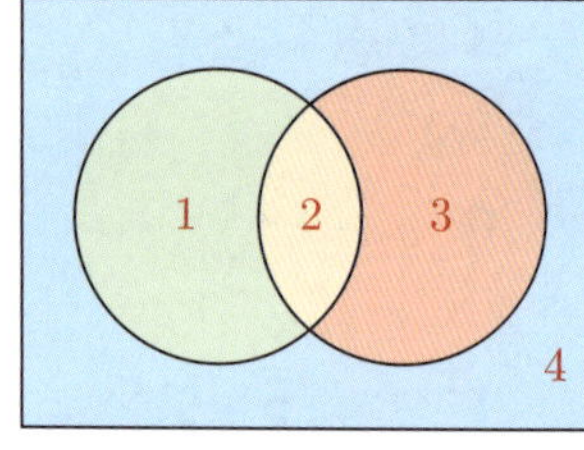

   a Find the maximum number of regions which can be formed using:

      i 1 circle   **ii** 3 circles   **iii** 4 circles.

    Illustrate each case.

   b Find the maximum number of regions which can be formed using 10 circles.

   c Kevin has used overlapping circles to form 155 regions. What is the minimum number of circles he could have used?

DISCUSSION

Most measurable quantities in the real world cannot have negative values. Therefore, it is often appropriate to talk about values of the variables for which a formula is valid.

Look back at the questions in the previous Exercise.

- In question **4**, the number of people p hiring a bus must be a *positive integer*.
- In question **7**, Jessica's mass x must be *positive*.

Discuss the variables in question **5**. What can we say about B and h?

D GEOMETRIC PATTERNS

Many problems can be solved by observing patterns. We first look at simple cases, and use them to find a **formula** connecting the variables. We then use the formula to solve the problem.

Example 3   ◀) **Self Tutor**

Sam builds steel fences in sections.

 is called a 1-section and is made from 4 lengths of steel.

is called a 2-section and is made from 7 lengths of steel.

Let the number of sections be n and the number of steel lengths be S.

a Find the formula which connects S and n.

b How many lengths of steel does Sam need to make a 63-section fence?

c How many sections of fence can be made with 400 lengths of steel?

---

a We sketch the first few cases to help establish a pattern.

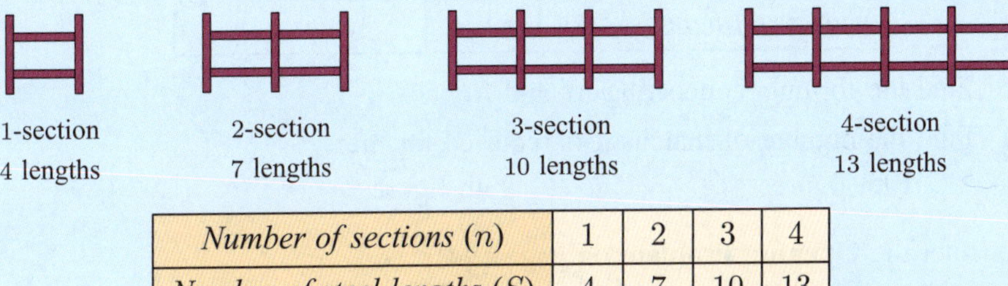

Number of sections (n)	1	2	3	4
Number of steel lengths (S)	4	7	10	13

$+3 \quad +3 \quad +3$

When n is increased by 1, S increases by 3.

We therefore compare $3n$ with S.

$3n$	3	6	9	12
S	4	7	10	13

The S values are always 1 more than the $3n$ values, so $S = 3n + 1$.

b When $n = 63$, $\quad S = 3 \times 63 + 1$

$$= 190$$

190 lengths of steel are needed to make a 63-section fence.

c When $S = 400$, $\quad 400 = 3n + 1$

$$\therefore \ 400 - 1 = 3n + 1 - 1$$

$$\therefore \ 399 = 3n$$

$$\therefore \ \frac{399}{3} = \frac{3n}{3}$$

$$\therefore \ n = 133$$

A 133-section fence can be made with 400 lengths of steel.

EXERCISE 10D

1 Consider the pattern:  ,  ,  , ...

   a Draw the next two figures in the pattern.

   b Copy and complete:

Figure number (n)	1	2	3	4	5
Number of matchsticks (M)	5				

   c Find the formula connecting M and n.

   d Find the number of matchsticks required for the 17th figure.

2 Consider the pattern: 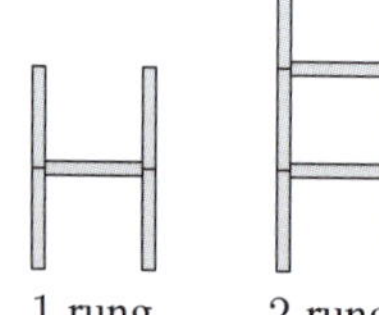 , 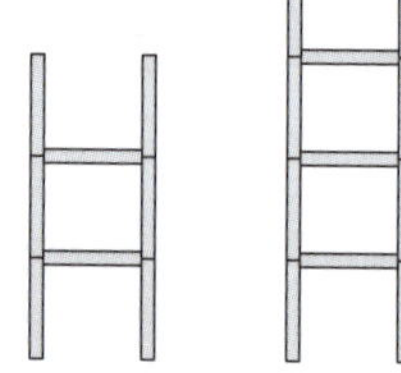 , ...

   a Draw the next two figures in the pattern.

   b Copy and complete:

Figure number (n)	1	2	3	4	5
Number of matchsticks (M)	3				

   c Find the formula connecting M and n.

   d Find the number of matchsticks required for the:

      i 10th figure   **ii** 25th figure.

3 Consider the **Opening Problem** on page **194**.

   a Copy and complete:

Number of rungs (n)	1	2	3	4	5
Number of metal pieces (M)	5	8			

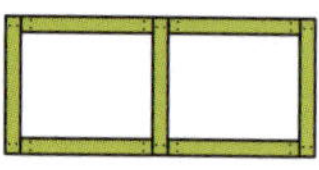
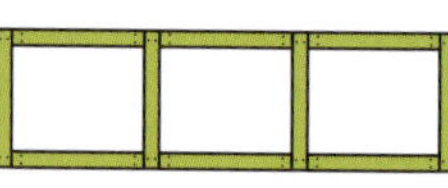

   b Find the formula connecting M and n.

   c How many metal pieces are required to build a ladder with 12 rungs?

   d Suppose Leonard has 50 pieces of metal. What size ladder can he build?

4 Mrs Moyle uses pieces of cardboard to make pigeonholes for her students.

   a Copy and complete the table.

   b Find the formula connecting C and p.

Pigeonholes (p)	1	2	3	4	5
Cardboard pieces (C)	4				

   c How many pieces of cardboard will Mrs Moyle need to construct 27 pigeonholes?

   d Suppose Mrs Moyle only has 60 pieces of cardboard. How many complete pigeonholes can she make?

5 Consider the pattern:

, ...

   a Draw the next two figures in the pattern.

   b Copy and complete:

Figure number (n)	1	2	3	4	5
Number of dots (D)					

   c Find a formula connecting D and n.

   d Find the number of dots in the 11th figure.

   e Which figure in the pattern has 85 dots?

6 Richard is experimenting with toothpicks to make tower designs.

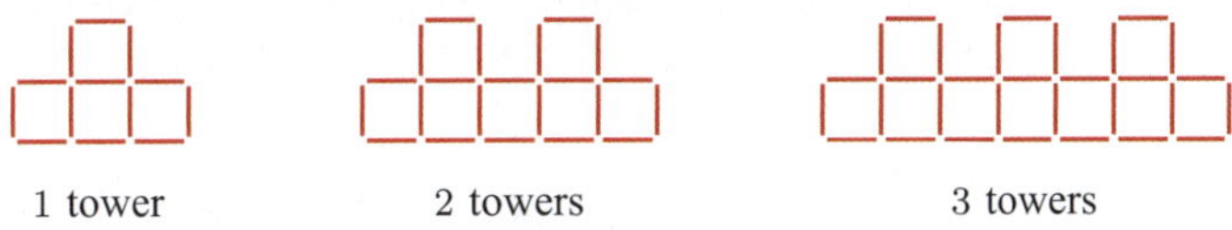

1 tower   2 towers   3 towers

   a Draw the toothpick diagram for 4 towers.

   b In a design with n towers, let H be the number of horizontal toothpicks, V be the number of vertical toothpicks, and T be the total number of toothpicks.

     Write a formula connecting:

     i H and n **ii** V and n **iii** T and n.

   c **i** Find the total number of toothpicks needed for a design with 10 towers.

     ii How many of these toothpicks are vertical?

   d Richard makes a design using a total of 220 toothpicks. How many of the toothpicks are horizontal?

INVESTIGATION   TRIANGULAR NUMBERS

Look at the pattern of dots below. The number of dots used to form each of the triangles in this pattern is called a **triangular number**.

1st   2nd   3rd   4th

1 dot   3 dots   6 dots   10 dots

What to do:

1 Construct the next three triangles in the pattern.

2 Copy and complete:

Triangle (n)	1st	2nd	3rd	4th	5th	6th	7th
Number of dots (D)							

3 Predict the number of dots in the 8th, 9th, and 10th triangles.

4 Copy and complete: $\dfrac{1 \times 2}{2} = \ldots\ldots,$ $\dfrac{2 \times 3}{2} = \ldots\ldots,$ $\dfrac{3 \times 4}{2} = \ldots\ldots,$ $\dfrac{4 \times 5}{2} = \ldots\ldots$

Hence write a formula connecting D and n.

5 Use your formula to find the number of dots in the 15th triangle.

6 When Alan, Brenton, Claude, and Daniel shake hands in all possible ways, 6 handshakes take place. We can represent these as AB, AC, AD, BC, BD, and CD, where AB represents the Alan - Brenton handshake.

a Copy and complete:

Number of people (p)	2	3	4	5	6
Number of handshakes (H)			6		

b Explain how the number of handshakes is related to the triangular numbers.

c Write a formula connecting H and p.

d How many handshakes can take place within a group of 20 people?

E   PRACTICAL PROBLEMS

The methods we have used for geometric patterns can also be used to solve more practical problems.

Example 4   ◄») **Self Tutor**

Peter is a plumber. He charges a \$50 call-out fee plus \$40 for each hour spent working.
  a What will t hours of labour cost?
  b Find the total charge C dollars for a job taking t hours.
  c Find the total charge for a job taking 5 hours.
  d Find the time taken for a job with total charge \$170.

---

a Each hour of labour costs \$40, so the cost of t hours of labour is \$$40t$.

b Total charge = cost of labour + call-out fee

$$\therefore \quad C = 40t + 50$$

c When $t = 5$, $C = 40 \times 5 + 50$
$$= 250$$

So, the total charge for a 5 hour job is \$250.

d When $C = 170$, $170 = 40t + 50$

$$\therefore \quad 170 - 50 = 40t + 50 - 50$$

$$\therefore \quad 120 = 40t$$

$$\therefore \quad \frac{120}{40} = \frac{40t}{40}$$

$$\therefore \quad t = 3$$

A job with total charge \$170 takes 3 hours.

EXERCISE 10E

1 Domenica's Pizza House charges $15 per pizza, with a $5 delivery fee.

   a What will p pizzas cost without delivery?

   b Find the total cost $\$C$ of having p pizzas delivered.

   c Find the total cost of having:

      i 3 pizzas delivered   **ii** 8 pizzas delivered.

2 Milan is stacking boxes of tiles onto a wooden pallet. The pallet has mass 30 kg, and each box of tiles has mass 20 kg.

   a Explain why the total mass of the pallet and b boxes of tiles is given by
$W = 20b + 30$ kg.

   b Find the total mass if 11 boxes are placed on the pallet.

   c The forklift used to lift the pallet and boxes cannot lift more than 800 kg. Find the maximum number of boxes which can be placed on the pallet.

3 Danielle inherits her great-grandmother's shell collection containing 200 shells. Each month she adds another 7 shells to her collection.

   a Write a formula for the number of shells S Danielle has in her collection after n months.

   b How many shells will Danielle have in her collection after:

      i 3 months   **ii** 1 year   **iii** $2\frac{1}{4}$ years?

   c How long will it take for Danielle's collection to reach 550 shells? Write your answer in years and months.

4 Mary has $5000 in her bank account. She only uses the account to pay rent, which costs her $75 per week.

   a How much money will Mary have withdrawn for rent after n weeks?

   b Write a formula for the amount $\$M$ that Mary will have left in her bank account after n weeks.

   c How much money will be in Mary's account after:

      i 8 weeks   **ii** 1 year?

   d How long will it take for the amount in Mary's account to fall to $2000?

5 300 people went to the cinema to see a new film. However, they were so bored by the film that 5 people left the cinema every 10 minutes.

   a How many people had left the cinema after t lots of 10 minutes?

   b Find the total number of people P who were still in the cinema after t lots of 10 minutes.

   c Find the number of people still at the film after:

      i 40 minutes   **ii** $1\frac{1}{2}$ hours.

6 A printer makes 12 000 greeting cards in the first year of production, and 8000 greeting cards each year after that.

   a In total, how many cards are printed after:

     i 2 years   **ii** 3 years   **iii** 4 years?

   b Write a formula for the total number of greeting cards C printed after n years.

   c Find the total number of greeting cards printed after 10 years.

   d How long will it take for 140 000 cards to be printed?

7 Patrick begins the year with \$8400 in his savings account. Each fortnight he is paid \$2300 in wages, and he spends \$$e$ in rent and other living expenses.

   a Write a formula for the amount \$$A$ in Patrick's account after n fortnights.

   b Find the value of A if $e = 1460$ and $n = 5$.

   c Find the value of e if $A = 14\,250$ and $n = 9$.

8 A street grid has n streets each way.

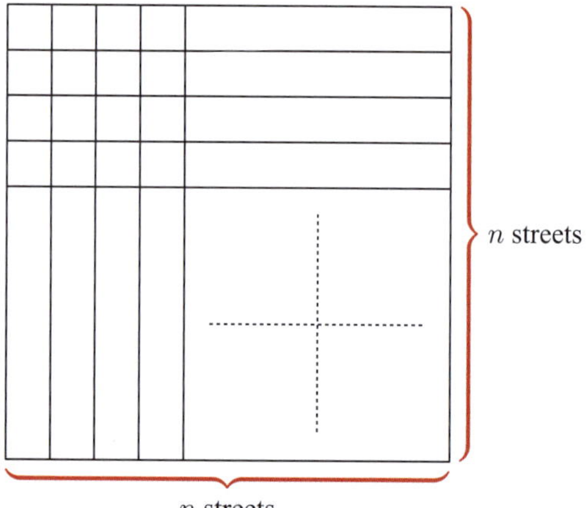

   a $\top$ is called a T-junction.

     Write a formula for the total number of T-junctions, T.

   b $+$ is called an intersection.

     Write a formula for the total number of intersections, I.

   c If $n = 9$, find the value of:

     i T **ii** I.

   d If $T = 36$, find the value of:

     i n **ii** I.

   e Suppose the total number of junctions $T + I$ is 140. Find the value of n.

MULTIPLE CHOICE QUIZ

QUICK QUIZ

REVIEW SET 10A

1 Consider the rule:

   Four times the input number, minus five.

   Complete the table by applying the rule to each input number.

Input	Output
2	
5	
8	
12	

2 For each table, write a formula connecting the variables.

a

n	1	2	3	4
M	5	9	13	17

b

x	1	2	3	4
y	5	12	19	26

3 Consider the formula $W = 6n - 13$.   Find the value of:

    a W when $n = 4$
    b n when $W = 47$.

4 Consider the formula $P = 4n^2 + 3$.

    a Find the value of P when:

        i $n = 2$
        ii $n = 6$.

    b Find the possible values of n when:

        i $P = 7$
        ii $P = 103$.

5 The cost of hiring a tennis court for t hours is given by $C = 15t + 8$ dollars.
Find the cost of hiring the tennis court for:

    a 12 hours
    b $4\frac{1}{2}$ hours.

6 Luke has a five dollar notes and b ten dollar notes in his wallet. The total value of the
notes is $V = 5a + 10b$ dollars.

    a Find the total value of the notes if $a = 3$ and $b = 4$.
    b Find b if $V = 80$ and $a = 6$.
    c Find a if $V = 75$ and $b = 3$.

7 Consider the pattern:

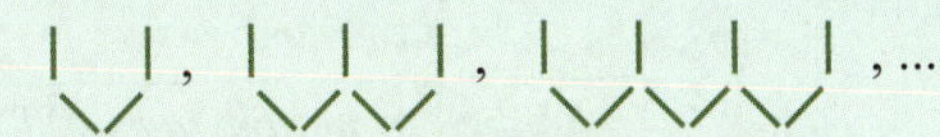

    a Draw the next two figures in the pattern.

    b Copy and complete:

Figure number (n)	1	2	3	4	5
Number of matchsticks (M)					

    c Find the formula connecting M and n.
    d How many matchsticks are required for the 25th figure?

8 Consider the pattern:

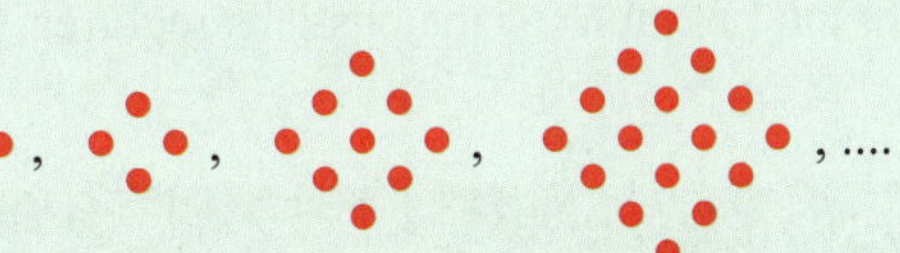

    a Draw the next figure in the pattern.

    b Copy and complete:

Figure number (n)	1	2	3	4	5
Number of dots (D)					

    c Find the formula connecting D and n.
    d Find the number of dots in the 12th figure.
    e Which figure in the pattern has 484 dots?

9 Josie has \$450 in her savings account. She is able to deposit \$25 per week into the account. She does not withdraw any money.

   a Find the amount of money that Josie will have deposited into the account after w weeks.

   b Find the *total* amount A dollars in the account after w weeks.

   c How much money will Josie have in her account after:

      i 4 weeks   **ii** 10 weeks?

10 Carl is buying concert tickets online. Each ticket costs \$19, and there is a fixed credit card fee of \$3.

   a Explain why the total cost of buying n tickets is $C = 19n + 3$ dollars.

   b Find the total cost of buying: **i** 2 tickets   **ii** 5 tickets.

   c Carl spent a total of \$79 on tickets. How many tickets did he buy?

11 There are initially 20 dog biscuits in a bowl. Each day, Claire puts 12 biscuits in the bowl, and her dog eats b biscuits.

   a Write a formula for the number of biscuits N after d days.

   b Find the value of N if $b = 15$ and $d = 3$.

   c Find the value of b if $N = 10$ and $d = 5$.

12 Consider this pattern: □ , □□ , □□□ , ...

   a Draw the next two figures in the pattern.

   b Copy and complete:

Figure number (n)	1	2	3	4	5
Number of matchsticks (M)					

   c Find the formula connecting M and n.

   d Find the number of matchsticks required for the 10th figure.

   e Which figure number contains 70 matchsticks?

   f The *total* number of matchsticks required for the first n figures is given by $T = 2n^2 + 4n$ matchsticks.

      i Check that this formula is correct for $n = 1, 2, 3$.

      ii Find the total number of matchsticks required for the first 20 figures.

REVIEW SET 10B

1 For each rule, write a formula connecting the input number n and the output number M.

   a The output number is the input number divided by 6.

   b The output number is four times the input number, plus 3.

2 For each table, write a formula connecting the variables:

a

p	1	2	3	4
Q	4	10	16	22

b

t	1	2	3	4
D	5	8	11	14

3 Consider the formula $L = 5n + 7$. Find the value of:

   a L when $n = 3$ **b** n when $L = 47$.

4 The momentum of an object with mass m kg and velocity v m/s is given by $p = mv$ kg m/s.

   a Find the momentum of a person with mass 60 kg moving at 5 m/s.

   b A truck with mass 3500 kg has momentum 70 000 kg m/s. Find the velocity of the truck.

5 Consider the pattern:

   a Draw the next two figures in the pattern.

   b Copy and complete:

Figure number (n)	1	2	3	4	5
Number of matchsticks (M)					

   c Find the formula connecting M and n.

   d Find the number of matchsticks required for the:

      i 15th figure   **ii** 42nd figure.

6 Consider the pattern:

   a Copy and complete:

Figure number (n)	1	2	3	4	5
Number of matchsticks (M)					

   b Find the formula connecting M and n.

   c Find the number of matchsticks required for the 8th figure.

   d Which figure number contains 94 matchsticks?

7 Consider the formula $P = 2n^3 - 5$.

   a Find the value of P when:

      i $n = 2$ **ii** $n = 4$ **iii** $n = 7$.

   b Find the value of n when:

      i $P = 245$ **ii** $P = 1995$ **iii** $P = 1019$.

8 The distance of a train from its destination after t hours of travel is given by the formula $D = 600 - 75t$ kilometres.

   a How far is the train from its destination after:

      i 2 hours   **ii** 5 hours?

   b How long will it take for the train to arrive at its destination?

9 The mass of a basket containing n plums is $M = 45n + 95$ grams.

   a Find the mass of the basket when empty.

   b Find the mass of a basket containing a dozen plums.

   c A basket with plums has total mass 500 grams. How many plums are in the basket?

10 A school fundraising committee decides to hold a bake sale.

They start with $60 in their money box, and they receive $1.50 for every cupcake they sell.

  a Explain why the total amount of money in the money box after selling c cupcakes is given by $A = 60 + 1.5c$ dollars.

  b How much money will be in the money box if they sell a total of:

    i 20 cupcakes   **ii** 50 cupcakes?

11 Before the school holidays, Pauline had read 150 pages of her book. She decided to read 20 pages of the book each day during the holidays.

  a Find the number of pages Pauline will read in n days of the holidays.

  b Find the total number of pages of the book Pauline will have read after n days of the holidays.

  c How many pages of the book will she have read after 7 days of the holidays?

  d The book is 510 pages long. How many days of the holidays will it take Pauline to finish the book?

12 When a circle is divided by n distinct lines, the maximum number of regions which can be formed is given by the formula $R = \frac{1}{2}n^2 + \frac{1}{2}n + 1$.

For example, when a circle is divided by 2 lines, a maximum of 4 regions can be formed, as shown alongside.

  a Find the maximum number of regions formed when a circle is divided by:

    i 3 lines   **ii** 4 lines.

  Illustrate each case.

  b Find the maximum number of regions formed when a circle is divided by 12 lines.

  c Find a formula for the *minimum* number of regions formed M when a circle is divided by n distinct lines.

Measurement: Length and area

Contents:

OPENING PROBLEM

Asha has bought an outdoor swimming pool. It is circular with diameter 5 m.

Asha is trying to work out the distance around the edge of the pool.

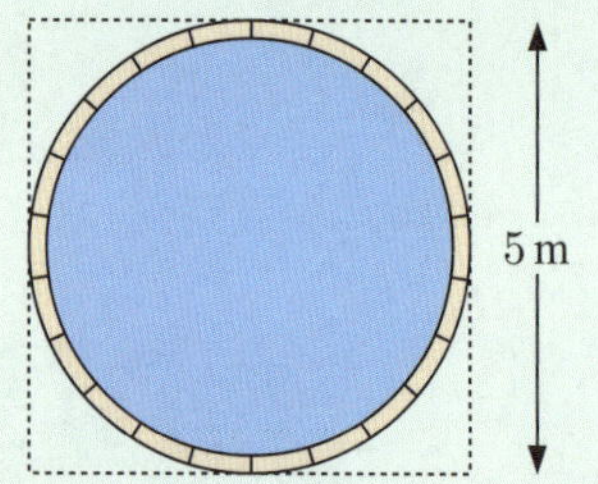

Things to think about:

a Can you use the diagram to explain why the distance around the edge of the pool must be:

    i greater than 10 m   **ii** less than 20 m?

b Using a piece of string and a tape measure, Asha finds that the distance around the edge of the pool is about 15.71 m.

    i How many times greater is the distance around the edge of the pool than the diameter of the pool?

    ii Can we use this information to find the distance around the edge of a pool with diameter 8 m?

c What is the *area* of the base of the pool?

Many people such as builders, engineers, landscapers, and surveyors rely on accurate **measurements** to perform their jobs.

In this Chapter we consider measurements of **length** and **area**.

A   LENGTH

Length is a measure of distance.

The **metre** (m) is the base unit for length in the metric system.

From this base unit, we can define the following related units for measuring smaller and larger distances:

- 1 **kilometre** (km) = 1000 metres

- 1 **centimetre** (cm) = $\dfrac{1}{100}$ metre

- 1 **millimetre** (mm) = $\dfrac{1}{1000}$ metre   or $\dfrac{1}{10}$ centimetre

LENGTH CONVERSIONS

1 km = 1000 m
1 m = 100 cm
1 cm = 10 mm

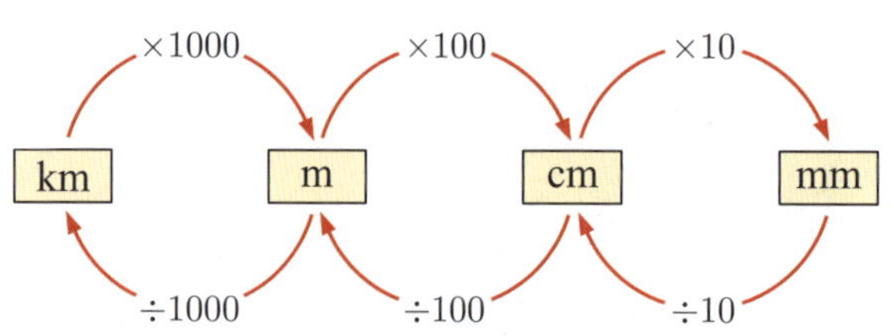

Example 1   ◀ϑ **Self Tutor**

Convert:

  a   25 km into m   **b**   200 cm into m.

  a   25 km
       $= 25 \times 1000$ m
       $= 25\,000$ m

  b   200 cm
       $= 200 \div 100$ m
       $= 2$ m

EXERCISE 11A

1 Convert:

  a 70 mm into cm   **b** 12 km into m   **c** 120 cm into m

  d 3.82 m into cm   **e** 1250 m into km   **f** 350 mm into cm

  g 43.4 cm into mm   **h** 3.2 m into mm   **i** 1240 mm into m.

2 A submarine dived to a depth of 356 m. Write this depth in:

  a kilometres   **b** centimetres.

3 Find the sum of:

  a 7 m + 35 cm + 21 mm   in centimetres   **b** 4.3 km + 520 m + 860 cm   in metres.

4 My rain gauge contained 1.63 cm of rain. A further 7.8 mm of rain fell last night. How many millimetres of rain is in the gauge now?

5 The world records for the triple jump were both set at the 1995 World Championships in Gothenburg:

Mens	Jonathon Edwards (GBR)	18.29 m
Womens	Inessa Kravets (UKR)	15.50 m

  a Write each record in cm.

  b How much farther, in cm, did Jonathon Edwards jump than Inessa Kravets?

6 How many 12 mm thick biscuits can be cut from a 54 cm roll of biscuit dough?

7 A Swiss 20 centimes coin is 1.65 mm thick.

  a How many cm high is a stack of eighteen 20 centimes coins?

  b How many 20 centimes coins would you need to build a stack 0.99 m high?

8 A snail and a caterpillar are 2.7 m apart. The snail moves 33 mm towards the caterpillar, and the caterpillar moves 19.2 cm towards the snail. Find the new distance between them, in centimetres.

9 An office building has 12 floors above ground level. There are 18 steps between each floor. Each step is 17.8 cm high. How many metres does a worker climb if he walks up from ground level to the top floor?

ACTIVITY 1   ESTIMATING LENGTH

Being able to estimate lengths is a useful skill. It allows you to communicate the size of objects to other people.

What to do:

1   Copy or print the table below:

Length	Units	Estimate	Measured length
length of your pen			
width of your exercise book			
height of your desk			
length of your whiteboard			
height of your teacher			
length of your classroom			
width of your school soccer field			

PRINTABLE TABLE

   a   Choose appropriate units for each measurement.

   b   Look at each thing and *estimate* the length.

   c   Check your estimates by *measuring* each length with a ruler, tape measure, or trundle wheel.

2   Discuss how you would estimate:

   a   the height of a building

   b   the distance between your school and your house

   c   the thickness of a sheet of paper.

HISTORICAL NOTE

In 1793, the **metre** was defined as one ten-millionth of the distance from the north pole to the equator along the line of longitude through Paris, France. After difficult and exhaustive surveys, a piece of platinum alloy was prepared to this length and called the **standard metre**.

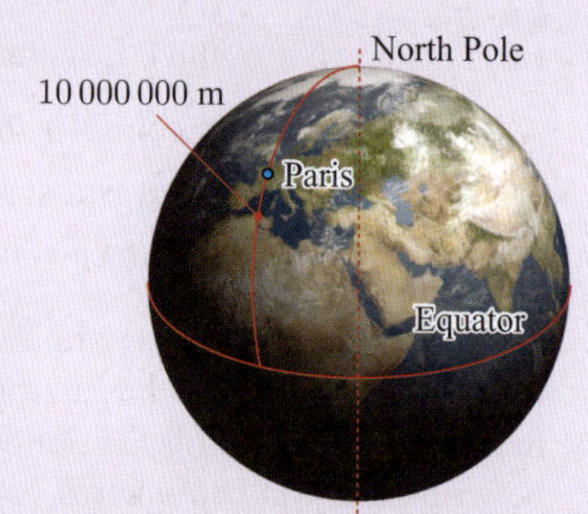

The standard metre was kept at the International Bureau of Weights and Measures at Sèvres, near Paris. However, this meant it was not easily accessible to scientists around the world.

From 1960 to 1983, the metre was defined as 1 650 763.73 wavelengths of orange-red light from the isotope Krypton 86, measured in a vacuum.

In 1983, the metre was redefined as the distance light travels in a vacuum in $\dfrac{1}{299\,792\,458}$ of a second.

DISCUSSION

Are the scales on all tape measures *exactly* the same?

How do you know whether your tape measure is accurate?

What is the purpose of a "standard metre"?

B   PERIMETER

The **perimeter** of a closed figure is the distance around its boundary.

The perimeter of a polygon is found by adding the lengths of its sides.

Example 2   ◀》 Self Tutor

Find the perimeter of each figure:

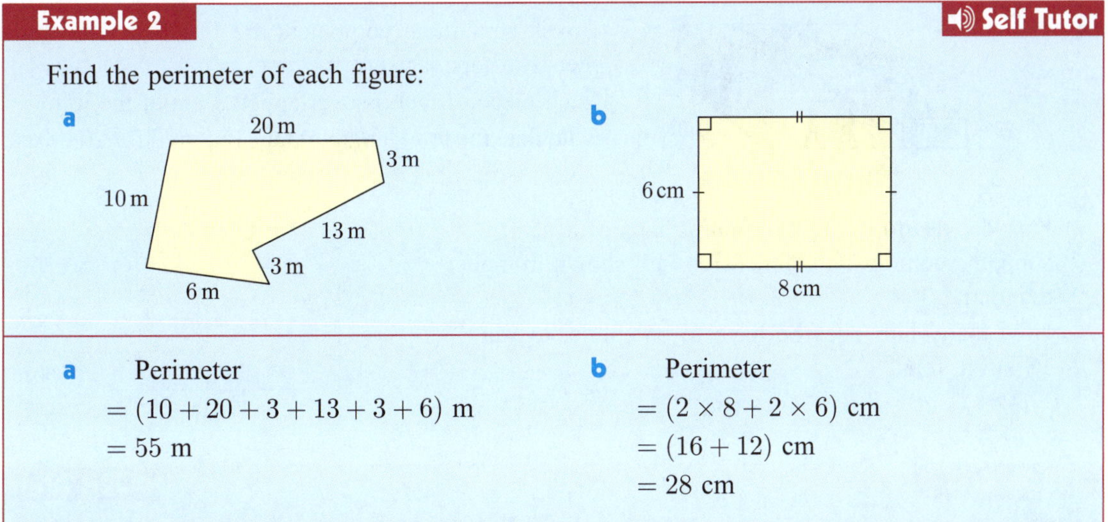

a   Perimeter
$= (10 + 20 + 3 + 13 + 3 + 6)$ m
$= 55$ m

b   Perimeter
$= (2 \times 8 + 2 \times 6)$ cm
$= (16 + 12)$ cm
$= 28$ cm

EXERCISE 11B

1   Find the perimeter of each figure:

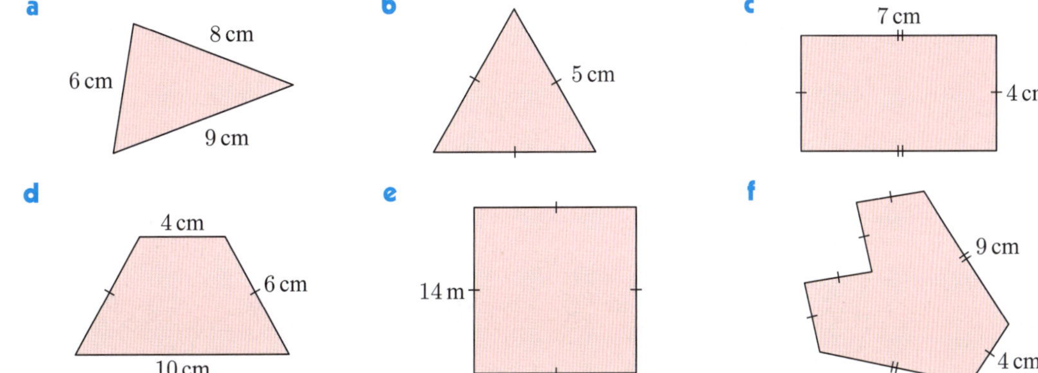

2 Find the perimeter of each figure:

a
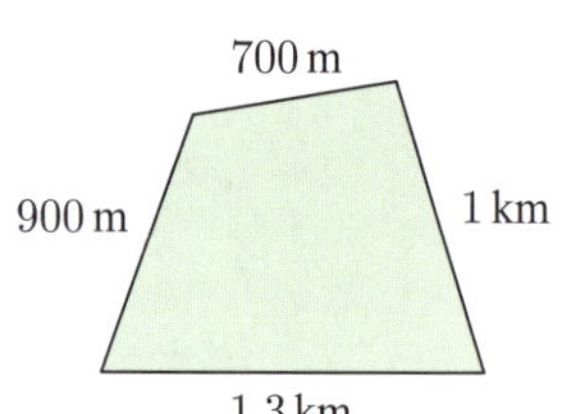

b
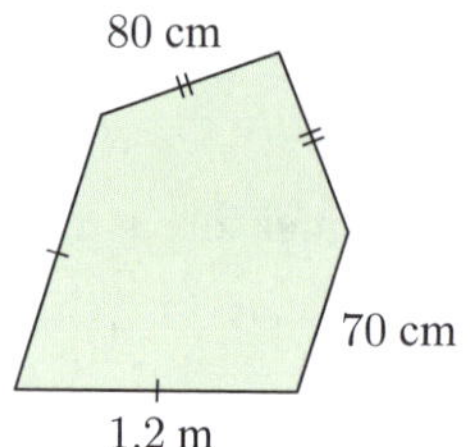

c
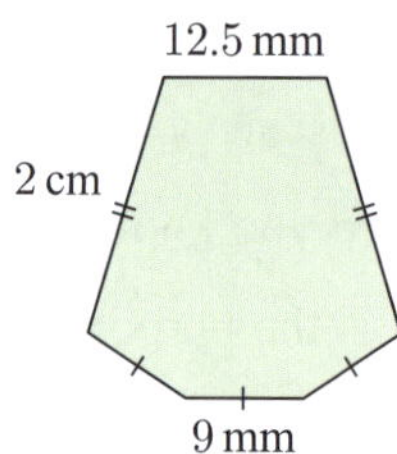

3 A farmer fences a 400 m by 350 m rectangular field with a 4-strand wire fence. Each strand costs \$38.20 per 100 metres. Find:

   a the perimeter of the field

   b the total length of wire required

   c the total cost of wire.

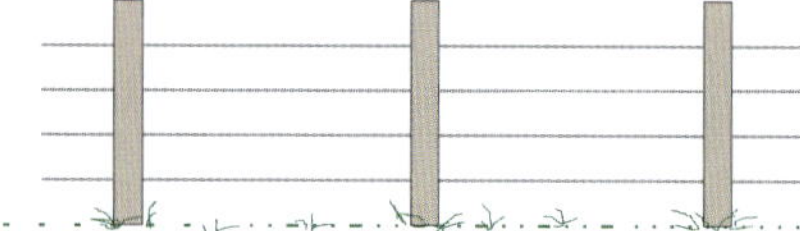

4

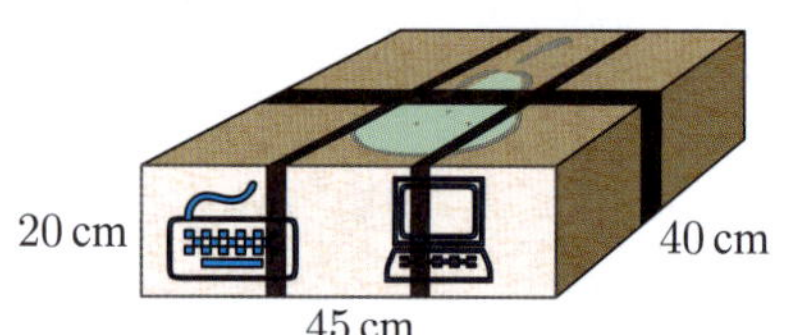

Boxes containing computers are fastened with three pieces of tape, as shown.

Each piece of tape is overlapped 5 cm at the join.

Calculate the total length of tape required for 20 boxes.

5 Kim decides to get fit by running laps of a nearby shopping complex. Its dimensions are shown in the diagram.

How many laps does he have to complete, to run 13 km in total?

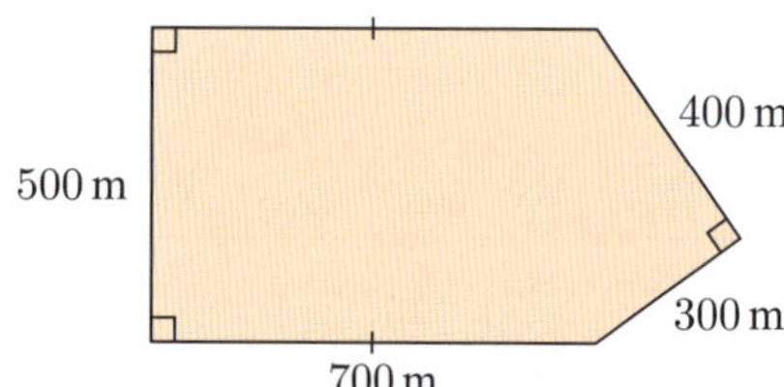

Example 3   ◀》 Self Tutor

Find, in simplest form, the perimeter P of each figure:

a
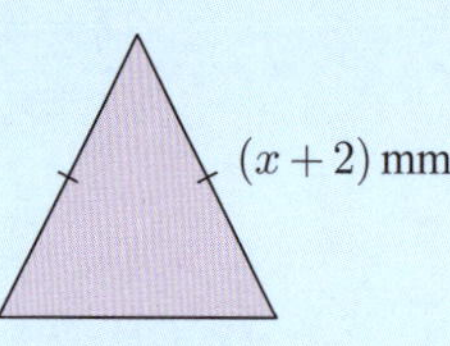

b
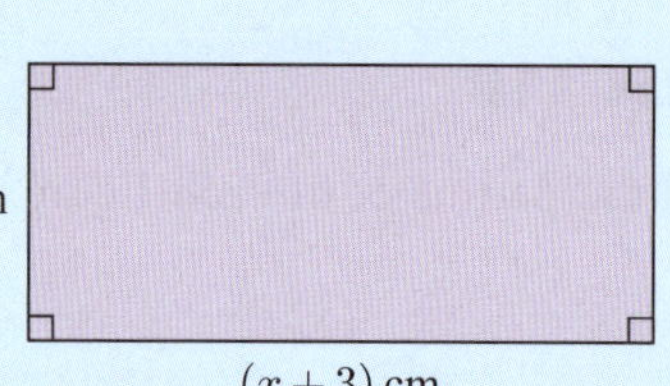

a $P = x \text{ mm} + 2(x + 2) \text{ mm}$
$ = (x + 2x + 4) \text{ mm}$
$ = (3x + 4) \text{ mm}$

b The opposite sides of a rectangle are equal in length.
$\therefore\ P = 2x \text{ cm} + 2(x + 3) \text{ cm}$
$ = (2x + 2x + 6) \text{ cm}$
$ = (4x + 6) \text{ cm}$

6 Find, in simplest form, the perimeter P of each figure:

a
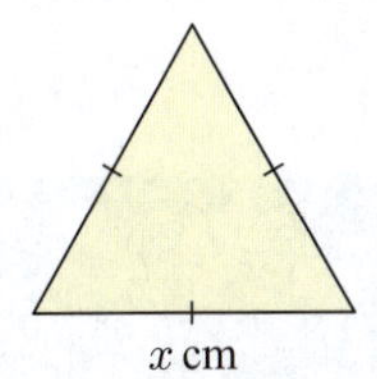

b
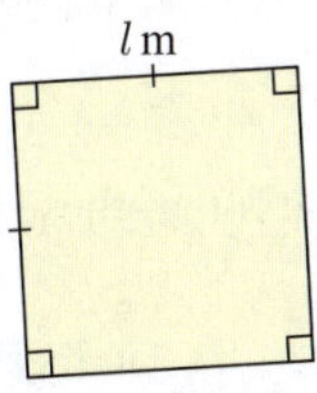

c
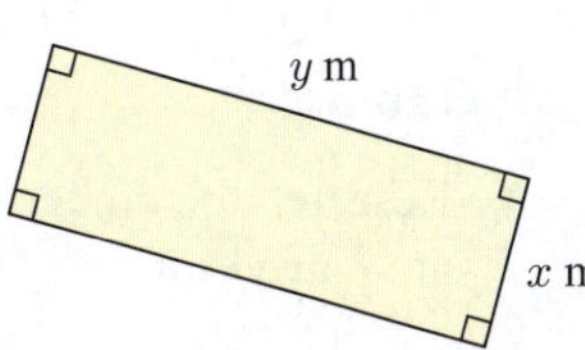

d
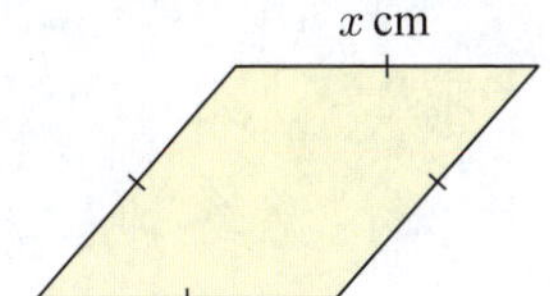

e
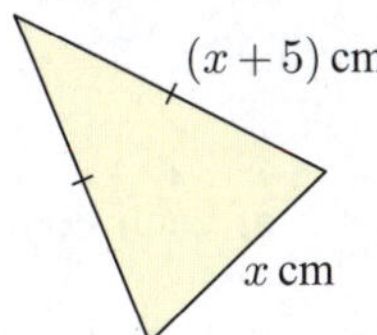

f
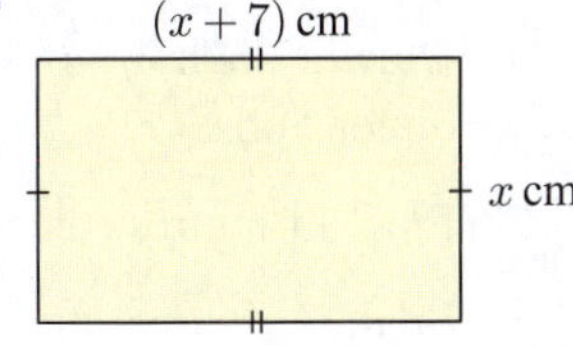

g
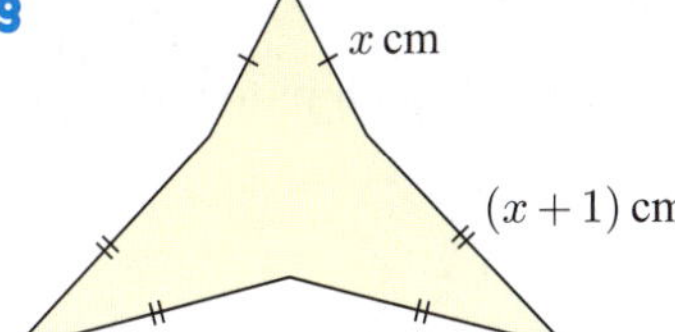

h
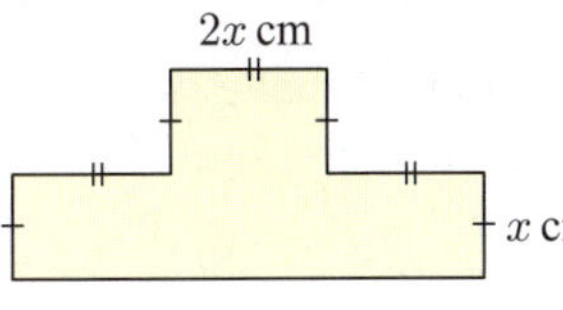

i
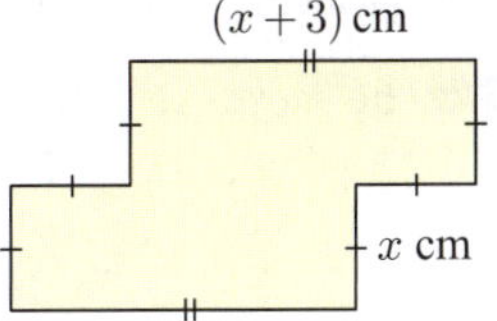

7   **a** Write a formula for the perimeter P of this triangle.

    b If $x = 3$, find the perimeter.

    c If the perimeter is 58 cm, find the value of x.

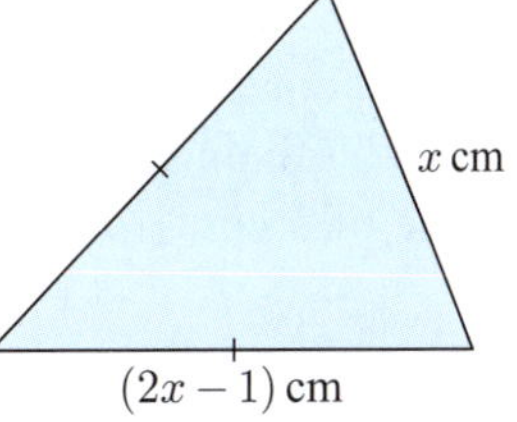

8   **a** Find the perimeter P of the figure alongside.

    b Suppose the perimeter of the figure is 43 m.

      i Find the value of x.

     ii Find the length of the longest side.

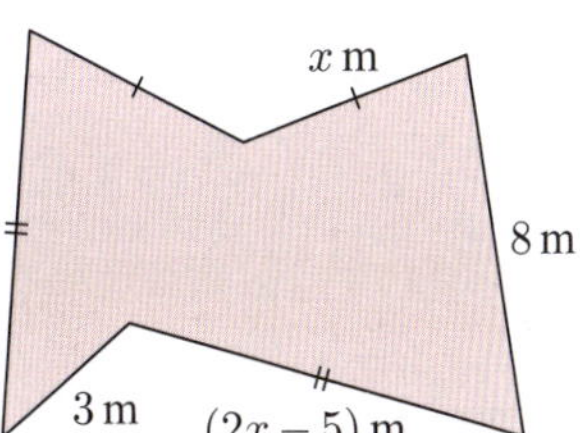

9   **a** Write, in simplest form, a formula for the perimeter P of this pentagon.

    b Suppose $x = 9$ and $y = 5$. Find the value of P.

    c Suppose $P = 23$ cm and $y = 3$. Find the value of x.

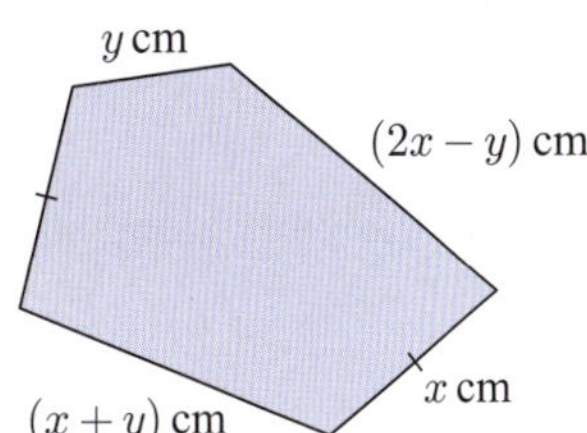

ACTIVITY 2 — COURT LINES

What to do:

1 Research the dimensions of the courts that these sports are played on:

   a badminton   **b** basketball   **c** netball

   d tennis   **e** volleyball.

Draw a diagram of each court, including all dimensions.

2 Find, in metres, the outer perimeter of each court in **1**.

3 Find, in metres, the total length of the lines of each court in **1**.

4 Write the courts in order from smallest to largest length of lines.

5 The new school gymnasium will have 1 basketball court, 4 badminton courts, and 2 volleyball courts. Find the total length of lines to be painted.

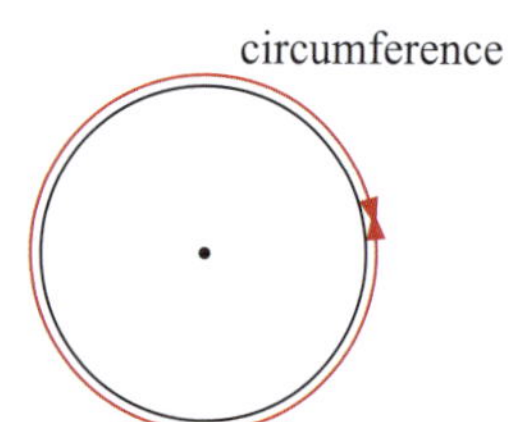

C   CIRCUMFERENCE

So far we have found the perimeters of shapes with straight edges. We will now look at finding the perimeter of a circle. This is called the **circumference** of a circle.

In the following **Investigation** we explore the relationship between the circumference and diameter of any circle.

INVESTIGATION 1 — CIRCUMFERENCE

You will need:

- a ruler
- a pencil
- some cylinders such as a soft drink can, a toilet roll, and a coin.

What to do:

1 Copy the table alongside:

Object	Diameter	Circumference	Circumference / Diameter
⋮	⋮	⋮	⋮

For each object:

   a Use a ruler to measure its diameter.

b Mark a point on the circumference of the object. Roll it along the floor for one complete revolution, as shown. The distance between the two marks is the circumference of your object.

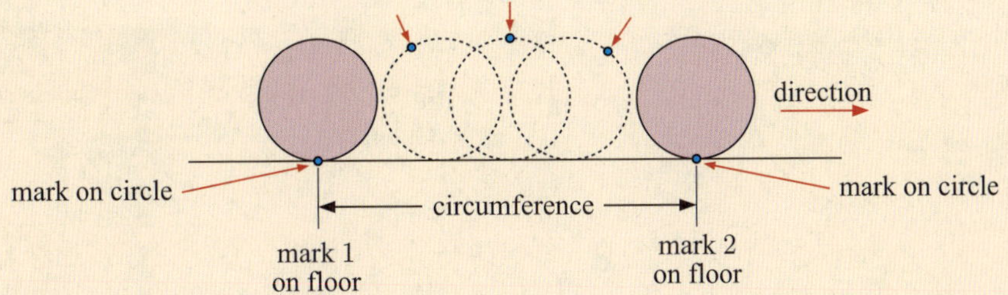

Alternatively, it may be more accurate to roll the object 10 times, then divide the total distance by 10 to get the circumference.

c Calculate the value $\dfrac{\text{circumference}}{\text{diameter}}$.

2 Compare your results and those of other students. What do you notice?

Whenever the circumference of a circle is divided by its diameter, the answer is always the same. In the **Investigation** above, you should have found that this value lies between 3.1 and 3.2.

The ratio $\dfrac{\text{circumference}}{\text{diameter}}$ is represented by the Greek letter π known as "pi".

Since $\dfrac{\text{circumference}}{\text{diameter}} = \pi$, we conclude that:

> The **circumference** of a circle with **diameter** d is $C = \pi d$.
>
> The **circumference** of a circle with **radius** r is $C = 2\pi r$.

The value of π correct to 30 decimal places is
$\pi \approx 3.141\,592\,653\,589\,793\,238\,462\,643\,383\,279$.

In practice we use $\pi \approx 3.14$, which is rounded to 3 significant figures, or else use the appropriate keys on our calculator. When using a calculator, we round the **final answer** only.

Example 4   ◀) **Self Tutor**

Use $\pi \approx 3.14$ to find the circumference of a circle with:
  a diameter 20 cm
  b radius 3.6 cm

a $C = \pi d$
  $\approx 3.14 \times 20$ cm
  ≈ 62.8 cm
  The circumference is about 62.8 cm.

b $C = 2\pi r$
  $\approx 2 \times 3.14 \times 3.6$ cm
  ≈ 22.6 cm
  The circumference is about 22.6 cm.

EXERCISE 11C

1 Use $\pi \approx 3.14$ to find the circumference of each circle:

a 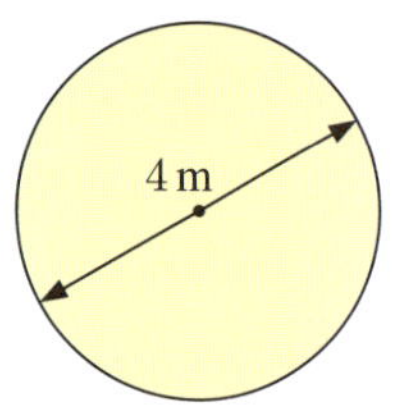

b 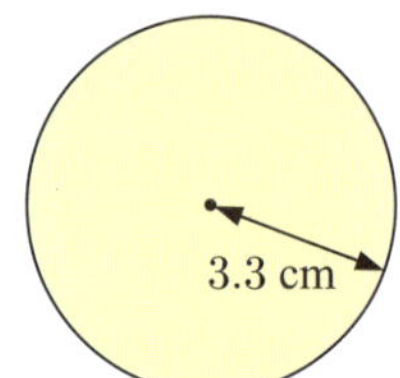

c 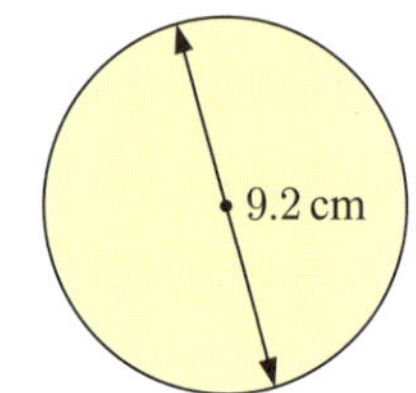

d

2 Use $\pi \approx 3.14$ to find the circumference of a circle with:

   a radius 164 mm   **b** diameter 28 cm

Example 5
◀⎞ **Self Tutor**

Use your calculator to find the circumference of a circle with radius 2.5 m. Round your answer to 2 decimal places.

The radius of the circle is 2.5 metres.

$C = 2\pi r$

$\quad = 2 \times \pi \times 2.5 \text{ m}$

$\quad \approx 15.71 \text{ m}$

The circumference is about 15.71 m.

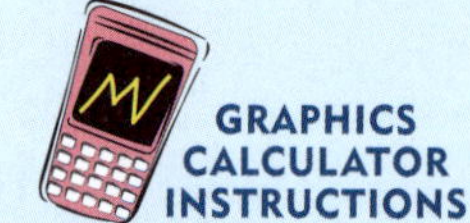

GRAPHICS CALCULATOR INSTRUCTIONS

---

3 Use your calculator to find the circumference of each circle. Round your answers to 2 decimal places.

a 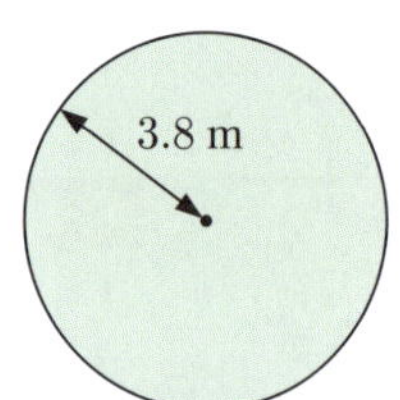

b 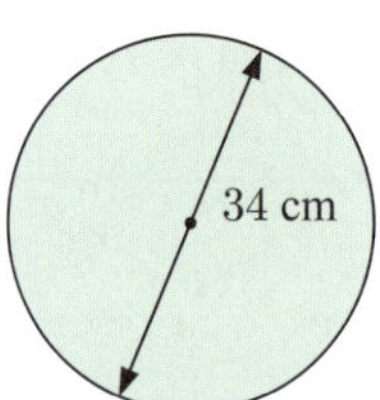

4 Use your calculator to find the circumference of a circle with:

   a radius 9 m   **b** diameter 16 cm   **c** radius 6.8 km

5 Find the circumference of a cylindrical deodorant can with base diameter 4 cm.

6 A circular pond has radius 1.5 m. Find the perimeter of the pond. Round your answer to 4 significant figures.

7 Natalie is sewing a fringe on a circular rug with the dimensions shown.

How many metres of fringe will Natalie need?

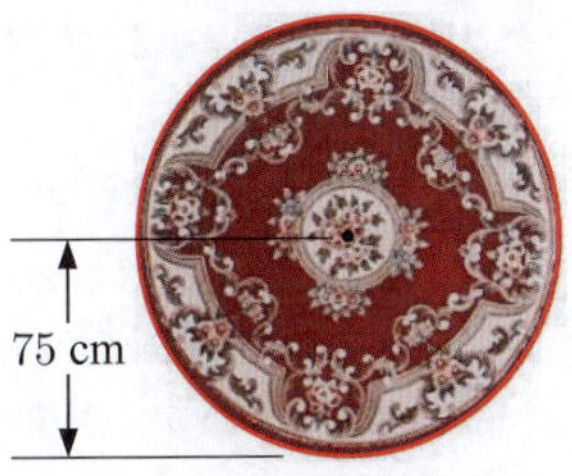

8 The minute hand of a clock is 6 cm long. How far does the tip travel between noon and 1:30 pm?

9 A car tyre has diameter 70 cm.

   a Find the circumference of the tyre.

   b How many kilometres are travelled if the tyre rotates 20 000 times?

   c How many times does the tyre rotate if the car travels 45 km?

10 A circle has the same perimeter as a square with side length 8 cm. Find the circle's diameter.

11 Which is the shorter path from A to B: around the 2 smaller semi-circles or around the large semi-circle?

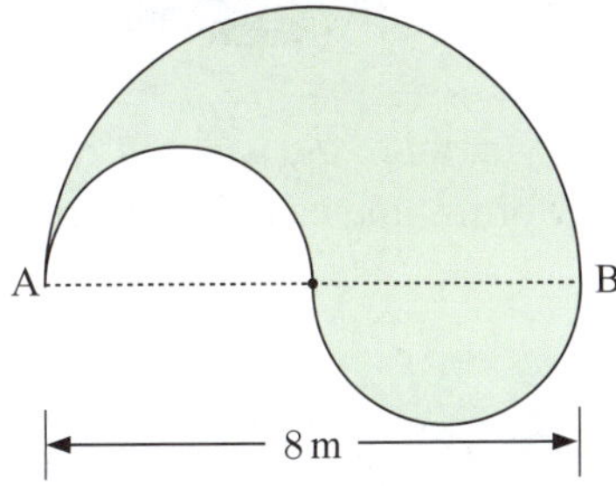

12 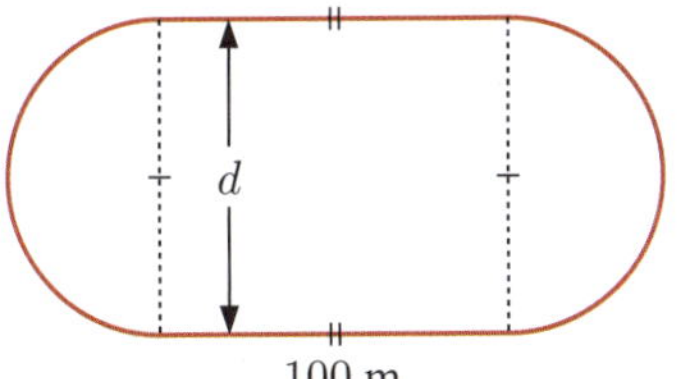

An athletics track has two "straights" measuring 100 m, and two identical semi-circular sections of 100 m each. Find the distance d between the two straight sections of track.

13 It is claimed that the circumference of a trundle wheel is 100 cm.

   a Find the radius that the wheel should have, correct to 6 significant figures.

   b The radius of the wheel is actually 0.1 mm too small. By what percentage is the circumference of the wheel too small?

14   **a** Write a formula for the perimeter P of this shape.

   b If $x = 4$, find the perimeter of the shape.

   c If the perimeter is 58.26 cm, calculate x to 3 decimal places.

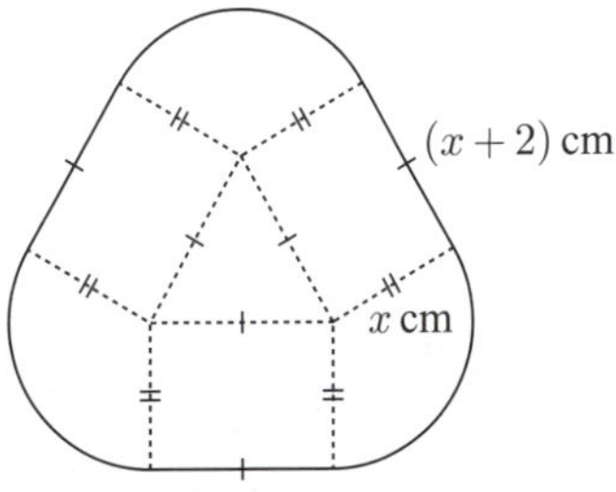

PUZZLE   ROPE AROUND THE EARTH

Consider these two scenarios:

1 A loop of rope is placed tightly around a circular table.
1 metre of rope is then added to the loop, and the rope is pulled
into a circle so there is an equal gap between the rope and the
table.

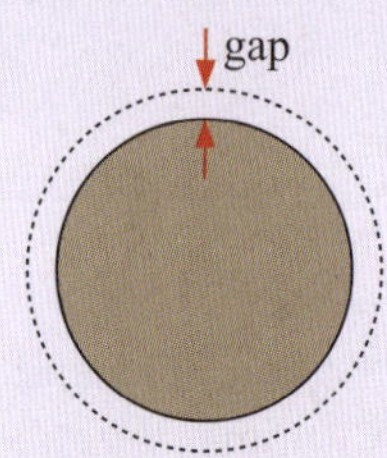

2

A loop of rope is placed tightly around the Earth.
Again, 1 metre of rope is then added to the loop, and the
rope is pulled into a circle so there is an equal gap between
the rope and the Earth.

Do you think the gap will be larger in scenario **1** or scenario **2**?
Perform some calculations to determine whether you are correct!

D   AREA

Area is the measurement of the size of a surface.

We can measure area using the following units:

1 square millimetre (mm^2) is the area enclosed by a square of side length 1 mm.

1 square centimetre (cm^2) is the area enclosed by a square of side length 1 cm.

1 square metre (m^2) is the area enclosed by a square of side length 1 m.

1 hectare (ha) is the area enclosed by a square of side length 100 m.

1 square kilometre (km^2) is the area enclosed by a square of side length 1 km.

AREA CONVERSIONS

We can use the length unit conversions to help us convert from one area unit to another.

For example, a square with sides of length 1 cm has area $1 \ cm^2$:

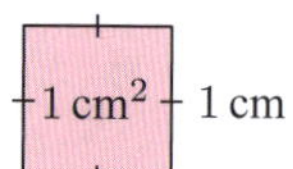

We could also measure the sides of this square in millimetres:

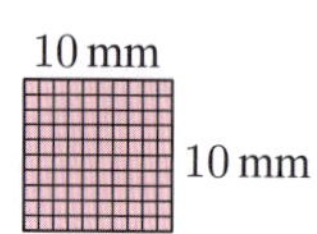

Each of the small squares has area $1 \ mm^2$.

There are $10 \times 10 = 100$ square millimetres in the square,
so $1 \ cm^2 = 100 \ mm^2$.

Likewise:

- $1 \text{ m}^2 = 1 \text{ m} \times 1 \text{ m}$
  $\phantom{1 \text{ m}^2} = 100 \text{ cm} \times 100 \text{ cm}$
  $\phantom{1 \text{ m}^2} = 10\,000 \text{ cm}^2$

- $1 \text{ ha} = 100 \text{ m} \times 100 \text{ m}$
  $\phantom{1 \text{ ha}} = 10\,000 \text{ m}^2$

- $1 \text{ km}^2 = 1 \text{ km} \times 1 \text{ km}$
  $\phantom{1 \text{ km}^2} = 1000 \text{ m} \times 1000 \text{ m}$
  $\phantom{1 \text{ km}^2} = 1\,000\,000 \text{ m}^2$
  $\phantom{1 \text{ km}^2} = 100 \times 10\,000 \text{ m}^2$
  $\phantom{1 \text{ km}^2} = 100 \text{ ha}$

$1 \text{ km}^2 = 100 \text{ ha}$
$1 \text{ ha} = 10\,000 \text{ m}^2$
$1 \text{ m}^2 = 10\,000 \text{ cm}^2$
$1 \text{ cm}^2 = 100 \text{ mm}^2$

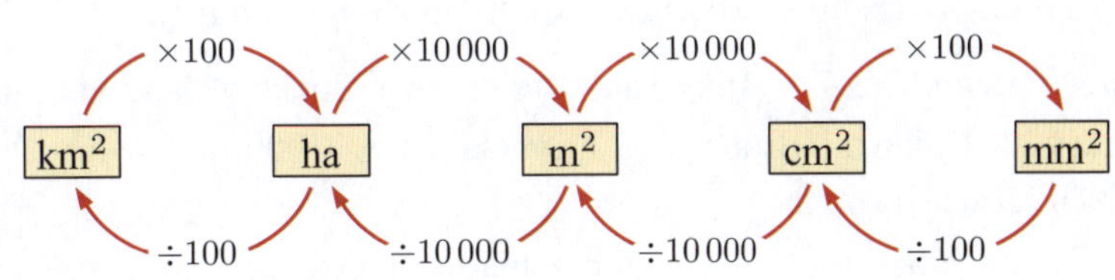

Example 6   ◀) **Self Tutor**

Convert:

a 2.2 ha into m^2

b 540 mm^2 into cm^2.

a   2.2 ha
$ = 2.2 \times 10\,000 \text{ m}^2$
$ = 22\,000 \text{ m}^2$

b   540 mm^2
$ = 540 \div 100 \text{ cm}^2$
$ = 5.4 \text{ cm}^2$

EXERCISE 11D

1 Choose the correct answer:

  a The area of a postage stamp is about:

    A 10 mm^2 **B** 10 cm^2 **C** 10 m^2 **D** 10 ha

  b The area of the floor of an elevator is about:

    A 3 mm^2 **B** 3 cm^2 **C** 3 m^2 **D** 3 ha

  c The area of a school is about:

    A 2 cm^2 **B** 2 m^2 **C** 2 ha   **D** 2 km^2

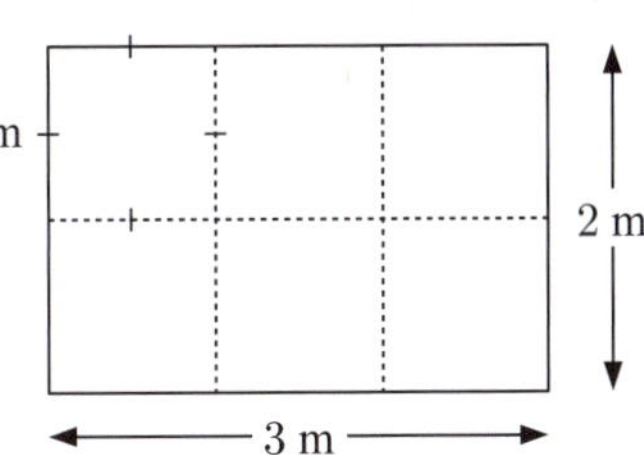

2 Erin has a rectangular vegetable garden measuring 2 m by 3 m.

  a Suppose the garden is divided into squares with side length 1 m.

    State the area of the garden in m^2.

  b Now suppose the garden is divided into squares with side length 1 cm.

    i State the number of rows and columns of squares.

    ii Hence find the area of the garden in cm^2.

3 Convert:

  a 5 cm^2 into mm^2
  b 2500 mm^2 into cm^2
  c 7 ha into m^2

  d 3.6 m^2 into cm^2
  e 0.4 km^2 into ha
  f 83 cm^2 into mm^2

  g 80 ha into m^2
  h $15\,600 \text{ cm}^2$ into m^2
  i 1200 ha into km^2

  j 900 mm^2 into cm^2
  k $76\,000 \text{ m}^2$ into ha
  l 2.8 cm^2 into mm^2

  m 0.25 km^2 into ha
  n 12.48 m^2 into cm^2
  o 0.0092 m^2 into mm^2.

4 A photograph has area 150 cm^2. Express this area in mm^2.

5 Three farmers Alessio, Bruno, and Carlos own blocks of land with the following areas:
Alessio 2.15 km^2,   Bruno 320 ha,   Carlos $640\,000 \text{ m}^2$.
Which farmer owns the:

  a largest block
  b smallest block?

6 The gardener of a school has 5.7 ha of lawn to manage. He must apply 60 g of fertiliser to every m^2 of lawn. He wants to buy fertiliser in 25 kg bags, each costing $\$19.85$.

  a How much fertiliser will be required?

  b How much will the gardener spend on fertiliser?

7 One hectare is about $2.471\,05$ acres. How many square metres are there in an acre?

E   AREA FORMULAE

In previous years we have seen formulae for the area of **rectangles**, **triangles**, **parallelograms**, and **trapezia**.

RECTANGLES

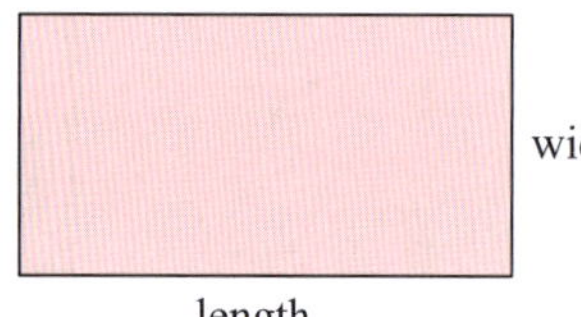

Area of rectangle $=$ length $\times$ width

TRIANGLES

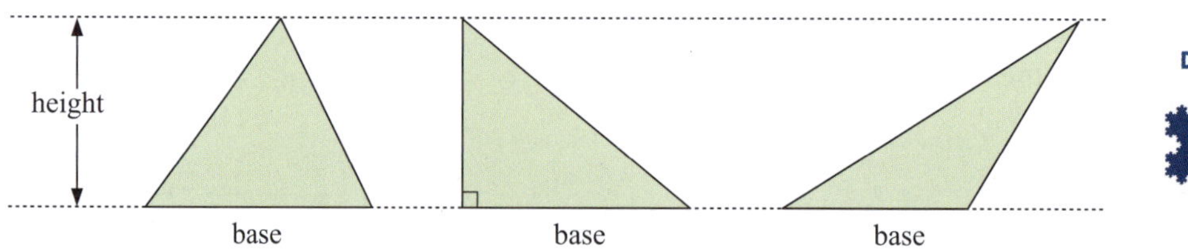

Area of triangle $= \frac{1}{2} \times$ base $\times$ height

PARALLELOGRAMS

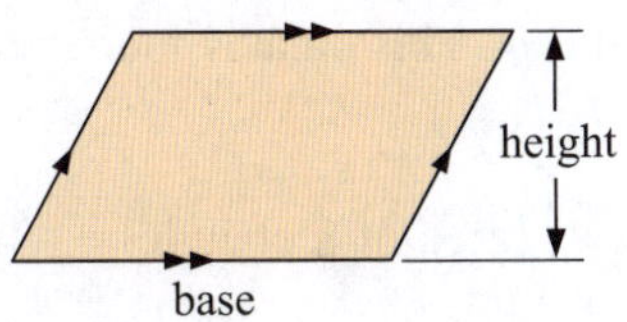

Area of parallelogram = base $\times$ height

TRAPEZIA

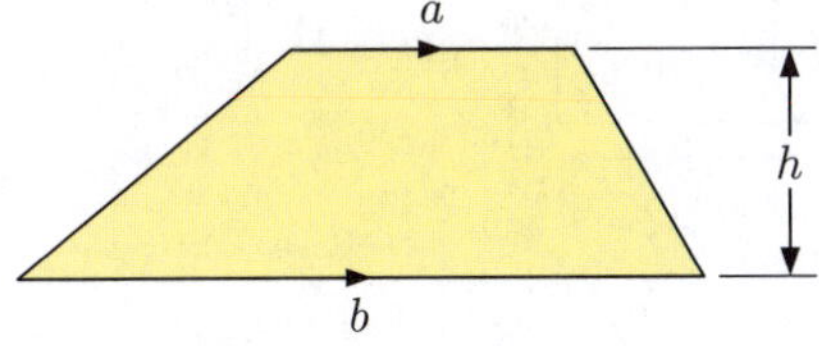

Area of trapezium = average length of parallel sides $\times$ distance between parallel sides

$$= \left(\frac{a+b}{2}\right) \times h$$

Example 7 ◆) **Self Tutor**

Find the area of:

a

5 m

12 m

b

5 cm

10 cm

c

11 m

4 m

16 m

a Area	**b** Area	**c** Area

a Area
$$= \tfrac{1}{2} \times \text{base} \times \text{height}$$
$$= \left(\tfrac{1}{2} \times 12 \times 5\right) \text{ m}^2$$
$$= 30 \text{ m}^2$$

b Area
$$= \text{base} \times \text{height}$$
$$= (10 \times 5) \text{ cm}^2$$
$$= 50 \text{ cm}^2$$

c Area
$$= \left(\frac{a+b}{2}\right) \times h$$
$$= \left(\frac{11+16}{2}\right) \times 4 \text{ m}^2$$
$$= 54 \text{ m}^2$$

KITES

Area of kite
$$= \tfrac{1}{2} \times \text{the product of the lengths of the diagonals}$$
$$= \tfrac{1}{2}(x \times y)$$

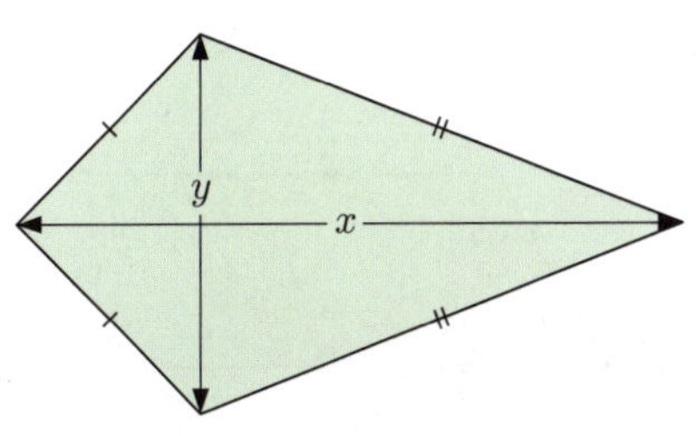

Proof:

Suppose we draw the kite within a rectangle with length x and width y. This creates four smaller rectangles, each divided in half by a side of the kite.

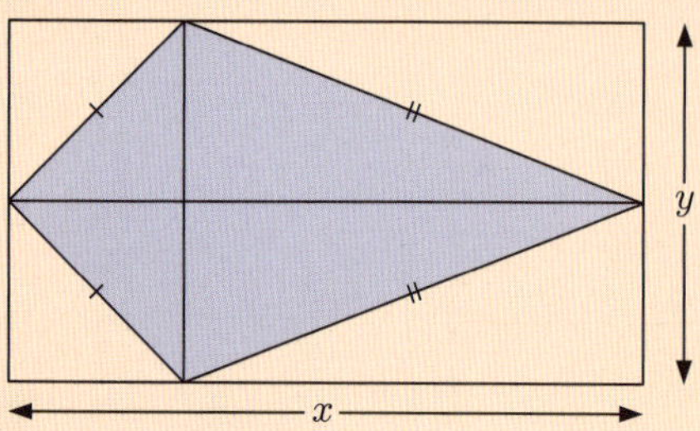

Area of kite $= \frac{1}{2}$(area of rectangle)

$= \frac{1}{2}(x \times y)$

DEMO

Example 8 — Self Tutor

Find the area of the kite:

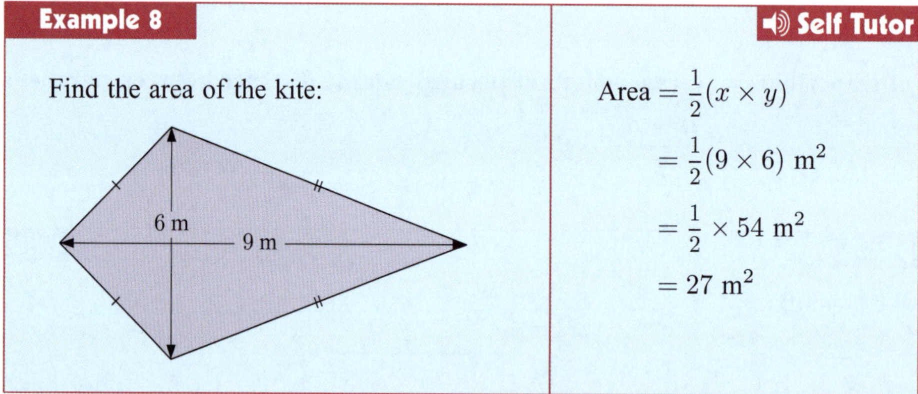

Area $= \frac{1}{2}(x \times y)$

$= \frac{1}{2}(9 \times 6)$ m^2

$= \frac{1}{2} \times 54$ m^2

$= 27$ m^2

EXERCISE 11E

1 Find the area of each polygon:

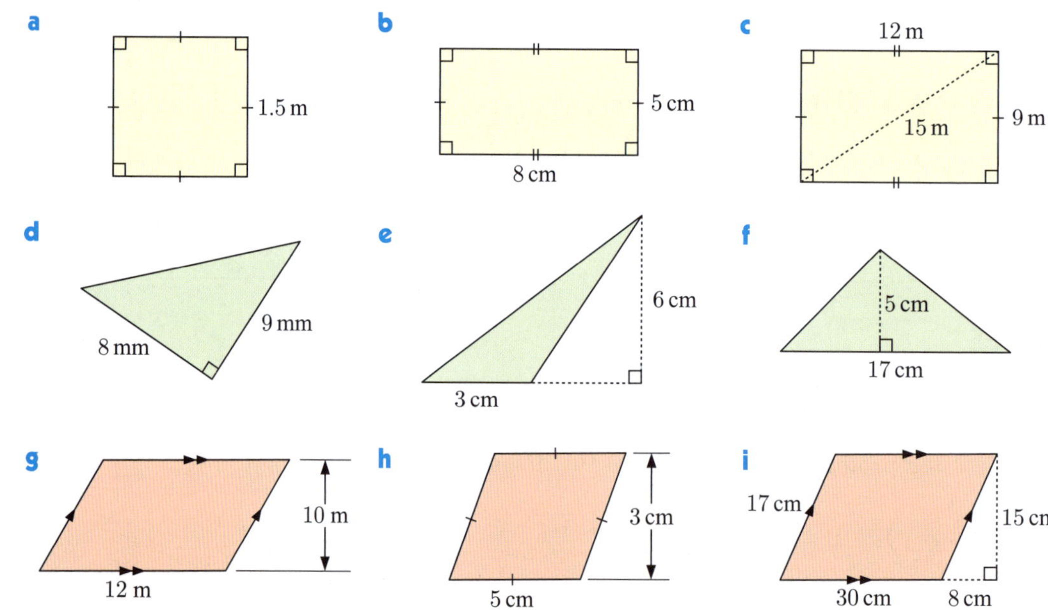

2 Find the area of each polygon:

a   **b**   **c**

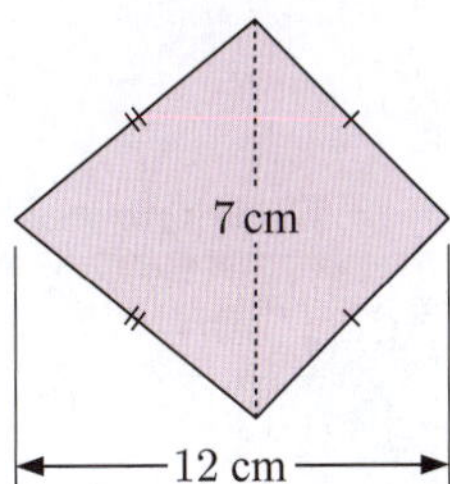
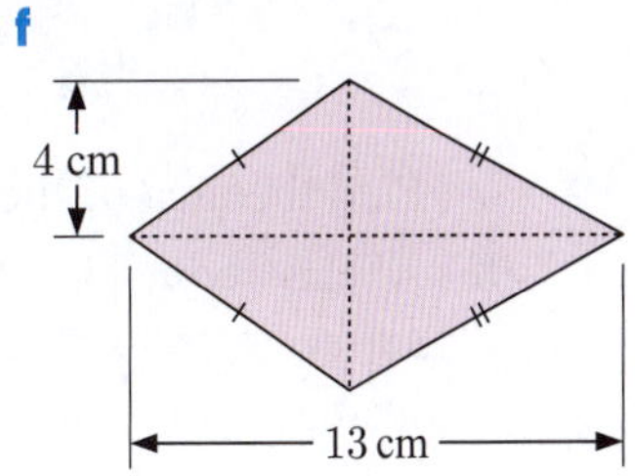

d   **e**   **f**

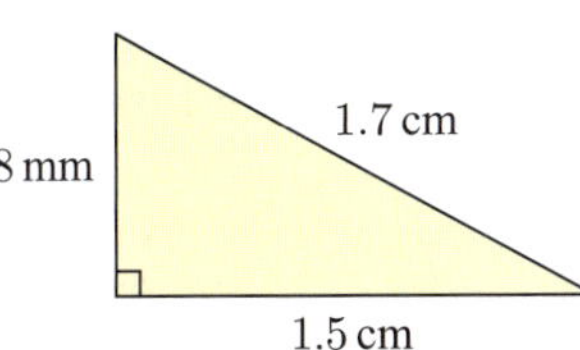
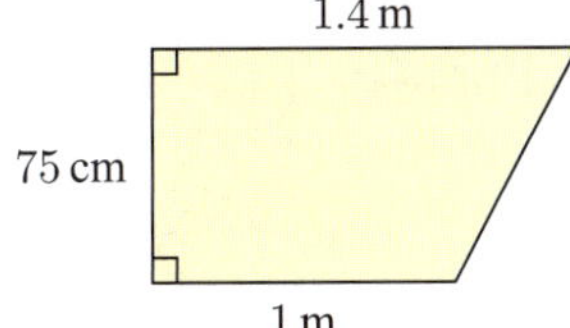
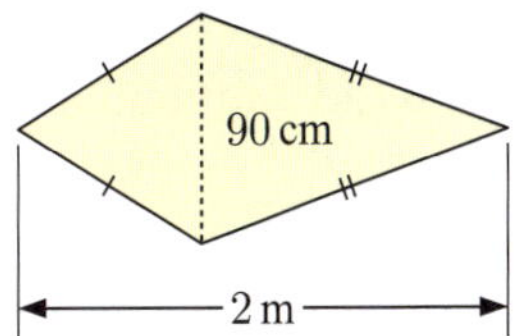

3 Find the area of each polygon:

a   **b**   **c**

4 A double bed sheet is 254 cm long and 228 cm wide. A queen bed sheet is 274 cm long and 245 cm wide.

  a Find the area of:

   i a double bed sheet   **ii** a queen bed sheet.

  b What percentage larger is the area of a queen bed sheet than the area of a double bed sheet?

5 **a** Find the amount of material required to make the sail shown.

  b The material costs \$19.50 per square metre. Find the cost of the sail.

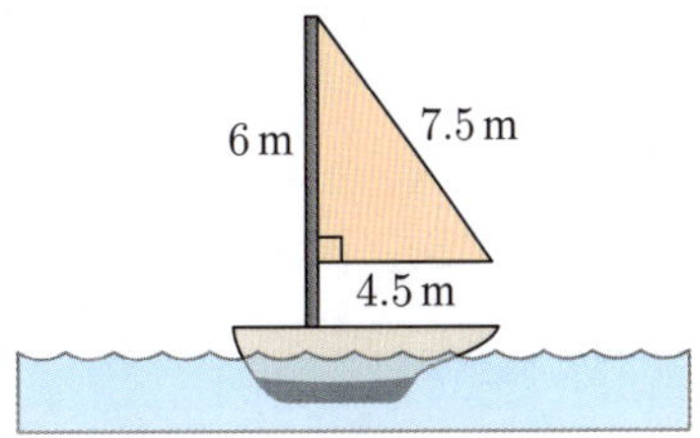

6 Find, in m², the total amount of material required to make 30 of these kites:

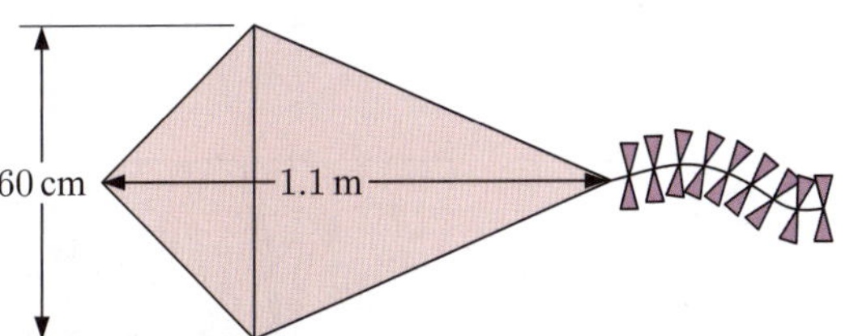

7 Anya's farm measures 800 metres by 1.2 km. 36 hectares are sown with wheat. Find the percentage of her farm that is sown with wheat.

8 A square paving brick has area 225 cm^2. If each brick fits tightly in place, how many bricks are needed to pave a 1.2 m by 12 m path?

9 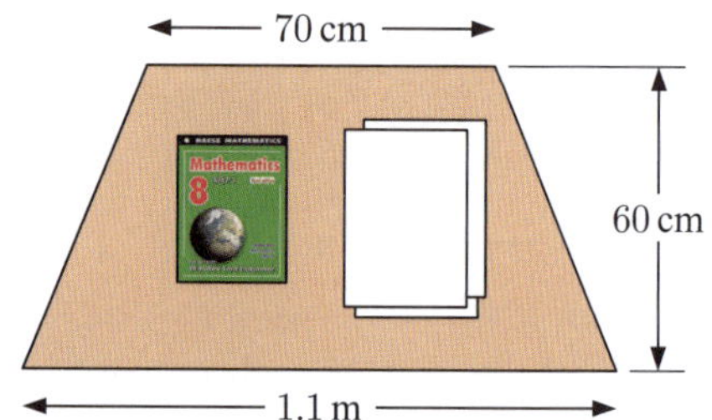

A feature wall at a movie theatre is made from green and red panels, as shown.

  a Find the area of the feature wall.

  b Find the area of each:

   i red panel   **ii** green panel.

10 **a** Find the area of the top of this desk.

  b A classroom has 25 of these desks. The top surface of each desk is to be repainted. Find the total area to be repainted, in m^2.

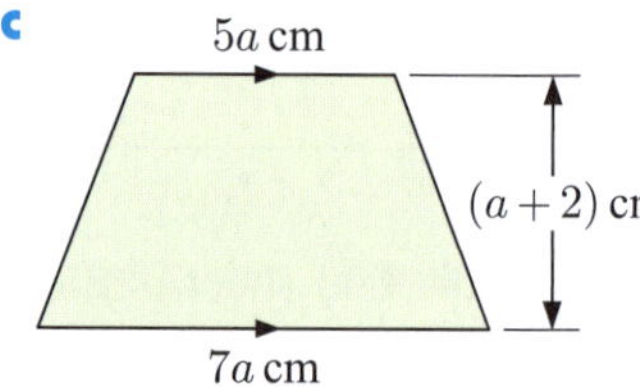

11 Write a formula for the area A of each shape, giving your answers in simplest form:

  a 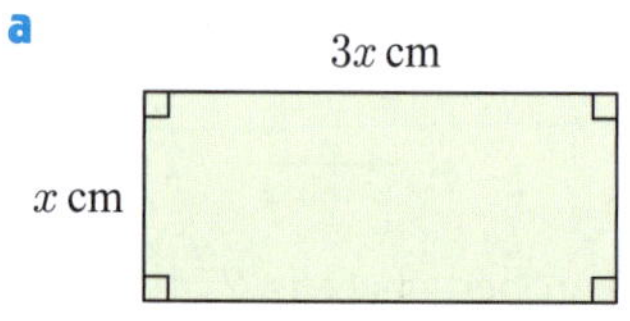

  b 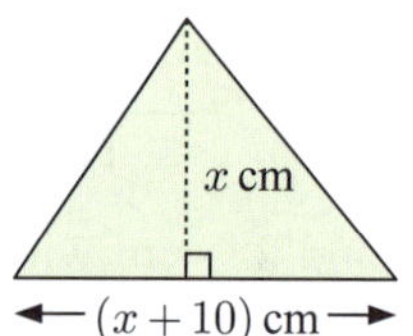

  c

12 The trapezium alongside has area 57 cm^2. Find the length of [AD].

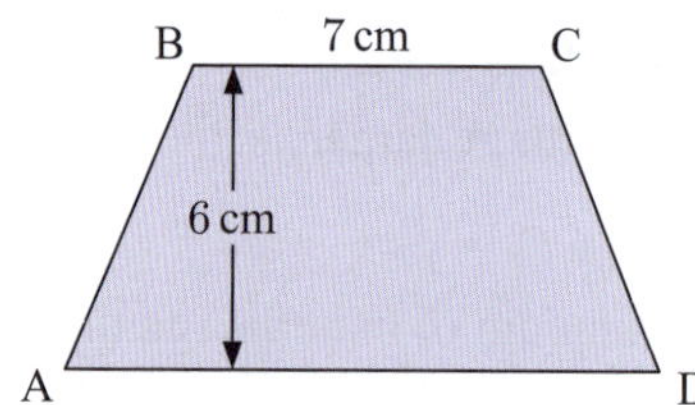

F

THE AREA OF A CIRCLE

Consider cutting a circle with radius r into 16 equal sectors and arranging them as shown:

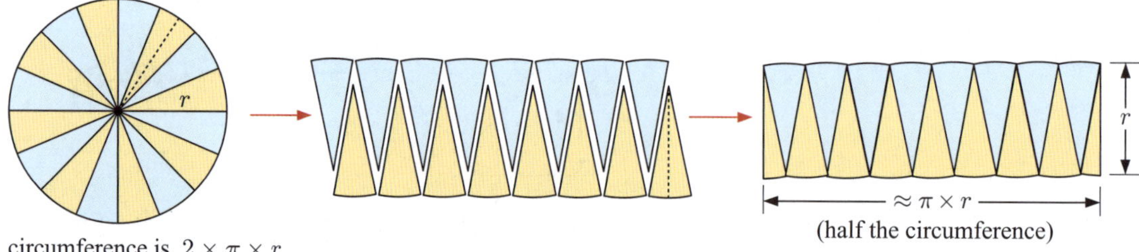

circumference is $2 \times \pi \times r$

The figure obtained closely resembles a rectangle.

- The height is approximately the radius of the circle.
- The top "edge" is approximately the sum of the arc lengths of the blue sectors. This is half the circumference of the circle, which is

$$\frac{1}{2} \times 2\pi r = \pi \times r.$$

Perform this demonstration for yourself. Click on the icon, then print the protractor and cut it into equal sectors.

If the original circle is cut into 1000 equal sectors and arranged in the same way, the resulting figure is indistinguishable from a rectangle.

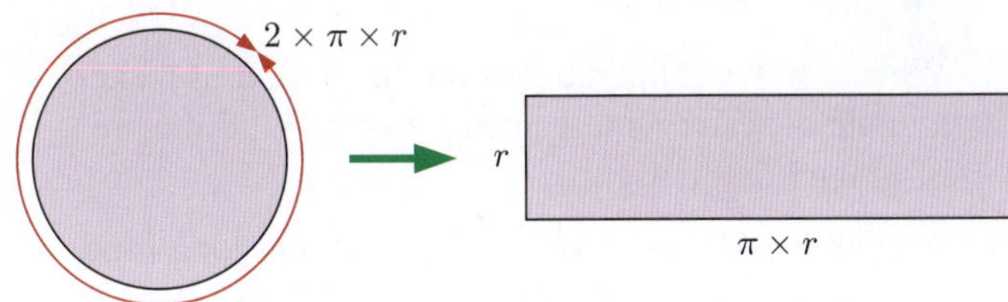

So, the area of the circle = length $\times$ width of the rectangle

$$= \pi \times r \times r$$
$$= \pi r^2$$

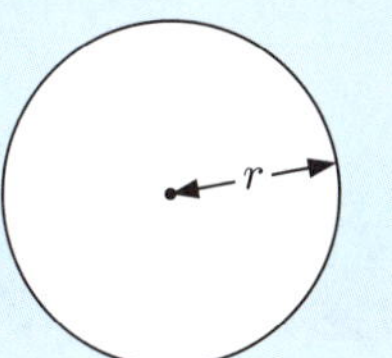

Area of circle $= \pi r^2$

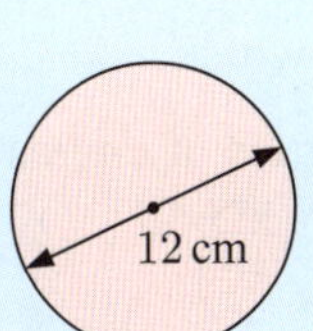

Example 9

Find the area of this circle.

Round your answer to 1 decimal place.

12 cm

Radius $r = \frac{1}{2} \times$ diameter

$$= \frac{1}{2} \times 12 \text{ cm} = 6 \text{ cm}$$

$$\therefore \quad \text{area} = \pi r^2 = \pi \times 6^2 \text{ cm}^2$$
$$\approx 113.1 \text{ cm}^2$$

EXERCISE 11F

1 Find the area of each figure. Round your answers to 2 decimal places.

a

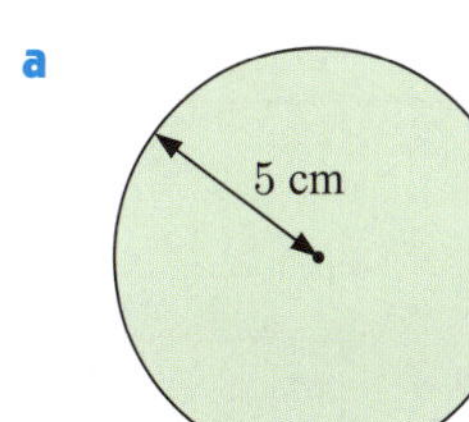

b

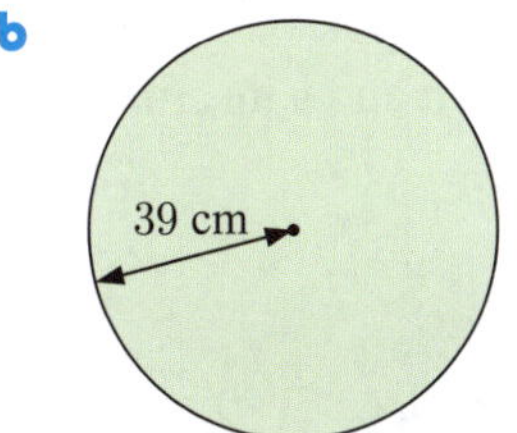

c

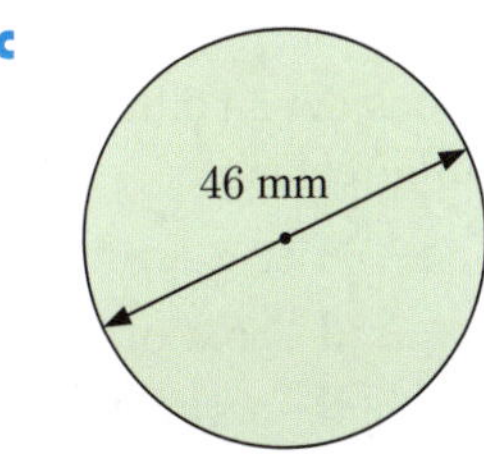

d 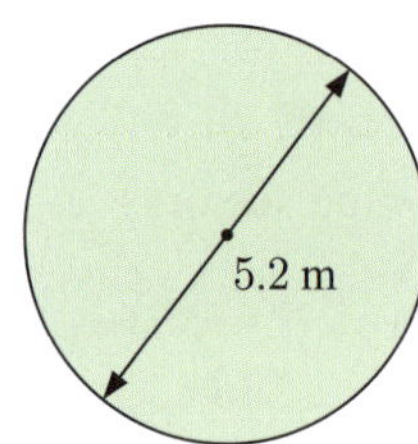

e 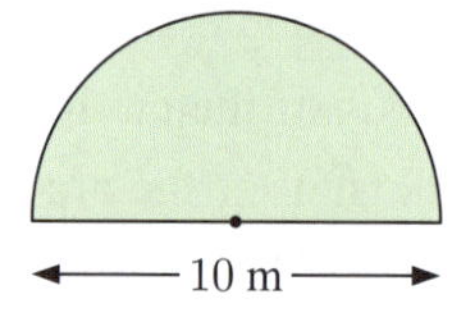

f 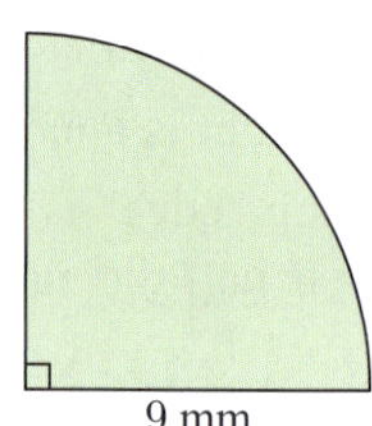

2 A circle has diameter 4.8 cm. Find, to 2 decimal places, its:

  a perimeter   **b** area.

3 When a water balloon bursts on the ground, it makes a circle of water with diameter 26 cm. Find the area of ground that is wet.

4 A donkey is tethered to a post by a 3.6 m long rope. Over what area can the donkey roam?

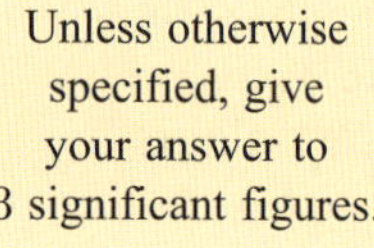

5 Look at the sector shown.

  a What fraction of a circle is it?

  b Write a formula for the area A of the sector.

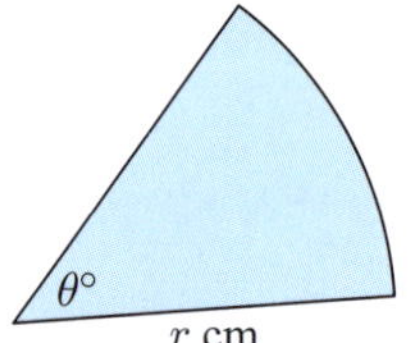

ACTIVITY 3   CIRCLES AND ELLIPSES

An ellipse is an oval-shaped figure, obtained by stretching a circle in one direction. The ellipse alongside has **semi-axes** a and b.

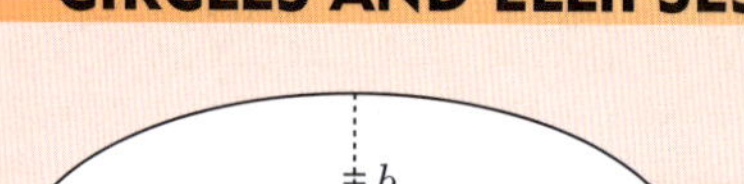

A circle is the special case of an ellipse for which $a = b$.

In this Activity we will see how the properties of circles are related to the properties of ellipses in general.

What to do:

1 In the diagram alongside, a circle with radius r is inscribed in a square.
Write in terms of π, the fraction:

  a $\dfrac{\text{area of circle}}{\text{area of square}}$ **b** $\dfrac{\text{perimeter of circle}}{\text{perimeter of square}}$

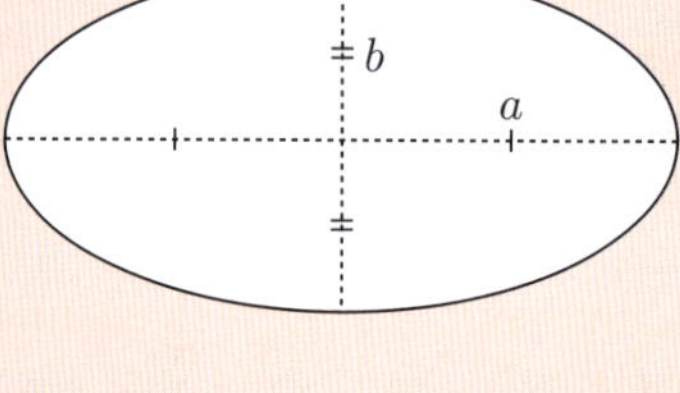

2 Consider an ellipse with semi-axes a and b inscribed in a rectangle.

  a Write a formula for the:

    i area of the rectangle

    ii perimeter of the rectangle.

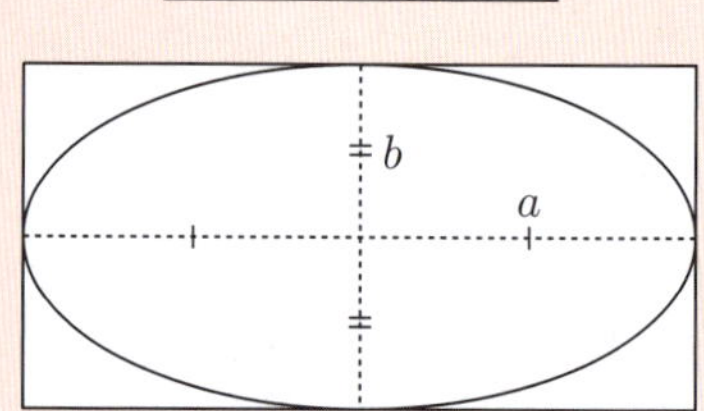

  b Explain why we might expect the area of the ellipse to be given by $A = \pi ab$.
Research online to confirm that this formula is indeed true for all ellipses.

  c Explain why we might expect the perimeter of the ellipse to be given by $P = \pi(a+b)$.
Research online to confirm that this formula is *not* true for ellipses in general.
Is it *possible* to write a formula for the perimeter of an ellipse?

3 Use the area formula $A = \pi ab$ to find the area of each ellipse:

a
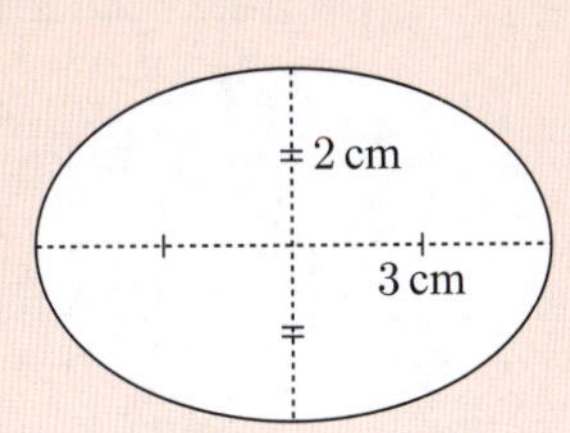

b
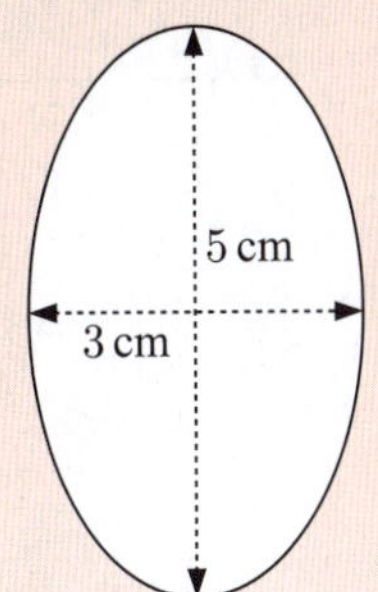

c
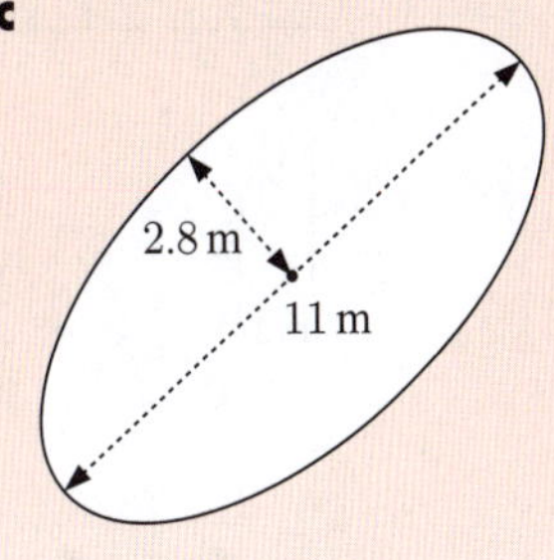

G   AREAS OF COMPOSITE FIGURES

Composite figures are figures made up of two or more basic shapes.

To calculate the area of a composite figure, we divide it into shapes that we are familiar with.

Example 10   ◄》 Self Tutor

Find the green shaded area:

a
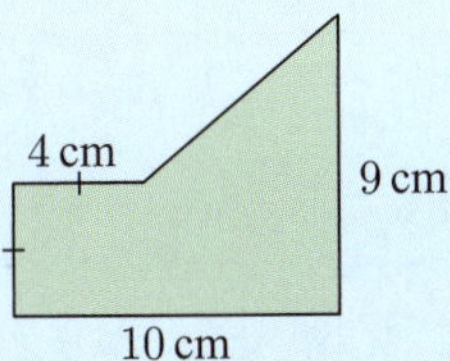

b

---

  a We divide the figure into a rectangle and a triangle as shown.

    Area = area of rectangle + area of triangle

$$= 10 \times 4 \text{ cm}^2 + \frac{1}{2} \times 6 \times 5 \text{ cm}^2$$

$$= (40 + 15) \text{ cm}^2$$

$$= 55 \text{ cm}^2$$

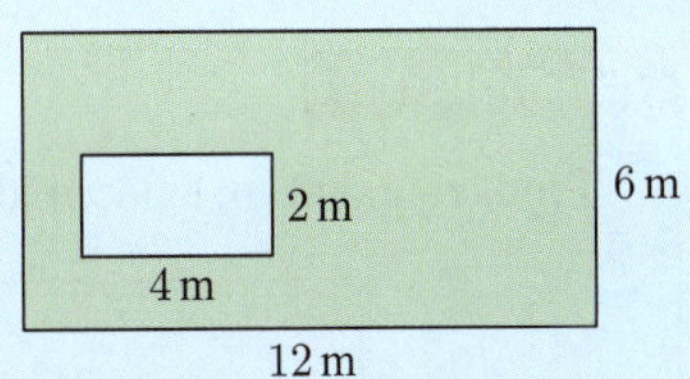

  b Area = area of large rectangle − area of small rectangle

$$= 12 \times 6 \text{ m}^2 - 4 \times 2 \text{ m}^2$$

$$= (72 - 8) \text{ m}^2$$

$$= 64 \text{ m}^2$$

EXERCISE 11G

1 Find the shaded area:

a
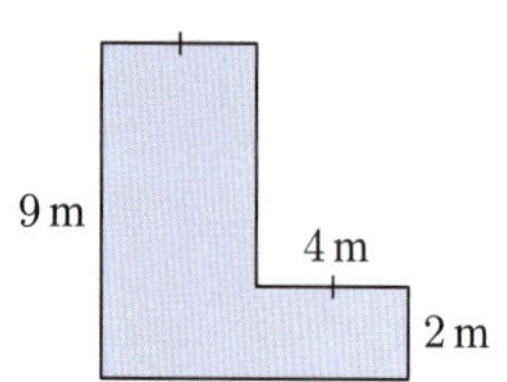

b
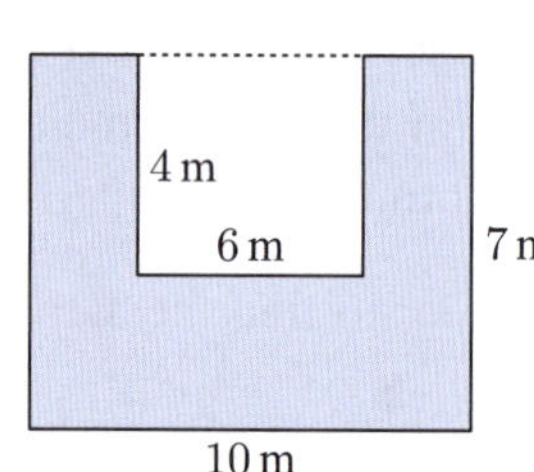

c
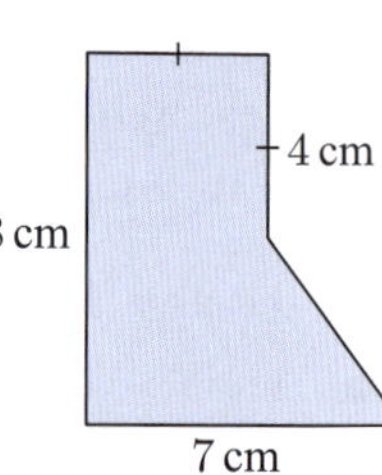

d
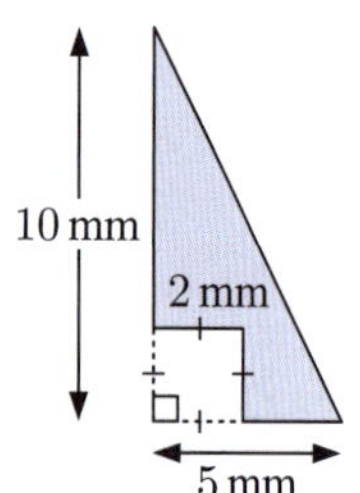

e
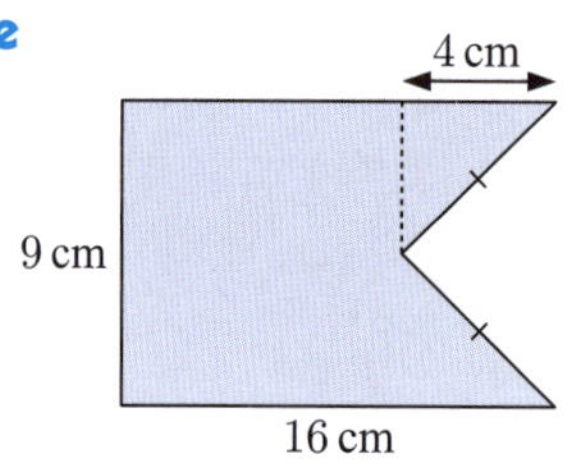

f
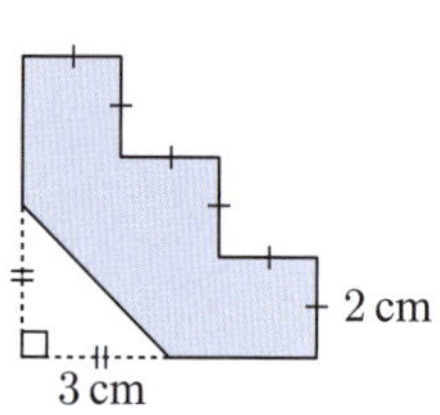

2 Find the shaded area:

a
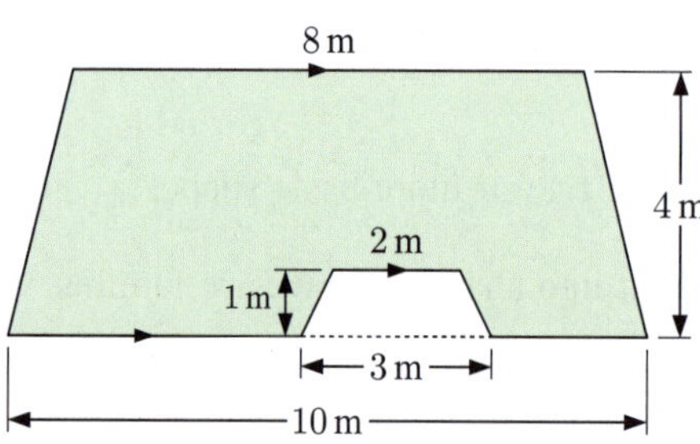

b
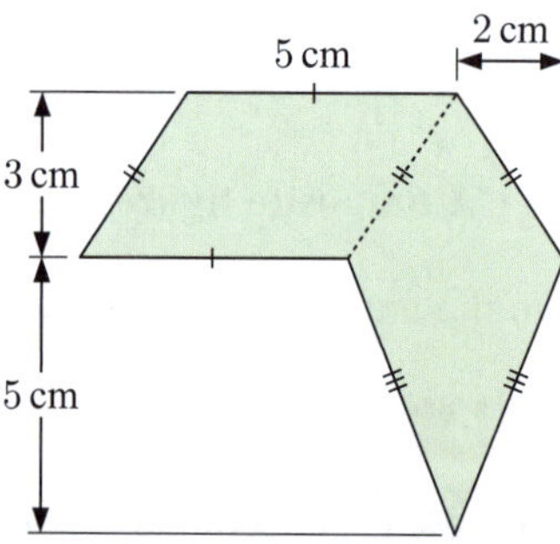

Example 11   ◀)) Self Tutor

Find, rounded to 1 decimal place, the pink shaded area:

a
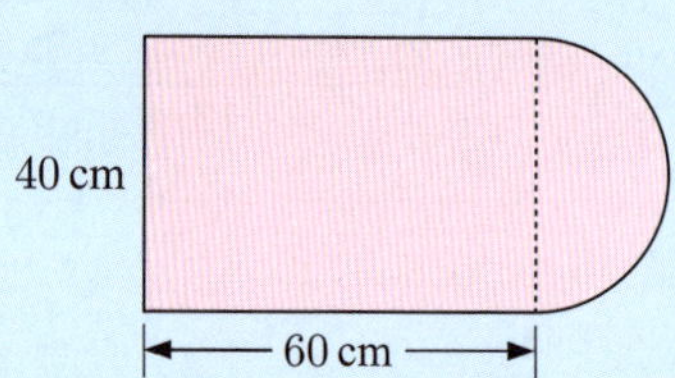

b
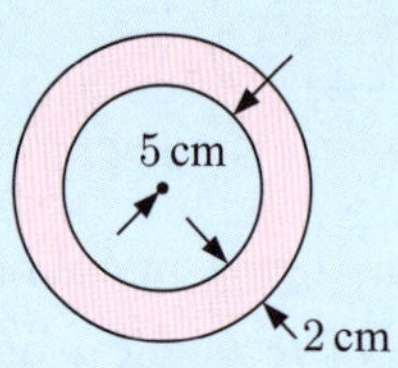

a Area = area of rectangle + area of semi-circle

$$= 40 \times 60 \text{ cm}^2 + \frac{1}{2} \times (\pi \times 20^2) \text{ cm}^2$$

$$\approx 3028.3 \text{ cm}^2$$

b Area = area of large circle − area of small circle

$$= \pi \times 7^2 \text{ cm}^2 - \pi \times 5^2 \text{ cm}^2$$

$$\approx 75.4 \text{ cm}^2$$

3 Find the area of the shaded region, rounding your answer to 1 decimal place:

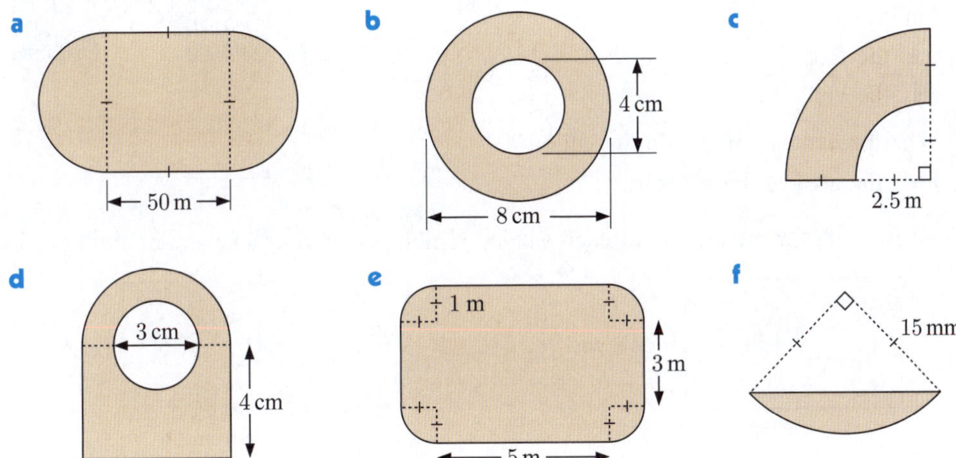

a 50 m

b 4 cm, 8 cm

c 2.5 m

d 3 cm, 4 cm, 5 cm

e 1 m, 3 m, 5 m

f 15 mm

4 The wall shown is to be tiled with 30 cm square tiles.
How many tiles are needed?

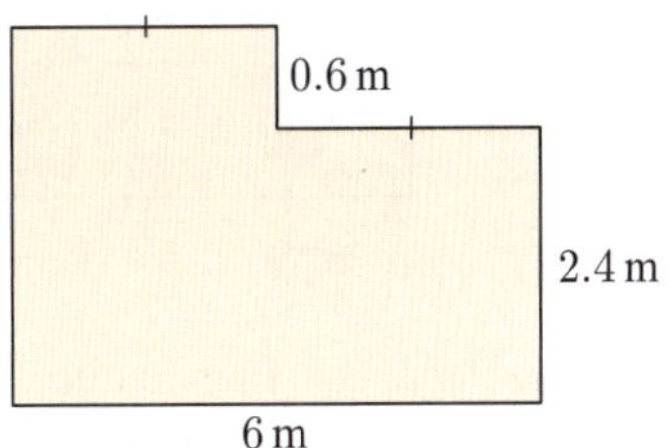
0.6 m, 2.4 m, 6 m

5 A plain metal washer is shown below.
Find the area of the top surface.

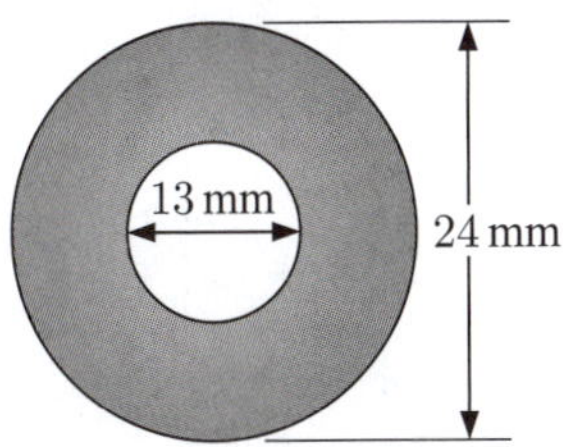
13 mm, 24 mm

6 Find the area of this stained glass window.

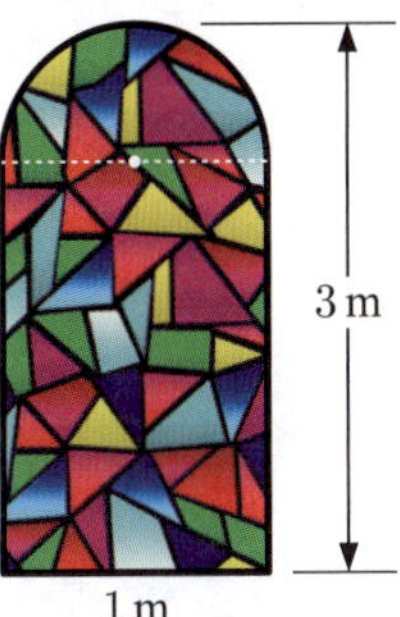
3 m, 1 m

7 Find the area of the jigsaw puzzle piece shown.
Give your answer to the nearest mm^2.

8 mm

8 Ian cuts 12 circles of pastry from a 30 cm by 40 cm sheet.
Find the area of pastry which remains.

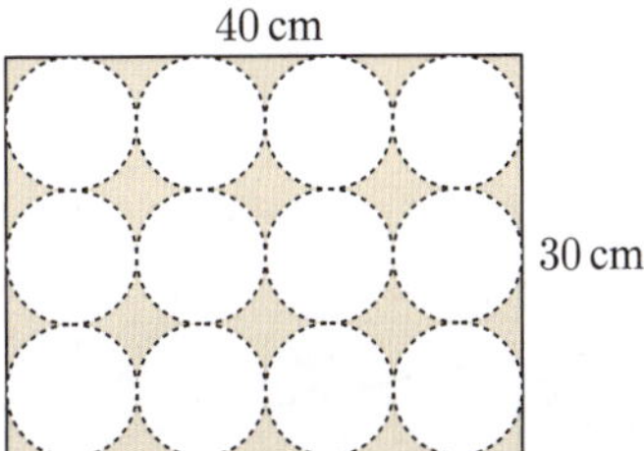
40 cm, 30 cm

9 An Australian Aboriginal flag is shown alongside.

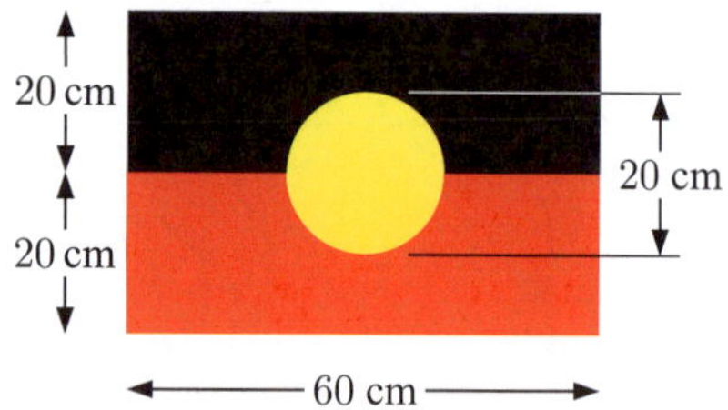

    a Find the area of:

        i the flag

        ii the yellow circle.

    b What percentage of the flag is yellow?

    c Find the area of the red part.

10 Write a formula for the area A of each shaded region. Give your answers in simplest form.

    a

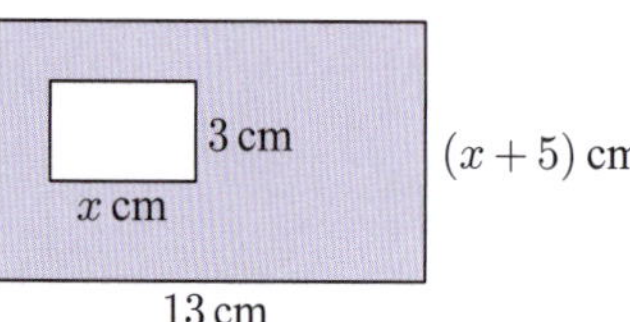

    b

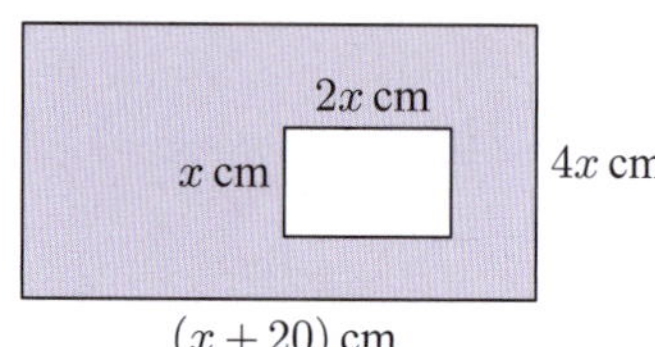

    c

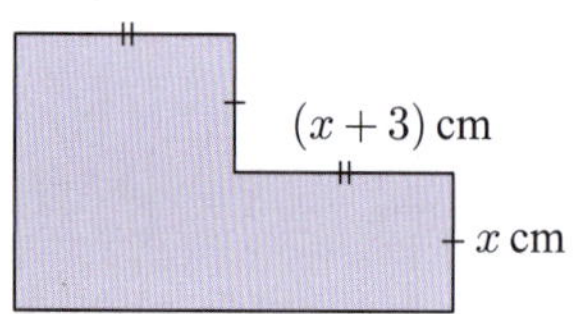

    d

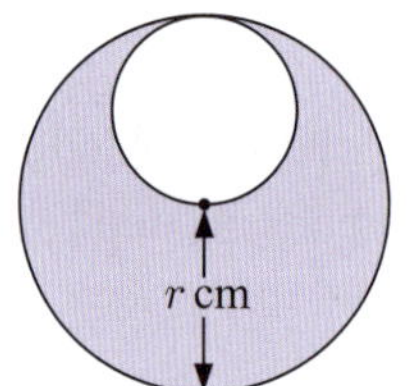

11 The Japanese flag features a red circular "sundisc" on a white rectangular "field".

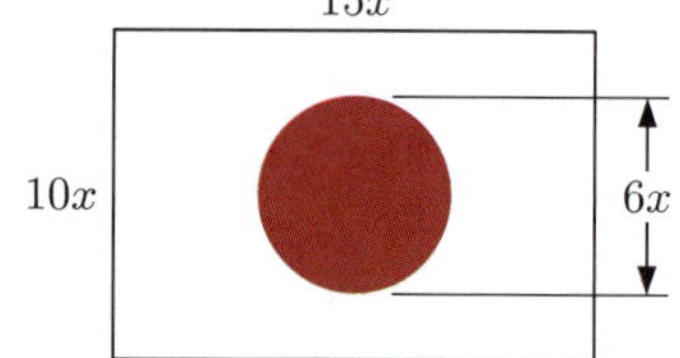

    a What fraction of the flag is taken up by the "sundisc"? Write your answer in terms of π.

    b What percentage of the flag is the "field"?

12 **a** Find, in simplest form, the area A of this figure.

    b Given that the figure has area 54 m^2, find the value of x.

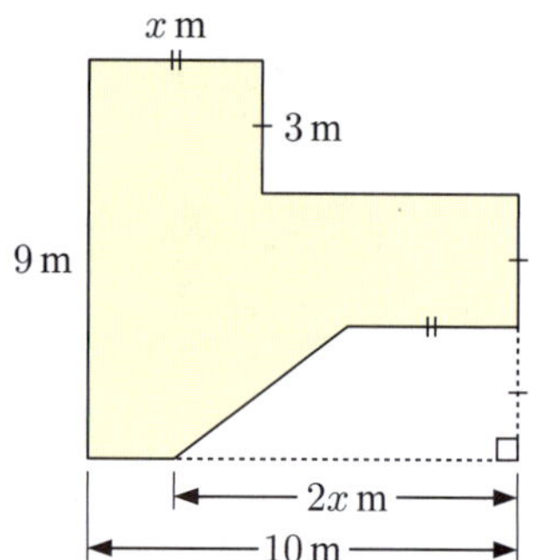

13 Find the area of the shaded region:

    a

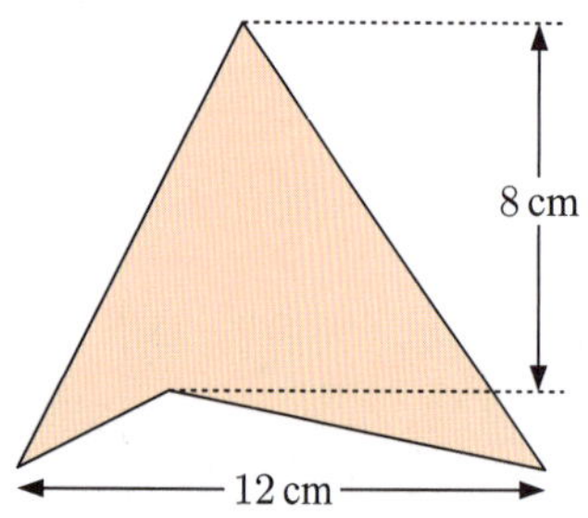

    b

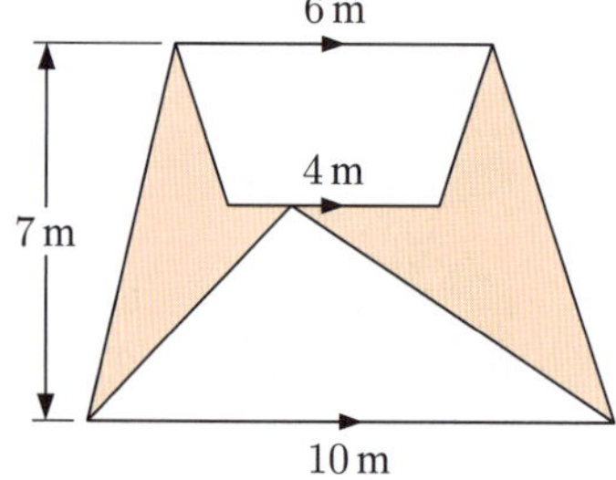

INVESTIGATION 2   PICK'S RULE FOR AREAS

This diagram shows a polygon whose **vertices** lie on **lattice points**.

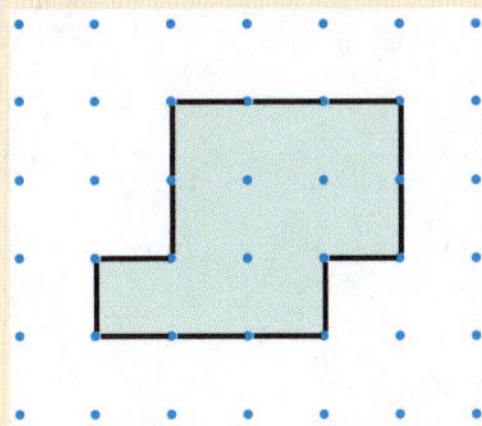

The lattice points which lie *on* the figure are called **boundary points**.

The lattice points which lie *inside* the figure are called **interior points**.

The figure shown has 14 boundary points and 3 interior points.

Suppose that for a given polygon:

$$B = \text{the number of boundary points}$$
$$I = \text{the number of interior points.}$$

What to do:

1 Copy the table below, then complete it using these figures:

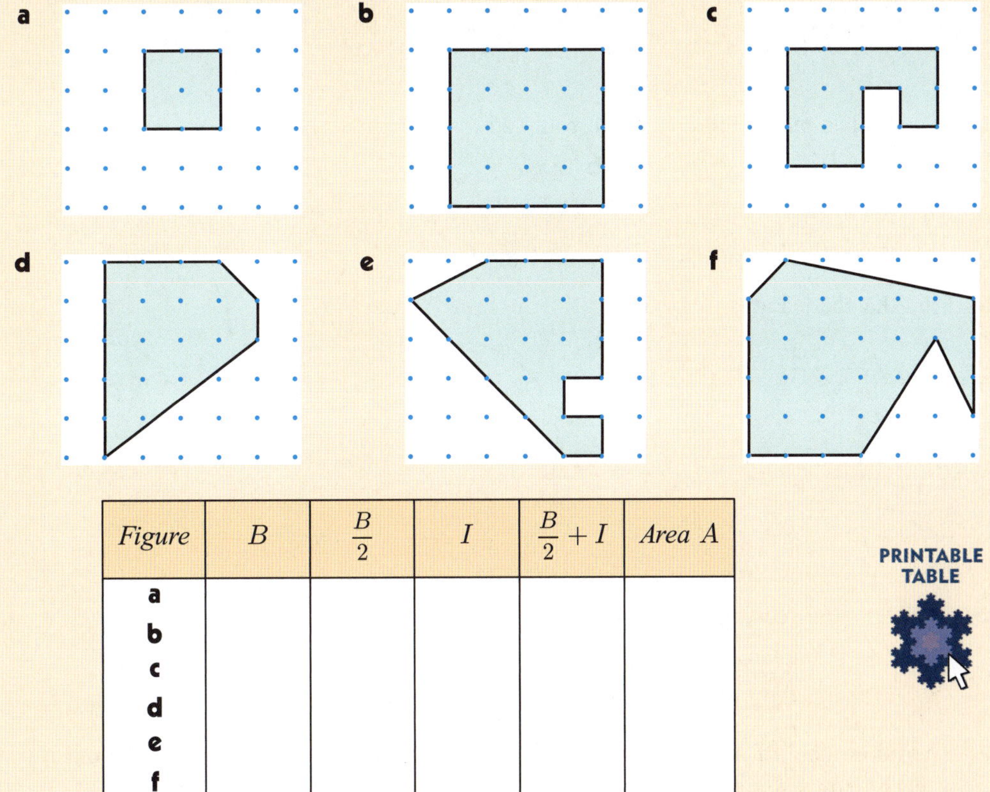

Figure	B	$\dfrac{B}{2}$	I	$\dfrac{B}{2} + I$	Area A
a					
b					
c					
d					
e					
f					

PRINTABLE TABLE

2 Suggest a rule connecting A, B, and I. This result is called **Pick's Rule**.

3 Draw several polygons of your own on grid paper, and check that your rule is correct.

PRINTABLE DOT GRID PAPER

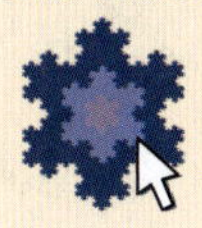

MULTIPLE CHOICE QUIZ

REVIEW SET 11A

1 Convert:

   a 95 mm into cm   **b** 4.8 km into m   **c** 270 cm into m.

2 Find the perimeter of each figure:

a
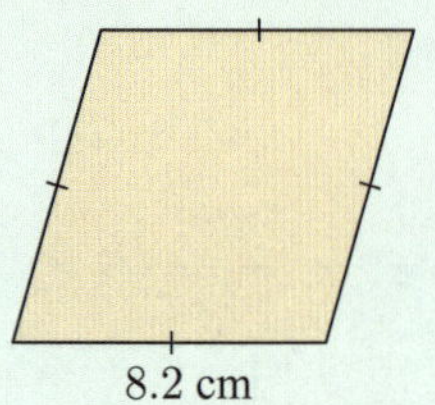

b
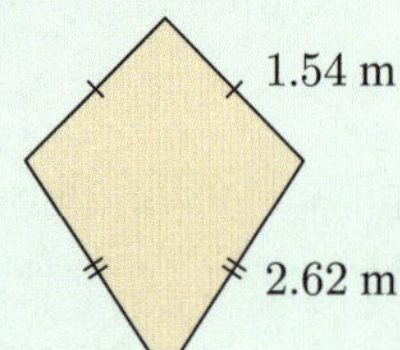

c
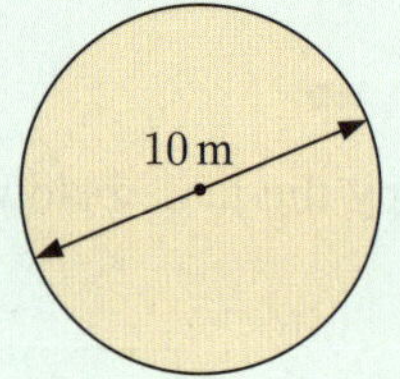

3 A circle has radius 5.6 cm. Find, rounded to 1 decimal place, its:

   a circumference   **b** area.

4 Convert:

   a $55\,000$ cm^2 into m^2 **b** 3.4 m^2 into cm^2 **c** $18\,500$ m^2 into ha.

5 Find the shaded area:

a
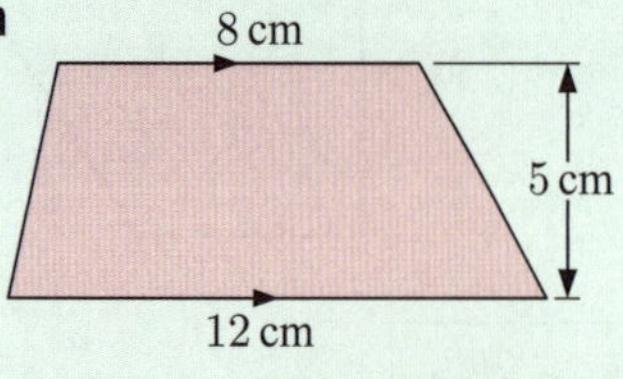

b
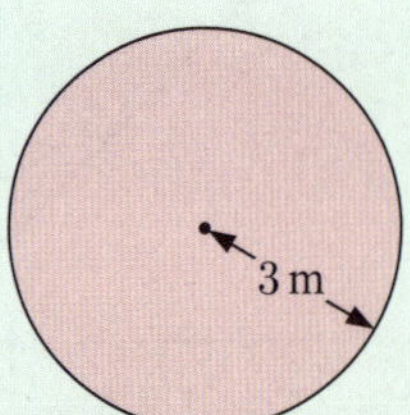

c
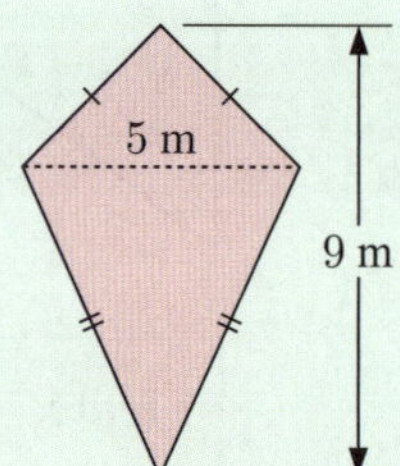

6 A rectangular recreation reserve is 800 m wide and 2.3 km long.

   a Find the area of the reserve in hectares.

   b If 20 trees are planted in each hectare, how many trees are planted in total?

7 The boom gate of a railway crossing contains a pattern of parallelograms, with a triangle at each end as shown.

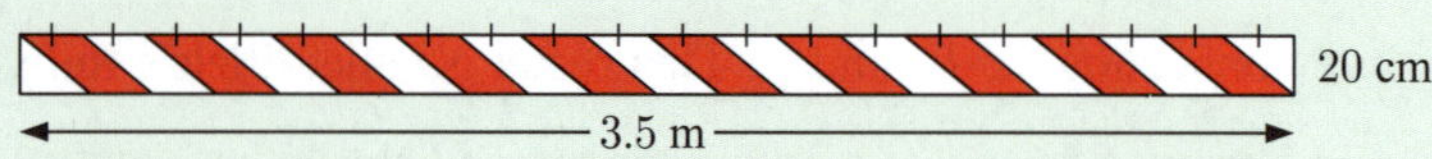

   a Find the area of each: **i** parallelogram   **ii** triangle.

   b Check your answers by adding the areas of the sections and comparing this with the total area of the rectangular boom gate.

8 Find the shaded area:

a
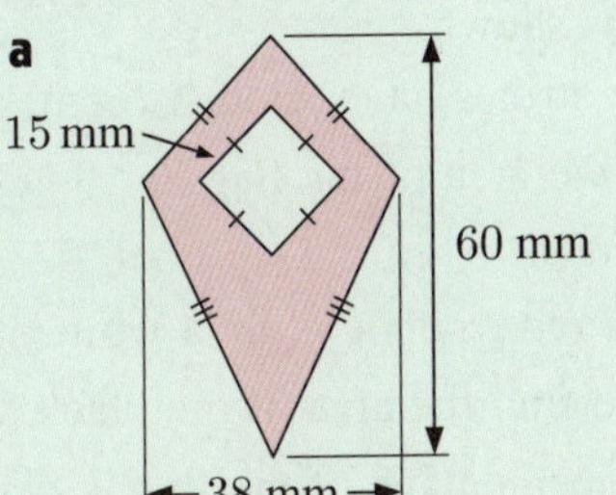

b
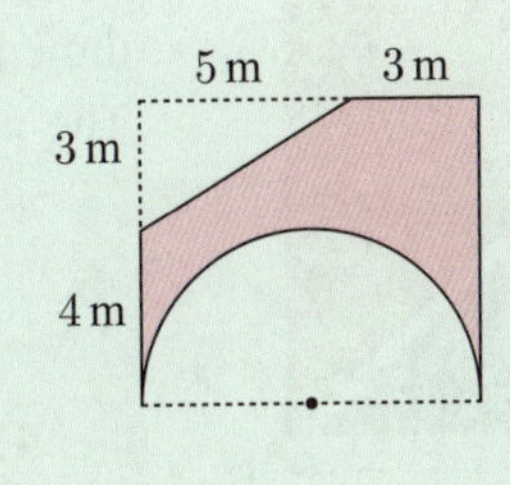

c
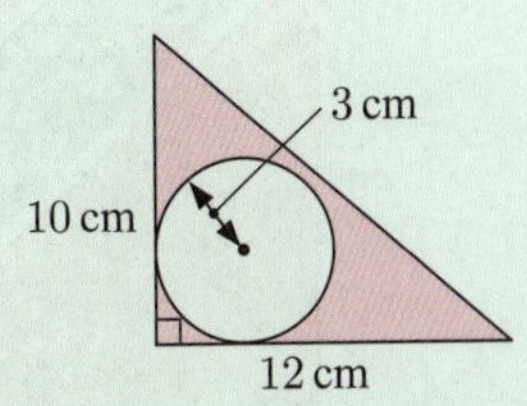

9 This door is fitted into a metal door frame. Find:

   a the length of metal required for the frame

   b the area of the door.

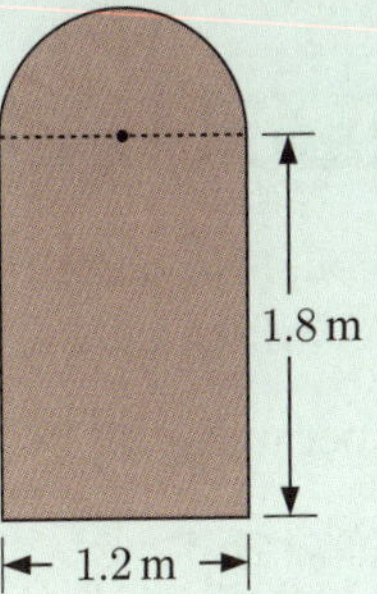

10 This circular pond has diameter 1.6 m.

   a Find the area of plastic needed to line the bottom of the pond.

   b A 10 cm wide strip filled with pebbles will surround the edge of the pond. What area will the pebbles cover?

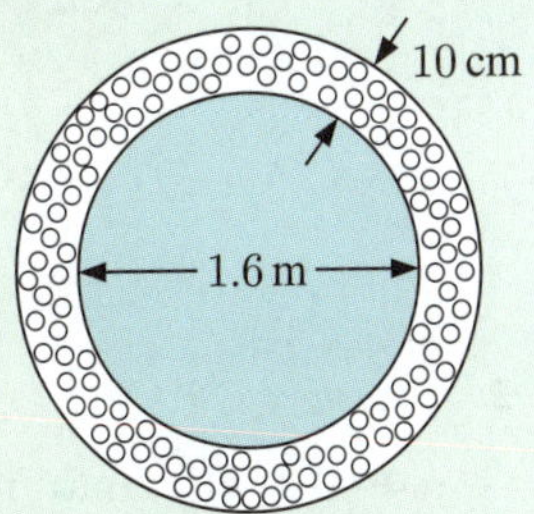

11 **a** Find, in simplest form, the area A of this figure.

   b Given that the figure has area 24 km^2, find the value of x.

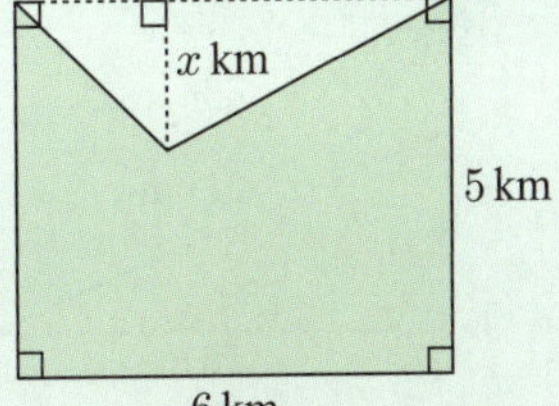

12 A "Narrow 5s" dartboard is shown alongside.

   a Find the:

     i circumference of the board

     ii area of the board.

   b The "double" ring has width 8 mm. Find, in mm:

     i the diameter of the inner boundary of the "double" ring

     ii the circumference of the inner boundary of the "double" ring.

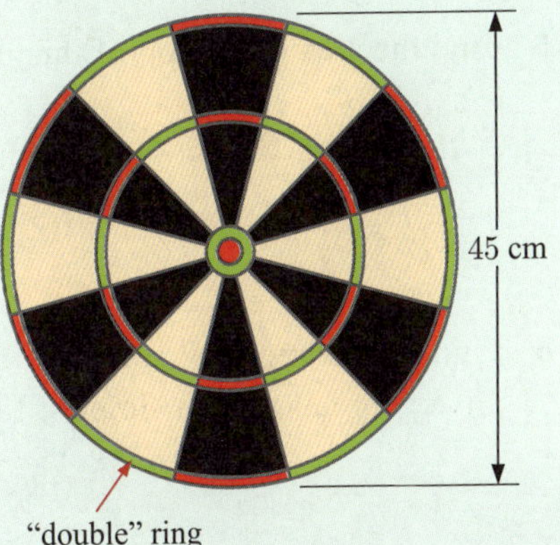

13

A flag of Trinidad & Tobago has the dimensions shown.

The white stripes have width 1.5 cm.

  a Find the area of the flag.

  b Find the area of:

    i a red triangle   **ii** a white stripe.

  c Hence find the area of the black stripe.

  d What percentage of the flag is black?

REVIEW SET 11B

1 Hayden bought a packet containing 60 m of dental floss. He uses 40 cm of floss each day. How long will he take to use the whole packet?

2 Find the perimeter of each figure:

  a

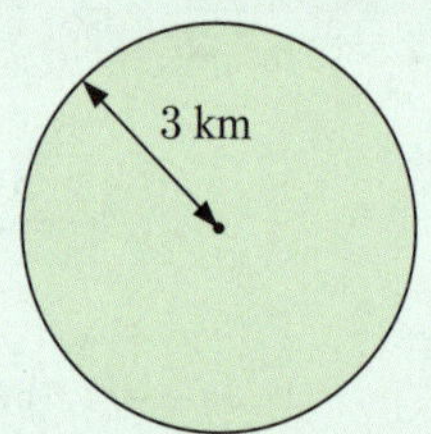

  b

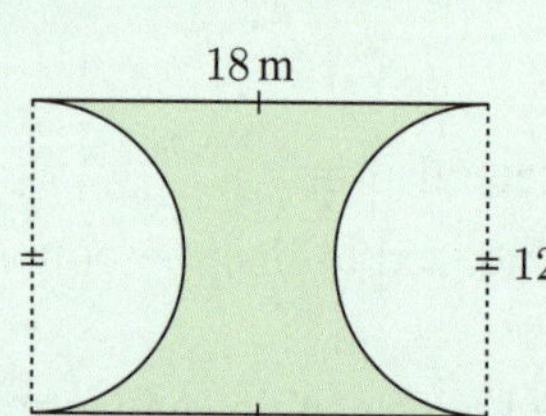

  c

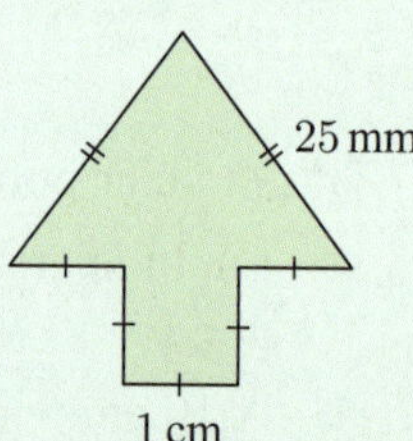

3 Convert:

  a 340 cm^2 into mm^2 **b** 3270 m^2 into ha.

4 Find the area of:

  a

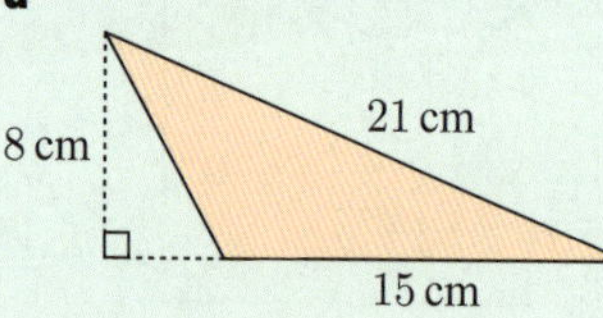

  b

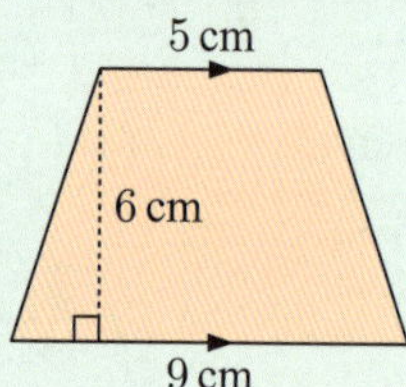

  c

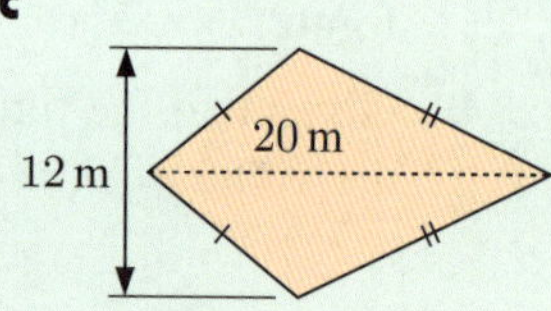

5 Find the area of the top of this piece of shortbread:

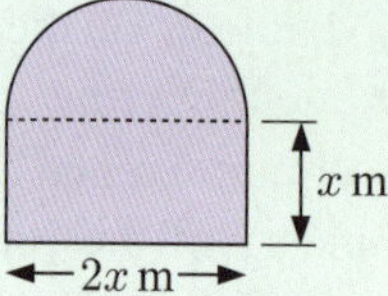

6 Find the perimeter P and area A of this figure. Give your answers in simplest form.

7 Whose pizza is larger?

Eliza's pizza

18 cm

Claire's pizza

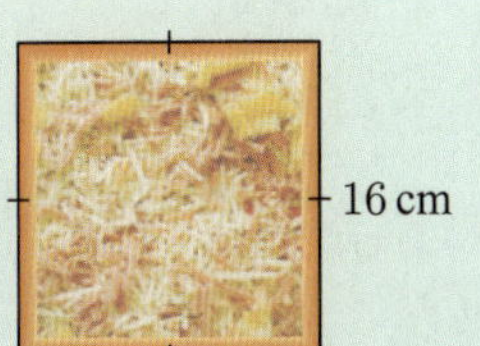

16 cm

8 The room shown is to be covered in floorboards. Each floorboard measures 20 cm by 80 cm. How many floorboards are needed?

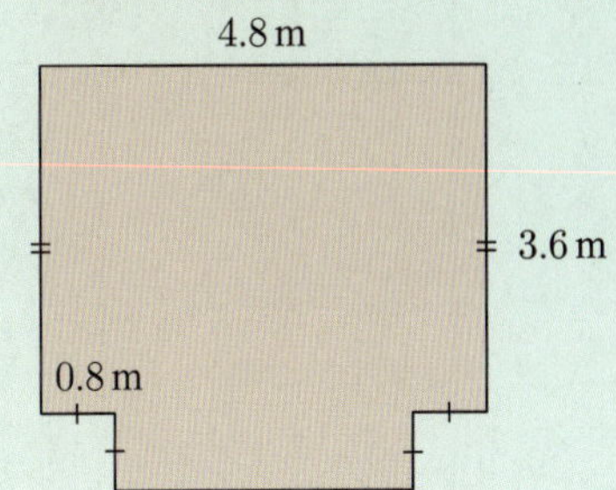

9 Find the pink shaded area:

a

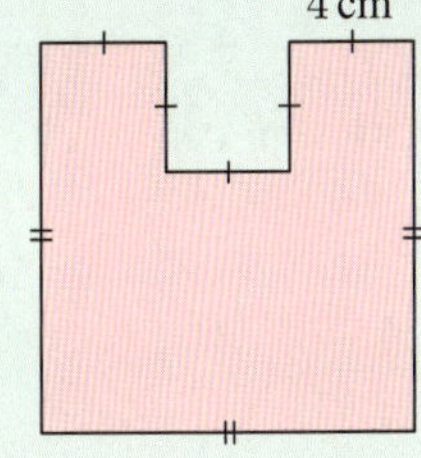

b

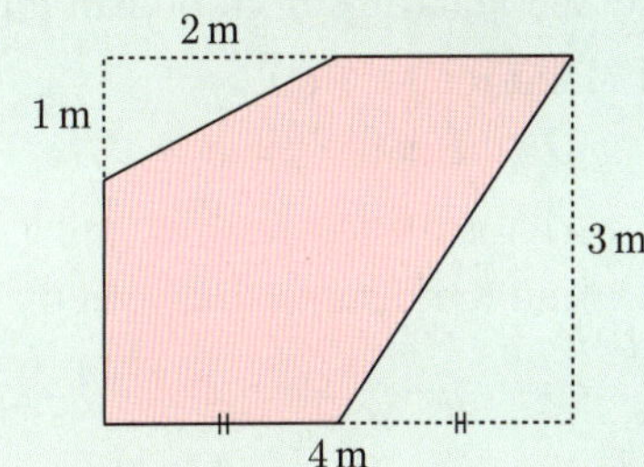

10 A rug measuring 3.5 m by 2.5 m was placed in a room 8.2 m long and 6.4 m wide. What area of floor is not covered by the rug?

11 The flag of Nepal is shown alongside. Find its area.

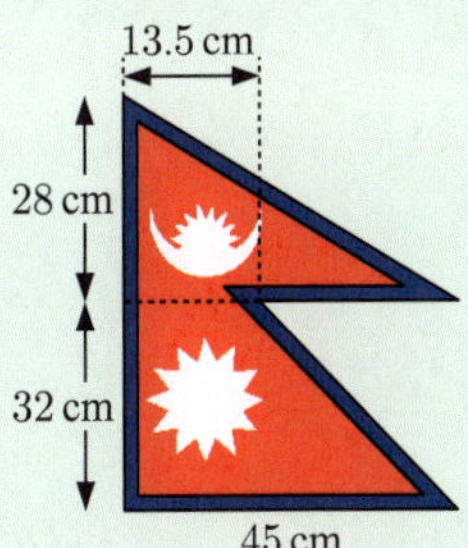

12 Mary is making a kite with the dimensions shown.

  a Find the total length of wood required for the crosspieces.

  b Mary wants to put edging around the perimeter of the kite. How many metres of edging will she need?

  c Find the total area of cloth Mary will need for her kite.

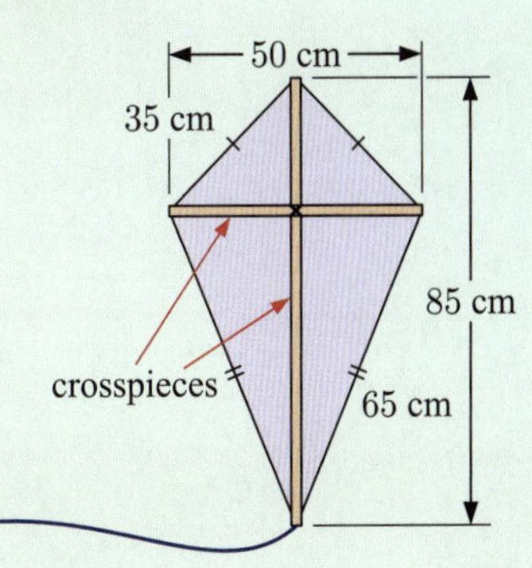

13

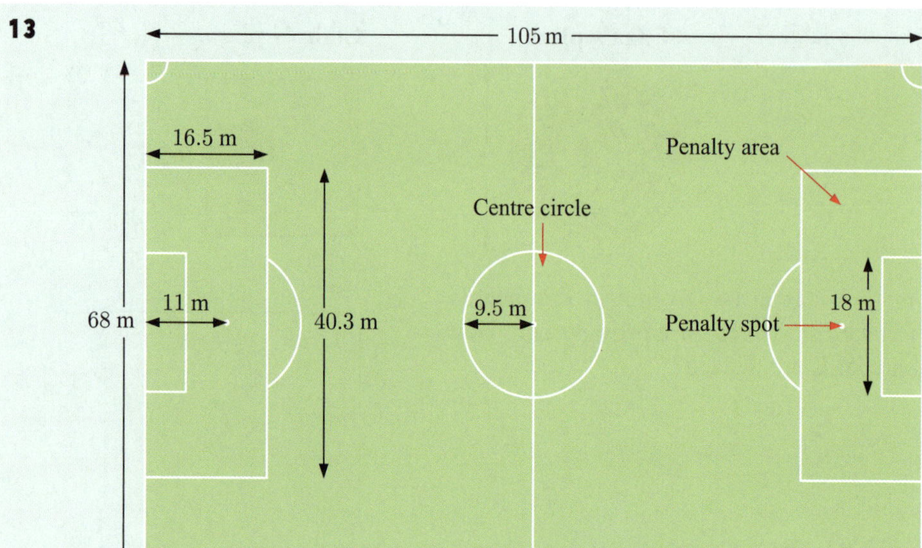

The line markings of a football pitch are shown above.

a Find the area of:

   i the entire pitch   **ii** each penalty area   **iii** the centre circle.

b What percentage of the pitch's area is the centre circle?

c Find the distance between the two penalty spots.

Measurement: Surface area, volume, and capacity

Contents:

OPENING PROBLEM

Silas is baking a cake. His baking tin is cylindrical with radius 10 cm and height 8 cm.

Silas has prepared 3 litres of cake mix.

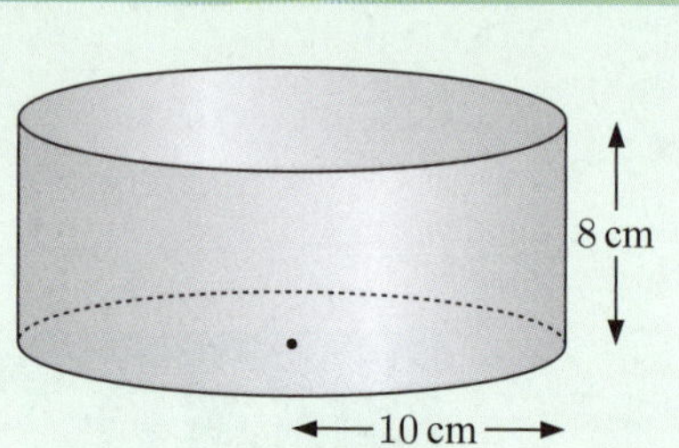

Things to think about:

a Which *two-dimensional* shapes of metal could be used to construct the tin?

b What *area* of baking paper is required to cover the bottom and sides of the tin?

c What *volume* of cake mix would completely fill the tin?

d Will Silas be able to pour all of his cake mix into the tin?

In this Chapter we will extend our study of measurement to consider 3-dimensional objects.

We will consider the **surface area** and **volume** of solids.

A   SURFACE AREA

The **surface area** of a solid is the sum of the areas of its surfaces.

To find the surface area of a solid, it is often helpful to draw the **net** of the solid.

Example 1   ◀)) Self Tutor

Find the surface area of this rectangular prism.

We first draw a net of the prism.

$A_1 = 4 \times 3 = 12$ cm^2

$A_2 = 4 \times 2 = 8$ cm^2

$A_3 = 3 \times 2 = 6$ cm^2

Surface area

$= 2 \times A_1 + 2 \times A_2 + 2 \times A_3$

$= 2 \times 12$ cm$^2 + 2 \times 8$ cm$^2 + 2 \times 6$ cm^2

$= 52$ cm^2

EXERCISE 12A

1 Find the surface area of each cube:

a
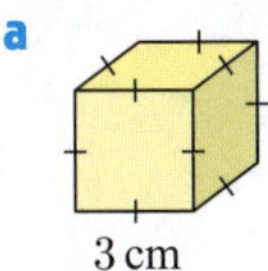
3 cm

b
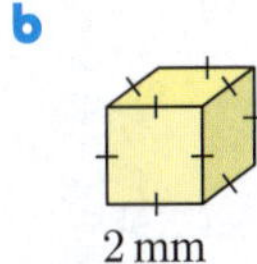
2 mm

c
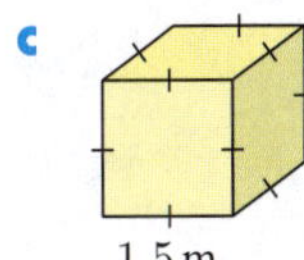
1.5 m

2 Find the surface area of each rectangular prism:

a
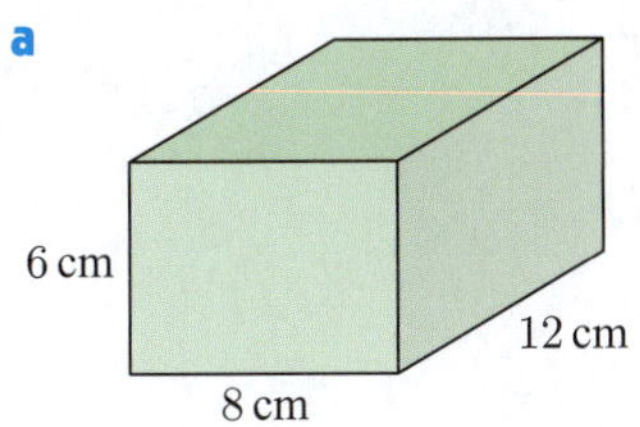
6 cm
12 cm
8 cm

b
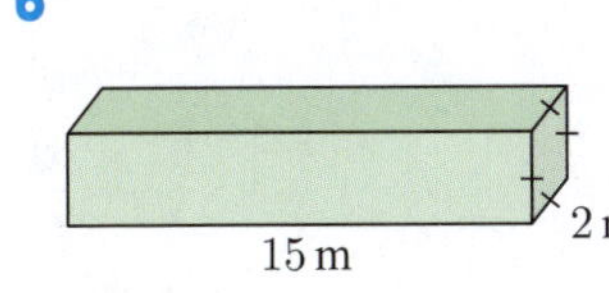
15 m
2 m

c
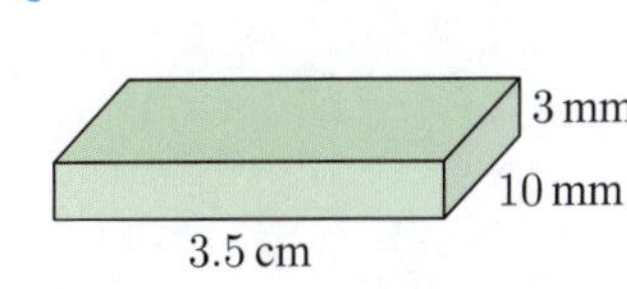
3 mm
10 mm
3.5 cm

3 Victor is erecting a 6 m by 4 m rectangular animal shelter that is 2 m high. The metal sheeting costs \$15 per square metre. Find the cost of the sheeting for the sides and roof of the shelter.

Example 2   ◆) **Self Tutor**

Find the total surface area of this triangular prism.

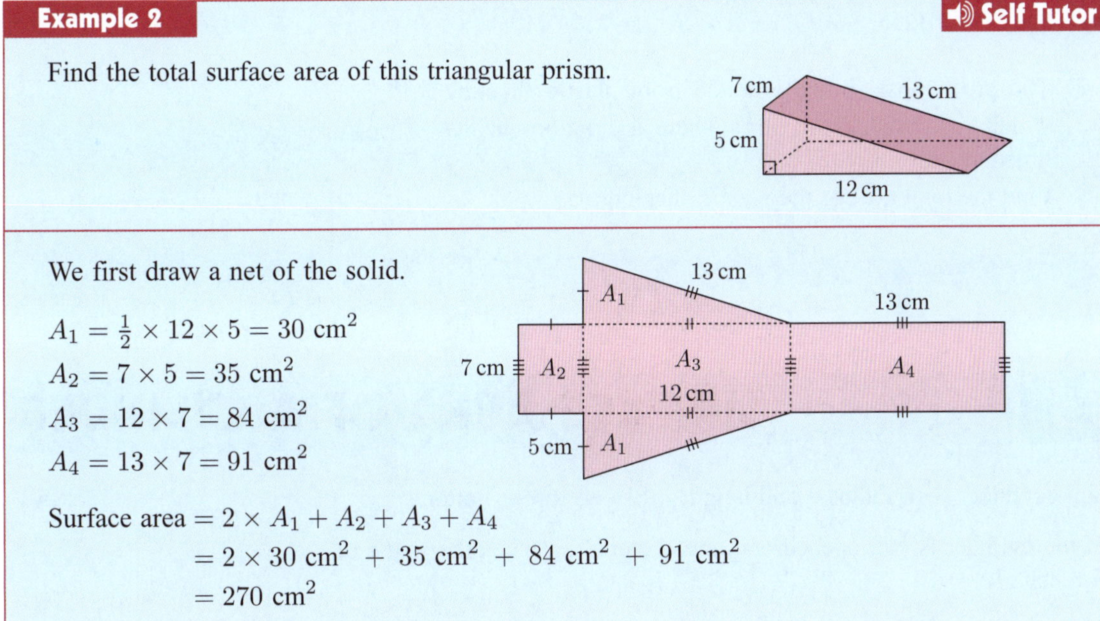

We first draw a net of the solid.

$A_1 = \frac{1}{2} \times 12 \times 5 = 30$ cm^2

$A_2 = 7 \times 5 = 35$ cm^2

$A_3 = 12 \times 7 = 84$ cm^2

$A_4 = 13 \times 7 = 91$ cm^2

Surface area $= 2 \times A_1 + A_2 + A_3 + A_4$

$\quad\quad\quad\quad\quad = 2 \times 30 \text{ cm}^2 + 35 \text{ cm}^2 + 84 \text{ cm}^2 + 91 \text{ cm}^2$

$\quad\quad\quad\quad\quad = 270 \text{ cm}^2$

---

4 Find the surface area of each triangular prism:

a
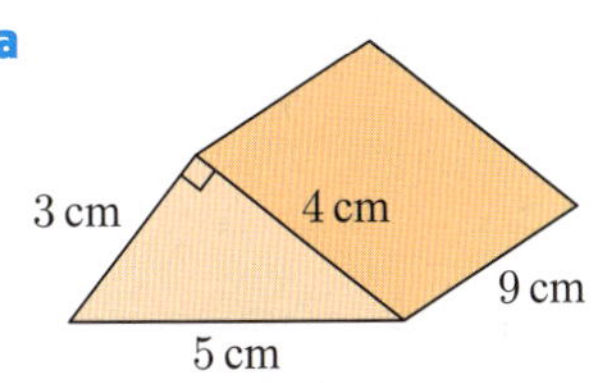
3 cm
4 cm
9 cm
5 cm

b
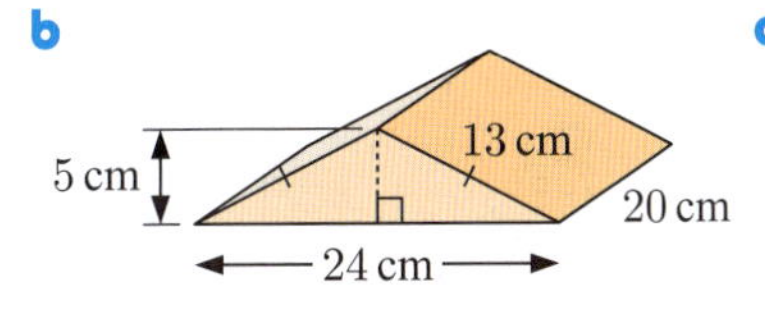
5 cm
13 cm
20 cm
24 cm

c
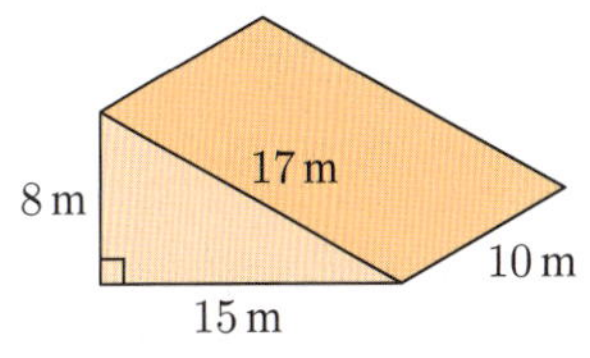
17 m
8 m
10 m
15 m

5 Find the surface area of each prism:

a
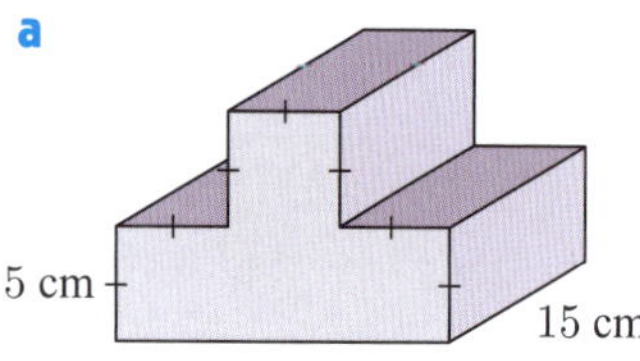

b
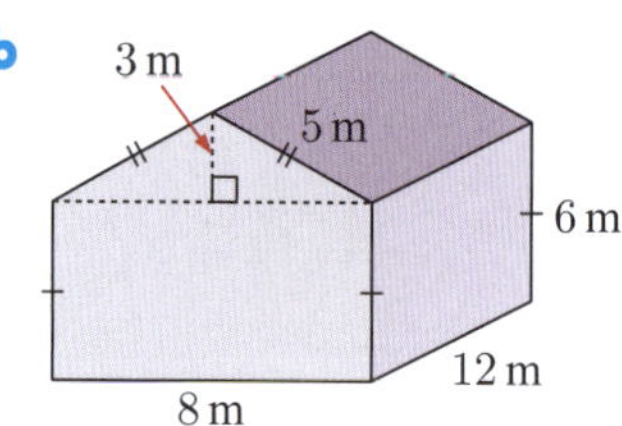

c
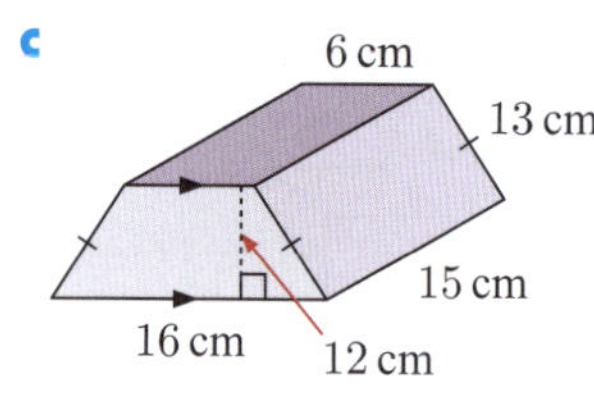

6 Find the surface area of each pyramid:

a
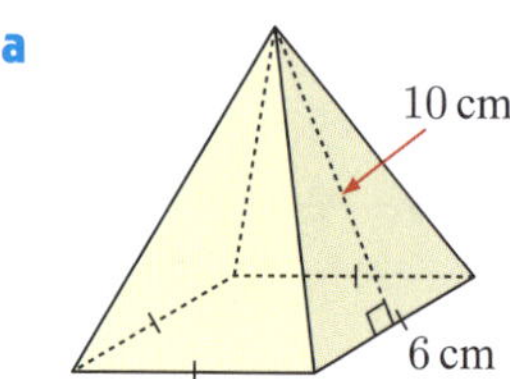

b
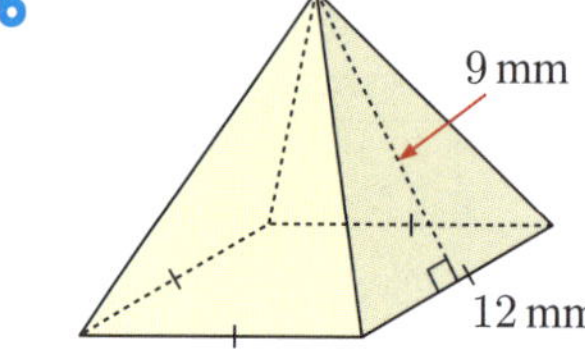

c
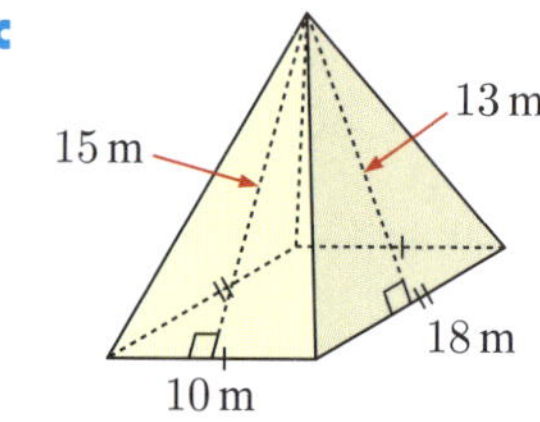

7
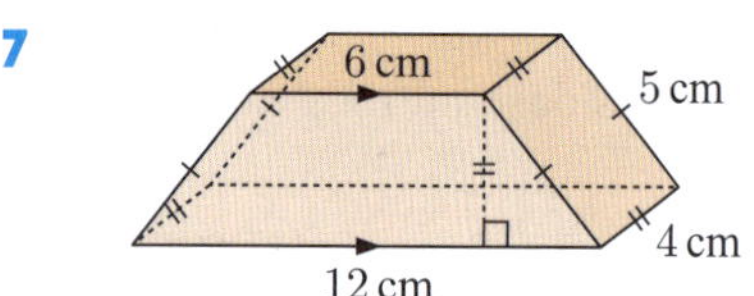

A box for a printer cartridge has the dimensions shown. Find the area of cardboard needed to make the box.

8 The greenhouse shown is made from plastic sheeting which costs \$24.50 per m^2. There is no floor in the greenhouse.
Find the total cost of the plastic sheeting.

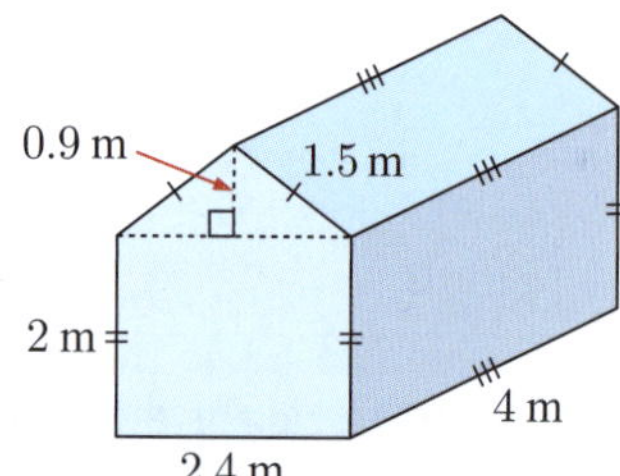

B   SURFACE AREA OF A CYLINDER

This cylinder with radius r and height h has no top or bottom.

If the cylinder is cut, opened out, and flattened, it takes the shape of a rectangle.

You can check this by peeling the label off a cylindrical can.

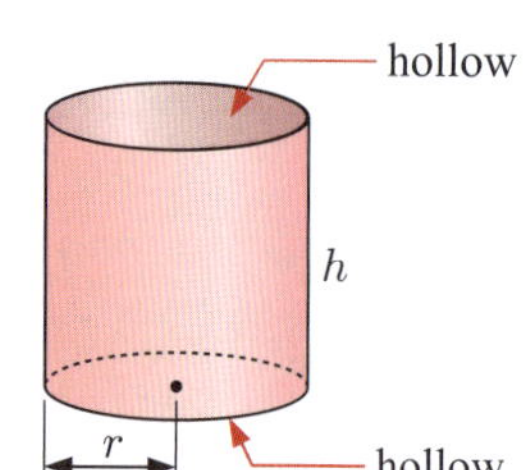

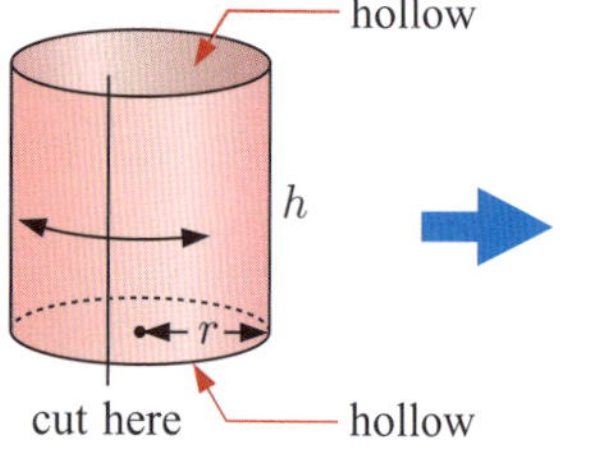

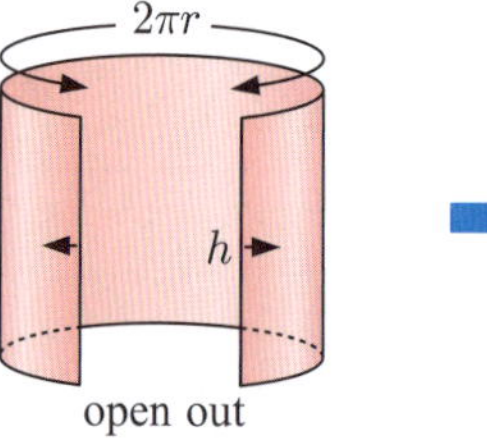

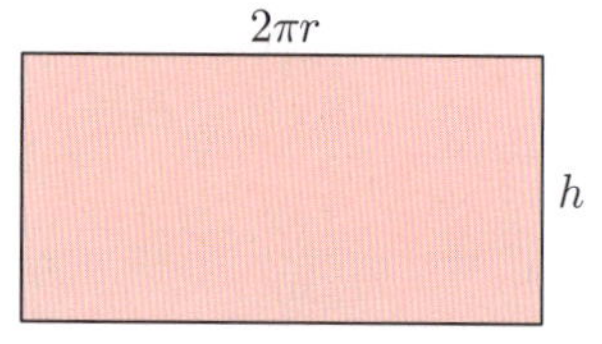

The length of the rectangle is the circumference of the cylinder, which is $2\pi r$.

So, the area of the *curved surface* of a cylinder is

$$\begin{aligned} A &= \text{area of rectangle} \\ &= \text{length} \times \text{width} \\ &= 2\pi r \times h \\ &= 2\pi rh \end{aligned}$$

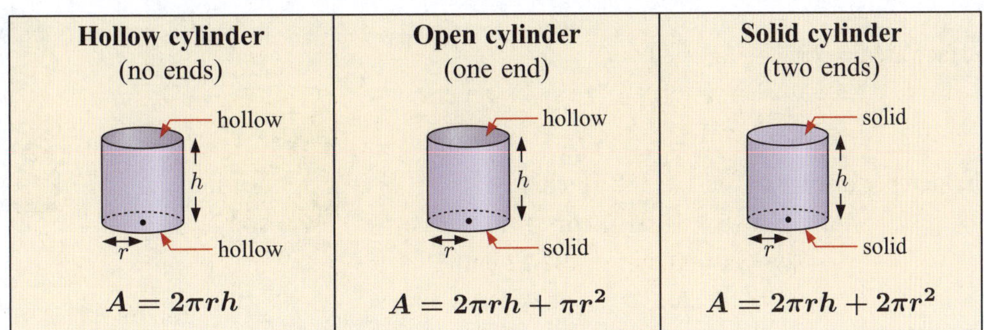

Example 3 ◀)) Self Tutor

Find the surface area of this open cylinder.

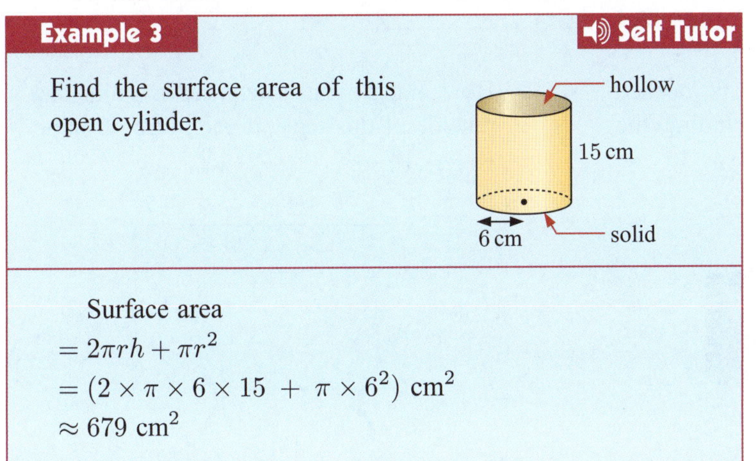

Surface area
$$\begin{aligned} &= 2\pi rh + \pi r^2 \\ &= (2 \times \pi \times 6 \times 15 + \pi \times 6^2) \text{ cm}^2 \\ &\approx 679 \text{ cm}^2 \end{aligned}$$

When we talk about the surface area of a hollow or open object, we are talking about the *outer surface* only.

EXERCISE 12B

1 Find the outer surface area of each cylinder. Round your answers to 3 significant figures.

a hollow throughout

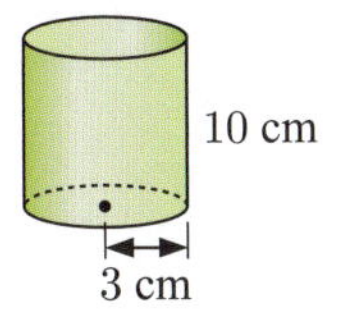

b solid

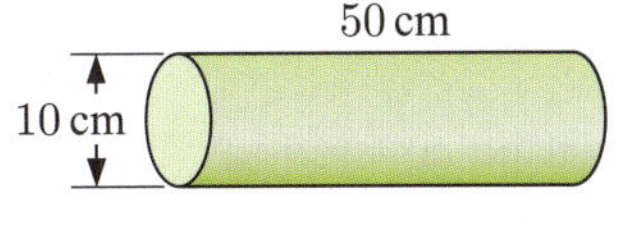

c can (no top)

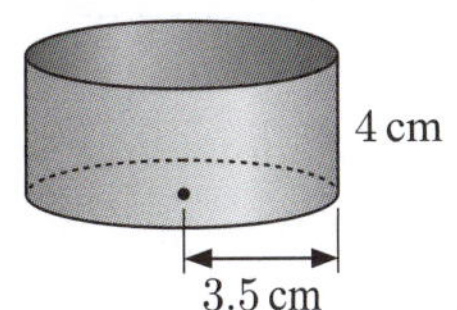

d solid

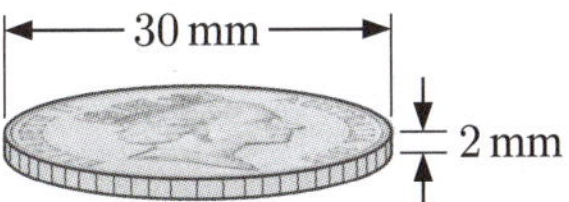

e tank (no top)

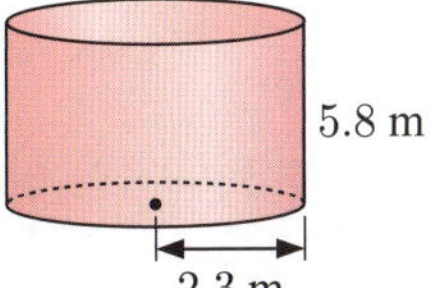

f hollow throughout

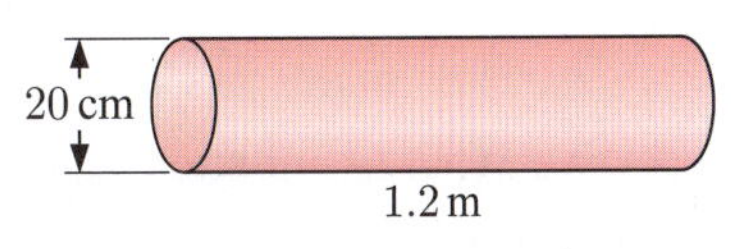

Example 4

Jane is painting cylindrical tin cans which have height 12 cm and diameter 8 cm. She paints the top, bottom, and sides of each can.

Jane has enough paint to cover 5 m^2. How many cans is she able to paint?

Surface area $= 2\pi rh + 2\pi r^2$

$\qquad\qquad = (2 \times \pi \times 4 \times 12 \ + \ 2 \times \pi \times 4^2)$ cm^2

$\qquad\qquad \approx 402.12$ cm^2

The number of cans ≈ 5 m$^2 \div 402.12$ cm^2

$\qquad\qquad\qquad \approx (5 \times 10\,000)$ cm$^2 \div 402.12$ cm^2

$\qquad\qquad\qquad \approx 50\,000$ cm$^2 \div 402.12$ cm^2

$\qquad\qquad\qquad \approx 124.34$

So, Jane can paint 124 cans.

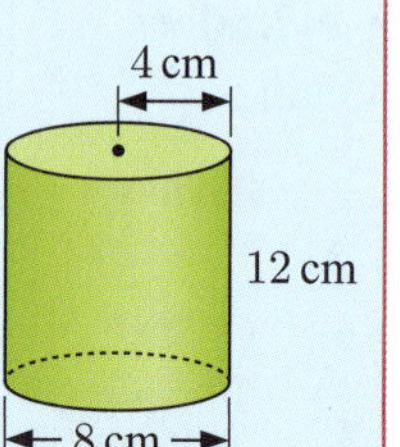

2 What area of sheet metal is needed to make this saucepan, excluding the handle? Round your answer to the nearest square centimetre.

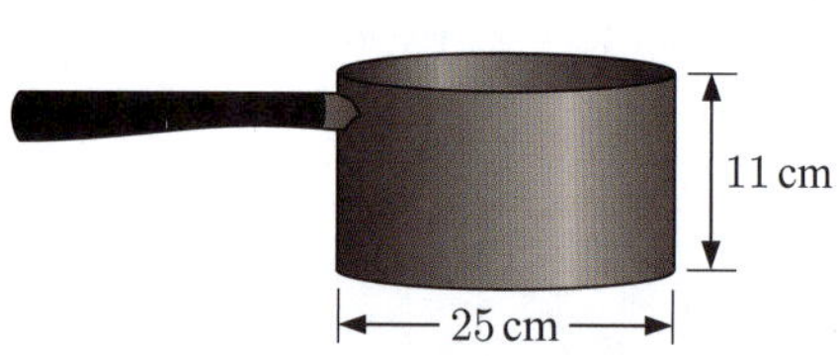

3 How much paint is required to paint the outside of this tank, if each litre of paint covers 15 m^2?

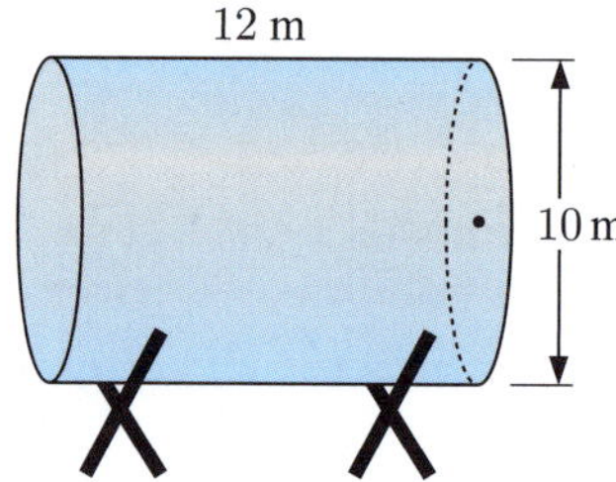

4 Andreas is painting poles for showjumping. He paints 12 poles with the colour scheme shown. The ends of each pole are left unpainted.

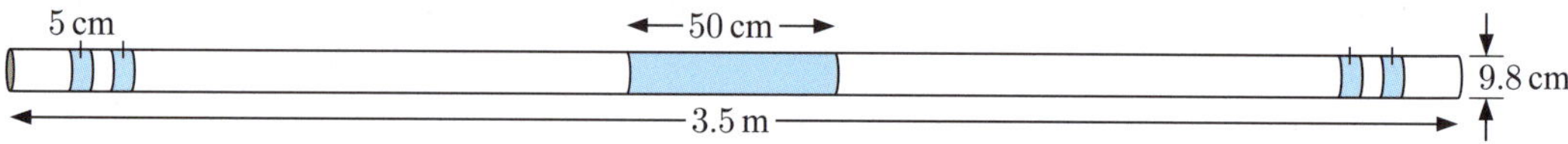

Find the total area which must be covered in:

  a blue paint   **b** white paint.

5 Fernandez has a collection of cylindrical jars with external radius 3 cm and height 10 cm. He needs to cover the curved side of each jar with a label. He has 2 m^2 of sticky paper to use for the labels.

  a How many jars can Fernandez cover?

  b What assumption have you made in your calculation in **a**?

C   SURFACE AREA OF A SPHERE

For each of the solids we have seen so far, we have either calculated the surface area directly from a net of the solid, or else used the net as the basis for a formula.

It is not possible to draw a net for a sphere, so we need to use a different method to find its surface area.

INVESTIGATION 1   SURFACE AREA OF A SPHERE

You will need:

a solid sphere such as an orange or a foam ball, cord, nail, scissors.

What to do:

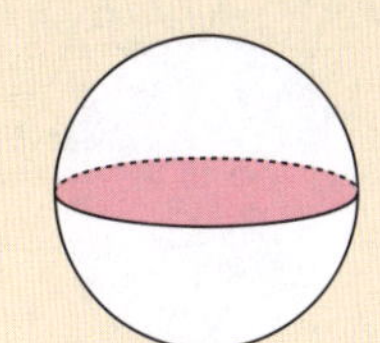

1 Cut your sphere in half to obtain two hemispheres.

2 Insert a nail in the centre of the flat circular surface of one hemisphere. Wind a length of cord around the nail in a spiral until the flat surface is completely covered. Cut the length of cord required and put it aside.

3 Now insert the nail in the centre of the curved surface of the hemisphere. Wind another length of cord around the nail until the curved surface of the hemisphere is completely covered.

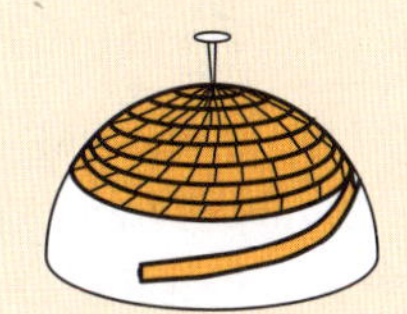

4 Compare the lengths of the cord in both cases. Is the area of the curved surface about twice the area of the flat circular surface?

Archimedes of Syracuse (287 BC - 212 BC) was first to prove a formula for the surface area of a sphere.

The surface area of a sphere is four times the area of a circle with the same radius.

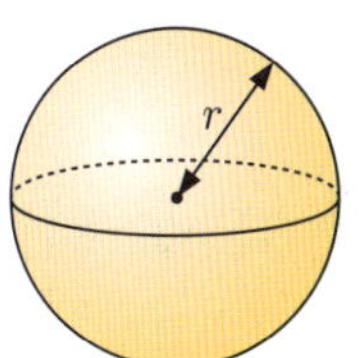

$$\text{Surface area of a sphere} = 4\pi r^2$$

Example 5	◀) Self Tutor

Calculate the surface area of the sphere:

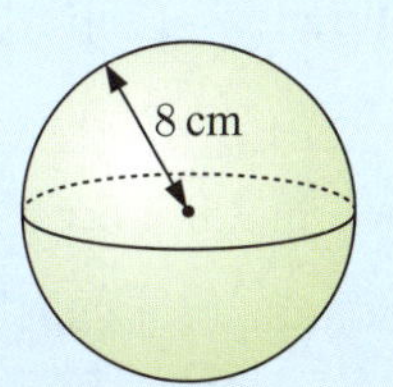

Surface area
$$= 4\pi r^2$$
$$= 4 \times \pi \times 8^2 \text{ cm}^2$$
$$\approx 804 \text{ cm}^2$$

EXERCISE 12C

1 Find the surface area of each solid. Round your answers to 3 significant figures.

a 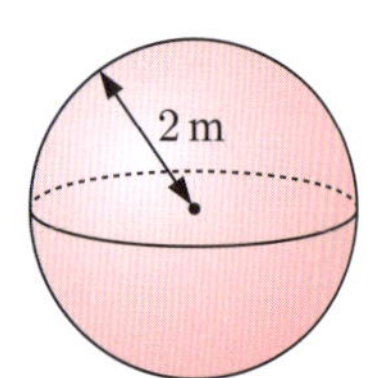

b 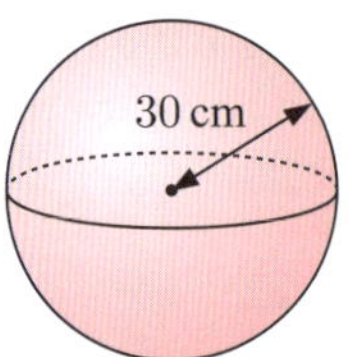

c 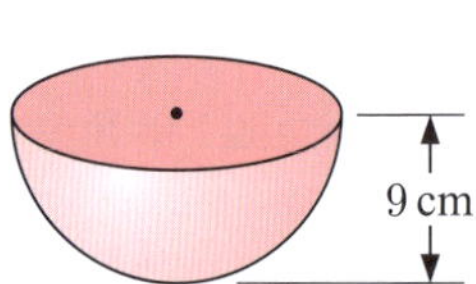

2 Find, in terms of π, the surface area of each solid:

a 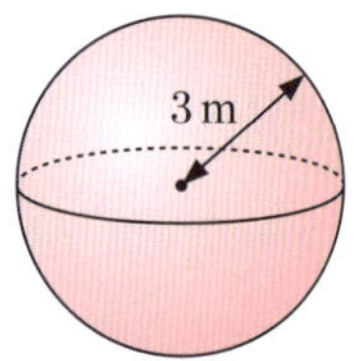

b 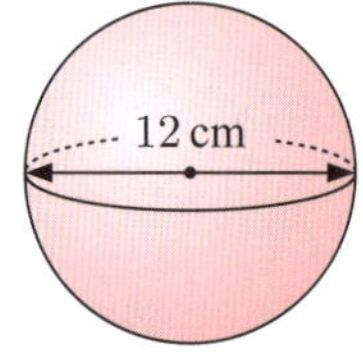

c 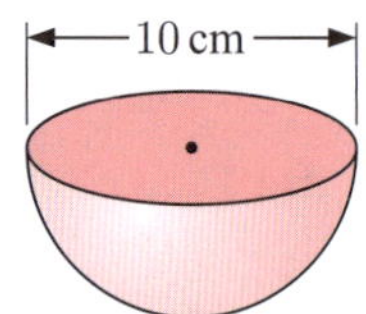

3 Find the surface area of a tennis ball with diameter 7 cm.

4 We often use a sphere to model Earth, even though our planet is not a *perfect* sphere. Earth has a radius of approximately 6400 km. Estimate the surface area of Earth.

5 Find the radius of a sphere with surface area 100 cm^2.

6 During the half-time break at the hockey game, my mum gave me an orange with radius 3.6 cm, sliced into six equal wedges.

   a Find the surface area of the peel on each wedge.

   b Find the *total* surface area of each wedge.

7 The surface area of the planet Mercury is about $74\,800\,000$ km^2. Estimate the diameter of Mercury.

8 A pot of paint will cover 40 m^2. How many spherical balls with diameter 15 cm can be painted?

D   VOLUME

> The **volume** of a solid is the amount of space it occupies.

We can measure volume in **cubic millimetres** (mm^3), **cubic centimetres** (cm^3), or **cubic metres** (m^3).

> 1 **cubic millimetre** is the volume of a cube with side length 1 mm.
>
> 1 **cubic centimetre** is the volume of a cube with side length 1 cm.
>
> 1 **cubic metre** is the volume of a cube with side length 1 m.

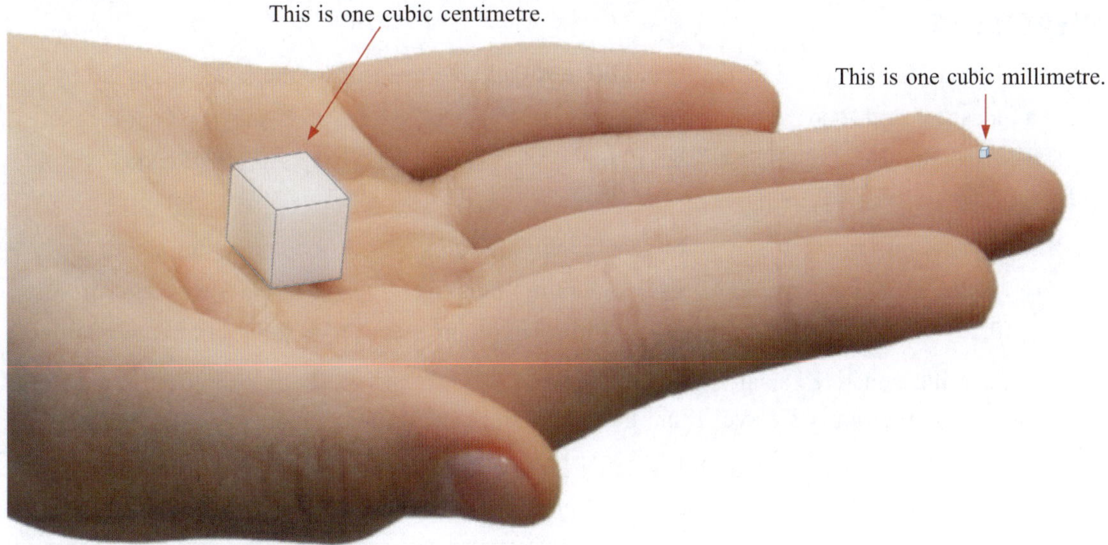

VOLUME CONVERSIONS

This cube with side length 1 cm has volume 1 cm^3.

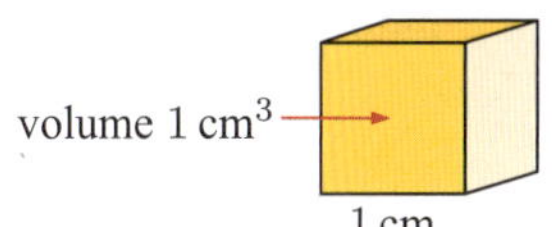

Since 1 cm $= 10$ mm, we can fit $10 \times 10 = 100$ cubic millimetres on the bottom surface of the cube.

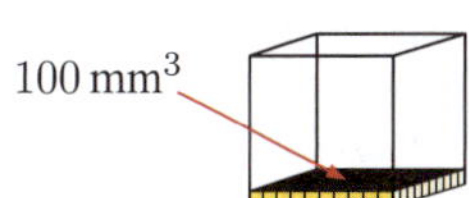

We can fit 10 of these layers in the cube, using $10 \times 100 = 1000$ cubic millimetres in total.

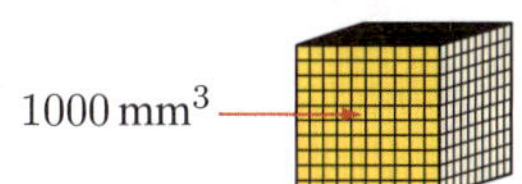

So, 1 cm$^3 = 10$ mm $\times 10$ mm $\times 10$ mm $= 1000$ mm^3.

Using the same principle, 1 m$^3 = 100$ cm $\times 100$ cm $\times 100$ cm $= 1\,000\,000$ cm^3.

$$1 \text{ m}^3 = 1\,000\,000 \text{ cm}^3$$
$$1 \text{ cm}^3 = 1000 \text{ mm}^3$$

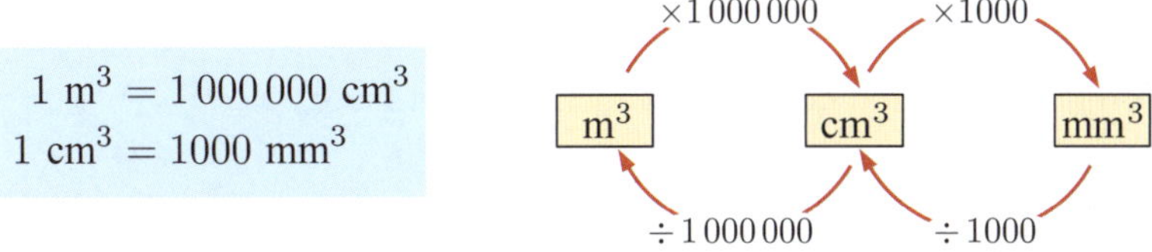

Example 6   ◀ Self Tutor

Convert:

a 4.56 cm^3 into mm^3

b $324\,000$ cm^3 into m^3.

a 4.56 cm^3
$= 4.56 \times 1000$ mm^3
$= 4560$ mm^3

b $324\,000$ cm^3
$= 324\,000 \div 1\,000\,000$ m^3
$= 0.324$ m^3

EXERCISE 12D

1 State the units of volume which would be most suitable to measure the space occupied by:

    a a can of dog food   **b** a house   **c** a stapler

    d a mountain   **e** a book   **f** a grain of sand.

2 Convert:

    a 48 cm^3 into mm^3 **b** $29\,000 \text{ cm}^3$ into m^3 **c** 1.2 m^3 into cm^3

    d $12\,485 \text{ mm}^3$ into cm^3 **e** $0.000\,45 \text{ m}^3$ into cm^3 **f** $14\,500 \text{ cm}^3$ into m^3

    g 0.295 cm^3 into mm^3 **h** 1.43 mm^3 into cm^3 **i** 0.0056 m^3 into cm^3.

3 To make the concrete for a path, Wendy mixed $50\,000 \text{ cm}^3$ of sand, $25\,000 \text{ cm}^3$ of cement powder, and 0.16 m^3 of gravel. Find the total volume of these components.

4 A slab of freeze dried coffee with volume 2000 cm^3 is broken into tiny pieces with average volume 10 mm^3. Find the total number of pieces.

5 The table alongside summarises the plastic bricks used to construct a department store display.

Find the total volume of the display, giving your answer in m^3.

Number of bricks	Volume of each brick (cm^3)
136 400	2.16
71 916	3.48
268 518	1.35
49 372	4.78

E   VOLUME OF A SOLID OF UNIFORM CROSS-SECTION

In the solid alongside, any vertical slice parallel to the front face will be the same size and shape as that face. Solids like this are called **solids of uniform cross-section**. In this case, the cross-section is a triangle.

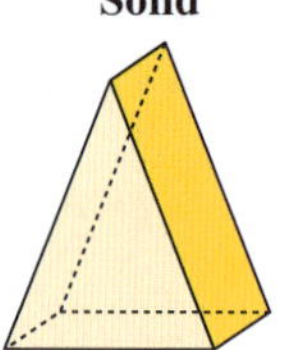

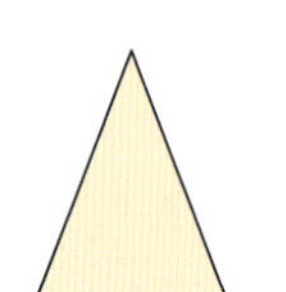

All solids of uniform cross-section contain two identical end faces.

> **Volume of solid of uniform cross-section**
> $=$ **area of end $\times$ length**

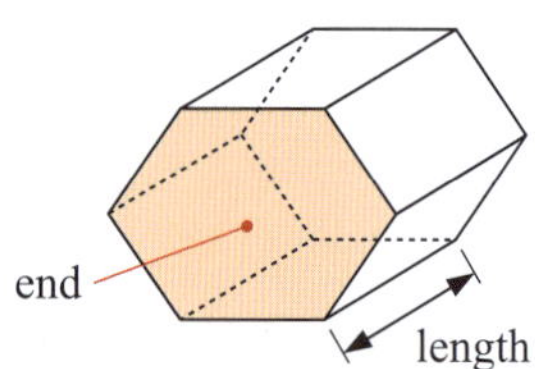

We can use more specific formulae for some special solids of uniform cross-section.

- For a rectangular prism, the cross-section is a rectangle.

> **Volume of rectangular prism**
> $=$ **length $\times$ width $\times$ height**

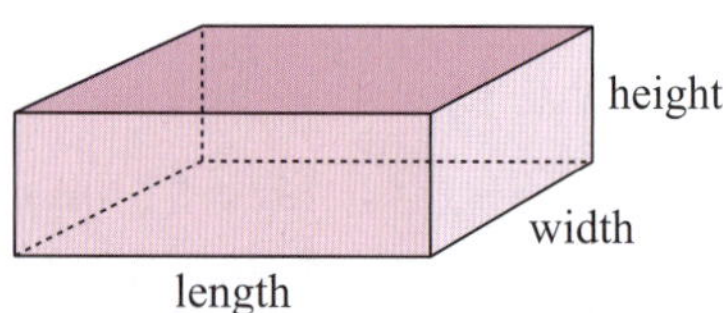

- For a cylinder, the cross-section is a circle.

  Volume = area of circle × length
  $$= \pi r^2 \times l$$

  > **Volume of cylinder** $V = \pi r^2 l$
  > or $V = \pi r^2 h$

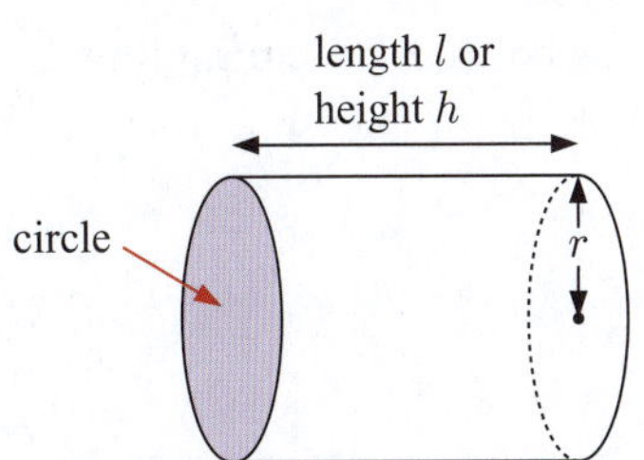

Example 7

Find the volume of each solid:

a
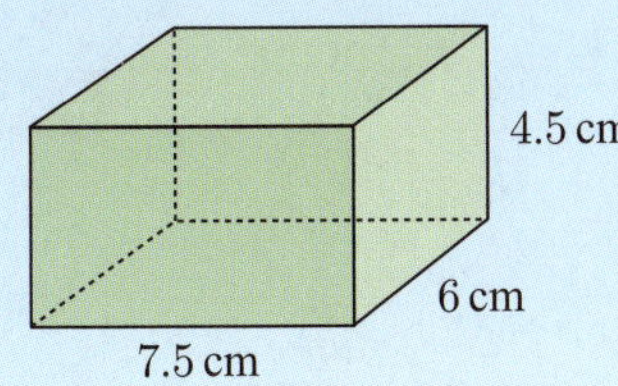

b
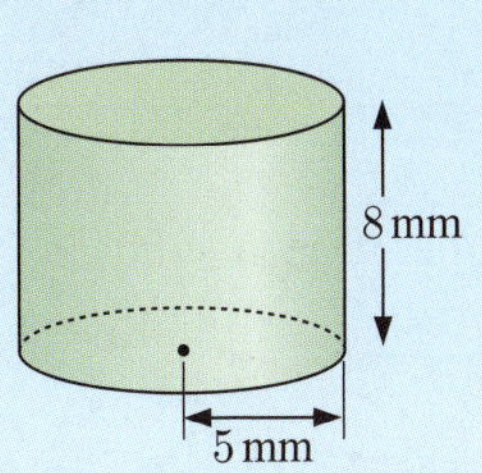

a Volume = length × width × height
$$= 7.5 \text{ cm} \times 6 \text{ cm} \times 4.5 \text{ cm}$$
$$= 202.5 \text{ cm}^3$$

b Volume $= \pi r^2 h$
$$= \pi \times 5^2 \times 8 \text{ mm}^3$$
$$\approx 628 \text{ mm}^3$$

EXERCISE 12E

1 Find the volume of each rectangular prism:

a
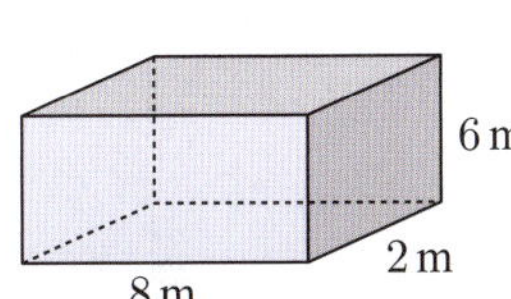

b
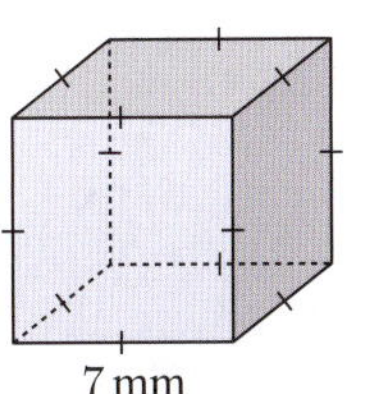

c
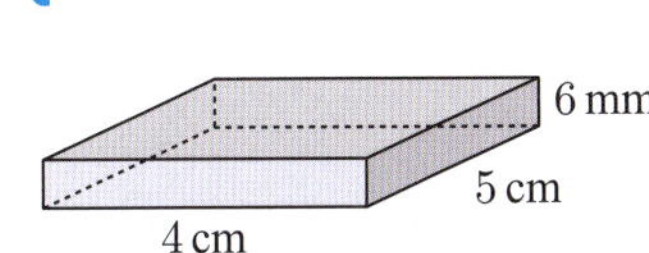

2 Find the volume of each cylinder:

a
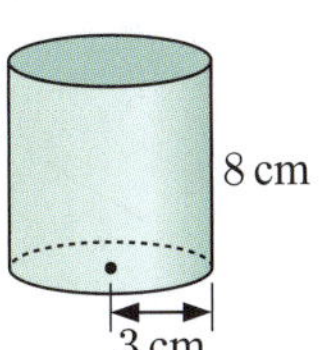

b
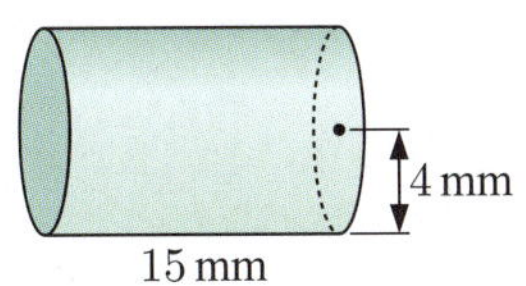

c
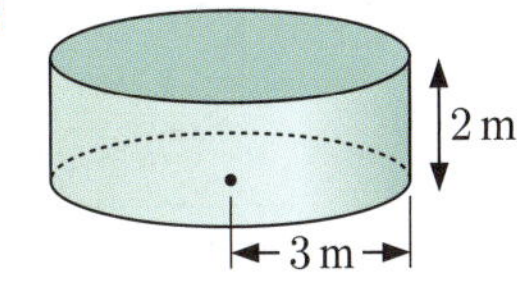

d
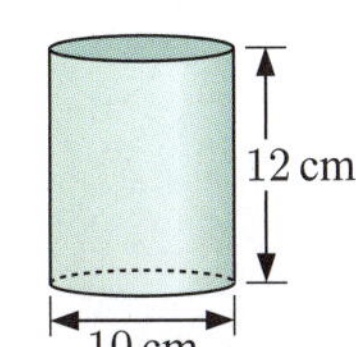

e
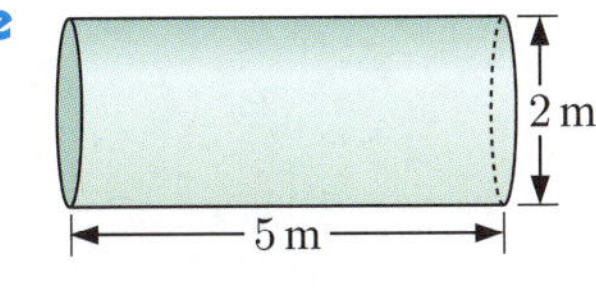

f
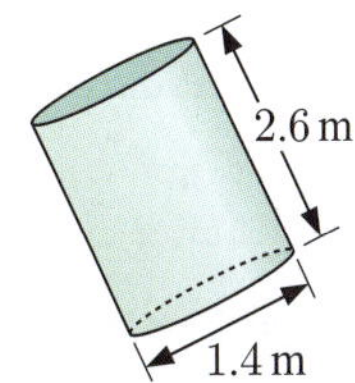

3 A round of cheese is 18 cm high and has radius 12 cm. Find the volume of cheese in this round.

4 A cylindrical steel rod is 2.2 m long and has diameter 5 cm. Find the volume of the rod in cm^3.

5 Find the side length of a cube with volume 2.744 cm^3.

6 Find the height of a cylinder with diameter 2.4 cm and volume 3.68 cm^3.

7 Find the radius of a cylindrical rod with length 1.3 m and volume 915 cm^3.

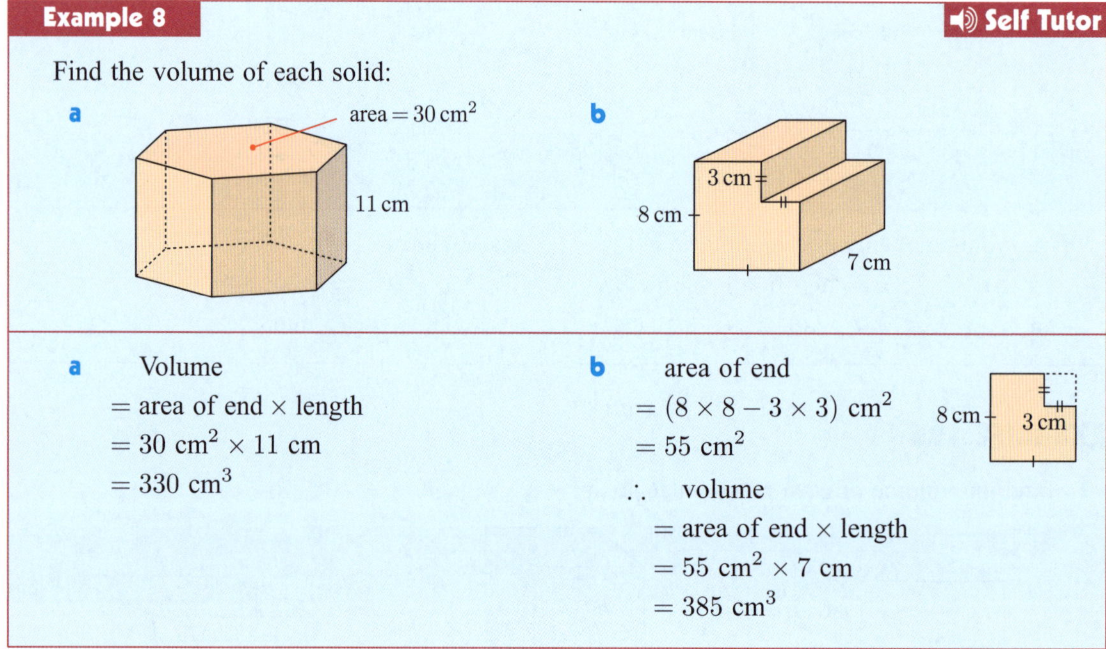

Example 8   ◀◉ Self Tutor

Find the volume of each solid:

a   area = 30 cm^2 11 cm

b   3 cm   8 cm   7 cm

a   Volume
= area of end × length
= 30 cm^2 × 11 cm
= 330 cm^3

b   area of end
= $(8 \times 8 - 3 \times 3)$ cm^2
= 55 cm^2

∴   volume
= area of end × length
= 55 cm^2 × 7 cm
= 385 cm^3

8 cm   3 cm

8 Find the volume of each solid:

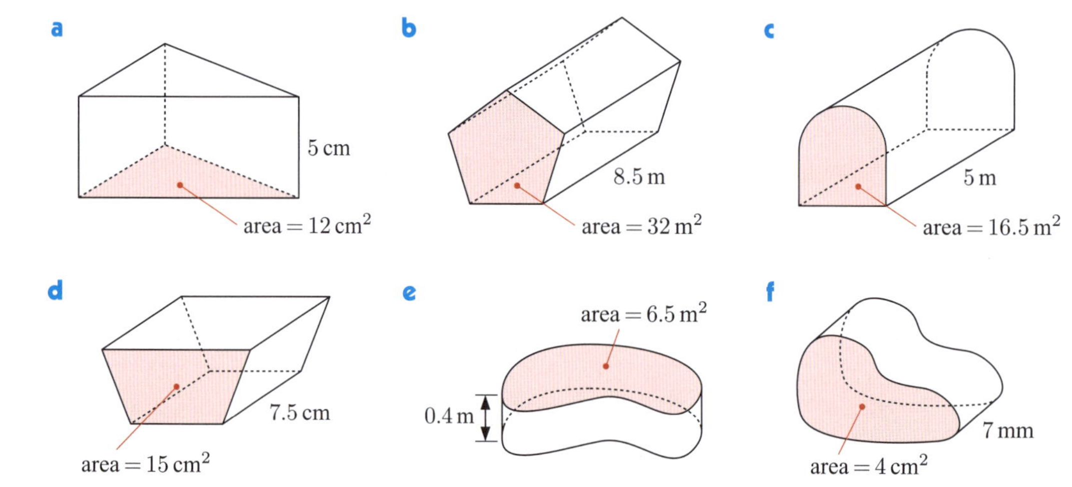

a   5 cm   area = 12 cm^2

b   8.5 m   area = 32 m^2

c   5 m   area = 16.5 m^2

d   7.5 cm   area = 15 cm^2

e   area = 6.5 m^2 0.4 m

f   7 mm   area = 4 cm^2

9 Find the volume of each solid:

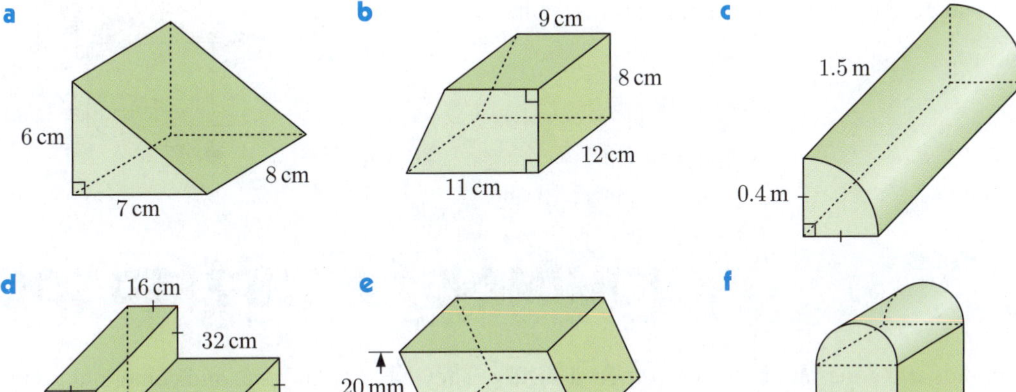

10 Find the volume of concrete required to make this ramp.

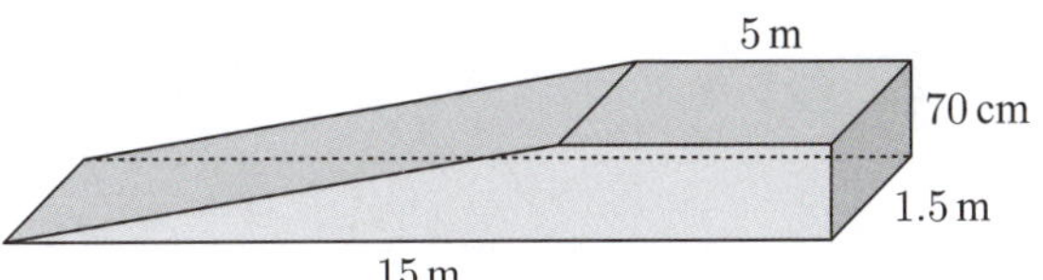

11 A speed bump across a 7.5 m wide road has the cross-section shown. Its volume is 0.81 m^3. Find the height of the speed bump.

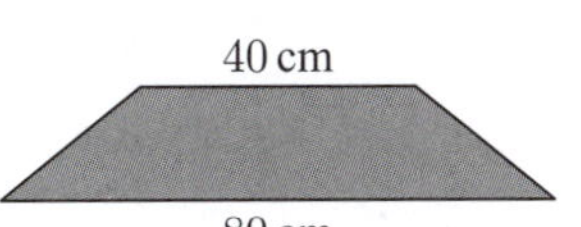

12
  a Find the volume of air inside this letter box.

  b Three rectangular envelopes with dimensions 23 cm by 12 cm by 5 mm are placed in the letter box. What volume of air remains inside it?

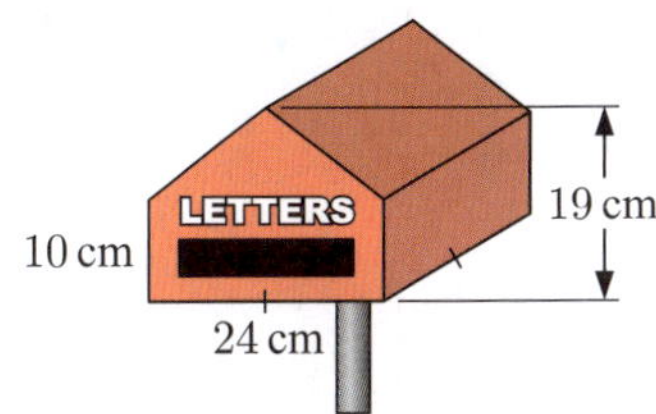

13 A gold ring has outer diameter 21 mm, inner diameter 20 mm, and is 4 mm thick.

  a Find the volume of the ring.

  b How many of these rings can be made from 8 cm^3 of gold?

  c 1 cm^3 of gold has mass 19.3 g. Find the mass of the ring.

  d 1 g of gold is worth $61.28. Find the value of the ring.

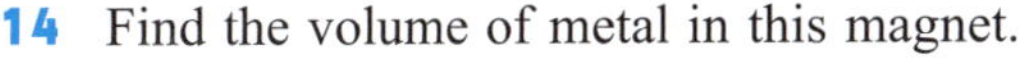

14 Find the volume of metal in this magnet.

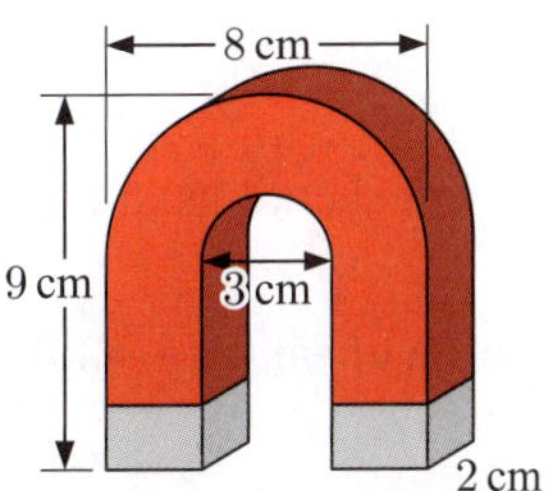

15 Three faces of a rectangular prism have areas 6 cm^2, 10 cm^2, and 15 cm^2. Find the volume of the prism.

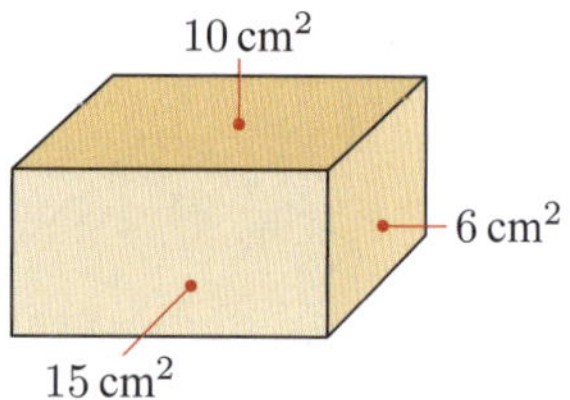

 # F   VOLUME OF A TAPERED SOLID

Pyramids and cones are known as **tapered solids**. They have a flat base, and come to a point called the **apex**.

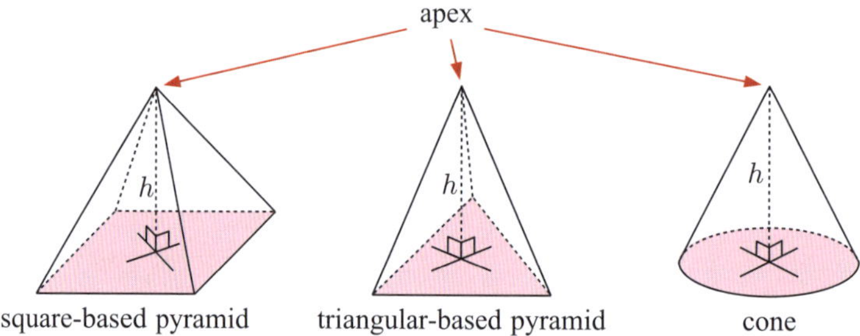

Tapered solids do *not* have uniform cross-sections. Their cross-sections always have the same *shape*, but not the same *size*.

For example, the cross-section of a cone parallel to the base is always a circle, but its radius decreases as we move up the cone.

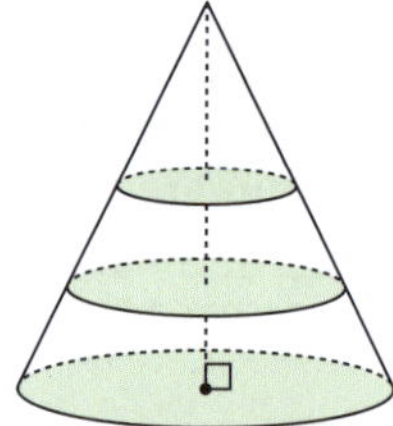

INVESTIGATION 2   VOLUME OF A TAPERED SOLID

In this practical Investigation, you will establish a formula for the volume of a tapered solid. To achieve this we compare tapered solids with solids of uniform cross-section that have the same base and height as the tapered solids.

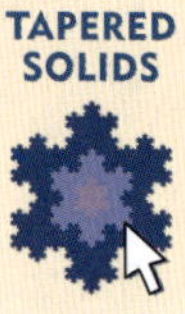

From the **Investigation**, you should have found that:

$$\text{Volume of tapered solid} = \tfrac{1}{3} \times \text{area of base} \times \text{height}$$

Example 9

Find the volume of each tapered solid:

a

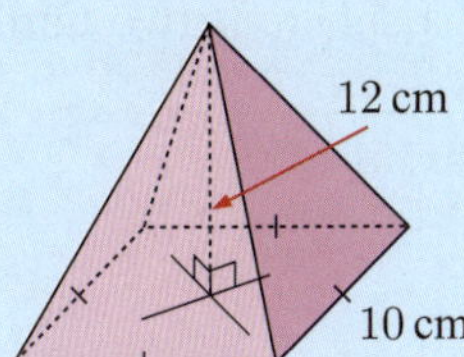

b

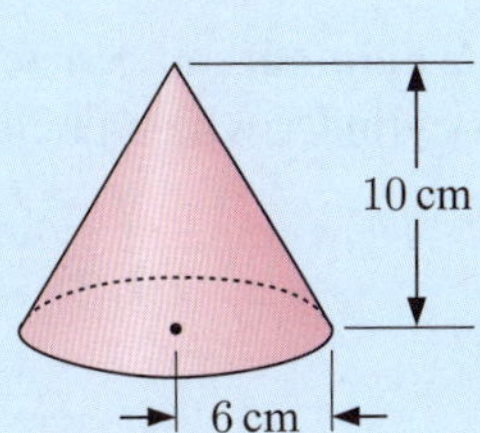

a   Volume

$= \frac{1}{3} \times$ area of base $\times$ height

$= \frac{1}{3} \times 10 \times 10 \times 12$ cm^3

$= 400$ cm^3

b   Volume

$= \frac{1}{3} \times$ area of base $\times$ height

$= \frac{1}{3} \times \pi \times 6^2 \times 10$ cm^3

≈ 377 cm^3

EXERCISE 12F

1   Find the volume of each tapered solid:

a

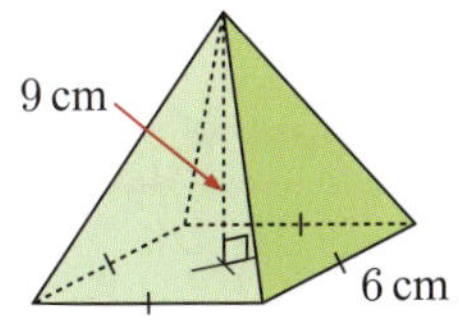

b

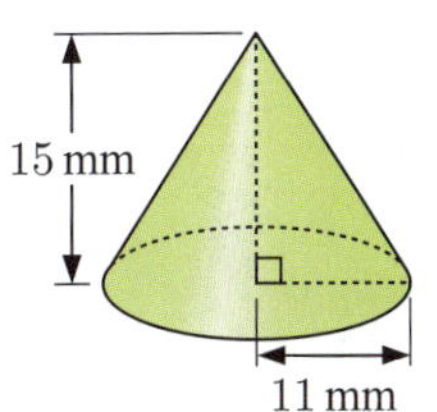

c

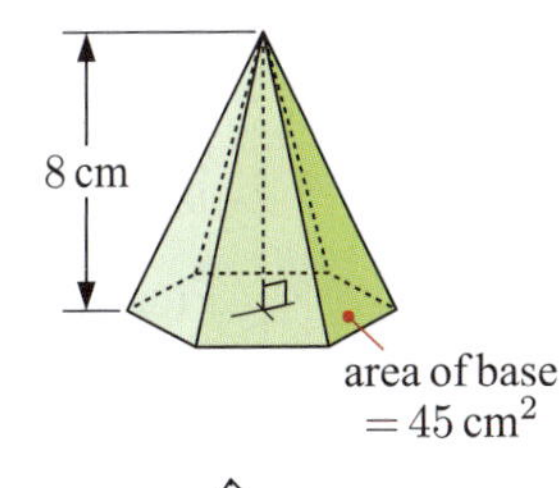

d

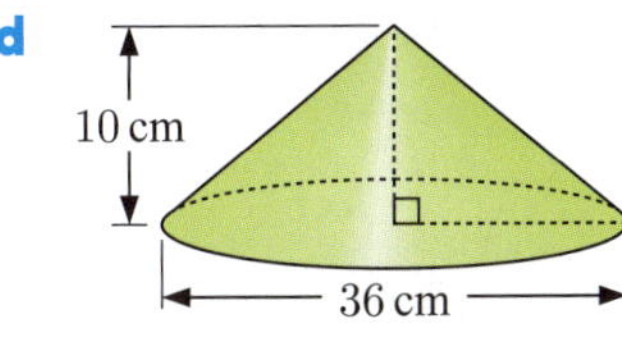

e

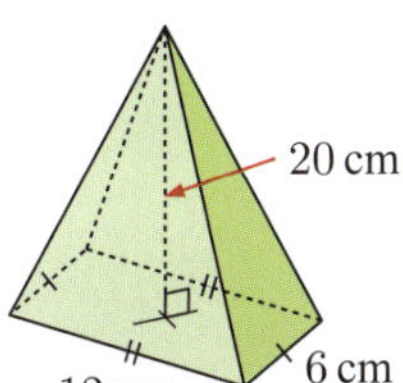

f

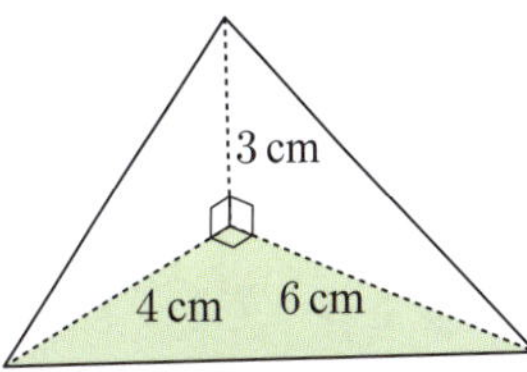

2   Sulphur dropped from a conveyor belt forms a conical heap on the ground.

a   The heap is 20 m in diameter and 8 m high. Find the volume of the sulphur.

b   A cubic metre of sulphur has mass 1830 kg. Find, in tonnes, the mass of the sulphur.

c   The sulphur is worth \$6750 per tonne. Find the total value of the sulphur.

3   Find, in terms of x, the volume of each solid:

a

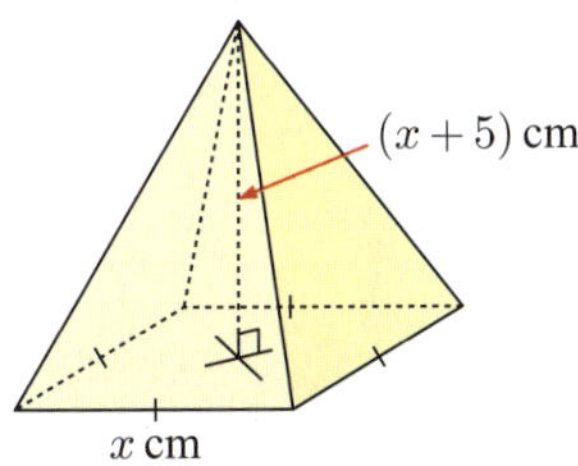

b

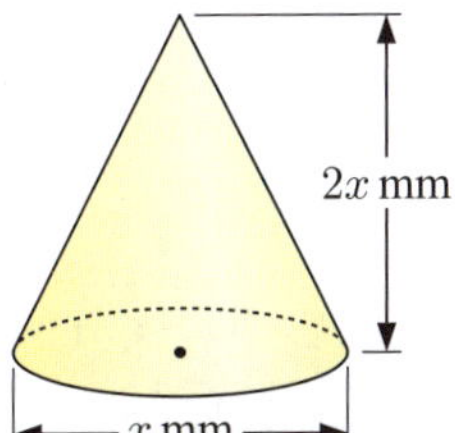

c

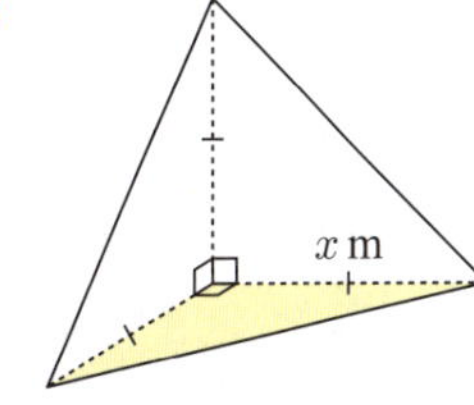

G  VOLUME OF A SPHERE

Another discovery by **Archimedes of Syracuse** was that the volume of a sphere is two thirds of the volume of the smallest cylinder which encloses it.

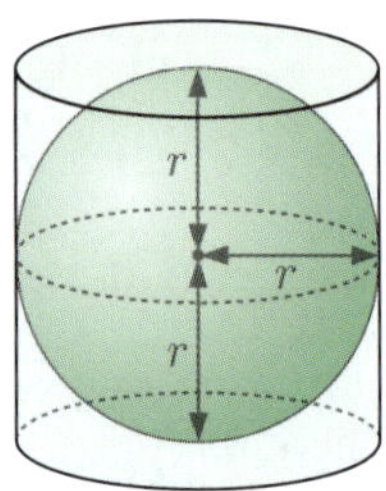

$$\text{Volume of cylinder} = \pi r^2 \times h$$
$$= \pi r^2 \times 2r$$
$$= 2\pi r^3$$
$$\therefore \quad \text{volume of sphere} = \tfrac{2}{3} \times \text{volume of cylinder}$$
$$= \tfrac{2}{3} \times 2\pi r^3$$
$$= \tfrac{4}{3}\pi r^3$$

Thus, the volume of a sphere with radius r is given by:

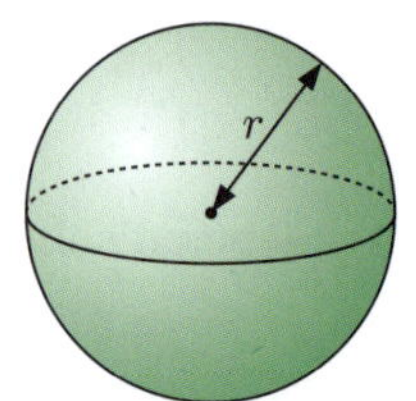

$$\textbf{Volume of a sphere} = \frac{4}{3}\pi r^3$$

Example 10 ◀) **Self Tutor**

Find the volume of this sphere:

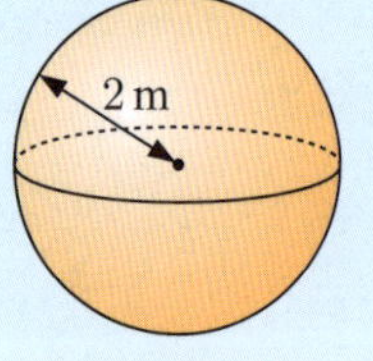

$$\text{Volume} = \tfrac{4}{3}\pi r^3$$
$$= \tfrac{4}{3} \times \pi \times 2^3 \text{ m}^3$$
$$\approx 33.5 \text{ m}^3$$

EXERCISE 12G

1 Find the volume of each solid:

a

6 cm

b
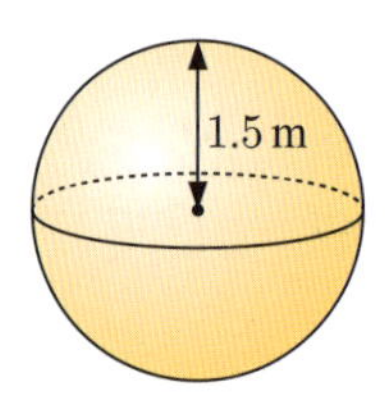
1.5 m

c

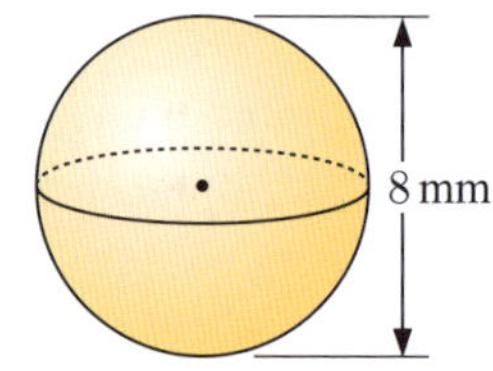
8 mm

d
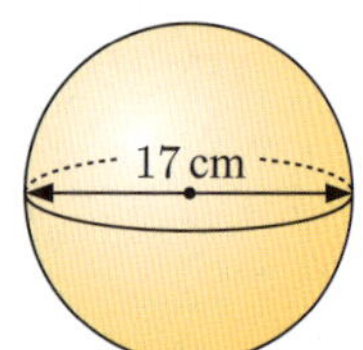
17 cm

e
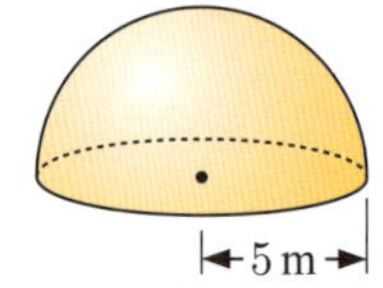
5 m

f
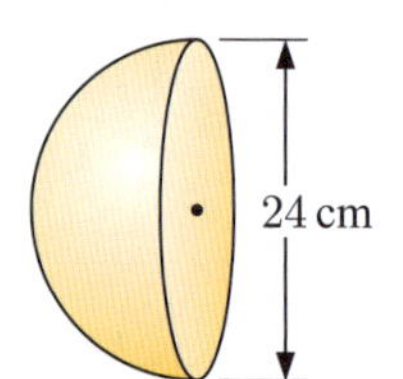
24 cm

2 A spherical balloon has radius 9 cm. Find, in terms of π, the volume of the balloon.

3 Lucy's necklace contains 15 spherical glass beads of radius 8 mm, and 12 spherical glass beads of radius 5 mm.

Find the total volume of glass in Lucy's necklace, giving your answer in:

  a mm^3 **b** cm^3.

4 **a** Jack shapes some clay into a sphere with diameter 9 cm. How much clay has Jack used?

    b If Jack decides to reshape this clay into a cube, what will the side lengths of the cube be?

H   CAPACITY

The **capacity** of a container is the amount of fluid it can contain.

The units for capacity are:

- **millilitres** (mL)

400 mL

- **litres** (L)

12 L

- **kilolitres** (kL)

40 kL

- **megalitres** (ML).

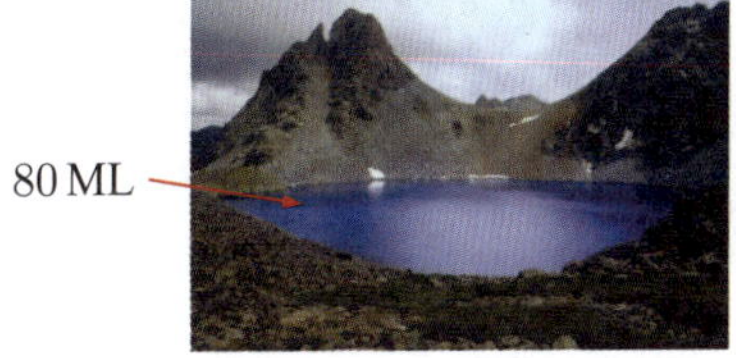

80 ML

CAPACITY CONVERSIONS

1 ML = 1000 kL
1 kL = 1000 L
1 L = 1000 mL

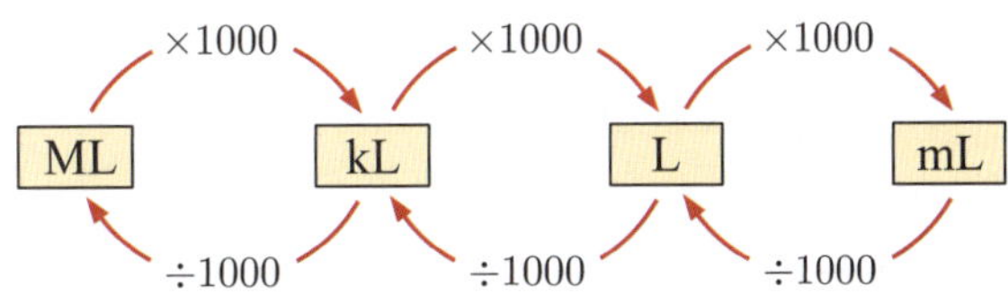

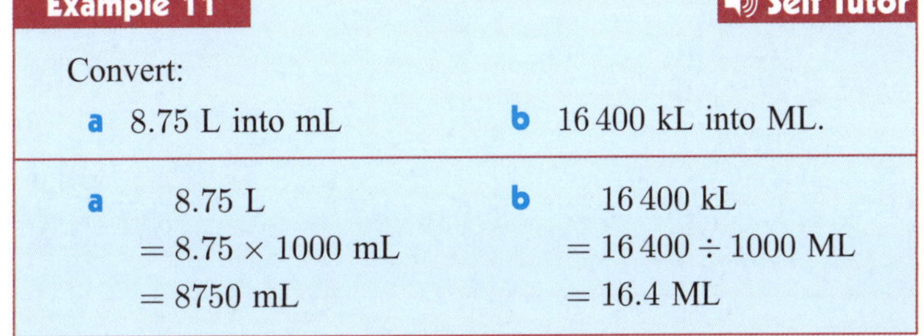

Example 11　　　　　　　　　　　　　　　　　　🔊 **Self Tutor**

Convert:

  a 8.75 L into mL   **b** 16 400 kL into ML.

a   8.75 L

     = 8.75 × 1000 mL

     = 8750 mL

b   16 400 kL

     = 16 400 ÷ 1000 ML

     = 16.4 ML

EXERCISE 12H

1 State the units you would use to measure the capacity of a:

    a test tube   **b** small water bottle   **c** reservoir

    d swimming pool   **e** bucket   **f** shipping container.

2 Convert:

    a 2960 mL into L   **b** 3.01 ML into kL   **c** 99.1 L into kL

    d 1800 kL into ML   **e** 6.6 kL into L   **f** 18.7 L into mL

    g 56 L into kL   **h** 0.35 kL into L   **i** 0.03 ML into L.

3 A fully stocked vending machine holds 144 cans with capacity 375 mL each, and 96 bottles with capacity 600 mL each. Find, in litres, the total capacity of drinks in the machine.

4 Water flows from Shelley's garden hose at 20 litres per minute. How long will it take to fill her outdoor spa with 1.5 kL of water?

5 One dose of a vaccine contains 0.12 mL of liquid. If a country is supplied with 1.428 kL of vaccine, how many doses are there?

I CONNECTING VOLUME AND CAPACITY

The units of **capacity** and the units of **volume** are closely related.

- 1 mL of fluid will fill a cube 1 cm × 1 cm × 1 cm, so $1 \text{ mL} \equiv 1 \text{ cm}^3$

- 1 L of fluid will fill a cube 10 cm × 10 cm × 10 cm, so $1 \text{ L} \equiv 1000 \text{ cm}^3$

- 1 kL of fluid will fill a cube 1 m × 1 m × 1 m, so $1 \text{ kL} \equiv 1 \text{ m}^3$

Example 12   ◀)) **Self Tutor**

Find the capacity of a cylindrical rainwater tank with height 3 m and diameter 4 m. Give your answer in kilolitres.

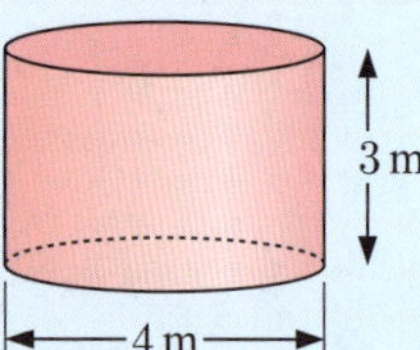

$$\text{Volume} = \pi r^2 h$$
$$= \pi \times 2^2 \times 3 \text{ m}^3$$
$$\approx 37.7 \text{ m}^3$$

Since $1 \text{ kL} \equiv 1 \text{ m}^3$, the capacity ≈ 37.7 kL.

EXERCISE 12I

1 Find the capacity of each container. Express each answer using appropriate units.

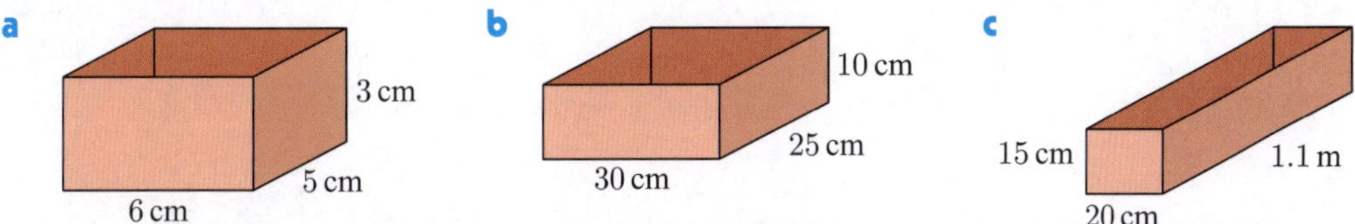

2 A refrigerator is 60 cm long, 50 cm wide, and 1.2 m high. Find the capacity of the refrigerator, in litres.

3 How many litres of oil can be stored in this drum?

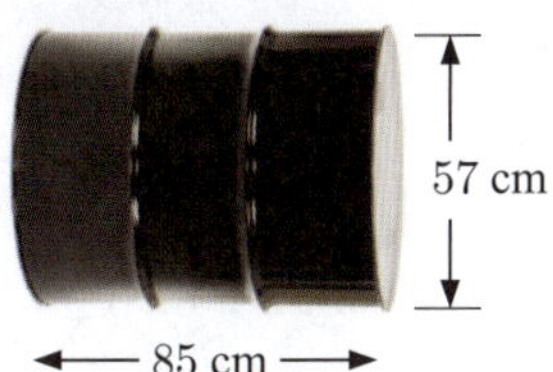

4 Find the number of kilolitres of water required to fill this swimming pool.

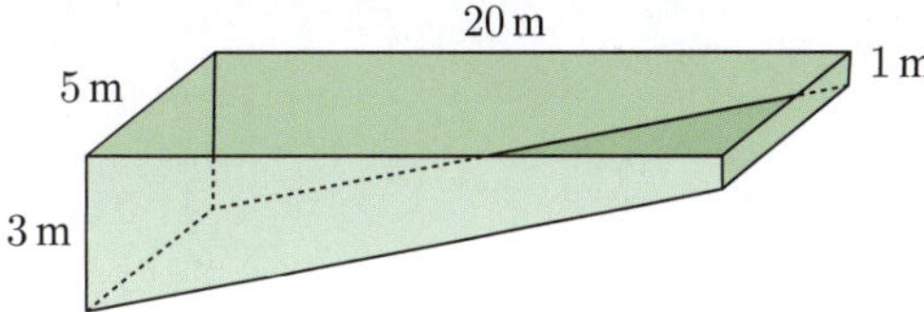

5 A children's lawn bowls set contains 8 spherical bowls of radius 6.1 cm, each filled with water. Find the total amount of water in the bowls.

6

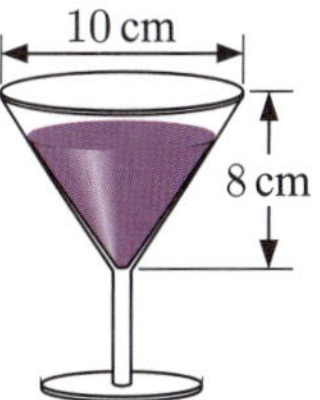

Find the amount of water in this aquarium, giving your answer in:

a kilolitres   **b** megalitres.

7 Mrs Foster has made $3\frac{1}{2}$ L of marmalade. She ladles it into cylindrical jars which are 12 cm high and have 7 cm internal diameter. How many jars can she completely fill?

8 From a 300 mL bottle of grape juice, the conical wine glass shown is filled to 80% of its capacity. The remaining juice is poured into a cylindrical glass with base radius 3 cm. To what height does the juice reach?

GLOBAL CONTEXT   **ICEBERGS**

GLOBAL CONTEXT

Global context: Globalisation and sustainability

Statement of inquiry: Taking measurements allows us to be more aware of changes to our natural resources.

Criterion: Applying mathematics in real-life contexts

MULTIPLE CHOICE QUIZ

REVIEW SET 12A

1 Find the surface area of each prism:

a 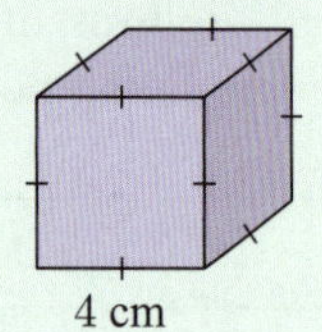

b 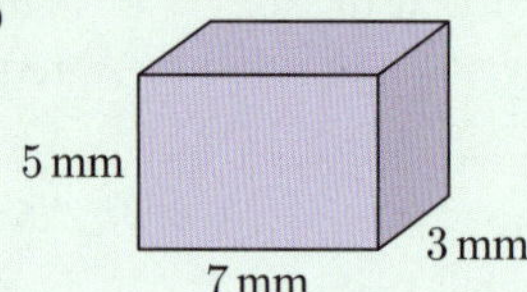

c

2 A garden shed is 3 m by 4 m by 2 m high. Find the cost of painting the outside walls and ceiling of the shed given that each litre of paint costs \$26 and covers 10 m^2.

3 Find the surface area of each solid:

a 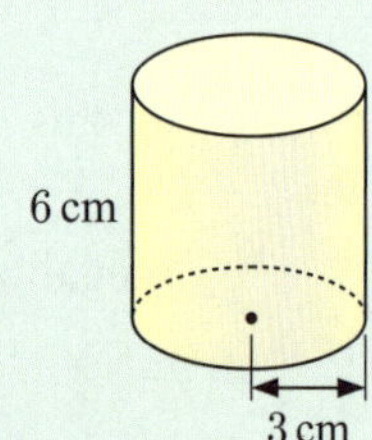

b 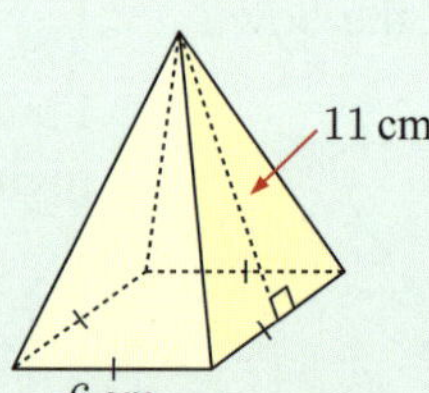

c 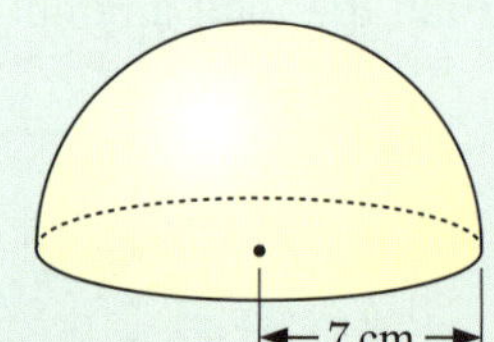

4 Convert:

    a 2.8 m^3 into cm^3 **b** 4 200 000 cm^3 into m^3 **c** 1.1 cm^3 into mm^3.

5 Find the volume of each solid:

a 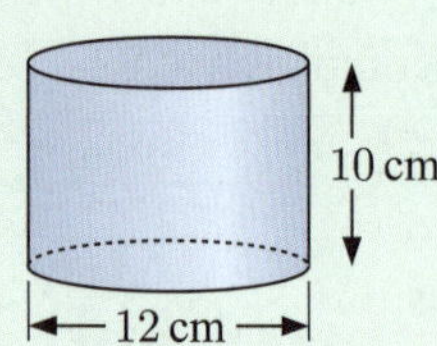

b 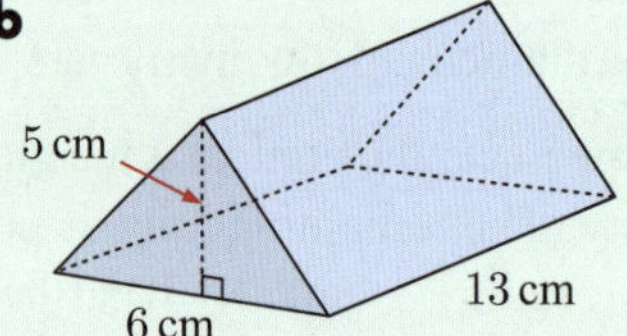

c 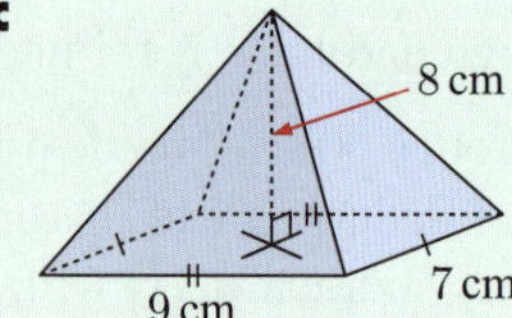

6 A semi-circular tunnel with the dimensions shown is made of concrete.

    a Find the cross-sectional area of the tunnel.

    b The tunnel is 200 m long, and concrete costs \$120 per cubic metre. Find the cost of the concrete for the tunnel.

7 A sphere has diameter 8 cm. Find, in terms of π, its:

    a surface area   **b** volume.

8 Tyson's Trophy Company makes wooden trophy bases with the dimensions shown.

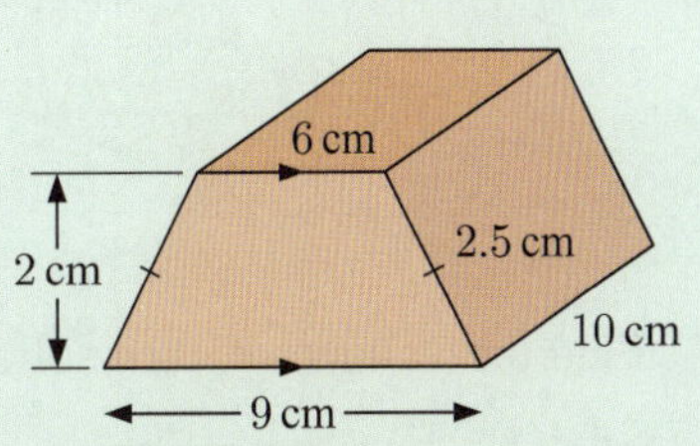

  a What volume of wood is required for each base?

  b Find the surface area of the trophy base.

  c Tyson paints his bases with black paint. He knows that one tin of paint covers 3 square metres. How many *complete* trophy bases can Tyson paint with one tin?

9 Ten glass bottles of lime cordial are lined up on a supermarket shelf. Marina accidentally knocks one off the shelf and it breaks. In total, 6.75 L of lime cordial remains on the shelf. Find the capacity of each bottle, in millilitres.

10 A petrol tank is a rectangular prism 80 cm by 120 cm by 50 cm. Find the capacity of the tank, in litres.

11 Find the capacity of this conical funnel, in kilolitres.

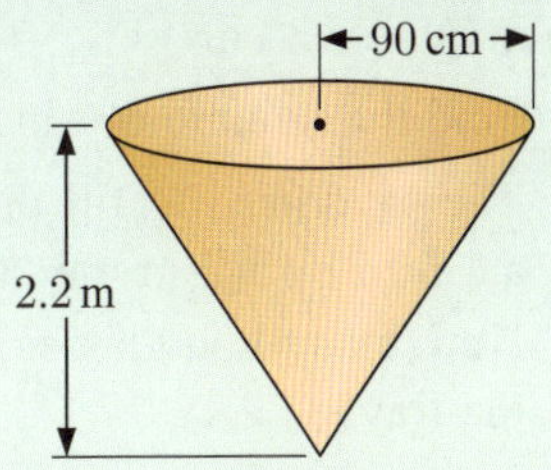

12 Answer the **Opening Problem** on page **242**.

REVIEW SET 12B

1 Find the surface area of each solid:

  a   **b**   **c**

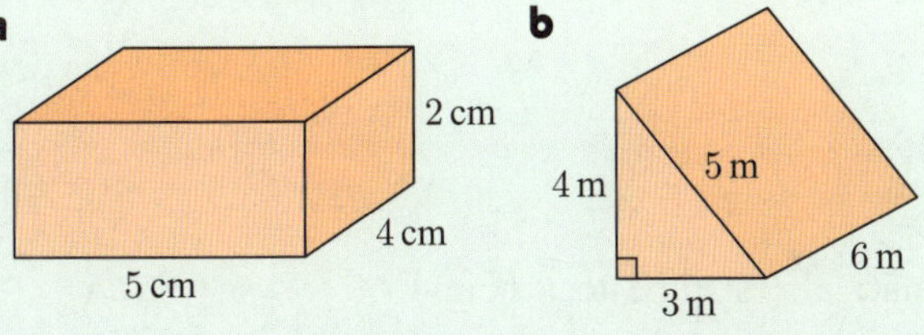

2 Find the surface area of this solid cylindrical rod.

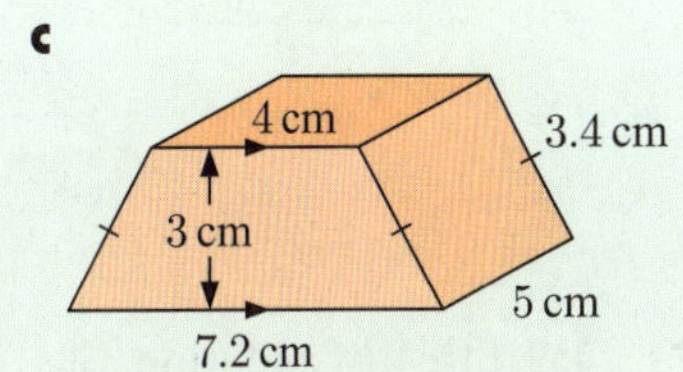

3 A company sells cylindrical vases with base radius 10 cm and height 50 cm. The curved surface of each vase must be wrapped with bubble wrap before it is shipped overseas. How many vases can be wrapped with 6 m^2 of bubble wrap?

4 A volleyball has surface area 1400 cm^2. Find the diameter of the volleyball.

5 How many cubic metres of molten metal are needed to fill 8000 moulds with 150 cm^3 of metal each?

6

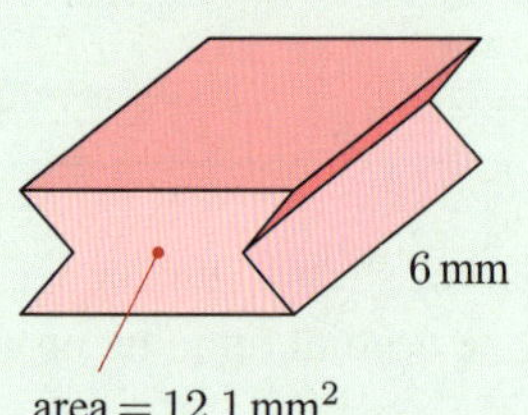

Find the volume of the wooden ramp illustrated.

7 Find the volume of each solid:

a

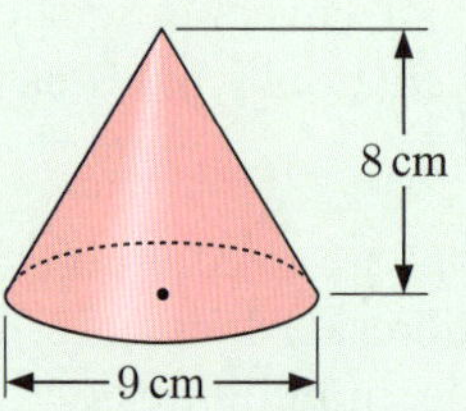

b

c

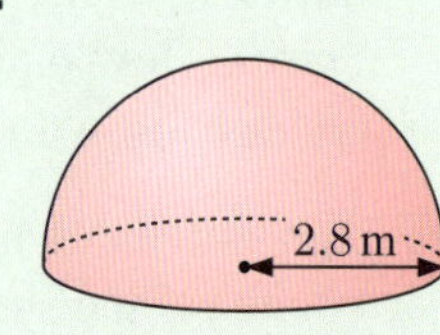

8 A factory produces hourglasses with the dimensions shown. The sand fills the cylindrical part of the hourglass to a height of 3 cm.

  a Find the volume of sand in each hourglass.

  b The sand used to fill the hourglasses is contained in a 1.5 m by 1.2 m tray, to a depth of 50 cm.
How many hourglasses can be filled with sand from the tray?

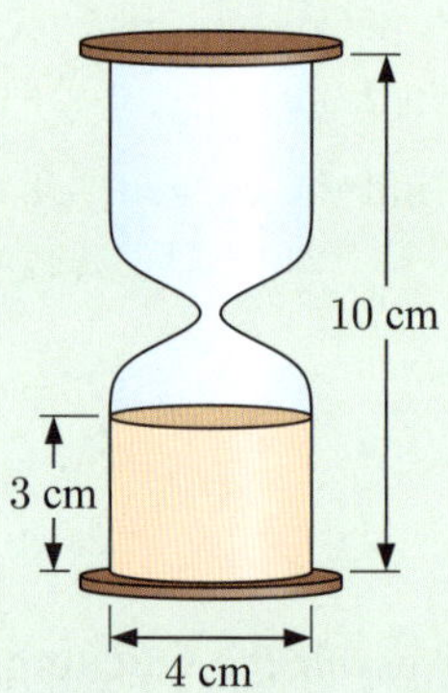

9 Convert:

  a 820 mL into L   **b** 3.07 ML into kL   **c** 0.72 kL into mL.

10

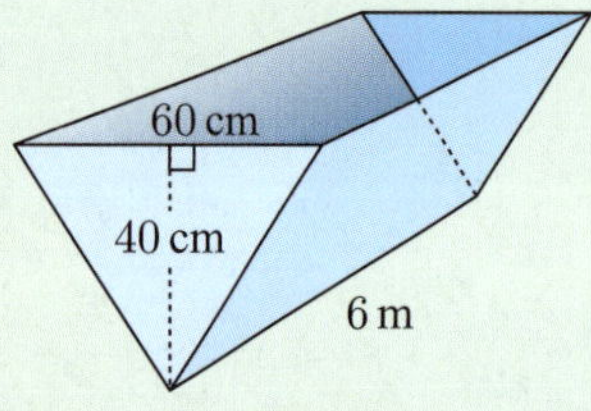

A water trough has the dimensions shown. Find the capacity of the trough in kilolitres.

11 1.2 litres of solution fills a cylindrical test tube to a height of 18 cm. Find the radius of the test tube.

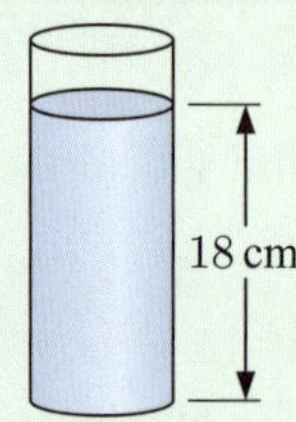

12 In the diagram alongside, a cone and a sphere are inscribed in a cylinder.

Show that the volume of the cylinder is equal to the sum of the volumes of the cone and the sphere.

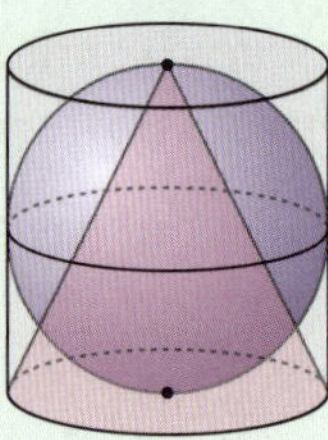

Time

Contents:

OPENING PROBLEM

Mila recently drove from Berlin to Vienna. The table alongside shows the cities that Mila stopped at, and the times when she arrived and departed.

City	Arrived	Departed
Berlin	-	09:00
Dresden	11:16	11:45
Prague	13:31	14:53
Brno	17:14	?
Vienna	19:24	-

Things to think about:

a Did Mila arrive in Vienna in the morning or evening?

b How long did Mila's trip take?

c Between which two cities was Mila driving at 3 pm?

d It took Mila 1 hour and 48 minutes to drive from Brno to Vienna. At what time did she depart from Brno?

e Which leg of Mila's trip was the shortest?

The measurement of **time** is a very important part of our lives. We encounter it in transport timetables, school schedules, sports, and appointments.

An understanding of time allows us to organise our day and schedule events.

A   UNITS OF TIME

The units of time we use today are based on the rotation of the Earth and its movement around the Sun.

The time taken for the Earth to complete one rotation about its axis is called a **day**.

The day is divided into **hours** (h), **minutes** (min), and **seconds** (s).

The time taken for the Earth to complete an orbit of the Sun is called a **year**.

TIME CONVERSIONS

1 minute = 60 seconds

1 hour = 60 minutes = 3600 seconds

1 day = 24 hours

1 week = 7 days

1 year $\approx 365\frac{1}{4}$ days

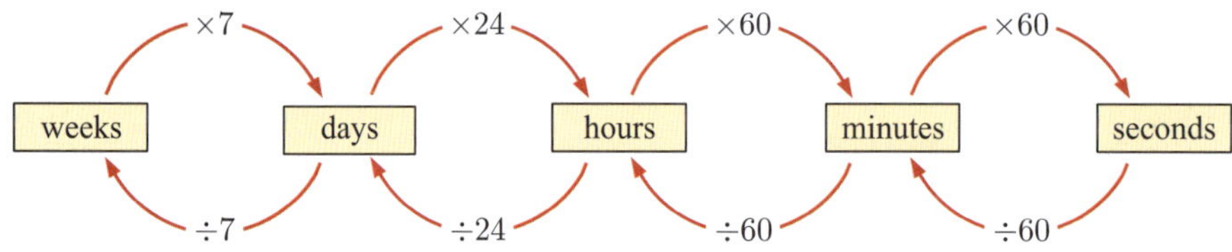

Example 1   ◀) **Self Tutor**

Write:

a 4 hours and 35 minutes in minutes only   b 90 000 seconds in hours.

a   4 hours + 35 minutes
   $= 4 \times 60$ min $+ 35$ min
   $= 275$ min

b   90 000 seconds
   $= 90\,000 \div 60$ min
   $= 1500$ min
   $= 1500 \div 60$ hours
   $= 25$ hours

EXERCISE 13A.1

1 Write in minutes:

  a 17 hours   b 1380 seconds   c 3 days

  d 4 hours and 28 minutes   e 11 h 33 min   f 2 days 5 hours

2 Convert into days:

  a 6 weeks   b 8 years   c 8640 minutes

  d 1152 hours   e 259 200 seconds

3 Convert into seconds:

  a 7 minutes   b 13 min 45 s   c 6 hours

  d 1 hour and 16 minutes   e 1 day   f 4 h 27 min

Example 2   ◀) **Self Tutor**

Write:

a 557 min in hours and minutes   b 200 h in days and hours.

a   9 h $= 9 \times 60$ min $= 540$ min
   $\therefore$ 557 min $= 540$ min $+ 17$ min
   $= 9$ h 17 min

b   8 days $= 8 \times 24$ h $= 192$ h
   $\therefore$ 200 h $= 192$ h $+ 8$ h
   $= 8$ days 8 h

---

4 Write in hours and minutes:

  a 107 minutes   b 282 minutes   c 492 min   d 737 min

5 Write in minutes and seconds:

  a 186 seconds   b 214 seconds   c 350 s   d 851 s

6 Write in days and hours:

  a 43 hours   b 61 hours   c 102 h   d 1500 min

7 Jeremy's geography class lasted 80 minutes. Write this in hours and minutes.

8 A film in the cinema has duration 2 hours and 14 minutes. Write this in minutes.

9   Marcia spends 45 minutes exercising every day. Calculate the total time she spends exercising in a year. Give your answer to the nearest day.

10   A cake decorator decorated 12 cakes in 25 hours and 24 minutes. On average, how long did it take to decorate each cake? Give your answer in hours and minutes.

11   The table alongside shows the times Marcel practised the drums during a week. Find, in hours and minutes:

    a   the total time Marcel spent practising during the week

    b   the average time Marcel practised each day.

Day	Time
Mon	20 min
Tue	1 h 20 min
Wed	55 min
Thu	30 min
Fri	1 h 10 min
Sat	2 h 35 min
Sun	45 min

12   The athletes in a 4×400 m relay team ran their legs in 48.67 s, 50.39 s, 49.68 s, and 48.12 s. Find the team's time for the relay, giving your answer in minutes and seconds.

13   The women's 400 m individual medley world record is held by Katinka Hosszú of Hungary with a time of 4 min 26.36 s. It took Katinka 1 min 0.91 s to swim 100 m butterfly, 1 min 7.48 s to swim 100 m backstroke, and 1 min 16.11 s to swim 100 m breaststroke. How long did it take Katinka to swim the final 100 m freestyle?

LARGE TIME UNITS

From the **year**, we can define very **large** units of time.

$$1 \ \textbf{decade} = 10 \text{ years}$$
$$1 \ \textbf{century} = 100 \text{ years}$$
$$1 \ \textbf{millennium} = 1000 \text{ years}$$

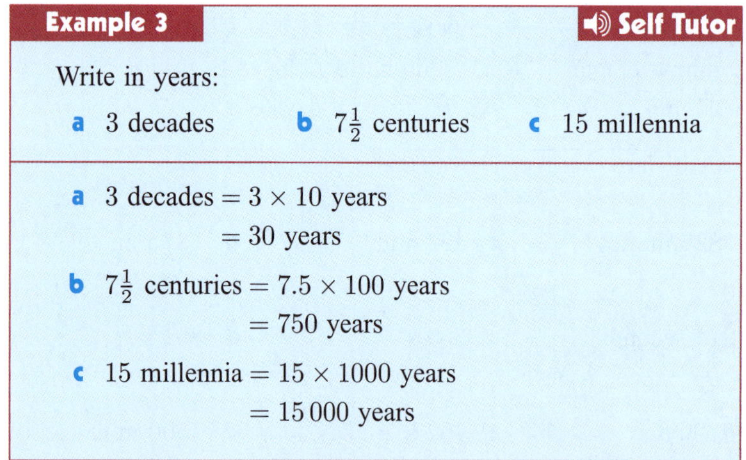

Example 3   ◀ﾘ **Self Tutor**

Write in years:

    a 3 decades   **b** $7\frac{1}{2}$ centuries   **c** 15 millennia

    a   3 decades $= 3 \times 10$ years
                  $= 30$ years

    b $7\frac{1}{2}$ centuries $= 7.5 \times 100$ years
                      $= 750$ years

    c   15 millennia $= 15 \times 1000$ years
                      $= 15\,000$ years

EXERCISE 13A.2

1 Write in years:

  a 7 decades
  b 12 centuries
  c 3 millennia

  d $4\frac{1}{2}$ centuries
  e $3\frac{1}{2}$ decades
  f $2\frac{1}{4}$ millennia

  g 2 centuries 3 decades
  h 5 millennia 4 centuries

2 Which is longer?

  a 18 decades or 2 centuries
  b 43 centuries or 4 millennia

3 Indigenous Australians have been living in Australia for at least 50 000 years.
Express this in:

  a centuries
  b millennia.

4 Dinosaurs first came into existence about 230 million years ago, and became extinct about 65 million years ago.
For how many millennia did dinosaurs exist?

RESEARCH   TIME IN DIFFERENT CULTURES

The units of time we have studied allow us to record events on a number line called a **time line**. Whether we are describing the schedule for our day or the passage of history over millennia, a time line gives a *linear* view of time from a starting point to a finishing point.

Some cultures and religions also consider a linear view of time from a definitive start of the universe to an end which is yet to come. Other cultures and religions have a more cyclic view of time which may be the repeating cycle of days, seasons, or years, or else repeating life cycles through reincarnation.

Use your library or the internet to research the following questions:

1 Which major religions have:

  a a linear view of time
  b a cyclic view of time?

2 What are the four great epochs in a Hindu cycle? How many millennia does each epoch correspond to?

3 In Buddhism, how long is a "kalpa"? What is supposed to happen in this length of time?

4 Is it reasonable to say that Christianity has two time lines? If so, how are they connected?

5 In the cultures and religions with a cyclic view of time, is there always a way to escape from the cycle?

6 Research the view of time of the indigenous people of your country. Discuss whether their view of time is linear or cyclic, and how their view of time has shaped their cultural traditions.

B TIME CALCULATIONS

We often need to do calculations with time. For example:

- we add or subtract hours and minutes to find when events start and finish
- to find the **duration** of an event, we calculate the difference between its starting and finishing times.

Example 4
◀)) **Self Tutor**

Find the time which is:

  a 4 hours 50 minutes after 11:20 am   **b** $2\frac{3}{4}$ hours before 5:10 pm.

  a   11:20 am $+$ 4 hours $=$ 3:20 pm

     3:20 pm $+$ 40 minutes $=$ 4:00 pm

     4:00 pm $+$ 10 minutes $=$ 4:10 pm

     The time 4 hours 50 minutes after 11:20 am is 4:10 pm.

  b $2\frac{3}{4}$ hours $=$ 2 hours 45 minutes

     5:10 pm $-$ 2 hours $=$ 3:10 pm

     3:10 pm $-$ 10 minutes $=$ 3:00 pm

     3:00 pm $-$ 35 minutes $=$ 2:25 pm

     The time $2\frac{3}{4}$ hours before 5:10 pm is 2:25 pm.

EXERCISE 13B

1 Find the time which is:

  a 6 hours after 2:12 pm   **b** 22 minutes after 6:50 am

  c 3 hours 5 minutes before 7:09 pm   **d** 2 hours 35 minutes after 1:44 am

  e $3\frac{1}{4}$ hours after 9:00 am   **f** $8\frac{1}{2}$ hours before 8:20 am

  g $2\frac{3}{5}$ hours before 10:20 am   **h** 11 hours after 5:30 pm.

2 Lauren drives from her house to her sister's house which is $2\frac{1}{4}$ hours away. If Lauren leaves her house at 11:30 am, at what time will she arrive at her sister's house?

3 A waiter finished his $6\frac{1}{2}$ hour shift at 11:15 pm. At what time did he start his shift?

4 Evan started a bushwalk at 8:20 am. He hiked for 5 hours and 40 minutes before arriving at his destination. At what time did he arrive?

5 Russ is making dinner for a party, and wants it to be ready for 7 pm. The dinner will take $1\frac{1}{2}$ hours to prepare, and 45 minutes to cook. At what time should Russ start making the dinner?

6 The table shows the duration of each quarter of an NFL football match.

There was a 2 minute break after the first and third quarters, and a 12 minute break at half-time. If the game started at 1:40 pm, at what time did it finish?

Quarter	1	2	3	4
Duration (min)	41	47	43	50

Example 5 ◀)) **Self Tutor**

What is the time difference between 8:45 am and 2:30 pm?

$$
\begin{aligned}
&\text{8:45 am to 1:45 pm} = 5 \text{ h} \\
&\text{1:45 pm to 2:00 pm} = \quad\ 15 \text{ min} \\
&\text{2:00 pm to 2:30 pm} = \quad\ 30 \text{ min} \\
\hline
&\therefore \ \text{the time difference is} \quad 5 \text{ h } 45 \text{ min}
\end{aligned}
$$

---

7 Find the time difference between:

   a 4:10 am and 8:35 am   **b** 10:33 am and 5:49 pm

   c 3:20 pm and 6:08 pm   **d** 9:52 am and 1:38 pm

   e 9:27 pm and 6:30 am the next day   **f** 7:45 am and 10:05 am the next day.

8 A theatre show started at 7:30 pm and finished at 9:12 pm. How long was the show?

9 Paula started a marathon at 8:50 am and finished at 1:13 pm. How long did she take to complete the marathon?

10 Tai went to sleep at 8:15 pm and woke up at 7:10 am the next morning. The school bus leaves at 8:02 am.

   a For how long did Tai sleep?

   b How long does Tai have to get ready?

11 A hairdresser works from 8 am to 6 pm. Her schedule for today is shown alongside.

   a Who has the longest appointment?

   b Who has the shortest appointment?

   c Another customer requests a 90 minute appointment. Will the hairdresser be able to see her today?

   d The hairdresser charges \$50 per hour. How much money will she earn today?

8:30 - 9:30	Tracey
10:15 - 10:30	Terry
10:30 - 11:00	Adrian
11:15 - 1:00	Deborah
1:30 - 2:15	Carol
3:30 - 4:00	Ihaka
4:00 - 4:30	Alex
4:30 - 6:00	Michelle

12 Anthea worked the following hours last week:

Monday	8:30 am - 4:15 pm
Tuesday	8:25 am - 4:30 pm
Wednesday	8:45 am - 4:45 pm
Thursday	9:00 am - 5:05 pm
Friday	8:40 am - 4:45 pm

a How many hours did Anthea work last week?

b Anthea's pay rate is $21 per hour. What was Anthea's income for last week?

13 The 2016 Newport Bermuda yacht race started at 5:30 pm on June 17. The winning yacht arrived in Bermuda at 4:22 am on June 19. Find the time taken for this yacht to complete the race. Give your answer in hours and minutes.

14 The daily schedule of performances at a marine park is shown below.

	Duration	*Times*
Diving Dolphins	25 min	9:50 am, 1:40 pm, 4:00 pm
Whale Mania	30 min	10:15 am, 12:45 pm, 3:15 pm
Seal of Approval	35 min	11:30 am, 2:00 pm
Otter Odyssey	20 min	10:00 am, 3:45 pm
3D Underwater World	40 min	10:10 am, 12:30 pm, 3:50 pm
Marine Park Parade	25 min	4:00 pm

a Which performance is shortest?

b Which performance finishes latest?

c At what time will the 1:40 pm dolphin performance end?

d Which performances will be in progress at 10:30 am?

e Justine arrives at the park $1\frac{3}{4}$ hours before the first seal performance starts. At what time does she arrive?

f Jim wants to see the 10:00 am otter performance, then the next available whale performance. How much time will he have to wait between the performances?

g Harriet wants to see all six performances in one day.

   i Find the total amount of time she will spend watching performances.

   ii Determine the time at which she will need to see each performance.

C · 24-HOUR TIME

24-hour time indicates the amount of time which has passed since midnight.

By using 24-hour time, we avoid the need for using am and pm to indicate morning and afternoon.

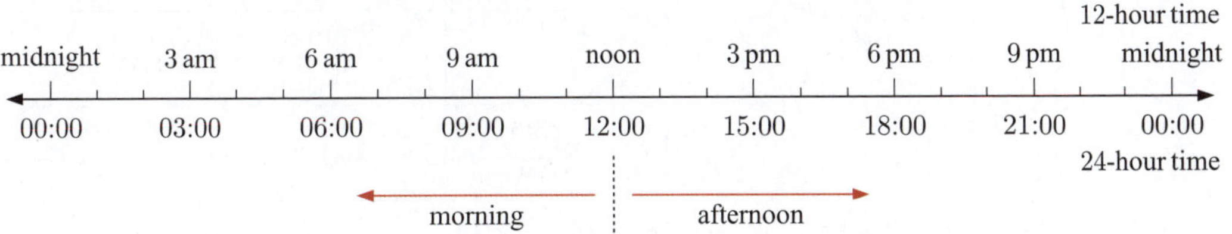

We always use *four* digits to write 24-hour time. For example:

- 5:27 am is 05:27
- 3:48 pm is 15:48

Notice that:

- Midnight is 00:00 not 24:00.
- Times from 00:00 to 11:59 are morning (am) times.
- Times from 12:00 to 23:59 are afternoon (pm) times.
- To convert times from 13:00 to 23:59 into 12-hour time, we subtract 12 hours.

Example 6   ◀)) **Self Tutor**

Write:

  a 7:42 am in 24-hour time   **b** 18:50 in 12-hour time.

  a 7:42 am is 07:42.

  b 18:50 is a pm time.
     18 hours − 12 hours = 6 hours
     So, 18:50 is 6:50 pm.

EXERCISE 13C

1 Write in 24-hour time:

  a 3:47 am   **b** 10:12 am   **c** 6 o'clock pm

  d 1:35 pm   **e** 8:41 am   **f** 9:25 pm

  g noon   **h** 6:39 pm   **i** 7 minutes past midnight

2 Write in 12-hour time:

  a 02:49   **b** 11:53   **c** 15:24   **d** 20:00

  e 09:00   **f** 12:34   **g** 23:23   **h** 00:20

3 Find the time difference between:

  a 08:30 and 11:50   **b** 09:45 and 14:50   **c** 12:40 and 22:25

  d 10:10 am and 14:47   **e** 13:18 and 9:05 pm   **f** 5:35 pm and 23:16.

4 Find the time difference between:

  a 21:40 and 08:23 the next day

  b 19:40 and 15:00 the next day

  c 16:30 Monday and 08:30 Wednesday.

5 Copy and complete the railway schedule alongside:

Departure	Travelling time	Arrival
07:25	2 h 15 min	
11:30		16:05
	29 min	11:08
	3 h 41 min	12:30
23:52		04:55 (next day)

6 Yesterday, sunrise was at 06:43 and sunset was at 19:31.

  a How long was it between sunrise and sunset?

  b The Sun was highest in the sky at the time halfway between sunrise and sunset. When did this occur?

7 Look at the schedule of arrivals into Cape Town airport.

  a How many of the listed flights arrive:

   i before 3:00 pm   **ii** after 6:00 pm?

  b How long after the Windhoek flight arriving will the Johannesburg flight arrive?

  c Kaya arrived at the airport at 5:20 pm to pick up her friend from Harare. How long will she need to wait?

  d The flight from East London was delayed, and did not arrive until 3:22 pm. By how long was it delayed?

  e The flight from Lanseria departed at 7:55 pm. How long was the flight?

ARRIVALS		
Flight	Origin	Arrival time
4Z601	Bloemfontein	09:55
WV912	Windhoek	10:30
BA6411	Johannesburg	11:55
ET847	Addis Ababa	13:45
BA6302	Durban	14:25
BA6322	East London	14:45
4Z644	Nelspruit	16:00
4Z383	Harare	17:50
4Z349	Walvis Bay	19:00
KL597	Amsterdam	21:25
FA317	Lanseria	22:00

DISCUSSION

What are the advantages and disadvantages of using 24-hour time instead of 12-hour time?

D — TIME ZONES

At any given time, different parts of the world are experiencing different phases of day and night. This means the time of day varies depending on where you are.

For example, when it is the middle of the day in New York, Bangkok is in complete darkness.

Until the 19th century, every city and town would calculate their own time by measuring the position of the Sun. In Europe, the time of day was broadcast to the people of the town using a bell on a church. This meant that cities that were only a short distance from each other would use slightly different times.

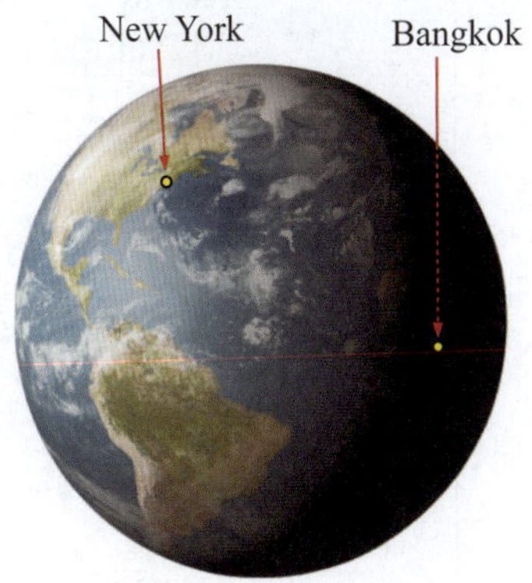

With the invention of railways, Great Britain began using electric telegraph signals to standardise time in the railway stations across the country. Other countries quickly adopted their own standardised times.

To account for the time of day varying according to location on the Earth, the world was divided into **time zones**.

STANDARD TIME ZONES

The **Prime Meridian** is the line with longitude $0°$, which passes through **Greenwich** near London. It is the starting point for 12 time zones west of Greenwich and 12 time zones east of Greenwich. Time along the Prime Meridian is called **Greenwich Mean Time (GMT)**.

Places which lie in the same time zone share the same **standard time**. Standard Time Zones are usually measured in 1 hour units, but there are also a few $\frac{1}{2}$ hour units around the world.

Places to the **east** of the Prime Meridian are **ahead** of GMT.

Places to the **west** of the Prime Meridian are **behind** GMT.

The map on page **274** shows the main time zones of the world. Their borders are not all straight like the lines of longitude, but rather follow the borders of countries or regions, and natural boundaries such as rivers and mountains. The numbers in the zones show how many hours have to be added or subtracted from Greenwich Mean Time to work out the standard time for that zone.

Example 7　　　　　　　　◀) **Self Tutor**

If it is 12 noon in Greenwich, what is the standard time in:

a Mumbai　　　　　**b** Los Angeles?

a Mumbai is in a zone marked $+5\frac{1}{2}$, so it is $5\frac{1}{2}$ hours ahead of GMT.
　∴　the standard time in Mumbai is 5:30 pm.

b Los Angeles is in a zone marked -8, so it is 8 hours behind GMT.
　∴　the standard time in Los Angeles is 4 am.

STANDARD TIME ZONES OF THE WORLD

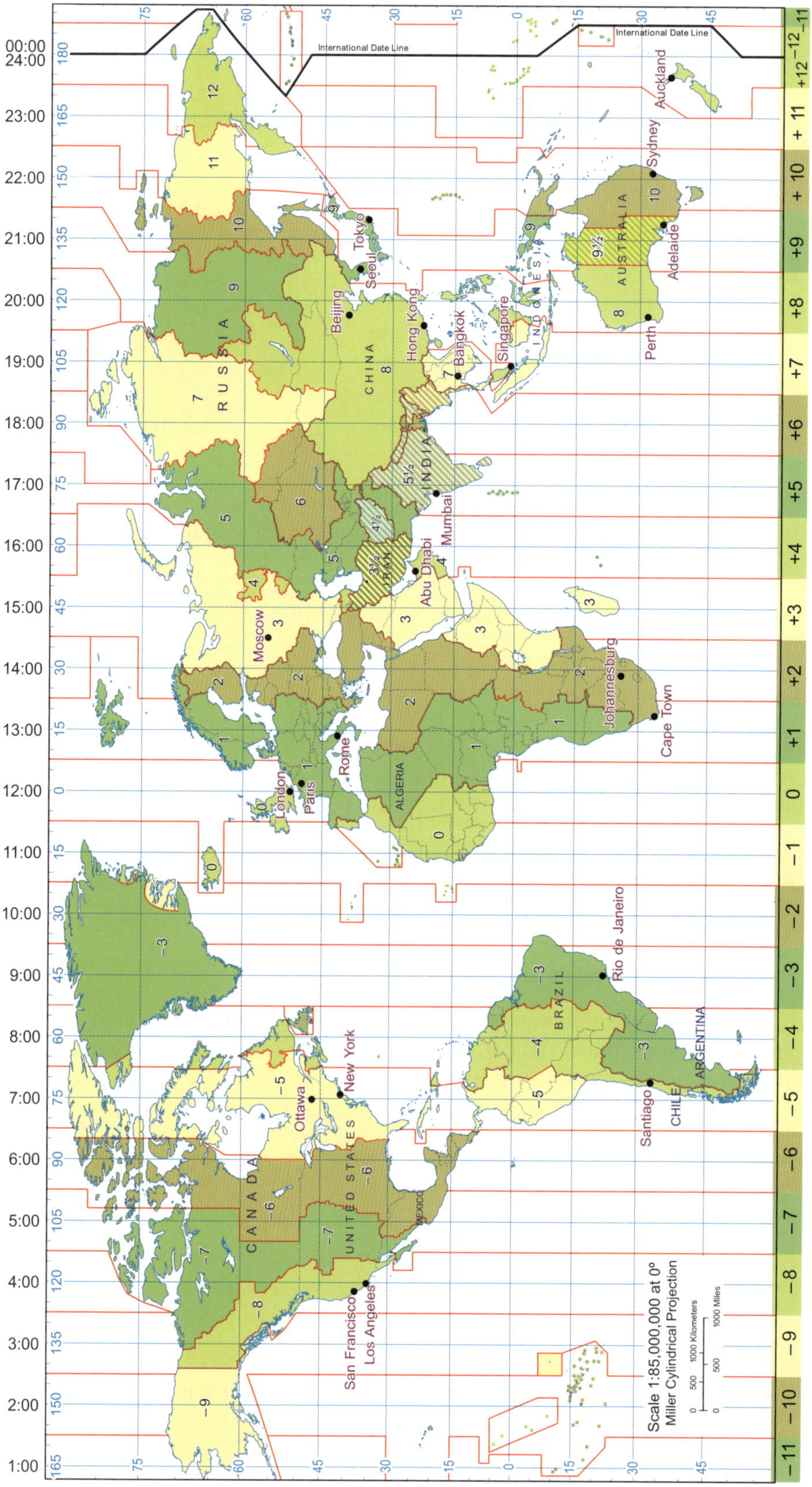

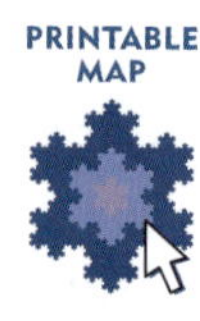

Adapted from *Standard Time Zones of the World* found at
https://www.cia.gov/the-world-factbook/maps/world-regional/

EXERCISE 13D

1   If it is 12 noon in Greenwich, what is the standard time in:

   a Ottawa   **b** Cape Town   **c** Hong Kong   **d** Perth?

2   If it is 11 pm on Tuesday in Greenwich, what is the standard time in:

   a Auckland   **b** Abu Dhabi   **c** Tokyo   **d** Santiago?

3   If it is 5 pm on Friday in Sydney, what is the standard time in:

   a Los Angeles   **b** Paris   **c** Moscow   **d** Beijing?

4   Clint lives in Ottawa. At 11 am Ottawa time, he called his brother Kirk, who lives in Rome. At what time did Kirk receive the call in Rome?

5   The first football match in the 2018 FIFA World Cup started at 6 pm in Moscow. What time did this match start for people watching in:

   a Paris   **b** Tokyo   **c** New York   **d** Rio de Janeiro?

Example 8   ◄)) **Self Tutor**

A flight from Perth to Sydney leaves at 7 am Perth time, and takes 4 hours. What is the local time when the plane arrives in Sydney?

The flight leaves at 7 am Perth time, and takes 4 hours.

∴   the plane arrives in Sydney at 11 am Perth time.

Now Perth is in a zone marked +8, and Sydney is in a zone marked +10.

∴   the standard time in Sydney is 2 hours ahead of Perth.

∴   the plane arrives in Sydney at 1 pm local time.

---

6   Lee-Yen takes a flight from Tokyo to Seoul. The flight takes 2 hours and 20 minutes, and arrives in Seoul at 10:15 pm. Find the time when the flight leaves Tokyo.

7   Noah's flight leaves Perth at 10:10 pm, and arrives in Hong Kong at 5:55 am the next day. How long was his flight?

8   Drew takes a 2:00 pm flight from Los Angeles to New York. The flight takes $5\frac{3}{4}$ hours. What is the time in New York when he arrives?

9   A flight from Mumbai to Johannesburg leaves at 11:50 pm. The flight takes 11 hours and 40 minutes. What is the local time in Johannesburg when the plane arrives?

10   Ali takes a flight from Abu Dhabi to Singapore. He leaves at 1:05 am Abu Dhabi time, and arrives at 12:35 pm Singapore time. How long was the flight?

11   Ji-hoon is flying from Seoul to Los Angeles. He leaves at 19:40 on Tuesday, and when he arrives in Los Angeles the local time is 14:50 on the same day. How long was Ji-hoon's flight?

12 Tomoko is travelling from Adelaide to Tokyo. She leaves Adelaide on a Tuesday, and must stop at Sydney and Hong Kong along the way. Her travel plan is shown alongside, with all times given as local times.

Adelaide → Tokyo	
8:00 am Adelaide	10:25 am Sydney
11:55 am Sydney	7:50 pm Hong Kong
1:05 am Hong Kong (Wed)	6:00 am Tokyo (Wed)

   a At what time does the flight from Sydney to Hong Kong arrive?

   b How long must Tomoko wait between flights at Sydney?

   c How long is the flight from:

      i Adelaide to Sydney   **ii** Hong Kong to Tokyo?

   d Find the total time between Tomoko leaving Adelaide and arriving in Tokyo.

RESEARCH

What is the **international date line**?

What is its purpose?

Why is it drawn so it only passes through ocean and a little of Antarctica?

MULTIPLE CHOICE QUIZ

QUICK QUIZ

REVIEW SET 13A

1 Convert:

   a 840 minutes into hours   **b** 7 minutes and 16 seconds into seconds only.

2 A ferry takes 12 minutes to cross a river, unload passengers, and load the next group. How many river crossings does the ferry make if it operates continuously for 6 hours?

3 Which is longer?

   a 32 decades or 3 centuries   **b** 210 decades or 2 millennia

4 Find the time which is:

   a 3 hours after 10:20 am   **b** $1\frac{1}{2}$ hours before 5:16 pm

   c 2 hours 27 minutes after 2:22 pm   **d** 4 hours 43 minutes before 12:14 pm.

5 Find the time difference between:

   a 5:15 pm and 10:25 pm   **b** 7:45 am and 1:00 pm

   c 11:24 and 14:54   **d** 20:20 and 23:57.

6 Daisy started work at 10:45 am and finished work at 2:30 pm. For how long did she work?

7 A train from Shanghai to Nanjing leaves at 11:45 am and takes 2 hours and 23 minutes. The cities are in the same time zone. At what time does the train arrive in Nanjing?

8 Write in 24-hour time:

   a 9:47 am   **b** 12:49 pm   **c** 7:17 pm   **d** 12:26 am

9 A meteorology website reported that, at a particular location, the high tides will occur at 06:17 and 18:40. A low tide will occur 6 hours and 47 minutes after the first high tide.

   a Find the time difference between the high tides.

   b At what time will the low tide occur?

10 Tokyo is in a time zone marked $+9$ and Rio de Janeiro is in a time zone marked -3. If it is 12 noon in Greenwich, what is the standard time in:

   a Tokyo   **b** Rio de Janeiro?

11 A furniture store has the opening hours shown alongside.

   a For how long is the store open on:

      i Wednesday   **ii** Saturday?

   b On which day is the store open the longest?

   c For how many hours does the store open each week?

OPENING HOURS	
Mon-Wed	9:00 am - 7:00 pm
Thursday	9:00 am - 9:00 pm
Friday	8:30 am - 6:00 pm
Saturday	10:00 am - 5:30 pm
Sunday	11:00 am - 5:00 pm

12 Answer the **Opening Problem** on page **264**.

REVIEW SET 13B

1 Write:

   a 457 minutes in hours and minutes   **b** 82 h in days and hours.

2 Reuben ran an 800 m race in 3 minutes and 14 seconds. Write this time in seconds.

3 The last ice age ended about 11 700 years ago. Express this in millennia.

4 How long is it between 1:49 pm and 6:53 am the next day?

5 Jess leaves the house at 7:20 am. She walks for 12 minutes to the bus stop, and waits 6 minutes for the bus. The ride to school takes 35 minutes. Jess then takes another 8 minutes to walk to her classroom. At what time does Jess arrive at class?

6 Find the time which is:

   a 5 h 29 min after 3:16 am

   b $2\frac{3}{4}$ hours before 1:07 pm.

7 Nadine is driving from Sydney to Canberra. The trip takes 3 hours and 26 minutes. Nadine wants to arrive in Canberra at 11:00 am. Given that the cities are in the same time zone, at what time should she leave Sydney?

8 Write in 12-hour time:

   a 07:12   **b** 15:02   **c** 21:59   **d** 03:48

9 Caleb's train leaves at 18:40. His watch currently reads 3:52 pm. How long does Caleb have before his train leaves?

10 Tom lives in Adelaide and wants to call his sister Anita in London. He wants to call Anita at 7 pm Friday London time. At what time in Adelaide should he make the call?

11 A local cinema is having a "Classic Films" festival. The screening timetable is shown below.

Movie	Duration	Screening times
The Italian Job	95 minutes	10:30 am, 12:30 pm, 2:30 pm, 5:00 pm
Gone With the Wind	222 minutes	9:00 am, 2:00 pm, 7:00 pm
Fantasia	125 minutes	9:15 am, 12:00 pm, 6:00 pm, 8:45 pm
The Sound of Music	174 minutes	2:30 pm, 6:00 pm, 9:30 pm
Singin' in the Rain	103 minutes	11:45 am, 3:15 pm
The Godfather	175 minutes	11:30 am, 3:00 pm, 6:30 pm

  a Write the duration of "Fantasia" in hours and minutes.

  b Which movie is the shortest?

  c Which movie finishes the latest? At what time does it finish?

  d Which movies are in progress at 4:15 pm?

  e Katie wants to see the first screening of "The Italian Job", then the next available screening of "The Godfather". How much time will she have to wait between movies?

12 Aroon is travelling from Bangkok to Johannesburg. His itinerary is shown below, with all times given as local times.

Bangkok → Johannesburg	
6:30 pm Bangkok (Tuesday)	9:55 pm Singapore
1:30 am Singapore (Wednesday)	6:10 am Johannesburg

  a If Aroon must arrive at the airport $2\frac{1}{2}$ hours before takeoff to check in, at what time must he be at the airport?

  b How long must Aroon wait between flights at Singapore?

  c How long is the flight from:

    i Bangkok to Singapore   **ii** Singapore to Johannesburg?

  d Find the total time between Aroon leaving Bangkok and arriving in Johannesburg.

14

Coordinate geometry

Contents:

OPENING PROBLEM

When Tiffany makes an international phone call, the *cost* of the call depends on the *time* that she spends on the call. This relationship is illustrated on the graph alongside.

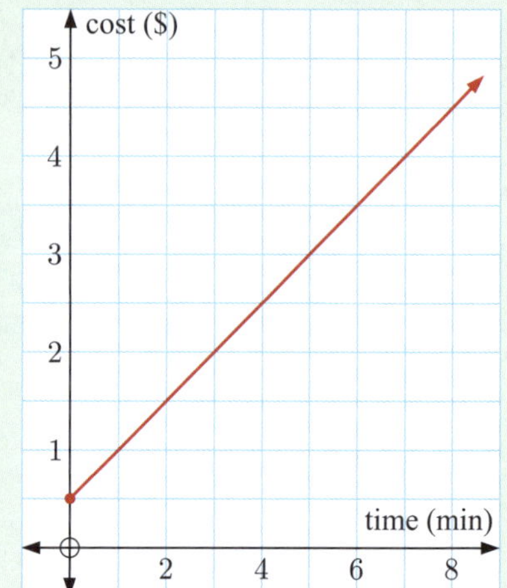

Things to think about:

a How can we describe the relationship between *cost* and *time*?

b What is the *connection fee* or fixed cost charged for connecting the phone call?

c By how much does the cost of the call increase for each minute spent on the call?

d Can you use this graph to determine:

  i the cost of a 3 minute call

  ii the length of a call which costs $4?

A   THE CARTESIAN PLANE

The number grid alongside is a **Cartesian plane**, named after **René Descartes**.

We start with a point of reference O called the **origin**.

Through the origin we draw two fixed number lines or **axes** which are perpendicular to each other.

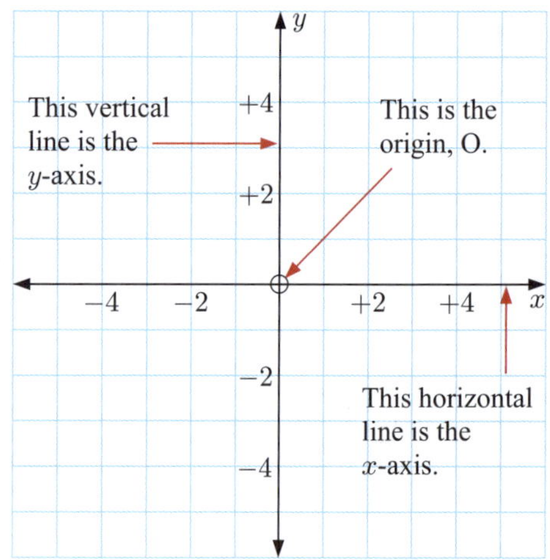

- The horizontal axis is called the **x-axis**. It is positive to the right of O and negative to the left of O.

- The vertical axis is called the **y-axis**. It is positive above O and negative below O.

QUADRANTS

The x and y-axes divide the Cartesian plane into four regions called **quadrants**. These quadrants are numbered in an **anticlockwise direction** as shown:

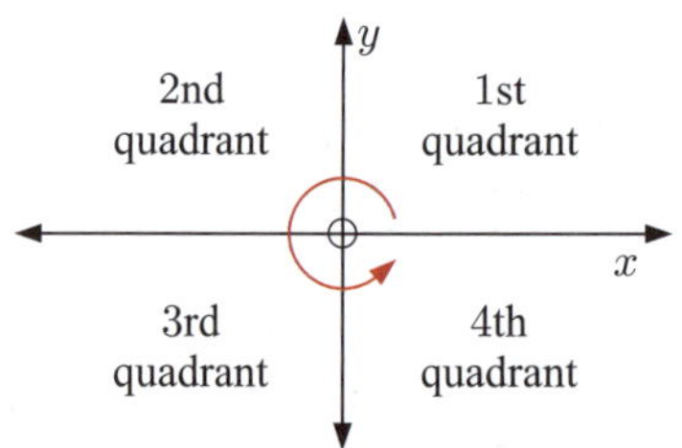

COORDINATES

We can describe any point on the Cartesian plane using an **ordered pair** of numbers called **coordinates**.

- The coordinates of a point are written (x-coordinate, y-coordinate).
- The x-coordinate gives the horizontal position along the x-axis.
- The y-coordinate gives the vertical position along the y-axis.

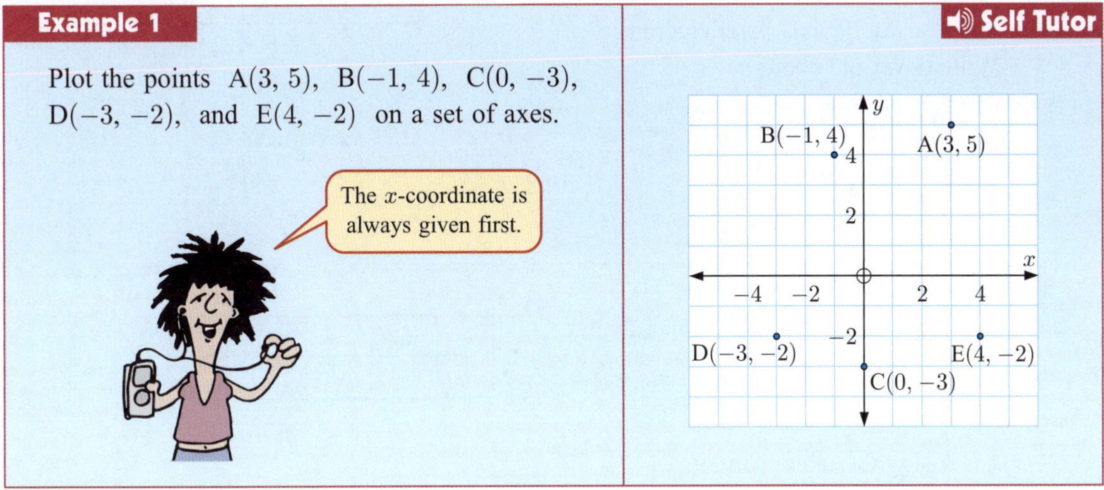

EXERCISE 14A

1 Look at the grid.

  a Write down the x-coordinate of:

  i A   **ii** E   **iii** D.

  b Write down the y-coordinate of:

  i C   **ii** B   **iii** A.

  c State the point which lies on the:

  i x-axis   **ii** y-axis.

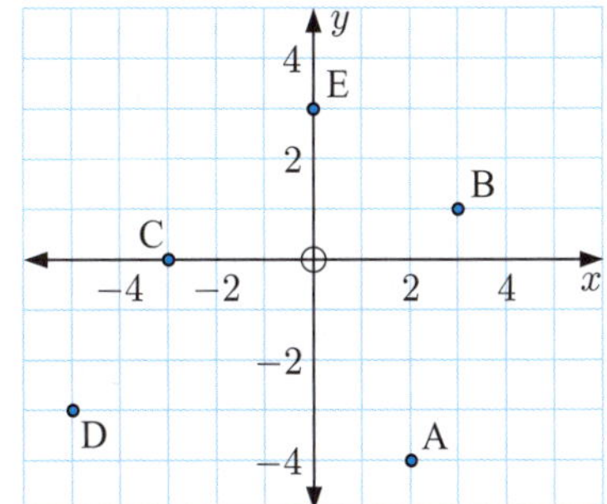

2 Look at the grid.

  a State the coordinates of the points P, Q, R, S, and T.

  b State the quadrant in which we find the point S.

  c State the point(s) in the 4th quadrant.

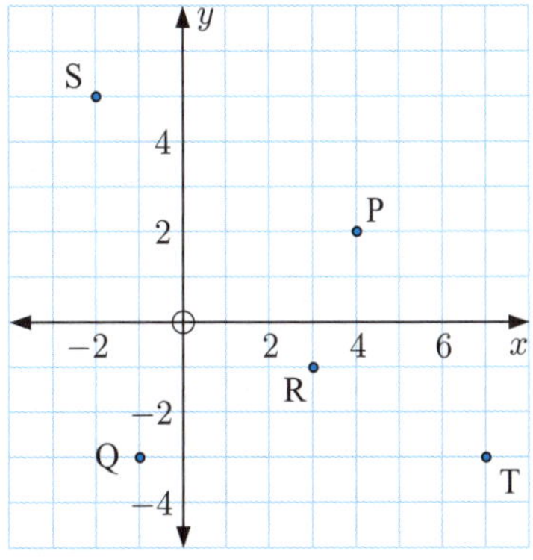

3 On the same set of axes, plot the points:

A(2, 5), B(4, −2), C(−5, 0), D(−1, −4), E(−2, 3), F(0, 3), G(5, 1), H(−5, −1).

4 State the quadrant in which you would find:

a (−2, −1)   **b** (3, 4)   **c** (−1, 5)   **d** (5, −2).

5 The points A, B, and C are shown on the grid.

  a Suppose ABCD is a square.
  Write down the coordinates of D.

  b Suppose ACBE is a parallelogram.
  Write down the coordinates of E.

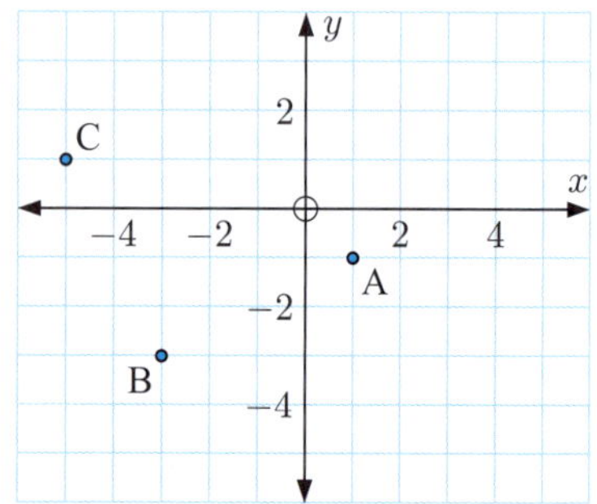

Example 2   ◀)) **Self Tutor**

Plot the points in this table of values on the same set of axes.

x	−2	−1	1	3	4
y	−1	0	1	2	3

Using the table we plot
(−2, −1), (−1, 0), (1, 1), (3, 2),
and (4, 3).

---

6 Plot the points in each table of values:

a

x	−3	−2	0	2	3
y	4	2	1	0	−1

b

x	−2	−1	0	1	3
y	−3	−1	0	3	2

7 The height of a seedling is measured each week, and the results recorded in the table shown.

Week number (x)	1	2	3	4	5
Height (y cm)	3	5	7	9	11

  a Plot these points on the same set of axes.

  b Do the points lie in a straight line?

  c Assuming this pattern continues, predict the height of the seedling after:

   i 6 weeks   **ii** 8 weeks.

8 The points in this table of values lie in a straight line.

x	1	3	5	
y	-2	2		12

   a Use the first two points to draw the line on a Cartesian plane.

   b Complete the table of values.

   c Identify the point on the line which lies on the:

      i x-axis   **ii** y-axis.

Example 3   ◀) Self Tutor

Use the equation $y = 2x - 1$ to construct a table of values for $x = -2, -1, 0, 1, 2$.

Plot the resulting points on a Cartesian plane.

When $x = -2$, $y = 2(-2) - 1 = -5$
When $x = -1$, $y = 2(-1) - 1 = -3$
When $x = 0$, $y = 2(0) - 1 = -1$
When $x = 1$, $y = 2(1) - 1 = 1$
When $x = 2$, $y = 2(2) - 1 = 3$

The table of values is:

x	-2	-1	0	1	2
y	-5	-3	-1	1	3

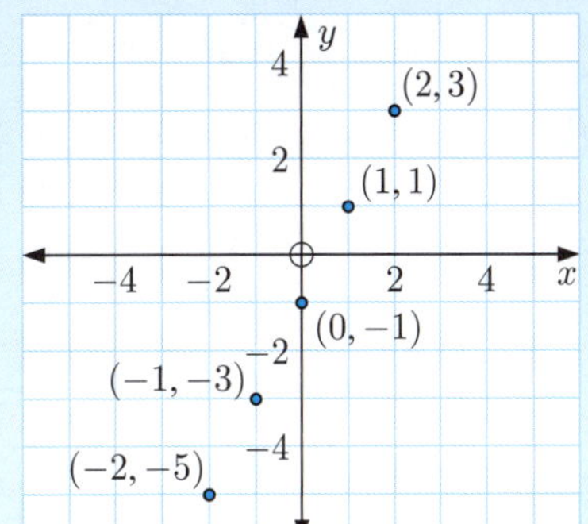

---

9 For each equation given, construct a table of values for $x = -2, -1, 0, 1, 2$. Plot the resulting points on a Cartesian plane.

   a $y = 3x$ **b** $y = x + 1$ **c** $y = 3 - x$

   d $y = x^2$ **e** $y = 1 - x^2$ **f** $y = x^3$

Example 4   ◀) Self Tutor

On a Cartesian plane, show all the points with positive x-coordinate and negative y-coordinate.

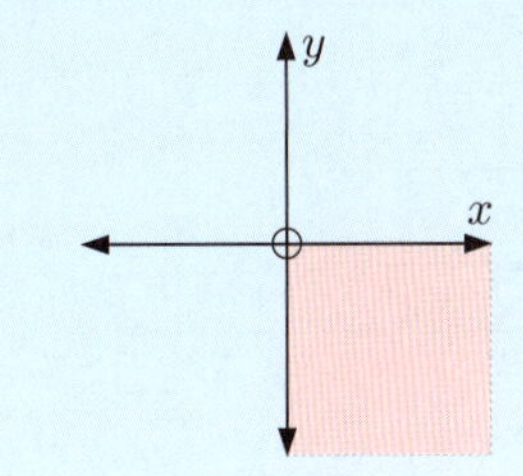

The points on the axes are not included.

---

10 On different sets of axes, show all the points with:

   a x-coordinate equal to -1 **b** y-coordinate equal to 3

   c x-coordinate equal to 0   **d** y-coordinate equal to 0.

11   On different sets of axes, show all the points with:

    a   negative x-coordinate   **b**   positive y-coordinate

    c   negative x and y-coordinates   **d**   negative x-coordinate and positive y-coordinate.

12   State the quadrants in which I would find points whose coordinates have:

    a   the same sign   **b**   different signs.

13   Make *two* copies of the grid alongside, including points A and B. Use one grid for each part of the question.

Sketch *all* the possible locations of point C such that triangle ABC is:

    a   isosceles   **b**   right angled.

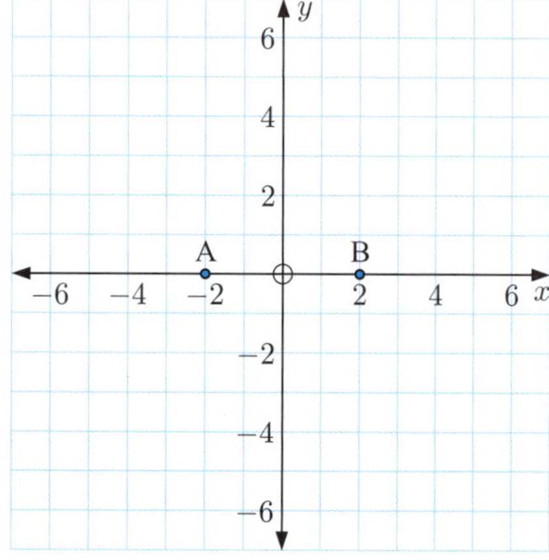

B   STRAIGHT LINES

A **straight line** is an infinite set of points in a particular direction.

To describe *all* of the points on a line, we need an **equation**.

INVESTIGATION 1   HORIZONTAL LINES

1   Plot the points in each table of values on a Cartesian plane:

    a

x	-2	-1	0	1	2
y	3	3	3	3	3

    b

x	-2	-1	0	1	2
y	-2	-2	-2	-2	-2

    Write down your observations.

2   Look at the line drawn on the number grid.

    a   What can you say about the y-coordinate of each point on the line?

    b   Copy and complete:

        The equation of the line is $y = \ldots\ldots$

3   What is the *form* of the equation of any horizontal line?

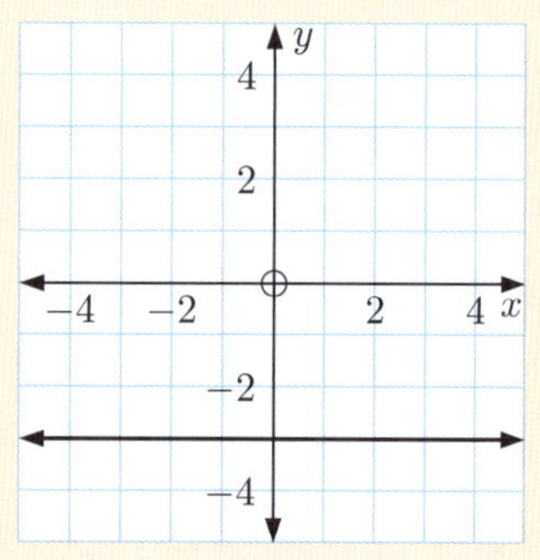

INVESTIGATION 2   VERTICAL LINES

1 Plot the points in each table of values on a Cartesian plane:

a

x	2	2	2	2	2
y	-2	-1	0	1	2

b

x	-3	-3	-3	-3	-3
y	-2	-1	0	1	2

Write down your observations.

2 Look at the line drawn on the number grid.

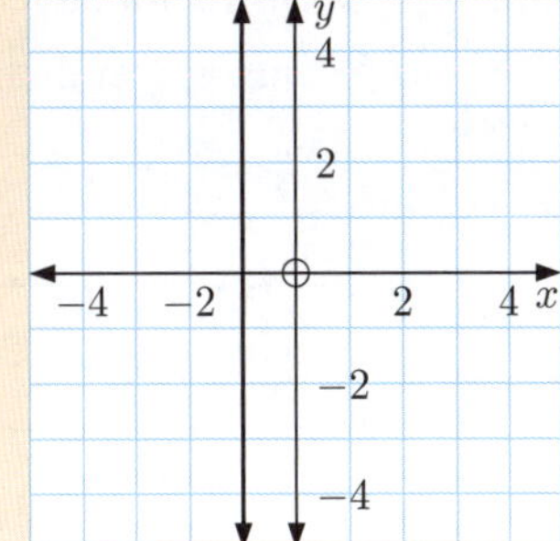

a What can you say about the x-coordinate of each point on the line?

b Copy and complete:
The equation of the line is $x = \ldots\ldots$

3 What is the *form* of the equation of any vertical line?

INVESTIGATION 3   OTHER STRAIGHT LINES

1 For each equation given, construct a table of values for $x = -2, -1, 1, 2$. Plot the resulting points on a Cartesian plane.

a $y = 2x$ **b** $y = x - 2$ **c** $y = x^2$

d $y = -x + 2$ **e** $y = -2x - 1$ **f** $y = -x$

g $y = -x^2$ **h** $y = x + 3$ **i** $y = \dfrac{1}{x}$

2 Which of the equations result in points in a straight line?

3 What is the *form* of the equations which result in straight lines?

All **horizontal lines** have equations of the form $y = c$ where c is a constant.

All **vertical lines** have equations of the form $x = a$ where a is a constant.

All other straight lines have equations of the form $y = mx + c$ where m and c are constants.

EXERCISE 14B

1 Draw the graph of the line with equation:

a $y = 1$ **b** $y = -4$ **c** $y = \dfrac{3}{2}$ **d** $y = 0$.

2 Draw the graph of the line with equation:

a $x = 2$ **b** $x = -3$ **c** $x = 0$ **d** $x = -\dfrac{1}{2}$.

3 **a** On the same set of axes, graph the lines $x = -2$, $x = 6$, $y = 5$, and $y = -6$.

   b Consider the quadrilateral formed by the intersection points of these lines.

   i What type of quadrilateral is it?   **ii** Find the area of the quadrilateral.

Example 5   ◆⟩ **Self Tutor**

Consider the equation $y = \frac{1}{2}x - 1$.

   a Construct a table of values using $x = -3, -2, -1, 0, 1, 2,$ and 3.

   b Hence graph the straight line.

a

x	-3	-2	-1	0	1	2	3
y	$-2\frac{1}{2}$	-2	$-1\frac{1}{2}$	-1	$-\frac{1}{2}$	0	$\frac{1}{2}$

b

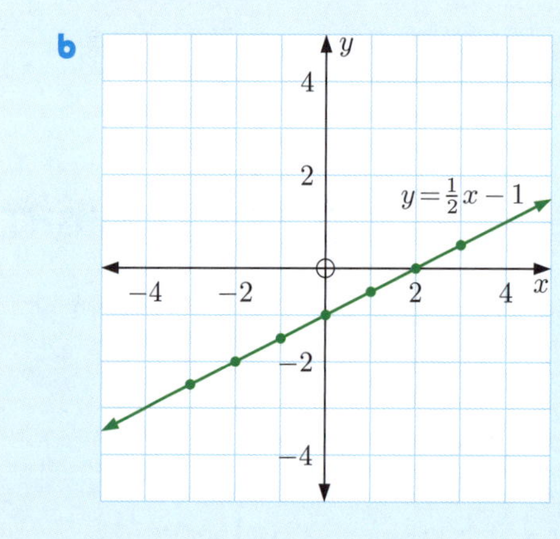

---

4 For each of the following equations:

   i Construct a table of values using $x = -3, -2, -1, 0, 1, 2,$ and 3.

   ii Draw the graph of the straight line.

a $y = \frac{1}{2}x$ 　　　　　　　**b** $y = -\frac{1}{2}x$ 　　　　　　　**c** $y = 3x + 2$

d $y = -3x + 2$ 　　　　　　**e** $y = 2x - 3$ 　　　　　　**f** $y = -2x - 3$

g $y = \frac{1}{4}x + 2$ 　　　　　　**h** $y = -\frac{1}{4}x + 2$ 　　　　　**i** $y = \frac{3}{2}x - 1$

j $y = -\frac{3}{2}x - 1$ 　　　　　**k** $y = 4x - 3$ 　　　　　　**l** $y = 4x + 3$

Use your **graphics calculator** or the **graphing package** to
check your answers.

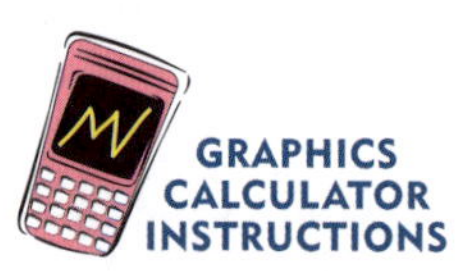

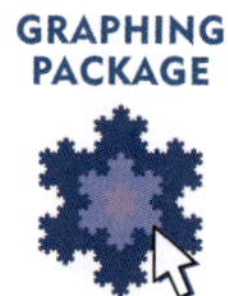

Example 6   ◀🔊 **Self Tutor**

By inspection only, find the equation of the straight line passing through the points in each table of values:

a

x	1	2	3	4	5
y	-3	-6	-9	-12	-15

b

x	1	2	3	4	5
y	5	4	3	2	1

a Each y-value is -3 times its corresponding x-value, so $y = -3x$.

b The sum of each x and y-value is always 6, so $x + y = 6$.

5 By inspection only, find the equation of the straight line passing through the points in each table of values:

a

x	1	2	3	4	5
y	4	8	12	16	20

b

x	2	3	4	5	6
y	4	5	6	7	8

c

x	0	1	2	3
y	0	-2	-4	-6

d

x	0	1	2	3
y	3	2	1	0

e

x	1	2	3	4	5
y	3	5	7	9	11

f

x	1	2	3	4	5
y	1	3	5	7	9

Example 7   ◀🔊 **Self Tutor**

Determine whether $(2, 5)$ and $(-1, -5)$ lie on the line with equation $y = 4x - 3$.

The point on the line with x-coordinate 2, has y-coordinate $= 4 \times 2 - 3$
$$= 8 - 3$$
$$= 5 \checkmark$$

So, $(2, 5)$ does lie on the line.

The point on the line with x-coordinate -1, has y-coordinate $= 4 \times (-1) - 3$
$$= -4 - 3$$
$$= -7$$

Since the y-coordinate $\neq -5$, $(-1, -5)$ does not lie on the line.

6 Determine whether the point:

    a $(3, 4)$ lies on the line $y = 2x - 2$ **b** $(-1, 4)$ lies on the line $y = x + 6$

    c $(-2, 10)$ lies on the line $y = -3x + 4$ **d** $\left(\frac{1}{2}, -6\right)$ lies on the line $y = 2x - 8$.

7 Consider the line $y = 2x - 3$.

    a Find the y-coordinate of the point on the line with x-coordinate 4.

    b Find the x-coordinate of the point on the line with y-coordinate -4.

8 Consider the line $y = 5 - 2x$.

    a Find the y-coordinate of the point on the line with x-coordinate -2.

    b Find the x-coordinate of the point on the line with y-coordinate -3.

C   GRADIENT

DISCUSSION   WHICH LINE IS STEEPER?

Look at the two lines alongside.

- What does it mean to say that one line is *steeper* than another?
- Which line do you think is steeper?
- How can we *measure* the steepness of a line?

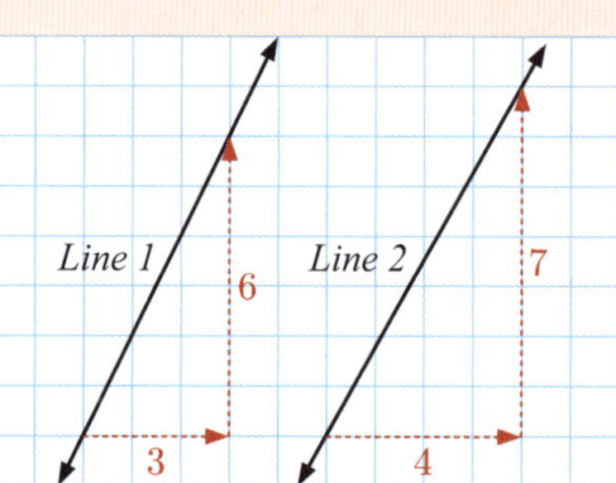

The **gradient** of a line is a measure of its steepness.

We calculate the rate at which a line rises or falls by choosing two points on the line, then dividing the vertical step between them by the horizontal step between them.

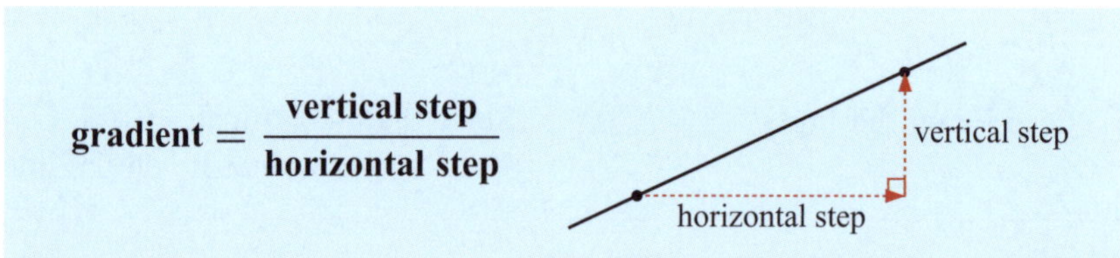

In the **Discussion** above:

- The gradient of *line 1* is $\frac{6}{3} = 2$. For every 1 unit moved horizontally, the line moves 2 units upwards.

- The gradient of *line 2* is $\frac{7}{4} = 1.75$.

- *Line 1* has a higher gradient than *line 2*, so *line 1* is *steeper* than *line 2*.

In general:

- An **upward sloping line** has a **positive gradient**.

- A **downward sloping line** has a **negative gradient**.

- For a **horizontal line**, the vertical step is 0, so the **gradient of a horizontal line is 0**.
- For a **vertical line**, the horizontal step is 0, so the **gradient of a vertical line is undefined**.

Example 8

Find the gradient of:

a [AB]   **b** [BC]

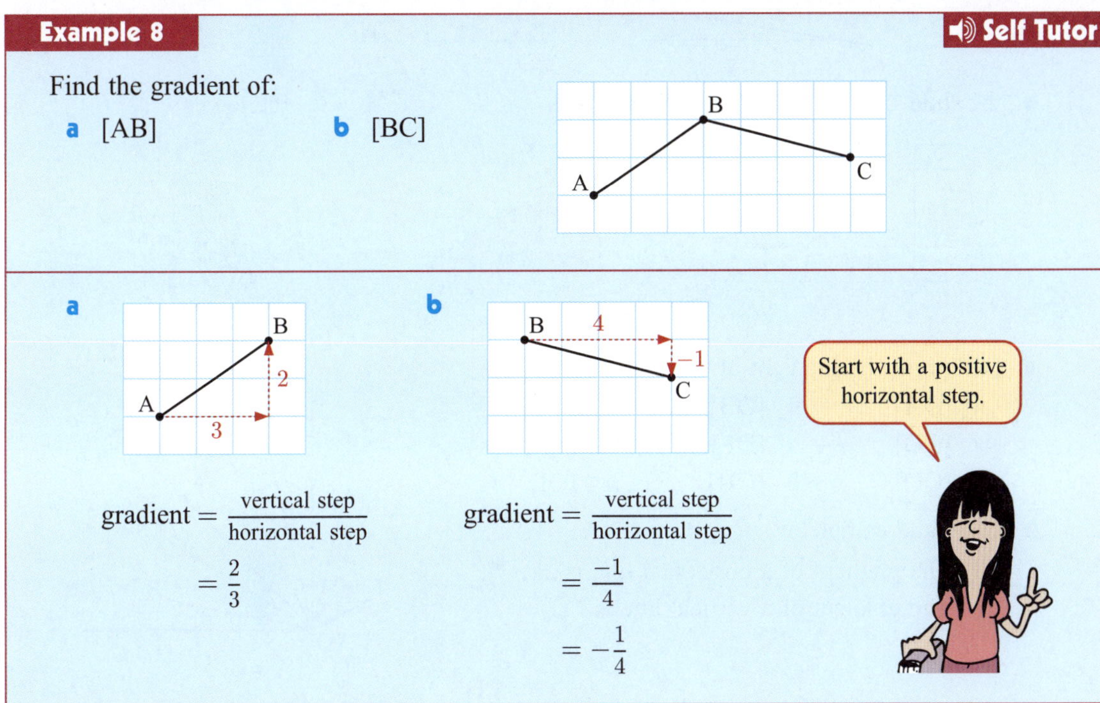

a

gradient = $\dfrac{\text{vertical step}}{\text{horizontal step}}$

$= \dfrac{2}{3}$

b

gradient = $\dfrac{\text{vertical step}}{\text{horizontal step}}$

$= \dfrac{-1}{4}$

$= -\dfrac{1}{4}$

EXERCISE 14C

1 Find the gradient of each line:

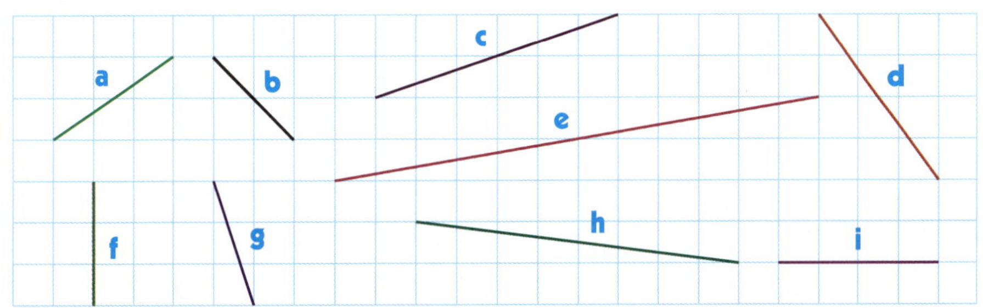

2 On grid paper, draw a line segment with gradient:

a $\dfrac{1}{2}$ **b** $\dfrac{1}{3}$ **c** $-\dfrac{3}{4}$ **d** 1 **e** 2 **f** -3

g $\dfrac{4}{3}$ **h** $-\dfrac{5}{2}$ **i** 0 **j** $1\dfrac{1}{5}$ **k** 5 **l** undefined.

3 On grid paper, draw a triangle which has:

a all three sides with positive gradients

b two sides with positive gradients and one side with negative gradient

c two sides with negative gradients and one side with positive gradient

d all three sides with negative gradients.

4 Find the gradient of:

a the road

b the slide   **c** the stairs   **d** the tower.

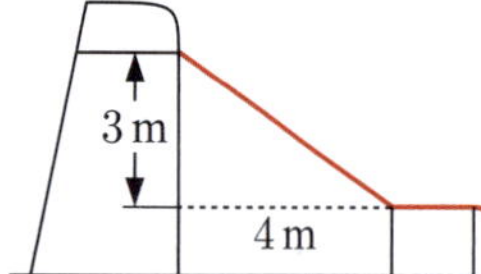

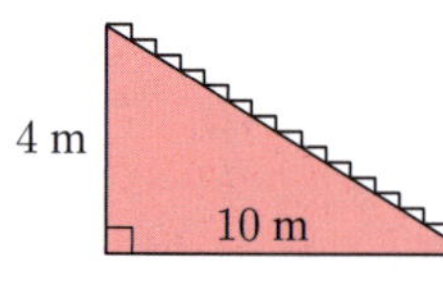

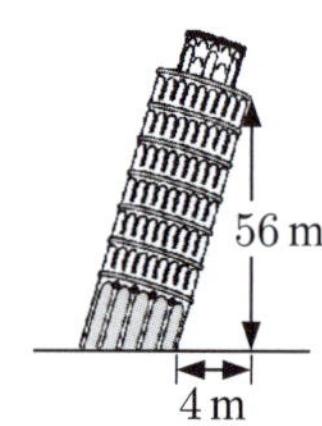

5 **a** Determine the gradient of:

   i [OA]   **ii** [OB]   **iii** [OC]

   iv [OD]   **v** [OE]   **vi** [OF]

   vii [OG]   **viii** [OH]   **ix** [OI]

b Copy and complete:

   i The gradient of a horizontal line is ......

   ii The gradient of a vertical line is ......

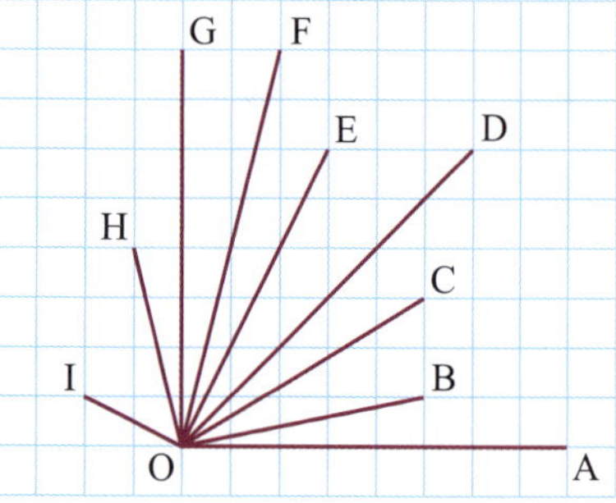

Example 9 🔊 **Self Tutor**

Draw a line through the point $(1, 1)$, with gradient $\dfrac{2}{3}$.

$$\text{gradient} = \dfrac{2}{3} \quad \begin{array}{l} \leftarrow \text{ vertical step} \\ \leftarrow \text{ horizontal step} \end{array}$$

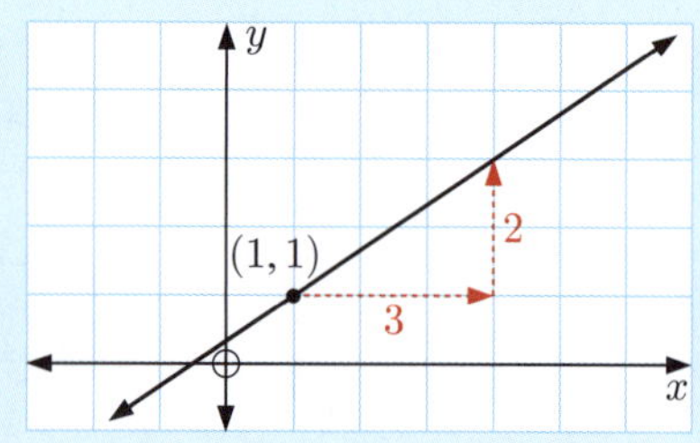

6 Draw a line through the point:

  a $(4, 1)$ with gradient $\dfrac{3}{2}$

  b $(3, -1)$ with gradient $\dfrac{1}{4}$

  c $(-2, 3)$ with gradient $-\dfrac{1}{5}$

  d $(0, 4)$ with gradient $-\dfrac{2}{3}$

  e $(0, -2)$ with gradient 3

  f $(-2, -1)$ with gradient -2

  g $(0, 6)$ with gradient $-\dfrac{5}{2}$

  h $(0, -3)$ with gradient $-\dfrac{3}{4}$.

7 By plotting the points on grid paper, find the gradient of the line segment joining:

  a $O(0, 0)$ and $A(2, 6)$

  b $O(0, 0)$ and $B(-4, 2)$

  c $G(0, -1)$ and $H(2, 5)$

  d $K(1, 1)$ and $L(-2, -2)$

  e $M(3, 1)$ and $N(-1, 3)$

  f $P(-2, 4)$ and $Q(2, 0)$.

8 Suppose l_1 is the line $x = 5$ and l_2 is the line $y = -3$.

  a Graph the lines on the same set of axes.

  b Suppose l_1 cuts the x-axis at A, and l_2 cuts the y-axis at B. Find:

   i the coordinates of A and B

   ii the gradient of [AB].

9 $A(-2, 3)$, $B(3, 6)$, $C(6, 2)$, and $D(-4, -4)$ are vertices of a quadrilateral. The sides [AB] and [CD] are parallel.

  a Plot quadrilateral ABCD on grid paper.

  b What type of quadrilateral is ABCD?

  c Find the gradient of: **i** [AB]   **ii** [BC]   **iii** [CD]   **iv** [DA].

  d What can we say about the gradients of parallel lines?

Example 10 ◀ᴺ **Self Tutor**

Consider the line with equation $y = 2x + 3$.

  a Find the points on the line with x-coordinates 1 and 3.

  b Plot the points on a number grid and hence draw the line.

  c Use the points to find the gradient of the line.

a When $x = 1$, $y = 2(1) + 3 = 5$.

When $x = 3$, $y = 2(3) + 3 = 9$.

$\therefore$ $(1, 5)$ and $(3, 9)$ lie on the line.

c The gradient of the line $= \dfrac{4}{2} = 2$.

b
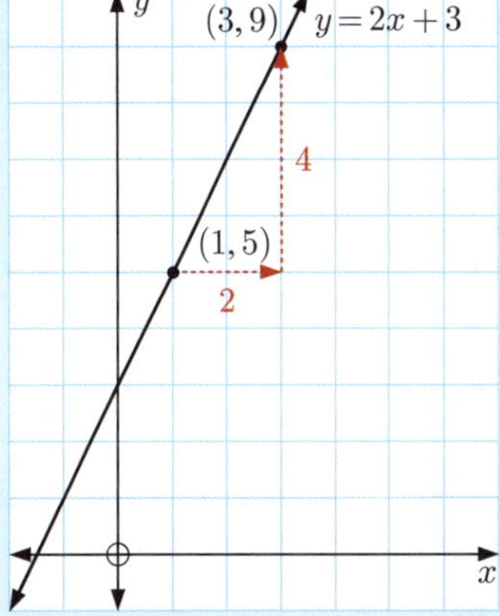

10 Consider the line with equation $y = x - 2$.

    a Find the points on the line with x-coordinates 1 and 3.

    b Plot the points on a number grid and hence draw the line.

    c Use the points to find the gradient of the line.

11 Consider the line with equation $y = -x + 3$.

    a Find the points on the line with x-coordinates 1 and 3.

    b Plot the points on a number grid and hence draw the line.

    c Use the points to find the gradient of the line.

12 Find the points on the given line with x-coordinates 0 and 2 and use them to find its gradient:

    a $y = x + 3$ **b** $y = -x + 1$ **c** $y = 2x - 1$

    d $y = -2x + 3$ **e** $y = \frac{1}{2}x$ **f** $y = -3x + 4$

    g $y = -\frac{3}{2}x + 1$ **h** $y = 4x - 3$

DISCUSSION

Look at your answers to questions **10** to **12** above.

What do you think is the gradient of a line with equation $y = mx + c$?

INVESTIGATION 4   THE GRADIENT FORMULA

We have seen how to calculate the gradient of the line segment between two points using the horizontal and vertical steps between them. In this Investigation we develop this method into a **formula**.

THE GRADIENT FORMULA

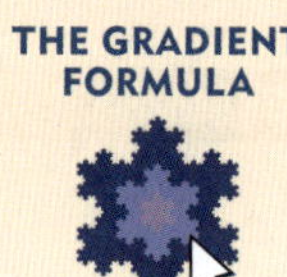

D   THE GRADIENT-INTERCEPT FORM OF A LINE

The **y-intercept** of a line is the y-coordinate of the point where the line cuts the y-axis.

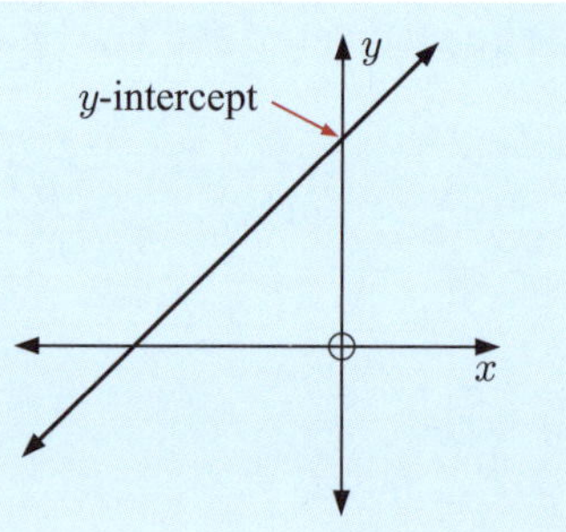

INVESTIGATION 5 — THE y-INTERCEPT OF A LINE

Since the y-axis is the line $x = 0$, we can find the y-intercept of a line by substituting $x = 0$.

What to do:

1 By substituting $x = 0$, find the y-intercept of the line with equation:

  a $y = 3x + 2$ **b** $y = 4 - x$ **c** $y = \frac{1}{2}x - 3$ **d** $y = -x - 1$.

2 By substituting $x = 0$, find the y-intercept of the line with equation $y = mx + c$.

We have already seen that the line $y = mx + c$ has gradient m, so we can now conclude that:

$y = mx + c$ is the equation of a straight line with gradient m and y-intercept c.

We call this the **gradient-intercept form** of a line.

Example 11 — ◀) Self Tutor

State the gradient and y-intercept of the line with equation:

  a $y = 3x - 2$ **b** $y = 7 - 2x$ **c** $y = 0$.

a $y = 3x - 2$ has $m = 3$ and $c = -2$
   $\therefore$ the gradient is 3 and the y-intercept is -2.

b $y = 7 - 2x$ can be written as $y = -2x + 7$, with $m = -2$ and $c = 7$
   $\therefore$ the gradient is -2 and the y-intercept is 7.

c $y = 0$ can be written as $y = 0x + 0$, with $m = 0$ and $c = 0$
   $\therefore$ the gradient is 0 and the y-intercept is 0.

EXERCISE 14D

1 State the y-intercept of each line:

  a 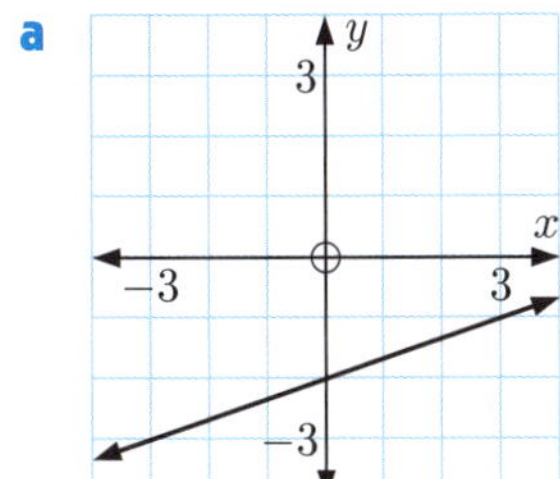   **b** 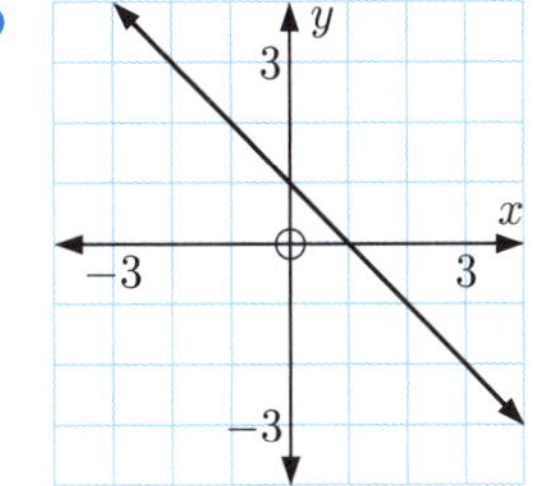   **c** 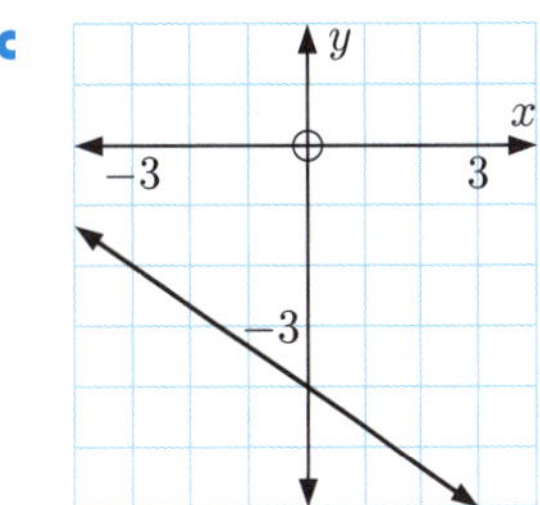

2 Write down the y-intercept of the line with equation:

  a $y = 2x - 1$ **b** $y = x + 4$ **c** $y = 3 + 2x$

  d $y = 5 - x$ **e** $y = -x - 3$ **f** $y = \frac{2}{3}x - 4$

3 State the gradient and y-intercept of the line with equation:

a $y = 4x + 8$

b $y = -3x + 2$

c $y = 6 - x$

d $y = -2x + 3$

e $y = -2$

f $y = 11 - 3x$

g $y = \dfrac{1}{2}x - 5$

h $y = 3 - \dfrac{3}{2}x$

i $y = \dfrac{2}{5}x + \dfrac{4}{5}$

j $y = \dfrac{x + 1}{2}$

k $y = \dfrac{2x - 10}{5}$

l $y = \dfrac{11 - 3x}{2}$

E │ GRAPHING A LINE FROM ITS GRADIENT-INTERCEPT FORM

To draw the graph of $y = mx + c$:

- Use the y-intercept c to plot the point $(0, c)$.
- Starting from $(0, c)$, use horizontal and vertical steps from the gradient m to locate another point on the line.
- Join the two points and extend the line in both directions.

Example 12   ◀) **Self Tutor**

Draw the graph of:

a $y = \dfrac{2}{3}x + 1$

b $y = -2x - 3$.

a For $y = \dfrac{2}{3}x + 1$:

- the y-intercept is $c = 1$
- the gradient is

$$m = \dfrac{2 \leftarrow \text{vertical step}}{3 \leftarrow \text{horizontal step}}$$

b For $y = -2x - 3$:

- the y-intercept is $c = -3$
- the gradient is

$$m = \dfrac{-2 \leftarrow \text{vertical step}}{1 \leftarrow \text{horizontal step}}$$

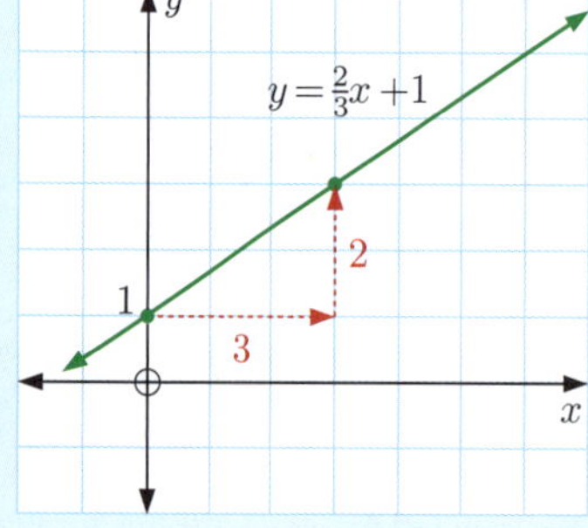

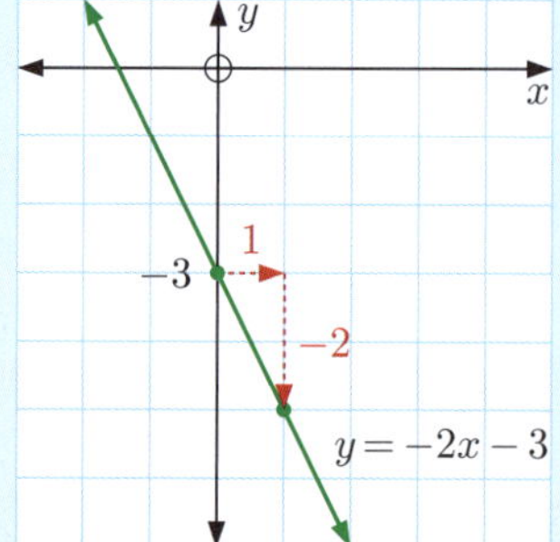

EXERCISE 14E

1 Draw the graph of:

   a $y = x + 3$ **b** $y = -x + 4$ **c** $y = 2x + 2$

   d $y = -3x - 2$ **e** $y = \frac{1}{2}x - 1$ **f** $y = \frac{2}{3}x + 4$

   g $y = 3x$ **h** $y = -\frac{1}{2}x$ **i** $y = -2x + 1$

   j $y = 3 - \frac{1}{3}x$ **k** $y = \frac{3}{4}x - 2$ **l** $y = -\frac{1}{4}x - 3$

2 Draw the graphs of each line on the same set of axes. Hence find the coordinates of any point where the lines intersect.

   a $y = x + 1, \quad y = -x + 3$ **b** $y = 2x + 1, \quad y = -\frac{1}{2}x - 4$

   c $y = 1, \quad y = \frac{1}{3}x - 2$ **d** $y = -3x, \quad y = -\frac{3}{2}x - 3$

   e $y = -2x + 4, \quad y = 2 - 2x$ **f** $x = 3, \quad y = \frac{2}{3}x + 2$

F   THE x-INTERCEPT OF A LINE

The **x-intercept** of a line is the x-coordinate of the point where the line cuts the x-axis.

Together, the x and y-intercepts are called the **axes intercepts** of the line.

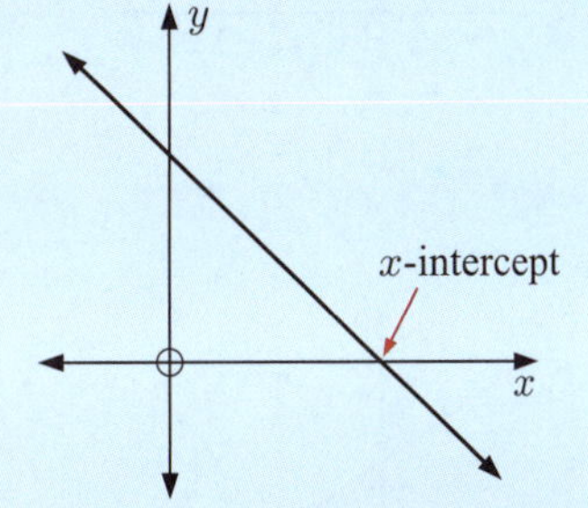

Example 13   ◀) Self Tutor

Find the x and y-intercept of the given line:

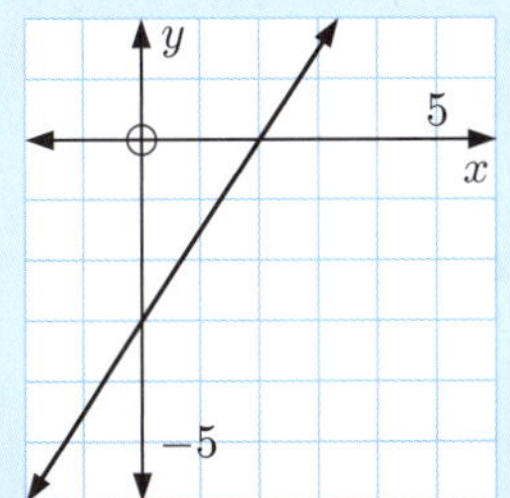

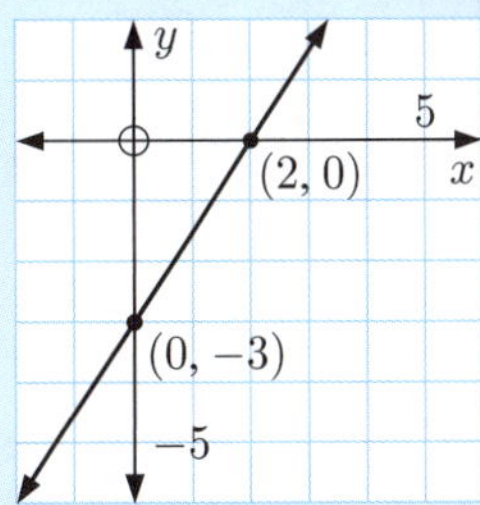

The line cuts the x-axis at $(2, 0)$, so the x-intercept is 2.

The line cuts the y-axis at $(0, -3)$, so the y-intercept is -3.

Since the x-axis is the line $y = 0$, we can find the x-intercept of a line by substituting $y = 0$.

Example 14   🔊 **Self Tutor**

Find the x-intercept of the line with equation $y = 4x - 3$.

Substituting $y = 0$, we find $4x - 3 = 0$

$$\therefore \ 4x - 3 + 3 = 0 + 3 \quad \text{\{adding 3 to both sides\}}$$
$$\therefore \ 4x = 3$$
$$\therefore \ \frac{4x}{4} = \frac{3}{4} \quad \text{\{dividing both sides by 4\}}$$
$$\therefore \ x = \frac{3}{4}$$

So, the x-intercept of the line is $\frac{3}{4}$.

EXERCISE 14F

1 Find the x and y-intercept of each line:

a 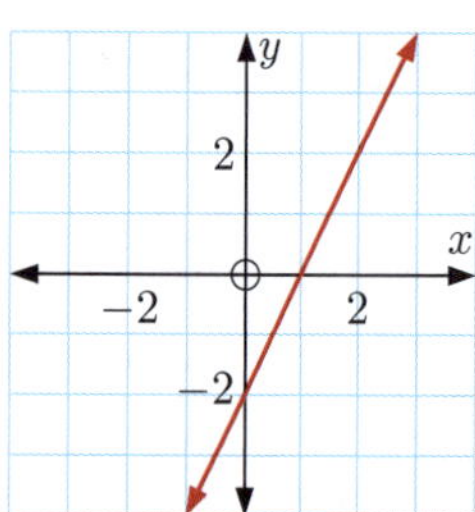

b 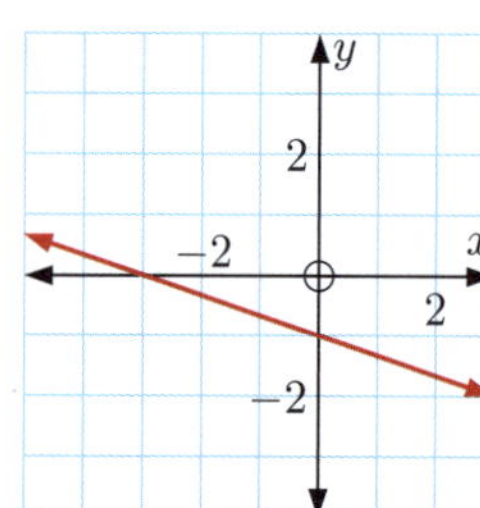

c 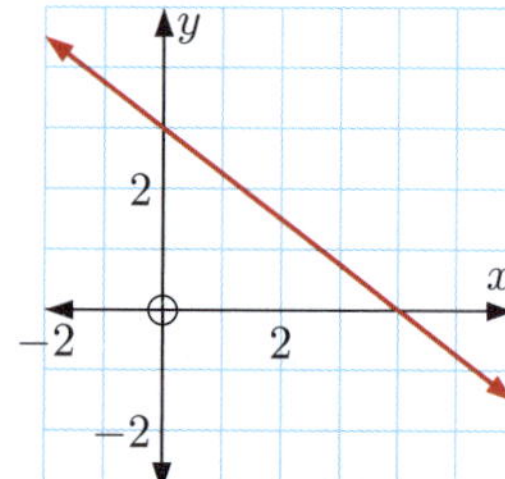

d 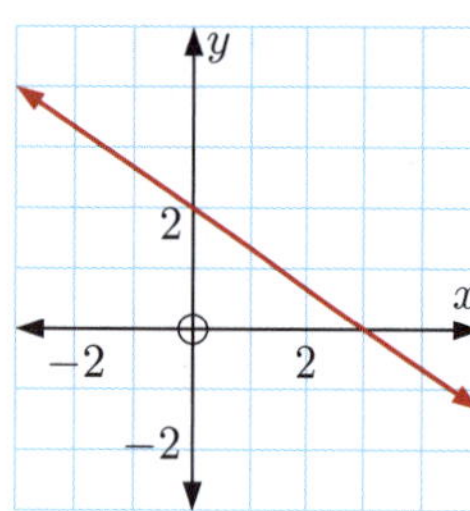

e 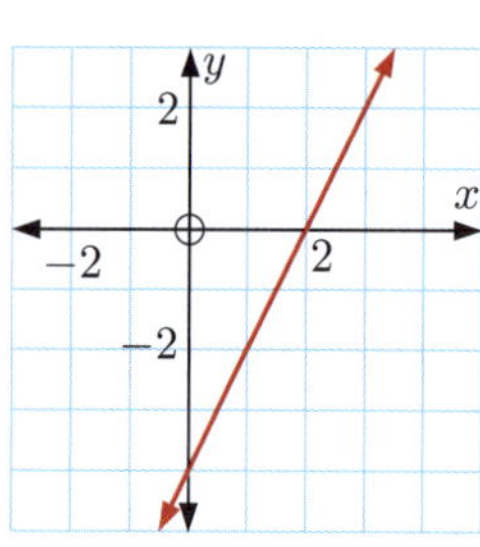

f 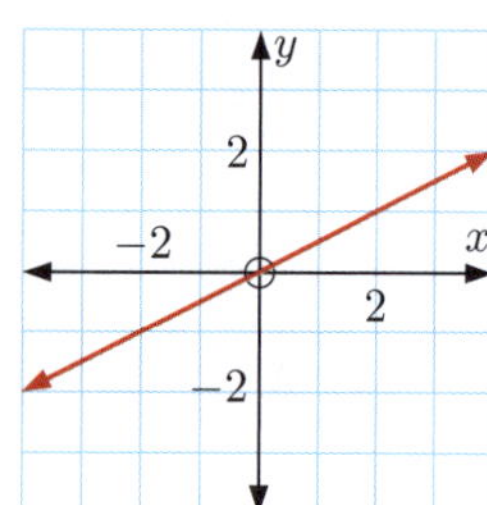

2 a Use the x and y-intercept of each line to find its gradient.

i 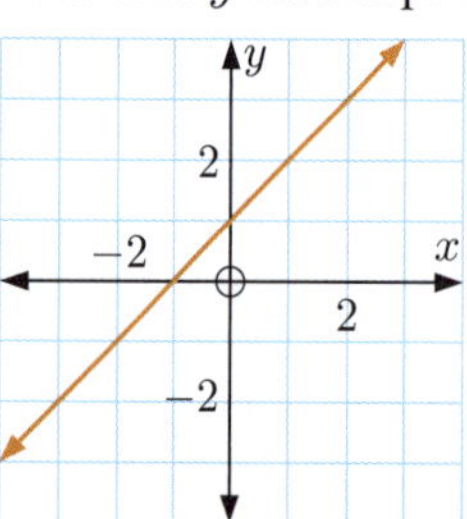

ii 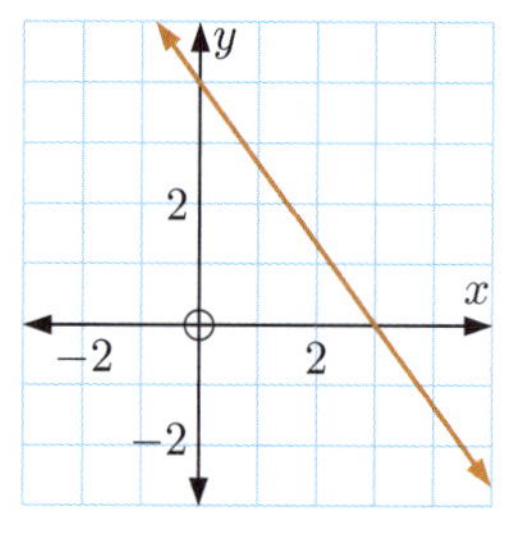

iii 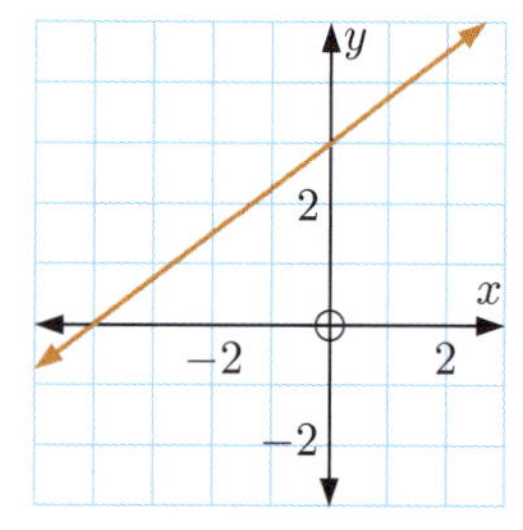

b Under what circumstance can we *not* determine the gradient of a line from the axes intercepts? Discuss your answer.

3 Find the x-intercept of the line with equation:

   a $y = 2x - 6$ **b** $y = 3x + 12$ **c** $y = -4x - 1$

   d $y = \dfrac{1}{2}x + 4$ **e** $y = -\dfrac{1}{3}x + 3$ **f** $y = 12 - \dfrac{2}{3}x$

4   **a** Find the x-intercept of the line with equation $y = mx + c$.

      b What assumption have you made in **a**?

      c Under what circumstances does the line with equation $y = mx + c$ have:

         i no x-intercept   **ii** infinitely many x-intercepts?

G   GRAPHING A LINE FROM ITS AXES INTERCEPTS

If the axes intercepts of a line are non-zero, we can use them to draw its graph.

Example 15   ◀) Self Tutor

Draw the graph of the line with x-intercept -2 and y-intercept -3.
Find the gradient of the line.

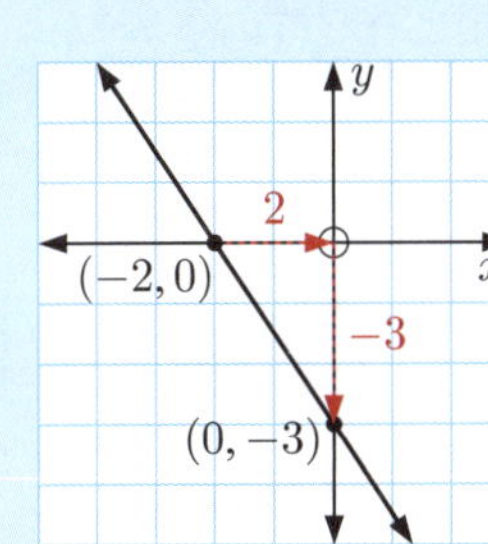

The line passes through $(-2, 0)$ and $(0, -3)$.

Its gradient is $\dfrac{-3}{2}$ or $-\dfrac{3}{2}$.

EXERCISE 14G

1 Draw the graph of the line with:

   a x-intercept 4 and y-intercept 3   **b** x-intercept -3 and y-intercept 5

   c x-intercept -2 and y-intercept -2 **d** x-intercept 1 and y-intercept -4.

2 Draw the graph of the line with these axes intercepts, and determine the line's gradient:

   a x-intercept -1 and y-intercept 2   **b** x-intercept 2 and y-intercept 6

   c x-intercept 3 and y-intercept -5 **d** x-intercept -4 and y-intercept -2

H   FINDING THE EQUATION FROM THE GRAPH OF A LINE

Given the graph of a line, we can use its y-intercept c and one other point on the line to find its gradient m.

We can then write down the equation of the line in the gradient-intercept form $y = mx + c$.

Example 16

For the graph alongside, find:

a the y-intercept

b the gradient of the line

c the equation of the line.

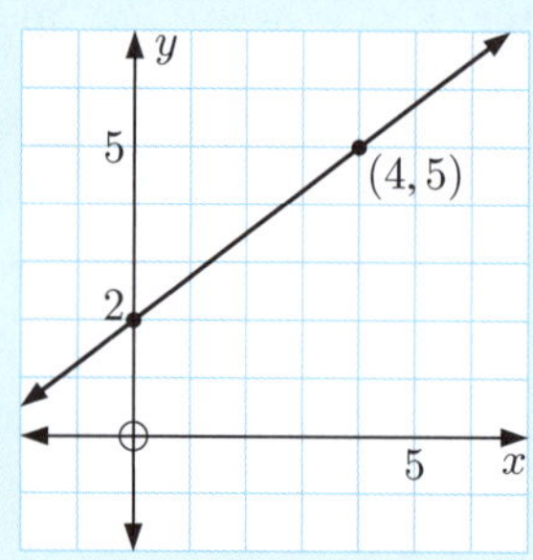

---

a The y-intercept $c = 2$.

b The gradient $m = \dfrac{\text{vertical step}}{\text{horizontal step}}$

$\qquad = \dfrac{3}{4}$

c The equation of the line is $y = \dfrac{3}{4}x + 2$.

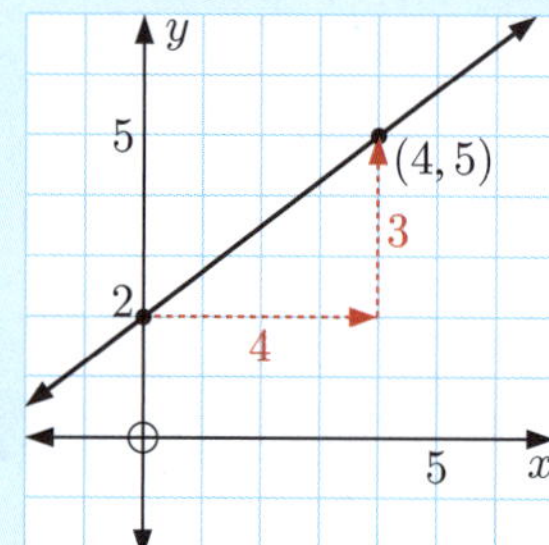

EXERCISE 14H

1 For each line, find:

i the y-intercept   **ii** the gradient   **iii** the equation of the line.

a
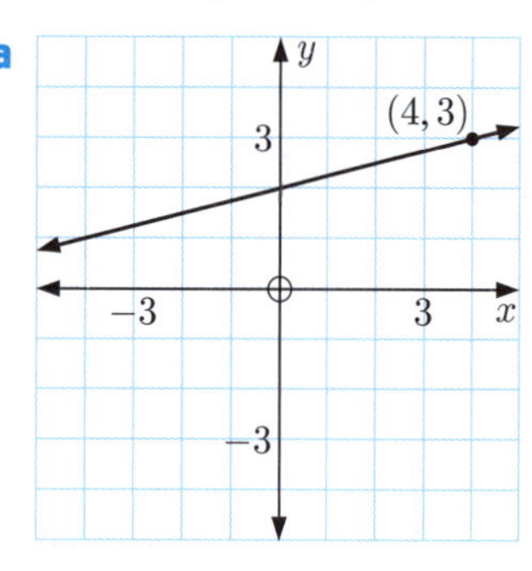

b
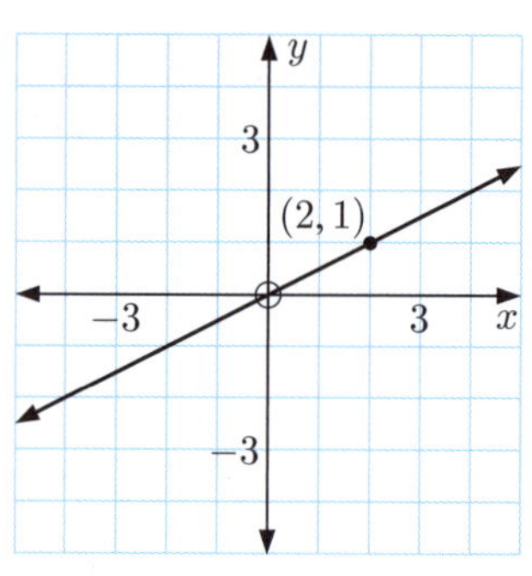

c
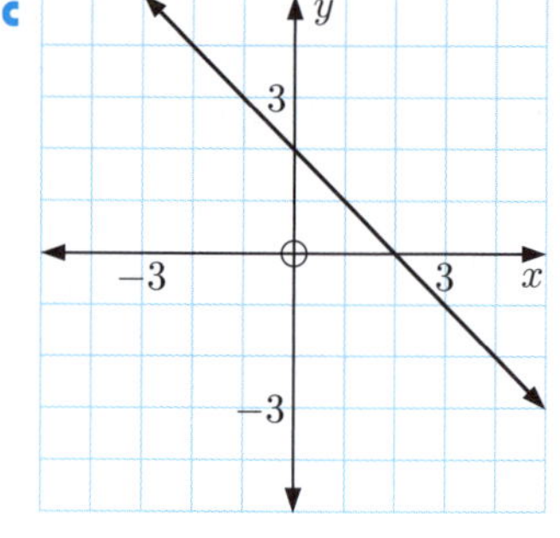

d
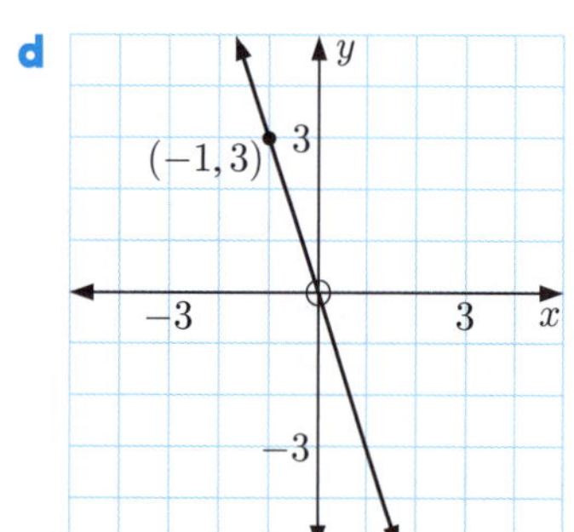

e
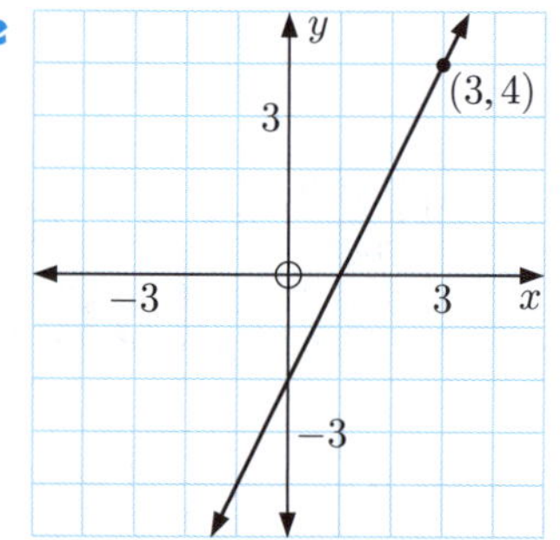

f
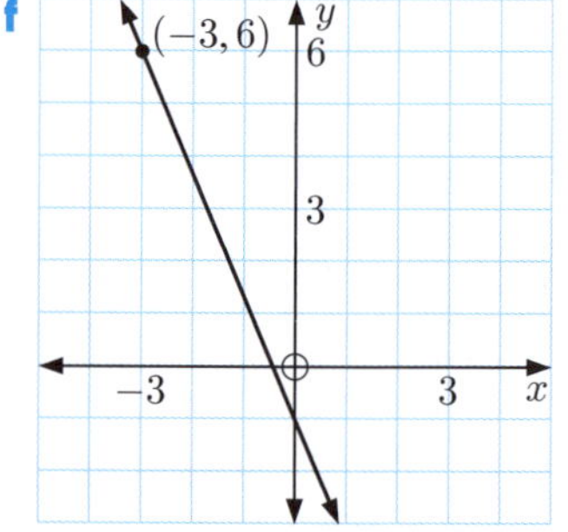

2 Look at each graph carefully. Describe the line and hence write down its equation:

a   **b** 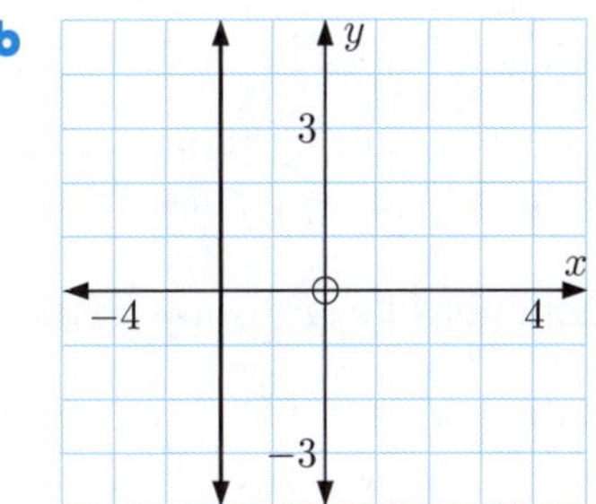

3 Find the equation of each line:

a 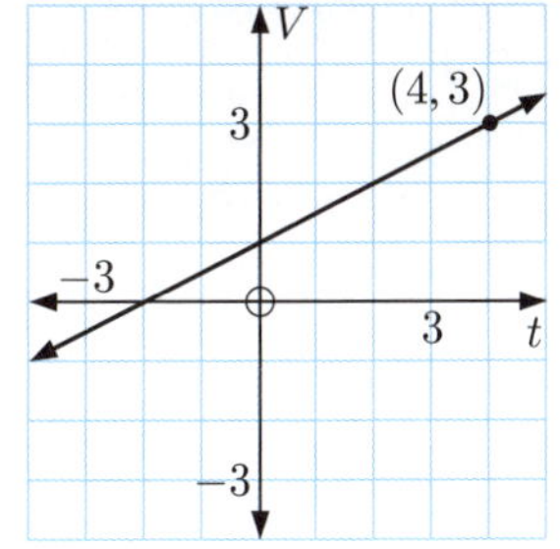   **b** 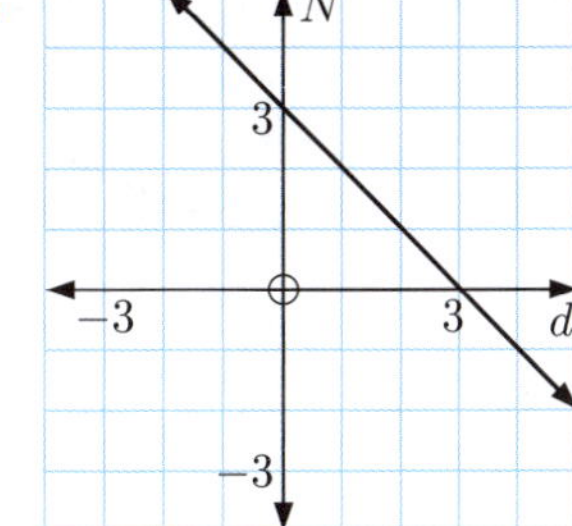   **c** 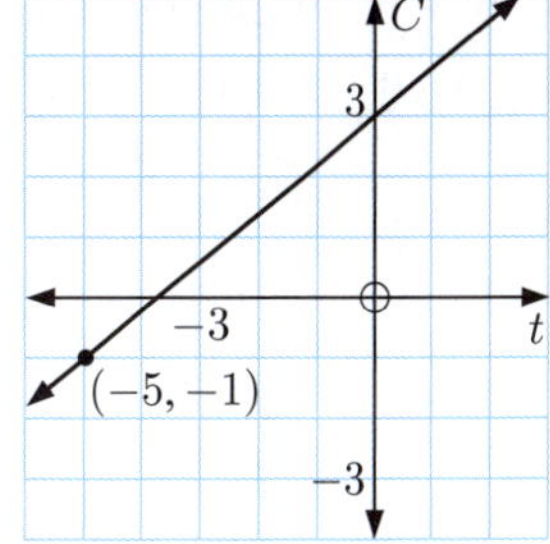

MULTIPLE CHOICE QUIZ

REVIEW SET 14A

1 **a** State the coordinates of the points A, B, C, D, and E.

   b State the quadrant in which we find the point D.

   c State the point which lies on the x-axis.

2 Lucy trains on the treadmill at a gym. Each day she records the time she spent running. Her results are shown in the table alongside.

Day (x)	1	2	3	4	5
Time (y min)	2	5	8	11	14

   a Plot these points on the same set of axes.

   b Do the points lie in a straight line?

   c Assuming this pattern continues, predict how long Lucy will spend running on:

   i day 7   **ii** day 8.

3 On a set of axes, show all the points which have equal x and y-coordinates.

4 For the equation $y = 4 - 3x$:

    a Construct a table of values using $x = -3, -2, -1, 0, 1, 2, 3$.

    b Draw the graph of the straight line.

5 On the same set of axes, draw graphs of the lines with equations $x = -2\frac{1}{2}$ and $y = -4$.

6 Consider the line $y = 6 - 4x$.

    a Find the y-coordinate of the point on the line with x-coordinate 3.

    b Find the x-coordinate of the point on the line with y-coordinate 4.

7 Find the gradient of each line:

  a 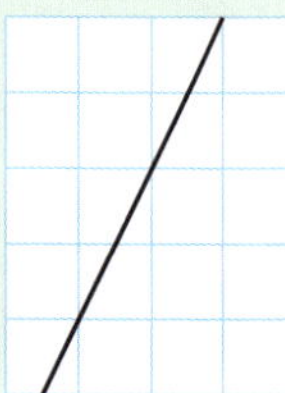   **b** 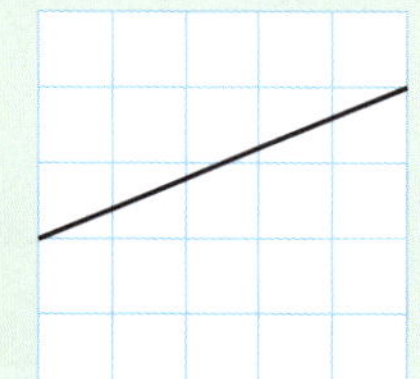   **c** 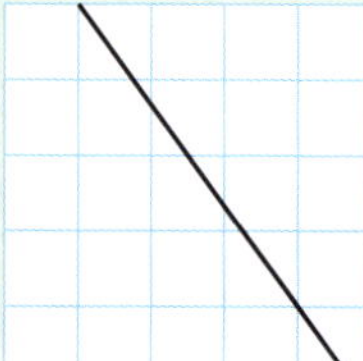

8 Draw a line through the point:

    a $(3, 5)$ with gradient $\frac{1}{4}$ **b** $(-2, 3)$ with gradient -2.

9 Find the x and y-intercept of each line.

  a 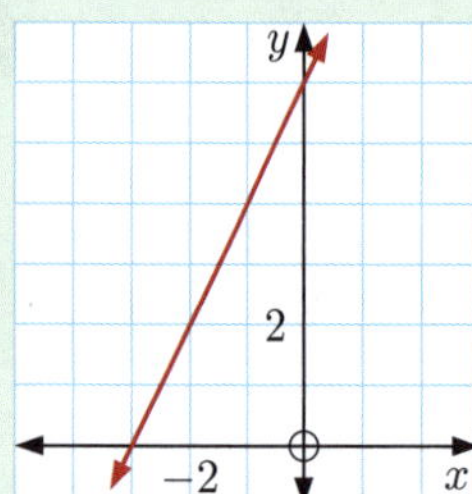   **b** 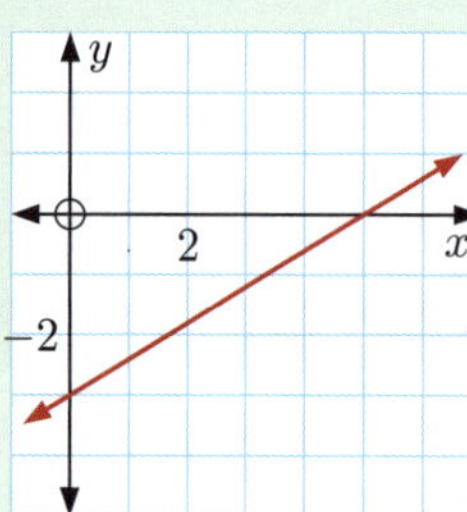   **c** 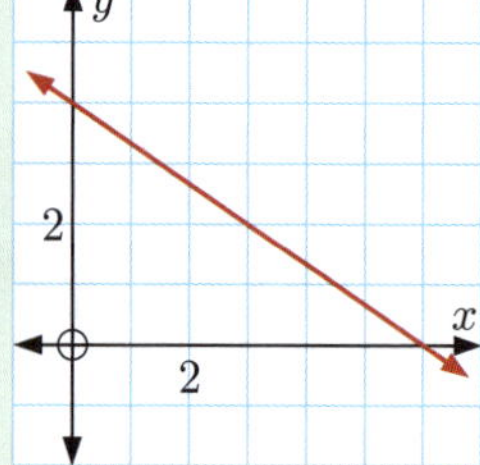

10 Write down the y-intercept of the line with equation:

    a $y = 5x - 2$ **b** $y = x + 6$ **c** $y = 1 - \frac{1}{3}x$

11 Consider the points $A(-5, 2)$ and $B(5, -3)$.

    a Name the quadrant in which each point lies.

    b Plot A and B on a set of axes, and draw the line L which passes through them.

    c Find the gradient of line L.

    d Use your graph to state the x-intercept of L.

12 Draw the graph of:

    a $y = 2x - 3$ **b** $y = \frac{1}{3}x + 1$ **c** $y = 8 - 3x$

13 Draw the graph of the line with:

    a x-intercept 5 and y-intercept 2   **b** x-intercept -1 and y-intercept -3.

14 For each line, find:

    i the y-intercept   **ii** the gradient   **iii** the equation of the line.

a 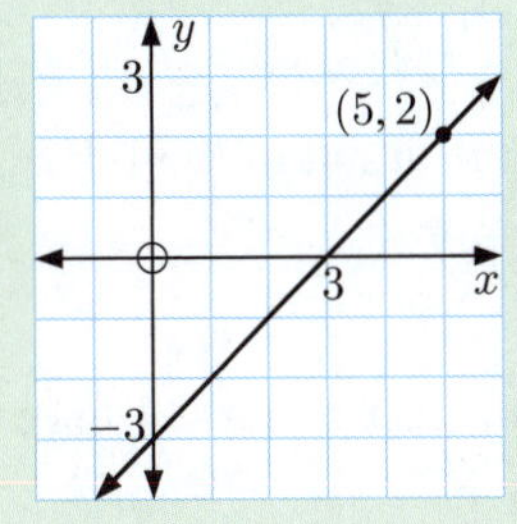   **b** 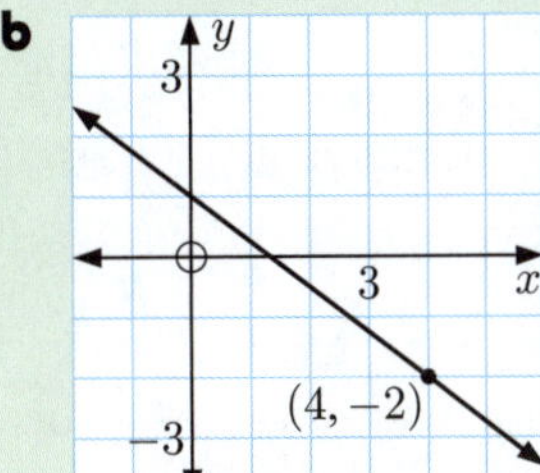   **c** 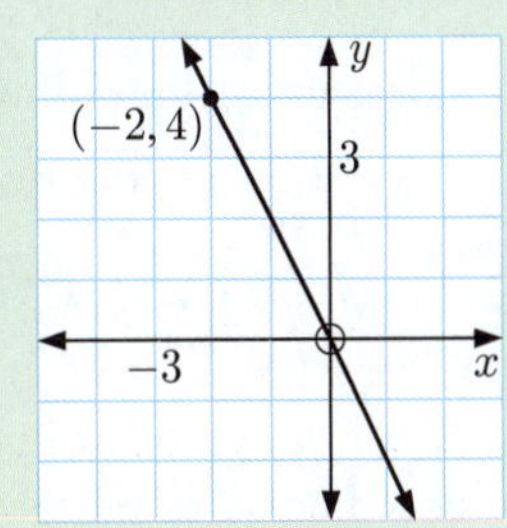

REVIEW SET 14B

1 Plot the points P(2, 5), Q(3, −3), R(−4, 0), and S(−2, −4) on a set of axes.

2 Find, by inspection, the equation of the straight line passing through these points:

x	0	1	2	3	4	5
y	5	4	3	2	1	0

3 Consider the equation $y = -\dfrac{5}{2}x + 4$.

    a Construct a table of values using $x = -3, -2, -1, 0, 1, 2, 3$.

    b Draw the graph of the straight line.

4 Graph the lines $x = -2$ and $y = 6$ on the same set of axes. Label the point where the lines intersect.

5 Determine whether the point $(-2, 5)$ lies on the line with equation $y = -2x + 1$.

6 On grid paper, draw a line with gradient:

    a $\dfrac{2}{5}$ **b** $-\dfrac{4}{3}$ **c** 0.

7 By plotting the points on grid paper, find the gradient of the line segment joining:

    a $(2, 1)$ and $(8, 13)$ **b** $(2, -3)$ and $(-7, 12)$.

8 For the graph alongside, find:

    a the x-intercept

    b the y-intercept

    c the gradient of the line.

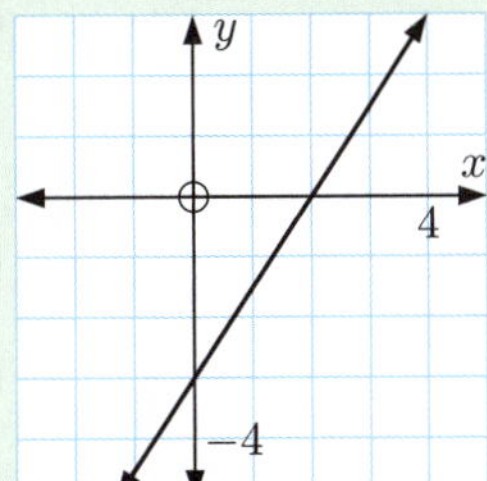

9 Consider the line with equation $y = \dfrac{1}{2}x + 2$.

    a Find the points on the line with x-coordinates 2 and 4.

    b Plot the points on a number grid and hence draw the line.

    c Use the points to find the gradient of the line.

10 Find the x-intercept of the line with equation:

    a $y = 3x - 15$ **b** $y = -2x - 7$ **c** $y = \frac{1}{4}x - 3$

11 For each of the following lines:

    i State the gradient and y-intercept.   **ii** Draw the graph of the line.

    a $y = -\frac{1}{3}x + 2$ **b** $y = \frac{3}{4}x - 3$

12 Draw the graphs of $y = 3x - 2$ and $y = x + 4$ on the same set of axes. Hence find the coordinates of the point where the lines intersect.

13 Find the equation of each line:

    a

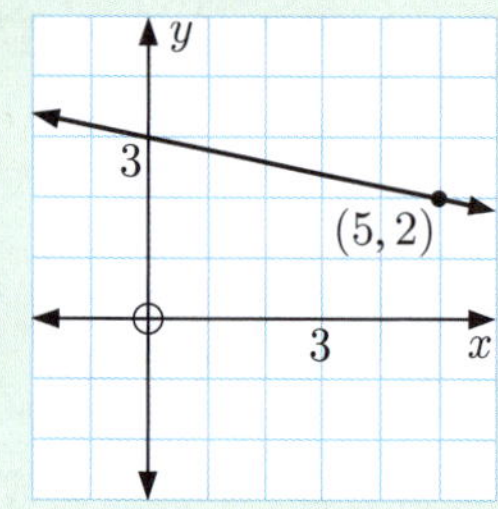

    b

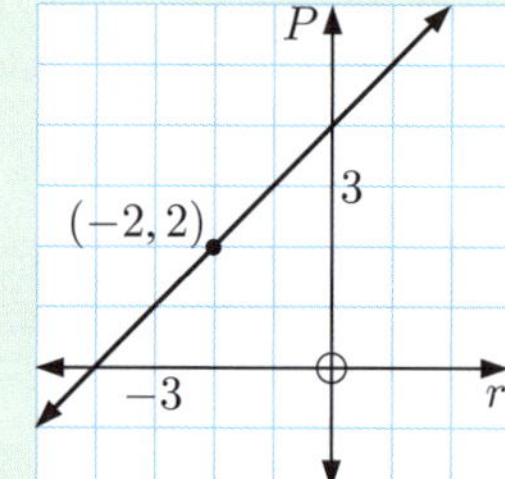

14 **a** Draw the graph of the line with these axes intercepts, and determine the line's gradient:

      i x-intercept 2 and y-intercept 3   **ii** x-intercept -4 and y-intercept 1

      iii x-intercept -3 and y-intercept -4 **iv** x-intercept 1 and y-intercept -2

    b Suppose a straight line has x-intercept a and y-intercept b, where $a, b \neq 0$. Write, in terms of a and b, the equation of the line.

Chapter 15

Ratio

Contents:

OPENING PROBLEM

Andrew wanted to make banana bread. Looking in his pantry, he discovered that he had run out of self-raising flour. His friend Marie told him he could make self-raising flour by mixing plain flour and baking powder "in the ratio $20 : 1$".

Things to think about:

a What does this ratio *mean*?

b Will the mixture contain more plain flour or baking powder?

c If Andrew uses 5 g of baking powder, how much plain flour should he use?

d How much baking powder does Andrew need to make 420 g of self-raising flour?

A   RATIO

In the **Opening Problem**, Andrew is mixing two ingredients together. The ingredients are both measured by their mass, so we say they are quantities of the *same kind*.

> A **ratio** is an ordered comparison of quantities of the **same kind**.

Ned wants to paint his fence green. He is told to mix "two parts blue paint with one part yellow paint".

We say that the ratio of blue paint to yellow paint is $2 : 1$ or "2 is to 1".

The actual quantities of blue paint and yellow paint used are not important. As long as the paints are combined in the correct *ratio*, Ned will get the correct shade of green paint.

Example 1	◀) Self Tutor

Write as a ratio: 9 mm is to 2 cm.

$9 \text{ mm is to } 2 \text{ cm} = 9 \text{ mm} : 20 \text{ mm}$
$= 9 : 20$

Ratios may involve more than two quantities.

For example, to make concrete a builder mixes 1 bucket of cement, 2 buckets of sand, and 4 buckets of gravel.

The ratio of cement to sand to gravel is $1 : 2 : 4$, which we read as "1 is to 2 is to 4".

EXERCISE 15A

1 Write as a ratio:

   a 6 mL is to 11 mL   **b** 5 kg is to 2 kg   **c** 12 m is to 25 m

   d 165 cm is to 152 cm   **e** $1 is to $3 is to $8   **f** 2 kg is to 5 kg is to 9 kg

2 Find the ratio of:

   a gold coins to silver coins   **b** pens to pencils

   c acute angles to obtuse angles   **d** knives to forks to spoons.

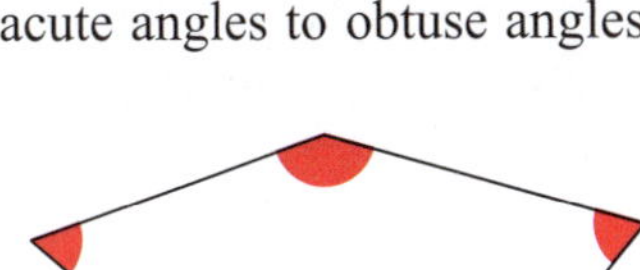

3 Find the ratio of vowels to consonants in the English alphabet.

4 Write as a ratio:

   a 93 cm is to 1 m   **b** 3 years is to 11 months

   c 14 mm is to 3.1 cm   **d** 4 hours is to 80 minutes

   e 23 seconds is to 2 minutes   **f** 225 kg is to 0.75 tonnes

5 Write as a ratio:

   a 1.6 kg is to 1807 g is to 2.12 kg   **b** 860 m : 1.28 km : 0.94 km

   c 105 seconds is to 2 minutes is to 134 seconds

Example 2   ◀)) **Self Tutor**

Write as a ratio:

   There are 5 adults for every 2 children at the restaurant.

   adults : children = 5 : 2

---

6 Write as a ratio:

   a A bakery sells 2 pastries for every 7 pies.

   b My lacrosse team wins 3 games for each game they lose.

   c There are 5 cats for every 4 dogs at the pound.

   d For every 2 hours of use, my laptop requires 37 minutes of recharging.

   e For every $10 Marcia spends on food, she spends $7 on clothes and $3 on books.

   f For every kilogram of tomatoes, add 250 g of onions and 400 g of carrots.

B   EQUAL RATIOS

An amusement park has three ticket options:

- an individual ticket gives 1 entrance pass and 4 ride passes
- a couples ticket gives 2 entrance passes and 8 ride passes
- a family ticket gives 4 entrance passes and 16 ride passes.

Notice that with every ticket type, for every entrance pass there are 4 ride passes.

The three ratios $1:4$, $2:8$, and $4:16$ are therefore **equal**.

We can write $1:4 = 2:8 = 4:16$.

Notice that we can:

- double each part of $2:8$ to obtain $4:16$

$$2:8 = 4:16$$

- halve each part of $2:8$ to obtain $1:4$

$$2:8 = 1:4$$

If we multiply or divide all parts of a ratio by the same non-zero number, we obtain an **equal ratio**.

Example 3	◀) Self Tutor

Write a ratio that is equal to $7:10$ by:

  a multiplying both parts by 3   **b** dividing both parts by 10.

a $\quad 7:10 = 7 \times 3 : 10 \times 3$
$\qquad\quad = 21:30$

b $\quad 7:10 = 7 \div 10 : 10 \div 10$
$\qquad\quad = 0.7:1$

EXERCISE 15B

1 Write a ratio equal to $6:15$ by:

  a multiplying both parts by 2   **b** dividing both parts by 3.

2 Write a ratio equal to $2:5$ by:

  a multiplying both parts by 4   **b** dividing both parts by 5.

3 Write a ratio equal to $2:3:8$ by:

  a multiplying all parts by 7   **b** dividing all parts by 4.

4 Explain why the ratios $4:5$ and $12:15$ are equal.

5 Explain why the ratios $12:9$ and $6:3$ are *not* equal.

6 Decide whether the ratios in each pair are equal:

    a $2:4$ and $8:16$ **b** $12:30$ and $3:10$

    c $48:42$ and $7:6$ **d** $72:40$ and $8:5$

    e $5:8$ and $1:\dfrac{4}{5}$ **f** $0.2:0.3$ and $4:6$

    g $3:6:7$ and $12:24:28$ **h** $2:3:5$ and $10:15:30$

7 Find a ratio equal to $5:7$, whose parts sum to 60.

8 The two flags below are both correct flags of the Czech Republic.

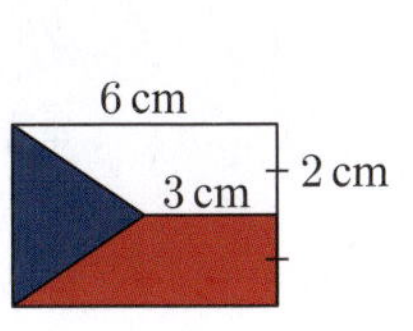

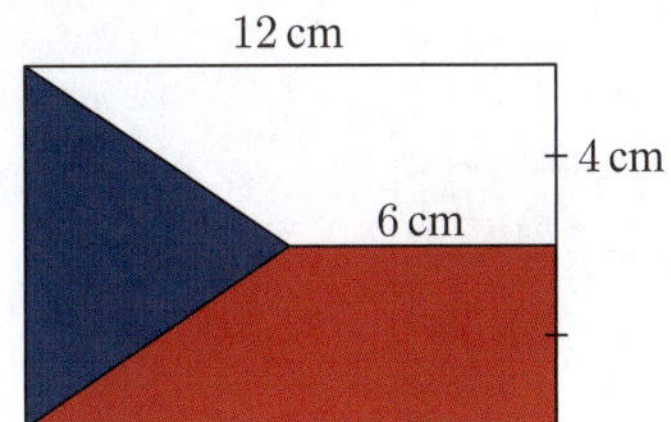

    a Write down the ratio height : width for each flag.
Show that these ratios are equal.

    b Write down the ratio of areas blue : white : red for each flag.
Show that these ratios are equal.

9 Suppose the side lengths of a square are doubled.

    a Write down the ratio new side length : old side length.

    b Write down the ratio new area : old area.

    c Are the ratios you have found equal? Explain your answer.

C   LOWEST TERMS

A ratio is in **lowest terms** or **simplest form** when it is written in terms of whole numbers with no common factors.

To write a ratio containing whole numbers in lowest terms, we divide the parts of the ratio by their **highest common factor**.

Example 4	◆) Self Tutor

Write in lowest terms:

    a $8:12$ **b** $18:48$

  a $8:12$

      $= 8 \div 4 : 12 \div 4$ {HCF of 8 and 12 is 4}

      $= 2:3$

  b $18:48$

      $= 18 \div 6 : 48 \div 6$ {HCF of 18 and 48 is 6}

      $= 3:8$

We are now able to define equal ratios more specifically:

Two ratios are **equal** if they have the same lowest terms.

EXERCISE 15C

1 Write in lowest terms:

a $3 : 12$ **b** $16 : 8$ **c** $5 : 30$ **d** $6 : 9$

e $35 : 15$ **f** $24 : 40$ **g** $63 : 49$ **h** $80 : 90$

i $600 : 800$ **j** $250 : 120$ **k** $132 : 84$ **l** $91 : 169$

2 Write in lowest terms:

a $6 : 8 : 14$ **b** $15 : 40 : 55$ **c** $42 : 12 : 21$

3 

A chessboard with all the pieces set up is shown alongside.

king   queen   rook   bishop   knight   pawn

Find, in lowest terms, the ratio of:

a pawns to kings   **b** pawns to rooks

c bishops to knights   **d** bishops to queens

e black pieces to white pieces.

Example 5 🔊 **Self Tutor**

Write in lowest terms:

a $\dfrac{2}{5} : \dfrac{3}{5}$ **b** $1\dfrac{1}{2} : 1\dfrac{2}{3}$

a $\dfrac{2}{5} : \dfrac{3}{5}$

$= \dfrac{2}{5} \times 5 : \dfrac{3}{5} \times 5$

$= 2 : 3$

b $1\dfrac{1}{2} : 1\dfrac{2}{3}$

$= \dfrac{3}{2} : \dfrac{5}{3}$ {convert to improper fractions}

$= \dfrac{3}{2} \times 6 : \dfrac{5}{3} \times 6$ {LCM of 2 and 3 is 6}

$= 9 : 10$

4 Write in lowest terms:

a $\dfrac{2}{3} : \dfrac{5}{3}$ **b** $\dfrac{5}{2} : \dfrac{1}{2}$ **c** $1 : \dfrac{1}{4}$

d $2 : \dfrac{2}{7}$ **e** $\dfrac{1}{8} : 3$ **f** $1\dfrac{1}{5} : 2\dfrac{1}{5}$

g $4\dfrac{1}{2} : 9$ **h** $\dfrac{5}{8} : 1\dfrac{1}{8}$ **i** $\dfrac{7}{3} : 3$

5 Write in lowest terms:

a $\dfrac{1}{4} : \dfrac{1}{2}$

b $\dfrac{1}{3} : \dfrac{1}{4}$

c $\dfrac{2}{3} : \dfrac{3}{2}$

d $1\dfrac{1}{4} : \dfrac{1}{3}$

e $\dfrac{2}{5} : 1\dfrac{1}{2}$

f $1\dfrac{3}{4} : 2\dfrac{1}{3}$

Example 6 ◀)) **Self Tutor**

Write in lowest terms:

a $0.3 : 1.7$

b $0.05 : 0.15$

a $\quad 0.3 : 1.7$
$= 0.3 \times 10 : 1.7 \times 10$
$= 3 : 17$

b $\quad 0.05 : 0.15$
$= 0.05 \times 100 : 0.15 \times 100$
$= 5 : 15$
$= 5 \div 5 : 15 \div 5 \quad$ {HCF of 5 and 15 is 5}
$= 1 : 3$

---

6 Write in lowest terms:

a $0.1 : 0.3$

b $0.4 : 0.7$

c $1.1 : 0.9$

d $0.4 : 0.8$

e $1.6 : 0.4$

f $1.2 : 2$

g $0.03 : 0.27$

h $0.08 : 0.24$

i $0.81 : 0.6$

7 Write as a ratio in lowest terms:

a 800 m is to 500 m

b 250 g is to 300 g

c \$3.20 is to \$2.40

d 1 L is to 600 mL

e 90 cm is to 2.1 m

f 600 kg is to 1.8 tonnes

g 45 minutes is to 4 hours

h 1.4 cm^2 is to 40 mm^2

i 3.6 kg is to 900 g

j 80 min is to 2 h 20 min

k 750 mL is to $2\dfrac{1}{2}$ L

l 6000 m^2 is to 2.8 ha

8 By writing each ratio in lowest terms, determine whether the ratios in each pair are equal:

a $6 : 8$ and $20 : 15$

b $6 : 16$ and $9 : 30$

c $\dfrac{3}{7} : \dfrac{5}{7}$ and $15 : 25$

d $50 : 125$ and $0.2 : 0.5$

e $\dfrac{1}{3} : \dfrac{1}{6}$ and $3 : 6$

f $0.35 : 0.6$ and $\dfrac{7}{6} : 2$

g $10 : 12 : 16$ and $25 : 30 : 40$

h $24 : 6 : 18$ and $36 : 12 : 48$

DISCUSSION

The ratio of the circumference of a circle to its diameter is $\pi : 1$.

Can this ratio be written in "lowest terms"?

D   PROPORTIONS

A **proportion** is a statement that two ratios are equal.

For example, the statement $6 : 15 = 12 : 30$ is a proportion.

Example 7

Find $\square$ to complete the proportion $3 : 5 = 18 : \square$.

$$3 : 5 = 18 : \square$$
$$\therefore \quad 3 : 5 = 18 : \square \quad (\times 6)$$
$$\therefore \quad \square = 5 \times 6 = 30$$

EXERCISE 15D

1   Find the unknown value $\square$ in each proportion:

a $3 : 4 = 6 : \square$

b $3 : 6 = 12 : \square$

c $2 : 5 = 4 : \square$

d $5 : 8 = \square : 40$

e $1 : 3 = \square : 27$

f $4 : 1 = 24 : \square$

g $30 : 18 = 5 : \square$

h $56 : 21 = 8 : \square$

i $36 : 12 = \square : 1$

j $3 : 4 = 1\frac{1}{2} : \square$

k $0.5 : 0.9 = \square : 9$

l $1 : 6 = \square : 1$

m $20 : \square = 4 : 3$

n $2 : \square = 6 : 9$

o $24 : \square = 8 : 11$

p $\square : 30 = 2 : 5$

q $\square : 12 = 8 : 3$

r $\square : 2\frac{1}{2} = 9 : 5$

2   Find the missing numbers $\square$ and $\triangle$ in each proportion:

a $5 : 6 : 7 = 10 : \square : \triangle$

b $2 : 5 : 9 = \square : 15 : \triangle$

c $7 : 3 : 8 = \square : \triangle : 40$

Example 8

The ratio of prefects to students on a grade 5 excursion is $3 : 10$.

12 prefects are going on the excursion. How many grade 5 students are going with them?

$$\text{prefects : students} = 3 : 10$$
$$\therefore \quad 3 : 10 = 12 : \square \quad (\times 4)$$
$$\therefore \quad \square = 10 \times 4 = 40$$

So, 40 grade 5 students are going on the excursion.

3 A group of people auditioning for a quiz show consists of men and women in the ratio $11 : 2$.

   a If there are 33 men auditioning, find the number of women.

   b If there are 12 women auditioning, find the number of men.

4 A street stall sells chicken kebabs and beef kebabs in the ratio $3 : 8$. If 96 beef kebabs were sold, how many chicken kebabs were sold?

5 Marion is training for a duathlon. Her time spent running and cycling is split in the ratio $3 : 2$. If she spends 10 hours cycling in one week, how long does she spend running?

6 Pedro makes drinks for his soccer teammates by mixing cordial and water in the ratio $2 : 9$. If he has 0.4 litres of cordial, how much water should he add?

7 In a biscuit recipe, flour and sugar are combined in the ratio $5 : 2$. If $\frac{1}{2}$ cup of sugar is used, how much flour should be added?

8 Sean has a choice of four different sizes to print out his photographs. The width : length ratio is the same in each case. Copy and complete the table showing the dimensions of each photograph size.

Size	Width	Length
small	6 cm	8 cm
regular	9 cm	
medium		16 cm
large	24 cm	

9 Rachel is baking a chocolate cake. She uses flour, sugar, and cocoa in the ratio $9 : 7 : 4$. If she uses 270 g of flour, what mass will she use of:

   a sugar   **b** cocoa?

10 Sue invested money in stocks, shares, and property in the ratio $6 : 4 : 5$. If she invested $36 000 in property, how much did she invest in the other two areas?

E   USING RATIOS TO DIVIDE QUANTITIES

Quantities can be divided in a particular ratio by considering the total number of parts the whole quantity is to be divided into.

Example 9   ◀)) **Self Tutor**

35 lollies are to be divided between Michelle and Wade in the ratio $3 : 4$. How many lollies will each person receive?

The ratio contains $3 + 4 = 7$ parts in total.

$\therefore$ Michelle receives $\frac{3}{7}$ of 35   and   Wade receives $\frac{4}{7}$ of 35

$$= \frac{3}{7} \times 35 \qquad\qquad\qquad = \frac{4}{7} \times 35$$

$$= 15 \text{ lollies} \qquad\qquad\quad = 20 \text{ lollies}$$

EXERCISE 15E

1 20 balloons are to be divided between Celena and Drew in the ratio 2 : 3.

    a What fraction of the balloons will be given to:

        i Celena   **ii** Drew?

    b How many balloons will be given to:

        i Celena   **ii** Drew?

2 At a pro-am tournament involving 60 golfers, the ratio of professionals to amateurs is 5 : 7. How many amateurs are playing in the tournament?

3 Divide \$450 in the ratio 5 : 4.

4 To make a ground cover for his pet tortoise, Chris mixes soil and sand in the ratio 7 : 3. How much of each will Chris need to make:

    a 20 kg   **b** 50 kg of ground cover?

5 A recipe says to mix lemon juice to lime juice in the ratio 5 : 2.

    a If a 70 mL mixture is made, how much is lime juice?

    b If 75 mL of lemon juice is used, how much lime juice should be added?

6 Ben and Alice share the cost of a laptop computer in the ratio 5 : 8.

    a If the laptop costs \$2600, how much will each person pay?

    b If Ben pays \$450, how much will Alice pay?

    c If Alice pays \$600, how much does the laptop cost in total?

Example 10   ◄⑴ **Self Tutor**

Clay pavers are made using clay, sand, and cement in the ratio 2 : 4 : 1.

I wish to make 14 tonnes of clay pavers. How much clay, sand, and cement must I purchase?

The ratio contains $2 + 4 + 1 = 7$ parts in total.

$\therefore$ I need $\dfrac{2}{7} \times 14$ tonnes $= 4$ tonnes of clay,

$\dfrac{4}{7} \times 14$ tonnes $= 8$ tonnes of sand,

and $\dfrac{1}{7} \times 14$ tonnes $= 2$ tonnes of cement.

---

7 Kurt is making three-cheese pasta. He needs to use ricotta, mozzarella, and parmesan cheese in the ratio 6 : 3 : 1. If Kurt uses 800 g of cheese in total, how much of each type of cheese does he use?

8 An alloy is made from tin, zinc, and lead in the ratio 15 : 4 : 1. How much tin is required to make 8 tonnes of the alloy?

9 A hotel receives local, interstate, and overseas guests in the ratio $1 : 4 : 3$. Last year the hotel had 12 000 guests. How many of the guests were from:

    a interstate   **b** overseas?

10 Steve owns two sports stores. Store A has 15 baseball bats in stock, and store B has 25.

Steve receives a shipment of 50 new baseball bats. How many of these bats should Steve send to each store so that the ratio of stock in store A to store B is $3 : 2$?

11 Leon has two buckets of lemons and limes, each containing 40 pieces in total. The first bucket contains 31 lemons. If the overall ratio of lemons to limes is $5 : 3$, how many limes are in the second bucket?

PUZZLE

One full glass contains vinegar and water in the ratio $1 : 3$. Another glass with twice the capacity of the first, has vinegar and water in the ratio $1 : 4$.

If the contents of both glasses are mixed together, what is the ratio of vinegar to water?

F   SCALE DIAGRAMS

When designing a house, it would be ridiculous for an architect to draw a full-size plan.

Instead, the architect draws a smaller diagram in which all lengths have been divided by the same **scale factor**.

We see that the lengths on the diagram are *in proportion* to the actual house.

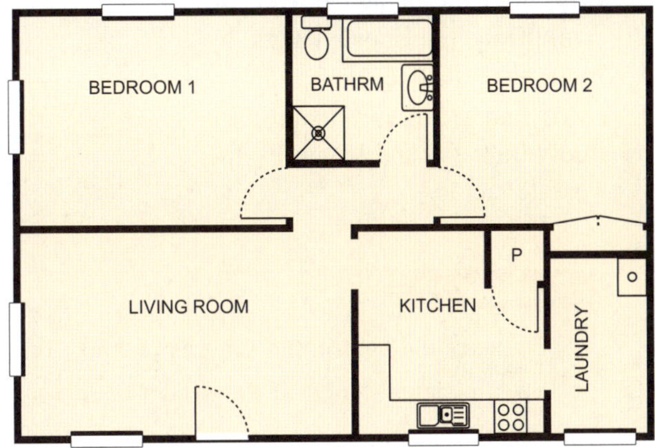

DISCUSSION

In a scale diagram, all lengths are divided by the same *scale factor*.

What happens to the angles?

To properly use a scale diagram, we need to know the **scale** used.

Scales are commonly given in several ways:

Scale: 1 cm represents 50 m.	This tells us that 1 cm on the scale diagram represents 50 m in reality.
Scale 0 50 100 150 200 250 metres	By measuring the scale, we see that 1 cm on the scale diagram represents 50 m in reality.
Scale: 1 : 5000	This ratio tells us that 1 unit on the scale diagram represents 5000 of the same units in reality. For example: • 1 cm would represent 5000 cm or 50 m • 1 mm would represent 5000 mm or 5 m.

Example 11   ◀ Self Tutor

On a scale diagram, 1 cm represents 20 m.

 a Write the scale as a ratio.   **b** State the scale factor.

a   1 cm : 20 m
   $= 1$ cm : 20×100 cm
   $= 1$ cm : 2000 cm
   $= 1 : 2000$

b The ratio simplifies to
   $1 : 2000$ so the scale
   factor is 2000.

- To find the actual length, we **multiply** the drawn length by the scale factor.
- To find the drawn length, we **divide** the actual length by the scale factor.

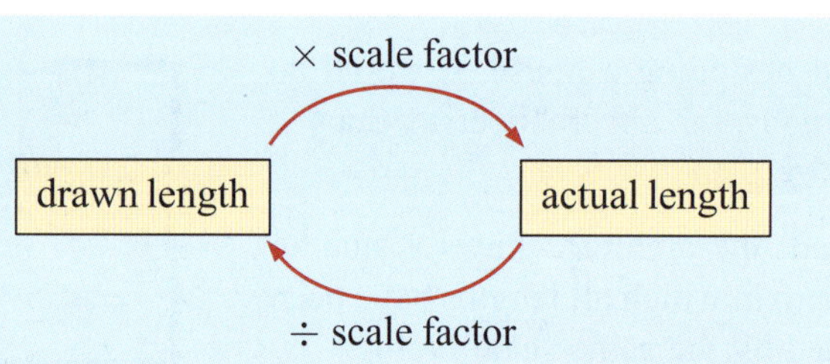

Example 12   ◀ Self Tutor

On a scale diagram, the scale is 1 : 50. Find:

 a the actual length if the drawn length is 15 cm

 b the drawn length if the actual length is 12 m.

a   actual length
   $=$ drawn length $\times$ scale factor
   $= 15$ cm $\times 50$
   $= 750$ cm
   $= 750 \div 100$ m
   $= 7.5$ m

b   drawn length
   $=$ actual length $\div$ scale factor
   $= 12$ m $\div 50$
   $= 0.24$ m
   $= 0.24 \times 100$ cm
   $= 24$ cm

EXERCISE 15F

1 Write each scale as a ratio, and state the scale factor:

   a 1 cm represents 1 m

   b 1 cm represents 1 km

   c 1 cm represents 30 m

   d 1 mm represents 2 km

   e 1 mm represents 250 m

   f 1 cm represents 200 km

2 For each scale, explain what 1 cm represents:

   a 1 : 250

   b 1 : 4000

   c 1 : 500

   d 1 : 25 000

   e 1 : 150 000

   f 1 : 22 000 000

3 Write each scale as a ratio:

   a
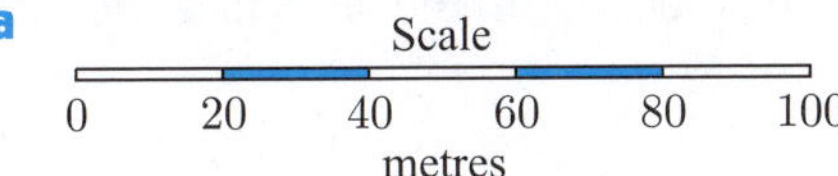

   b
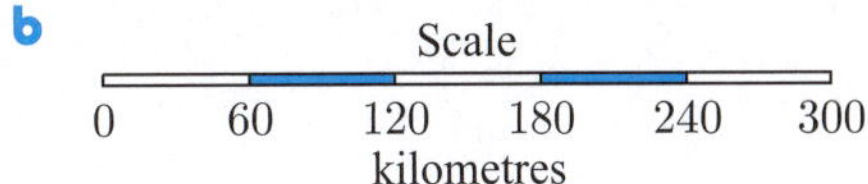

4 Consider the scale 1 : 10 000. Find the actual length if the drawn length is:

   a 3.5 mm   **b** 16 cm   **c** 5.2 cm   **d** 6.4 mm.

5 Consider the scale 1 : 20. Find the drawn length if the actual length is:

   a 10 m   **b** 2.6 m   **c** 480 cm   **d** 5.6 m.

6 Consider the scale diagram of a rectangle.

   a Find the actual dimensions of the rectangle.

   b Which of the following could the rectangle represent?

     A a bank note   **B** a domino

     C a tennis court   **D** a chopping board

Scale: 1 : 800

7
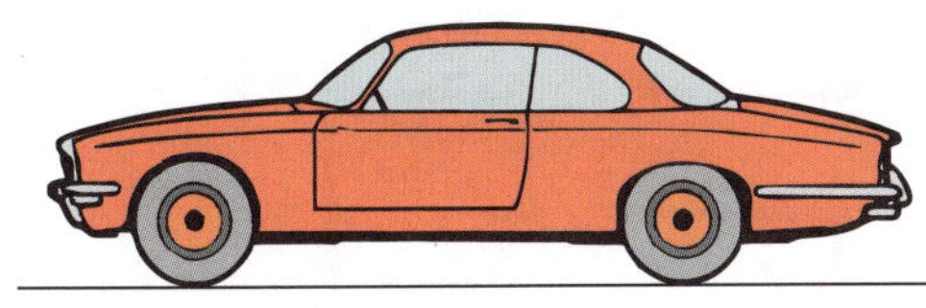

Scale: 1 : 80

Use this scale diagram to find:

   a the length of the vehicle

   b the diameter of a tyre

   c the height of the top of the vehicle above ground level

   d the width of the bottom of the door.

8 Use this scale diagram to find:

   a the total length of the ship

   b the height of the taller mast

   c the distance between the masts.

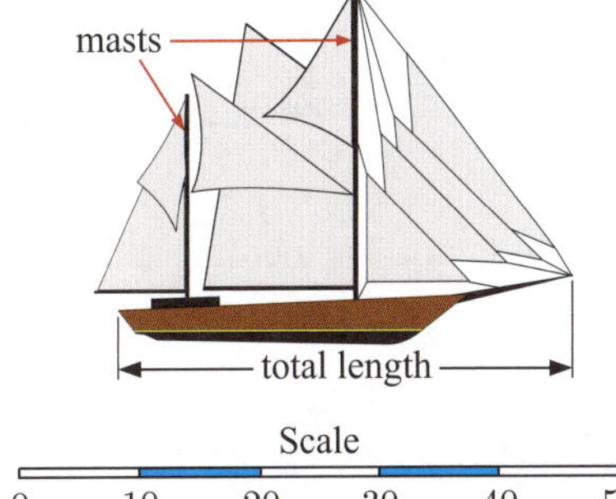

9 Draw a scale diagram of:

   a a square with sides 25 m using the scale 1 cm represents 10 m

   b a rectangle 3 km by 6 km using the scale 1 cm represents 2 km

   c a triangle with sides 10 m, 24 m, and 26 m using the scale 1 : 500

   d a circle with diameter 5 km using the scale 1 : 250 000.

10 Choose an appropriate scale and draw a scale diagram of:

   a a rectangular house block 13 m by 29 m

   b a garage door 4.6 m by 2.2 m

   c a triangular park with sides 45 m, 60 m, and 75 m.

11

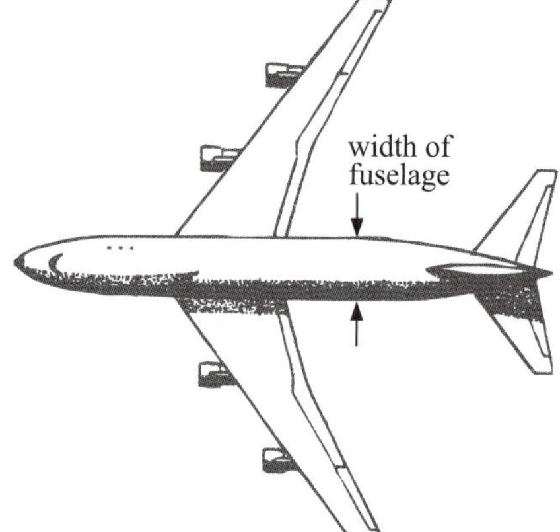

The actual length of the aeroplane in the scale diagram is 70 m.

Find:

   a the scale used in the drawing

   b the actual wingspan of the aeroplane

   c the actual width of the fuselage.

12 The actual dimensions of the verandah in this scale diagram are 3.24 m by 1.68 m.

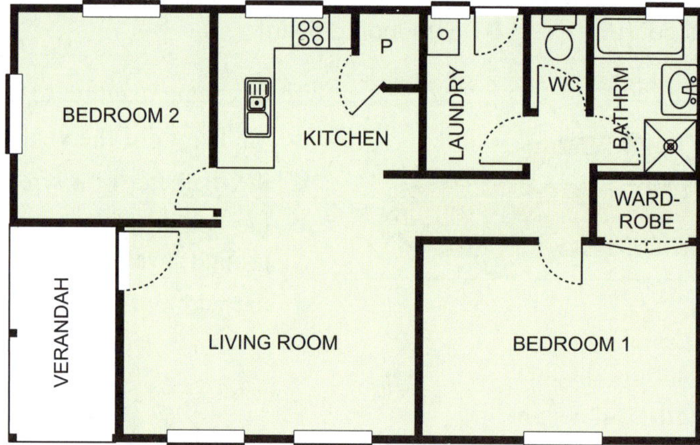

Find:

   a the scale factor for the diagram

   b the external dimensions of the house, including the verandah

   c the cost of covering the bedroom floors with wooden floorboards costing $127.50 per square metre.

13 Use this map of the USA to find the actual distance between:

    a New York and New Orleans   **b** El Paso and Miami   **c** Seattle and Denver.

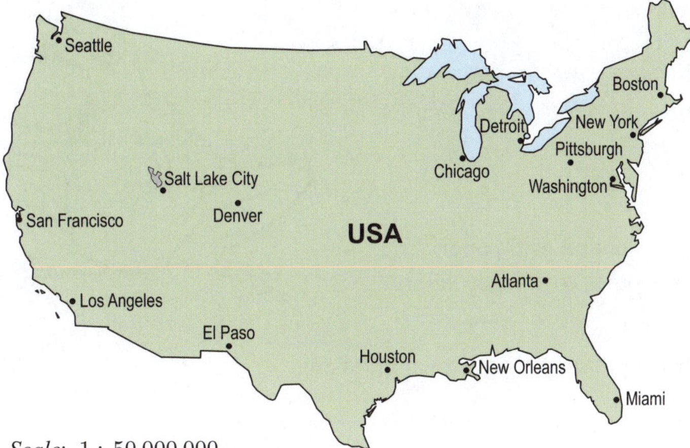

Scale: 1 : 50 000 000

14 This diagram is an enlargement of a mosquito drawn with the scale $1 : 0.25$. Find the actual lengths of the:

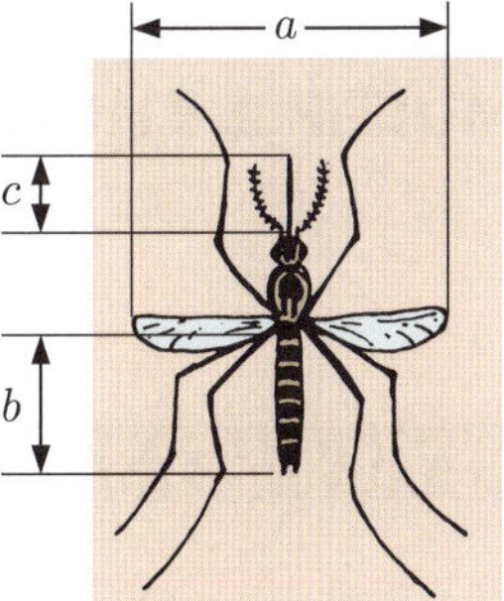

    a wingspan, a

    b abdomen length, b

    c proboscis length, c.

15 The diagram given shows a microscopic organism enlarged using the scale $8000 : 1$.

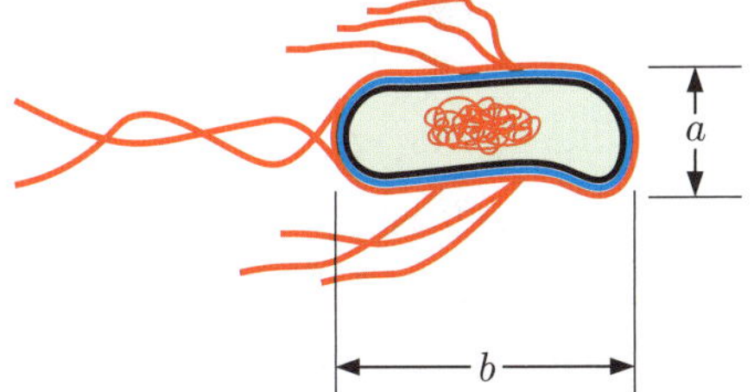

    a Find the scale factor for this diagram.

    b Find the actual length of the:

       i cell width, a **ii** cell length, b.

<table><tr><td>ACTIVITY 1</td><td align="right">HOUSE PLANS</td></tr></table>

What to do:

1   **a** Use a measuring tape or ruler to find the dimensions of the rooms in your house.

    b Choose an appropriate scale then draw a plan of your house like the one shown on page **313**. Do not forget to include the scale on your plan.

2 Look up some house plans on the internet from real estate sites.

    a Check that the length and width of each room are in the same ratio as the dimensions given.

    b Why do you think it is important that real estate plans are to scale?

ACTIVITY 2   SCALE DIAGRAMS

One way to make a scale drawing is to draw a grid over the picture to be enlarged or reduced. We then copy the picture one square at a time onto a larger or smaller grid.

What to do:

Click on the icon to obtain grid paper. You could use a photocopier to further enlarge or reduce it.

Copy the frog, or a picture of your own choosing.

MULTIPLE CHOICE QUIZ

REVIEW SET 15A

1   Write as a ratio:

   a 5 cm is to 9 cm
   b 2 kg is to 839 g

   c 141 minutes is to 3 hours is to 53 minutes.

2   Write a ratio equal to $10 : 30 : 25$ by:

   a multiplying all parts by 4
   b dividing all parts by 5.

3   Write in lowest terms:

   a $24 : 54$
   b $\frac{3}{5} : 2\frac{1}{5}$
   c $3 : 2.2$

4   By writing each ratio in lowest terms, determine whether the ratios in each pair are equal:

   a $45 : 60$ and $39 : 52$
   b $\frac{2}{5} : 4$ and $4 : 20$
   c $48 : 32$ and $0.45 : 0.3$

5   Find the unknown value $\square$ in each proportion:

   a $5 : 8 = 15 : \square$
   b $72 : \square = 9 : 4$
   c $5 : 7 = 2\frac{1}{2} : \square$

6   A choir has baritone, soprano, and alto singers in the ratio $2 : 5 : 3$. If there are 15 alto singers, find the number of:

   a baritone singers
   b soprano singers.

7   Divide \$3500 in the ratio $5 : 2$.

8   Three families share the cost of a 120 kg side of beef in the ratio $4:5:3$. How much meat should each family receive?

9   **a**   For the scale given in this scale diagram, explain what 1 cm represents.

    b   Find the actual:

      i   height of the building

      ii   width of the building.

Scale: $1:400$

10   Draw a scale diagram of a triangular block of land with side lengths 18 m, 24 m, and 30 m. Use the scale $1:500$.

11   Answer the **Opening Problem** on page **304**.

12   Housemates Izaak, Penny, and Sean share the cost of their internet bills in the ratio $2:4:3$.

    a   What fraction does each person pay?

    b   To set up the account, Izaak paid \$30. How much was paid by:

      i   Penny   **ii**   Sean?

    c   The total bill in one month was \$72. How much did each person pay?

    d   When Sean moved out of the house, Izaak and Penny agreed to pay half of Sean's share of the cost. Find, in lowest terms, the ratio in which Izaak and Penny now share the cost.

REVIEW SET 15B

1   Write as a ratio:

    a   A gardener plants 5 roses for every 3 tulips.

    b   For every hour Sara spends running, she spends 45 minutes cycling and 18 minutes swimming.

2   Decide whether the ratios in each pair are equal:

    a $\frac{1}{2}:3$ and $4:24$ **b** $4:5:8$ and $16:20:48$

3   Write in lowest terms:

    a $20:55$ **b** $0.28:0.1$ **c** $1\frac{1}{4}:\frac{2}{3}$

4   Write as a ratio in lowest terms:

    a   800 kg is to 1.4 tonnes   **b**   2.8 cm^2 is to 80 mm^2

5   Find the missing numbers $\square$ and $\triangle$ in the proportion $6:3:7 = \square:18:\triangle$.

6   In a school the ratio of students playing soccer and basketball is $8:5$. If 96 students play soccer, how many students play basketball?

7 A 26 hectare vineyard grows vines for white and red grapes in the ratio 4 : 9. How many hectares of each type of grape are planted?

8 Each year, Greg donates money to his three favourite charities in the ratio 7 : 6 : 3. If he donated a total of $320 this year, how much did each charity receive?

9 A 1 : 3000 scale model of the Sydney Harbour Bridge is being built for a museum. The actual bridge is 1149 metres long. Find the length of the model.

10 The rectangle is drawn with the scale 1 : 10.

a Find the actual dimensions of the rectangle.

b Which of the following could the rectangle represent?

A a room   **B** a stamp   **C** a paver

11 The actual length of the bus in the scale diagram is 10 m.

a Find the scale used for the diagram.

b Hence find:

i the actual height of the windows

ii the actual height of the bus.

12 Mr Brown is designing a garden. He draws the scale diagram shown.

a Find the actual dimensions of:

i the chicken pen

ii the fruit tree area.

b Find the actual area of the lawn.

c Mr Brown wants to add another vegetable bed between the two existing beds. He wants the bed to be 2 m wide by 5 m long.

i How big would this new bed be on the diagram?

ii Will this new bed fit? Explain your answer.

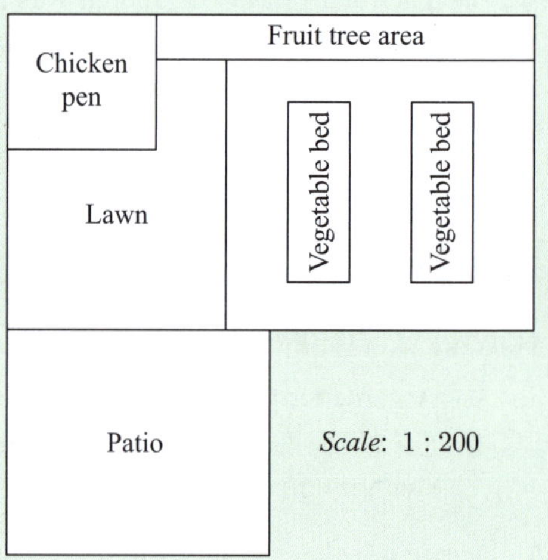

Rates and line graphs

Contents:

OPENING PROBLEM

As Lily looks out the car window, she sees that the car is exactly level with a train. Both the car and the train are travelling at constant speeds.

In the next 2 minutes, the car travels 2600 m and the train travels 3.8 km.

Things to think about:

a   What do we mean by *speed*? What does it measure?

b   Which is travelling faster, the car or the train?

c   Can you write the speed of:

    i   the car in metres per second   ii   the train in kilometres per hour?

d   Can you graph the distance travelled by the car and the train against time on a set of axes?

e   How are the speeds of the car and train observed on the graph?

A   RATES

We have seen that a **ratio** is an ordered comparison of quantities of the **same** kind.

For example, we can have a ratio of lengths or a ratio of times.

> A **rate** is an ordered comparison of quantities of **different** kinds.

For example, a person's *heart rate* is a comparison between a *number of heart beats* and the *time taken* for them to happen.

DISCUSSION

1   Think about quantities which you can *measure*, such as mass, area, and volume. Discuss scenarios where we may be interested in the rate of change of these quantities with respect to time.

2   What rates are you familiar with, which are *not* rates of change with respect to time?

When we write a rate, we do not use a ratio sign ":", but instead we divide one quantity by another.

Since we are comparing quantities of *different* kinds, units are very important. We must always include units in our answer. We use the word *per* which means "for every", or a slash "/" to separate the units.

For example, if a person's heart beats 65 times every minute, we write their heart rate as 65 beats per minute or 65 beats/minute.

ACTIVITY 1   MEASURING YOUR HEART RATE

What to do:

1 Find your pulse on your wrist or the side of your neck.

2 Count how many times you can feel your pulse in one minute. This is your heart rate in beats per minute.

3 Compare your heart rate with those of your classmates.

4 What happens to your heart rate when you exercise?

5 Find out how your heart rate can be used to measure your fitness.

Example 1   ◀)) **Self Tutor**

Three weeks ago, baby Hazel weighed 12.5 kg. She now weighs 13.1 kg. Find Hazel's rate of weight gain in grams per week.

In 3 weeks, Hazel's weight gain $= 13.1 \text{ kg} - 12.5 \text{ kg}$
$$= 0.6 \text{ kg}$$
$$= 600 \text{ g}$$

Hazel's rate of weight gain $= \dfrac{600 \text{ g}}{3 \text{ weeks}}$
$$= 200 \text{ g per week}$$

EXERCISE 16A

1 Suggest units which could be used to measure each rate:

   a a person's rate of pay   **b** an aeroplane's speed

   c the price of petrol   **d** a person's typing speed

   e the rate at which a car's temperature increases on a hot day.

2 Jennifer's heart beats 375 times in 5 minutes. Express this as a rate in beats per minute.

3 The Peterson household used 1170 megajoules of gas during April. Express this as a rate in megajoules per day.

4 In 2015, a tree was 19.5 m tall. By 2021 it had grown to 26.7 m tall. Find its rate of growth in metres per year.

5 A 2.8 kg durian costs $18.60. Express this as a rate in $/kg.

6 The mass of a baby mouse is recorded at birth and at the end of each week thereafter.

Birth	Week 1	Week 2	Week 3
1.38 g	4.67 g	8.14 g	11.09 g

   a Find the rate of change in mass over the 3 week period.

   b In which week was the growth fastest?

7 The table below shows the populations of several cities in the years 2009 and 2019:

City	2009 population	2019 population	Population growth	Population growth per year	% population growth
Sao Paulo	19 377 000	21 847 000	2 470 000	247 000	12.7%
Lagos	10 115 000	13 904 000			
Cairo	16 539 000	20 485 000			
Mexico City	19 958 000	21 672 000			
Beijing	15 685 000	20 035 000			
Manila	11 622 000	13 699 000			

  a Copy and complete the table.

  b Which city had the highest population growth per year over this period?

  c Which city had the highest *percentage* population growth rate over this period?

Example 2 ◀)) **Self Tutor**

Henry the elephant eats 240 peanuts every 3 minutes.

  a Find Henry's rate of eating peanuts.

  b How many peanuts will Henry eat in 10 minutes?

  a Henry's rate of eating peanuts $= \dfrac{240 \text{ peanuts}}{3 \text{ minutes}}$

  $= 80$ peanuts per minute

  b In 10 minutes Henry will eat $10 \times 80 = 800$ peanuts.

---

8 A family of four uses 2800 litres of water each week.

  a Find the rate of water usage in litres per day.

  b How much water will the family use in 20 days?

9 Judy works part-time at a local café. She earned $86.40 for working 4 hours last week.

  a Find Judy's rate of pay.

  b This week Judy worked 19 hours. How much will she earn this week?

10 A milk truck takes 5 minutes to discharge 6750 litres of milk.
  At this rate, how much milk would the truck discharge in 18 minutes?

11 A bus can travel 92 km on 8 litres of diesel.

  a How far could the bus travel on 28 litres of diesel?

  b How many litres of diesel are required to travel 690 km?

12 It costs $640 to buy a 32 m length of fibre optic cable.

  a Find the cost of each metre of cable.

  b Find the cost of a 27 m length of cable.

  c Find the length of cable that could be bought for $4400.

13 Leo's car uses 28 litres of petrol to travel 518 km.

    a Find the rate at which the petrol is used in:

        i km per litre   **ii** litres per 100 km.

    b Leo is driving 1480 km on vacation.

        i How many litres of petrol will he need?

        ii If petrol costs \$1.35 per litre, how much will he spend on petrol?

14 26.5 tonnes of gravel costs \$1030.85. Find the cost of 42.8 tonnes of gravel.

ACTIVITY 2   POPULATION DENSITY

Population density is a rate which compares the population of a region with the area of that region.

Click on the icon to obtain this Activity.

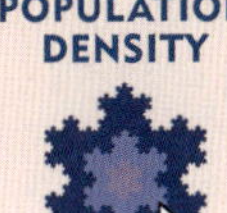

B   SPEED

The most common rate that we use is **speed**, which is a comparison between the *distance travelled* and the *time taken*.

When we go on a car journey, we do not always travel at a constant speed. We need to slow down for other cars, and stop at traffic lights.

We can therefore consider speed in two different ways.

- The **instantaneous speed** of an object is the rate at which it is travelling at a particular point in time.
  For example, this speedometer tells us the instantaneous speed of the car is 50 km per hour.

- The **average speed** of an object during its journey is given by

$$\text{average speed} = \frac{\text{total distance travelled}}{\text{total time taken}}$$

This formula can be rearranged into:

$$\text{distance} = \text{speed} \times \text{time}$$

$$\text{or}\quad \text{time} = \frac{\text{distance}}{\text{speed}}$$

You can use the triangle alongside to help you remember these. Click on the icon for a demonstration.

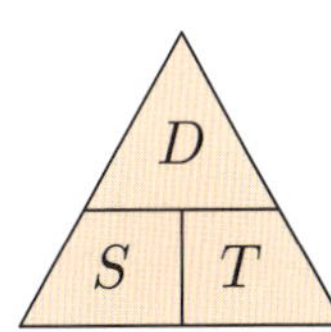

DISCUSSION

Discuss whether you would be more interested in *instantaneous speed* or *average speed* when:

- you are driving past a police officer with a radar gun
- you are planning a holiday road trip
- you are driving to your best friend's wedding
- you are driving past a school.

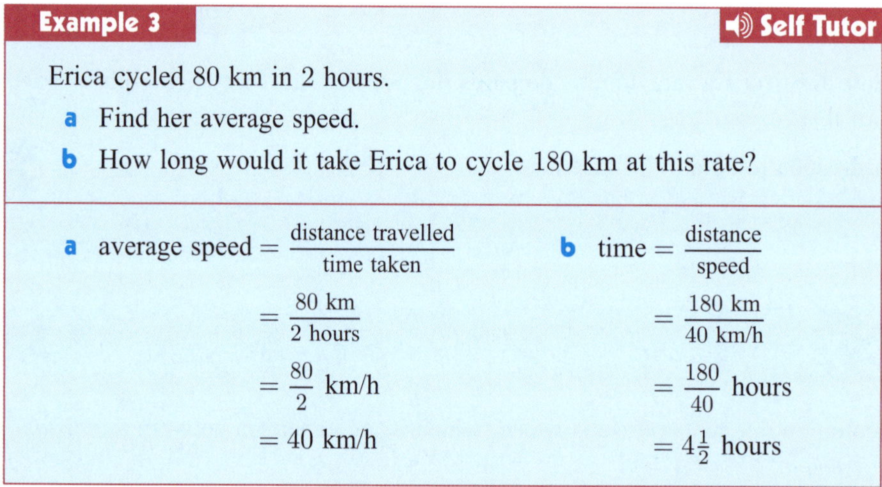

Example 3   ◀) **Self Tutor**

Erica cycled 80 km in 2 hours.

a Find her average speed.

b How long would it take Erica to cycle 180 km at this rate?

a $\text{average speed} = \dfrac{\text{distance travelled}}{\text{time taken}}$

$= \dfrac{80 \text{ km}}{2 \text{ hours}}$

$= \dfrac{80}{2} \text{ km/h}$

$= 40 \text{ km/h}$

b $\text{time} = \dfrac{\text{distance}}{\text{speed}}$

$= \dfrac{180 \text{ km}}{40 \text{ km/h}}$

$= \dfrac{180}{40} \text{ hours}$

$= 4\tfrac{1}{2} \text{ hours}$

EXERCISE 16B

1 Find, in kilometres per hour, the average speed of:

a a cyclist who rides 100 km in 4 hours

b a jet boat which travels 150 km in 5 hours

c an athlete who runs 18 km in 1.5 hours

d an aeroplane which takes 50 minutes to fly 750 km.

2 The speed limit on a freeway is 100 km/h. Jason drives 210 km along the freeway in 2 hours. Has he broken the law?

3 Bernadette drives her car at an average speed of 72 km/h.

a If Bernadette drives for 3 hours, how far does she travel?

b How long would it take Bernadette to travel 54 kilometres at this speed?

4 A model train travels around a circular track with radius 5 m. The train takes 20 seconds to complete a lap of the track. Find the average speed of the train, giving your answer in metres per second correct to 1 decimal place.

5 **a** Liam's aeroplane flew at an average speed of 210 km/h for 1 hour and 40 minutes. How far did Liam fly?

  b When Liam makes the return journey, he is now flying against the wind, and his plane averages only 175 km/h. How long does the return flight take him?

6 Yiren walks 60 metres in 22.5 seconds, while Sean walks 150 metres in 1 minute.

  a Find the average speed of each person.

  b Yiren and Sean each walk 2000 m at their normal speed. Who will finish first, and by how much?

7 A truck travels 90 km at 90 km/h, then another 90 km at 60 km/h. Find the average speed of the truck for the entire journey.

C   DENSITY

Which is heavier:
1 tonne of lead or 1 tonne of feathers?

The objects both have mass 1 tonne. However, many people guess that the lead is heavier, since any given *volume* of lead will be much heavier than the *same volume* of feathers. They have in fact compared the lead and feathers using a rate called **density**.

> The **density** of an object is its mass per unit of volume.
>
> $$\text{density} = \frac{\text{mass}}{\text{volume}}$$

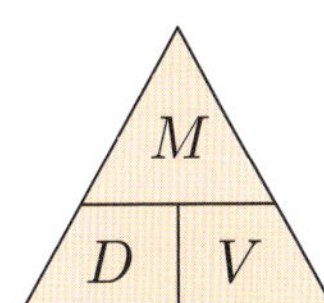

Density is usually measured in grams per cubic centimetre.

For example, the density of pure gold is 19.30 grams per cm^3. This means that every cubic centimetre of pure gold has mass 19.30 grams.

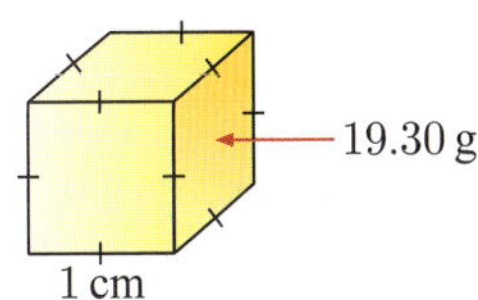

THE DENSITY OF WATER

We have previously seen that 1 mL or 1 cm^3 of pure water at 4°C weighs 1 gram.

> The density of pure water is 1 gram per cm^3.

If an object has density less than 1 gram per cm^3, then it will float on water.

If its density is greater than 1 gram per cm^3, then it will sink.

These tables list some common densities in grams per cm^3.

Material	Density
carbon dioxide	0.002
petrol	0.70
ice	0.92
water	1.00
milk	1.03

Material	Density
aluminium	2.70
iron	7.87
lead	11.35
gold	19.30
platinum	21.45

Example 4 — ◀)) Self Tutor

Find the density of a piece of timber which is 60 cm by 10 cm by 3 cm and weighs 1.62 kg.

The timber has mass $= 1.62$ kg   and volume $= 60 \times 10 \times 3$ cm^3
$\qquad\qquad\qquad = 1620$ g $\qquad\qquad\qquad\qquad = 1800$ cm^3

$$\therefore \quad \text{density} = \frac{\text{mass}}{\text{volume}}$$

$$= \frac{1620 \text{ g}}{1800 \text{ cm}^3}$$

$$= 0.9 \text{ g per cm}^3$$

EXERCISE 16C

1 Find the density of:

  a an object with mass 20 g and volume 5 cm^3

  b a metal disc which weighs 1.13 kg and has volume 50 cm^3

  c a block of ebony which is 1.1 m $\times$ 3 cm $\times$ 4 cm and weighs 1.4 kg.

2 How many times more dense is:

  a lead than water   **b** platinum than aluminium   **c** milk than petrol?

3 A pair of cubic dice and their weights are shown alongside.
Which die is more dense?

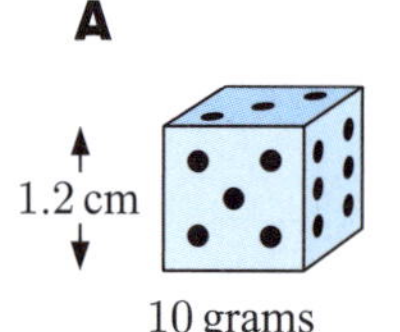

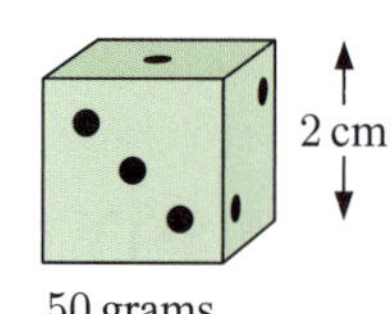

4

Petrol and water are *immiscible* which means they do not mix. If the two liquids are poured into a container, they will separate into two layers. Which is the upper layer? Explain your answer.

5 The doorstop shown weighs 200 grams. If it is dropped into water, will it sink or float? Explain your answer.

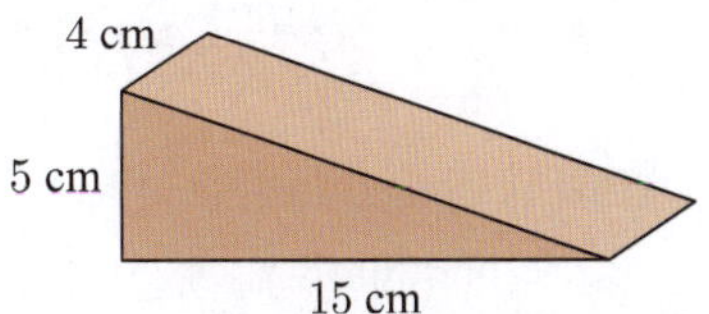

6 A cube of solid silver has side length 2 cm and mass 84 g.

   a Find the density of solid silver.

   b Find the mass of a 5 cm by 6 cm by 10 cm block of silver.

7 A cylindrical iron bar has radius 2 cm and length 1.2 m. Find the mass of the bar in kilograms.

8 Belinda has a square-based pyramid made of crystal with mass 18.3 g and dimensions as shown. Calculate the density of the crystal.

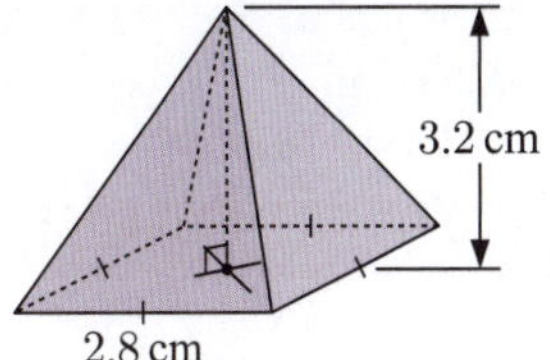

9 Calculate the average density of a tennis ball with diameter 66 mm and mass 57 g.

ACTIVITY 3   WILL IT SINK OR FLOAT?

You will need: ruler, set of scales, container of water.

What to do:

1 Gather several solid waterproof objects from around your classroom or your home. The objects should have a shape that you can calculate the volume of, such as a rectangular prism or a cylinder.

2 Copy this table and list your objects in it:

Object	Prediction	Mass	Volume	Density	Result
⋮					

3 Predict whether each object will sink or float when placed in the container of water.

4 Measure the mass of each object.

5 Measure the dimensions of each object, and hence calculate its volume.

6 Calculate the density of each object. If you wish, you may now change your predictions about whether each object will sink or float.

7 Place each object in the container of water. Record the result.

DISCUSSION

- How is density affected by:
    - temperature
    - the *state* of a substance (solid, liquid, or gas)?
- What is special about the solid and liquid states of water?

D   CONVERTING RATES

It is sometimes useful to convert a rate into different units so it is easier to understand in the context of the situation.

DISCUSSION

2 metres per second is equivalent to 7.2 kilometres per hour.

Which units are more helpful if you are:

- walking 300 m to the bus stop
- hiking for 5 hours?

Example 5   ◄) Self Tutor

A petrol bowser pumps petrol at the rate of 600 L per hour. Write this rate in L per minute.

In 1 hour, the bowser pumps 600 L.

$\therefore$ in 1 minute, the bowser pumps $\dfrac{600}{60} = 10$ L.   {1 h = 60 min}

This is a rate of 10 L per minute.

SPEED CONVERSIONS

Roger rides his bicycle at $36 \text{ km/h} = \dfrac{36 \text{ km}}{1 \text{ hour}}$

$\qquad\qquad\qquad\qquad = \dfrac{36\,000 \text{ metres}}{3600 \text{ seconds}}$ {1 h = 60 min = 60 × 60 s}

$\qquad\qquad\qquad\qquad = 10$ m/s

So, travelling at 10 m/s is the same as travelling at 36 km/h.

Notice that travelling at 1 m/s is the same as travelling at 3.6 km/h.

- To convert m/s into km/h we **multiply by 3.6**.
- To convert km/h into m/s we **divide by 3.6**.

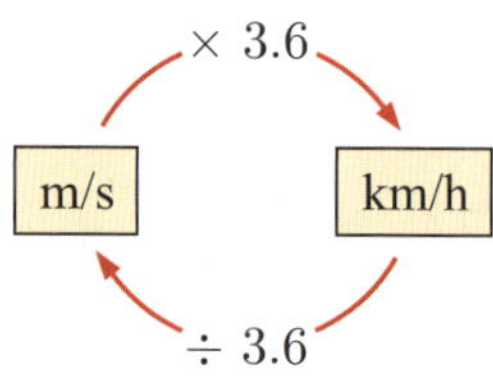

Example 6   🔊 **Self Tutor**

a A sprinter runs at 11 m/s. Convert this speed into km/h.

b An aeroplane travels at 900 km/h. Convert this speed into m/s.

a 11 m/s	**b** 900 km/h
$= 11 \times 3.6$ km/h	$= 900 \div 3.6$ m/s
$= 39.6$ km/h	$= 250$ m/s

EXERCISE 16D

1 A fire hose discharges water at the rate of 180 litres per minute. Write this rate in L per hour.

2 Kelly's heart rate is 60 beats per minute. Write her heart rate in:

    a beats per second   **b** beats per hour   **c** beats per day.

3 A shower head has a flow rate of 7.5 L per minute. Write this rate in:

    a L per hour   **b** mL per second.

4 A bamboo plant grows 18 m in 60 days. Write this growth rate in:

    a m per day   **b** m per hour   **c** mm per hour.

5 Reg eats 175 g of potato chips per day. Write this rate in:

    a grams per week   **b** kilograms per week.

6 The density of a metal is 6.8 g per cm^3. Write this density in kg per m^3.

7 A helium balloon with volume 2000 cm^3 contains 358 mg of helium.

    a Find the density of helium in mg per cm^3.

    b The density of air is approximately 0.0013 g per cm^3. Is helium more dense or less dense than air?

8 Convert into km/h:

    a 200 m/s   **b** 45 m/s   **c** 27 m/s   **d** 800 m/s

9 Convert into m/s:

    a 50 km/h   **b** 110 km/h   **c** 21 km/h   **d** 540 km/h

10 In 2009, Usain Bolt achieved a world record time of 19.19 seconds for the 200 metre sprint. Write his average speed, correct to 2 decimal places, in:

    a m/s   **b** km/h.

11 Write the following speeds in km/h:

    a A sprinter runs 100 m in 9.7 seconds.

    b A greyhound races 500 m in 29 seconds.

    c A horse gallops 2 km in 2 min 10 seconds.

    d A swimmer travels 1.5 km in 15 minutes.

    e A cheetah sprints 380 m in 12.9 seconds.

12 A hypersonic jet can travel at Mach 5, which is 5 times the speed of sound. Given that the speed of sound is approximately 340 m/s, find the time it would take the jet to fly 12 066 km from Sydney to Los Angeles. Give your answer in hours and minutes.

E   LINE GRAPHS

One way to *visualise* the relationship between two quantities is to draw a graph of one quantity against the other.

To determine which quantity to place on which axis, we consider whether the value of one variable *depends* on the value of the other.

The **dependent variable** *depends* on the **independent variable**.

We place the independent variable on the horizontal axis and the dependent variable on the vertical axis.

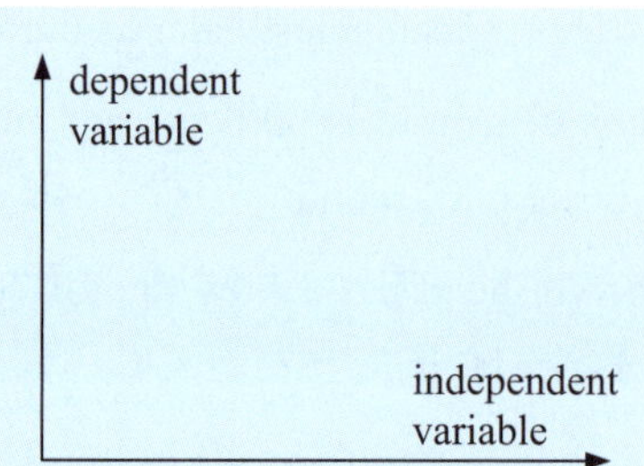

For example, a **travel graph** for a journey shows the relationship between *distance* travelled and the *time* taken. The *distance* travelled depends on the *time* taken, so *time* is on the horizontal axis and *distance* is on the vertical axis.

Example 7	◀◈ **Self Tutor**

This travel graph shows the progress of a train travelling between cities.

a How far does the train travel in the first 3 hours?

b Find the speed of the train for the first 3 hours.

c Find the speed of the train during the final hour of the journey.

d Find the average speed of the train for the entire journey.

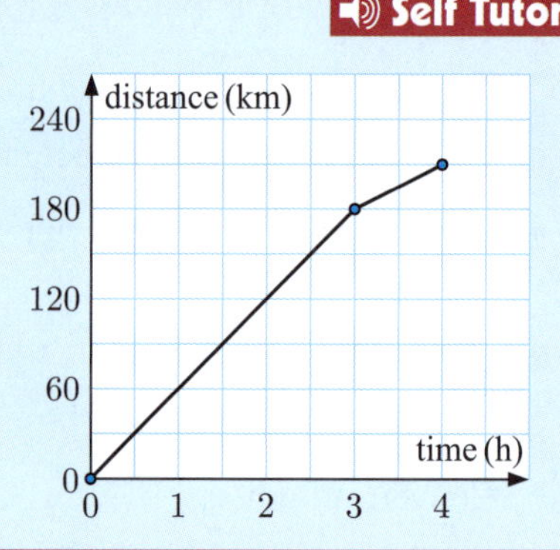

a The train travels 180 km in the first 3 hours.

b $\text{Speed} = \dfrac{\text{distance travelled}}{\text{time taken}} = \dfrac{180 \text{ km}}{3 \text{ hours}} = 60 \text{ km/h}$

c $\text{Speed} = \dfrac{\text{distance travelled}}{\text{time taken}} = \dfrac{30 \text{ km}}{1 \text{ hour}} = 30 \text{ km/h}$

d $\text{Average speed} = \dfrac{\text{total distance travelled}}{\text{total time taken}}$

$= \dfrac{210 \text{ km}}{4 \text{ hours}}$

$= 52.5 \text{ km/h}$

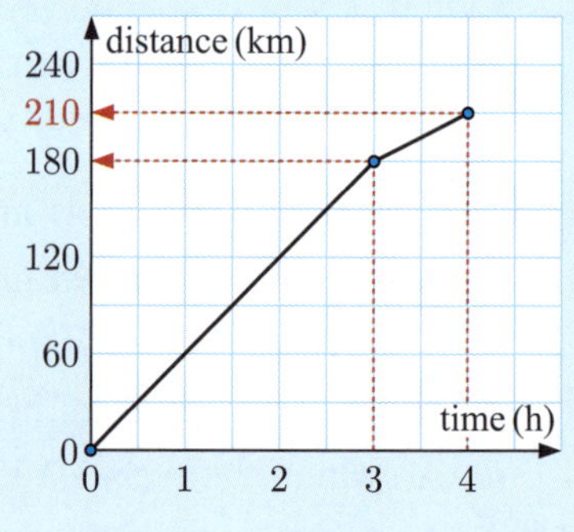

EXERCISE 16E

1 This graph shows the progress of a secretary typing a document.

a How many words did the secretary type in the first 2 minutes?

b How long did the secretary take to type 450 words?

c Find the rate at which the secretary can type.

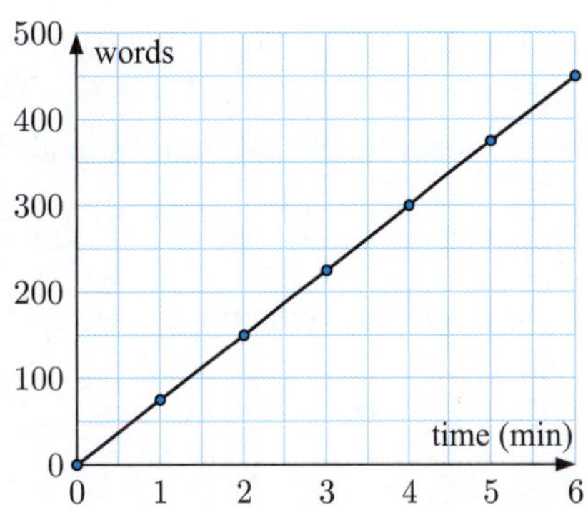

2 This travel graph shows the progress of a car travelling between towns A, B, and C.

a How far is it from A to B?

b Find the speed of the car between A and B.

c How far is it from B to C?

d Find the speed of the car between B and C.

e Find the total distance from A to C.

f Find the *average speed* of the car between A and C.

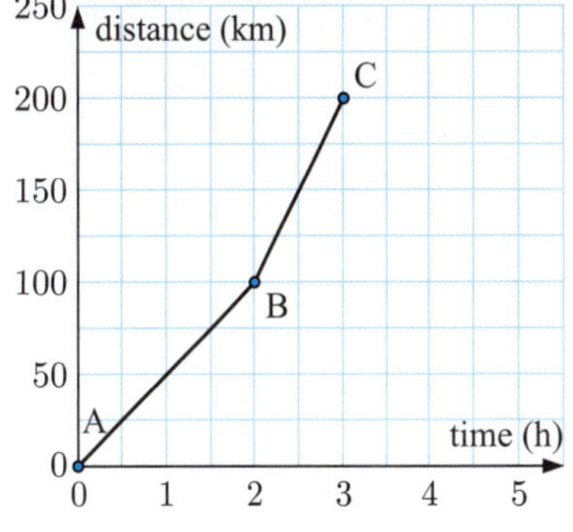

3 This graph shows how much a student earns for working at a service station.

a How much does the student receive for working 6 hours?

b Find the student's rate of pay for the first 6 hours.

c How much does the student receive for working 8 hours?

d Find the student's rate of pay for each additional hour after the first 6 hours.

e Find the student's *average* rate of pay for working 8 hours.

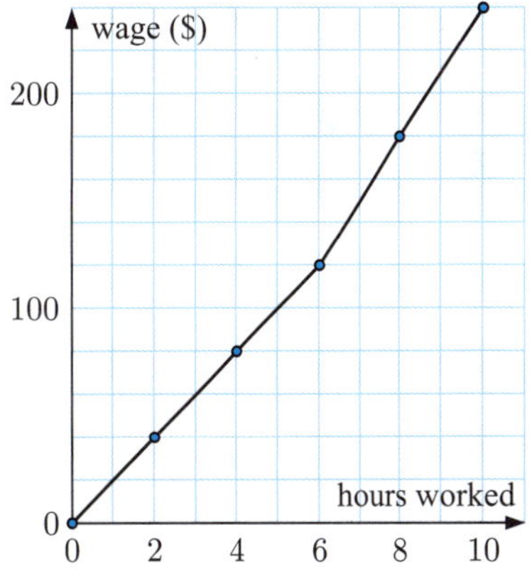

4 This travel graph shows the progress of a cyclist riding along a flat road.

a Calculate the speed of the cyclist.

b Calculate the gradient of the straight line. Comment on your answer.

c Explain the significance of the graph being a straight line.

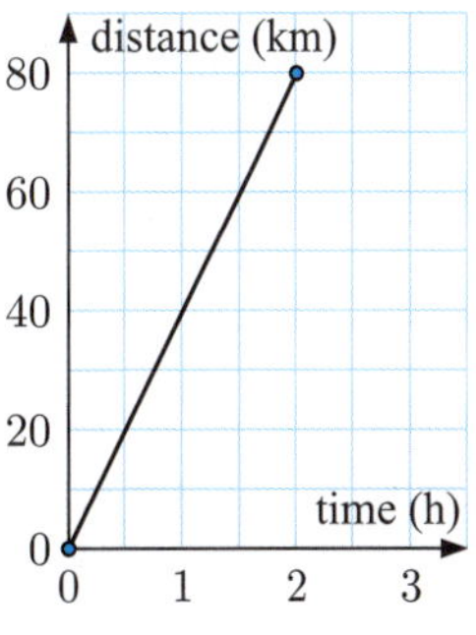

Example 8   ◄) **Self Tutor**

Max has 10 litres of fuel left in his car's petrol tank. When he fills it, petrol is pumped into the tank at 15 litres per minute. The petrol tank can hold 70 litres.

a Identify the independent and dependent variables.

b For every time increase of 1 minute, what is the change in the amount of petrol in the tank?

c Copy and complete:

Time (min)	0	1	2	3	4
Amount (litres)					

d Draw a line graph to display the relationship between the variables.

e Use your graph to find:

   i the number of litres of petrol in the tank after 1.5 minutes

   ii the time when there are 50 litres of petrol in the tank.

---

a The *amount of petrol* in the tank depends on the *time* it has been filling.

∴ *time* is the independent variable and the *amount of petrol* is the dependent variable.

b For every time increase of 1 minute, the amount of petrol in the tank increases by 15 litres.

c

Time (min)	0	1	2	3	4
Amount (litres)	10	25	40	55	70

d

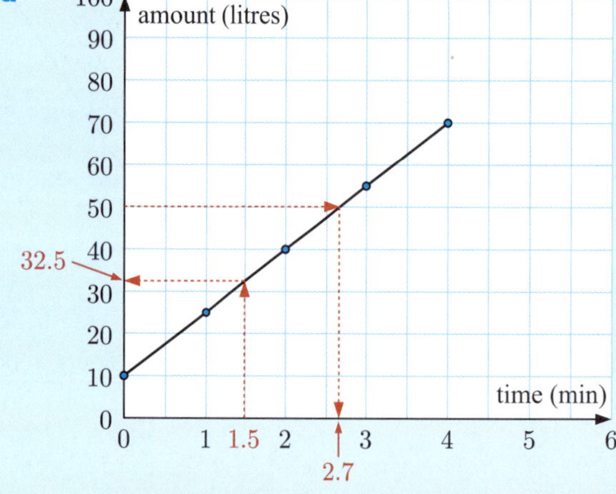

e **i** After 1.5 minutes there are 32.5 litres of petrol in the tank.

    ii There are 50 litres of petrol in the tank after about 2.7 minutes.

---

5 Identify the independent and dependent variables when considering:

a the mass of a kitten and its age in weeks

b the score you achieve for a test and the time you spent studying for it

c the amount of water given to a plant and the amount it grows.

6 Chickpeas can be bought for $6 per kilogram.

   a Copy and complete:

Weight of chickpeas (kg)	0	1	2	3	4	5
Cost ($)						

   b Identify the independent and dependent variables.

   c Draw a line graph to display the relationship between the variables.

   d Use your graph to find:

      i the cost of 1.5 kg of chickpeas

      ii the weight of chickpeas which can be bought for $15.

   e Find the gradient of the straight line. Explain your answer.

7 A cargo ship travels between two ports at a constant speed of 40 km/h.

   a Copy and complete:

Time (h)	0	1	2	3	4	5
Distance (km)						

   b Identify the independent and dependent variables.

   c Draw a travel graph to display the relationship between the variables.

   d How far will the ship travel in 2.5 hours?

   e How long will the ship take to travel 150 km?

   f Find the gradient of the straight line. Explain your answer.

8 A Formula 1 car uses 30 litres of fuel to travel 40 km.

   a Copy and complete:

Fuel used (L)	0	30	60	90
Distance (km)				

   b Identify the independent and dependent variables.

   c Draw a line graph to display the relationship between the variables.

   d Hence find the distance the car would travel on 75 litres of fuel.

   e How many litres of fuel would be needed to complete a race of 160 km?

   f Find the rate of fuel consumption in kilometres per litre.

9 When Chris took his phone off the charger, it had 80% charge. His phone lost charge at 5% per hour for the next 3 hours. Chris then used his phone to watch a movie, and his phone lost charge at 10% per hour for the next two hours.

   a Copy and complete:

Time (hours)	0	1	2	3	4	5
Charge (%)						

   b Identify the independent and dependent variables.

   c Draw a line graph to display the relationship between the variables.

   d Find the charge on the phone after 2.5 hours.

   e How long did it take for the charge to fall to 50%?

   f Find the *average* rate at which the charge decreased during the five hours.

GLOBAL CONTEXT   RENEWABLE ENERGY

Global context:	Globalisation and sustainability
Statement of inquiry:	Mathematics is an important tool for making informed decisions about global issues.
Criterion:	Applying mathematics in real-life contexts

MULTIPLE CHOICE QUIZ

REVIEW SET 16A

1 At a local market it costs \$5.10 to buy 0.6 kg of rhubarb.

  a Find the price per kilogram of the rhubarb.

  b How much would it cost to buy 2.5 kg of rhubarb?

2 Water from a tap will fill a 9 L watering can in 45 seconds.

  a Find the rate at which water flows from the tap, in litres per minute.

  b How long will it take to fill a 120 L pond?

3 A freight train travels 770 km in 8 hours, while a truck on the highway travels 120 km in 85 minutes. Which mode of transport travels faster?

4 A plane travels at 600 km/h for 1 hour, then 750 km/h for 3 hours. Find the total distance travelled by the plane.

5 Find the density of a 420 g paperweight with volume 150 cm^3.

6 A cube of iridium has side length 6 cm. Given that the density of iridium is 22.56 g/cm^3, find the mass of the cube in kilograms.

7 A cat has a resting heart rate of 150 beats per minute. Write this heart rate in:

  a beats per second   **b** beats per hour.

8 This graph shows the energy consumed by a space heater.

  a How much energy did the heater consume in the first 4 minutes?

  b How long did it take for the heater to consume 750 kJ of energy?

  c Find the rate at which the heater consumes energy.

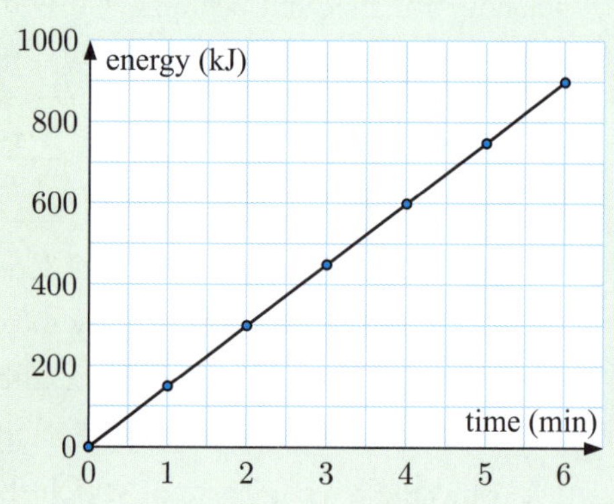

9 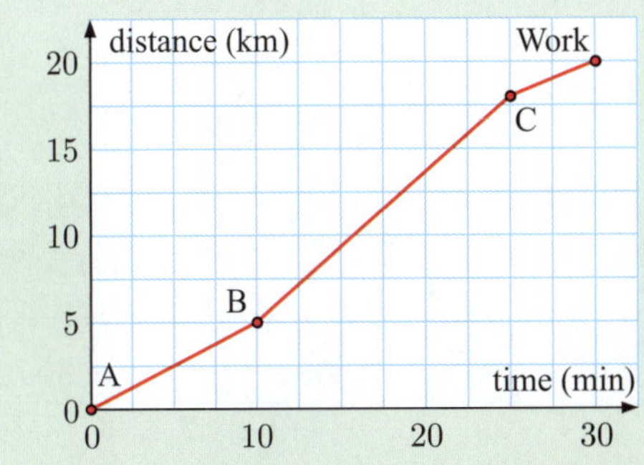

This travel graph shows Sylvie's progress when driving her car from home to work.

a How far is it from Sylvie's home to her work?

b How long did it take Sylvie to get to work?

c Find Sylvie's average speed for the whole journey.

d Find Sylvie's speed between B and C.

10 A particular fabric can be bought for $4 per metre.

a Copy and complete:

Length of fabric (m)	0	1	2	3	4	5
Cost ($)						

b Identify the independent and dependent variables.

c Draw a line graph to display the relationship between the variables.

d Use your graph to find:

    i the cost of 2.5 m of fabric   **ii** the length of fabric which costs $14.

11 Answer the **Opening Problem** on page **322**.

REVIEW SET 16B

1 In 3 hours on his motorcycle, Trent travels 198 km and uses 11 litres of fuel.

a Find Trent's average speed in km/h.

b Find the fuel consumption of the motorcycle in km/L.

2 In 2013, a house was valued at $580 000. In 2021, it was valued at $676 000. Find the rate of increase in value, in dollars per year.

3 Alex drove 200 km in 4 hours.

a Find his average speed.

b Driving at this speed, how long would it take Alex to drive 325 km?

c Write Alex's average speed in metres per second.

4 A jogger ran at an average speed of 12 km/h for 35 minutes. How far did she travel?

5 Foam has density 0.03 g/cm^3. Find the mass of a block of foam with volume 24 000 cm^3.

6 A cylinder 10 cm long has diameter 6 cm and mass 200 g. If it is dropped into water, will it sink or float?

7 A tree grows at a rate of 15 cm per year. Convert this rate into mm per month.

8 A soft drink manufacturer makes 144 000 L of drinks per day. Write this rate in:

a L per hour   **b** kL per week.

9 This graph shows the progress of a car as it travels between cities.

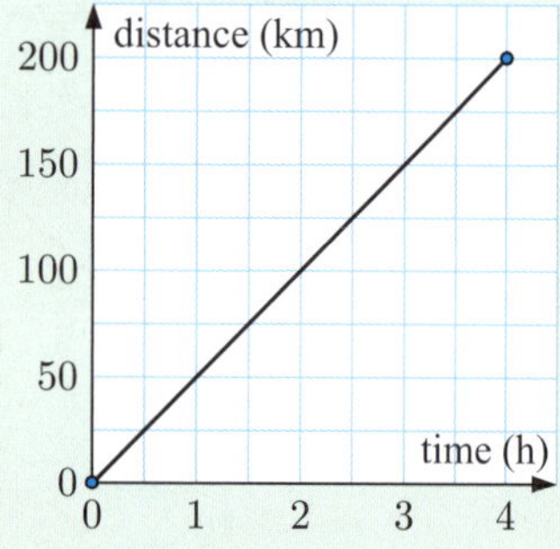

  a How far does the car travel in 3 hours?

  b How long does it take for the car to travel 100 km?

  c Find the speed of the car.

10 A satellite orbits the Earth at a constant speed of 3 km/s.

  a Copy and complete:

Time (s)	0	5	10	15	20	25
Distance (km)						

  b Draw a travel graph to display the relationship between the variables.

  c How far will the satellite travel in 7 seconds?

11 Adam is classifying metal samples.

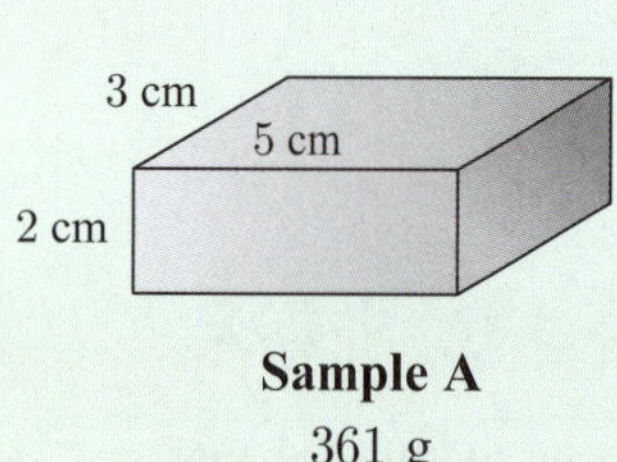

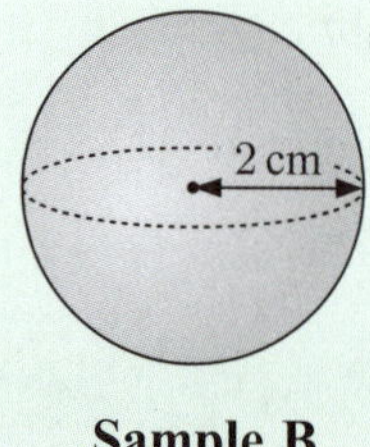

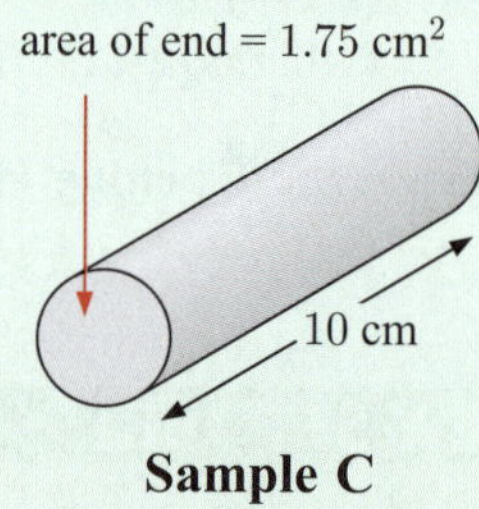

Sample A	**Sample B**	**Sample C**
361 g	247 g	184 g

  a Calculate the density of each sample.

  b **i** Use the table alongside to find the most likely metal of each sample.

   ii Explain why the densities do not match exactly.

  c Adam's assistant Jamie gives him a 0.36 cm^3 lump of metal. Jamie says it is platinum.

   i If the metal is platinum, what should its mass be?

   ii The actual mass of the sample is 8.12 g. Which metal is it?

Metal	*Density* (g/cm^3)
Tin	7.37
Iron	7.87
Silver	10.49
Aluminium	2.70
Iridium	22.56
Palladium	12.02
Platinum	21.45

Probability

Contents:

OPENING PROBLEM

Of the 20 students in Class 8A, 8 have a food allergy.

Of the 25 students in Class 8B, 9 have a food allergy.

Things to think about:

a More students have a food allergy in Class 8B than Class 8A. Does this mean that if a student is randomly selected from each class, then it is more likely that the student from Class 8B has a food allergy?

b What is the probability that a randomly selected student has a food allergy if chosen from:

    i Class 8A   **ii** Class 8B?

c Suppose one student is randomly chosen from each class. What is the probability that *both* selected students have a food allergy?

d How can we *estimate* the probability that a randomly chosen student from the *whole school* has a food allergy?

A   PROBABILITY

The **probability** of an event is the chance or likelihood of it occurring.

An **impossible** event has 0% chance of happening, and is assigned the probability 0.

A **certain** event has 100% chance of happening, and is assigned the probability 1.

All other events have probability between 0 and 1.

Probabilities may be given as decimals, proper fractions, or percentages.

This number line shows how we could interpret different probabilities:

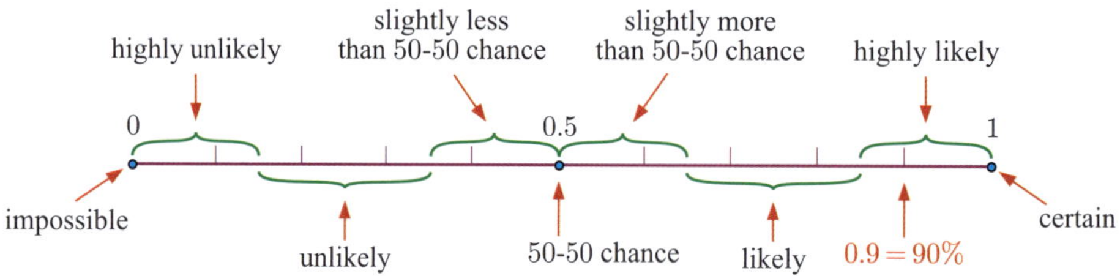

For example, if the weather forecast says there is a 90% chance of rain tomorrow, we would say it is *highly likely* that it will rain tomorrow.

In probability, we often use capital letters to represent events.

For example, we could let E be the event that it will rain tomorrow.

The probability of event E occurring is written $P(E)$, so we write $P(E) = 0.9$.

The **complement** of an event E is the event that E does *not* occur. The complement of E is written E'.

For any event E, $P(E') = 1 - P(E)$.

If E is the event that it will rain tomorrow, then E' is the event that it will *not* rain tomorrow, and $P(E') = 1 - 0.9 = 0.1$.

EXERCISE 17A

1 Use a word or phrase to describe the probability of each event:

   a There is a 25% chance that Ella will score a goal in her next football match.

   b There is a 60% chance that the restaurant will be booked out on Saturday night.

   c There is a 5% chance that William will forget to take his lunch to school tomorrow.

2 The schedule for the next round of baseball matches is given alongside, as well as the probability of each team winning.

32%	Cubs	vs Lions	68%
59%	Wildcats	vs Flames	41%
73%	Eagles	vs Angels	27%
20%	Pumas	vs Strikers	80%

   a Which team is most likely to win the game between the Cubs and the Lions?

   b Which team is most likely to win their game?

   c What is the probability that the Wildcats will lose their game?

   d True or false?   "It is *likely* that the Angels will beat the Eagles."

3 Five students are competing in a long distance race. Each student's probability of winning the race is given alongside.

Julie	20%
Edward	22%
Rob	7%
Tran	15%
Patricia	36%

   a Who is most likely to win the race?

   b Who is least likely to win the race?

   c Find the sum of the probabilities given. Explain your result.

   d Find the probability that either Julie or Tran will win the race.

   e Describe in words, the probability that Rob will win the race.

   f Let E be the event that Edward will win the race.

      i Find $P(E)$.   **ii** State the complementary event E'.   **iii** Find $P(E')$.

4 A bag is filled with balls. One ball is to be chosen at random. Use a word or phrase to describe the probability of choosing a *red* ball, if the bag contains:

   a 1 red ball and 1 blue ball   **b** 5 red balls

   c 2 blue balls and 3 green balls   **d** 1 red ball and 10 blue balls.

5 Suppose S is the event that it will snow tomorrow, and $P(S) = 0.03$.

   a State the complementary event S'.   **b** Find $P(S')$.

6 Suppose R is the event it will rain tomorrow, and S is the event I will have soup for dinner.
$P(R) = 0.2$ and $P(S) = 0.8$.

Are R and S complementary events? Explain your answer.

B   SAMPLE SPACE

Before we can calculate the probability of obtaining a particular result in an experiment, we must first understand what **outcomes** can occur.

> The **sample space** of an experiment is the set of its possible outcomes.

For example, suppose five cards numbered 1 to 5 are placed in a hat.
One card is drawn out at random.

The sample space is $\{1, 2, 3, 4, 5\}$.

There are 5 possible outcomes.

2-DIMENSIONAL GRIDS

If an experiment involves *two* operations, we can use a **2-dimensional grid** to display the set of possible outcomes.

For example, the grid alongside displays the sample space for spinning these spinners:

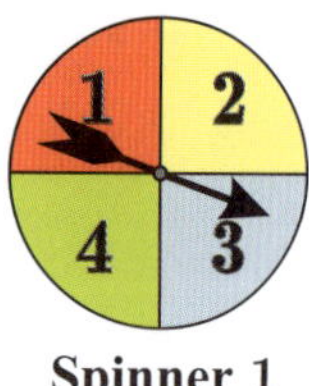

Spinner 1

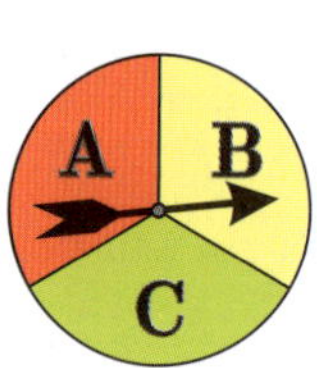

Spinner 2

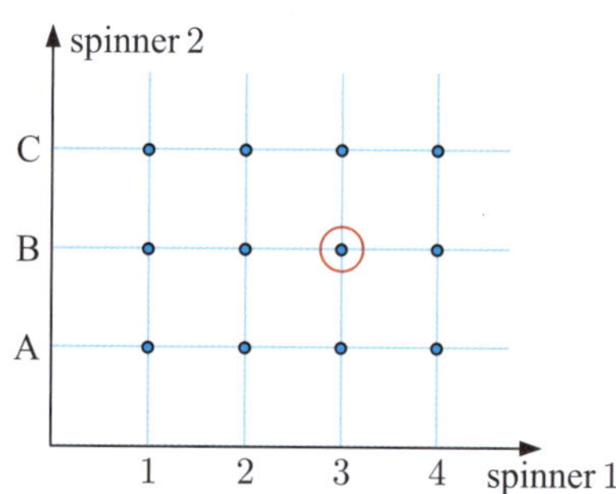

The four possibilities for spinner 1 are listed on the horizontal axis, and the three possibilities for spinner 2 are listed on the vertical axis.

Each point on the grid represents one of the 12 possible outcomes of the experiment.

For example, the circled point represents spinning 3 with spinner 1 and B with spinner 2.

EXERCISE 17B

1 List the sample space, and state the number of possible outcomes for:

   a rolling an ordinary 6-sided die

   b twirling a spinner with 3 segments marked A, B, and C

   c choosing a season of the year

   d taking a card from a pack of playing cards and looking at its suit

   e choosing a two-digit number which is less than 20

   f choosing a factor of 200.

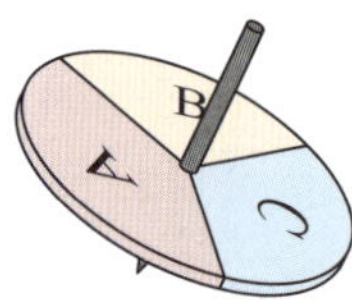

2 A Year 8 class contains 9 twelve year olds, 15 thirteen year olds, and 2 fourteen year olds. 22 of the students are right-handed, and the rest are left-handed.

A student from the class is randomly selected.

  a State the number of possible outcomes.

  b How many of the outcomes are a student who is:

    i at least thirteen years old   **ii** left-handed?

3 For each experiment below, draw a 2-dimensional grid to display the sample space, and state the number of possible outcomes:

  a tossing a coin and spinning a spinner marked with green, orange, and purple sectors

  b rolling a die and spinning a spinner marked 1, 2, and 3

  c selecting one consonant and one vowel from the letters in the name SINGAPORE.

4 **a** Draw a 2-dimensional grid to display the sample space for rolling a die and spinning a spinner marked A, B, C, and D.

  b How many possible outcomes are there?

  c Draw a red circle around the outcome of rolling a 5 and spinning C.

  d Draw green circles around the outcomes of rolling an even number and spinning A.

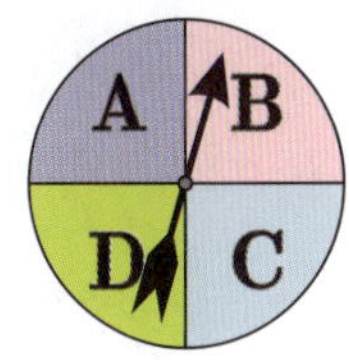

5 **a** Draw a 2-dimensional grid to display the sample space when selecting a ticket from each of these boxes.

  b State the number of possible outcomes.

  c Draw a circle around the outcome of choosing two tickets marked B.

  d Draw a square around the outcomes which include *exactly one* ticket marked B.

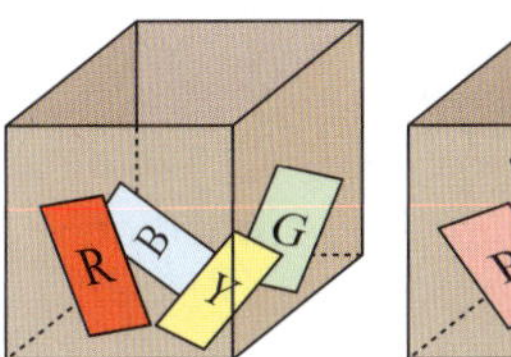
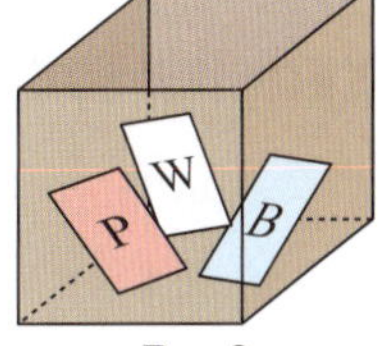

<table><tr><td style="background:#1e90ff;color:white;font-weight:bold;">C</td><td></td></tr></table>

THEORETICAL PROBABILITY

The sample space for spinning this octagonal spinner is $\{1, 2, 3, 4, 5, 6, 7, 8\}$. There are 8 possible outcomes.

Since the spinner is a *regular* octagon, the outcomes are *equally likely* to occur.

> If the possible outcomes of an experiment are equally likely, the theoretical probability of an event E is given by
>
> $$P(E) = \frac{\text{number of outcomes in } E}{\text{total number of possible outcomes}}.$$

Example 1

A ticket is randomly selected from a basket containing 3 green, 4 yellow, and 5 blue tickets. Determine the probability of getting:

  a a green ticket

  b a green or yellow ticket

  c an orange ticket

  d a green, yellow, or blue ticket.

There are $3 + 4 + 5 = 12$ tickets which could be selected with equal chance.

a $\text{P(green)} = \dfrac{3}{12} = \dfrac{1}{4}$

b $\text{P(a green or a yellow)} = \dfrac{3+4}{12} = \dfrac{7}{12}$

c $\text{P(orange)} = \dfrac{0}{12} = 0$

d $\text{P(green, yellow, or blue)} = \dfrac{3+4+5}{12} = \dfrac{12}{12} = 1$

EXERCISE 17C

1 An ordinary die is rolled once. Determine the probability of rolling:

  a a 4

  b an odd number

  c a number greater than 1

  d a multiple of 3.

2 Ten cards with the numbers 1 to 10 written on them are placed in a bag. A card is chosen at random. Find the probability that the number chosen is:

  a 7

  b 9 or 10

  c a multiple of 3

  d a number greater than 4

  e a prime number

  f *not* a prime number.

3 A standard set of balls for a pool table is shown alongside.

If a ball is chosen at random, determine the probability of choosing:

  a the white ball

  b a green ball

  c a number less than 8.

4 A carton contains eight brown and four white eggs. Find the probability that an egg selected at random is:

  a brown

  b white.

5 A ticket is randomly selected from a box containing 3 yellow, 4 green, and 8 blue tickets. Find the probability that the ticket is:

    a yellow   **b** green   **c** blue

    d red   **e** not yellow   **f** yellow or blue.

6 For each of the following spinners:

    i Find the probability of spinning *red*.

    ii Decide whether it is more likely that red will be spun or that blue will be spun.

a 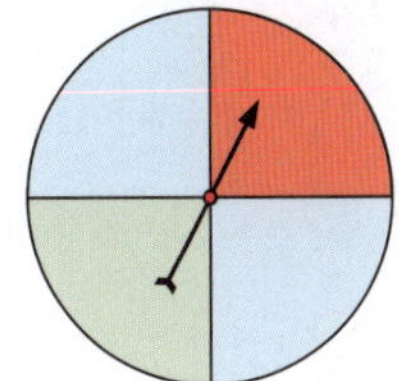   **b** 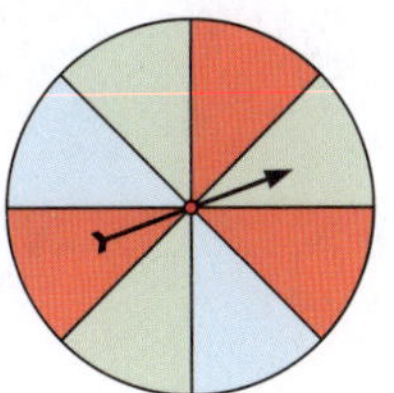   **c** 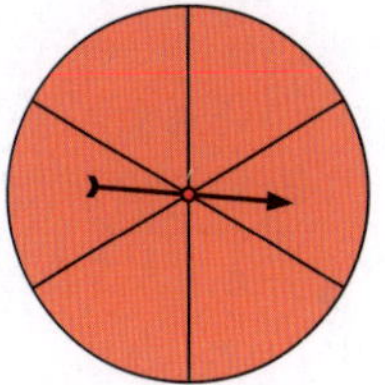

7 At the local cinema, one movie session sells 20 adult tickets, 16 concession tickets, and 9 child tickets. One person from the audience is selected at random. Find the probability that this person has:

    a a concession ticket   **b** an adult ticket

    c either a child or a concession ticket.

8 A jigsaw puzzle for a small child is shown alongside.
If a piece of the puzzle is chosen at random, determine the probability of choosing:

    a a corner piece

    b an edge piece

    c a corner or an edge piece

    d a piece which is neither a corner nor an edge.

9 The layout of seats on an aeroplane is given alongside. The seats already occupied are shaded red.

If you are allocated an unoccupied seat at random, determine the probability of getting:

    a a front row seat

    b a seat in the emergency row

    c a window seat

    d an aisle seat.

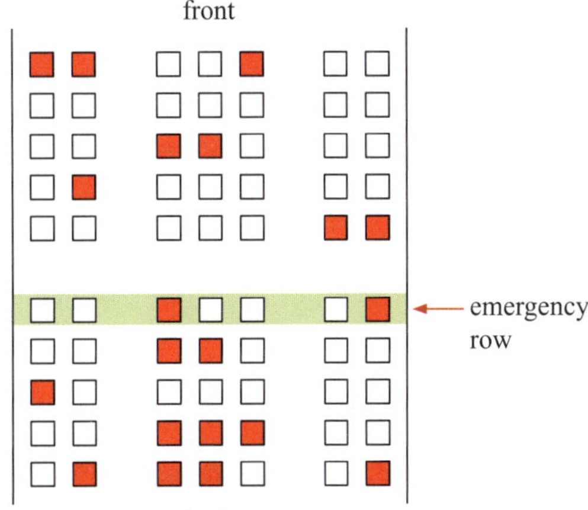

Example 2

A coin is tossed and a die is rolled simultaneously.

a Use a grid to illustrate the sample space.

b Find the probability of getting:

    i a tail and a 5   **ii** a head and a number less than 3   **iii** a tail or a 5.

a 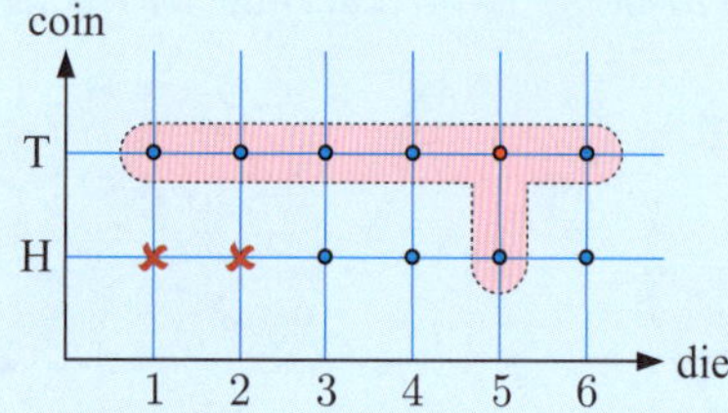

b The sample space has 12 outcomes, each of which is equally likely.

    i P(a tail and a 5) $= \dfrac{1}{12}$ $\{\bullet\}$

    ii P(a head and a number less than 3)

$$= \dfrac{2}{12} = \dfrac{1}{6} \quad \{\times\}$$

    iii P(a tail or a 5)

$$= \dfrac{7}{12} \quad \{\text{the enclosed points}\}$$

---

10 Two coins are tossed simultaneously.

    a Use a grid to illustrate the sample space.

    b How many possible outcomes are there?

    c Find the probability of getting:

        i two heads   **ii** exactly one head   **iii** at least one head.

11 A coin is tossed, and a spinner with 3 equal sectors A, B, and C is twirled.

    a Use a grid to illustrate the sample space.

    b How many possible outcomes are there?

    c Determine the probability of getting:

        i an A and a head   **ii** a head but *not* an A

        iii an A   **iv** an A or a B, and a tail

        v an A, or, a B and a tail.

12 A disc is taken at random from each of the bags shown.

    a Draw a grid to illustrate the sample space.

    b Hence find the probability of getting:

        i two red discs

        ii two discs the same colour

        iii a white disc

        iv a blue disc or a green disc

        v two discs which are different in colour.

Bag A

Bag B

13 Two spinners with equal sectors 1, 2, and 3 are twirled simultaneously.

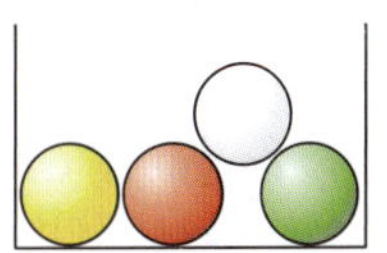 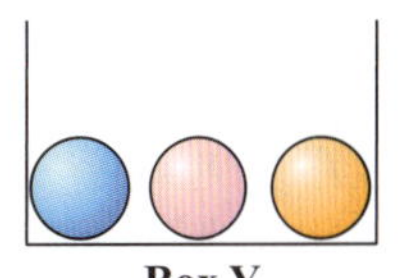

  a Draw a grid to display the sample space.

  b Find the probability of getting:

    i two 2s
    ii two odd numbers
    iii a 2 and a 3
    iv two numbers whose sum is 3
    v two identical numbers
    vi two numbers whose sum is 3 or 5.

14 Erika has separated the integers from 2 to 10 into primes and composites. She now selects one number from each group at random.

  a Draw a grid to display the sample space.

  b Find the probability that Erika selects:

    i 3 and 8
    ii two odd numbers
    iii two even numbers
    iv two numbers whose sum is 13
    v consecutive numbers
    vi two numbers with HCF > 1.

D INDEPENDENT EVENTS

When we calculate the probability of two or more events occurring, we are calculating the probability of **compound events** or **combined events**.

Suppose a ball is randomly selected from each of these boxes.

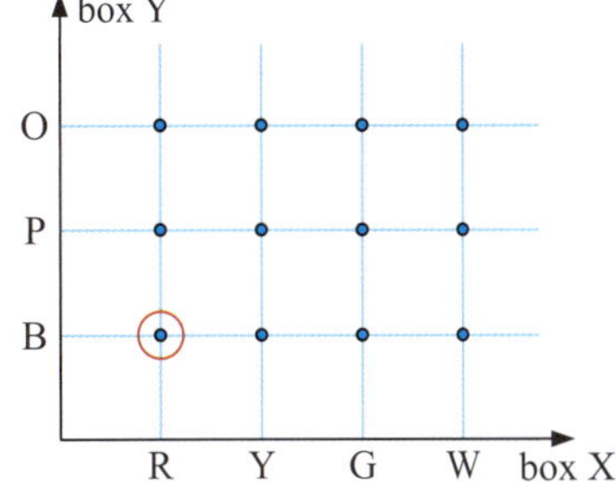

We have seen that we can illustrate the sample space on a 2-dimensional grid, and use this grid to calculate probabilities.

For example, only 1 of the 12 outcomes is a red from X and a blue from Y, so

$$P(\text{red from X } and \text{ blue from Y}) = \frac{1}{12}.$$

For a compound event like this, we can also calculate the probability *without* using a grid.

The selection from each box does not affect the selection from the other. We say the events are *independent*.

Notice that $P(\text{red from X}) = \frac{1}{4}$, $P(\text{blue from Y}) = \frac{1}{3}$,

and that $P(\text{red from X } \textbf{and } \text{blue from Y}) = P(\text{red from X}) \times P(\text{blue from Y})$.

Two events are **independent** if the occurrence of each event does not affect the occurrence of the other.

If two events A and B are **independent** then $P(A \text{ and } B) = P(A) \times P(B)$.

Example 3   ◀) **Self Tutor**

The two spinners alongside are spun. Find the probability that a "green" and a "3" result.

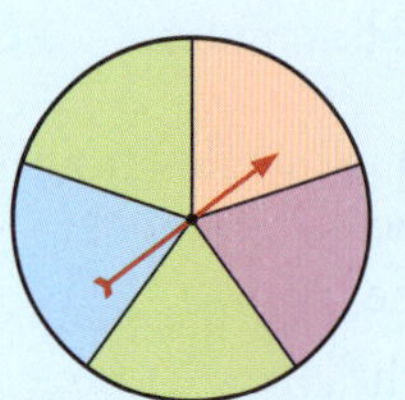

The outcome from one spinner does not affect the outcome from the other, so the events "spinning a green" and "spinning a 3" are independent.

$$\therefore \quad P(\text{a green and a 3}) = P(\text{a green}) \times P(\text{a 3})$$
$$= \frac{2}{5} \times \frac{1}{7}$$
$$= \frac{2}{35}$$

EXERCISE 17D

1 A coin and a pentagonal spinner with edges marked A, B, C, D, and E are tossed and twirled simultaneously. Find the probability of getting:

   a a head and a D
   b a tail, and either an A or a D.

2 A coin and an ordinary die are tossed and rolled simultaneously. Find the probability of getting:

   a a tail and a 3
   b a head and an even number.

3 The spinners alongside are twirled simultaneously. Find the probability of getting:

   a a red and a green
   b a blue and a purple
   c a yellow and a green
   d a red and a purple.

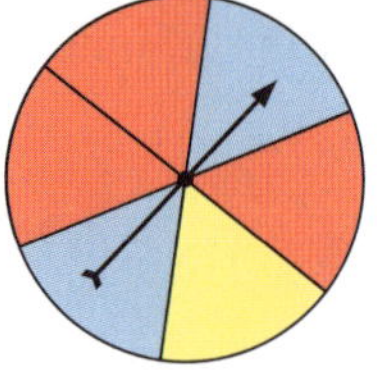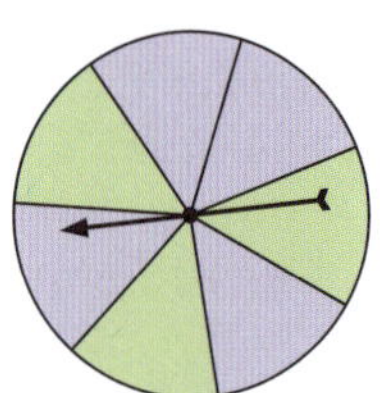

4 Janice and Lee take set shots at a netball goal. From past experience, Janice scores a goal on average 2 times in every 3 shots, whereas Lee scores a goal 4 times in every 7 shots. If the two girls both shoot for goals, find the probability that:

   a both score
   b both miss
   c Janice scores but Lee misses.

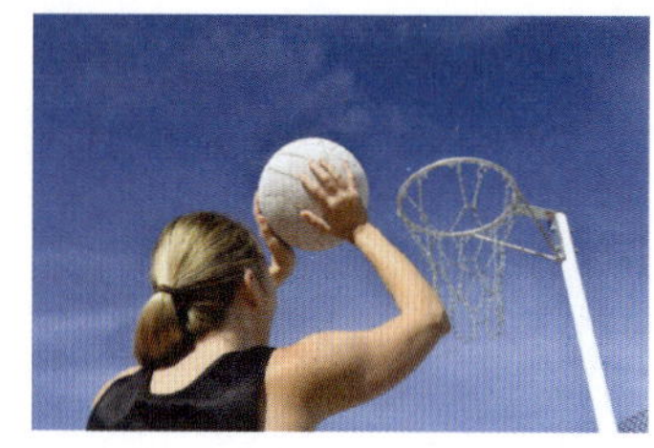

5 Tei has probability $\frac{1}{3}$ of hitting a target with an arrow, while See has probability $\frac{2}{5}$.

   a If they both fire at the target, determine the probability that:

     i both hit   **ii** both miss

     iii Tei hits but See misses   **iv** Tei misses but See hits.

   b Which of the outcomes in **a** is most likely?

   c Find the sum of the probabilities in **a**. Explain your answer.

6 Vicky and Paul are taking their driving test tomorrow. Vicky has 70% chance of passing the test, while Paul has 60% chance. Find the probability that:

   a they both pass the test

   b at least one of them fails the test.

DISCUSSION

Are two events always independent?

Can you think of a situation where the outcome of one event *depends* on the outcome of another?

ACTIVITY   DEPENDENT EVENTS

Cards numbered 1 to 5 are placed in a hat. One card is selected from the hat, and is placed to one side. A second card is then selected from the hat from the cards which remain.

Let A be the event that the *first* card is odd, and B be the event that the *second* card is *even*.

Mitch says that $P(A) = \frac{3}{5}$ and $P(B) = \frac{2}{5}$, so $P(A \text{ and } B) = P(A) \times P(B) = \frac{3}{5} \times \frac{2}{5} = \frac{6}{25}$.

What to do:

1 List the sample space of possible outcomes for this experiment. Circle the outcomes which satisfy A and B.

2 Use your sample space to find $P(A \text{ and } B)$. Is Mitch correct?

3 Are A and B independent events?

4 *Given that* event A occurs, what is the probability that event B occurs?

5 How should the rule $P(A \text{ and } B) = P(A) \times P(B)$ be adjusted in the case that A and B are *dependent* events?

E   EXPERIMENTAL PROBABILITY

In situations where we cannot calculate theoretical probabilities, we can *estimate* the probability of an event by performing many trials of the experiment. This estimate is known as an **experimental probability**.

In a probability experiment:

- the **frequency** of a particular event is the number of times that this event is observed
- the **relative frequency** of an event is the frequency of that event divided by the total number of trials
- the relative frequency is the **experimental probability** of the event.

For example, suppose a small plastic cone was tossed into the air 300 times. It fell on its *side* 203 times and on its *base* 97 times. We say that:

- the number of trials is 300
- the possible outcomes are *side* and *base*
- the frequency of *side* is 203
- the frequency of *base* is 97
- the relative frequency of $side = \dfrac{203}{300} \approx 0.677$
- the relative frequency of $base = \dfrac{97}{300} \approx 0.323$.

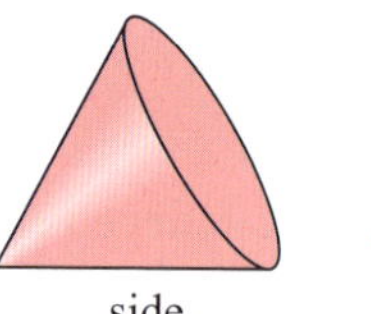
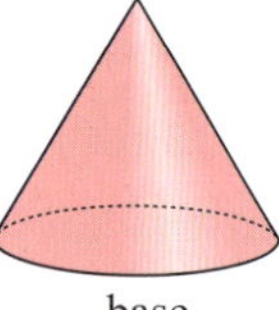

In the absence of any further data, the relative frequency of each outcome is our best estimate of the probability of it occurring.

We write $P(side) \approx 0.677$, $P(base) \approx 0.323$.

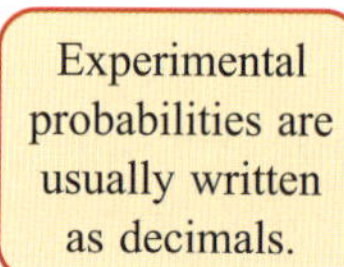

The accuracy of an experimental probability is improved by increasing the number of trials of the experiment.

DISCUSSION

Why is the accuracy of an experimental probability improved by increasing the number of trials of the experiment?

EXERCISE 17E

1 A batch of 145 paper clips was dropped onto 6 cm by 6 cm squared paper. 113 clips fell completely inside squares, and 32 finished on a grid line.

Estimate, to 2 decimal places, the probability of a clip falling:

    a inside a square   **b** on a grid line.

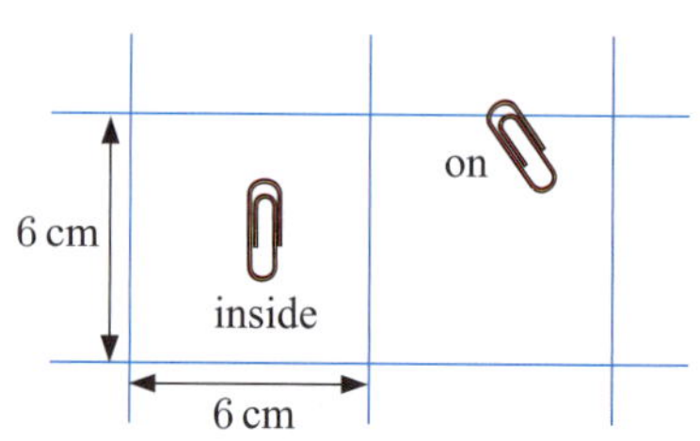

2 Connor visits the same bakery each day for lunch. On 13 out of his last 40 visits, the bakery had sold out of pies. Estimate the probability that on his next visit, the bakery:

    a will have sold out of pies   **b** will still have some pies remaining.

3 During one day at a carnival, 247 people played the "Coconut Shy" game. 175 people did not win a prize, 64 people won a minor prize, and the remainder won a major prize. Estimate the probability that the next player will win:

    a a major prize   **b** any prize.

4 When a nut was tossed 400 times, it finished on its edge 84 times and on its side for the rest.

    a Find the experimental probability that when this nut is tossed, it will finish on its edge.

edge   side

    b Hence estimate the probability that when two identical nuts are tossed:

        i they both fall on their edges   **ii** they both fall on their sides.

5 Suzanne and Tasha are surveying customers at a supermarket to see whether they buy brand X cheese.

Out of 54 shoppers Suzanne surveys, 23 buy brand X cheese.

Out of 67 shoppers Tasha surveys, 31 buy brand X cheese.

    a Find the experimental probability that a randomly selected shopper will buy brand X cheese using:

        i Suzanne's results   **ii** Tasha's results   **iii** the combined results.

    b Which of the experimental probabilities you have calculated do you expect to be most accurate? Explain your answer.

INVESTIGATION   SECRET SANTA

Many workplaces and families hold a **Secret Santa** or **Kris Kringle** at Christmas time.

The name of each person is placed in a hat. Each person selects a name from the hat and will buy a present for that person. Obviously, it is important that nobody draws their own name.

In this Investigation we will estimate the probability that, for a Secret Santa involving 8 people, nobody draws their own name.

What to do:

1 Write the letters A, B, C, D, E, F, G, and H on 8 identical pieces of paper. The letters represent the 8 participants. Place the pieces of paper in a hat or bag.

2 Randomly draw names from the hat, one at a time. The first name drawn represents the person selected by A, the second represents the person selected by B, and so on.

3   Determine whether any of the participants have drawn their own name.

4   Perform 50 trials of this experiment. For each trial, record whether any of the participants drew their own name.

5   Use your results to estimate the probability that, in a Secret Santa involving 8 participants, nobody will draw their own name.

6   Combine your results with your classmates to obtain a better estimate.

7   Click on the icon to open a simulation of this experiment. Use the simulation to perform:

    a   500 trials of the experiment   **b**   5000 trials of the experiment.

Compare the results with the results you obtained in your own experiment.

F   PROBABILITIES FROM TABLED DATA

When data is collected, it is often summarised and displayed in a frequency table. We include a column for the *relative frequency* of each result, which gives us the experimental probability.

Example 4   ◀) Self Tutor

A marketing company surveys 50 randomly selected people to discover what brand of toothpaste they use. The results are given in the table.

Brand	Frequency	Relative frequency
Shine	10	
Starbright	14	
Brite	4	
Clean	12	
No Name	10	
Total	50	

a   Copy and complete the table.

b   Estimate the probability that a randomly selected person uses:

    i   Starbright   **ii**   Clean.

---

a

Brand	Frequency	Relative frequency
Shine	10	$\frac{10}{50} = 0.2$
Starbright	14	$\frac{14}{50} = 0.28$
Brite	4	$\frac{4}{50} = 0.08$
Clean	12	$\frac{12}{50} = 0.24$
No Name	10	$\frac{10}{50} = 0.2$
Total	50	1.00

b   **i**   P(Starbright) ≈ 0.28

    ii   P(Clean) ≈ 0.24

EXERCISE 17F

1 A corner shop sells shortbread and chocolate chip biscuits. In one week 187 biscuits are sold.

  a Copy and complete the table.

  b Estimate the probability that the next customer will buy:

    i a shortbread biscuit   **ii** a chocolate chip biscuit.

Type	Frequency	Relative frequency
Shortbread	62	
Chocolate chip		
Total	187	

2 A football club has three levels of membership, as described in the table.

  a Copy and complete the table.

  b How many members does the football club have?

  c Estimate the probability that the next membership sold is:

    i silver   **ii** not platinum.

Level	Frequency	Relative frequency
Platinum	419	
Gold	628	
Silver	921	
Total		

3 A café sells four types of hot drinks. Its sales over a one week period are shown in the table.

  a Copy and complete the table.

  b Estimate the probability that the next customer will buy:

    i hot chocolate

    ii an espresso or tea.

Type	Frequency	Relative frequency
Espresso	49	
Milk coffee	187	
Tea	150	
Hot chocolate	113	
Total		

4 Passengers on a Baltic cruise ship boarded the ship in one of four cities, as shown in the table.

  a Copy and complete the table.

  b Find the probability that a randomly chosen passenger:

    i boarded in Gdańsk

    ii boarded in Tallinn or Klaipėda

    iii did not board in Tallinn.

City	Frequency	Relative frequency
Helsinki		0.3
Tallinn		0.24
Klaipėda	400	
Gdańsk	750	0.3
Total	2500	

G   PROBABILITIES FROM TWO-WAY TABLES

Sometimes data is categorised by not just one, but two variables. The data is represented in a **two-way table**.

In a two-way table we include the totals of each row and each column. This makes it easier for us to calculate probabilities.

Example 5

Self Tutor

60 students at a school were randomly selected and asked whether they studied Physics and English.

Physics

		Yes	No	*Total*
English	Yes	15	12	
	No	13	20	
	Total			

a Copy and complete the table by finding the total of each row and column.

b What does the 13 in the table represent?

c Estimate the probability that a randomly selected student studies English but not Physics.

d Estimate the probability that a randomly selected student who studies English, also studies Physics.

a

Physics

		Yes	No	*Total*
English	Yes	15	12	27
	No	13	20	33
	Total	28	32	60

b 13 of the selected students study Physics but do not study English.

c 12 out of the 60 students study English but not Physics.

$\therefore$ P(student studies English but not Physics) $\approx \dfrac{12}{60} \approx 0.2$

d Of the 27 students who study English, 15 also study Physics.

$\therefore$ P(student who studies English also studies Physics) $\approx \dfrac{15}{27} \approx 0.556$

EXERCISE 17G

1 A sample of Year 8 students at a school were asked whether they completed their homework for English and Mathematics.

Mathematics

		Yes	No	*Total*
English	Yes	29	6	
	No	7	12	
	Total			

a Copy and complete the table.

b How many students were surveyed?

c What does the 7 indicate?

d Estimate the probability that a randomly chosen student completed their Mathematics homework.

e Estimate the probability that a randomly chosen student who completed their English homework, did not complete their Mathematics homework.

2 200 randomly chosen people were surveyed to determine whether they enjoyed a new television show. The results are categorised by age.

Preference

Age		Like	Dislike	Undecided	*Total*
	Under 40	29	49	26	
	Over 40	48	14	34	
	Total				

a Copy and complete the table.

b What does the 49 indicate?

c Estimate the probability that a randomly chosen person:

    i will dislike the new television show

    ii is over 40 and undecided.

3 A sample of teenagers were asked whether they had ever learnt to play a musical instrument.   The results were further categorised by gender as shown.

Gender

Instrument		Male	Female	*Total*
	Yes	39	14	
	No	17	2	
	Total			

a Copy and complete the table.

b Estimate the probability that a randomly chosen female teenager has learnt to play an instrument.

c Estimate the probability that a randomly chosen male teenager has never learnt to play an instrument.

d Which of the estimates from **b** and **c** is more likely to be accurate? Explain your answer.

4 40 Year 8 students were asked what grades they achieved for Mathematics and Science last term.

Mathematics

Science		A	B	C	*Total*
	A	7	5	3	
	B	6	8	4	
	C	4	1	2	
	Total				

a Copy and complete the table.

b Estimate the probability that a randomly selected student scored:

    i an A for Mathematics and a B for Science

    ii an A for at least one of the subjects

    iii the same grade for both subjects.

c Estimate the probability that a student who scored an A for Science, scored a C for Mathematics.

5 A sample of 13 year olds were surveyed to find out whether they wanted to study at university, and whether they wanted children one day. Some of the data is shown in the table.

When a 13 year old is randomly chosen and it is found that they want to study at university, the probability that they also want children is 0.64 .

University

Children		Yes	No	*Total*
	Yes			101
	No		14	
	Total	125		

a Copy and complete the table.

b Estimate the probability that a randomly chosen 13 year old does not want to go to university nor have children.

H   PROBABILITIES FROM VENN DIAGRAMS

Venn diagrams give us an alternative way to show the number of individuals in a sample which have particular properties. We can therefore use Venn diagrams to find probabilities.

Example 6   ◀⑴ **Self Tutor**

30 tourists on a bus are discussing the countries they have visited. 18 of them have visited Japan, and 11 have visited Singapore. 6 have visited both countries.

a Represent this information on a Venn diagram.

b Find the probability that a randomly chosen tourist on the bus has visited:
  i Japan but not Singapore   **ii** at least one of the countries.

c Given that a tourist has visited Singapore, find the probability that they have not visited Japan.

---

a Let J represent the tourists who have visited Japan, and S represent the tourists who have visited Singapore.

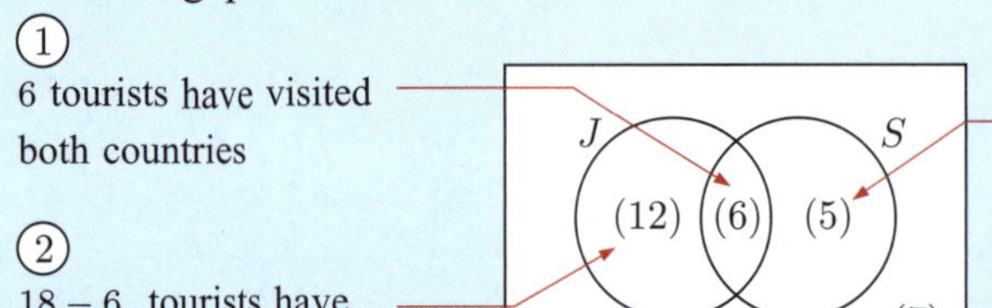

① 6 tourists have visited both countries

② $18 - 6$ tourists have visited only Japan

③ $11 - 6$ tourists have visited only Singapore

④ $30 - 12 - 6 - 5$ tourists have visited neither country

b **i** Of the 30 tourists on the bus, 12 have visited Japan but not Singapore.

$\therefore$ P(has visited Japan but not Singapore) $= \dfrac{12}{30} = \dfrac{2}{5}$.

ii Of the 30 tourists on the bus, $12 + 6 + 5 = 23$ have visited at least one of the countries.

$\therefore$ P(has visited at least one of the countries) $= \dfrac{23}{30}$.

c Of the 11 tourists who have visited Singapore, 5 have not visited Japan.

$\therefore$ P(has not visited Japan given they have visited Singapore) $= \dfrac{5}{11}$.

EXERCISE 17H

1 This Venn diagram shows dance lessons attended by students of a dance studio. Each dot represents one student.

J represents the students who attend jazz lessons.

B represents the students who attend ballet lessons.

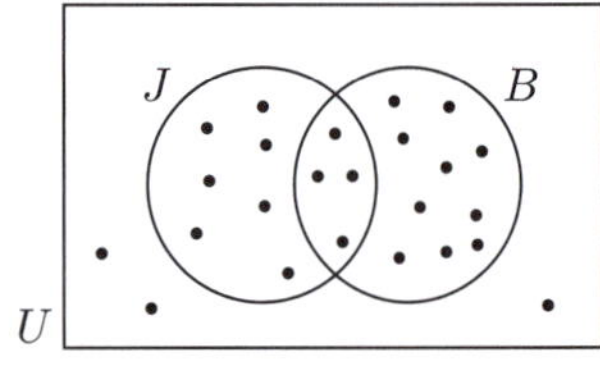

a How many students does the dance studio have?

b Find the probability that a randomly chosen student attends:
  i both ballet and jazz lessons
  ii either ballet or jazz lessons, but not both.

2 Of the 25 four year olds in a kindergarten class, 15 have swum in a pool, 7 have swum at the beach, 4 have done both, and 7 have done neither.

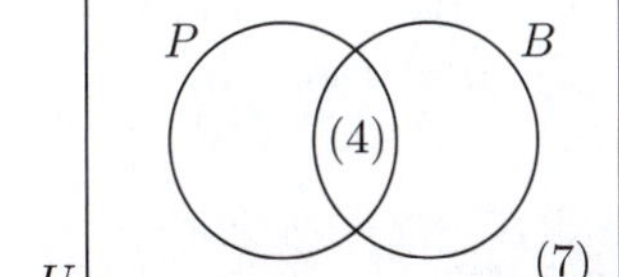

a Copy and complete this Venn diagram.

b Find the probability that a randomly selected class member has swum at the beach or in a pool.

3 There are 15 different pizza choices on a menu. 11 contain red meat, 5 contain white meat, and 3 contain both.

a Place this information on a Venn diagram.

b Forgetting that his friend Simon is a vegetarian, Joshua randomly chooses a pizza from the menu. Find the probability that Simon will be able to eat some.

4 Nathaniel's mum packs his lunch for him every day. During a 50 day term, Nathaniel finds fruit in his lunch box on 37 days, a sandwich on 35 days, and both a sandwich and fruit on 22 days.

a Place this information on a Venn diagram.

b Find the probability that on a randomly selected day, Nathaniel's lunch box contains fruit but not a sandwich.

c On a particular day, Nathaniel finds a sandwich in his lunch box. Find the probability that his lunch box does not also contain fruit.

5 75 people went through the zoo gates in one hour. 59 bought something from the café, 24 bought something from the gift shop, and 17 bought something from both the café and the gift shop.

a Copy and complete this Venn diagram to illustrate the information.

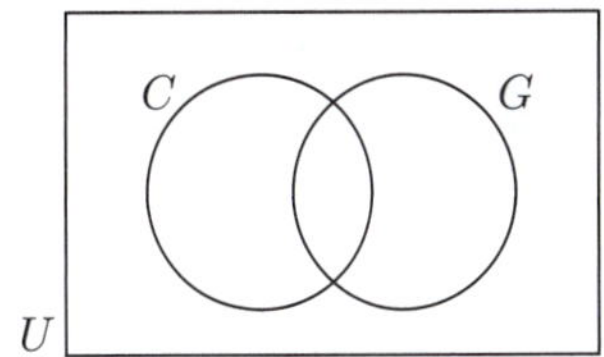

b Find the probability that a randomly selected person at the zoo bought something from the café or gift shop, but not both.

c Given that a person does not buy something from the gift shop, find the probability that they buy something from the café.

EXPECTATION

If the probability of an event is known, it can be used to predict the number of times the event will occur in a given number of trials.

For example, suppose a die is rolled 120 times. We know that on each roll, there is a $\frac{1}{6}$ probability of rolling a "5".

Since $\frac{1}{6}$ of 120 is 20, we would *expect* 20 of the 120 rolls to be a "5".

If there are n trials of an experiment, and the probability of an event occurring in each trial is p, then the **expectation** of the occurrence of that event is np.

Example 7

🔊 **Self Tutor**

In a special promotion, every bottle of iced tea comes with a 1 in 6 chance of winning a prize. In one month, the school canteen sells 539 bottles of iced tea. How many prizes would you expect to be won?

$p = \text{P(bottle wins a prize)} = \dfrac{1}{6}$

$\therefore \quad np = 539 \times \dfrac{1}{6}$

$\qquad \approx 89.8$

We would expect about 90 prizes to be won.

DISCUSSION

1 Look at the answer in **Example 7**.

Is it always *possible* to obtain exactly the "expected" number of occurrences of an event?

2 Is it reasonable to say that the expectation of an event is its *average* expected occurrence?

EXERCISE 17I

1 If a coin is tossed 500 times, how many times would you expect it to land heads up?

2 Each time Darren takes the train, he has probability $\dfrac{5}{8}$ of getting a seat. In 120 train trips, how many times can Darren expect to get a seat?

3 Turtle hatchlings only have probability 0.01 of surviving to adulthood. If 500 000 eggs hatch in one year, how many of the hatchlings are expected to survive to adulthood?

4 A library has found that the probability of a borrowed book being returned on time is 0.68. In one day the library lent 837 books. How many of those books can the library expect to be returned on time?

5 If the spinner alongside is spun 200 times, how many times would you expect the result to be:

  a blue   **b** red

  c yellow   **d** green?

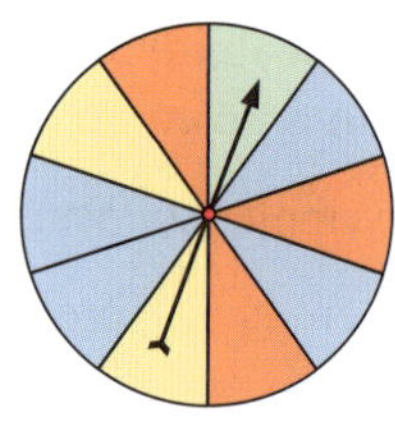

6 If a die is rolled 600 times, how many times would you expect to get:

  a a 3   **b** a prime number   **c** a composite number?

7 If a pair of dice is rolled 900 times, how many times would you expect to get:

    a a pair of 6s   **b** a 5 and a 2

    c two odd numbers   **d** two prime numbers

    e two numbers whose sum is 10   **f** two numbers whose product is 12?

GLOBAL CONTEXT   ROLE-PLAYING GAMES

Global context:	Personal and cultural expression
Statement of inquiry:	Using dice in different ways can dramatically change how a game is played.
Criterion:	Investigating patterns

MULTIPLE CHOICE QUIZ

REVIEW SET 17A

1 List the sample space and state the number of possible outcomes for:

    a selecting a letter from the alphabet

    b spinning the spinner shown.

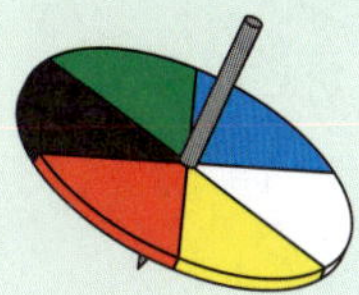

2 Suppose E is the event that Ned's bike will get a puncture next week, and $P(E) = 0.13$.

    a State the complementary event E'.   **b** Find $P(E')$.

3 One lollipop is chosen at random from a tub of lollipops. Find the probability of picking an orange lollipop if the tub contains:

    a 2 orange, 5 strawberry, and 1 lemon lollipop

    b 3 apple and 7 blackberry lollipops

    c 5 watermelon, 10 blackberry, 2 strawberry, and 3 orange lollipops.

4 A farmer fences his rectangular property into 9 rectangular paddocks as shown. A paddock is selected at random. Find the probability that it has:

    a no fences on the boundary of the property

    b one fence on the boundary of the property

    c two fences on the boundary of the property.

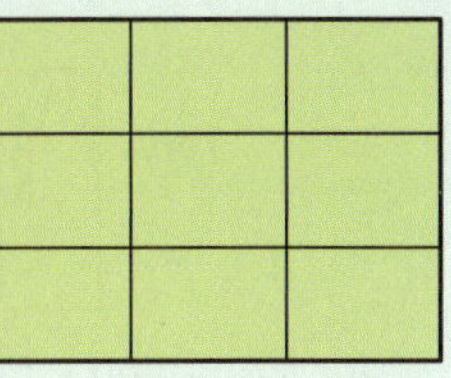

5 A coin is tossed, and a spinner with equal sectors numbered 1, 2, 3, 4, and 5 is spun.

    a Use a 2-dimensional grid to show the sample space of possible outcomes.

    b Find the probability of getting: **i** a head *and* a 5   **ii** a head *or* a 5.

6   The spinner alongside is spun twice.

   a   Use a grid to illustrate the sample space.

   b   Find the probability of spinning:

      i   two prime numbers

      ii   a 3 and a 5   **iii**   at least one 1

      iv   two numbers whose sum is 6 or 8.

   c   Which pair of events in **b** is complementary?

7   Jeff and Santi are baking a cake in a cooking class. Jeff has a 20% chance of burning his cake, and Santi has a 15% chance of burning his cake. Find the probability that:

   a   they both burn their cakes   **b**   neither burns their cake

   c   Jeff burns his cake, but Santi does not.

8   Carly has completed 13 of her last 20 half-marathons in less than two hours. Estimate the probability that:

   a   Carly will take more than two hours to complete her next half-marathon

   b   Carly will complete *both* of her next two half-marathons in less than two hours.

9   A marketing company surveyed teenagers about the main reason they used their home computers. The results of the survey are shown below.

Reason	Frequency	Relative frequency
Homework	29	
Social networking	43	
Playing games	15	
Downloading music	69	
Total		

   a   Copy and complete the table.

   b   Estimate the probability that a randomly selected teenager mainly uses their computer:

      i   for homework   **ii**   for something other than downloading music.

10   The cars serviced by a mechanic this month are summarised in this two-way table.

Age

Type		Less than 3 years old	3 - 10 years old	More than 10 years old	Total
	Automatic	9	17	9	
	Manual	1	7	6	
	Total				

   a   Copy and complete the table.

   b   Estimate the probability that the next car to be serviced will be:

      i   automatic   **ii**   manual and at least 3 years old.

   c   Estimate the probability that the next automatic car to be serviced will be more than 10 years old.

11 90 students from one school took part in a Geography competition and a History competition. 12 of the students won a prize in the Geography competition, and 17 won a prize in the History competition. 6 students won prizes in both competitions.

    a Display this information on a Venn diagram.

    b Find the probability that a randomly chosen student won at least one prize.

    c Given that a student won a prize in the History competition, find the probability that the student did not win a prize in the Geography competition.

12 When a high jump bar is set at 1.90 metres, Steve has probability 0.26 of successfully completing the jump. If Steve attempts 50 jumps at this height, how many times would you expect him to be successful?

REVIEW SET 17B

1 The probability that Lionel will pass his English exam is 0.93.

    a Use a word or phrase to describe this probability.

    b Find the probability that Lionel will *not* pass his English exam.

2 List the sample space for selecting:

    a a disc from a bag containing red, pink, white, and brown discs

    b a 2-digit multiple of 5.

3 A variety box contains 6 packets of plain chips, 6 packets of salt and vinegar chips, 4 packets of chicken chips, and 4 packets of barbecue chips. A packet is chosen at random from the box. Find the probability it contains:

    a barbecue chips   **b** chicken chips or plain chips.

4 Are you more likely to select a prime number by randomly choosing a whole number from 1 to 20, or by randomly choosing a whole number from 1 to 40?

5 The two spinners alongside are twirled simultaneously.

    a Draw a 2-dimensional grid to illustrate the sample space.

    b Find the probability of spinning:

      i the same number with both spinners

      ii two numbers whose sum is 4.

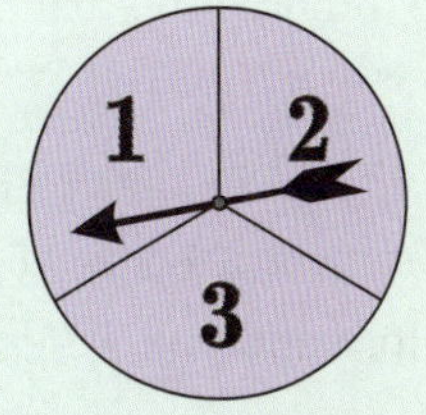
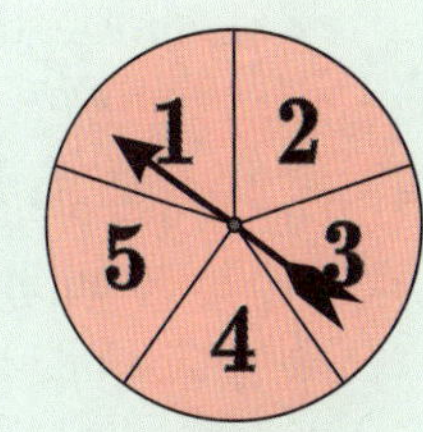

6 A card is randomly selected from each bag.

    a Find the probability that:

      i the card from bag A is green and the card from bag B is not green

      ii the card from bag A is not green and the card from bag B is green.

    b Are the events in **a** complementary? Explain your answer.

Bag A

Bag B

7 Evan has answered 7 of the last 12 phone calls he has received. Estimate the probability that he will *not* answer the next call he receives.

8 The numbers of middle school students sent to detention last term are shown in the table.

Year level	Frequency	Relative frequency
7	38	
8	29	
9	57	
Total		

a Copy and complete the table.

b Estimate the probability that the next student to arrive at detention is:

  i a Year 9 student **ii** not a Year 8 student.

9 Answer the **Opening Problem** on page **340**.

10 150 randomly selected people were asked what radio station they listen to. The results were categorised by age, and are shown in the table.

Radio station

		Mash	Planet	Gold	Total
Age	Under 30	43	14	24	
	Over 30	12	35	22	
	Total				

a Copy and complete the table.

b Estimate the probability that a randomly chosen person:

  i is under 30 and listens to Gold **ii** listens to Mash

  iii does not listen to Planet.

c Estimate the probability that a randomly chosen person who listens to Mash, is over 30.

11 The 40 people living on one street were asked if they played an instrument or sport. 16 played an instrument, 24 played a sport, and 9 played both.

a Place this information on a Venn diagram.

b Find the probability that a randomly selected person living on the street played an instrument, but not a sport.

12 A ball is removed from each of these boxes.

a Find the probability that both balls removed are yellow.

b Suppose this experiment is performed 60 times. How many times would you expect both balls removed to be yellow?

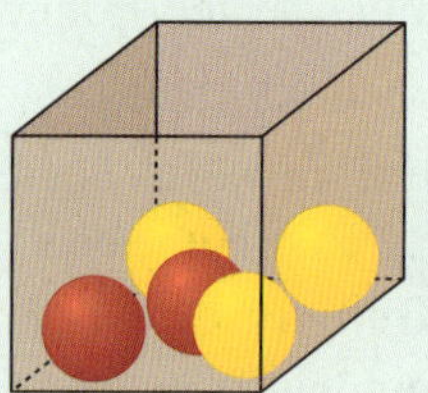
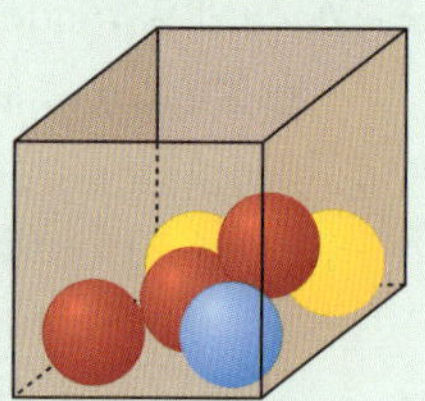

Chapter 18

Statistics

Contents:

OPENING PROBLEM

Chan-juan wants to find the mean number of people living in the apartments on her block. Her block is quite large, so instead of surveying every apartment, she only visits the apartments in her own building. She obtains the following results for the number of people living in each apartment:

1 2 2 4 3 2 1 3 5 2 1 4 1 2 2 3 2 4 1 2
2 3 1 2 5 1 2 4 1 3 5 2 8 4 1 3 4 2 3 1

Things to think about:

a What is the most commonly occurring number of people in an apartment?

b Can you find the *average* number of people per apartment?

c What things will Chan-juan need to consider if she is to use her data to predict the mean number of people per apartment for the whole block?

Facts or pieces of information are called **data**. The singular of data is one *datum*.

Statistics is the study of collecting and analysing data.

The data we collect describes a particular feature or **variable** which differs between the people or objects we are studying.

The process of **statistical enquiry** or **investigation** includes the following steps:

Step 1: Decide on a topic you wish to investigate. Determine what data you will need to collect and how you will collect it.

Step 2: Collect the data.

Step 3: Organise the data.

Step 4: Summarise and display the data.

Step 5: Analyse the data and make a conclusion in the context of your investigation.

Step 6: Write a report to describe what you have done.

A   DATA COLLECTION

For any statistical problem there is a target **population**. This is the group of things or people we are interested in finding information about.

For example, we may want to know:

- the colours of cats in a particular animal shelter
- the ages of people who live in a suburb
- how much money people borrow when buying a house.

CENSUS OR SAMPLE

When we collect data, we can either perform a **census** or a **sample**.

> A **census** involves collecting data about *every* individual in the whole population.

For example, if we wanted to know the colours of cats in a particular animal shelter, we could visit the shelter and record the colour of every cat.

Analysing the data obtained would provide exact information, such as the exact percentage of cats that are black.

> A **sample** involves collecting data about a *part* of the population only.

For example, if we wanted to know the colours of cats in a city, it would be very difficult and expensive to collect information from the whole population. We may choose instead to take a **sample** from the population.

Conclusions based on data from samples always involves some error. However, we can use the properties of the sample to **estimate** the properties of the population.

We can make our estimate more reliable by choosing a sample that is **unbiased** and **sufficiently large**.

BIAS IN SAMPLING

We can only reliably use the properties of a sample to estimate the properties of the whole population if the sample is **representative** of the population.

For example, if you sampled a selection of students leaving a high school in South Africa, the information you would obtain would not be representative of the whole population of South Africa. We would call this a **biased sample**.

To obtain more representative data, we could choose our sample from a location where there is less age bias, such as a shopping centre.

SAMPLE SIZE

For the results from a sample to be reliable, the sample must also be **sufficiently large**. For example, if we were to estimate the average age of all South Africans by surveying only five people, we would not have a very reliable estimate.

EXERCISE 18A

1 State whether a census or a sample would be used to investigate:

   a the country of origin of the parents of students in your class

   b the proportion of Canadians who are concerned about global warming

   c opinions about the public transport system in Lisbon

   d the heights of trees in a garden.

2 Explain any bias in the following samples:

    a To determine the proportion of Californians who can swim, Bill surveys a selection of people at Santa Monica State beach.

    b To determine the average height of plants in her wheat crop, Jill measures the plants nearest the barn.

    c To determine the average time workers in England take to travel to work, Gary surveys a group of workers in Liverpool.

    d To determine how many meals the average adult eats out per week, a "Healthy Eating" group takes a sample of 30 adults who all live in the central business district.

3 Maura is interested in finding the average membership size for gyms around New Zealand. She uses the membership numbers from four randomly selected gyms for her estimate. Explain why Maura's estimate is likely to be unreliable.

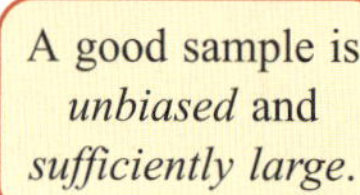

4 Cindy wants to know what proportion of her youth group prefers drinking tea to coffee. She randomly selects 80 fellow group members to survey. 56 prefer tea and 24 prefer coffee.

    a Estimate the proportion of Cindy's youth group members who prefer tea.

    b Do you think this estimate is reliable? Explain your answer.

ACTIVITY 1   RANDOM SAMPLING

The best way to avoid bias when selecting a sample is to make sure the sample is **randomly selected**. This means that each member of the population has the same chance of being selected in the sample.

In this Activity, we will consider two different sampling methods:

- **simple random sampling** where individuals are selected completely randomly
- **systematic sampling** where individuals are selected according to a rule or system.

What to do:

1 Identify the sampling method used in each scenario:

    a The names of club members are placed in a barrel. 10 members are selected by drawing names from the barrel.

    b The first name on every 3rd page of a list of school students is chosen.

    c Every 25th person leaving a supermarket is asked about changes to the deli section.

    d 6 tickets are chosen in a lottery draw.

    e Every Wednesday, a librarian records the number of books borrowed.

    f The first 100 people leaving a stadium are asked about proposed changes to surrounding car parks.

    g A handful of paperclips are taken from a well-mixed bin of paperclips.

2 How could you randomly select:

    a one ticket out of 5 tickets   **b** one of the letters A or B

    c one of the numbers 1, 2, 3, 4, 5, or 6   **d** one card from a pack of 52?

3 How could you select a random sample of:

    a 400 adults   **b** bottles of soft drink at a factory

    c 30 students at a school   **d** words from the English language?

4 Is it always *practical* to select a simple random sample?

5 Will a systematic sample always be biased? You can use the scenarios in **1** to help discuss your answer.

6 Research how **stratified sampling** works and discuss its advantages and disadvantages.

B   CATEGORICAL DATA

A **categorical variable** describes a particular quality or characteristic of the individuals in a population.

Categorical data is data which can be placed in categories.

Examples of categorical variables are:

- *Method of getting to school*:   the categories could be train, bus, car, and walking.
- *Colour of eyes*:   the categories could be blue, brown, hazel, green, and grey.

ORGANISING CATEGORICAL DATA

We can organise categorical data using a **tally and frequency table**.

- The **tally** is used to count the data in each category.
- The **frequency** gives the total number in each category.

The **mode** of a set of categorical data is the category that occurs most frequently.

Example 1   ◀) **Self Tutor**

Jordan wanted to know which season his classmates liked most. He collected the following categorical data from the 30 students in his class. The categories are autumn (A), winter (W), spring (Sp), and summer (Su).

Sp	A	A	Su	W		Su	Sp	A	Sp	W		A	Sp	W	Su	A
A	Sp	Su	W	Sp		Su	W	Sp	A	Sp		Sp	A	Sp	Sp	Su

    a Draw a tally and frequency table for the data.

    b Find the mode of the data. Explain what it means.

    c What fraction of the students chose summer as their favourite season?

a	Season	Tally	Frequency
	Autumn	$\parallel\parallel\parallel\parallel\; \parallel\parallel\parallel$	8
	Winter	$\parallel\parallel\parallel\parallel$	5
	Spring	$\parallel\parallel\parallel\parallel\; \parallel\parallel\parallel\parallel\; \parallel$	11
	Summer	$\parallel\parallel\parallel\parallel\; \parallel$	6
		Total	30

b Spring has the highest frequency so it is the mode.

The most popular season is spring.

c The fraction of students choosing summer as their favourite season is $\frac{6}{30} = \frac{1}{5}$.

DISPLAYING CATEGORICAL DATA

Having collected and organised data, we often want to display it visually using a **graph**. This is not only attractive, but often helps us to identify trends and *understand* the data. Graphs are often published by governments, companies, and the media to convey information to the public.

There are several types of graph we can use to display categorical data:

- **vertical column graph**

- **horizontal bar chart**

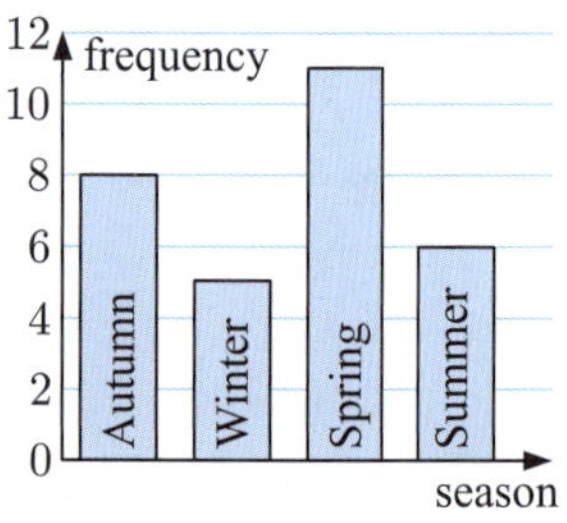
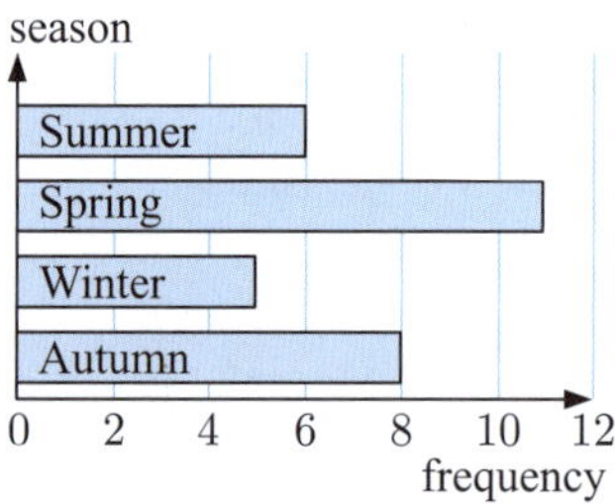

- In a **pie chart**, the angle of each sector illustrates the percentage of data in that category.

  For Jordan's seasons data, each student represents $\frac{1}{30}$ of the pie chart, which is $\frac{1}{30} \times 360° = 12°$.

  So, the sector for autumn is $8 \times 12° = 96°$, and so on.

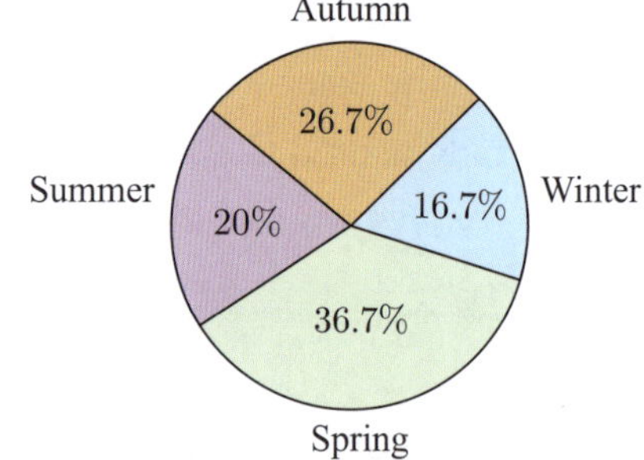

DISCUSSION

Add up the percentages in the pie chart above.

Why do they not add up to exactly 100%?

In the following Exercise you can draw the graphs either by hand or using the **statistics package**.

STATISTICS PACKAGE

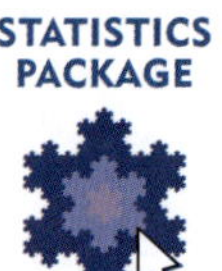

EXERCISE 18B

1 Write down possible categories for each categorical variable:

   a flavours of ice cream

   b methods of transport

   c families of musical instruments in a band

   d methods of communication.

2 50 students were asked their favourite way of spending time with their friends.
The responses are given below, using the categories beach (B), movies (M), park (P), shopping (S), and video games (V).

   V V S B P   M S B V S   M M M B S   M S M B B   V P V S M
   P S S M M   V S V M S   B B S V M   B S M P P   M V S M P

   a Construct a tally and frequency table for this data.

   b How many students chose shopping?

   c Find the mode of the data.

   d What fraction of the students chose the beach?

3 Guests of a hotel in Paris were asked which country they lived in. The results are shown in the vertical column graph.

   a How many guests were surveyed?

   b Find the mode of the data. Explain what it means.

   c How many more guests lived in China than in Canada?

   d What percentage of guests lived in Spain?

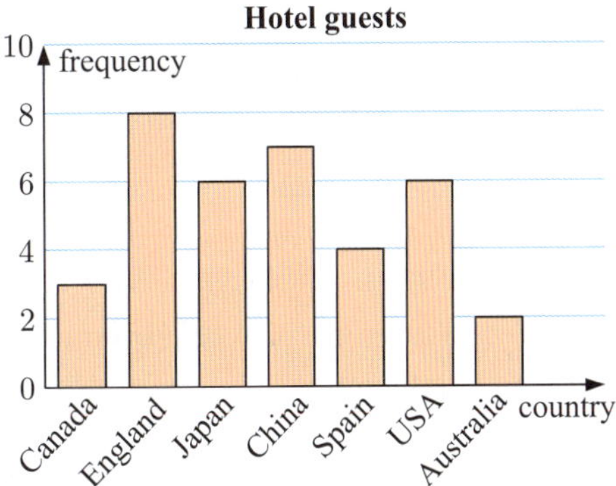

4 A class of 25 students is having a party for multicultural week. The class captain conducts a survey to determine which cuisines are most desired for the party. He records the results using the categories Chinese (C), Greek (G), Indian (In), Italian (It), and Thai (T).

   In It It G It   C T G G It   T G C In G
   G It It C T   It It T In It

   a Draw a tally and frequency table for the data.

   b Construct a column graph to display the data.

   c The three most popular cuisines will be used for the party. Which cuisines should the class captain organise?

5 A member of each household in one street was asked which brand of washing machine they own. The data collected is summarised alongside.

   a How many households were surveyed?

   b Find the mode of the data. Explain what it means.

   c Construct a horizontal bar chart for the data.

   d What percentage of households own a brand E washing machine?

Brand	Frequency
A	11
B	5
C	13
D	8
E	7
F	6

6 The horizontal bar chart shows the median house prices for four cities in 2020.

   a Find the median price in 2020 in:
      i Liverpool   **ii** Edinburgh.

   b How much higher was the median price in Edinburgh than in Glasgow?

   c By what percentage was the median price in Liverpool lower than the median price in Birmingham?

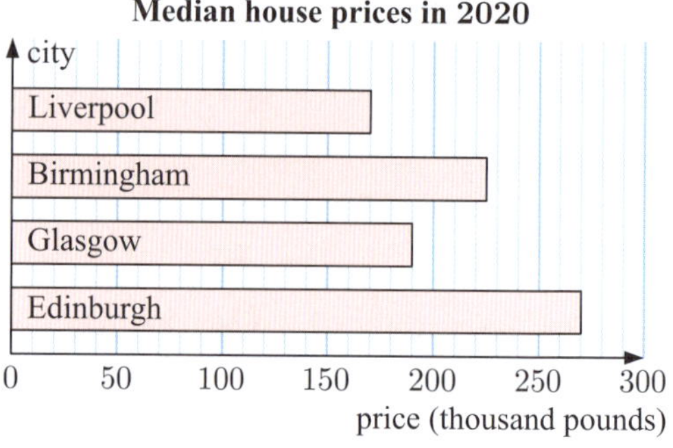

7 The pie chart alongside shows the distribution of supermarkets in a particular country.

   a Which supermarket chain accounted for the most supermarkets?

   b If there are 12 582 supermarkets in total, find the number of supermarkets which are:
      i Tudor's Foods
      ii Food City
      iii Independent.

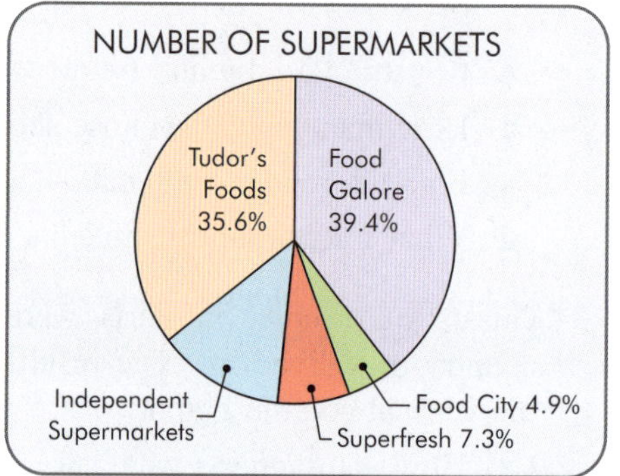

8 The graph displays the results of a national poll where children aged six to sixteen were asked what they do at home between 4 pm and 6 pm.

   a Estimate the percentage of children who:
      i watch TV, videos, play with computers, or listen to music
      ii spend time with friends, siblings, or pets.

   b 240 children were surveyed in the poll.
      i How many do schoolwork?
      ii How many more spend time on the phone than do chores?

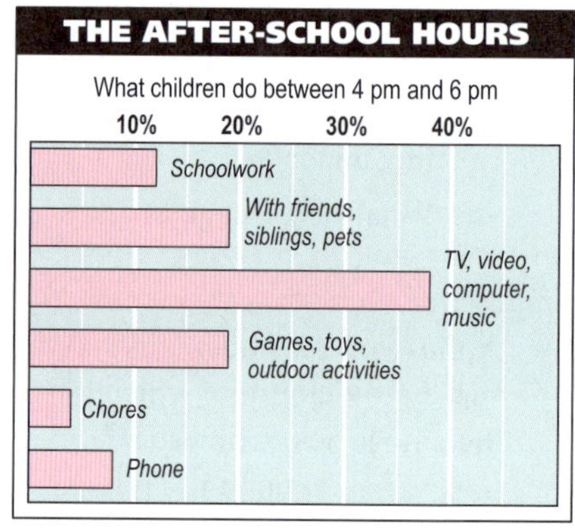

9 The graph alongside indicates the number of ticketed arts performances in a city from 2007 to 2017. Find:

   a the number of ticketed arts performances in 2007

   b the increase in ticketed arts performances from 2016 to 2017

   c the percentage decrease in ticketed arts performances from 2012 to 2013

   d the year which showed the greatest increase in performances from the previous year.

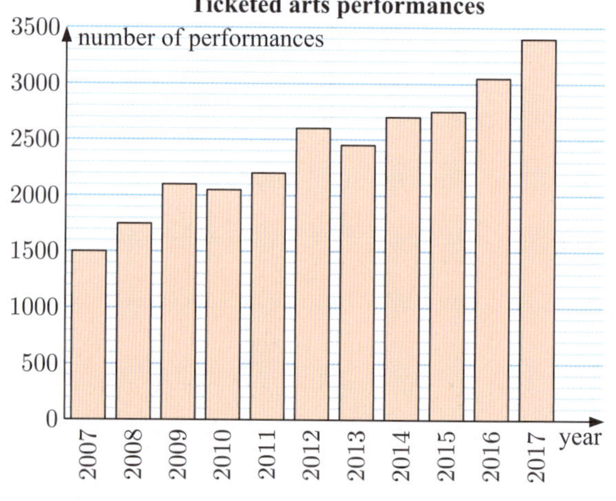

10 The pie chart alongside shows how the health budget will be allocated.

  a How much money was allocated to the health care budget in total?

  b What percentage of the budget will be allocated to mental health services?

  c Which service will receive the most money?

  d How much money will be spent on:

   i hospital support services

   ii other programs?

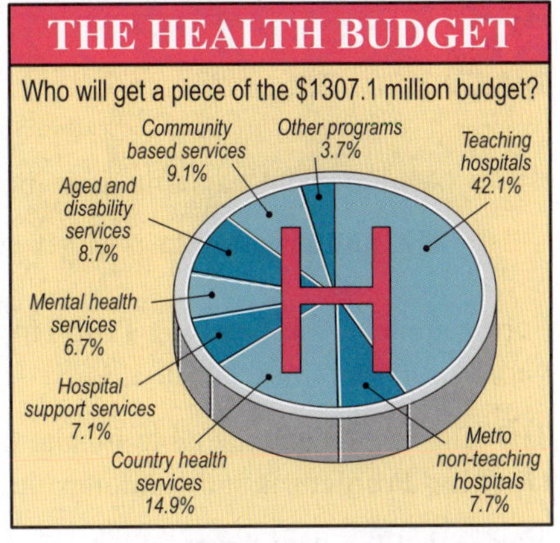

11 Sixty students were asked which elective subject they would like to learn next semester. Their first preferences are given in the table alongside.

  a What percentage of the students want to learn Art?

  b What fraction of the students want to study Textiles? Give your answer in lowest terms.

  c Draw a pie chart to display this data.

Subject	Frequency
Art	11
Design	8
Music	17
Performing Arts	15
Textiles	9

12 The graph shows the number of students studying nursing over a period of 4 years.

  a Estimate the total number of nursing students in 2018.

  b Estimate the percentage of the total nursing students in 2018 who were male.

  c In which year was the proportion of male students the highest?

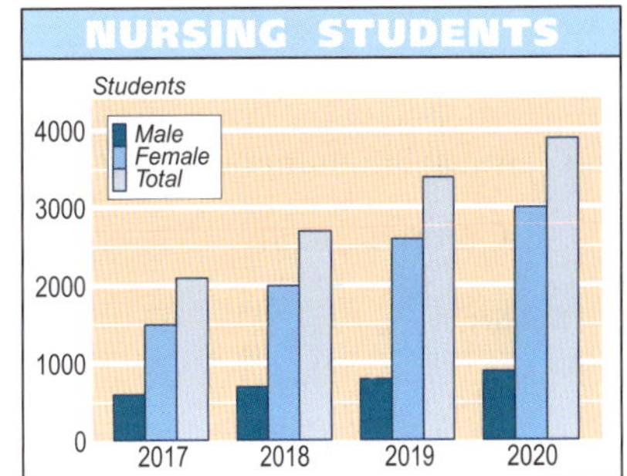

13 143 people voted on who they think is the greatest poet ever. The results are shown in the table.

  a What percentage of people voted for:

   i Shakespeare   **ii** Goethe?

  b Draw a pie chart to display the data.

Poet	Frequency
Yeats	34
Whitman	11
Shakespeare	56
Goethe	23
Dante	19

GLOBAL CONTEXT   MISLEADING GRAPHS

Global context: Fairness and development

Statement of inquiry: Knowing the ways in which statistical graphs can be misleading can help us to better interpret data.

Criterion: Communicating

C   NUMERICAL DATA

A **numerical variable** describes a quantity which takes a value which is a number.

The **numerical data** is either counted or measured.

For example in the **Opening Problem**, the *number of people living in an apartment* is a numerical variable. There could be 0, 1, 2, 3, 4, ... people living in the apartment.

The tally and frequency table for the data in the **Opening Problem** is shown alongside.

Notice that the data value "8" is separated from the rest of the data. We call this data value an **outlier**.

Number of people	Tally	Frequency
1	ⵑⵑⵑ ⵑⵑⵑ	10
2	ⵑⵑⵑ ⵑⵑⵑ |||	13
3	ⵑⵑⵑ ||	7
4	ⵑⵑⵑ		6
5					3
6		
7		
8			1
	Total	40

We can display numerical data using a **column graph** or a **dot plot**.

Column graph

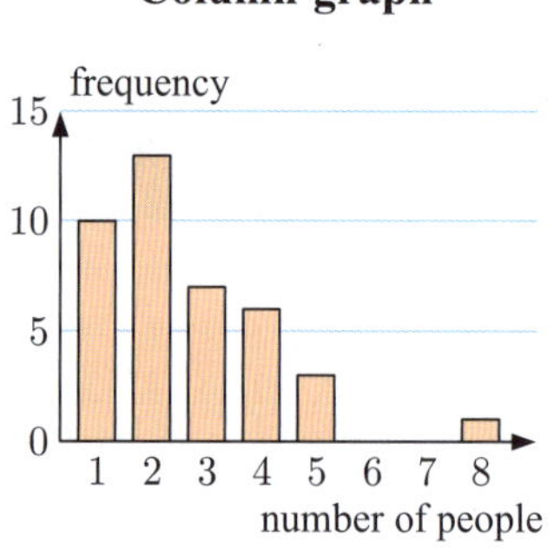

Dot plot

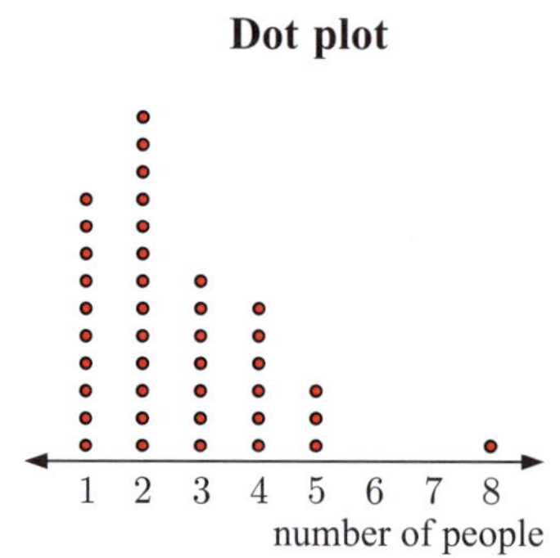

Example 2   ◀)) **Self Tutor**

Every time Ashley buys something at the canteen, he has to wait in line. On his last 20 visits to the canteen, he counted the number of people in front of him when he arrived:

7   10   6   5   5   8   10   7   9   7   8   6   1   7   9   11   5   8   9   8

a Draw a dot plot of the data.

b Are there any outliers in the data?

c On what percentage of visits were there more than 8 people in front of Ashley?

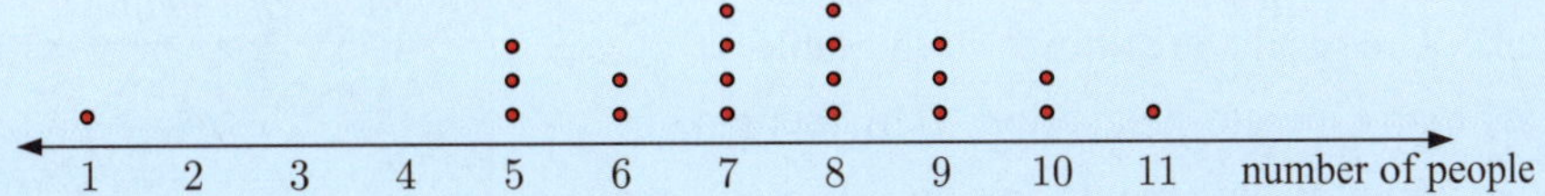

b Yes, the data value "1" is an outlier.

c There were more than 8 people in front of Ashley on $3 + 2 + 1 = 6$ occasions.
This is $\frac{6}{20} \times 100\% = 30\%$ of his visits.

You can use the **statistics package** or your **calculator** to graph numerical data.

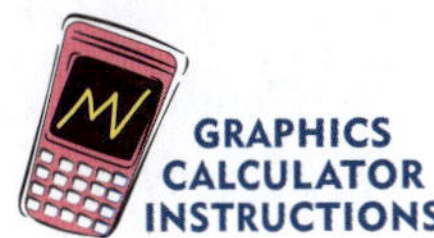

EXERCISE 18C

1 Classify each variable as either categorical or numerical:

a the number of hospitals in capital cities

b the places where you access the internet

c the brands of breakfast cereal

d the heights of students in your class

e the number of road fatalities each day

f the breeds of horses

g the number of hours you sleep each night.

2 A randomly selected sample of teenagers was asked "How many times a week do you eat meat?" A column graph has been constructed from the results.

a How many teenagers answered the survey?

b How many of the teenagers eat meat at least once a week?

c What fraction of the teenagers eat meat twice a week?

d What percentage of the teenagers eat meat more than twice a week?

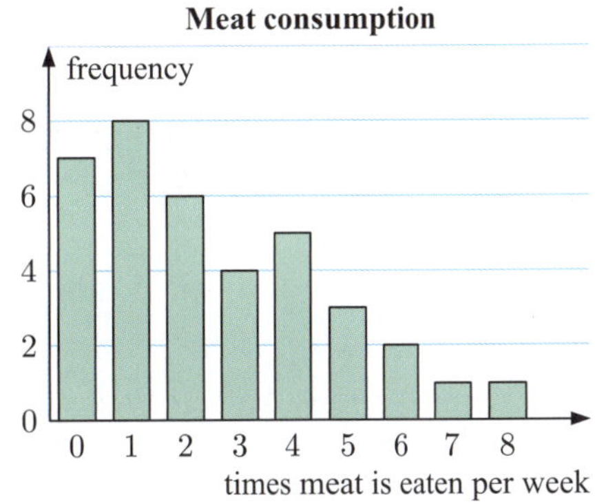

3 25 randomly selected residents of a retirement village were asked "How many great-grandchildren do you have?" Their responses are recorded in the dot plot below.

a How many residents do not have any great-grandchildren?

b What percentage of residents have more than one great-grandchild?

c How would you describe the data value "9"?

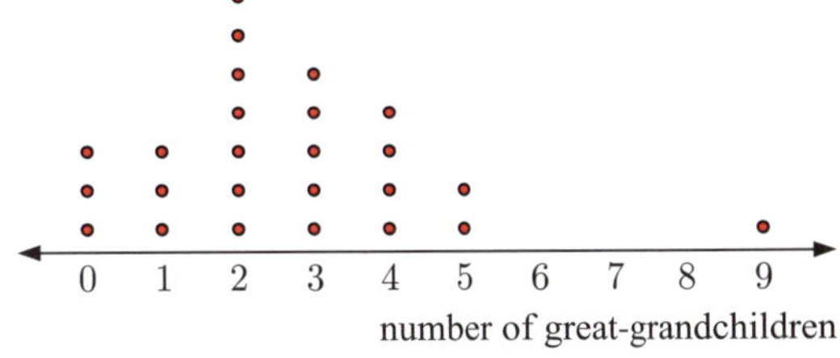

4 Bernadette counts the number of tomatoes on each of her plants. The results are shown in the table.

Number of tomatoes	Frequency
5	2
6	7
7	8
8	5
9	3
10	1

  a How many tomato plants does Bernadette own?

  b What fraction of plants produced more than 7 tomatoes?

  c Draw a column graph to display the data.

  d Are there any outliers?

5 Rosie records the number of doughnuts she sells each day in her bakery:

$$15 \quad 14 \quad 12 \quad 8 \quad 16 \quad 17 \quad 12 \quad 14 \quad 13 \quad 18 \quad 12 \quad 11 \quad 13 \quad 15$$
$$14 \quad 16 \quad 17 \quad 11 \quad 18 \quad 12 \quad 15 \quad 13 \quad 15 \quad 16 \quad 12 \quad 15 \quad 11 \quad 14$$

  a Construct a tally and frequency table for this data.

  b Display the data using a column graph.

  c On what percentage of days were 15 or more doughnuts sold?

  d State any outliers in this data.

6 30 customers in a supermarket were asked how many litres of milk they bought each week. The following data was collected:

$$3 \quad 2 \quad 1 \quad 4 \quad 5 \quad 0 \quad 2 \quad 6 \quad 3 \quad 2 \quad 1 \quad 1 \quad 2 \quad 2 \quad 3$$
$$2 \quad 4 \quad 0 \quad 1 \quad 1 \quad 2 \quad 3 \quad 5 \quad 0 \quad 1 \quad 2 \quad 4 \quad 2 \quad 10 \quad 2$$

  a Construct a dot plot to display the data.

  b Are there any outliers in the data?

  c What percentage of the customers do not buy milk?

  d What percentage of the customers buy 3 or more litres of milk each week?

D   GROUPED DATA

If there are a large number of different data values, it may be appropriate to *group* the data into **class intervals**.

For example, a local kindergarten was concerned about the number of vehicles passing by between 8:45 am and 9:00 am. They recorded the following data over 30 consecutive weekdays:

$$27 \quad 30 \quad 17 \quad 13 \quad 46 \quad 23 \quad 40 \quad 28 \quad 38 \quad 24 \quad 23 \quad 22 \quad 18 \quad 29 \quad 16$$
$$35 \quad 24 \quad 18 \quad 24 \quad 44 \quad 32 \quad 52 \quad 31 \quad 39 \quad 32 \quad 9 \quad 41 \quad 38 \quad 24 \quad 32$$

In this case we use class intervals of length 10 to construct a tally and frequency table:

The **modal class** is the class interval with the highest frequency. In this case the modal class is 20 to 29 cars.

Number of cars	Tally	Frequency
0 to 9	|	1
10 to 19	⦀⦀	5
20 to 29	⦀⦀ ⦀⦀	10
30 to 39	⦀⦀ ||||	9
40 to 49	||||	4
50 to 59	|	1
	Total	30

We can display grouped data using a **column graph** as before, except the columns correspond to the class intervals, rather than individual values.

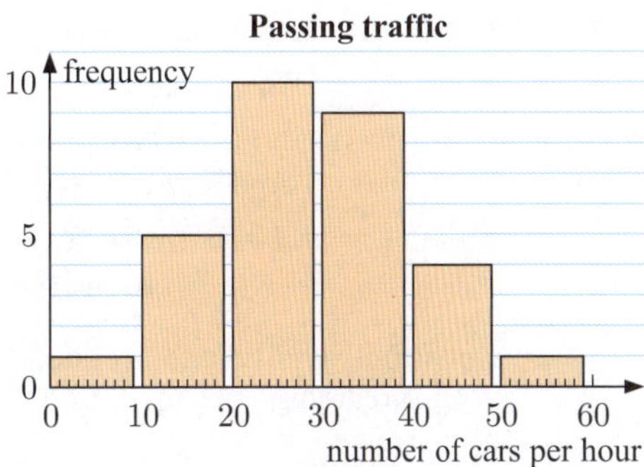

EXERCISE 18D

1 Dave plays golf almost every day. The scores for his last 40 rounds were:

$$
\begin{array}{cccccccc}
72 & 81 & 78 & 77 & 82 & 78 & 89 & 82 \\
85 & 76 & 75 & 81 & 74 & 87 & 84 & 77 \\
80 & 83 & 79 & 82 & 75 & 74 & 74 & 71 \\
75 & 79 & 72 & 76 & 71 & 75 & 82 & 78 \\
76 & 76 & 81 & 77 & 72 & 84 & 87 & 83
\end{array}
$$

   a Construct a tally and frequency table for this data using the class intervals 70 to 74, 75 to 79, 80 to 84, and 85 to 89.

   b How many of Dave's scores were more than 84?

   c What percentage of Dave's scores were less than 75?

   d Copy and complete: More scores were in the interval ...... than in any other interval.

2 An English teacher asked her students to write a brief description of themselves using 70 to 100 words. The teacher recorded the number of words each student used, and displayed the results in the column graph below.

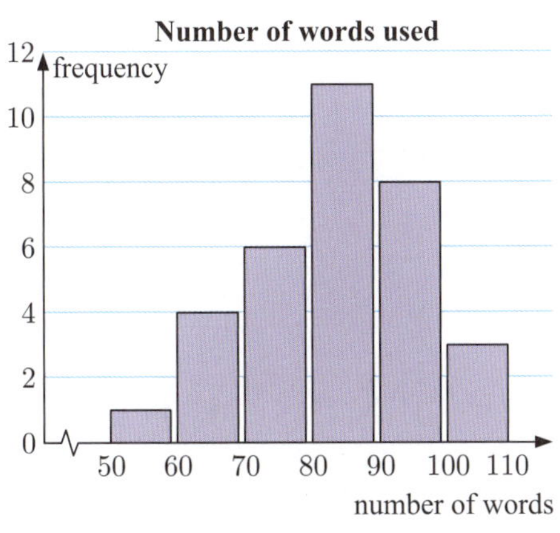

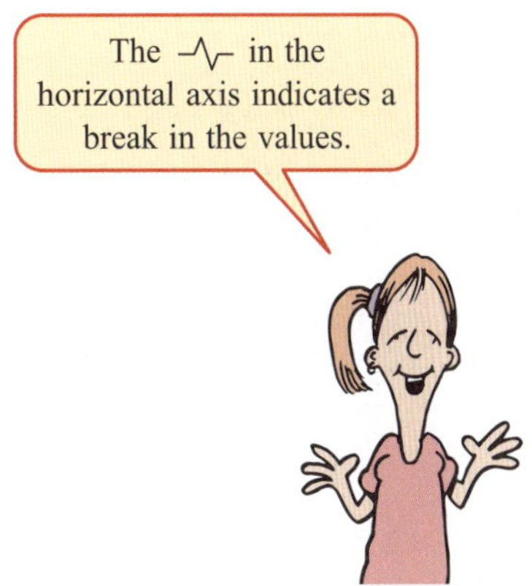

   a How many students are in the class?

   b Find the modal class of the data.

   c How many students did not reach the minimum word limit?

   d How many students exceeded the maximum word limit?

   e Is it possible to determine the highest number of words used?

3 A delicatessen records how many customers it has each day for 50 days. The results are:

$$\begin{array}{cccccccccc}
81 & 82 & 75 & 83 & 104 & 92 & 112 & 124 & 110 & 92 \\
111 & 105 & 78 & 87 & 99 & 86 & 129 & 115 & 121 & 76 \\
109 & 75 & 104 & 96 & 74 & 104 & 98 & 114 & 106 & 109 \\
101 & 110 & 92 & 86 & 122 & 118 & 115 & 117 & 84 & 71 \\
73 & 116 & 119 & 122 & 94 & 81 & 107 & 109 & 115 & 127
\end{array}$$

a Construct a tally and frequency table for this data using the class intervals 70 to 79, 80 to 89, 90 to 99, 100 to 109, 110 to 119, and 120 to 129.

b Draw a column graph for the data.

c Find the modal class of the data.

d On what percentage of the days did the delicatessen have at least 100 customers?

DISCUSSION

- If we are given a set of *raw* or unordered data, how can we efficiently find the lowest and highest data values?
- If the data values are grouped in class intervals on a frequency table or column graph, do we still know what the highest and lowest values are?
- Is there another way to display data in groups without "losing" information about the highest and lowest values?

E STEM-AND-LEAF PLOTS

A **stem-and-leaf plot** is a method of writing data in groups without losing the actual data values.

For each data value, the last digit is used as a **leaf**, and the digits before it determine the **stem** on which the leaf is placed.

We include a **scale** to tell us the place value of each leaf.

Once a stem-and-leaf plot has been constructed, we can write the leaves in ascending order to form an **ordered stem-and-leaf plot**.

Example 3   ◀ Self Tutor

The ages of guests at a 21st birthday party were:

$$\begin{array}{cccccccccccc}
22 & 20 & 18 & 27 & 35 & 21 & 48 & 19 & 21 & 47 & 17 & 26 & 49 \\
16 & 21 & 48 & 19 & 22 & 31 & 20 & 18 & 21 & 19 & 45 & 22
\end{array}$$

a Construct a stem-and-leaf plot for the data.

b Hence construct an *ordered* stem-and-leaf plot for the data.

c How many guests attended the party?

d How old was the youngest guest?

e How many guests were at least 30 years old?

f What percentage of guests were in their twenties?

a The stem-and-leaf plot is:

```
1 | 8 9 7 6 9 8 9
2 | 2 0 7 1 1 6 1 2 0 1 2
3 | 5 1
4 | 8 7 9 8 5
```

Scale: 1 | 8 means 18

b The ordered stem-and-leaf plot is:

```
1 | 6 7 8 8 9 9 9
2 | 0 0 1 1 1 1 2 2 2 6 7
3 | 1 5
4 | 5 7 8 8 9
```

Scale: 1 | 8 means 18

c 25 guests attended the party.

d The youngest guest was 16 years old.

e 7 guests were at least 30 years old.

f 11 guests were in their twenties.

So, the percentage of guests who were in their twenties $= \frac{11}{25} \times 100\% = 44\%$.

EXERCISE 18E

1 For the data in this stem-and-leaf plot, find:

a the value of the circled leaf

b the maximum value

c the number of data values greater than 50

d the total number of data values.

```
2 | 5 6 9
3 | 1 2 7 7
4 | 0 3 4 (6) 9
5 | 2 5 6
```

Scale: 2 | 6 means 26

2 For the stem-and-leaf plot given, find:

a the minimum value

b the maximum value

c the number of data values greater than 90

d the number of data values of at least 100

e the percentage of data values less than 75.

```
 7 | 3 4 6 7 8 9 9
 8 | 0 0 1 1 2 3 3 5 5 6 7 8 8
 9 | 1 1 2 4 4 5 8 9
10 | 2 3 7
11 |
12 | 2
```

Scale: 9 | 1 means 91

3 **a** Construct an unordered stem-and-leaf plot for the following data:

$$27 \quad 34 \quad 19 \quad 36 \quad 52 \quad 34 \quad 42 \quad 51 \quad 18 \quad 48$$
$$29 \quad 27 \quad 33 \quad 30 \quad 46 \quad 19 \quad 35 \quad 24 \quad 21 \quad 56$$

b Hence construct an ordered stem-and-leaf plot for the data.

c State the modal class for the data.

4 Mandy's netball team went undefeated for the season. The winning margins for each game were:

$$32 \quad 20 \quad 1 \quad 7 \quad 5 \quad 13 \quad 14 \quad 9 \quad 23 \quad 12 \quad 6 \quad 55$$
$$14 \quad 23 \quad 30 \quad 16 \quad 6 \quad 13 \quad 24 \quad 9 \quad 4 \quad 19 \quad 8 \quad 27$$

a Construct an unordered stem-and-leaf plot for the data.

b Order the stem-and-leaf plot.

c Find the highest and lowest winning margins.

d Find the modal class of the data.

e In how many games was the winning margin less than 20 points?

f Are there any outliers in this data set?

5 The number of students absent from City Bay High School was recorded for 40 days during winter:

$$21 \quad 16 \quad 23 \quad 24 \quad 18 \quad 29 \quad 35 \quad 37 \quad 41 \quad 38 \quad 33 \quad 28 \quad 29 \quad 34$$
$$25 \quad 23 \quad 17 \quad 26 \quad 34 \quad 33 \quad 10 \quad 27 \quad 49 \quad 52 \quad 43 \quad 41 \quad 31$$
$$24 \quad 20 \quad 25 \quad 23 \quad 19 \quad 22 \quad 12 \quad 16 \quad 17 \quad 15 \quad 14 \quad 14 \quad 32$$

  a Construct a stem-and-leaf plot for the data.

  b Redraw the stem-and-leaf plot so it is ordered.

  c Find the highest and lowest number of absences for the period.

  d On what percentage of the days were there 40 or more students absent?

  e Attendance is described as "good" if there are fewer than 25 absences per day in winter. On what percentage of the days was the attendance "good"?

F   MEASURES OF CENTRE AND SPREAD

We can gain a better understanding of a data set by locating the **middle** or **centre** of the data, and by measuring its **spread**. Knowing one of these without the other is often of little use.

MEASURING THE CENTRE

There are three statistics that are used to measure the **centre** of a data set:

- **mean**
- **median**
- **mode**.

THE MEAN

The **mean** of a data set is the *arithmetic average* of the data values.

$$\text{mean} = \frac{\text{sum of data values}}{\text{number of data values}}$$

The mean is not necessarily a member of the data set.

For example, suppose the *mean* success rate of a basketball team from the free throw line is 78%. It is likely that several of the players score less than 78%, and several score more than 78%. It is not necessary for any of the players to score exactly 78%.

THE MEDIAN

The **median** of a data set is the *middle* value of the *ordered* set of data values.

The median splits the data in two halves. Half of the data are less than or equal to the median, and half are greater than or equal to the median.

For example, suppose the *median* success rate of a basketball team from the free throw line is 78%. This means that half of the team scores less than or equal to 78%, and half scores greater than or equal to 78%.

If there is an *odd* number of data values, there is one middle value which is the median.

If there is an *even* number of data values, there are two middle values. The median is the average of these values.

> If there are n data values, the median is the $\left(\dfrac{n+1}{2}\right)$th data value.

For example:

- When $n = 9$, $\dfrac{n+1}{2} = 5$, so the median is the 5th ordered data value.

- When $n = 12$, $\dfrac{n+1}{2} = 6.5$, so the median is the average of the 6th and 7th ordered data values.

THE MODE

> The **mode** is the most frequently occurring value in the data set.

If there are two data values which are the most frequently occurring, both values are modes and we say the data is **bimodal**.

If there are more than two data values which are the most frequently occurring, we say the mode is **undefined** and do not use it.

MEASURING THE SPREAD

> The **range** of a data set is the difference between the **maximum** or largest data value, and the **minimum** or smallest data value.
>
> $$\text{range} = \textbf{maximum value} - \textbf{minimum value}$$

Example 4

The numbers of games won by a hockey team over the last 12 seasons have been:

$$5 \quad 7 \quad 3 \quad 6 \quad 5 \quad 9 \quad 8 \quad 7 \quad 5 \quad 7 \quad 8 \quad 6$$

For this data set, find the:

a mean   **b** median   **c** mode   **d** range.

a mean $= \dfrac{5+7+3+6+5+9+8+7+5+7+8+6}{12}$ ← sum of data values ← number of data values

$= \dfrac{76}{12}$

≈ 6.33 wins

b The ordered data set is: $3 \; 5 \; 5 \; 5 \; 6 \; 6 \; 7 \; 7 \; 7 \; 8 \; 8 \; 9$ $\{n = 12, \; \dfrac{n+1}{2} = 6.5\}$

$\therefore$ the median $= \dfrac{6+7}{2} = 6.5$ wins

c 5 and 7 are the values which occur most often.

$\therefore$ the data set is bimodal with modes 5 and 7 wins.

d The lowest value is 3 and the highest value is 9.

$\therefore$ the range $= 9 - 3 = 6$ wins.

Example 5

Find the median and mean of the data set illustrated:

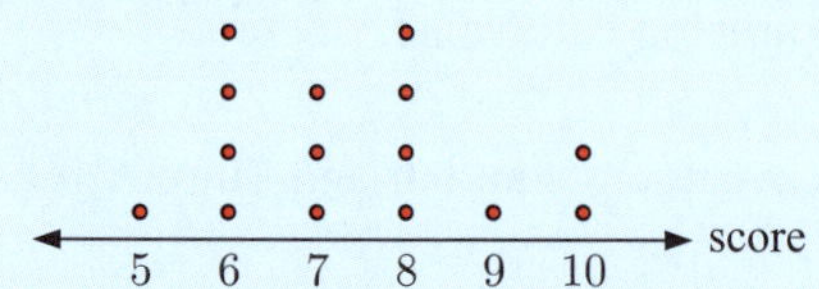

In order, the data values are:

$$5, 6, 6, 6, 6, 7, 7, 7, 8, 8, 8, 8, 9, 10, 10$$

Now $n = 15$ so $\dfrac{n+1}{2} = 8$

$\therefore$ the median is 7.

$$\text{The mean} = \frac{5 + 6 + 6 + ... + 9 + 10 + 10}{15}$$

$$= \frac{111}{15} = 7.4$$

You can use the **statistics package** or your **calculator** to find the measures of centre and spread.

EXERCISE 18F

1   Find the mean, median, mode, and range of each data set:

  a   1, 3, 4, 5, 9, 9, 11

  b   10, 12, 12, 15, 15, 17, 18, 18, 18, 19

  c   8, 4, 17, 11, 10, 10, 12, 11, 9, 18, 11, 6, 17, 7, 8

  d   3.8, 3.6, 4.1, 3.9, 3.6, 4.0, 3.7, 3.8, 4.0, 3.9, 3.7, 4.3

  e   6.8, 8.3, 7.9, 11.0, 9.2, 8.6, 10.1, 9.7, 8.3, 9.4, 6.9, 9.5, 8.9

  f   127, 123, 115, 105, 145, 133, 142, 115, 135, 148, 129, 127, 103, 130, 146, 140, 125, 124, 119, 128, 141, 116

2   The number of bedrooms in the houses on a street are listed below:

2   3   1   3   2   2   4   3   2   2   5   1   2   1   3   1

For this data set, find the:

  a   mean   **b**   median   **c**   mode   **d**   range.

3   In a survey, 25 randomly selected women were asked to give their shoe size. The results were:

7.5   7   6.5   8.5   9   10   11   6.5   8.5   9   8   8   9.5
8.5   7.5   5   7   9   8.5   8   8   9.5   10   6   7.5

  a   Find the range of the data. Explain what it means.

  b   Find the median of the data.

  c   Find the mode of the data. Explain what it means.

  d   If you were a shoe store owner, would you stock the same number of shoes for each size? Explain your answer.

4 At the supermarket checkout, the items in Anita's shopping basket cost:

$$\$2.30, \quad \$3.45, \quad \$1.98, \quad \$5.27, \quad \$0.95, \quad \$5.45, \quad \$3.99, \quad \$2.89,$$
$$\$1.45, \quad \$6.12, \quad \$12.99, \quad \$6.59, \quad \$9.10, \quad \$3.95, \quad \$1.85, \quad \$6.45.$$

  a Find the mean price of the items.

  b Find the median price of the items.

  c Does the data set have a mode? Explain your answer.

  d Find the range of the data.

5 Find the median and mean of the data set:

a 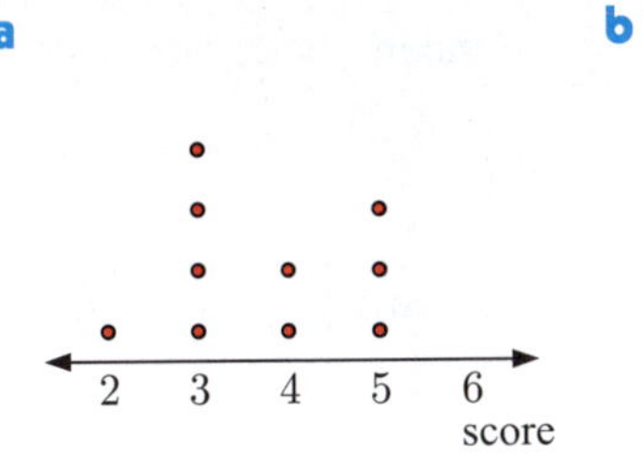

b 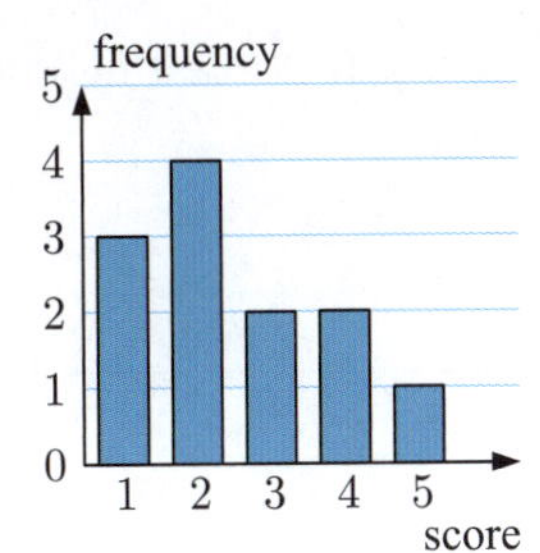

c
$$
\begin{array}{c|l}
2 & 1\ 3 \\
3 & 2\ 5\ 6 \\
4 & 3\ 4\ 4\ 5\ 8 \\
5 & 1\ 5\ 7 \\
\end{array}
$$

Scale: $2\,|\,1$ means 21

Example 6

🔊 **Self Tutor**

A set of 7 data values has mean 9. Find the sum of the data values.

$$\text{mean} = \frac{\text{sum of data values}}{\text{number of data values}}$$

$$\therefore\ 9 = \frac{\text{sum of data values}}{7}$$

$\therefore$ the sum of the data values $= 9 \times 7 = 63$

---

6 A set of 12 data values has a mean of 6. Find the sum of the data values.

7 A set of data values has sum 143 and mean 13. How many data values are in the set?

8 Huang does casual work at a petrol outlet. On average, he has earned \$214.50 per week for the last 6 weeks. How much has he earned in total?

9 It started snowing on Sunday January 21st. 11 cm of snow fell that day. On the next 5 days 24 cm, 28 cm, 22 cm, 16 cm, and 13 cm of snow fell.

  a Find the mean snowfall from Sunday to Friday.

  b How much snow would need to fall on Saturday for the mean to rise to 20 cm?

  c If 10 cm of snow fell on Saturday, calculate the mean snowfall for the week.

10 Helen rolled a die eight times. The numbers she rolled had mean 4, median 3.5, mode 3, and range 4. List the results of Helen's rolls, from smallest to largest.

ACTIVITY 2   DICE GAME FOR 2 PLAYERS

In this game players compare statistics from their dice rolls to earn points.

INVESTIGATION   THE EFFECT OF OUTLIERS

In this Investigation we will examine the effect of an outlier on the mean, mode, and median of a data set.

What to do:

1 **a** Copy and complete:

Data set	Mean	Mode	Median
4, 5, 6, 6, 6, 7, 7, 8, 9, 10			
4, 5, 6, 6, 6, 7, 7, 8, 9, 10, 100			

  b Comment on the effect that the outlier had on the:

   i mean   **ii** mode   **iii** median.

  c Which measure of centre was most affected by the inclusion of the outlier?

2 The prices of 20 cars in a used car dealership are:

$4800	$7900	$12 950	$3000	$4950	$9800	$12 000
$11 000	$6200	$5250	$8750	$6900	$9900	$7500
$8250	$6900	$5800	$10 000	$8500	$7600	

  a Calculate the mean and median prices.

  b A luxury car is now being sold at the dealership for $45 000.
     Find the new mean and median prices.

  c Which measure of centre was most affected by the inclusion of the outlier?

  d Which measure of centre gives a better indication of the price you would expect to pay for a car at this dealership?

3 Under what circumstances do you think the mean or the median would be more appropriate to use as a measure of the centre of a data set? Explain your answer.

G   MEASURES OF CENTRE AND SPREAD FROM A FREQUENCY TABLE

We can find the measures of centre and spread directly from a frequency table without having to list the individual data values.

To do this we add:

- a *product* column which records the sum of scores with each particular value
- a *cumulative frequency* column which records the number of data values less than or equal to that particular score.

We use the product column to help calculate the mean, and the cumulative frequency column to help calculate the median.

Example 7   ◄» Self Tutor

A class of 20 students take a spelling test. Their results out of 10 are shown in the table.

For these scores, calculate the:

a mode   **b** median

c mean   **d** range.

Score	Number of students
5	1
6	2
7	4
8	7
9	4
10	2
Total	20

Score	Number of students	Product	Cumulative frequency
5	1	$5 \times 1 = 5$	1
6	2	$6 \times 2 = 12$	$1 + 2 = 3$
7	4	$7 \times 4 = 28$	$3 + 4 = 7$
8	7	$8 \times 7 = 56$	$7 + 7 = 14$
9	4	$9 \times 4 = 36$	$14 + 4 = 18$
10	2	$10 \times 2 = 20$	$18 + 2 = 20$
Total	20	157	

a The score "8" had the highest frequency, so the mode $= 8$.

b $n = 20$ so $\dfrac{n+1}{2} = \dfrac{21}{2} = 10.5$

The 10th and 11th scores are both 8, so the median $= 8$.

c Mean $= \dfrac{\text{sum of scores}}{\text{number of scores}} = \dfrac{157}{20}$

$= 7.85$

d The lowest score is 5 and the highest score is 10, so the range $= 10 - 5 = 5$.

EXERCISE 18G

1 A class of 28 students was asked how many televisions they had in their home. The results are shown in the table.

For this data, calculate the:

a mode   **b** median

c mean   **d** range.

Number of televisions	Frequency
0	1
1	8
2	12
3	7
Total	28

2 Sarah recorded the number of occupants in each car that drove down her street in one hour. The results are shown in the column graph.

a Construct a frequency table from the graph.

b How many cars did Sarah record data for?

c For this data, find the:

    i mode   **ii** median

    iii mean   **iv** range.

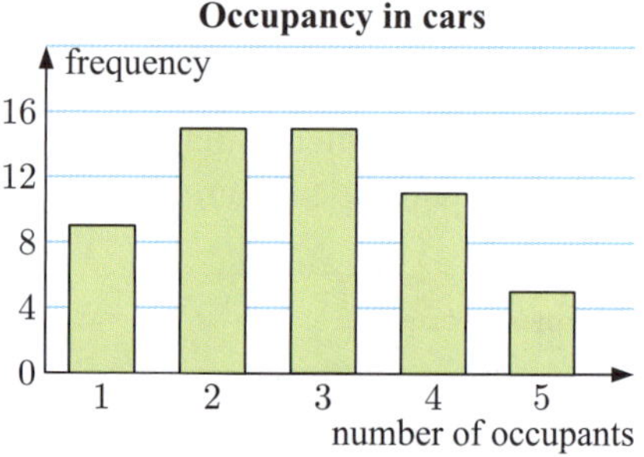

3 60 students in the Netherlands were asked how many times they had travelled to another country for a holiday. The results are given in the frequency table.

a For this data, find the:

    i mode   **ii** median

    iii mean   **iv** range.

b Construct a column graph of the data. Show the positions of the mode, median, and mean on the horizontal axis.

Number of international holidays	Frequency
0	8
1	9
2	11
3	12
4	8
5	5
6	3
7	2
8	2

4 The horizontal bar chart alongside shows the ages of students participating in an international school mathematics competition.

a How many students participated in the competition?

b Find the mode and explain what it means.

c Find the median age of the students.

d Find the mean age of the students.

e Compare the median and mean ages. Explain how you can tell which is greater by looking at the chart.

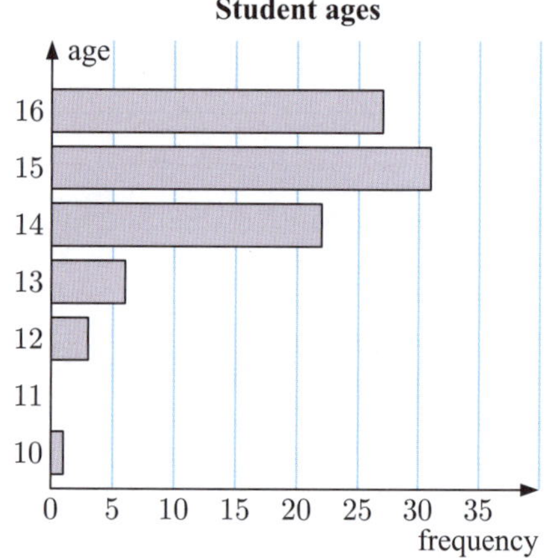

GLOBAL CONTEXT

THE AGE OF A POPULATION

Global context:	Identities and relationships
Statement of inquiry:	Calculating statistics and constructing graphs can help us understand the characteristics of and relationships within a population.
Criterion:	Communicating

PUZZLE

A MEAN GAME

Use the bridges to move from the Start island to the Finish island. The mean of the numbers you have landed on must never drop below 5 or rise above 10. You may not land on the same island more than once.

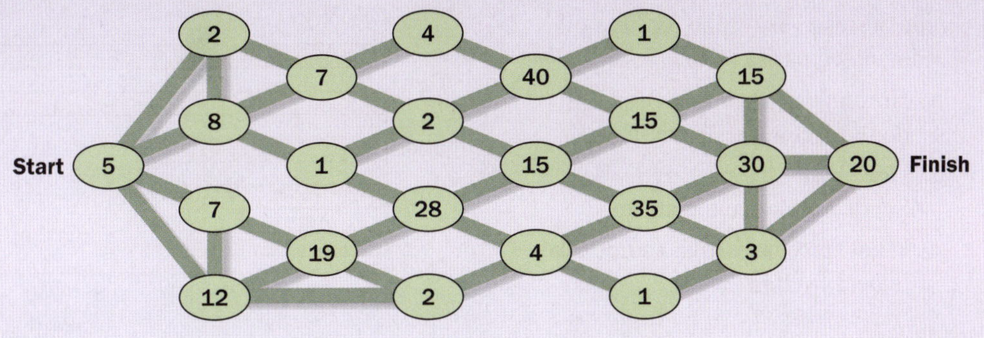

MULTIPLE CHOICE QUIZ

QUICK QUIZ

REVIEW SET 18A

1 Classify each variable as categorical or numerical:

  a the area of each block of land in a town

  b the marital status of people in a suburb

  c the number of matches in a match box.

2 The dot plot shows the number of goals scored by a group of netballers in 10 attempts.

  a How many netballers are in the group?

  b Are there any outliers?

  c Find the median number of goals scored.

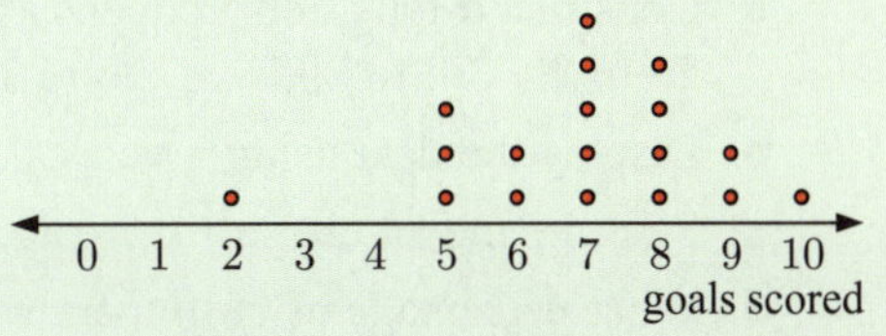

3 Of the passengers on a bus, 15 were children, 11 were students, 22 were adults, and 12 were pensioners.

   a Find the mode of the data.

   b What percentage of the passengers were pensioners?

   c What fraction of the passengers were children?

   d Display this data on a pie chart.

4 Explain any bias in the following samples:

   a To find the favourite TV show of high school students, the Year 8 students at the local high school are surveyed.

   b To determine the most popular dish at a restaurant, the diners at Sunday lunch are surveyed.

5 This horizontal bar chart shows the crude oil reserves for the top five oil producing countries in the world.

   a How many barrels of crude oil do these countries have in total?

   b How many more barrels does Saudi Arabia have than Iraq?

   c What percentage of the crude oil is held by Iran?

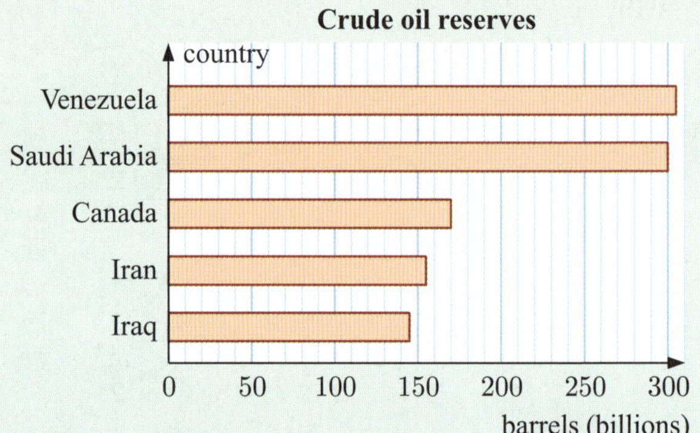

6 The results of the university students studying Quantum Physics were graded as high distinction (HD), distinction (D), credit (C), pass (P), or fail (F). The grades obtained were:

C P F P D   D C P F P   C C P D P
C F P P C   HD D C P F   C P F P

   a Draw a tally and frequency table to organise the data.

   b Construct a column graph to display the data.

   c State the mode of the data. Explain what it means.

   d What percentage of the students failed the course?

7 A customer satisfaction survey produced a rating out of 100 for each customer's overall experience. The results were:

62 85 100 57 12 74 63 91 54 58 100 96 80 86 98
100 82 66 92 75 89 68 85 73 82 95 95 61 100 56

   a Construct a tally and frequency table for this data using the class intervals 10 to 19, 20 to 29, 30 to 39, ..., 90 to 100.

   b Display the data using a vertical column graph.

   c Find the modal class of the data.

   d What percentage of customers were at least 80% satisfied?

8 The test scores out of 40 marks were recorded for a group of 30 students:

25	18	35	32	34	28	24	39	29	33	22	34	39	31	36
35	36	33	35	40	26	25	20	18	9	40	32	23	28	27

   a Construct a stem-and-leaf plot for this data using 0, 1, 2, 3, and 4 as the stems.

   b Redraw the stem-and-leaf plot so it is ordered.

   c What advantage does the stem-and-leaf plot have over a column graph?

   d Find the range of the marks scored for the test.

   e An "A" was awarded to students who scored 36 or more for the test. What percentage of students scored an "A"?

9 Damien bought his lunch each day for four days, spending an average of $6.20 each day.

   a Find the total amount of money that Damien has spent on lunches.

   b The next day, Damien pays $8.40 for his lunch. Find Damien's new average lunch cost.

10 A cinema wanted to find the average age of their patrons. At the end of a children's movie, 20 randomly selected audience members were asked their age. The responses were:

4	29	6	8	42	9	10	61	31	5	7	42	6	51	9	11	33	8	28	12

   a Find the mean of the sample.

   b Find the median of the sample.

   c Explain why neither the mean nor the median will be good estimates of the average age of the cinema's patrons.

11 For the data in this stem-and-leaf plot, find the:

   a mode   **b** median

   c mean   **d** range.

```
3 | 8 9
4 | 2 2 5
5 | 0 1 1 7 9
6 | 1 3 4
7 | 0
```

Scale: 3 | 8   means 3.8

12 Bernie asked 25 randomly selected adults how many hot beverages they drank per day. The results are listed below:

4	2	1	0	0	3	5	4	3	2	2	2	1
1	10	1	2	4	2	0	1	1	3	4	2	

   a Draw a tally and frequency table for the data.

   b Find the:

      i mode   **ii** median   **iii** mean.

   c Construct a column graph to display the data. Include the mode, median, and mean on the horizontal axis.

   d Identify the outlier in the data.

   e Which measure of centre is most affected by the outlier? Explain your answer.

   f What fraction of the adults drink less than 3 hot beverages per day?

13 The staff at a cinema asked customers to state their favourite type of film. The results are shown in the pie chart alongside.

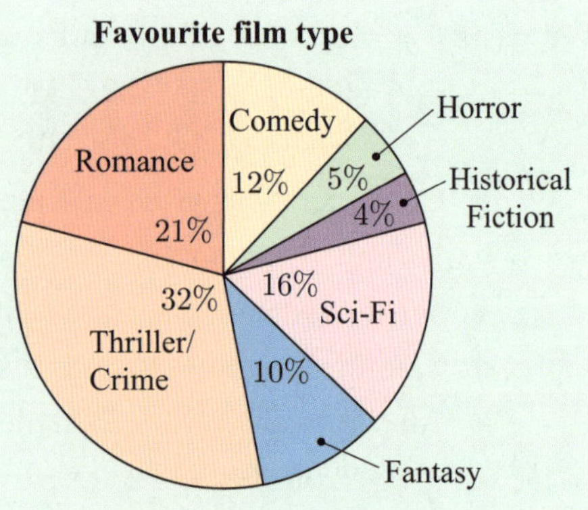

a What is the most popular film type?

b What percentage of customers chose Sci-Fi or Fantasy?

c 20 of the customers surveyed chose Horror. How many customers:

   i were surveyed

   ii chose Romance?

REVIEW SET 18B

1 Decide whether a census or sample would be appropriate for finding:

a the average number of books owned by members of a book club

b the median height of people living in Tokyo.

2 To investigate the public's opinion on whether money should be spent upgrading the local library, a questionnaire is placed in the library for people to answer.

Explain why this sample is likely to be biased.

3 The students in a class recorded how many siblings they have. The results were:

   0 1 2 1 1   0 1 3 1 0   0 1 2 2 1   1 0 1 2 4   1 2 0 0 1   2 1 0

a How many students are in the class?

b Draw a dot plot to display the data.

c Are there any outliers in the data? Explain your answer.

d What percentage of children in the class have no siblings?

4 An English class was asked how many pages they had read so far of their class novel. The results are given alongside.

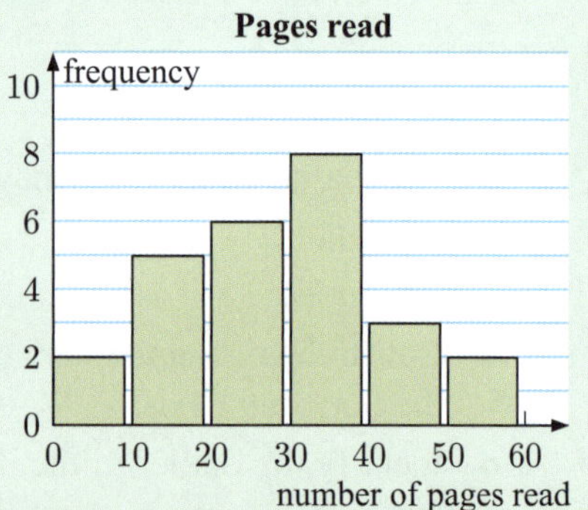

a What is the modal class for the data?

b Draw a frequency table for this data.

c What percentage of students have read 40 or more pages?

d The class was supposed to read the first 30 pages for homework. How many students did not complete their homework?

e Can you find the range of the data? Explain your answer.

5 Forty soccer players at a training camp were asked what position they played. The results were recorded as striker (S), midfielder (M), defender (D), or goalkeeper (G).

```
S  D  M  M  S   M  D  S  G  M
M  D  S  D  M   D  S  M  D  G
S  G  M  D  S   D  M  D  S  M
D  G  M  S  M   D  G  M  M  S
```

a Summarise the data in a tally and frequency table.

b Find the mode of the data. Explain what it means.

c What fraction of the players were midfielders?

d What percentage of the players were defenders or goalkeepers?

e Display the data on a horizontal bar chart.

6 An office recorded the number of phone calls it received each day for 30 days:

```
18  32  25  37  24  28  35  42  31  16  28  38  30  41  27
31  46  17  35  26  32  14  44  37  48  23  19  33  15  36
```

a Organise the data into a tally and frequency table, using the class intervals 10 to 19, 20 to 29, 30 to 39, and 40 to 49.

b Display the data on a column graph.

c Find the modal class of the data.

d On what percentage of days did the office receive at least 30 calls?

7 This stem-and-leaf plot shows the weights of Year 8 students in a class. Find:

a the minimum weight

b the maximum weight

c the median weight

d the range of weights

e the number of students who weigh at least 70 kg

f the percentage of students who weigh less than 48 kg.

```
3 | 2 4 8
4 | 0 4 4 7 9 9 9
5 | 0 0 1 2 2 3 3 5 5 6 8 8
6 | 0 1 2 4 4 5 7 9
7 | 0 2 6
8 | 4
9 | 1
```

Scale: 9 | 1 means 91 kg

8 The lengths of a group of cars were measured. The results in metres were:

```
4.46  4.58  4.37  4.21  4.19  4.28  4.75  4.54  4.64
4.68  4.27  4.38  4.35  4.48  4.58  4.52  4.43  4.24
```

a Draw an ordered stem-and-leaf plot to display this data.

b Find the range of the data.

9 For the data in the dot plot, find the:

a mode   **b** median

c mean   **d** range.

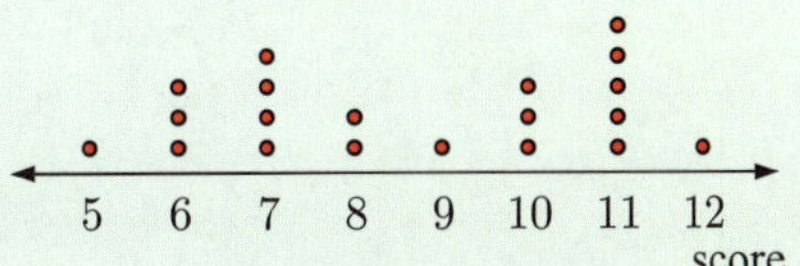

10

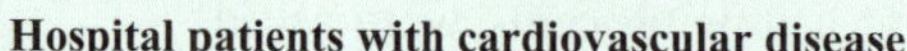

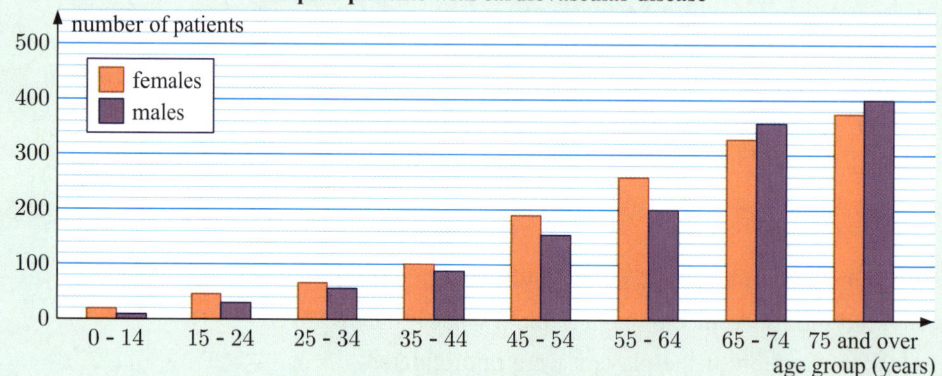

This graph shows the number of cardiovascular patients admitted to a hospital over 3 months.

  a How many 55 - 64 year old males were admitted to hospital?

  b Find the percentage of the total patients aged 45 - 54 who were female.

  c Which age groups had more male patients than female patients?

11 The table shows the scores in a fitness test for a class of 13-year-old Vietnamese students.

  a Calculate the median score.

  b Calculate the mean score.

  c The average score for all 13-year-old Vietnamese students was 7.2. How does this class compare with the national average?

Score	Frequency
6	2
7	4
8	7
9	12
10	5
Total	30

12 The column graph alongside shows the number of assists per game in a basketball player's international career.

  a Construct a frequency table for the data.

  b Determine the total number of assists made by the player.

  c For this data, find the:

   i mode   **ii** median

   iii mean   **iv** range.

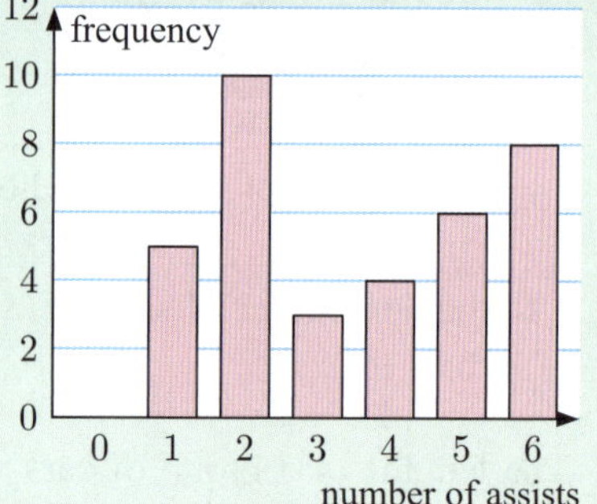

Congruence and similarity

Contents:

OPENING PROBLEM

Jane cut two triangular slices of cheesecake, and gave one to her brother Nathan.

"That's not fair!", Nathan said, "Your slice is bigger than mine!"

Jane used a ruler to measure the sides of each slice. "See, my slice has sides 5 cm, 6 cm, and 7 cm, and so does yours. That means the slices are the same size."

"Not necessarily", said Nathan, "the slices might have the same sides, but the angles might be different."

Things to think about:

a Who do you think is correct?

b What mathematical argument can you use to justify your answer?

In previous years we have studied several **transformations**.

When we perform a **translation**, **rotation**, or **reflection**, the image is identical in size and shape to the original object. We say that the object and image are **congruent**.

When we perform an **enlargement** or **reduction**, the image is identical in shape to the original object, but its size is different. We say that the object and image are **similar**.

In this Chapter we will study **congruence** and **similarity**, and see how these principles are used in practical problems.

A   CONGRUENCE

Congruent figures are identical in size and shape. They do not need to have the same position or orientation.

Example 1   ◄》 Self Tutor

Decide whether these pairs of figures are congruent:

a

b

c

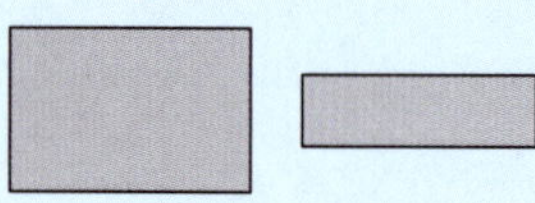

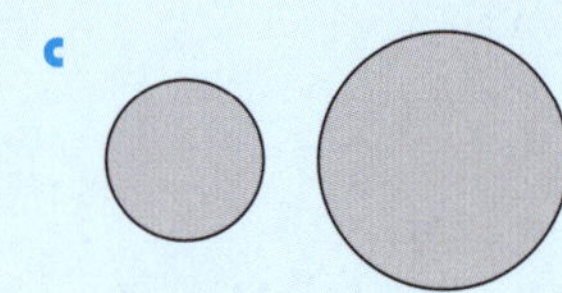

a The figures do not have the same shape, so they are *not congruent*.

b The figures are identical in size and shape even though one is rotated. They are therefore *congruent*.

c The figures have the same shape, but they are not the same size. They are therefore *not congruent*.

EXERCISE 19A

1 Decide whether these pairs of figures are congruent:

a

b

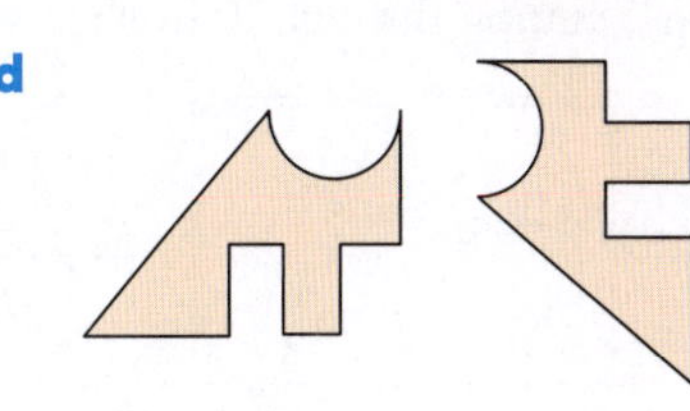

c

d

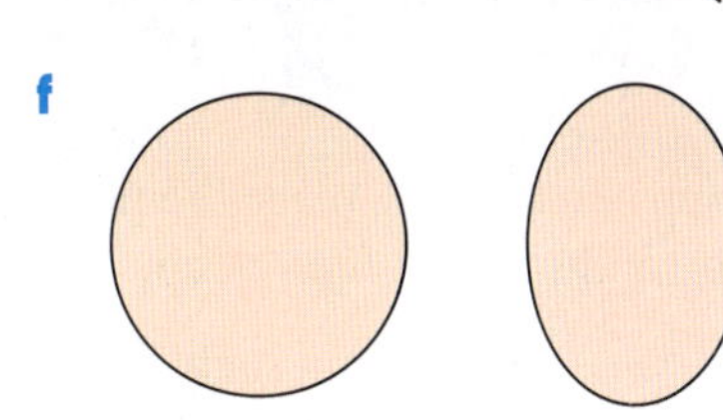

e

f

2 Which *two* of these figures are congruent?

A    **B** 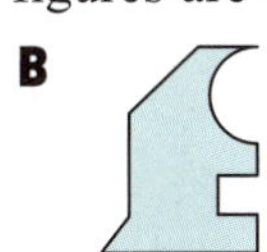   **C**    **D** 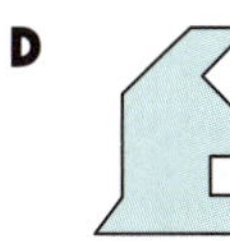   **E** 

3 Which *three* of these figures are congruent?

A 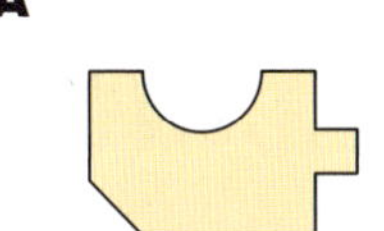   **B** 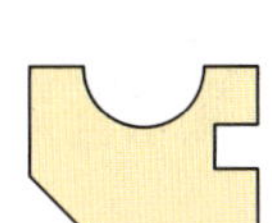   **C** 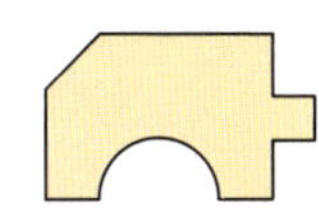   **D** 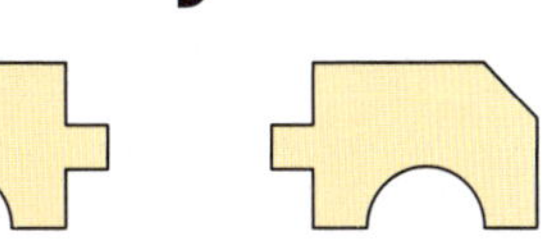   **E** 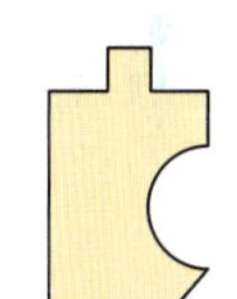

4 Quadrilaterals ABCD and EFGH are congruent.
Determine the:

  a length of side [EF]

  b size of $\hat{FGH}$

  c perimeter of EFGH.

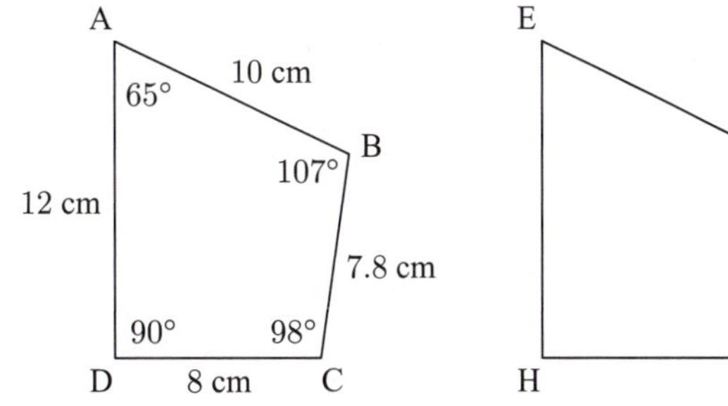

5 Triangles ABC and PQR are congruent.
Determine the:

  a length of side [RQ]

  b size of $\hat{PQR}$

  c area of PQR.

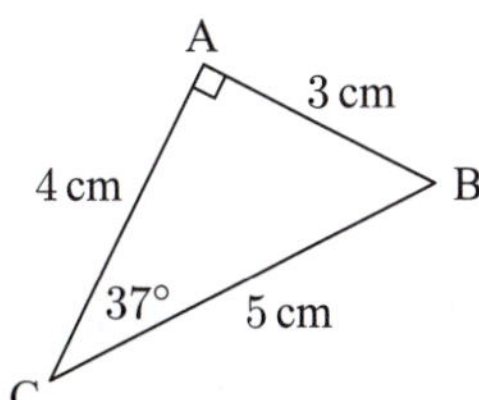
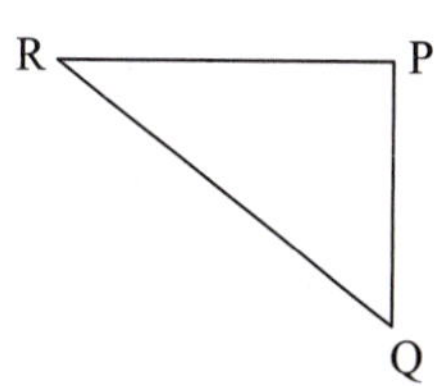

INVESTIGATION 1   CONGRUENCE

You will need: Two sheets of card, scissors.

What to do:

1   Draw a shape on one of the sheets of card.

2   Place the second sheet of card behind it, and hold them together tightly. Carefully cut out
the shape, cutting through both sheets of card. This will give you two congruent figures.

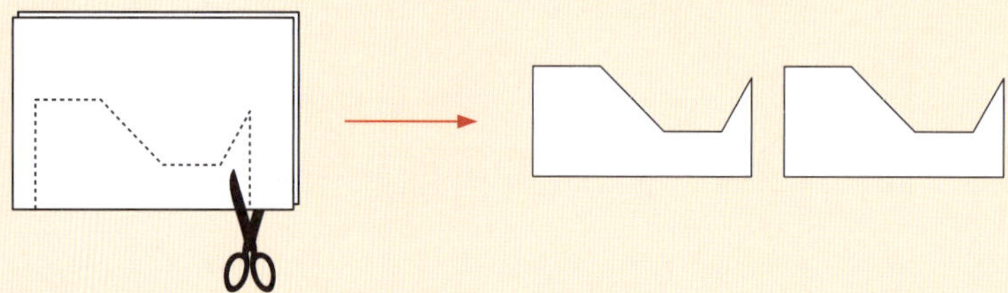

3   In a group or as a class, place the figures from each student
in a box, and mix the figures up.

4   Try to pair up the congruent figures.   Discuss what
transformations you apply to the figures in order to test
whether two figures are congruent.

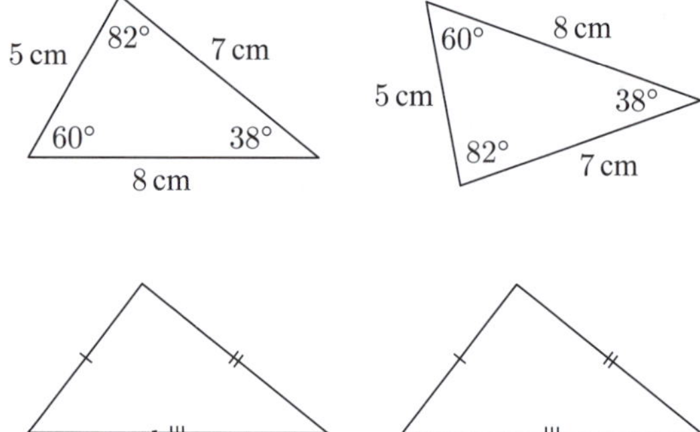

5   Can you *define* congruence in terms of transformations?
Write your answer in the form:

*Two figures are congruent if one can be placed exactly
on top of the other using a combination of ......*

B   CONGRUENT TRIANGLES

The triangles alongside have identical side
lengths and angles, so the triangles are
congruent.

However, we do not necessarily need to know
all of the information given to conclude that
the triangles are congruent.

For example, if we know that two triangles
have the same side lengths, this information is
sufficient to conclude that the triangles *must* be
congruent.

Click on the icon to explore this property for yourself.

DEMO

INVESTIGATION 2 — CONSTRUCTING TRIANGLES

In this Investigation we will discover other conditions which allow us to conclude that two triangles are congruent.

You will need: Paper, ruler, protractor.

What to do:

1   Two sides and an included angle

Draw a triangle with two side lengths 8 cm and 12 cm, and with an angle measuring 25° between these sides.

How many different triangles can be constructed?

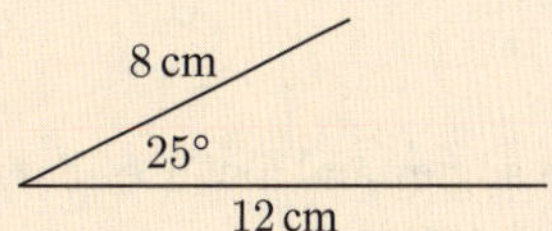

2   Two angles and a corresponding side

Draw a triangle with two angles measuring 70° and 45°, and with a side 10 cm long between these angles.

How many different triangles can be constructed?

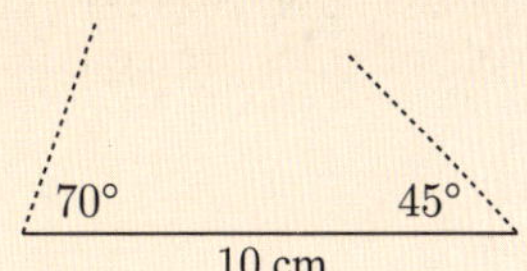

3   Right angle, hypotenuse, and a side

Draw a right angled triangle with hypotenuse 10 cm, and one other side 6 cm long. How many different triangles can be constructed?

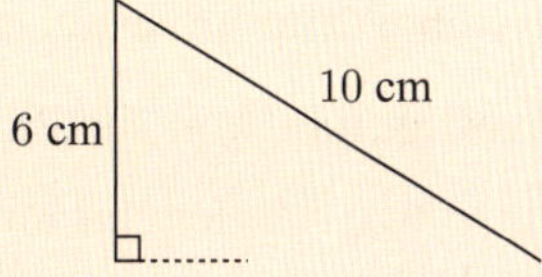

4   Two sides and a non-included angle

Draw a triangle with two side lengths 8 cm and 12 cm, and with an angle measuring 25° between the 12 cm side and the third side.

How many different triangles can be constructed?

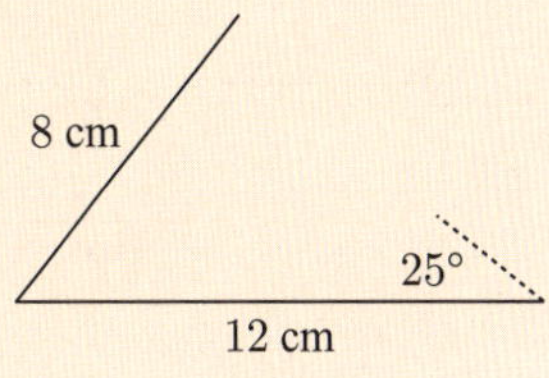

5   Three angles

Draw a triangle with angles measuring 50°, 60°, and 70°.

How many different triangles can be constructed?

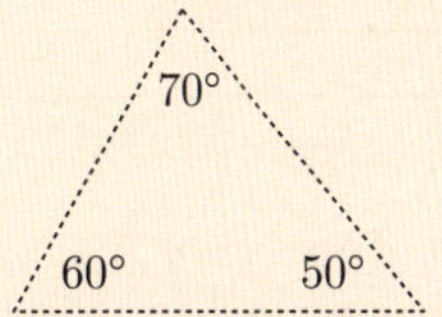

From the **Investigation** you should have made the following discoveries:

Two triangles are **congruent** if any one of the following is true:

- All corresponding sides are equal in length. (**SSS**)

- Two sides and the **included angle** are equal. (**SAS**)

- Two angles and a pair of **corresponding sides** are equal. (**AAcorS**)

- For right angled triangles, the hypotenuses and one pair of sides are equal. (**RHS**)

When we use one of these reasons to conclude that two triangles are congruent, we use the appropriate abbreviation to record our reason.

We use the symbol $\cong$ to indicate congruence.

From the **Investigation**, you should also have found that:

- If we know two side lengths and a non-included angle, there may be two ways to construct the triangle. We therefore *cannot* conclude that the two triangles are congruent.
- If we know all angles of a triangle, the triangle may still vary in size. We therefore *cannot* conclude that the two triangles are congruent.

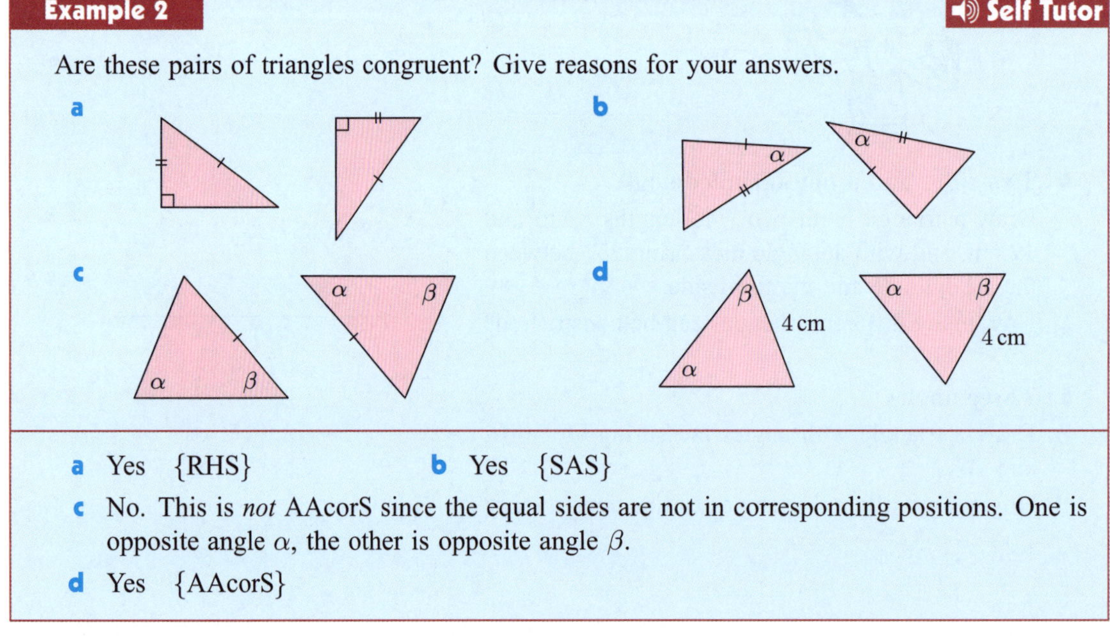

Example 2   ◀)) **Self Tutor**

Are these pairs of triangles congruent? Give reasons for your answers.

a

b

c

d

a Yes {RHS}   **b** Yes {SAS}

c No. This is *not* AAcorS since the equal sides are not in corresponding positions. One is opposite angle α, the other is opposite angle β.

d Yes {AAcorS}

Once we have established that two triangles are congruent, we can deduce that the remaining corresponding sides and angles of the triangles are equal.

Example 3

a Show that these triangles are congruent.

b What can be deduced from this congruence?

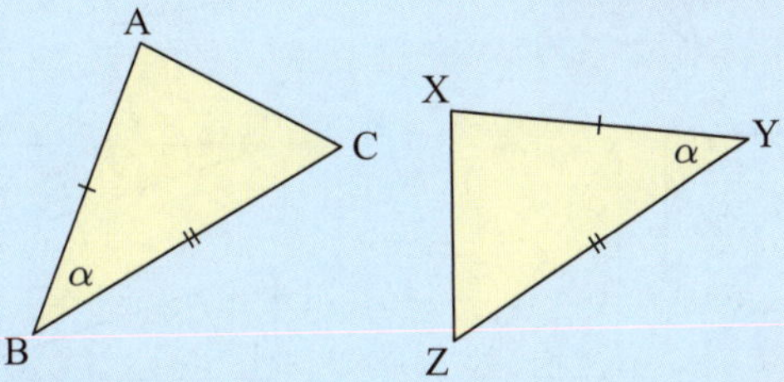

a In △s ABC and XYZ:

- $AB = XY$
- $BC = YZ$
- $\widehat{ABC} = \widehat{XYZ}$
- $\therefore \triangle ABC \cong \triangle XYZ$ {SAS}

b We can deduce that:

- $AC = XZ$
- $\widehat{BAC} = \widehat{YXZ}$ {β}
- $\widehat{ACB} = \widehat{XZY}$ {θ}

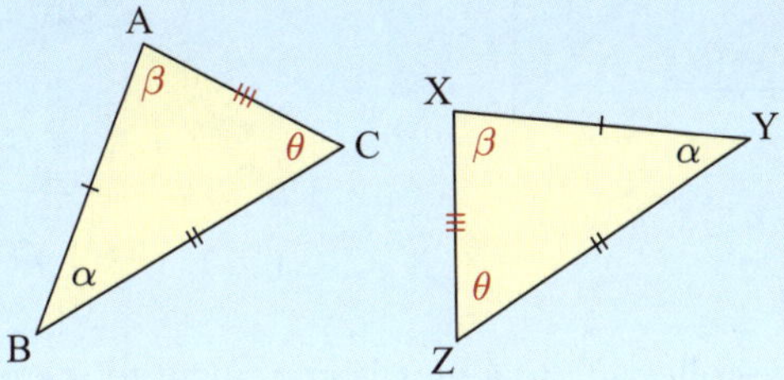

When we describe congruent triangles, we label the vertices that are in corresponding positions in the same order. In the previous Example, we therefore write $\triangle ABC \cong \triangle XYZ$,   not $\triangle ABC \cong \triangle YZX$.

EXERCISE 19B

1 Explain why we *cannot* conclude that these triangles are congruent:

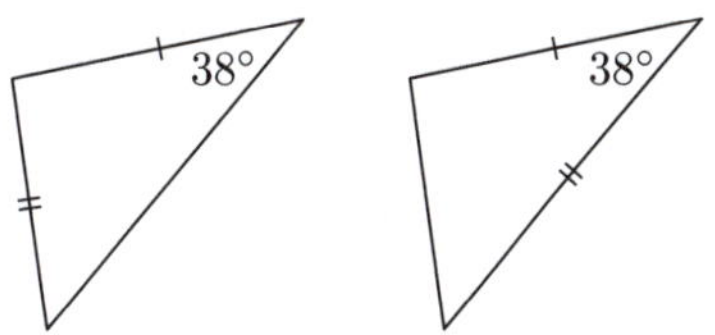

2 Decide whether these pairs of triangles are congruent, giving reasons for your answers:

a

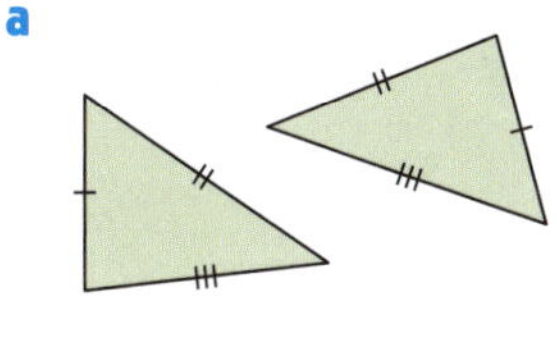

b

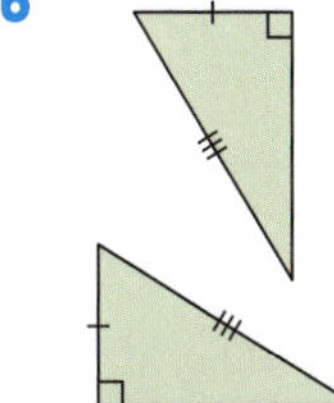

c

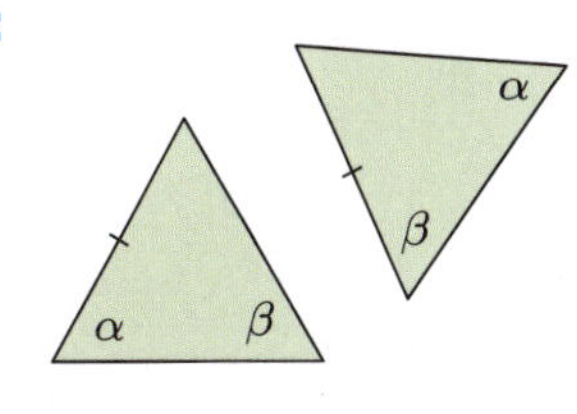

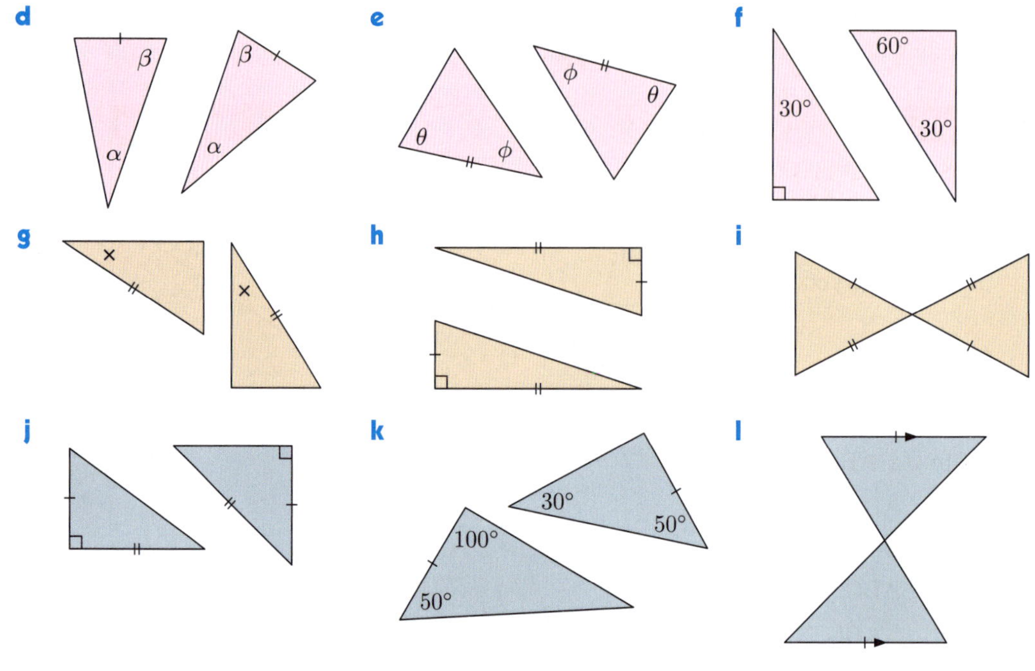

3 Which of the following triangles is congruent to the one alongside?

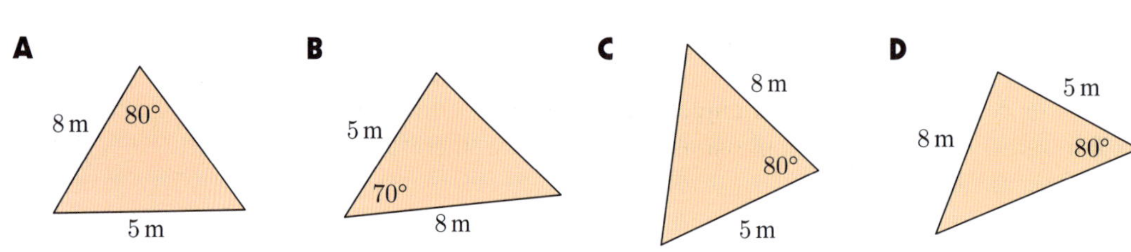

4 Which of these triangles are congruent to each other?

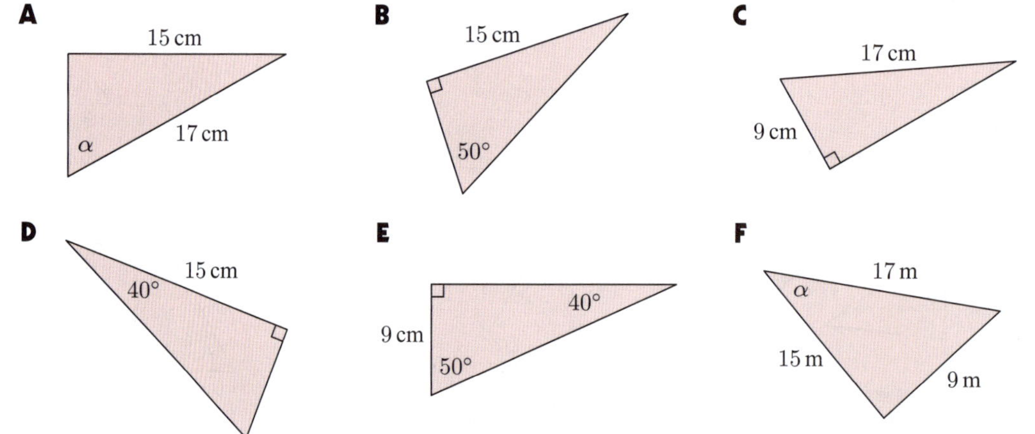

5 The following pairs of triangles are not drawn to scale, but the information on them is correct.

   i Determine whether the triangles are congruent.

   ii If the triangles are congruent, state what else we can deduce about them.

a
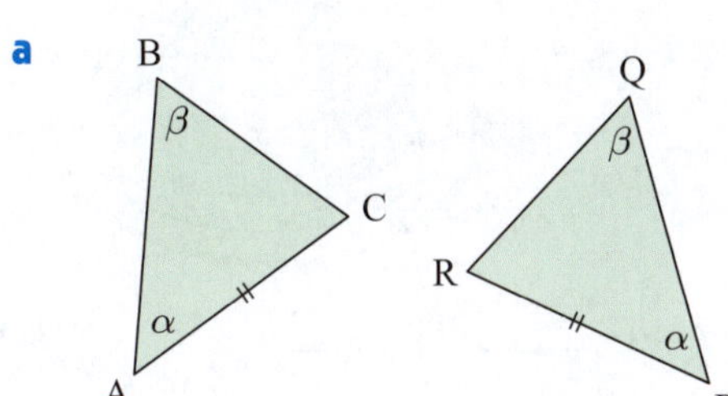

b
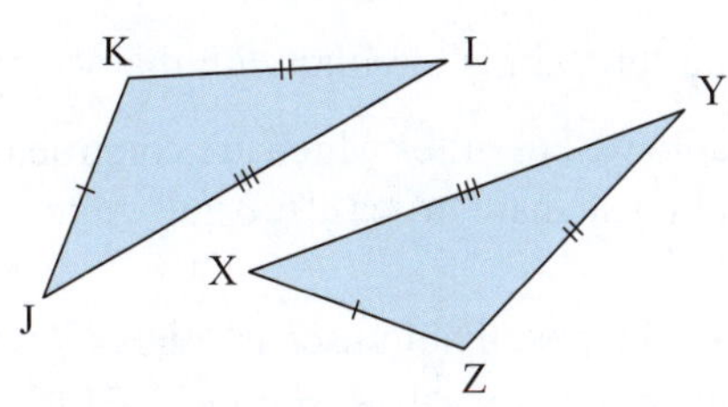

c
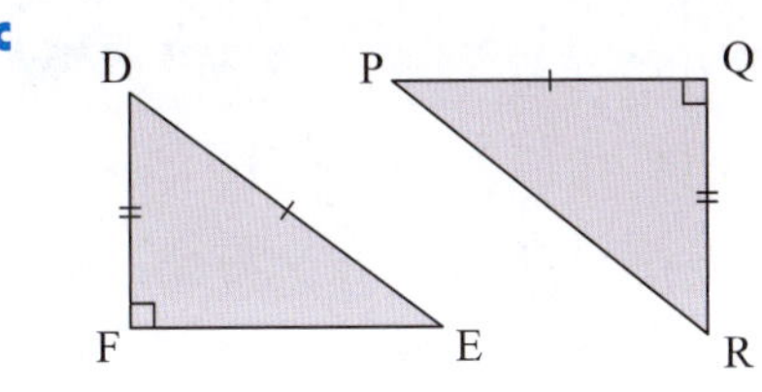

d
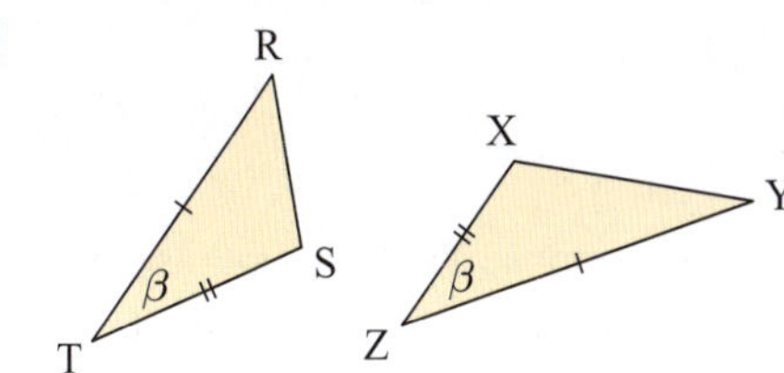

e
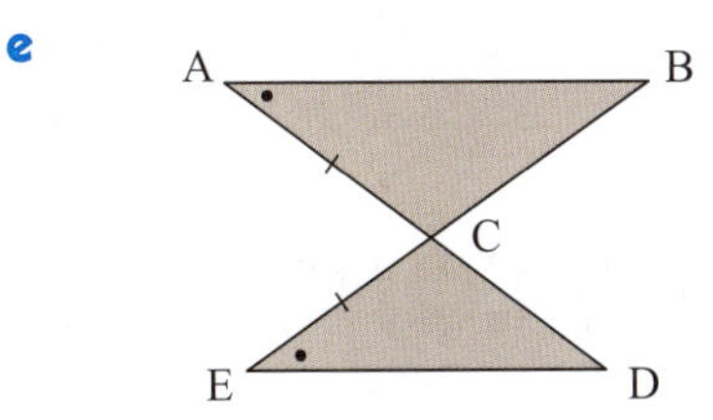

f
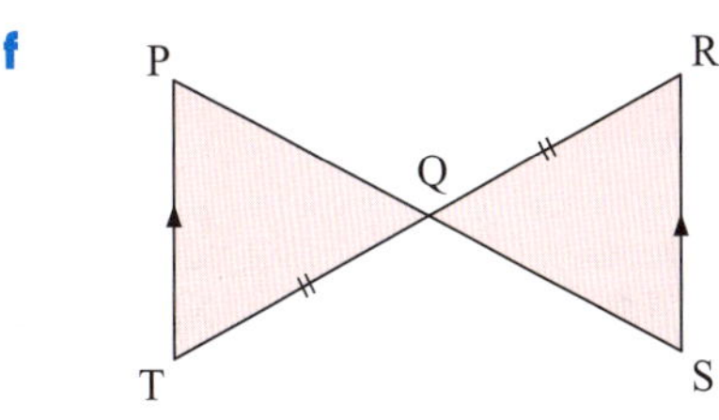

g
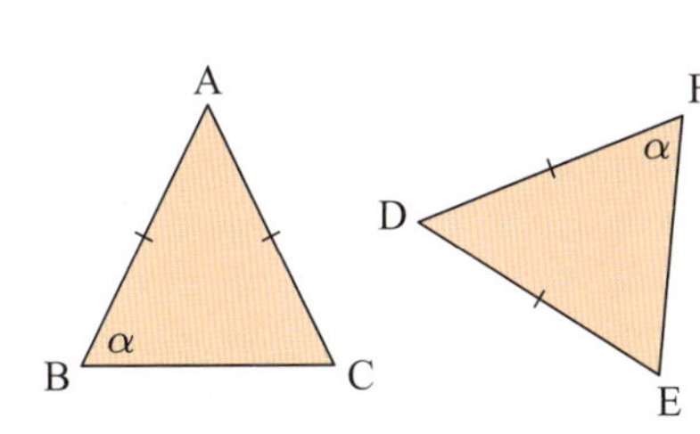

h
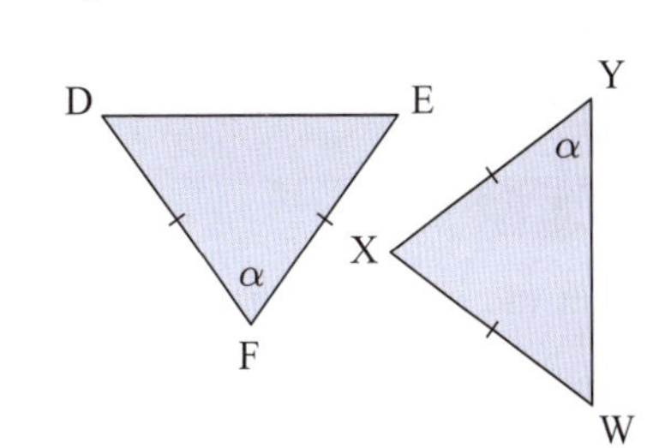

DISCUSSION

We have seen that if two *triangles* have equal corresponding sides, then they are congruent.

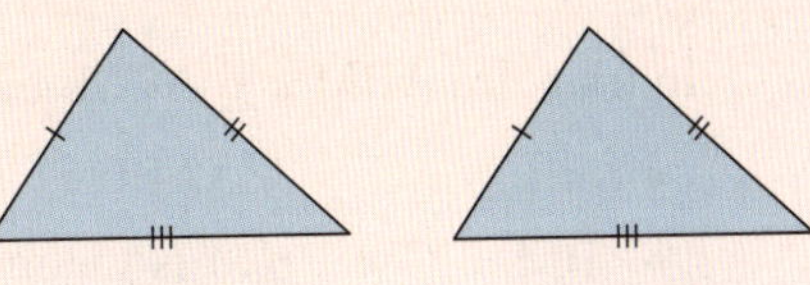

Is the same true for *quadrilaterals*? Can we say that the quadrilaterals alongside are congruent?

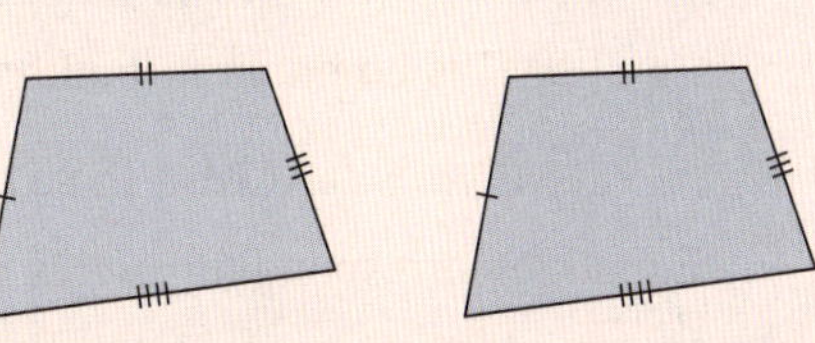

PUZZLE

Pablo wants to build a geometric path which completes a lap around his fountain.

The fountain is circular with diameter 1 m.

Pablo will use tiles which are congruent triangles. Every tile must share at least *two* full edges with other tiles.

1 Suppose Pablo uses the tiles alongside. What is the least number of tiles that Pablo needs? Illustrate your answer.

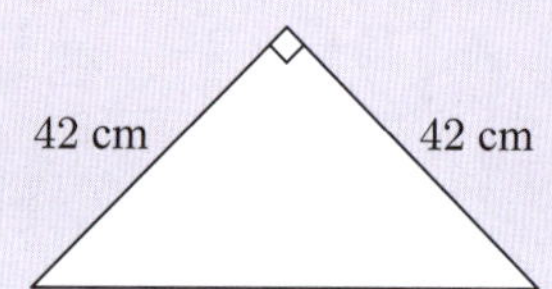

2 Now suppose Pablo uses these tiles instead. What is the least number of tiles that Pablo needs? Illustrate the solution which fits most tightly around the fountain.

C   PROOF USING CONGRUENCE

In **Chapter 9**, we studied the properties of isosceles triangles and special quadrilaterals. We can use congruence to prove many of these properties.

Example 4   ◀) Self Tutor

Consider the isosceles triangle ABC.

M is the midpoint of [BC].

a Use congruence to show that $B\hat{A}M = C\hat{A}M$.

b What property of isosceles triangles has been proven?

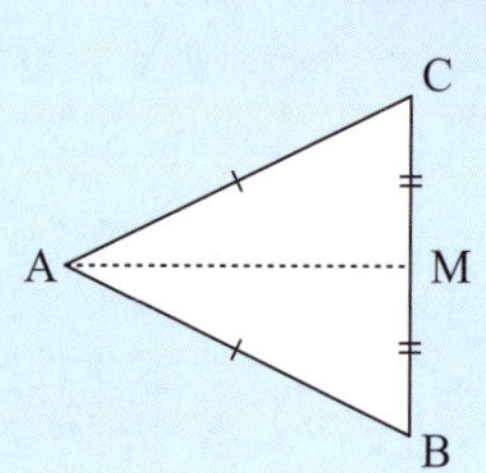

---

a In triangles ABM and ACM:

- $AB = AC$ {△ABC is isosceles}
- $BM = CM$ {M is the midpoint of [BC]}
- [AM] is common to both triangles.

$\therefore$ △ABM $\cong$ △ACM   {SSS}

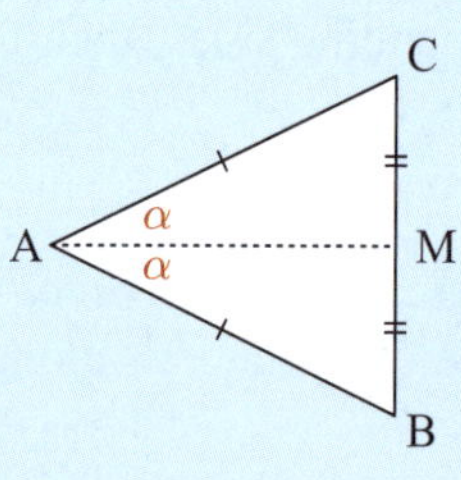

Equating corresponding angles, $B\hat{A}M = C\hat{A}M$.

b In any isosceles triangle, the line joining the apex to the midpoint of the base bisects the angle at the apex.

EXERCISE 19C

1 Consider the parallelogram ABCD.

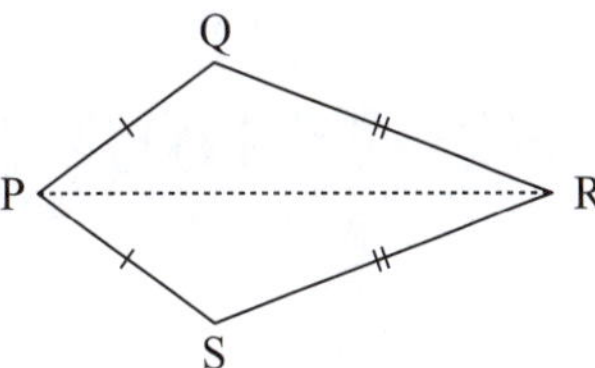

  a Copy and complete:

    In triangles ABD and CDB:

- $A\hat{D}B = $ ......   {equal alternate angles}
- $A\hat{B}D = $ ......   {equal alternate angles}
- [BD] is common to both triangles.

$\therefore$ $\triangle ABD \cong \triangle CDB$ {......}

    Equating corresponding sides,   AB = ...   and   AD = ...

  b What property of parallelograms has been proven in **a**?

2 Consider the kite PQRS.

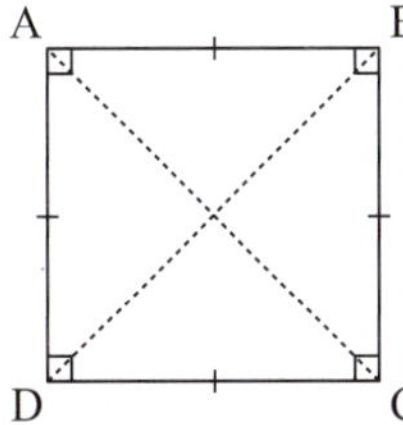

  a Use congruence to show that $Q\hat{P}R = S\hat{P}R$ and $Q\hat{R}P = S\hat{R}P$.

  b What property of kites has been proven?

3 Consider the square ABCD.

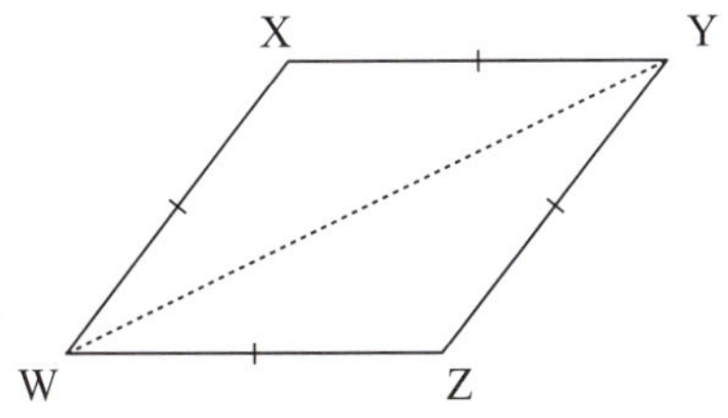

  a Show that $\triangle ABC \cong \triangle DAB$.

  b Hence show that AC = DB.

  c What property of squares has been proven?

4

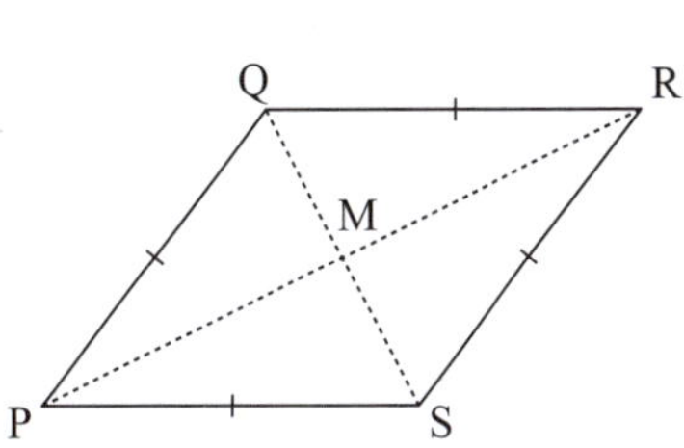

Consider the rhombus WXYZ.

  a Show that $\triangle WXY \cong \triangle YZW$.

  b Hence show that:

    i $X\hat{Y}W = Z\hat{W}Y$

    ii [XY] is parallel to [WZ]

    iii [XW] is parallel to [YZ].

  c What property of rhombuses has been proven?

5 The diagonals of rhombus PQRS meet at M.

  a Show that $\triangle PSQ \cong \triangle RSQ$.

  b Hence show that $P\hat{S}Q = R\hat{S}Q$.

  c What property of rhombuses has been proven?

  d Explain why $\triangle PSM \cong \triangle RSM$.

  e Hence show that PM = RM.

  f Find the sizes of $S\hat{M}P$ and $S\hat{M}R$.

  g Hence show that $\triangle SMP \cong \triangle QMR$ and therefore SM = QM.

  h What property of rhombuses has been proven in **e**, **f**, and **g**?

6 Use congruence to show that:

    a the base angles of an isosceles triangle are equal

    b the diagonals of a kite intersect at right angles.

7 In the diagram alongside:

- $\triangle ABC$ is right angled at C
- AMLC and CBNK are squares
- [MX] $\perp$ (AB) and [NY] $\perp$ (AB)
- A, B, X, and Y are collinear.

Prove that MX + NY = AB.

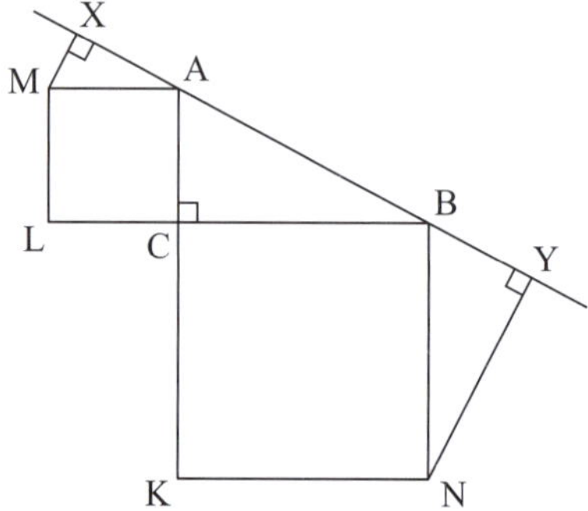

<table>
<tr><td>D</td><td></td></tr>
</table>

ENLARGEMENTS AND REDUCTIONS

The illustrated rectangles A, B, and C are not the same *size*, but they have the same *shape*. Their side lengths are in the same ratio.

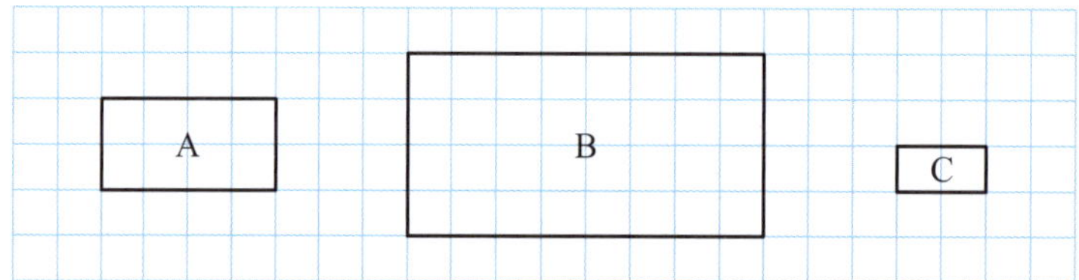

When an object is enlarged or reduced, all of its lengths are enlarged or reduced by the same **scale factor**.

For example:

- The side lengths in figure B are all twice the corresponding side lengths in figure A. We say that B is an **enlargement** of A with scale factor 2.
- The side lengths in figure C are all half the side lengths in figure A. We say that C is a **reduction** of A with scale factor $\frac{1}{2}$.

> If the scale factor is greater than 1, we have an **enlargement**.
>
> If the scale factor is less than 1, we have a **reduction**.

Example 5   ◀》 Self Tutor

Enlarge or reduce the given figure with scale factor:

a 3   **b** $\frac{1}{2}$

a For a scale factor of 3, the figure is enlarged and all lengths are *tripled*.

b For a scale factor of $\frac{1}{2}$, the figure is reduced and all lengths are *halved*.

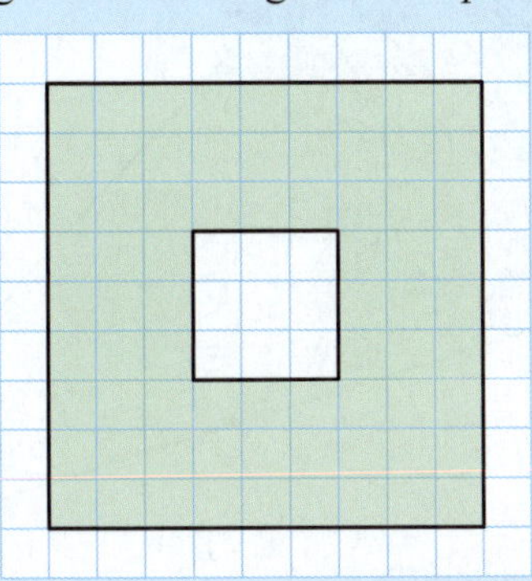

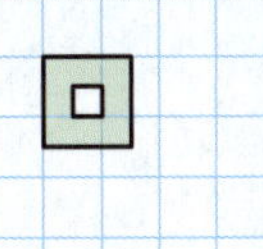

EXERCISE 19D

1 For each transformation from A to B, state whether the transformation is an enlargement or a reduction, and find the scale factor.

a
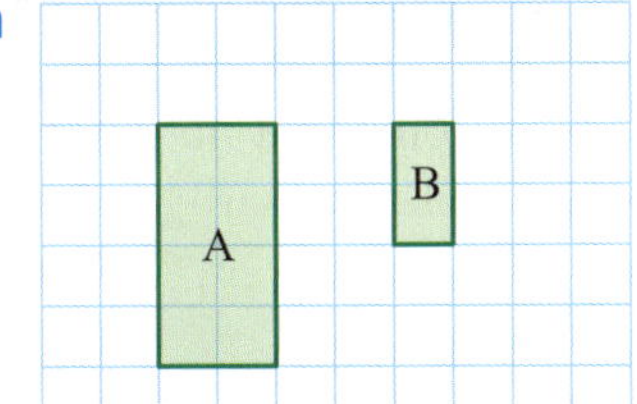

b
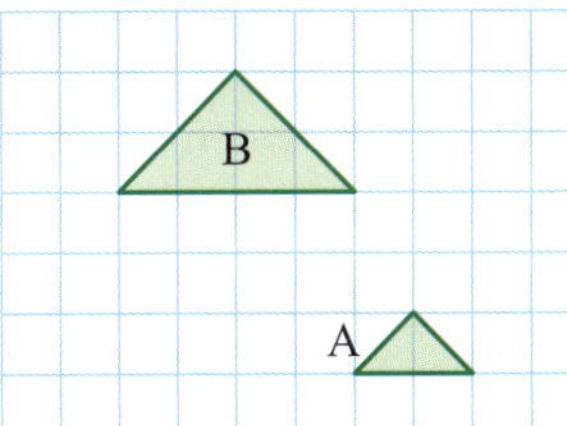

c
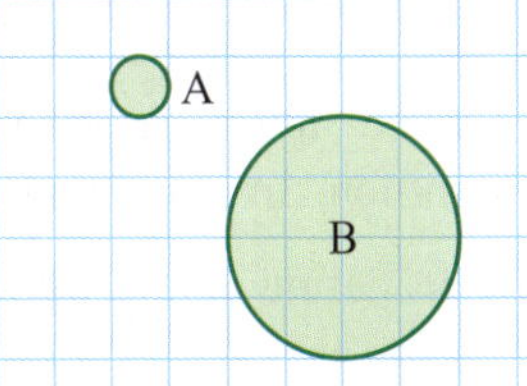

d
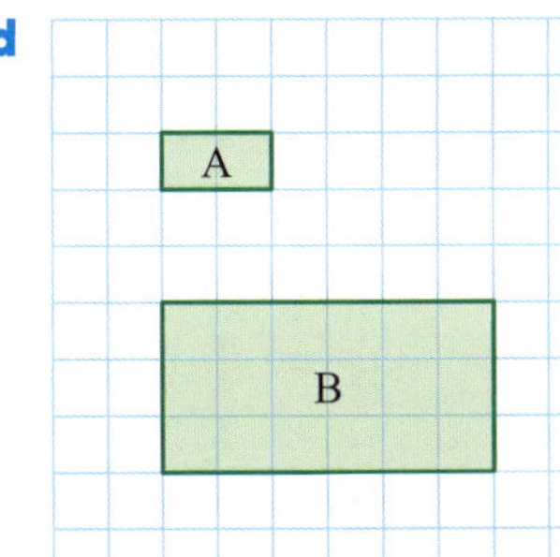

e
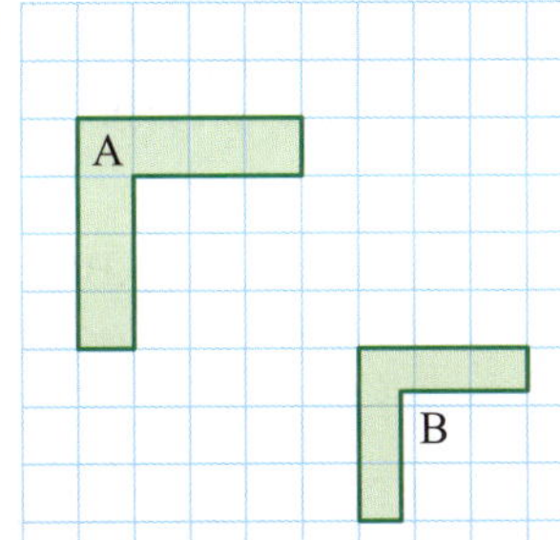

f
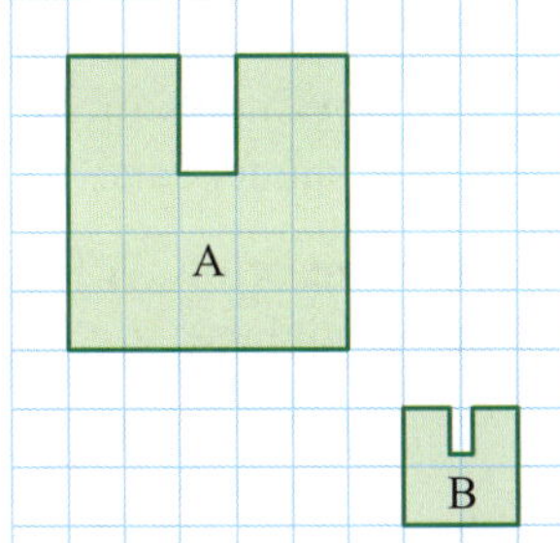

2 Enlarge or reduce the given figure with scale factor:

a 2

b $\frac{1}{2}$

c $\frac{3}{2}$

d $\frac{1}{4}$

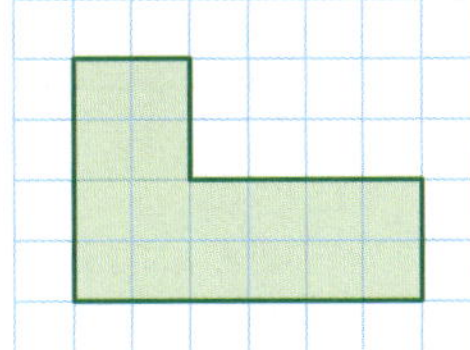

3 An object is reduced with scale factor k.

a What values could k have?

b What needs to be done to the image to return it to the original size?

4 Triangle B is an enlargement of triangle A.

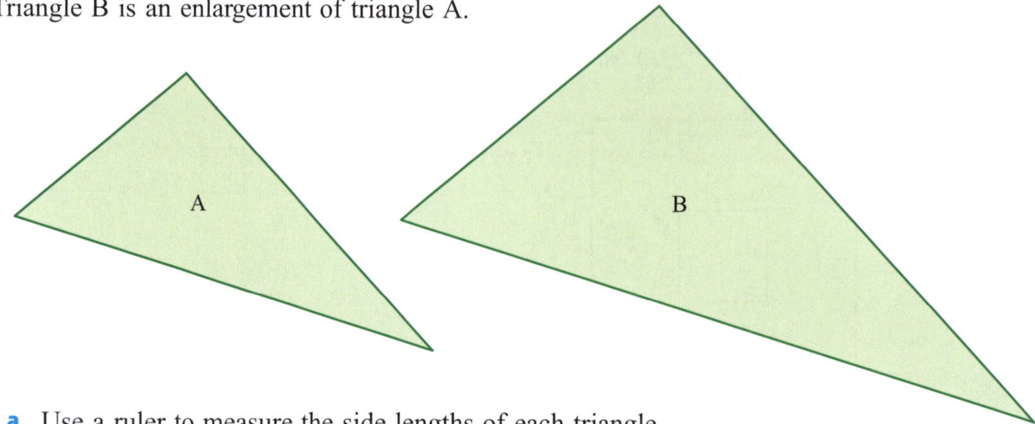

a Use a ruler to measure the side lengths of each triangle.
Hence find the scale factor for the enlargement.

b Use a protractor to measure the angles of each triangle. Comment on your results.

E   SIMILARITY

The word *similar* suggests a comparison between objects which have some, but not all, properties in common. In mathematics, similar figures have the same *shape*, but not necessarily the same *size*.

We can define similarity in terms of transformations:

> Two figures are **similar** if one can be placed exactly on top of the other using a combination of a translation, a rotation, a reflection, and an enlargement or reduction.

Common examples of similar figures include television images, photo enlargements, house plans, maps, and model cars.

In the figure below, A′B′C′D′ is an enlargement of ABCD with scale factor 3. The two figures are therefore similar.

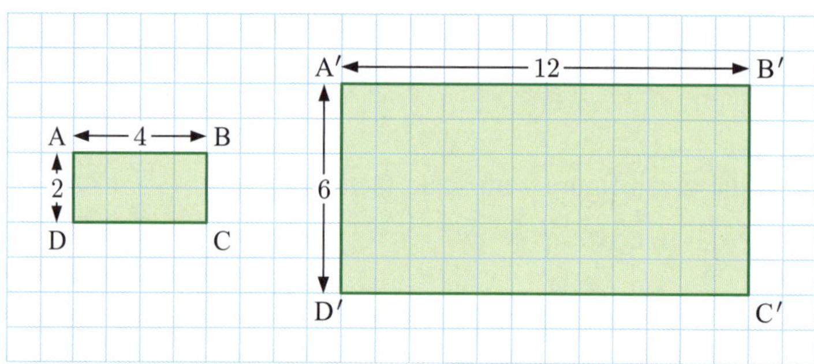

Notice that $\dfrac{A'B'}{AB} = \dfrac{B'C'}{BC} = \dfrac{C'D'}{CD} = \dfrac{D'A'}{DA} = 3,$ so the corresponding sides are in the **same ratio**.

When a figure is enlarged or reduced, the sizes of its angles do not change. We say that the figures are **equiangular**.

Two figures are **similar** if:

- the figures are **equiangular** *and*
- the corresponding side lengths are in the **same ratio**.

Example 6

Determine whether each pair of figures is similar:

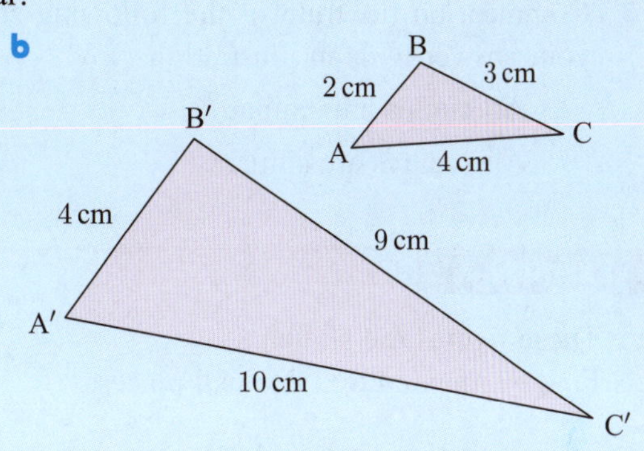

a $\dfrac{A'B'}{AB} = \dfrac{6}{4} = \dfrac{3}{2}$ and $\dfrac{B'C'}{BC} = \dfrac{3}{2}$

$\therefore$ the corresponding side lengths are in the same ratio.

The figures are also equiangular, so the figures are similar.

b $\dfrac{A'B'}{AB} = \dfrac{4}{2} = 2$ and $\dfrac{B'C'}{BC} = \dfrac{9}{3} = 3$

$\therefore$ the corresponding side lengths are *not* in the same ratio.

$\therefore$ the figures are not similar.

EXERCISE 19E

1   Determine whether each pair of figures is similar:

a
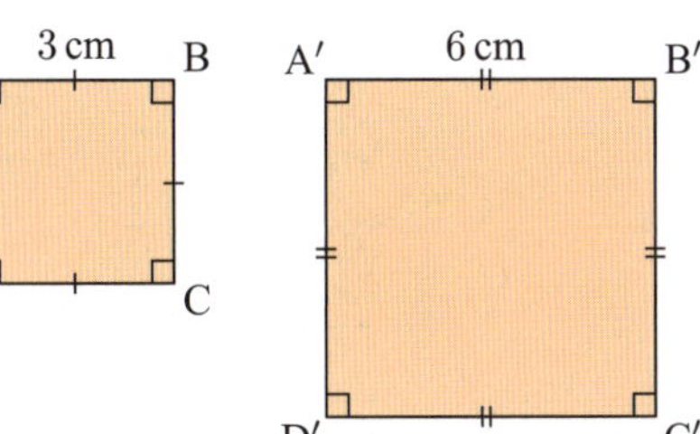

b
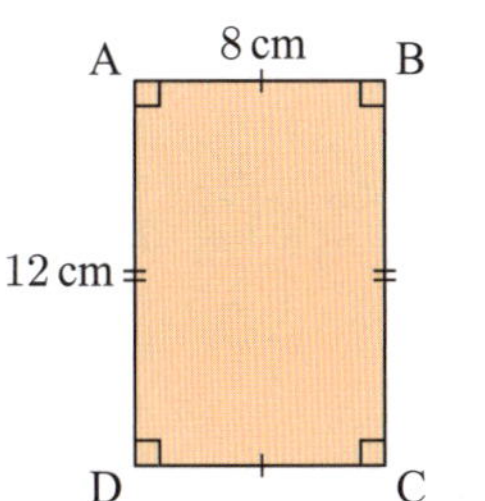

c
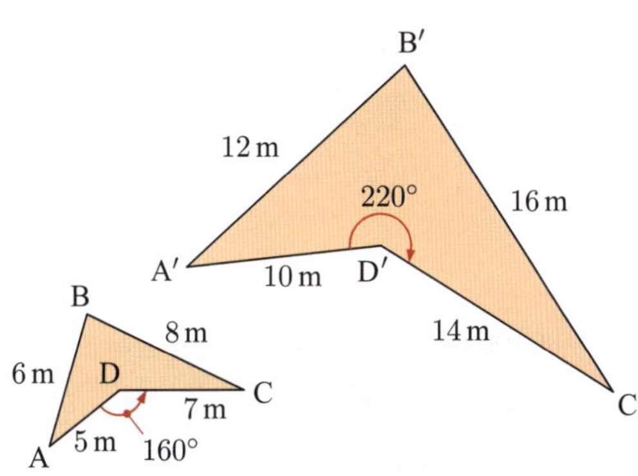

d
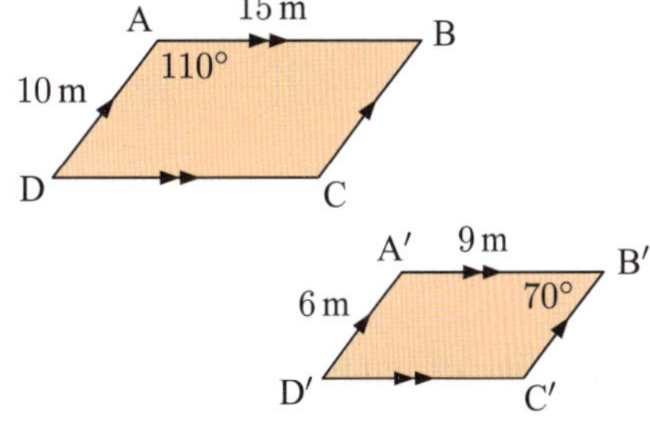

2

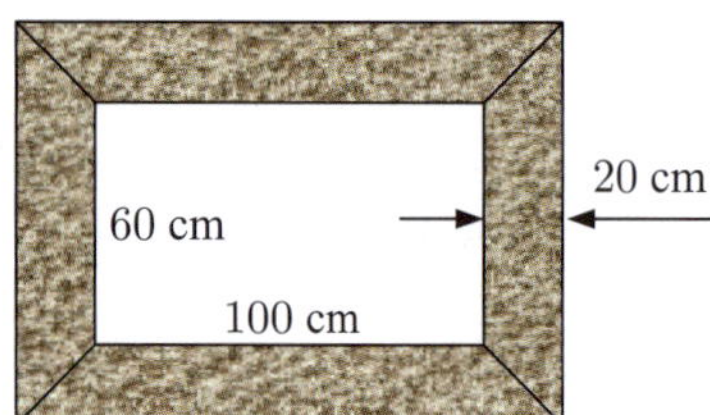

A 20 cm wide picture frame surrounds a painting which is 100 cm by 60 cm.

Are the two rectangles shown here similar?

3 Comment on the truth of the following statements. For any statement which is false, justify your answer with an illustration.

   a All circles are similar.   **b** All parallelograms are similar.

   c All squares are similar.   **d** All rectangles are similar.

Example 7   ◀ŷ **Self Tutor**

These figures are similar.
Find x, rounded to 2 decimal places.

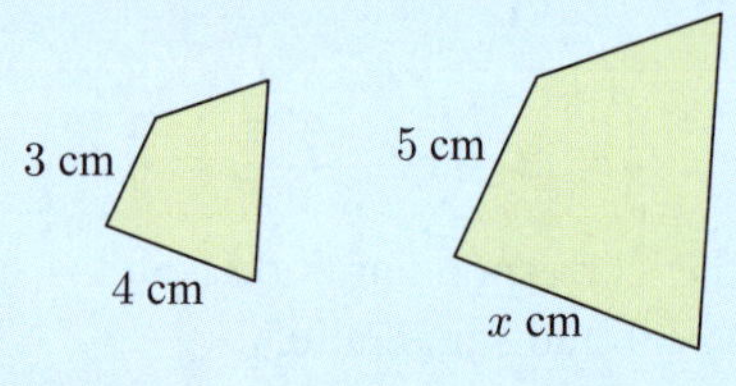

Since the figures are similar, their corresponding sides are in the same ratio.

$$\therefore \quad \frac{x}{4} = \frac{5}{3}$$

$$\therefore \quad x = \frac{5}{3} \times 4$$

$$\therefore \quad x = \frac{20}{3}$$

$$\therefore \quad x \approx 6.67$$

---

4 These figures are similar. Find the value of x, rounded to 2 decimal places if necessary:

a

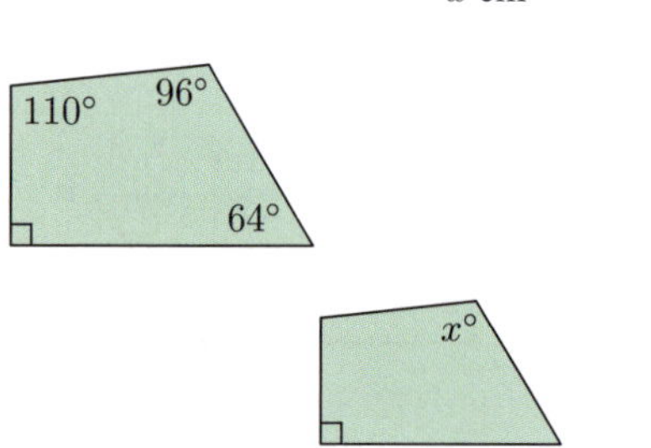

b

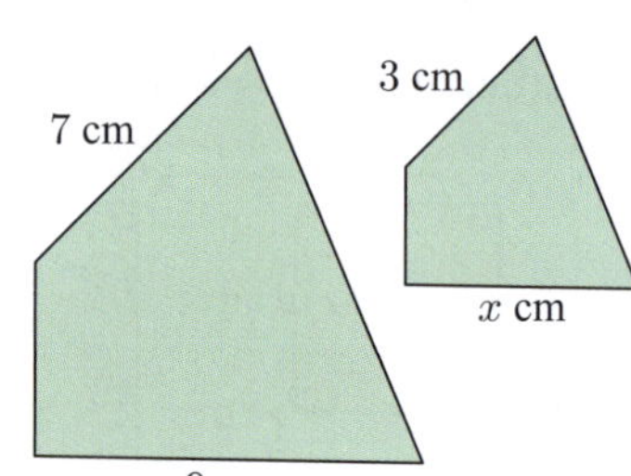

c

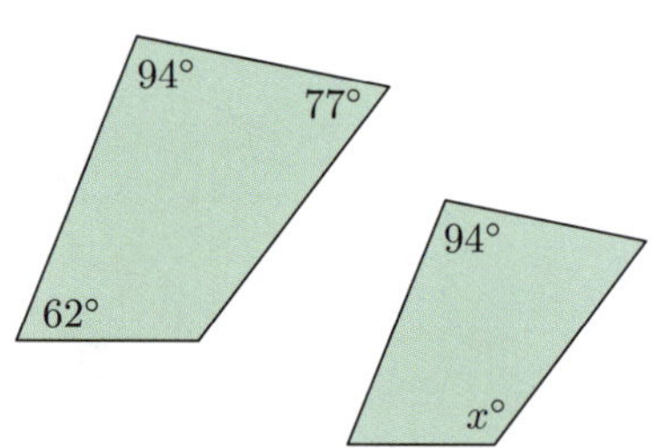

d

5 If a line of length 2 cm is enlarged with scale factor 3, find its new length.

6 A 3 cm length has been enlarged to 4.5 cm. Find the scale factor k.

7 Narelle is drawing a scale diagram of her bedroom, which is a rectangle 4.2 m long by 3.6 m wide. On her diagram she draws her room 7 cm long.

   a How wide will her bedroom be on her diagram?

   b What is the scale factor for the diagram?

8 Rectangles ABCD and EFGH are similar. Find the length of [EF].

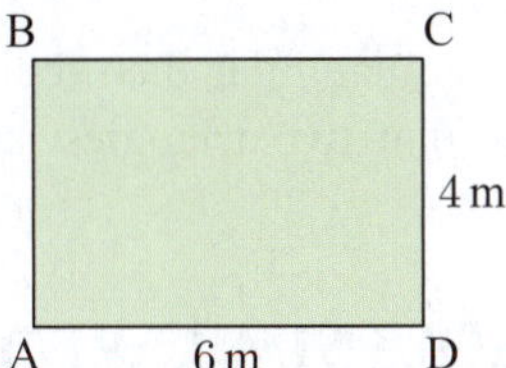
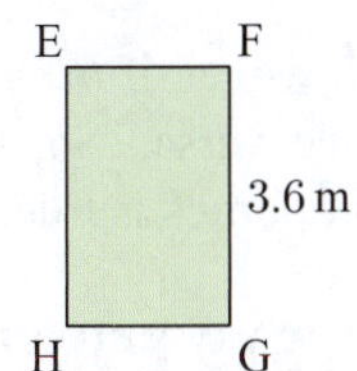

9 Find the value of x given that triangle ABC is similar to triangle A′B′C′:

  a

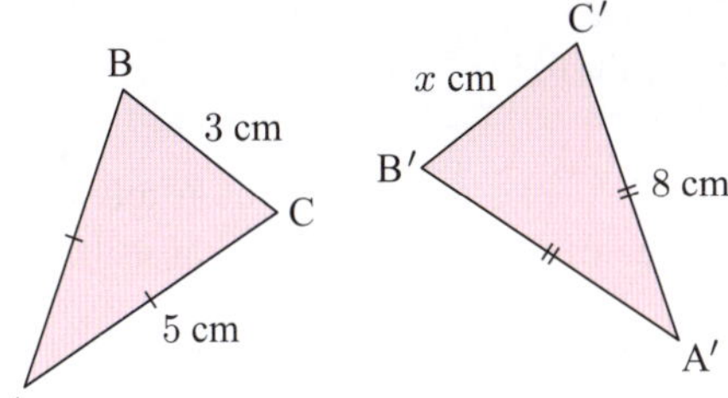

  b

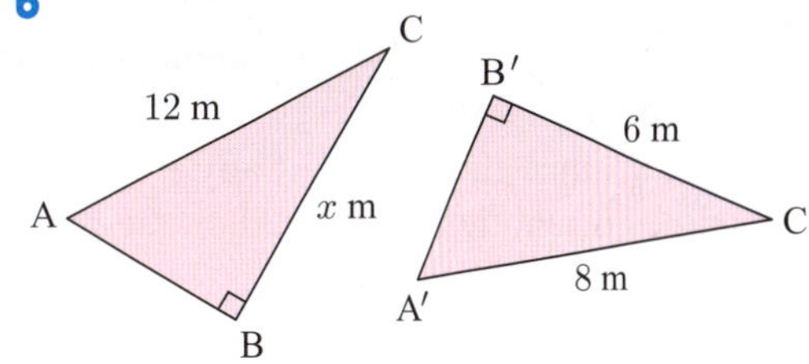

  c

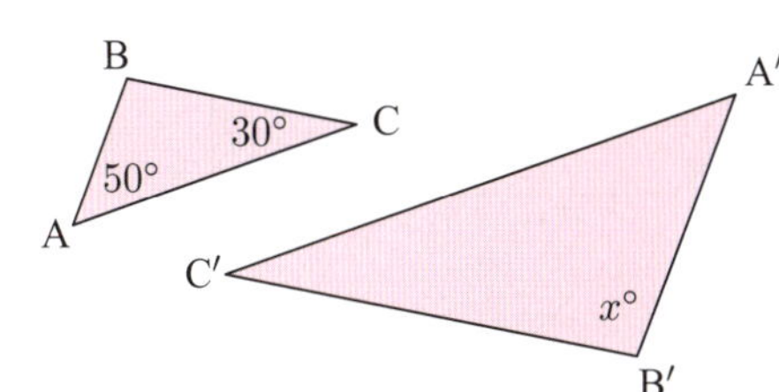

  d

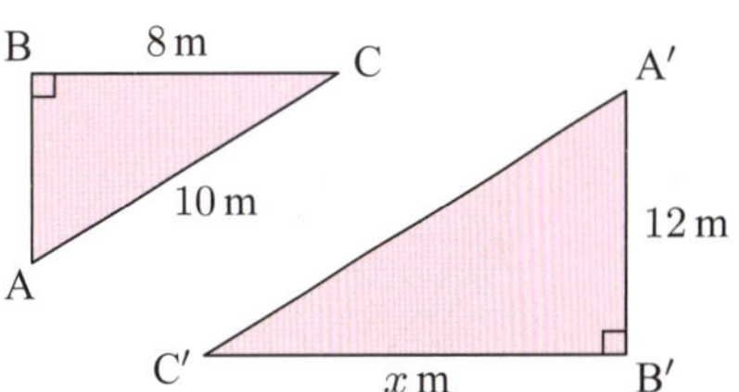

10 Sketch two quadrilaterals that:

   a are equiangular but not similar

   b have sides in proportion, but are not similar.

11 Can you draw two triangles which are equiangular but not similar?

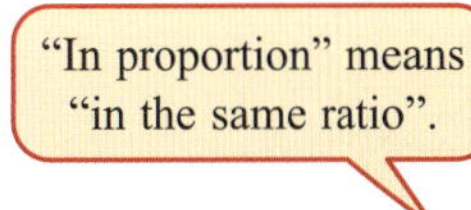

12

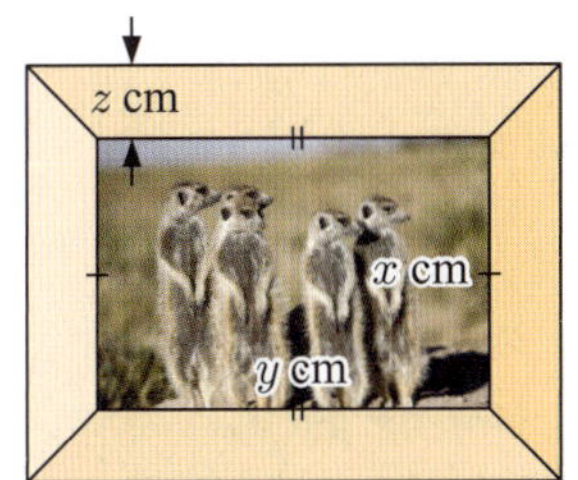

A picture x cm $\times$ y cm is surrounded by a frame z cm wide, as illustrated.

Prove that the rectangles in the figure are similar only if the picture is a square.

SIMILAR TRIANGLES

In the previous Exercise we saw that quadrilaterals which are equiangular are not necessarily similar, and quadrilaterals with sides in proportion are not necessarily similar.

However, if *triangles* are equiangular then their corresponding sides *must* be in the same ratio, and vice versa. So, to show that two triangles are similar, we only need to show that *one* of these properties is true.

TESTS FOR TRIANGLE SIMILARITY

Two triangles are similar if *either*:

- they are equiangular *or*
- their side lengths are in the same ratio.

Notice that since the angles of any triangle sum to $180°$, if two angles of one triangle are equal to two angles of another triangle, then the remaining angles of the triangles must also be equal.

Example 8

◀)) **Self Tutor**

Show that these figures possess similar triangles:

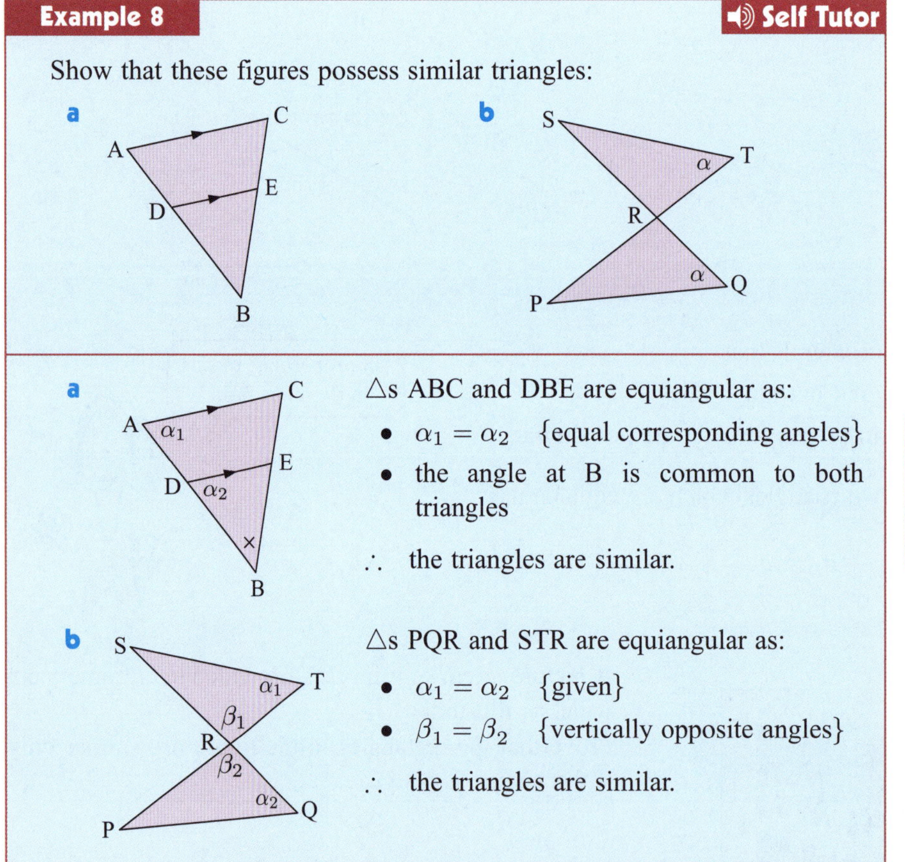

a $\triangle$s ABC and DBE are equiangular as:

- $\alpha_1 = \alpha_2$ {equal corresponding angles}
- the angle at B is common to both triangles

$\therefore$ the triangles are similar.

b $\triangle$s PQR and STR are equiangular as:

- $\alpha_1 = \alpha_2$ {given}
- $\beta_1 = \beta_2$ {vertically opposite angles}

$\therefore$ the triangles are similar.

If two triangles are similar, we list corresponding vertices in the same order.

EXERCISE 19F.1

1 Show that these figures possess similar triangles:

a 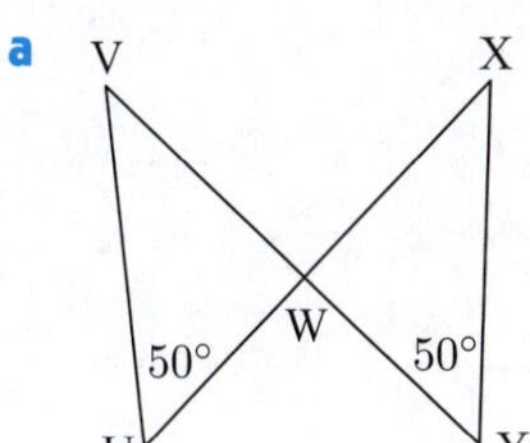   **b** 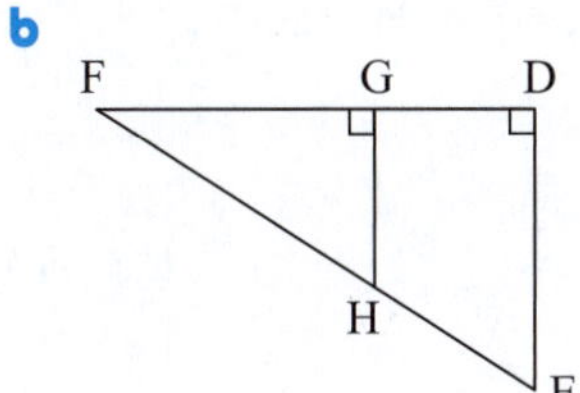   **c** 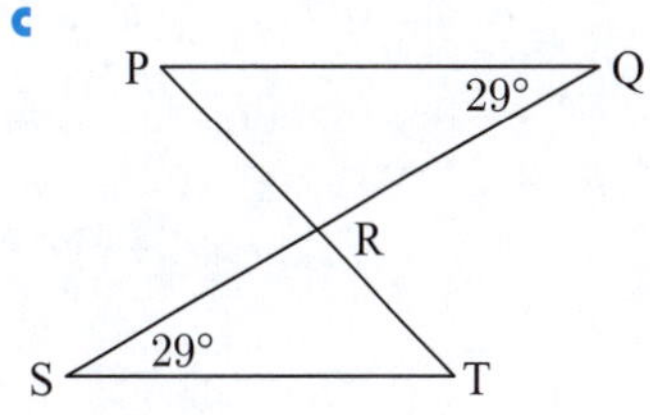

d 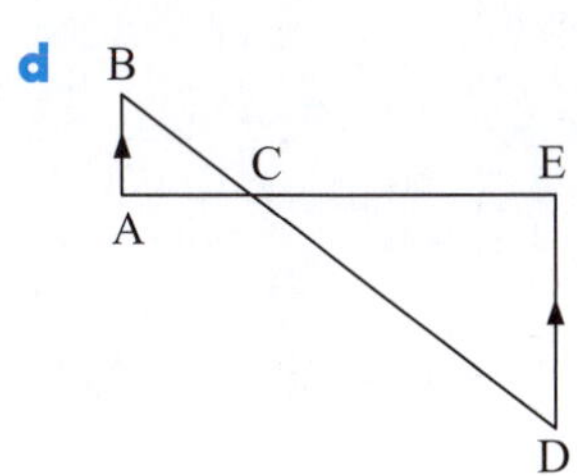   **e** 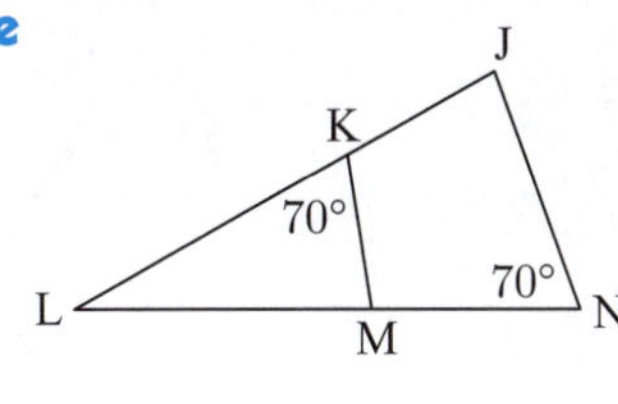

2 Explain why these triangles are similar:

a 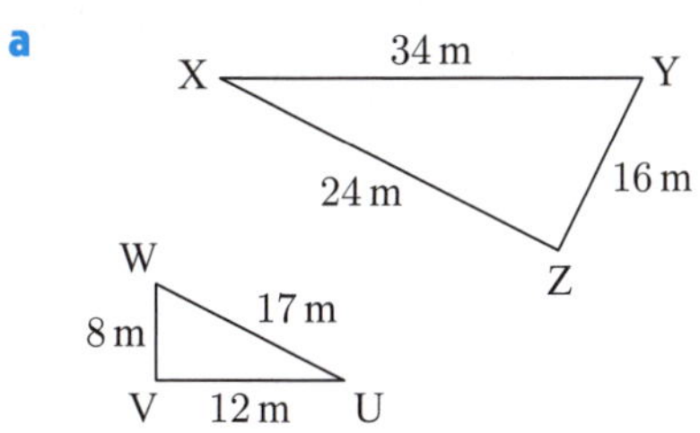   **b** 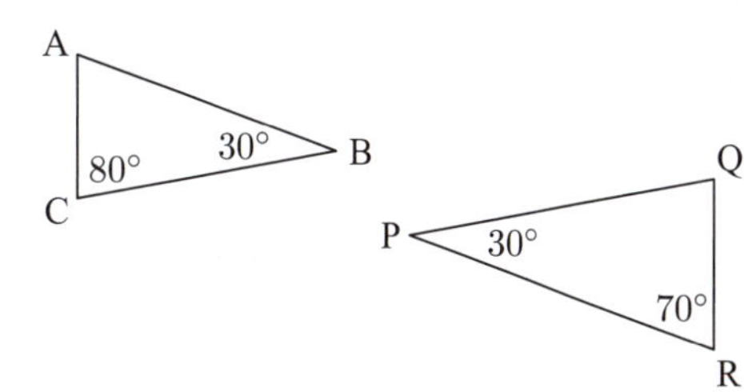

3 **a** Show that $A\widehat{C}B = \alpha$.

    b Hence show that the three triangles in the given figure are all similar to each other.

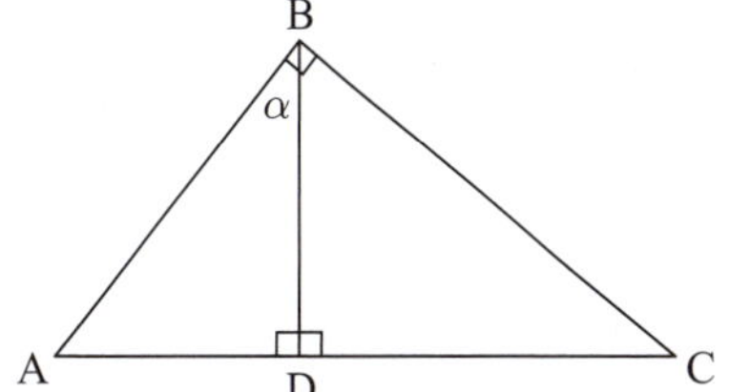

FINDING SIDE LENGTHS

Once we have shown that two triangles are similar, we can use the fact that corresponding sides are in the same ratio to find unknown lengths.

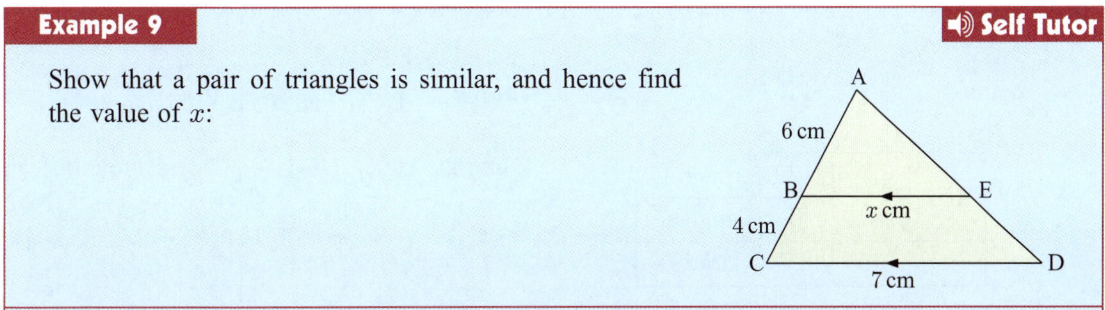

Example 9   ◄ Self Tutor

Show that a pair of triangles is similar, and hence find the value of x:

$\triangle$s ABE and ACD are equiangular as:

- $\alpha_1 = \alpha_2$ {equal corresponding angles}
- $\beta_1 = \beta_2$ {equal corresponding angles}

$\therefore$ the triangles are similar.

$\therefore$ $\dfrac{BE}{CD} = \dfrac{AB}{AC}$ {same ratio}

$\therefore$ $\dfrac{x}{7} = \dfrac{6}{6+4}$

$\therefore$ $x = \dfrac{6}{10} \times 7 = 4.2$

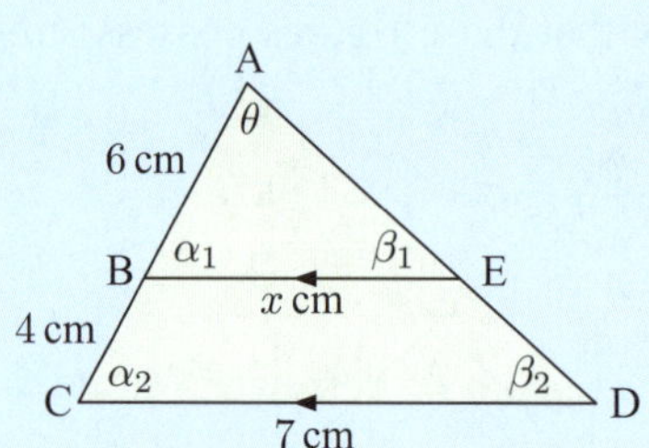

When solving similar triangle problems, it may be useful to use a table.

Step 1:   Label equal angles.

Step 2:   Show that the triangles are equiangular, and hence similar.

Step 3:   Put the information in a table, showing the equal angles and the side lengths *opposite* these angles.

Step 4:   Use the columns to write down the equation for the ratio of the corresponding sides.

Step 5:   Solve the equation.

For the triangles in **Example 9**, we would write:

α	β	θ	
-	6	x	small $\triangle$
-	10	7	large $\triangle$

$\therefore$ $\dfrac{6}{10} = \dfrac{x}{7}$

$\therefore$ $x = 4.2$

Example 10 ◀)) **Self Tutor**

Show that a pair of triangles is similar, and hence find the value of x:

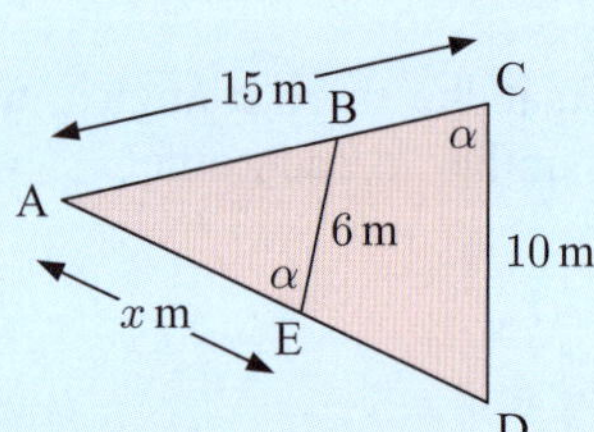

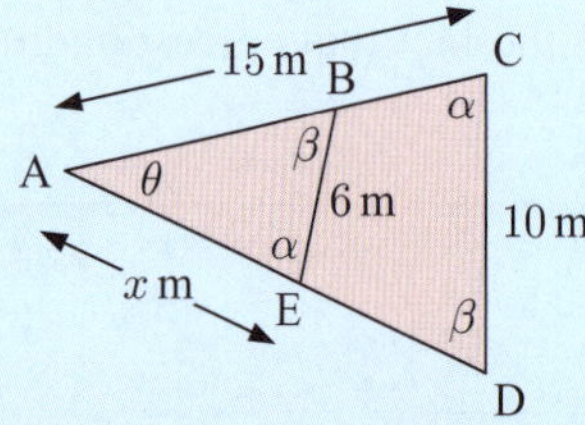

α	β	θ	
-	x	6	small $\triangle$
-	15	10	large $\triangle$

$\triangle$s ABE and ADC are equiangular since:

- $A\widehat{E}B = A\widehat{C}D$ {given}
- the angle at A is common

$\therefore$ the triangles are similar.

$\therefore$ $A\widehat{B}E = A\widehat{D}C$,   and we call this angle β.

Using the table, $\dfrac{x}{15} = \dfrac{6}{10}$ {same ratio}

$\therefore$ $x = 15 \times \dfrac{6}{10}$

$\therefore$ $x = 9$

EXERCISE 19F.2

1 In each figure, show that a pair of triangles is similar. Hence find the value of x.

a 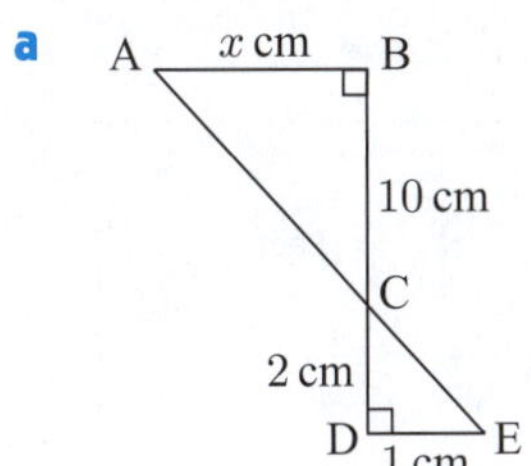

b 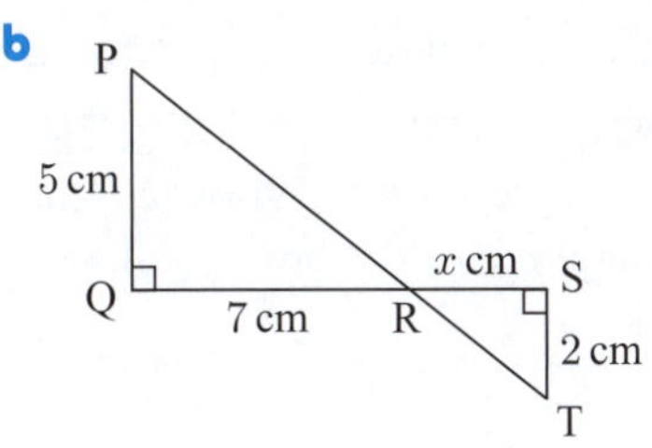

c 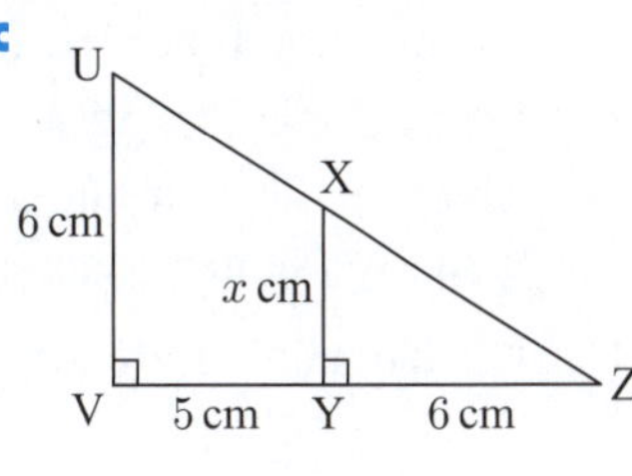

d 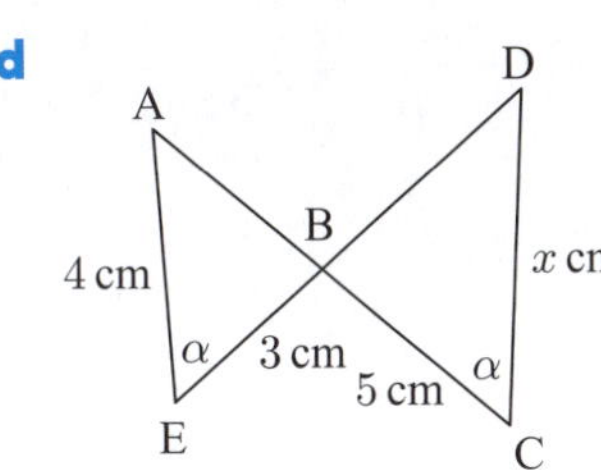

e 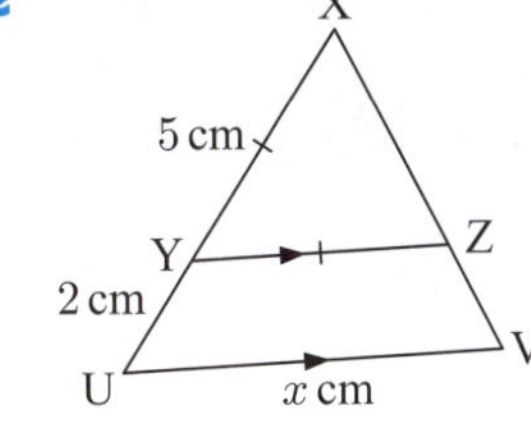

f 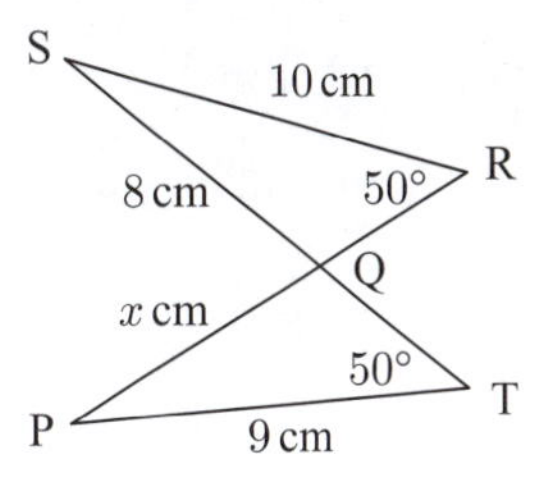

2 Given the figure alongside, find the length of [BF].

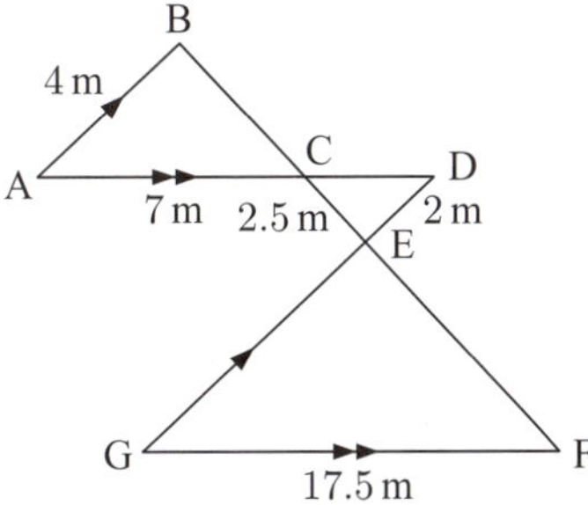

G   PROBLEM SOLVING

The properties of similar triangles have been known since ancient times. However, even with the technologically advanced measuring instruments available today, similar triangles are still important for finding heights and distances which would otherwise be difficult to measure.

Step 1:   Read the question carefully. Draw a diagram showing all of the given information.

Step 2:   Introduce a variable such as x, for the unknown quantity to be found.

Step 3:   Show that a pair of triangles are similar, and hence write an equation involving the variable.

Step 4:   Solve the equation.

Step 5:   Answer the question in a sentence.

Example 11   ◀)) Self Tutor

When a 30 cm stick is stood vertically on the ground, it casts a 24 cm shadow.
At the same time a man casts a shadow of length 152 cm. How tall is the man?

The sun shines at the same angle on both the stick and the
man. We suppose this is angle $\alpha°$ to the horizontal.

Let the man be h cm tall.

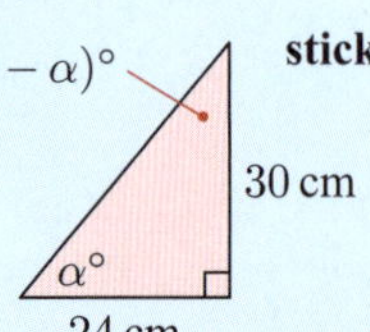

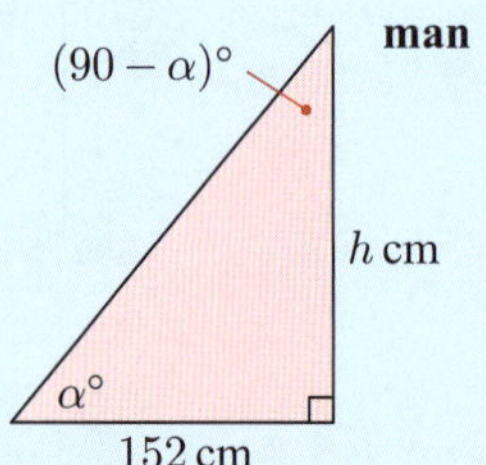

The triangles are equiangular and therefore similar.

$$\therefore \quad \frac{h}{30} = \frac{152}{24} \quad \text{\{same ratio\}}$$

$$\therefore \quad h = \frac{152}{24} \times 30$$

$$\therefore \quad h = 190$$

$\alpha°$	$(90 - \alpha)°$	$90°$	
h	152	-	large △
30	24	-	small △

The man is 190 cm tall.

EXERCISE 19G

1 Find the height of each pine tree:

a

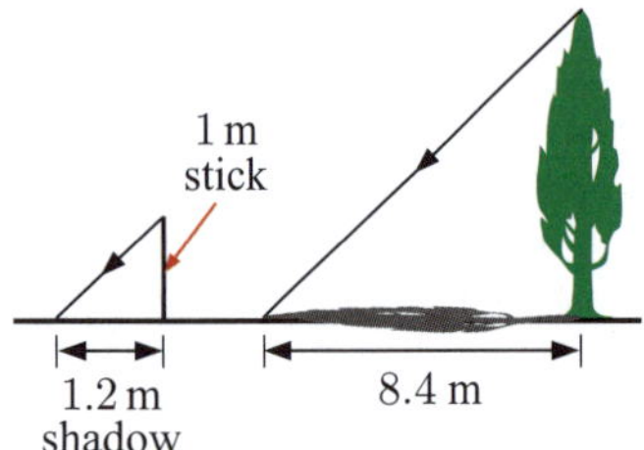

b

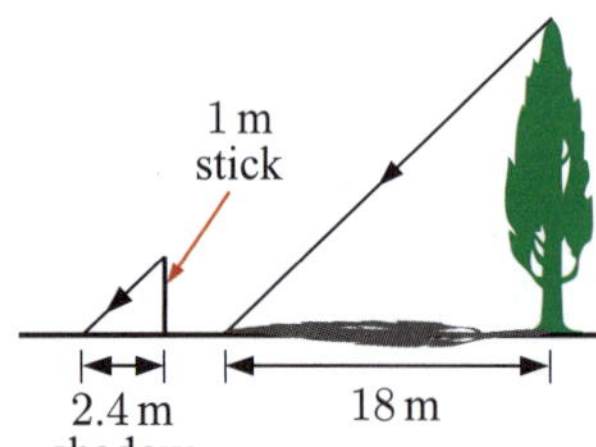

2 A ramp is built to enable wheelchair access to a building that is 24 cm above ground level. For
every 15 cm horizontally, the ramp rises 2 cm. Calculate the length of the base of the ramp.

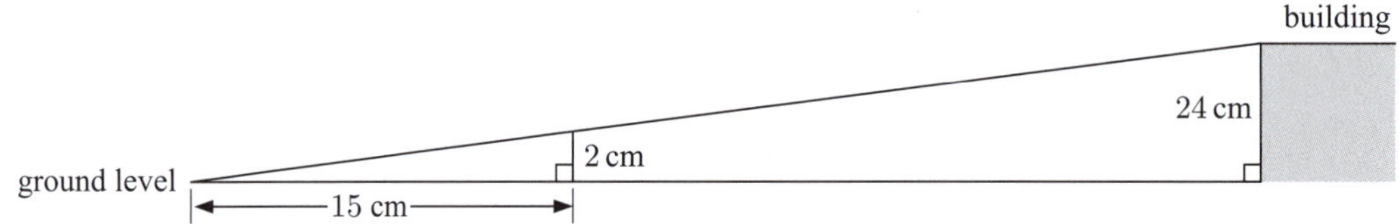

3 A piece of timber rests against both the top of a fence
and the wall behind it, as shown.

    a Find how far up the wall the timber reaches.

    b Find the length of the timber.

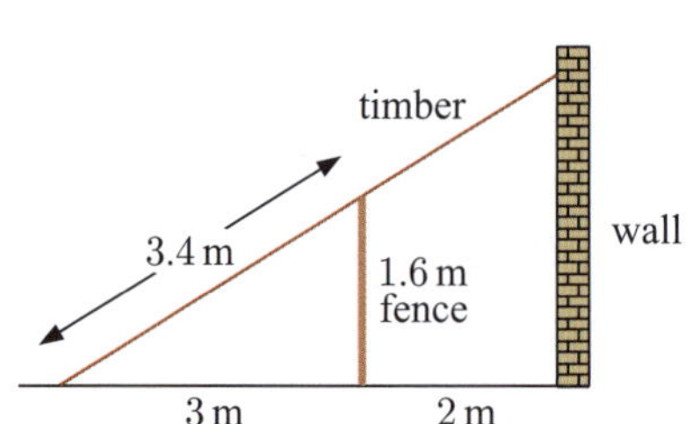

4 A pinhole camera displays an image on a screen as shown. The monument is 21 m tall, and its image is 3.5 cm high. The distance from the pinhole to the image is 6.5 cm.

How far is the pinhole from the monument?

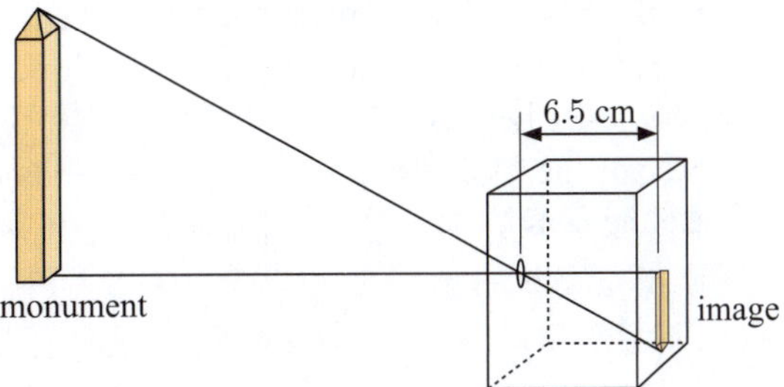

5 Ryan is standing on the edge of the shadow cast by a 5 m tall building. Ryan is 4 m from the building, and is 1.8 m tall. How far must Ryan walk towards the building to be completely shaded from the sun?

6 A swimming pool is 1.2 m deep at one end, and 2 m deep at the other end. The pool is 25 m long. Isaac jumps into the pool 10 metres from the shallow end. How deep is the pool at this point?

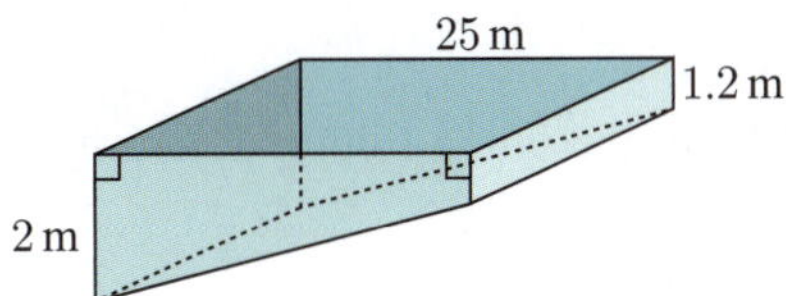

7

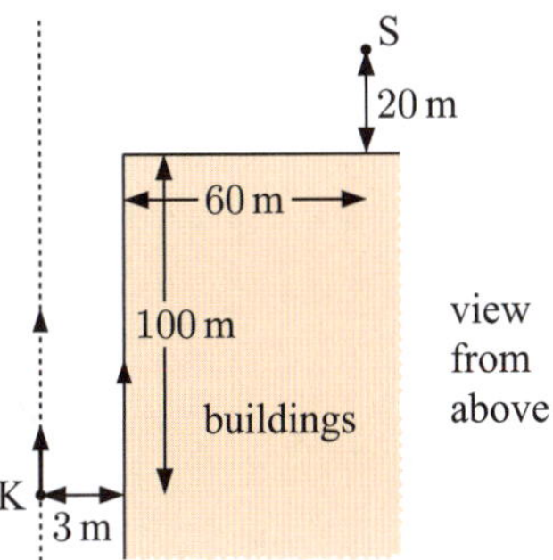

Kalev is currently at K, driving with speed 10 m/s along a street parallel to a line of buildings. A statue is located at S. How long will it be before Kalev will be able to see the statue?

8 A, B, C, and D are pegs on the bank of a canal which has parallel straight sides. C and D are directly opposite each other. AB = 30 m and BC = 140 m.

When I walk from A directly away from the bank, I reach a point E, 25 m from A, where E, B, and D are collinear. How wide is the canal?

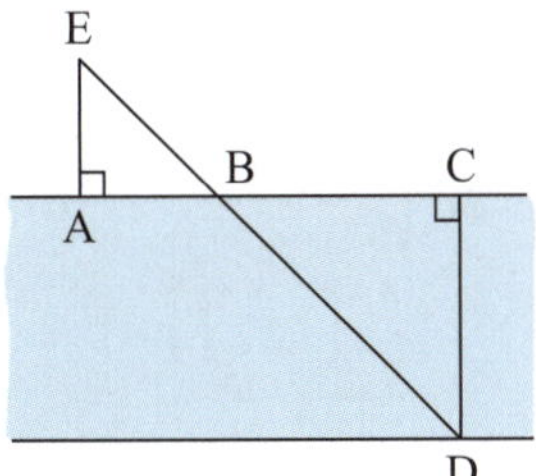

9 Two spheres with diameter 6 cm and 4 cm touch each other at A as they rest on a horizontal table. How high is A above the table?

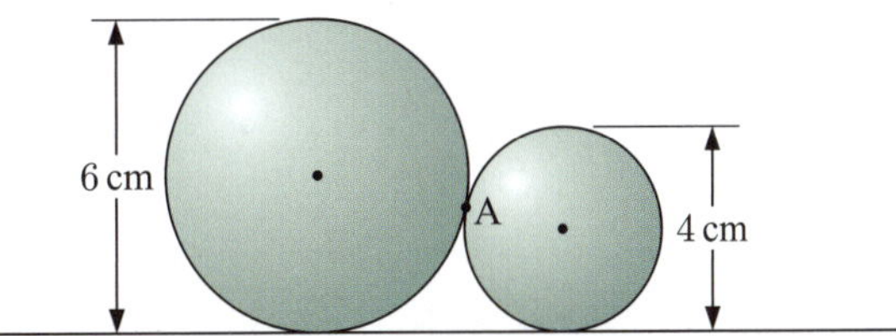

10 The dimensions of a tennis court are given alongside.
Samantha hits a shot from the baseline corner at S. The ball
passes over her service line at T such that UT = 1.92 m.
The ball then travels over the net and lands on the opposite
baseline [AD].

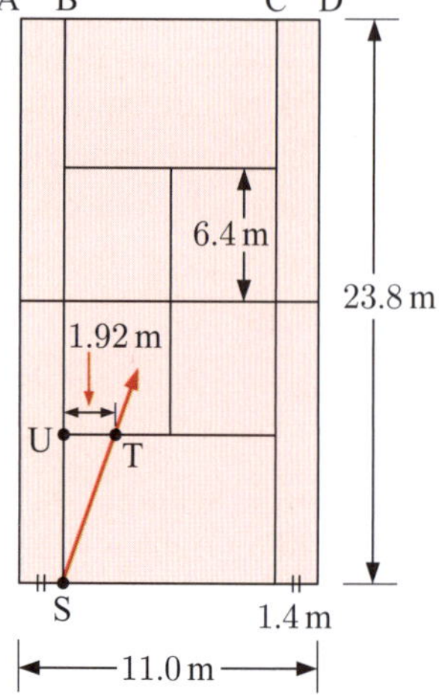

a Find the length of:

   i [SU]   **ii** [BC]

b A ball landing on the baseline is "in" if it lands between
points B and C.

Assuming the ball continues along the same trajectory,
will it land "in"?

11 In the given figure, determine the ratio BN : CM.

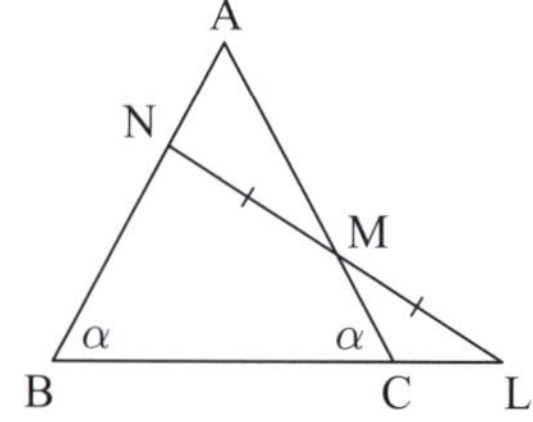

QUICK QUIZ

MULTIPLE CHOICE QUIZ

REVIEW SET 19A

1 Which two of these figures are congruent?

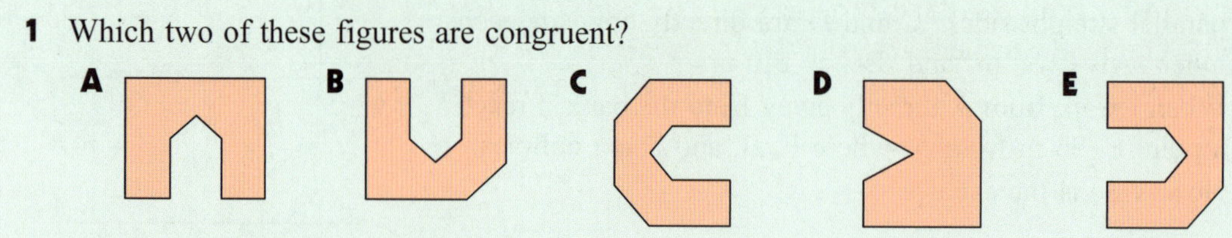

2 Which of the following triangles is congruent to the one
alongside?

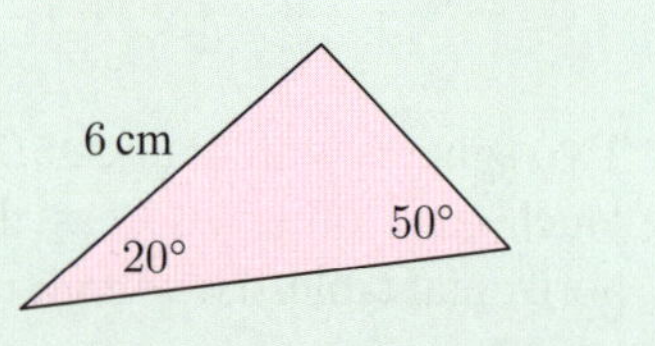

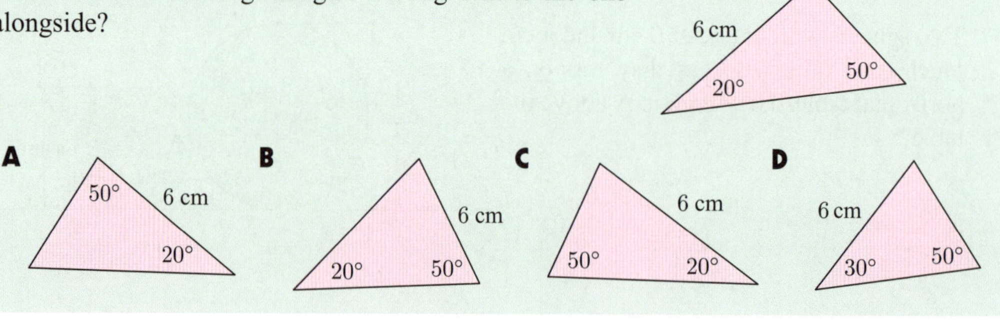

3 The following pairs of triangles are not drawn to scale, but the information on them is correct.

    i Determine whether the triangles are congruent.

    ii If the triangles are congruent, state what else we can deduce about them.

a
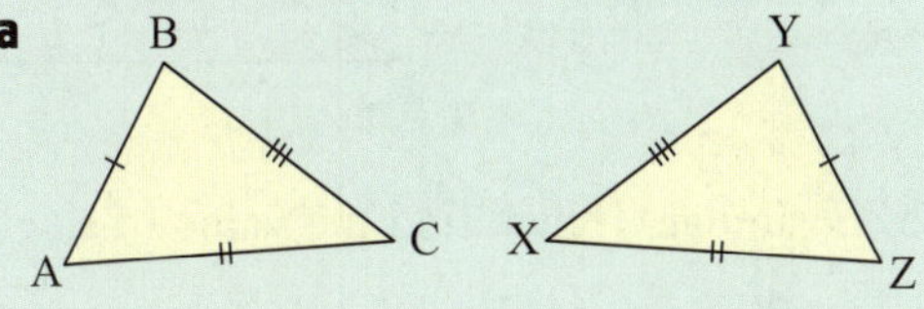

b
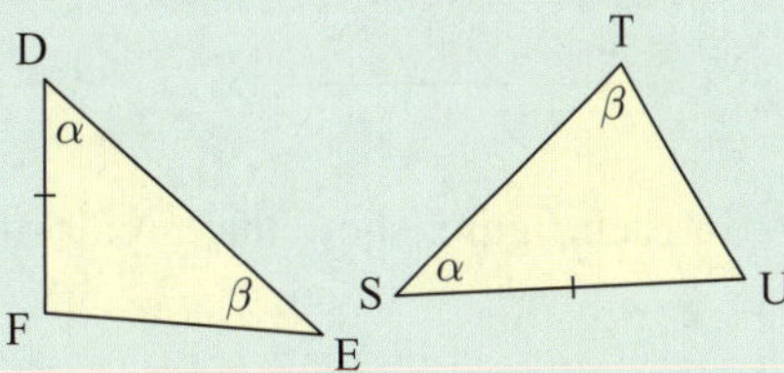

4 Consider the kite ABCD alongside.

  a Use congruence to show that $\widehat{BAC} = \widehat{DAC}$.

  b Hence show that $\triangle ABX \cong \triangle ADX$.

  c Show that [BX] and [DX] have the same length.

  d What property of kites has been proven in **c**?

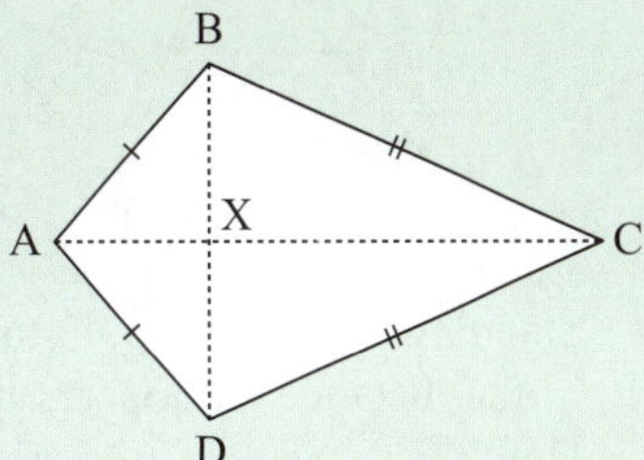

5 Use congruence to prove that, in an isosceles triangle, the line joining the apex to the midpoint of the base meets the base at right angles.

6 For each transformation from A to B, state whether the transformation is an enlargement or a reduction, and find the scale factor.

a
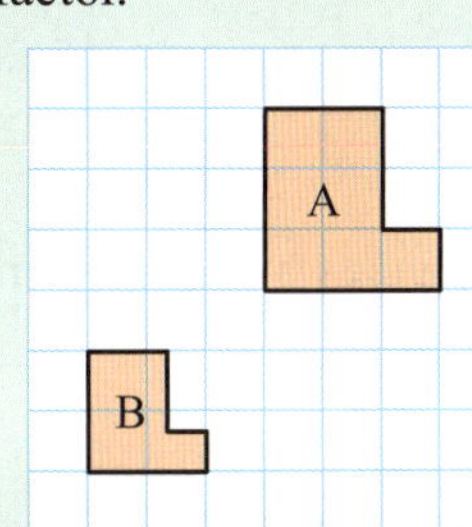

b
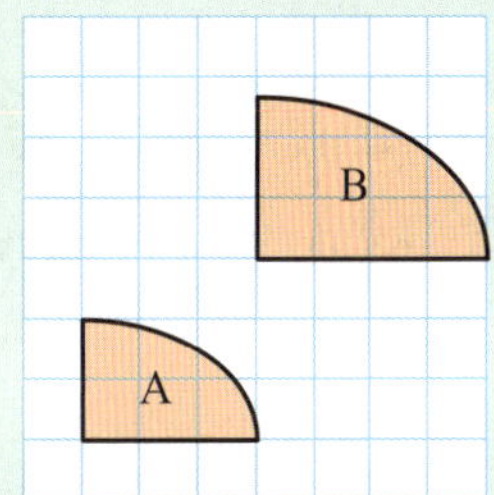

c

7 Determine whether each pair of figures is similar:

a
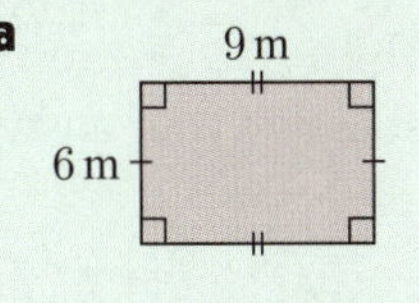
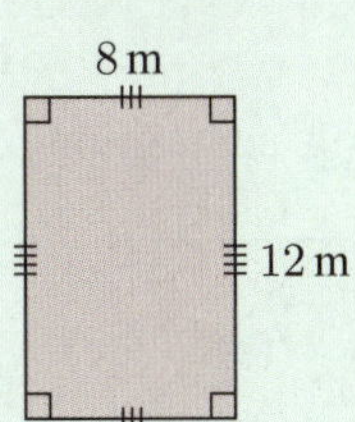

b
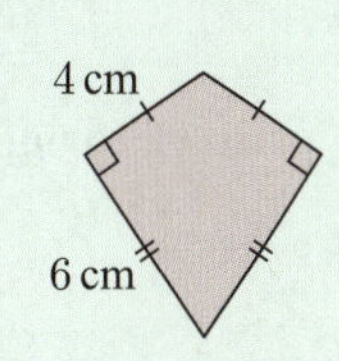
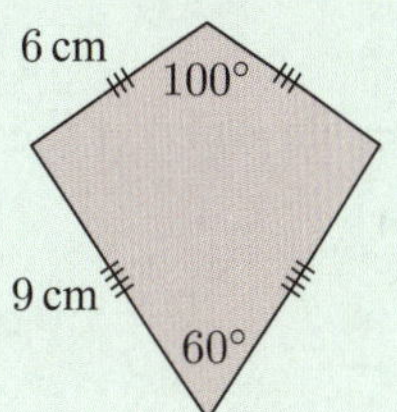

8 These figures are similar. Find the value of x:

a
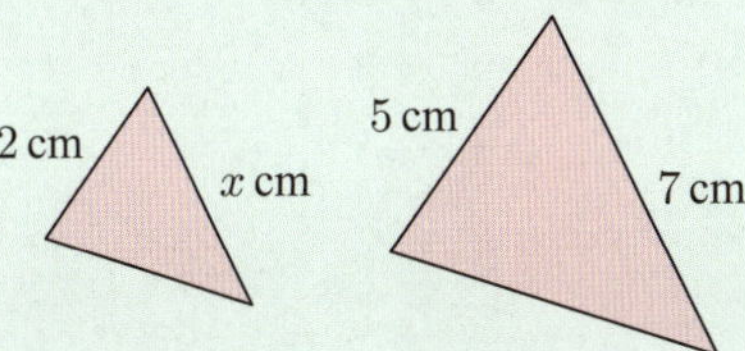

b

9 Show that these figures possess similar triangles:

a 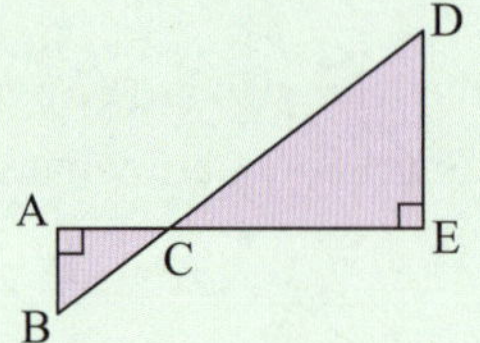

b 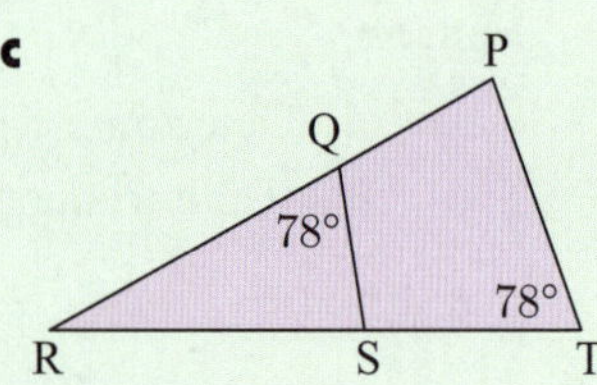

c

10 In each figure, show that a pair of triangles is similar. Hence find the value of x.

a 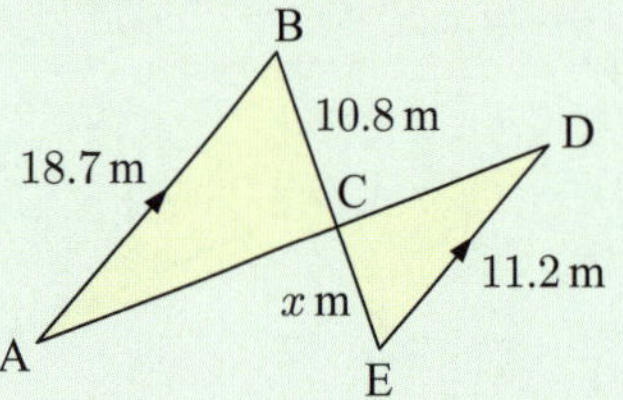

b 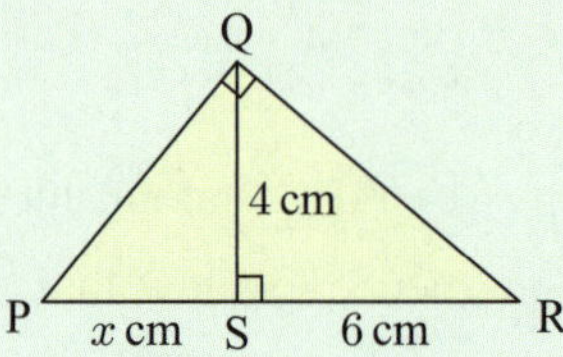

11 Standing on the beach, Francis holds a ruler 50 cm away from his eye, and measures a ship in the distance. It appears to be 1.6 cm high. Francis knows the ship is really 20 m high. How far is the ship from Francis?

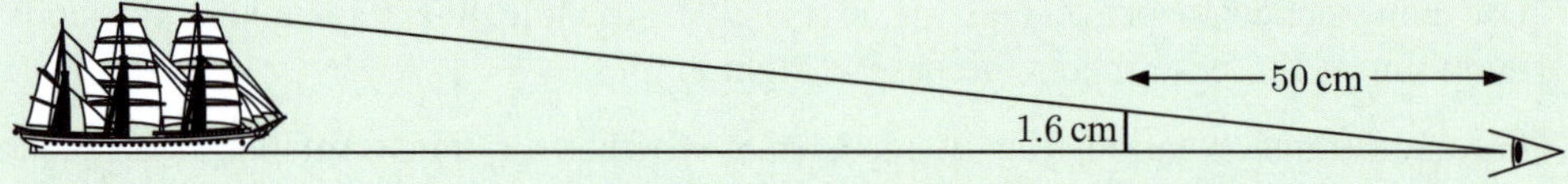

REVIEW SET 19B

1 Decide whether these pairs of figures are congruent:

a 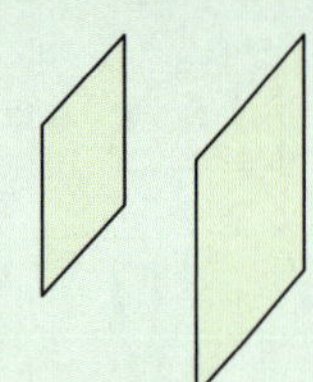

b 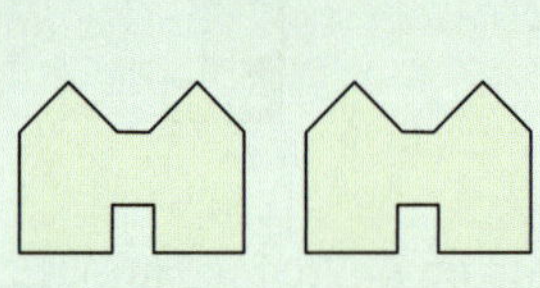

c 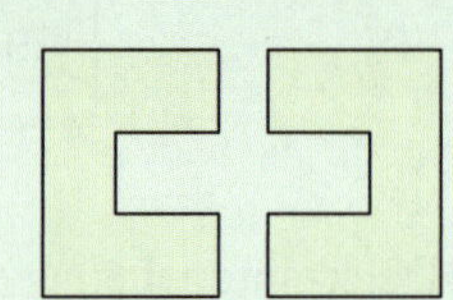

2 Decide whether these pairs of triangles are congruent, giving reasons for your answers:

a 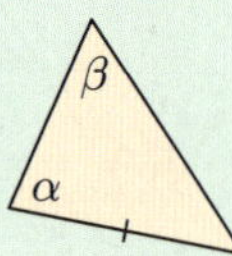

b 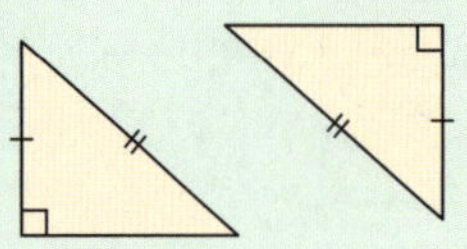

c 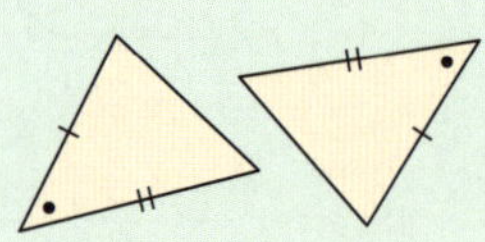

3 Consider the kite ABCD.

a Use congruence to show that $\widehat{ABC} = \widehat{ADC}$.

b What property of kites has been proven?

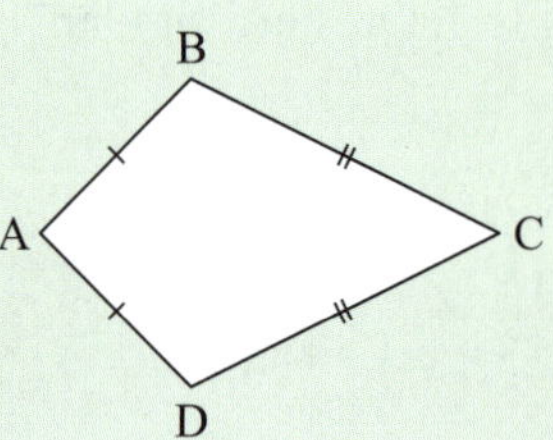

4   A square piece of paper is divided into four triangles as shown.

   a   Show that triangles A and D are congruent.

   b   Hence show that triangles B and C are congruent.

   c   Find the area of each triangle.

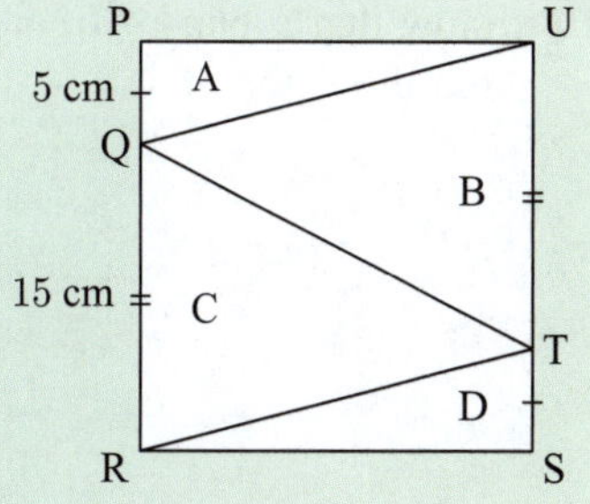

5   Enlarge or reduce the given figure with the scale factor:

   a $\frac{5}{2}$ **b** $\frac{3}{4}$

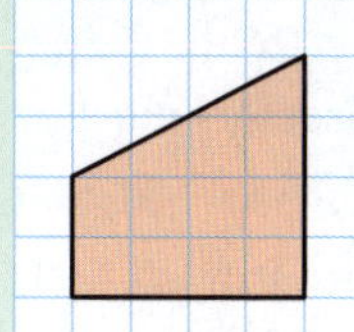

6   Are all rhombuses similar? Explain your answer.

7   The shaded rectangles in this diagram are similar. Show that the rectangle ABCD is also similar to the shaded rectangles.

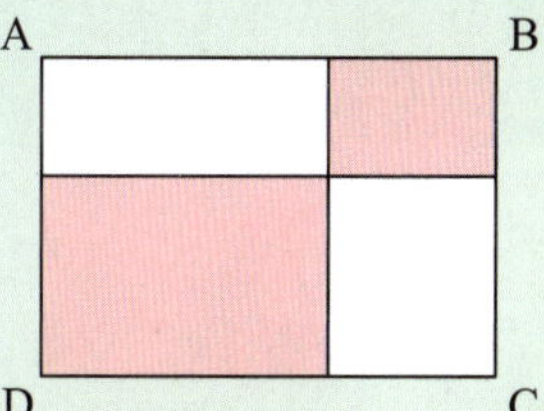

8   Show that these figures possess similar triangles:

   a

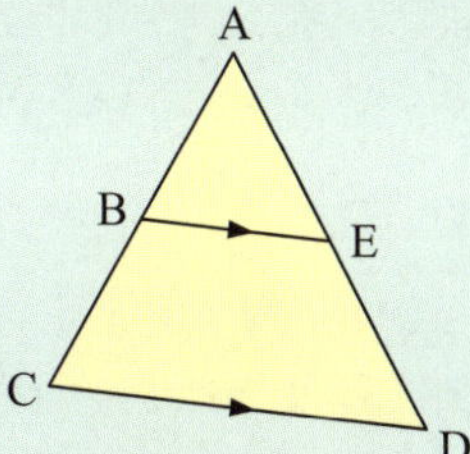

   b

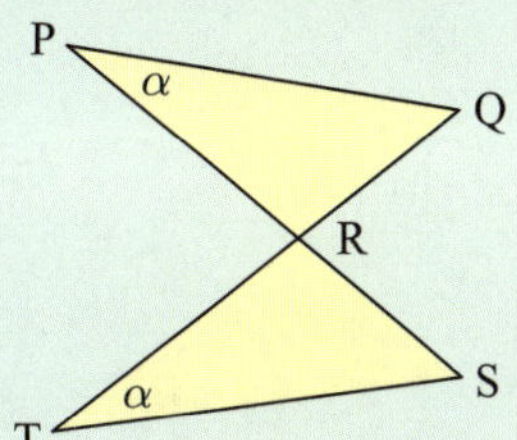

9   P and Q are markers on the bank of a canal which has parallel sides. R and S are telegraph poles which are directly opposite each other.

When I walk 20 m from P directly away from the bank, I reach the point T such that T, Q, and S are collinear.

How wide is the canal?

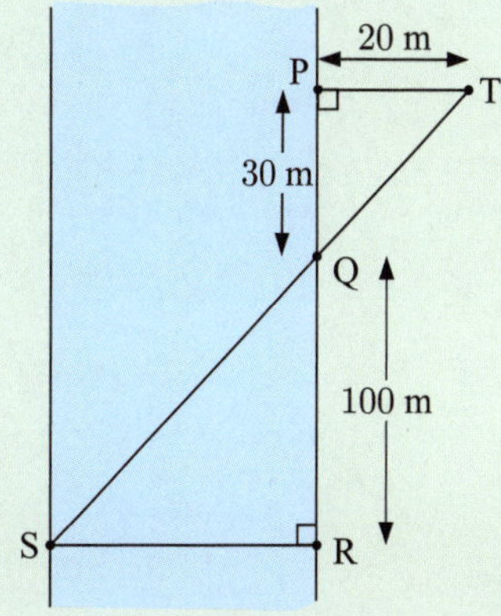

10 Show that a pair of triangles is similar, and hence find the value of x:

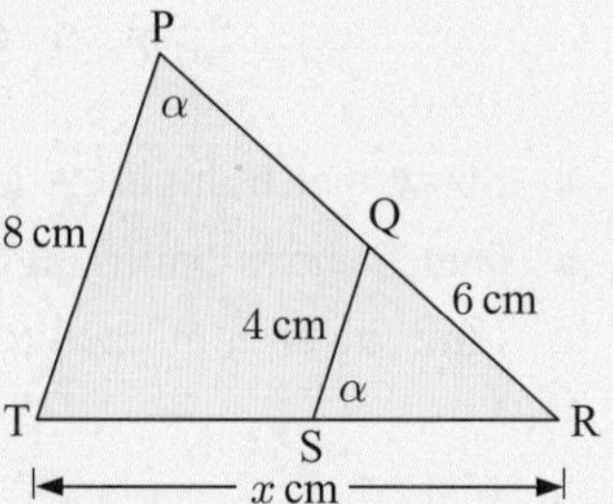

11 In the figure alongside, each angle of the isosceles triangle ABC is *trisected* into 3 equal angles. The angle trisectors meet at D, E, and F as shown.

a Show that $BF = CF$.

b Show that $\triangle ABD \cong \triangle ACE$, and hence that $BD = CE$.

c Show that $\triangle BDF \cong \triangle CEF$.

d Hence show that triangle DEF is also isosceles.

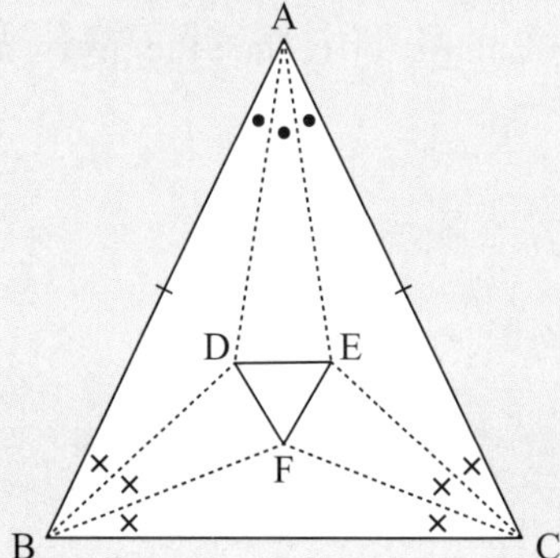

Chapter 20

Pythagoras' theorem

Contents:

OPENING PROBLEM

Phil is concerned that the street lamp outside his house is not quite at right angles to the ground.

He marks a point A on the lamp 1.5 m from its base, and a point B on the ground 2 m from the lamp's base. Using a tape measure, he finds that the distance between A and B is 2.40 m.

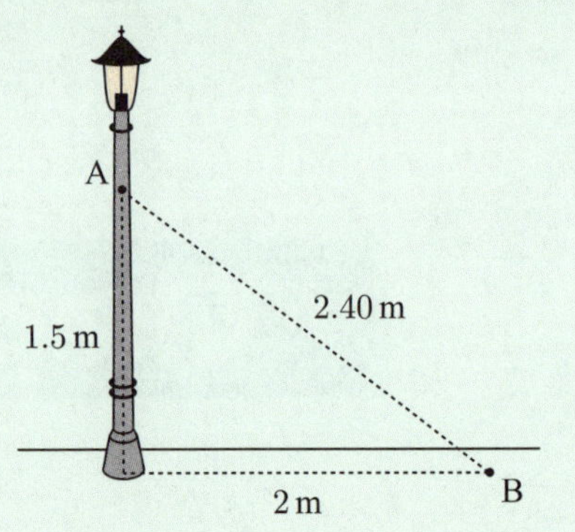

Things to think about:

a What assumptions has Phil made?

b Are the measured lengths sufficient information to determine whether the street lamp is at right angles to the ground?

A **right angled triangle** is a triangle which has a right angle as one of its angles.

The side **opposite** the right angle is called the **hypotenuse**. It is the **longest** side of the triangle.

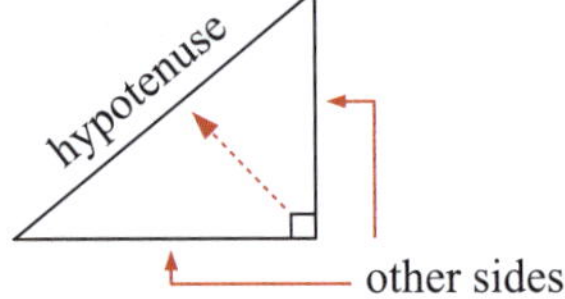

Right angled triangles are frequently observed in the real world. Some examples are shown below.

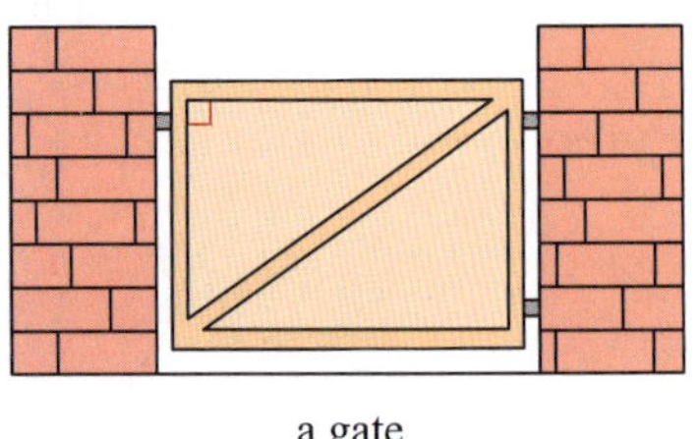

a gate

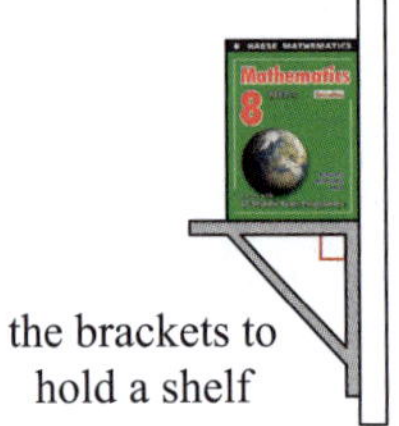

the brackets to hold a shelf

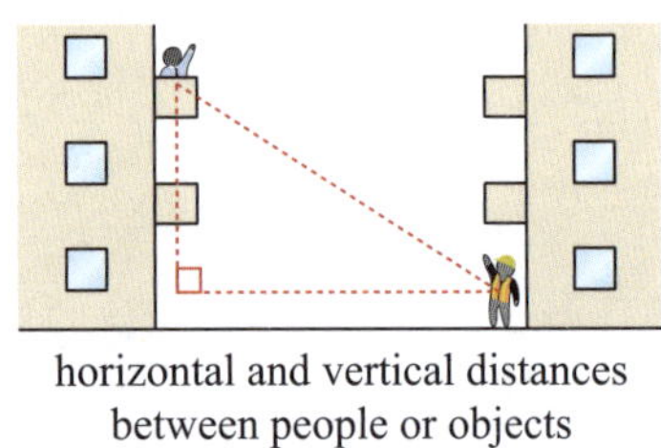

horizontal and vertical distances between people or objects

HISTORICAL NOTE

For many centuries people have used right angled corners to construct buildings and to divide land into rectangular fields. They have done this quite accurately by relatively simple means.

Over 3000 years ago the Egyptians knew that a triangle with sides in the ratio 3 : 4 : 5 was right angled. They used a loop of rope with 12 knots equally spaced along it to make corners in their building construction.

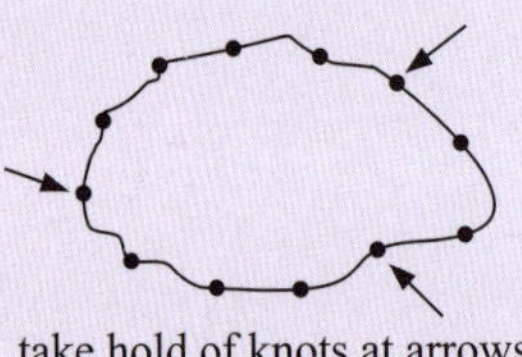

take hold of knots at arrows

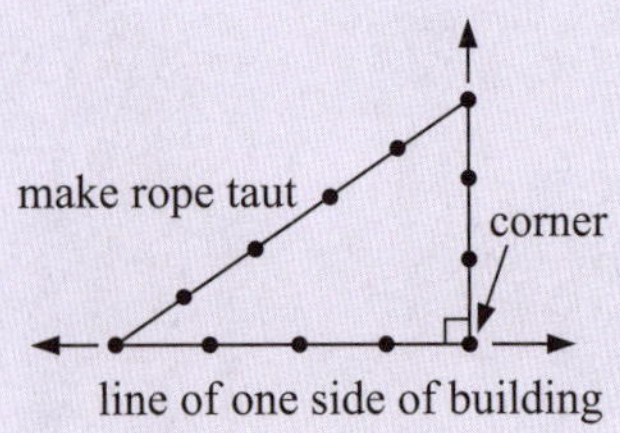

line of one side of building

Around 500 BC, the Greek mathematician **Pythagoras of Samos** proved a rule which connects the sides of a right angled triangle. According to legend, he discovered the rule while studying the tiled palace floor as he waited for an audience with the ruler Polycrates.

A PYTHAGORAS' THEOREM

DISCOVERING PYTHAGORAS' THEOREM

Consider a right angled triangle with hypotenuse of length c cm, and other side lengths a cm and b cm.

Your task is to find an equation which connects a, b, and c.

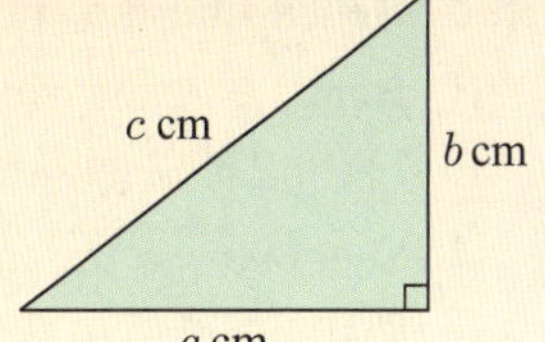

What to do:

1 Copy or print the table alongside.

PRINTABLE TABLE

a	b	c	a^2	b^2	c^2	$a^2 + b^2$
4	3		16			
6	8					
5	12					
7	4					

2 Draw a horizontal line of length $a = 4$ cm. At one end, draw a vertical line of length $b = 3$ cm. The horizontal and vertical lines meet at right angles.

Complete a right angled triangle by drawing in the hypotenuse. Find c by measuring this hypotenuse with a ruler.

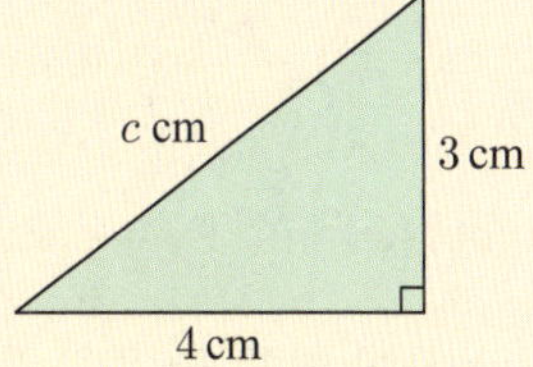

Write your answer in the table and hence complete the first row.

3 Repeat the process in **2** for the values of a and b given in the next three rows of the table.

4 Repeat the process in **2** for two more right angled triangles of your choosing. Write your results in the last two rows of the table.

5 By comparing your results for c^2 and $a^2 + b^2$, write down **Pythagoras' theorem**, which is an equation connecting a, b, and c.

6 In the era of Pythagoras, there was very little algebra. It is therefore likely that he discovered his theorem using geometry.

Suppose we start with a square. We mark a point along each side which divides the side into lengths a and b, then connect these points as shown.

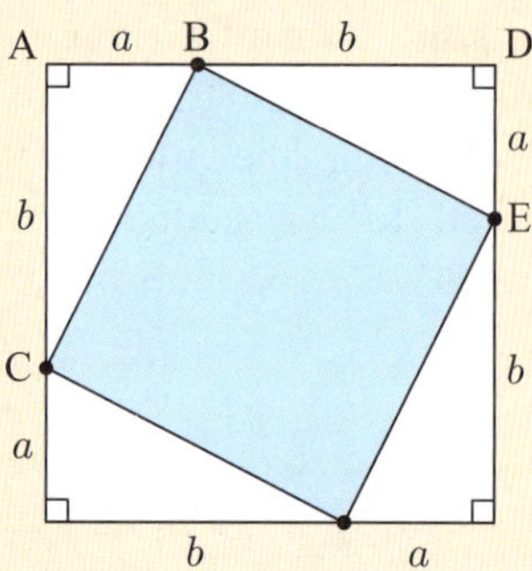

a Explain why the four right angled triangles formed are congruent.

b Find the sum of $\widehat{ABC}$ and $\widehat{ACB}$.

c Hence find the sum of $\widehat{ABC}$ and $\widehat{DBE}$.

d Hence explain why the shaded region is a square.

e Let the side length of the shaded region be c. Write down its area.

f Suppose we rearrange the four right angled triangles within the original square as shown. Write down the sum of the shaded areas.

g Use your results to prove Pythagoras' theorem.

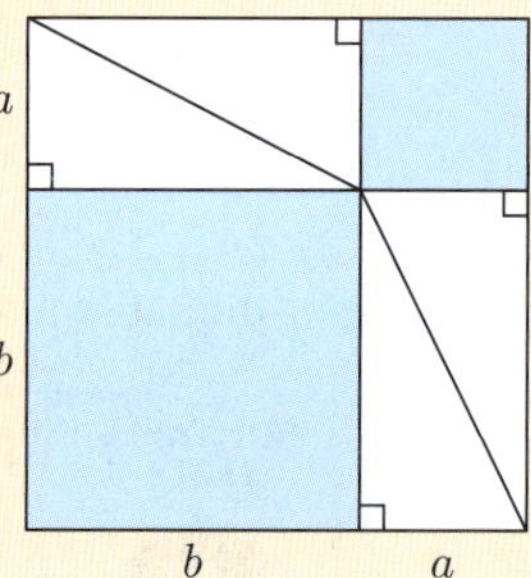

PYTHAGORAS' THEOREM

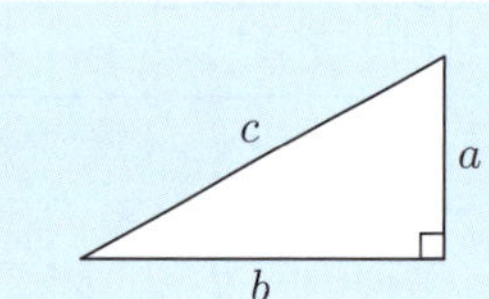

In a right angled triangle with hypotenuse of length c, and other sides of length a and b,

$$c^2 = a^2 + b^2.$$

In geometric form, Pythagoras' theorem states:

In any right angled triangle, the area of the square on the hypotenuse is equal to the sum of the areas of the squares on the other two sides.

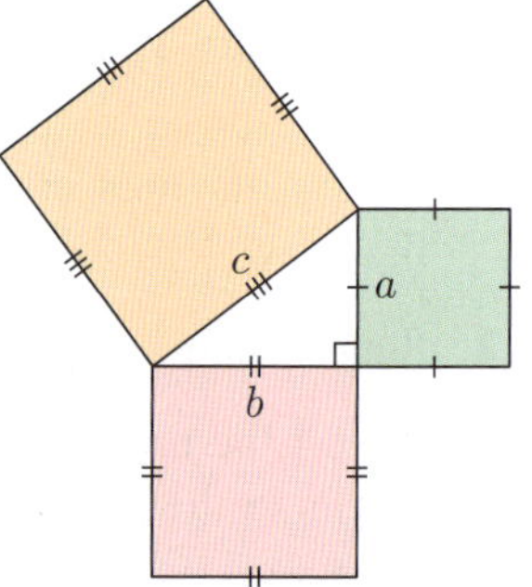

Look back at the tile pattern in the **Historical note** on page **421**.
Can you see the figure alongside in the pattern?

To find side lengths in triangles using Pythagoras' theorem, we need to solve a power equation. Since side lengths must be positive, we can always reject the negative solution.

Example 1 ◀》 Self Tutor

Find the length of the hypotenuse in this right angled triangle:

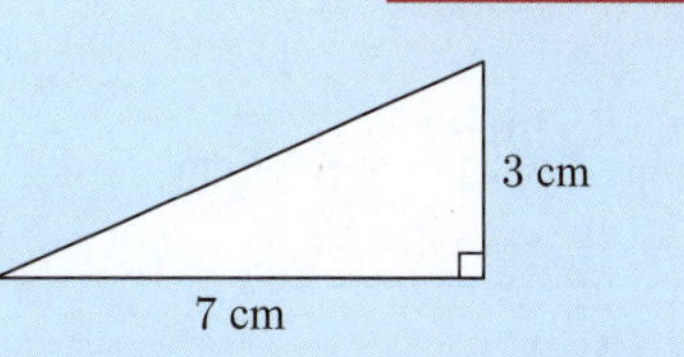

Let the hypotenuse have length x cm.

$\therefore \ x^2 = 7^2 + 3^2$ {Pythagoras}

$\therefore \ x^2 = 49 + 9$

$\therefore \ x^2 = 58$

$\therefore \ x = \sqrt{58}$ {since $x > 0$}

$\therefore \ x \approx 7.62$

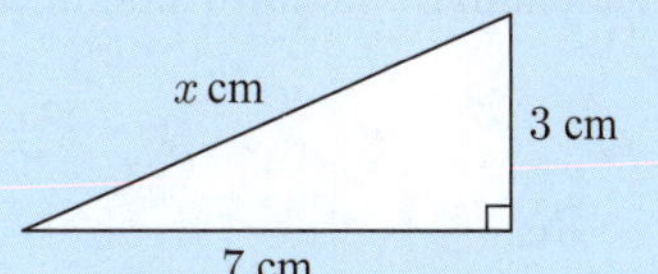

The hypotenuse is about 7.62 cm long.

EXERCISE 20A

1 Find the length of the hypotenuse in each right angled triangle:

a
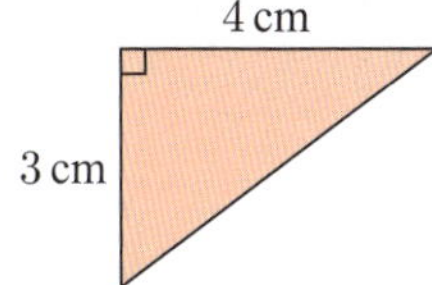

b
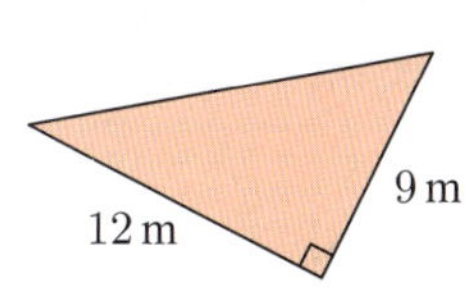

c
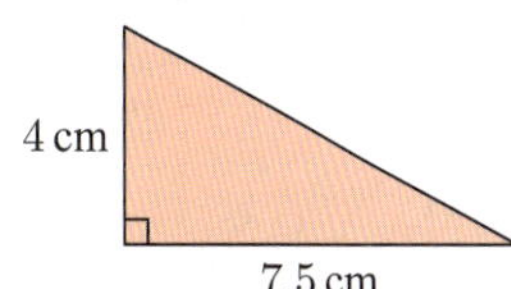

2 Find the length of the hypotenuse in each right angled triangle, rounding your answer to 2 decimal places:

a
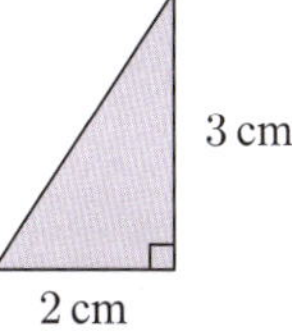

b
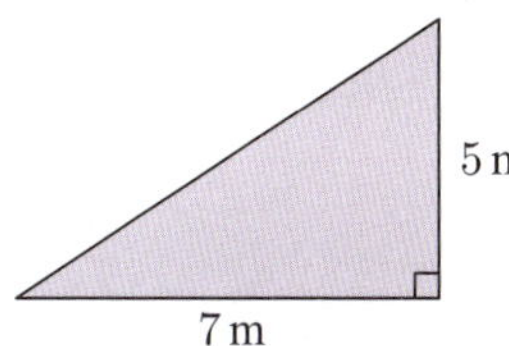

c
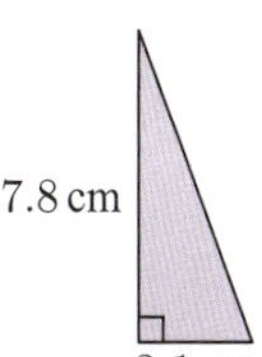

Example 2 ◀》 Self Tutor

Find the length of the third side of this right angled triangle:

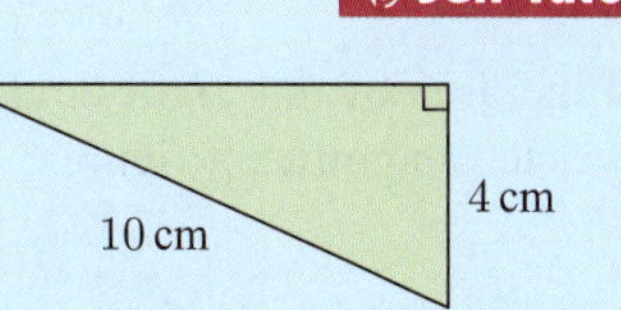

Let the unknown side have length x cm.

$\therefore \ x^2 + 4^2 = 10^2$ {Pythagoras}

$\therefore \ x^2 + 16 = 100$

$\therefore \ x^2 = 84$

$\therefore \ x = \sqrt{84}$ {since $x > 0$}

$\therefore \ x \approx 9.17$

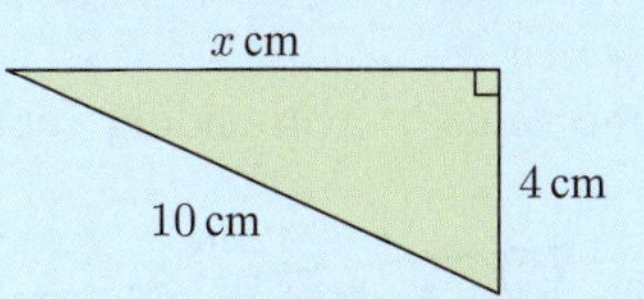

The third side is about 9.17 cm long.

3 Find the length of the third side in each right angled triangle:

a 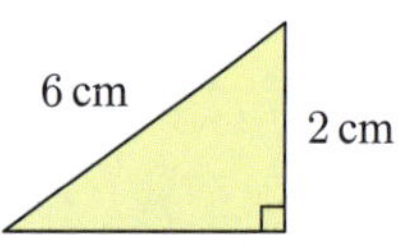

b 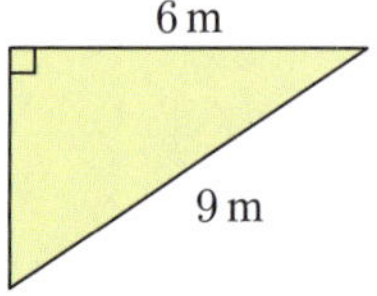

c 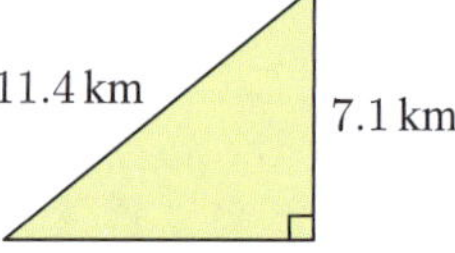

4 Find the length of the third side in each right angled triangle, rounding your answer to 2 decimal places:

a 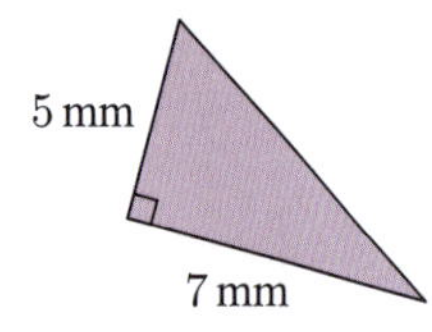

b 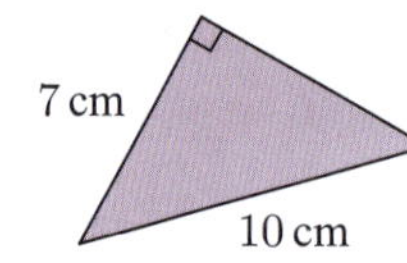

c 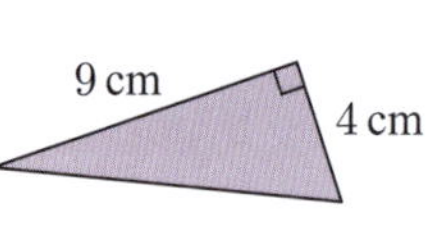

5 Find the length of the third side in each right angled triangle. Round your answer to 3 significant figures if necessary.

a 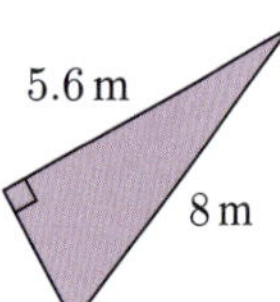

b 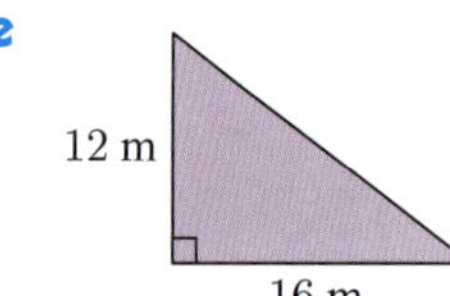

c 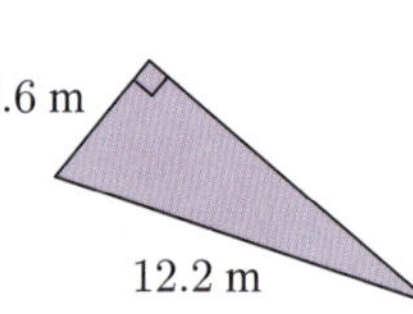

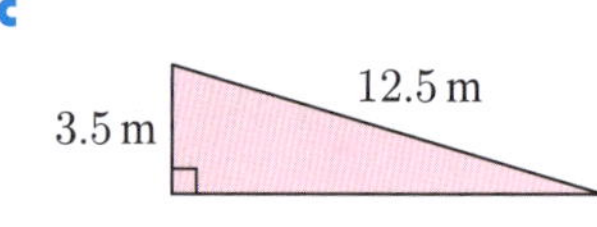

d 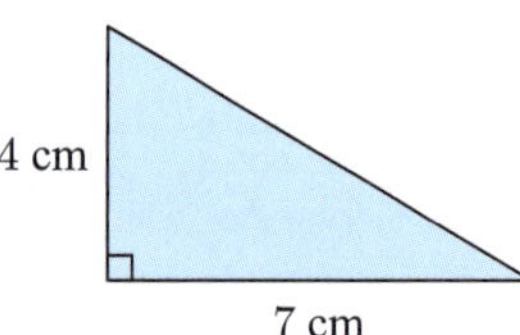

e

f 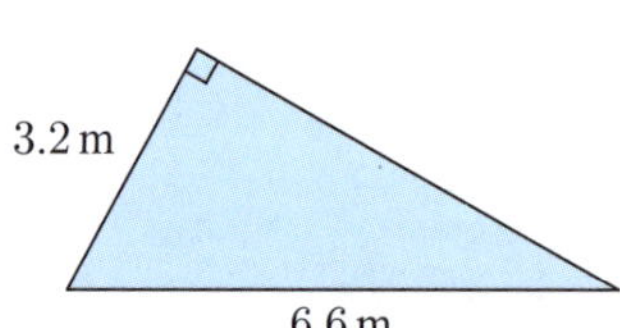

6 Find the perimeter of each right angled triangle, rounding your answer to 2 decimal places:

a

b

7 Find the area of this right angled triangle, rounding your answer to 3 significant figures:

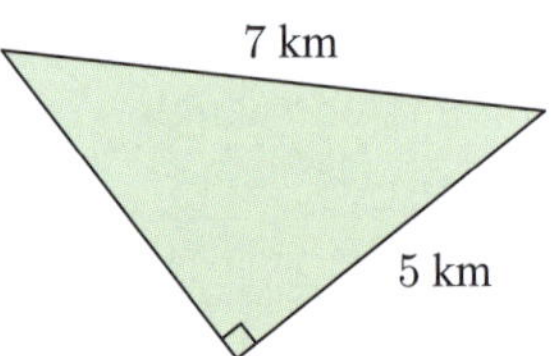

8 Find the value of x, rounding your answer to 3 significant figures:

a 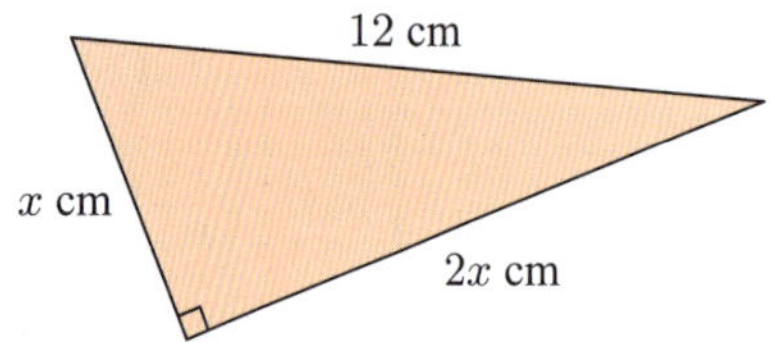

b 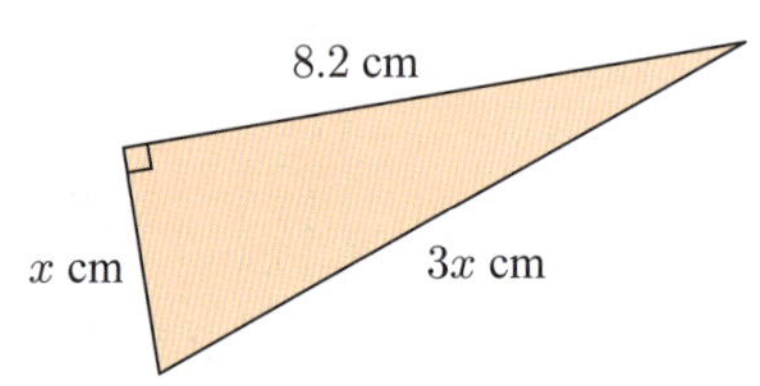

B   PROBLEM SOLVING

Right angled triangles occur in many practical problems. In these situations we can apply Pythagoras' theorem to help find unknown side lengths.

The problem solving approach involves the following steps:

Step 1:   Draw a neat, clear diagram of the situation.

Step 2:   Mark known lengths and right angles on the diagram.

Step 3:   Use a variable such as x to represent the unknown length.

Step 4:   Write down Pythagoras' theorem for the given information.

Step 5:   Solve the equation.

Step 6:   Where necessary, write your answer in sentence form.

Example 3   ◀ Self Tutor

Joe runs 6 km west and then 4 km south. How far is he now from his starting point?

Suppose Joe is x km from his starting point.

$\therefore \ x^2 = 4^2 + 6^2$ {Pythagoras}

$\therefore \ x^2 = 16 + 36$

$\therefore \ x^2 = 52$

$\therefore \ \ x = \sqrt{52}$ {since $x > 0$}

$\therefore \ \ x \approx 7.21$

Joe is about 7.21 km from his starting point.

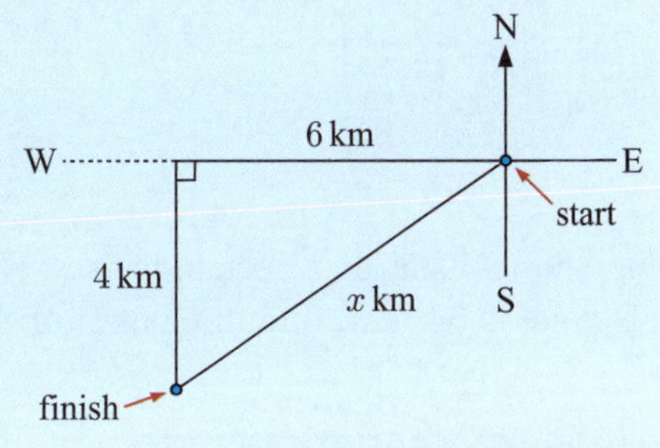

EXERCISE 20B

1   This rectangular gate has a diagonal strut for support and strength. How long is the strut?

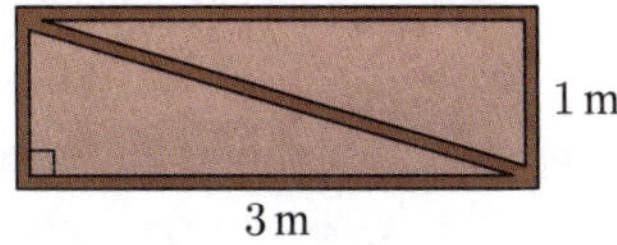

2   Mary walks 9 km north and then 3 km east. How far is she now from her starting point?

3

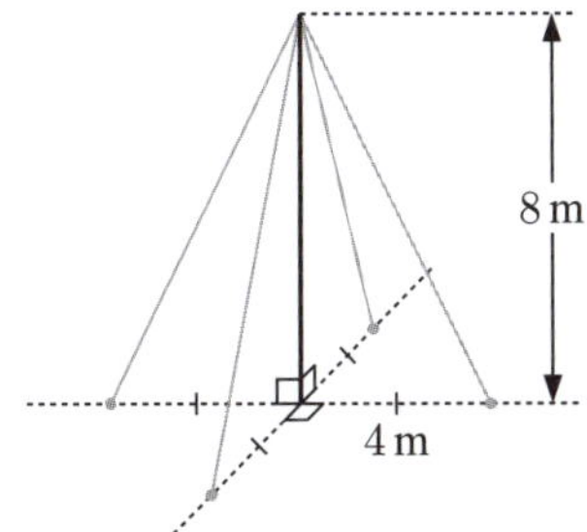

An 8 m high pole is held vertical by 4 cables. Each cable is anchored 4 m from the base of the pole. Find the total length of the cables.

4 An orienteer runs 540 m in a straight line, then turns at right angles and runs a further 380 m.

   a How far is the orienteer from his starting point?

   b The orienteer then turns and runs straight back to his starting point. How far has he run in total?

5 Find, to 3 significant figures, the length of a diagonal of a square courtyard with side length 4.6 m.

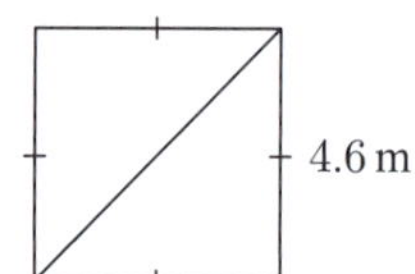

6 One end of a 38 m long zip-line is attached to a tree. The other end is fixed to the ground 32 m from the base of the tree. How high up the tree is the zip-line attached?

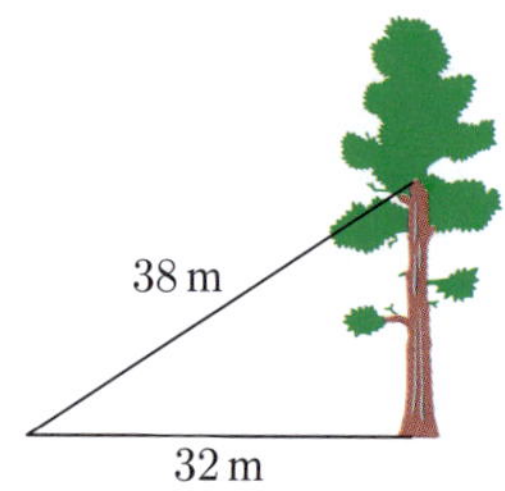

7 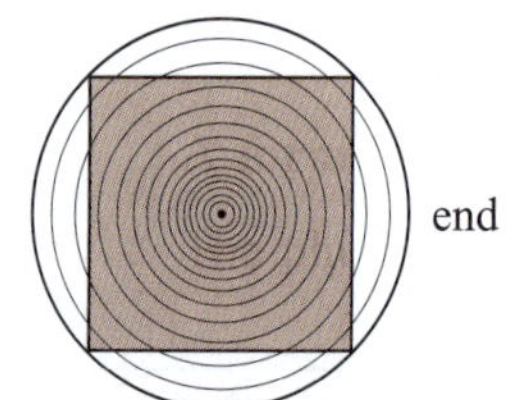

A log is 60 cm in diameter. Find the dimensions of the largest square section beam which can be cut from the log.

8 The shorter side of a rectangle is 40% shorter than the diagonal. If the longer side of the rectangle is 64 mm, find the length of the shorter side.

9 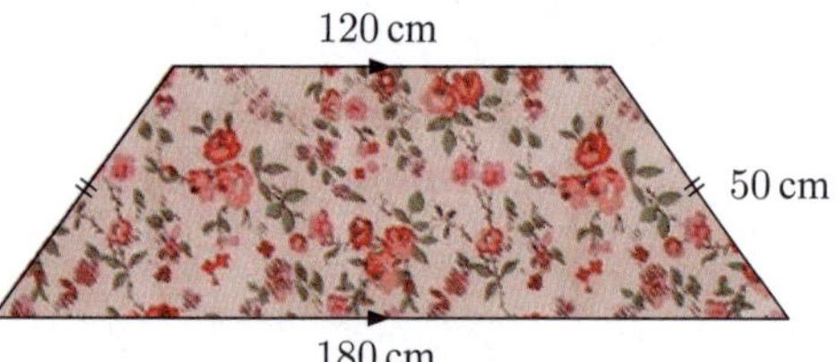

Vanessa is using the material alongside to make a skirt.

Find the area of the material.

10 The diagram shows the roof structure of a house. Find the length of the beam [AB].

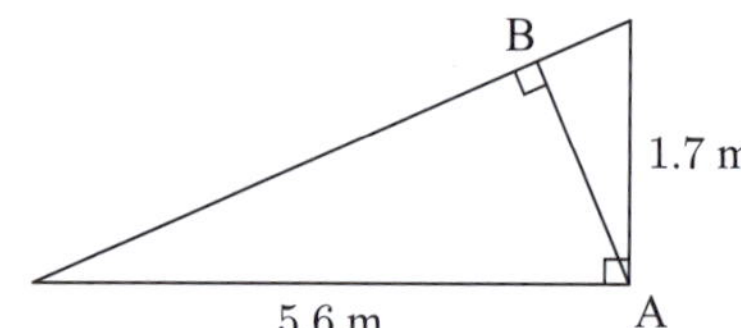

11 An aquarium has the dimensions as shown. Find the length of the diagonal [XY].

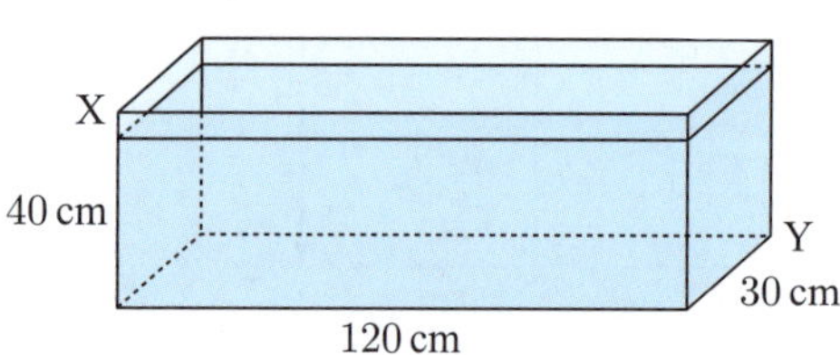

12

A huge concrete dam is used to control water flowing down a river. It has the cross-section shown, with area 177.31 m^2.

Find the length of:

  a the sloping face of the dam, [AB]

  b the sloping bottom of the dam, [AC].

C   THE CONVERSE OF PYTHAGORAS' THEOREM

If we are given the lengths of three sides of a triangle, the **converse of Pythagoras' theorem** gives us a simple **test** to determine whether the triangle is right angled.

THE CONVERSE OF PYTHAGORAS' THEOREM

> If a triangle has sides of length a, b, and c units where $a^2 + b^2 = c^2$, then the triangle is right angled.

Example 4   ◀) Self Tutor

A triangle has side lengths 9 cm, 15 cm, and 12 cm.

Is the triangle right angled?

The two shorter sides have lengths 9 cm and 12 cm.

Now $9^2 + 12^2 = 81 + 144 = 225$

and $15^2 = 225$

$\therefore 9^2 + 12^2 = 15^2$

$\therefore$ the triangle is right angled.

EXERCISE 20C

1 The following figures are not drawn to scale. Determine whether each triangle is right angled.

a

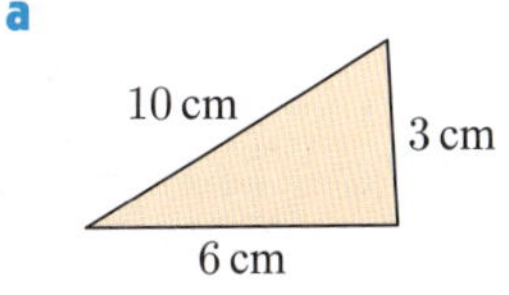

b

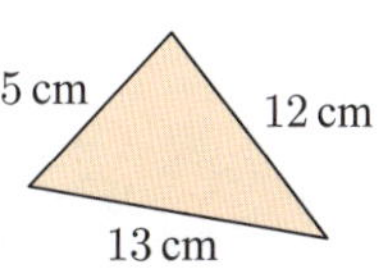

c

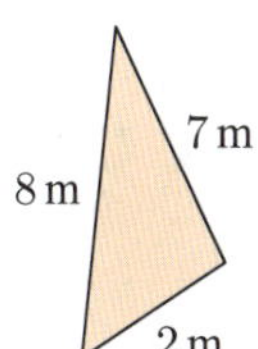

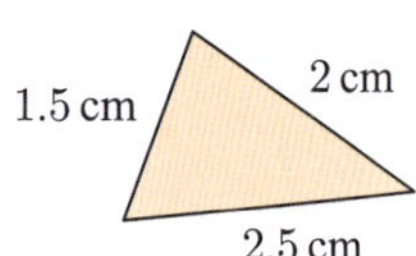

d

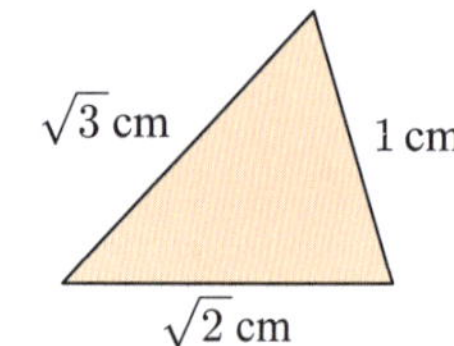

e

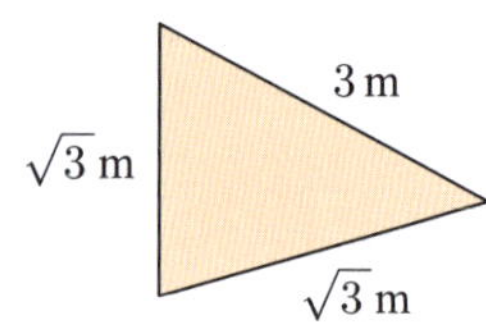

f

2 Answer the **Opening Problem** on page **420**.

3 Use the converse of Pythagoras' theorem to explain why this isosceles triangle cannot be right angled:

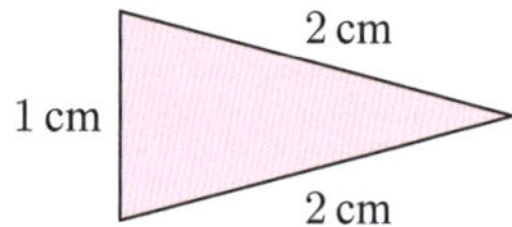

4 Consider the right angled isosceles triangle alongside.

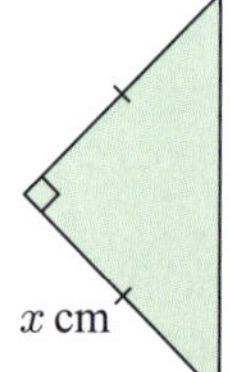

  a Use Pythagoras' theorem to find the length of the hypotenuse in terms of x.

  b Copy and complete:

     In any right angled isosceles triangle, the hypotenuse is always ...... times the length of the other sides.

ACTIVITY   TESTING FOR RIGHT ANGLES

What to do:

1 Select some objects around your school which appear to be at right angles to the ground. You could use a goal post, a wall, or a lamp post.

2 Use the converse of Pythagoras' theorem to determine whether each object is at right angles to the ground.

3 Write a short report to summarise your findings. Include your measurements, calculations, and conclusions. Discuss the accuracy of your results.

GLOBAL CONTEXT   PYRAMIDS

Global context: Personal and cultural expression

Statement of inquiry: Investigating the monuments of ancient civilisations can help us to understand their cultures.

Criterion: Communicating

MULTIPLE CHOICE QUIZ

REVIEW SET 20A

1 Find the length of the hypotenuse in each right angled triangle:

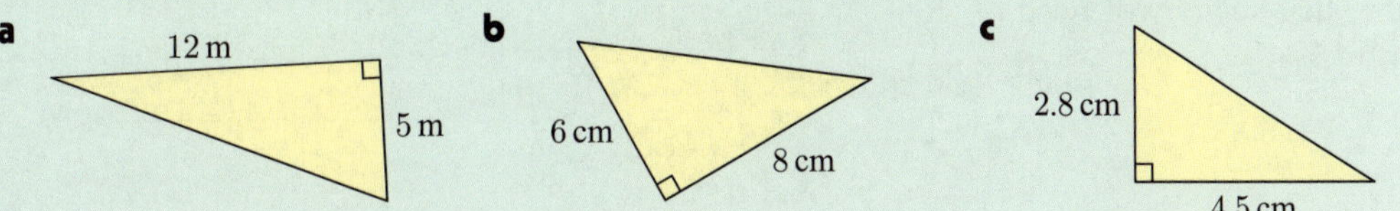

2 Find the length of the third side, rounding your answer to 2 decimal places:

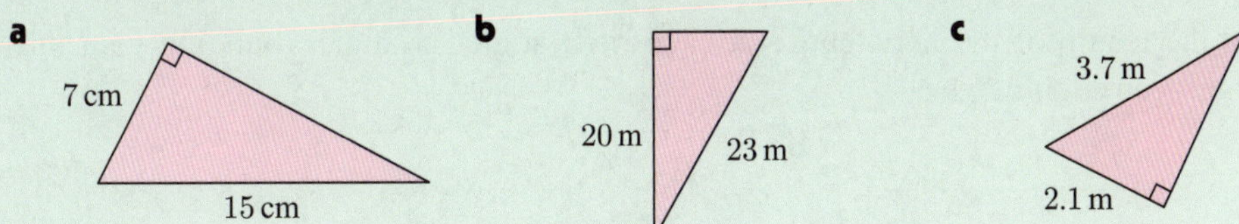

3 Find the perimeter of each triangle, rounding your answer to 2 decimal places:

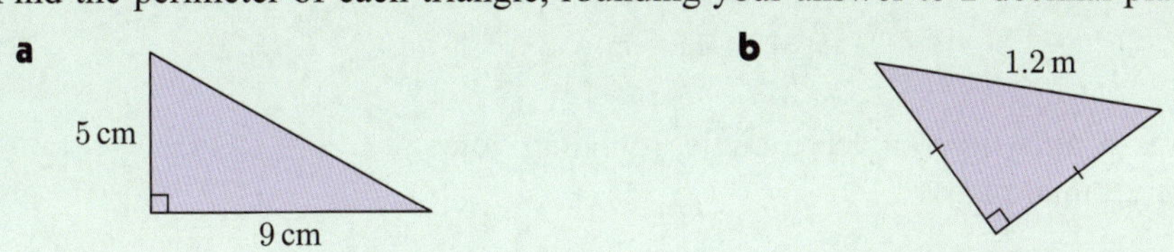

4 Find the length of the illustrated brace.

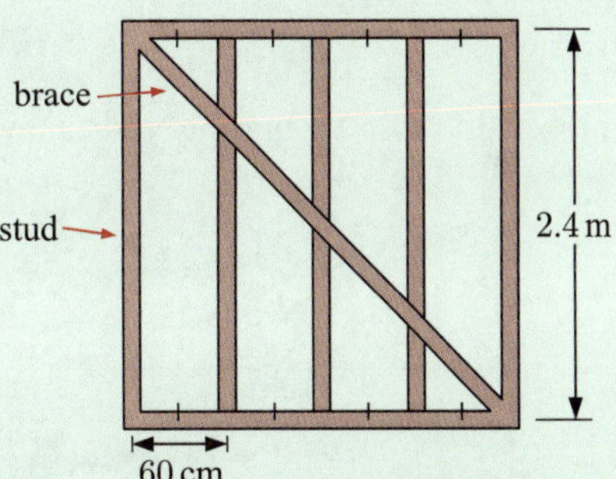

5 The following figures are not drawn to scale. Determine whether each triangle is right angled.

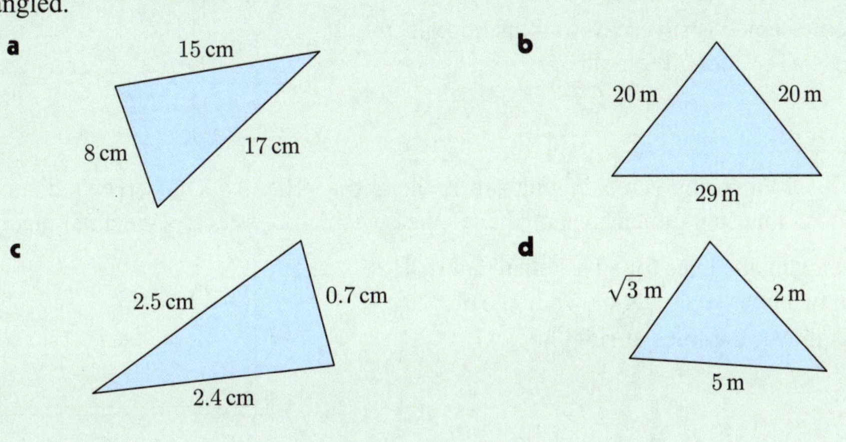

6 Jenny made a kite with the dimensions shown. She wants to decorate the edges of the kite with ribbon. Find the total length of ribbon that Jenny will need.

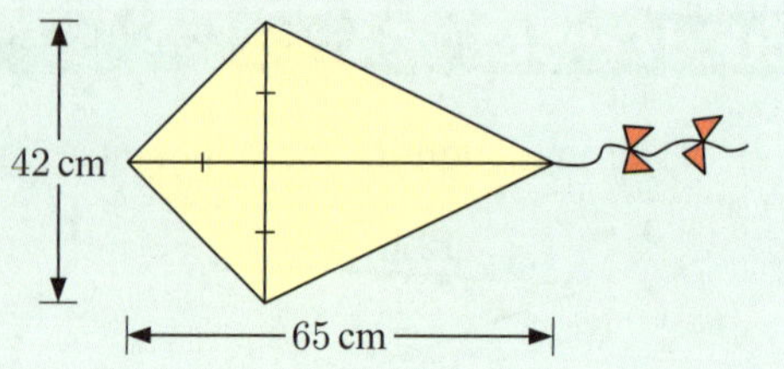

REVIEW SET 20B

1 Find the length of the hypotenuse in each right angled triangle, rounding your answer to 3 significant figures:

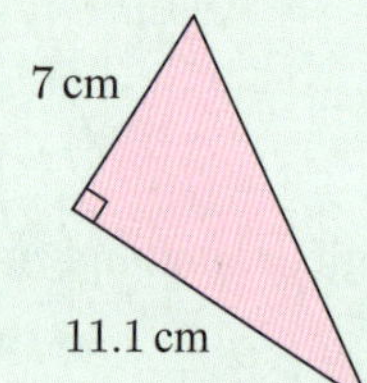
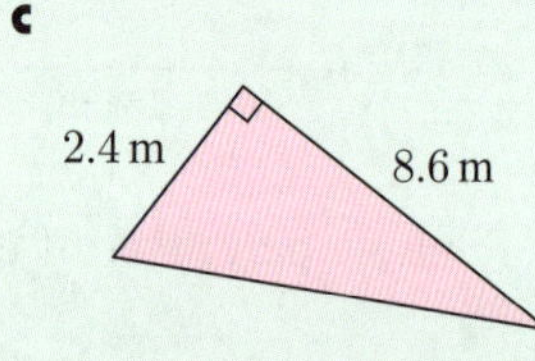

2 Find the area of this right angled triangle, rounding your answer to 2 decimal places.

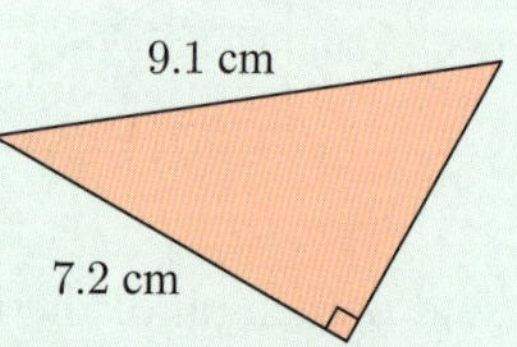

3 Find the value of x, rounding your answer to 3 significant figures:

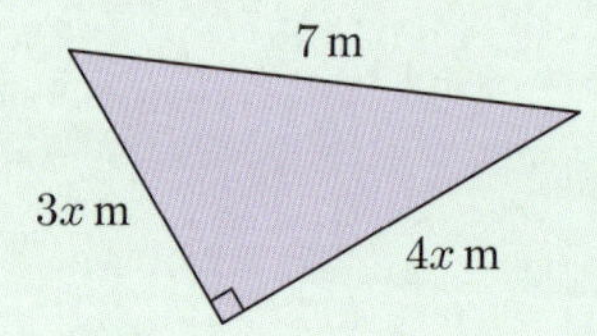

4 Two roads intersect at right angles. Point X is 5 km from the intersection of the two roads, whereas point Y is 3 km from the intersection.

How much distance is saved by walking straight from X to Y instead of along the roads?

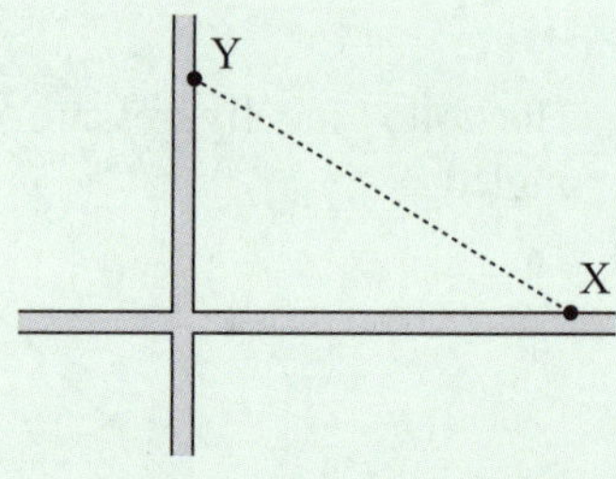

5 The ratio of a television screen's width to its height is $16 : 9$. The screen's diagonal is 127 cm long. Find the dimensions of the television screen correct to 1 decimal place.

6 Louis is drawing the lines for a basketball court. He has drawn two lines so far, as shown in the diagram. Has Louis drawn the lines at right angles?

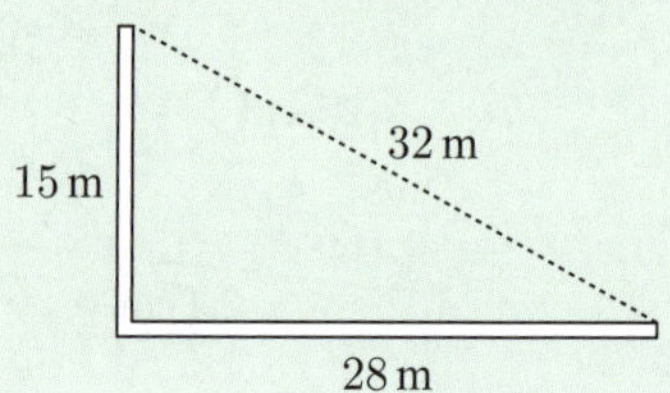

Problem solving

Contents:

OPENING PROBLEM

Kevin, Mai, and Bruce had breakfast at an all-you-can-eat pancake parlour. Bruce ate twice as many pancakes as Kevin. Mai ate one more pancake than Kevin. Between them they ate 13 pancakes.

Things to think about:

a How can we represent this problem using algebra?

b How many pancakes did Kevin eat?

Many problems we encounter are presented in sentences rather than symbols.

In this Chapter we will examine some techniques for solving these problems, including:

- writing the problem as an algebraic equation
- solution by working backwards.
- solution by search

A   WRITING PROBLEMS AS EQUATIONS

When solving problems presented in sentence form, it is important to be able to write the problem as a mathematical equation.

Example 1   ◀ Self Tutor

"I start with a number, add 4, then multiply by 3. The result is 33."

Write this statement as an algebraic equation.

Let the unknown number be x.

Start with a number	x
add 4	$x + 4$
multiply by 3	$3(x + 4)$
the result is 33	$3(x + 4) = 33$

The algebraic equation which represents the problem is $3(x + 4) = 33$.

EXERCISE 21A

1 Starting with the number x, rewrite each statement as an algebraic equation:

a I think of a number and subtract 7 from it. The result is 11.

b I think of a number and divide it by 8. The result is 5.

c I think of a number, multiply it by 3, then add 5. The result is 23.

d I think of a number, add 1 to it, then divide the result by 4. The final result is 10.

e I think of a number, subtract 3, then multiply the result by 9. The final result is 36.

f I think of a number, divide it by 4, then add the original number to the result. The final result is 20.

g I think of a number, multiply it by 5, then subtract 8. The result is 4 more than the original number.

2 Summarise the information as an equation:

a Clint has x sons and 3 daughters. He has 5 children in total.

b Beth took x minutes to walk to the shops and the same amount of time to walk back. She spent 15 minutes at the shops, and the trip took 37 minutes in total.

c Stan has mass x kg. Ted is 6 kg heavier than Stan. Their combined mass is 100 kg.

d Sylvia has $\$x$ in her bank account. Depositing $\$100$ into her account would triple the amount of money in the account.

e Jacinta is x years old. She is presently three times her brother's age. Four years from now, she will be twice her brother's age.

B   PROBLEM SOLVING WITH ALGEBRA

To solve problems presented in sentence form, we follow these steps:

Step 1:   Select a letter such as x to represent the unknown quantity to be found.

Step 2:   Write an equation using the information in the question.

Step 3:   Solve the equation. Check your solution.

Step 4:   Answer the question in a sentence.

Example 2   ◀)) **Self Tutor**

I think of a number, double it, then subtract 7. The result is 10. Find the original number.

Let x be the number, so $2x$ is the number doubled.

$\therefore \ 2x - 7$ is the number doubled minus 7.

$$\therefore \ 2x - 7 = 10$$
$$\therefore \ 2x - 7 + 7 = 10 + 7$$
$$\therefore \ 2x = 17$$
$$\therefore \ \frac{2x}{2} = \frac{17}{2}$$
$$\therefore \ x = 8\tfrac{1}{2}$$

Check: LHS $= 2 \times 8\tfrac{1}{2} - 7 = 17 - 7 = 10 =$ RHS ✓

The original number was $8\tfrac{1}{2}$.

EXERCISE 21B

1 When a number is doubled, the result is 28. Find the number.

2 I think of a number, treble it, then subtract 8. The result is -5. Find the original number.

3 When 3 is subtracted from a certain number and the result is divided by 4, the final result is 10. What was the original number?

4 When 4 is added to a certain number and the result is doubled, the final result is 6 more than the original number. What was the original number?

5 When a number is halved and then 5 is added, the result is 8 more than the original number. Find the original number.

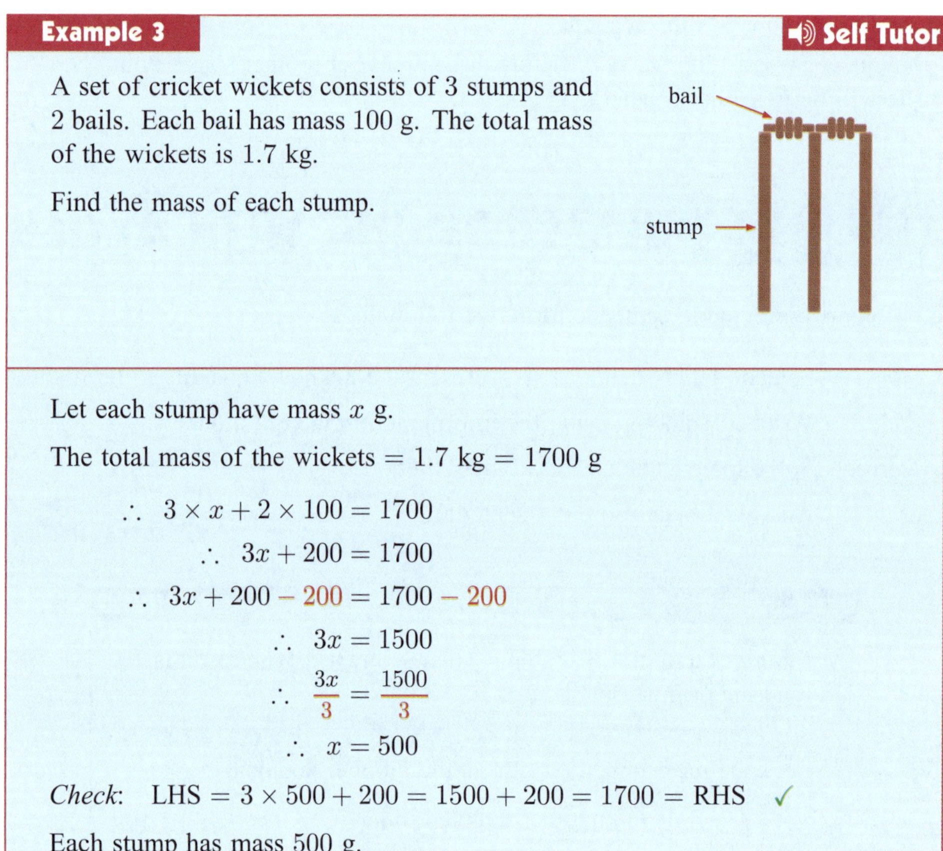

Example 3

A set of cricket wickets consists of 3 stumps and 2 bails. Each bail has mass 100 g. The total mass of the wickets is 1.7 kg.

Find the mass of each stump.

Let each stump have mass x g.

The total mass of the wickets $= 1.7$ kg $= 1700$ g

$$\therefore \quad 3 \times x + 2 \times 100 = 1700$$
$$\therefore \quad 3x + 200 = 1700$$
$$\therefore \quad 3x + 200 - 200 = 1700 - 200$$
$$\therefore \quad 3x = 1500$$
$$\therefore \quad \frac{3x}{3} = \frac{1500}{3}$$
$$\therefore \quad x = 500$$

Check: LHS $= 3 \times 500 + 200 = 1500 + 200 = 1700 =$ RHS ✓

Each stump has mass 500 g.

6 A market stall has 13 bags of potatoes, and 11 loose potatoes on display. In total there are 141 potatoes at the stall. How many potatoes are in each bag?

7 A packet of lollies is shared equally among 5 friends. Layla eats two of her lollies and has five left. How many lollies were in the packet?

8 Leanne cuts some roses in her front garden, and another 20 roses in her back garden. She uses one quarter of her roses to make a bouquet of a dozen flowers. How many roses did Leanne cut in her front garden?

Example 4 ◀)) Self Tutor

I have one rake and two identical shovels for sale. Each shovel costs $7 more than the rake. The total value of the items is $59. Find the cost of the rake.

Let the rake cost r, so each shovel costs $(r+7)$.

$$\therefore \ r + 2(r+7) = 59 \qquad \{\text{the total value is \$59}\}$$
$$\therefore \ r + 2r + 14 = 59$$
$$\therefore \ 3r + 14 = 59$$
$$\therefore \ 3r + 14 - 14 = 59 - 14$$
$$\therefore \ 3r = 45$$
$$\therefore \ \frac{3r}{3} = \frac{45}{3}$$
$$\therefore \ r = 15$$

The rake costs $15.

9 The Firebirds won last night's netball match, scoring 12 more goals than the Swifts. In total, 108 goals were scored in the match. How many goals did the Firebirds score?

10 Class A had 33 students and class B had 21 students. A number of students were moved from class A to class B so the classes now have the same number of students. How many students were moved?

11 Mandy bought some apples with mass 100 g each, and some oranges with mass 150 g each. In total she bought 10 pieces of fruit, with total mass 1.35 kg. How many apples did Mandy buy?

12 When the Norton family went out for dinner, they paid for their meal with a $100 note. The change they received was one seventh of the cost of the meal. Find the cost of the meal.

Example 5 ◀)) Self Tutor

A rectangle is 2 cm longer than it is wide. The perimeter is 52 cm. Find the width of the rectangle.

Let the width of the rectangle be x cm, so the length is $(x+2)$ cm.

Since the perimeter is 52 cm,

$$x + x + (x+2) + (x+2) = 52$$
$$\therefore \ 4x + 4 = 52$$
$$\therefore \ 4x + 4 - 4 = 52 - 4$$
$$\therefore \ 4x = 48$$
$$\therefore \ \frac{4x}{4} = \frac{48}{4}$$
$$\therefore \ x = 12$$

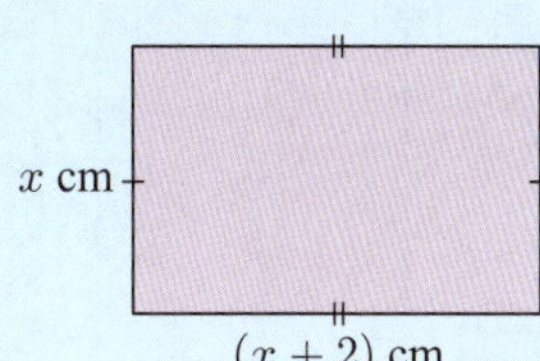

The width of the rectangle is 12 cm.

13   The length of a rectangle is double its width. The perimeter is 54 cm. Find the width of the rectangle.

14   A rectangle is 7 cm longer than it is wide. The perimeter is 74 cm. Find the width of the rectangle.

15   The perimeter of a square is 24 cm. Find its area.

16   The equal sides of an isosceles triangle are twice the length of the base. The perimeter of the triangle is 60 cm. Find the length of the base.

17   The parallelogram ABCD has perimeter 56 m.
The length AD is twice the height, and the length CD
is 1 m more than the height.
Find the area of the parallelogram.

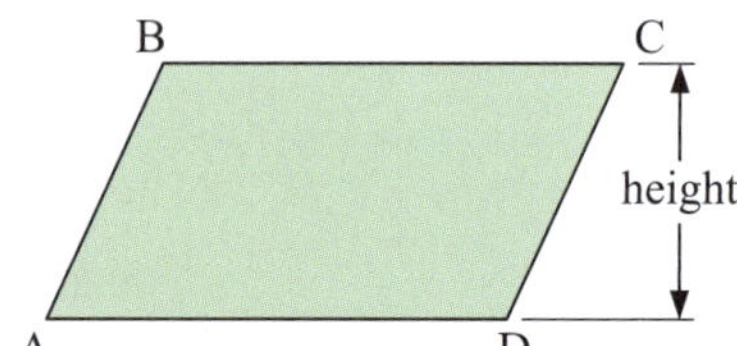

C   SOLUTION BY SEARCH

With some problems it is either difficult to write down an equation, or the equation may be difficult to solve. In some cases the equation may have more than one solution.

If the number of possibilities for the solution is small, it may be quickest to search through *all* of them to find the solutions which work. We call this a *proof by exhaustion*.

In other cases, a search through particular values may reveal a pattern which tells us what the correct solution must be.

Example 6   ◀)) Self Tutor

What rectangle with perimeter 24 cm has the greatest area?

The perimeter is 24 cm, so $2 \times (\text{length} + \text{width}) = 24$ cm

$\therefore \quad \text{length} + \text{width} = 12$ cm

We construct a table with some possible lengths and widths:

Length (cm)	Width (cm)	Area (cm^2)
1	11	$1 \times 11 = 11$
2	10	$2 \times 10 = 20$
3	9	$3 \times 9 = 27$
4	8	$4 \times 8 = 32$
5	7	$5 \times 7 = 35$
6	6	$6 \times 6 = 36$
7	5	$7 \times 5 = 35$
8	4	$8 \times 4 = 32$

We observe symmetry either side of the area 36 cm^2.
So, we conclude that the area is largest when the rectangle is a 6 cm by 6 cm square.

EXERCISE 21C

1 Consider the equation $3x + 4y = 30$ where x is a positive integer and y is positive.

   a Explain why x cannot be greater than 9.

   b Copy and complete the following table for all of the possible values of x and y.

x	1	2	3	4	5	6	7	8	9
y	$6\frac{3}{4}$								

For example, when $x = 1$, $\quad 3(1) + 4y = 30$

$$\therefore \ 4y = 27$$
$$\therefore \ y = \frac{27}{4} = 6\frac{3}{4}$$

   c Find the solutions to $3x + 4y = 30$ for which x and y are *both* positive integers.

2 Find positive integers x and y which satisfy $9x + 6y = 90$.

3 **a** Find the smallest positive number greater than 3 which, when divided by 4 and 5, leaves a remainder of 3. Try 4, 5, 6, ...

   b Describe a quicker way to find this number.

4 A rectangle has perimeter 32 cm and area 48 cm^2. Find the dimensions of the rectangle.

5 Find all positive integers less than 40 which *cannot* be written in the form $a^2 + b^3$, where a and b are natural numbers.

Hint: Be systematic in your approach.

$$\text{Calculate} \quad 0^2 + 1^3, \quad 1^2 + 0^3, \quad \text{and so on.}$$
$$0^2 + 2^3, \quad 1^2 + 1^3,$$
$$0^2 + 3^3, \quad 1^2 + 2^3,$$
$$0^2 + 4^3, \quad 1^2 + 3^3,$$

6 Find the smallest integer greater than 1 which is both a perfect square *and* a perfect cube.

7 Three sides of a rectangle have integer lengths which add to 32 cm.

   a Copy and complete this table:

Width (cm)	Length (cm)	Area (cm^2)
1	30	30
2	28	56
$\vdots$	$\vdots$	$\vdots$
10	12	120

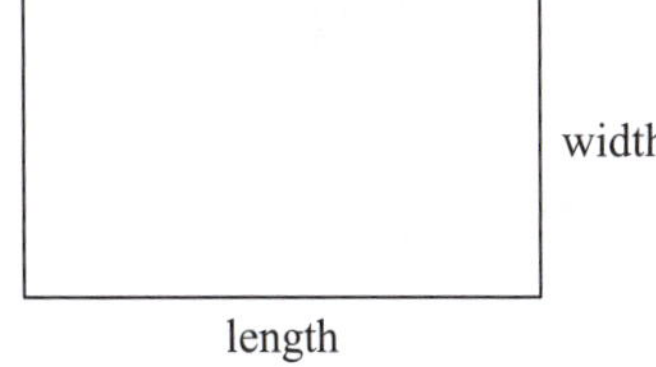

   b Hence find the dimensions of the rectangle which maximise its area.

8 A number is called **perfect** if the sum of its factors, excluding itself, is equal to itself. For example, 6 is a perfect number because its factors are 1, 2, 3, and 6, and $1 + 2 + 3 = 6$. Find the only other perfect number less than 40.

9 Leslie has bought ten presents for her two sons. The price of each present is shown below.

$20 $18 $10 $24 $5 $1 $8 $35 $15 $2

Leslie wants to divide the presents into two groups of five. The total value of the presents in each group must be the same. Explain how Leslie can do this.

10 Suppose a and b are positive integers such that $2a + b = 40$. Find the greatest possible value of ab^2.

D   SOLUTION BY WORKING BACKWARDS

In mathematics, we usually begin with an initial condition and work systematically towards an end point. However, if we know sufficient details about the final situation, we can **work backwards** to determine what the situation was initially.

Example 7   ◀) **Self Tutor**

My friend lost $5 from her wallet, but she then doubled what remained by selling a book to me. She finished with $50 in her wallet. How much did she have in her wallet initially?

My friend doubled her money to reach $50.

Double means multiply by 2, so to work backwards we divide by 2.

∴   she had $50 ÷ 2 = $25 in her wallet before selling the book to me.

Lose $5 means subtract $5, so to work backwards we add $5.

∴   before losing the $5, she had $25 + $5 = $30.

My friend had $30 in her wallet initially.

EXERCISE 21D

1 I think of a number. I multiply it by 7, add 1, divide by 8, subtract 2, then divide by 3. The result is 2. What was my original number?

2 A breeder sold half of his horses, then another two. He then sold half of his remaining horses, then another two. He then only had his one favourite horse left. How many horses did he have initially?

3 Xuen and Xia are sisters. In 2019, Xia grew 4 cm and Xuen grew 3 cm more than Xia. In 2020 Xuen grew 3 cm and Xia grew 2 cm. When measured at the end of 2020, both girls were 140 cm tall. What was the height of each girl at the end of 2018?

4 A teacher offers members of the class a peppermint for every problem they get right, but he takes back *two* for each incorrect answer.

Barry stored up his peppermints over a week. On Thursday he got 8 questions right and 2 wrong. On Friday he got 9 right and 1 wrong. He finished the week with 16 peppermints. How many did he have at the start of Thursday?

5

In the game "101", students sit in a circle. Going around in a clockwise direction, students call out the prime numbers in increasing order, starting with "2", "3", "5", and so on. The student who calls out "101" is the winner.

Fran will decide who should start the game. Which player should Fran choose so that Fran herself ends up winning?

Example 8   ◀)) **Self Tutor**

Sally cycles $\frac{1}{3}$ of the way to school, then is driven the remaining 6 km by a friend's family. How far does she live from her school?

She cycles $\frac{1}{3}$ of the way and is driven $\frac{2}{3}$ of the way.

$\frac{2}{3}$ of the way is 6 km.

$\therefore$ $\frac{1}{3}$ of the way is 3 km.

$\therefore$ Sally lives 3 km + 6 km = 9 km from her school.

---

6 Joe and Jerry went for a bike trek during the holidays.

On the first day they rode $\frac{1}{3}$ of the total distance.

On the second day they were tired and only rode $\frac{1}{4}$ of the remaining distance.

On the third day they rode half of the remaining distance.

The last day they rode 18 km.

How far did they ride altogether?

7 Anders gave $\frac{1}{3}$ of his lollies to Bree, then Bree gave $\frac{1}{4}$ of her lollies to Charlie. Each child now has 12 lollies. How many lollies did each child have originally?

8 Two bottles X and Y each contain some water.

From bottle X we pour into bottle Y twice as much water as bottle Y already contains.

We then pour from bottle Y into bottle X as much water as bottle X now contains.

Both bottles now contain 200 mL.

How much water was in each bottle to start with?

 E   MISCELLANEOUS PROBLEMS

In this Exercise we consider various problems which can be solved using the techniques in this Chapter. You should aim to choose the most appropriate technique for each problem.

EXERCISE 21E

1   I am thinking of two numbers. One of them is 5 more than the other, and their sum is 53. What are the numbers?

2   One number is three times as large as another, and a third number is 13 less than the smaller one. The sum of the three numbers is 42. Find the largest number.

3   Tim bought 4 packets of nails from a hardware store. He used 10 nails to fix his gate, and put the remaining 50 nails in his toolbox. How many nails were there in each packet?

4   Colleen has three daughters. The sum of her daughters' ages is 13, and the product of their ages is 48. Find the ages of the daughters.

5   Ashley brought some biscuits for morning tea at work. Her colleagues ate $\frac{7}{9}$ of them, then her boss came by and ate two biscuits. When Ashley went to the container, there were only four left. How many biscuits did she bring for morning tea?

6   The equal sides of an isosceles triangle are 6 cm longer than the third side. The perimeter of the triangle is 33.6 cm. Find the length of the third side.

7   A television costs \$200 more than a digital radio. Together, they cost \$378. Find the cost of the television.

8   When Samuel was 7, Claire was 34. Claire's age is now double Samuel's age. How old will Samuel be in 10 years' time?

9   In a multiple choice test there are 20 questions. A correct answer to a question earns 2 marks. An incorrect answer results in a deduction of 1 mark. Sarah answered every question and achieved a score of 19. How many questions did she answer correctly?

10   If x can take any whole number value from 1 to 10, find the smallest possible value of $x^2 - 8x + 25$.

11   A Russian Blue kitten costs \$150 more than a Manx kitten. 9 Russian Blue kittens and 8 Manx kittens cost \$14 100 in total. Find the cost of a Russian Blue kitten.

12   One half of an unknown number is equal to the number subtracted from forty eight. What is the number?

13   Each day, Sam the dog drinks half of the water remaining in his bowl. His owner adds another 200 mL of water to the bowl at the end of the day.
There was 500 mL of water in the bowl at the end of Thursday. How much water was in the bowl at the start of Wednesday?

14   Nine more than a number is the same as twenty four less a quarter of the number. What is the number?

15 Jim wrote down a sequence of numbers. After the first two numbers, each number in the sequence is equal to the sum of the previous two numbers.
The first number is 3 and the tenth number is 199. What is the fifth number in the sequence?

16 A pair of shoes costs $5 more than twice the cost of a tie. If Grant buys two pairs of shoes and three ties, the total cost is $255. Find the cost of:

    a a tie   **b** a pair of shoes.

17 Juan has two sets of metallic balls. The mass of each red ball is double the mass of each green ball. He knows that two red balls plus a five gram weight balance three green balls plus an eight gram weight. Find the mass of each red ball.

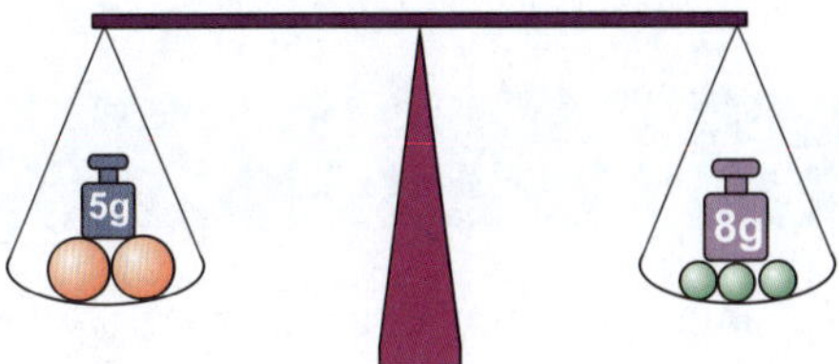

18 A 2-digit number is four times the sum of its digits.
Show that there are four possible numbers with this property.
Hint: The number "xy" with x "tens" and y "units" has value $10x + y$.

19 A 6-digit number begins with a 1. However, if the 1 is put at the other end of the number instead of at the start, the resulting 6-digit number is 3 times the original number. Find the number.

20 A motorcyclist and a cyclist leave simultaneously from A at 12 noon. They travel the same road to B. The motorcyclist arrives at B at 1:30 pm, and the cyclist arrives at 2 pm. At what time was the cyclist twice as far from B as the motorcyclist was?

INVESTIGATION   THE DIGITS PROBLEM

Using three different digits from 1 to 9, we can write six possible two-digit numbers.

For example, using the digits 2, 7, and 8 we can write 27, 28, 72, 78, 82, and 87.

In this Investigation we explore a property of the *sum* of the six two-digit numbers.

What to do:

1 Find the sum of 27, 28, 72, 78, 82, and 87, and write this sum in prime factored form.
Record your answers in a table like this:

Digits	Numbers	Sum of numbers	Sum in prime factored form
2, 7, 8	27, 28, 72, 78, 82, 87		

2   **a** Choose any three digits from 1 to 9, and write them in the next line of the table.

    b Write down the six possible two-digit numbers which can be formed from them.

    c Find the sum of these six numbers, and write the sum in prime factored form.

3 Repeat the process in **2** to complete the table.

4 Look at the sums in prime factored form. Find the HCF of all of the sums.

5 Prove that you will always obtain this HCF for all possible choices of 3 different digits.
Hint: The number "xy" with x "tens" and y "units" has value $10x + y$.

6 Is there a similar pattern when *four* different digits are chosen and we look at the HCF of the sum of all of the two-digit numbers that can be formed? Prove your answer.

F LATERAL THINKING

For some problems in mathematics, we need to allow ourselves to think differently.

The problems in this Section will require you to carefully analyse the information given to you, and think logically about what you need to do.

EXERCISE 21F

1 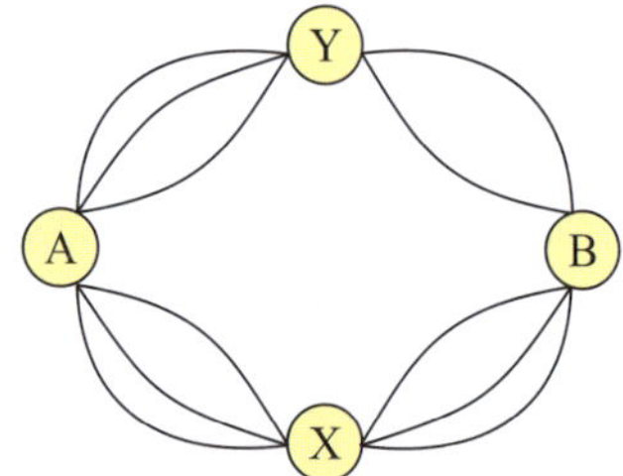

Two towns A and B are connected by the series of roadways passing through two other towns X and Y.

a How many different paths are there from A to B which pass through:

   i X   **ii** Y   **iii** X or Y?

b If X and Y were connected by a road, how many different paths would there now be from A to B?

2 How many positive integers less than 150 have an odd number of factors?

3 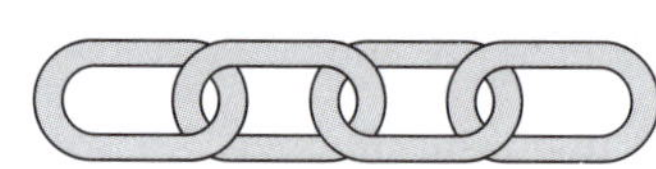

You have 6 chains, each containing 4 links as shown. You wish to make one chain of 24 links. It costs \$2 to cut each link and \$5 to weld a link. Describe the most economical way to make the chain.

4 Find the missing digits A, B, C, and D in the addition:

$$
\begin{array}{r}
A\ 8\ 7 \\
7\ 4\ B \\
+\ 2\ C\ 3 \\
\hline
2\ D\ 0\ 5
\end{array}
$$

5 This diagram shows 5 squares built from 16 matches. By moving 2 matches only, form 4 squares equal in area to one of the original squares. Each square must have a single match for each of its sides. You must use all of the matches.

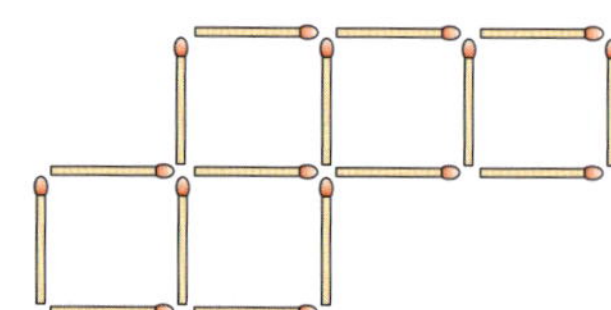

6 Is it possible to draw 4 straight lines through all of the 9 points without taking your pen off the paper? If so, sketch a possible solution.

7 Many years ago, MENSA inserted the notice alongside in a newspaper. This is really a mathematical problem. Can you see what the problem is? What is the solution to the problem?

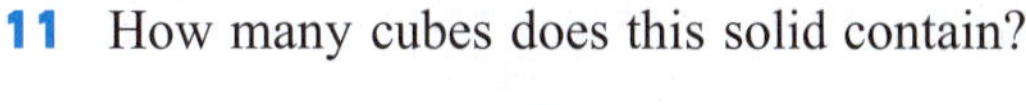

MENSA's contribution to Road Safety Week

C R O S S

R O A D S

D A N G E R

given that   **CAR = 956**

8 On a clear day, sound travels 335 m every second. Pam shouts across a valley and notices that her echo returns in 6 seconds. How far is it across the valley?

9 Two men dig a trench 30 m long in 3 days. How long would it take three men to dig a trench 15 m long working at the same rate?

10 How many triangles does this figure contain?

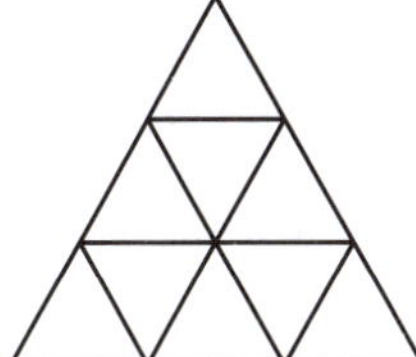

11 How many cubes does this solid contain?

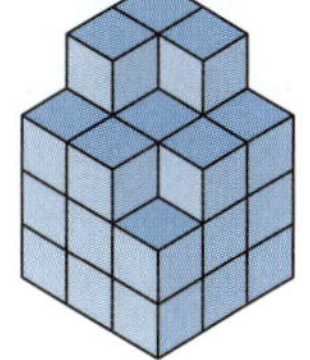

12 S, H, and E are different digits. What are they?

$$\begin{array}{r} H \ E \\ \times \ H \ E \\ \hline S \ H \ E \end{array}$$

13 King Arthur, Sir B, Sir C, and Sir D sit around a round table. Show that there are 6 different possible seating arrangements.

14 A cord fishing net is 100 m long and 4 m wide. The cord lengths are fixed 5 cm apart so that 5 cm by 5 cm holes are created. How many holes does the net have?

15 9 sheep are located in a square paddock, as shown. Draw 2 squares of fencing so that each sheep will be in a pen by itself.

16 Suppose you have a set of scales, and single masses of 9 kg, 3 kg, and 1 kg. Explain how you could measure the mass of any whole number from 1 kg to 13 kg.

17 In an enclosure at the zoo there are ostriches and giraffes. In total there are 17 heads and 50 legs. How many of each animal are present?

18 In how many different ways can 5 letters A, B, C, D, and E be mailed into 2 mail boxes?

19 Two wheels touch at their blue marks. The large wheel has diameter 20 cm and the small wheel has diameter 10 cm. The large wheel moves clockwise, and no slipping between the wheels occurs.

   a In which direction does the smaller wheel move?

   b After how many revolutions of the small wheel will the wheels once again touch at their blue marks?

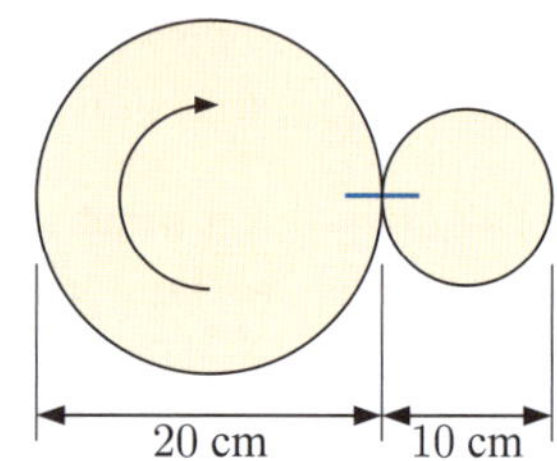

20 Using exactly four 4s and any of the operations $+$, $-$, $\times$, $\div$, find expressions for as many of the integers from 1 to 20 as you can.

$$\frac{4}{4} \div \frac{4}{4} = 1$$

$$\frac{4}{4} + \frac{4}{4} = 2$$

$$\frac{4 + 4 + 4}{4} = 3$$

21 Store A reduces its prices by 10%. By what percentage must the store increase its prices by so that its prices are back to what they were originally?

22 A log is sawn into 7 pieces in 7 minutes. How long would it have taken to saw the same log into 4 pieces?

23 The diagram below shows the start of a sequence of piles of tennis balls.

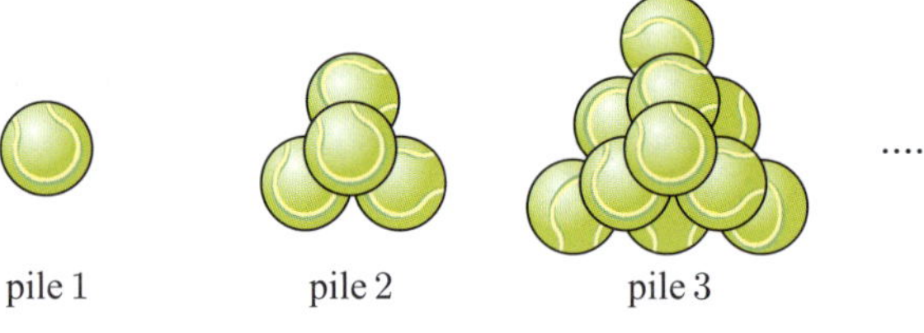

How many balls will be in pile: **a** 4 **b** 8?

24 Pete takes 3 days to dig a well working at a constant rate. It would take Matt 5 days to dig the same sized well, also working at a constant rate.

If they work together to dig the well, each working at their usual rates, how long will they take to dig the well?

25 The current in a river flows at 5 km per hour. When a boat travels at maximum speed against the current, it reaches its destination in 2 hours. On the return journey it takes just $1\frac{1}{2}$ hours.

   a What is the boat's maximum speed?

   b How far does the boat travel against the current?

26 Two numbers differ by 3, and the sum of their squares is 8. Find the product of the numbers.

27 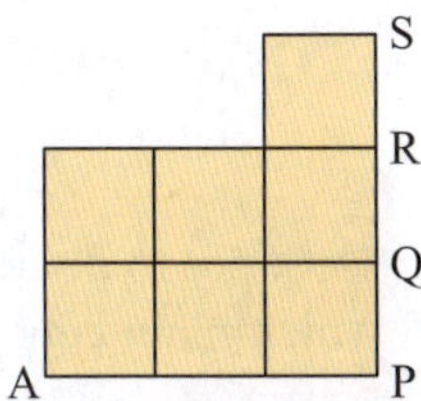 Where should point B be located along [PS] such that [AB] divides the figure into equal areas?

28 Nine people stand in a 3 by 3 grid with one person in each square.
Of the 3 people who are the *tallest* in each *row*, Celine is the *shortest*.
Of the 3 people who are the *shortest* in each *column*, Dianne is the *tallest*.
Who is taller, Celine or Dianne?

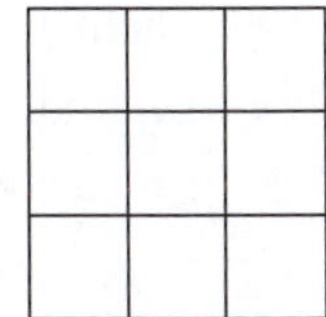

29 To pick all of the oranges in an orchard, it should take 14 people a total of 10 days. If 3 of the workers are sick and cannot work, and 2 others can only work at half the usual rate, how long will it take to pick all of the oranges?

30 Two consecutive perfect squares differ by 71. Find the smaller of the two numbers.

31 Show how this figure can be cut into four congruent pieces. 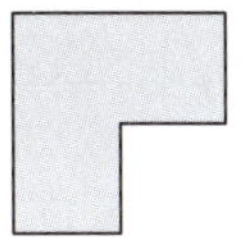

32 100 marbles are placed on a table. Adam and Belinda take turns removing marbles from the pile, starting with Adam. During their turn, a player can choose to remove any number from 1 to 10 marbles. The player who removes the last set of marbles wins the game.
Describe a strategy Adam can use to win the game.

PUZZLE   THE EMPEROR'S PROPOSITION

A prisoner has been sentenced to death. However, the emperor decides to give the prisoner a chance to live.

The prisoner is given 50 black marbles, 50 white marbles, and two empty bowls. The prisoner is told that he must divide the marbles between the bowls. He may place the marbles in the bowls in any combination he likes, but all of the marbles must be used, and neither bowl may be empty.

Once the prisoner has divided the marbles between the bowls, he will be blindfolded. The bowls will be placed on a table in front of the prisoner so he does not know which is which. The prisoner must choose a bowl at random, and then select one marble from that bowl. If the marble is black, the prisoner will be executed. If the marble is white, the prisoner will be allowed to live.

1 How should the prisoner divide the marbles between the bowls, to maximise the probability that he will live?

2 What is the probability that the prisoner will live in this case?

ACTIVITY 1   LOGIC PUZZLES

Use the clues given to solve these puzzles.

1 Two married couples, Louisa and James, and Damien and Brigitte, went out to dinner. Use the following information to determine where each person sat and which main course they ate:

- The married couples sat opposite one another.
- Brigitte, who sat in position A, is married to the man who ordered beef.
- Damien sat immediately to the left of the person who ordered fish.
- James, who enjoyed his chicken, did not sit in either position A or C.
- Louisa did not sit in position D.
- The person in position B did not order pork.

Name: ________________
Food: ________________

Name: ________________   A   Name: ________________
Food: ________________   D   B   Food: ________________
                                    C

Name: ________________
Food: ________________

2 Three people were discussing their cars, their pets, and the colour of their houses. Use the following information to identify the car, pet, and house colour for each person:

- Lisa owns a cat.
- Anne does not drive a Nissan, and does not live in the green house.
- Neither Sue, nor the person who drives a Honda, owns a dog.
- The person who owns the parrot lives in the blue house.

	Toyota	Nissan	Honda	Cat	Parrot	Dog	Green	Blue	White
Lisa									
Anne									
Sue									
Green									
Blue									
White									
Cat									
Parrot									
Dog									

PRINTABLE PUZZLES

3 A cross-country running race has just concluded. When the top 5 athletes were interviewed, each athlete told one truth and one lie. Their statements were:

   Allen: Dillon came second and I finished in third place.

   Billy: I won and Chris came second.

   Chris: I was third and Billy came last.

  Dillon: I finished second and Evan was fourth.

   Evan: I came fourth and Allen won.

In what order did the athletes finish the race?

ACTIVITY 2   VANISHING SHAPES

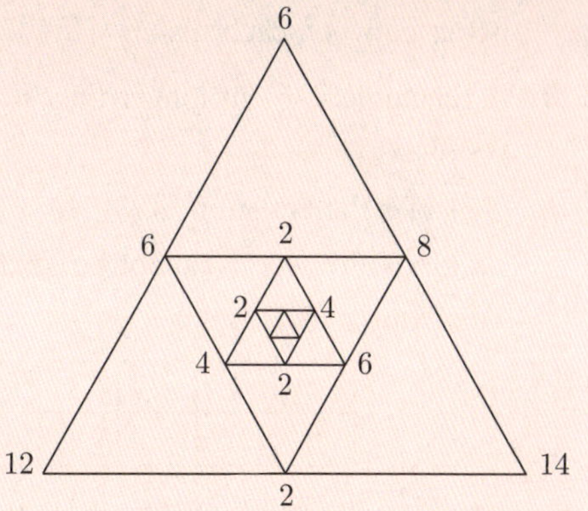

Draw an equilateral triangle and write a positive integer at each vertex. At the midpoint of each side, write the difference between the integers at the ends of each side. Join these midpoints, and repeat the procedure on the newly formed triangle.

For example, suppose the numbers 6, 12, and 14 are chosen. After three steps you should generate the figure alongside.

The procedure continues until either the numbers created are identical in consecutive triangles, or a pattern emerges.

What to do:

1 Copy the given example, making sure your original triangle is large enough. Continue the procedure until there is a clear pattern.

2 Repeat the procedure using triangles with the following combinations of integers:

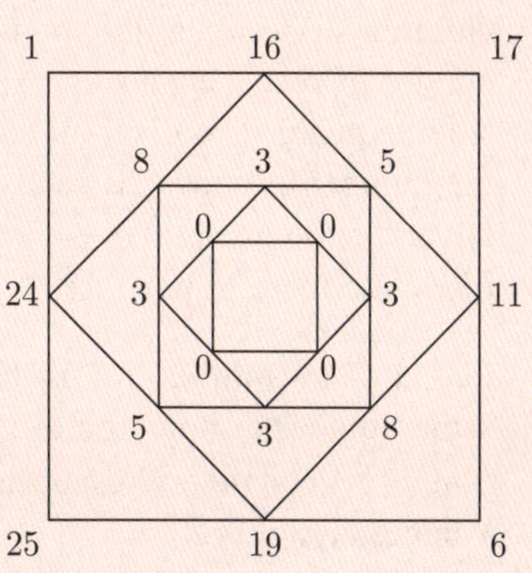

  a three even integers
       b two even integers

  c one even integer
        d no even integers.

Can you draw any general conclusions?

3 Try the same procedure starting with a square instead of an equilateral triangle.

For example, if the numbers 1, 17, 6, and 25 are chosen, the square alongside is produced.

Can you suggest a general result for the case of a square?

MULTIPLE CHOICE QUIZ

REVIEW SET 21A

1 **a** When a certain number is halved, the answer is 19. Find the number.

 b I think of a number, multiply it by 7, then subtract 2. The final result is 26. What was the original number?

2 Anita umpired 4 hockey matches between 6:00 pm and 9:30 pm. There was a 10 minute break between each match. How long did each match last?

3 A rectangle is 5 cm longer than it is wide. The perimeter is 50 cm. Find the length of the rectangle.

4 Consider the equation $3x + 4y = 40$ where x is a positive integer and y is positive.

 a Explain why x cannot be greater than 13.

 b Copy and complete:

x	1	2	3	4	5	6	7	8	9	10	11	12	13
y													

 c Find the solutions to $3x + 4y = 40$ for which x and y are *both* positive integers.

5 George has three times as many marbles as Steph. If George gives 5 of his marbles to Steph, George will have twice as many marbles as Steph. How many marbles does Steph have?

6 An isosceles triangle has integer side lengths, and a base which is 10 cm longer than its height. Its area is 48 cm^2 and its perimeter is 36 cm.

 a Find the length of the base of the triangle.

 b Find the lengths of the equal sides of the triangle.

7 There are 5 bowls of cupcakes on a table. Four children sit between the bowls as shown. Each child can reach cupcakes in the bowls either side of them. The diagram alongside indicates the number of cupcakes each child can reach. Given that there are 6 cupcakes in bowl 5, how many are there in bowl 1?

8 Leonard stays awake for 15 hours each day, then goes to sleep. Each night he sleeps the same number of hours before and after midnight.

Suppose Leonard wakes up at 5 am Friday morning. At what time did he go to sleep Wednesday night?

9 Reiko and Josie are buying food for their party. They find that a doughnut costs 80 cents more than a lamington.

   a If a doughnut costs $\$x$, write an expression for the cost of a lamington.

   b The girls buy 8 doughnuts and 12 lamingtons, spending a total of $25.40. Find the cost of each item.

   c If the girls had only bought lamingtons, how many could they have bought with $25.40?

10 A triangle has perimeter 21 cm, and the length of each side is a whole number of centimetres.

   a Explain why no side of the triangle can be longer than 10 cm.

   b Find all the possible dimensions of the triangle.

   c Find the dimensions of:

   i the isosceles triangle with the longest possible base

   ii the scalene triangle with the shortest possible side.

REVIEW SET 21B

1 When half of a number is subtracted from 20, the result is 14. Find the original number.

2 A necklace costs twice as much as a bracelet. I bought 4 necklaces and 3 bracelets for a total of $121. Find the cost of each item.

3 The sum of three consecutive numbers is 54. Find the middle number.

4 Karen takes 4 minutes longer to make a pizza than she does to make a pasta dish. She can make 4 pasta dishes in the time it takes to make 3 pizzas. How long does Karen take to make a pasta dish?

5 Suppose a and b are positive integers such that $3a + 4b = 22$. Find the greatest possible value of ab.

6 A number is **semiprime** if it is the product of two prime numbers. Find all of the 2-digit semiprimes.

7 Jo spent half of the money in her wallet on a new cardigan. She then spent $4.50 on a cup of coffee. After drinking her coffee, she spent three quarters of the money remaining in her wallet on a book. If Jo is left with $8 in her wallet, how much did she have in her wallet initially?

8 Brothers Jake and Phil have a mother named Sue. A year ago, Sue was three times as old as Jake. Four years ago, Sue was four times as old as Phil. If Phil is 13 years old, how old is Jake?

9 The total area of the kite is 6 times the area A.
Find the ratio $k : x$.

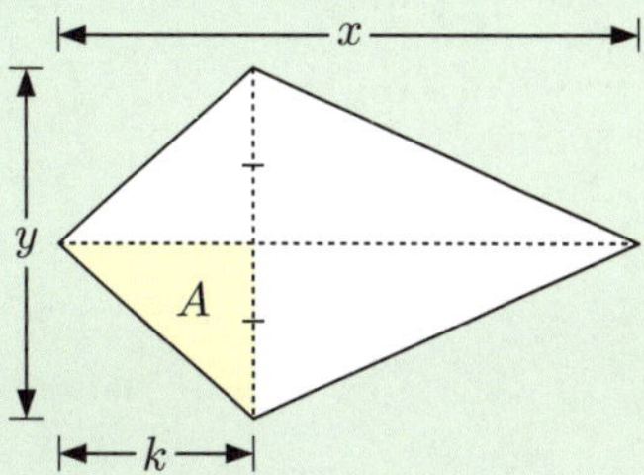

10 Tommy has some 20 cent and 50 cent coins in his pocket. He has 3 more 20 cent coins than 50 cent coins. Suppose Tommy has x 20 cent coins.

  a Write an expression for:

    i the total number of coins in Tommy's pocket

    ii the total value of coins in Tommy's pocket.

  b With the coins in his pocket, Tommy can afford a pie costing $4.30, but not a quiche costing $5.10.

    i How many of each coin does Tommy have?

    ii Find the exact value of Tommy's coins.

11 Ange, Bill, Chad, Dean, and Eve are kicking a football amongst themselves.

  Ange only kicks the ball to Dean.

  Bill only kicks the ball to Eve.

  Chad only kicks the ball to Ange.

  Dean kicks the ball to either Bill or Eve.

  Eve kicks the ball to either Ange or Chad.

  a Suppose Chad currently has the ball. Who will have the ball after two kicks?

  b Suppose Bill currently has the ball. Who had the ball two kicks ago?

  c Suppose Eve currently has the ball. Name the people who could have had the ball three kicks ago.

ANSWERS

EXERCISE 1A

1 **a** 9   **b** 9   **c** 19   **d** 19   **e** -9 **f** -9
   g -19 **h** -9

2 **a** 40   **b** -14 **c** -14 **d** 40   **e** -40 **f** 14
   g -40 **h** 14

3 **a** -9 **b** -7 **c** 7   **d** -9 **e** 15   **f** -1

4 **a** 9   **b** 21   **c** 21   **d** 9

5 **a** 36   **b** -36 **c** -36 **d** 36   **e** 33   **f** -33
   g -33 **h** 33   **i** 56   **j** -56 **k** 56   **l** -56

6 **a** $\square = 6$ **b** $\square = -8$ **c** $\square = 7$ **d** $\square = 3$
   e $\square = 11$ **f** $\square = -4$

7 **a** 4   **b** -4 **c** -4 **d** 4   **e** 5   **f** -5
   g -5 **h** 5   **i** 9   **j** -9 **k** -9 **l** 9

8 **a** $\square = -8$ **b** $\square = -20$ **c** $\square = -3$ **d** $\square = 55$
   e $\square = -9$ **f** $\square = -48$

9 **a** $\square = -10$ **b** $\square = -1$ **c** $\square = -16$ **d** $\square = -5$
   e $\square = -5$ **f** $\square = 10$ **g** $\square = -40$ **h** $\square = -2$
   i $\square = 18$ **j** $\square = 72$ **k** $\square = 10$ **l** $\square = -9$

EXERCISE 1B

1 **a** $2^1 = 2,\ 2^2 = 4,\ 2^3 = 8,\ 2^4 = 16,\ 2^5 = 32,\ 2^6 = 64$
   b $3^1 = 3,\ 3^2 = 9,\ 3^3 = 27,\ 3^4 = 81$
   c $5^1 = 5,\ 5^2 = 25,\ 5^3 = 125,\ 5^4 = 625$
   d $7^1 = 7,\ 7^2 = 49,\ 7^3 = 343$

2 **a** 18   **b** 60   **c** 108   **d** 225
   e 2250   **f** 1176   **g** 825   **h** 52 000

3 **a** $2^2 \times 3$ **b** $2 \times 3^2 \times 5$ **c** $3^2 \times 5^2$
   d $2^3 \times 5^2$ **e** $3^2 \times 5 \times 7^2$ **f** $2^4 \times 3^3 \times 5$

4 **a** 1   **b** -1 **c** -1 **d** 1   **e** -16 **f** 16
   g -16 **h** -25 **i** 125   **j** -125 **k** 81   **l** -49

5 **a** -12 **b** 270   **c** 2250

6 **a** 2^2 **b** 2^3 **c** 2^5 **d** 2^7

7 **a** 5^2 **b** 5^3 **c** 5^4

8 $3^1 = 3,\ 3^2 = 9,\ 3^3 = 27,\ 3^4 = 81,\ 3^5 = 243,\ 3^6 = 729,$
$3^7 = 2187,\ 3^8 = 6561,\ ....$
The last digit follows the pattern: 3, 9, 7, 1, 3, 9, 7, 1, ...
$\therefore$ the last digit of 3^{20} is 1.

9 **a** 8   **b** 1   **c** 7

EXERCISE 1C

1 **a** yes   **b** no   **c** no   **d** yes   **e** no   **f** yes

2 **a** 1, 2, 3, 6   **b** 1, 3, 5, 15   **c** 1, 2, 3, 4, 6, 8, 12, 24
   d 1, 3, 7, 9, 21, 63   **e** 1, 23   **f** 1, 5, 25

3 **a** 1 and 45, 3 and 15, 5 and 9
   b 1 and 48, 2 and 24, 3 and 16, 4 and 12, 6 and 8
   c 1 and 72, 2 and 36, 3 and 24, 4 and 18, 6 and 12,
     8 and 9
   d 1 and 100, 2 and 50, 4 and 25, 5 and 20, 10 and 10

4 **a** $2 \times 10,\ 4 \times 5,\ 2 \times 2 \times 5$
   b $2 \times 18,\ 3 \times 12,\ 4 \times 9,\ 6 \times 6,\ 2 \times 2 \times 9,\ 2 \times 3 \times 6,$
     $3 \times 3 \times 4,\ 2 \times 2 \times 3 \times 3$
   c $2 \times 32,\ 4 \times 16,\ 8 \times 8,\ 2 \times 2 \times 16,\ 2 \times 4 \times 8,\ 4 \times 4 \times 4,$
     $2 \times 2 \times 2 \times 8,\ 2 \times 2 \times 4 \times 4,\ 2 \times 2 \times 2 \times 2 \times 4,$
     $2 \times 2 \times 2 \times 2 \times 2 \times 2$

EXERCISE 1D

1 2, 3, 5, 7, 11, 13, 17, 19, 23, 29, 31, 37, 41, 43, 47

2 71, 73   **3** 28

4 **a** 2^4 **b** 3^4 **c** 2^6 **d** 3^5 **e** 13^2 **f** 2^8

5 **a** $24 = 2^3 \times 3$ **b** $40 = 2^3 \times 5$
   c $36 = 2^2 \times 3^2$ **d** $54 = 2 \times 3^3$
   e $66 = 2 \times 3 \times 11$ **f** $84 = 2^2 \times 3 \times 7$
   g $100 = 2^2 \times 5^2$ **h** $132 = 2^2 \times 3 \times 11$
   i $320 = 2^6 \times 5$ **j** $324 = 2^2 \times 3^4$
   k $588 = 2^2 \times 3 \times 7^2$ **l** $945 = 3^3 \times 5 \times 7$

6 **a** $1960 = 2^3 \times 5 \times 7^2$
   b $2^3 \times 5 \times 7^2 = (2^2 \times 7) \times (2 \times 5 \times 7)$

7 **a** 8   **b** 48   **c** 160

8 $a = 2,\ b = 5$ or $a = 5,\ b = 2$

EXERCISE 1E

1 **a** 4   **b** 3   **c** 14   **d** 8   **e** 12   **f** 5
   g 11   **h** 6   **i** 16

2 15 m $\times$ 15 m   **3** **a** 4   **b** 6   **c** 4

4 **a** 4   **b** 4   **c** 12   **d** 13   **e** 14   **f** 28

5 28 tables   **6** **a** ab **b** ab^2c **c** a^2b^2c

EXERCISE 1F

1 **a** 7, 14, 21, 28, 35   **b** 11, 22, 33, 44, 55
   c 18, 36, 54, 72, 90   **d** 21, 42, 63, 84, 105
   e 30, 60, 90, 120, 150

2 **a** 405   **b** 798   **3** 60, 75, 90

4 **a** 40   **b** 12   **c** 40   **d** 90   **e** 60   **f** 140
   g 108   **h** 630

5 **a** 12   **b** 70   **c** 24   **d** 360   **e** 120   **f** 180
   g 126   **h** 300

6 90 m   **7** 180 minutes   or   3 hours

8 **a** 24 days   **b** 120 days   **9** **a** a^2bc **b** $a^2b^2c^3$

10 The HCF is the largest factor common to both numbers.
The LCM is the product of the prime factors that appear in either number.
   $\therefore$ the LCM is equal to the HCF multiplied by the prime factors exclusive to each number.
   $\therefore$ LCM $\times$ HCF $=$ (HCF $\times$ factors exclusive to 1st number)
                     $\times$ (HCF $\times$ factors exclusive to 2nd number)
                $=$ product of numbers

EXERCISE 1G

1 **a** 5   **b** 9   **c** 7   **d** -11 **e** 1   **f** 9
   g 20   **h** -41 **i** 13   **j** 5   **k** 39   **l** 7
   m 15   **n** 16   **o** -23

2 **a** 1   **b** -7 **c** 1   **d** 2   **e** 4   **f** -15
   g 16   **h** 8   **i** 4

3 **a** 3   **b** 6   **c** 81   **d** 88   **e** 18   **f** 14
   g 27   **h** 7   **i** 21   **j** 22   **k** 44   **l** -54

4 **a** 11   **b** 4   **c** 17   **d** 0   **e** 9   **f** 69

5 **a** 438   **b** 874   **c** 33   **d** 8   **e** -83 **f** 28

6 **a** 60   **b** 39   **c** 23   **d** 4   **e** 8   **f** 13

7 **a** $7 + 3 - 4 = 6$ **b** $4 \times 6 - 3 = 21$
   c $12 \div 4 \times 3 = 9$ **d** $3 + 5 - 6 - 4 = -2$
   e $3 + 8 \div 2 - 5 = 2$ or $3 - 8 \div 2 + 5 = 2$

f $9 \div 3 + 2 \times 4 = 11$ or $9 + 3 \times 2 - 4 = 11$

8 a $(9 - 7) \times 4 = 8$ **b** $80 \div (8 \times 2) = 5$
c $80 \div 8 \times 2 = 20$ **d** $4 \times 8 - (7 - 1) = 26$
e $4 \times (8 - 7) - 1 = 3$ **f** $4 \times (8 - 7 - 1) = 0$
g $(5 + 2) \times 6 - 3 = 39$ **h** $5 + 2 \times (6 - 3) = 11$
i $(5 + 2) \times (6 - 3) = 21$

EXERCISE 1H

1 a $5 \times 10 + 3 \times 20 + 50$ **b** $160
2 a $4000 - (2 \times 900 + 5 \times 250)$ **b** 950 mL
3 a $(5 + 3) \times 12 \div 2$ **b** $48 each
4 a $45 - 20 \times 2$ **b** 5 km
5 a $(225 + 200 + 175 + 4 \times 60) \div 8$ **b** 105 g each
6 a $8 + 8^2$ **b** 72 kittens
7 a $196 \times 4 + (140 + 84) \times 5$ or $196 \times 4 + 140 \times 5 + 84 \times 5$
b $1904
8 a $(4^2 + 6^2) \div (7 + 8 - 2)$ **b** 4 blocks each

REVIEW SET 1A

1 a -3 **b** -72 **c** 36 **d** -9
2 a 200 **b** 350 **c** -125
3 a $1, 3, 7, 21$ **b** $1, 2, 4, 8, 16, 32$ **c** $1, 37$
4 $41, 43, 47$ **5 a** $450 = 2 \times 3^2 \times 5^2$ **b** $212 = 2^2 \times 53$
6 a 30 **b** 44 **c** 40 **7 a** 2 **b** 8 **c** 9
8 a 16 **b** -13 **c** 30 **d** 6 **e** 11
9 104 minutes   **10** 18 nectarines
11 a $3 \times 5 - 4 = 11$ **b** $8 - 6 \div 3 = 6$
12 a $200 \times (20 + 4) + 150 \times (8 + 4)$ **b** $6600

REVIEW SET 1B

1 a 24 **b** 4 **c** -7
2 a $\square = -9$ **b** $\square = 12$ **c** $\square = -11$
3 a 3^2 **b** 3^3 **c** 3^4 **4** $8, 16, 24, 32, 40$
5 $54 = 2 \times 3^3$ **6** $91, 93, 95$ **7 a** 6 **b** 90
8 729 and 784 **9 a** -2 **b** 8 **c** 4
10 a $12 \div (6 - 2) = 3$ **b** $(6 + 4) \div (2 + 3) = 2$
c $18 \div (1 + 2 \times 4) = 2$
11 60 lollies
12 a $(16\,000 - 8 \times 1200) \div (12 - 8)$ **b** 1600 votes each
c 200 minutes or 3 hours 20 minutes

EXERCISE 2A

1 a {Sunday, Monday, Tuesday, Wednesday, Thursday, Friday, Saturday}
b {F, O, T, B, A, L}   **c** {2, 3, 5, 7, 11, 13, 17, 19}
d {a, e, i, o, u}   **e** {1, 3, 7, 21}   **f** {white, black}
g {43, 44, 45, 46, 47}
h {Asia, Africa, North America, South America, Antarctica, Europe, Australia}
2 a **i** $n(S) = 6$ **ii** $n(T) = 4$
b **i** true   **ii** true   **iii** false   **iv** true   **c** no
3 a **i** $A = \{1, 4, 9\}$
  ii $B = \{4, 6, 8, 9, 10, 12, 14, 15, 16, 18\}$
  iii $C = \{1, 2, 3, 4, 6, 9, 12, 18, 36\}$
b **i** $n(A) = 3$ **ii** $n(B) = 10$ **iii** $n(C) = 9$
c **i** false   **ii** true
4 a $S = \{1, 2, 3, 6\},\ n(S) = 4$

b $S = \{6, 12, 18, 24, 30, 36\},\ n(S) = 6$
c $S = \{1, 17\},\ n(S) = 2$
d $S = \{1, 4, 9, 16, 25, 36, 49\},\ n(S) = 7$
e $S = \{2, 3, 5, 7, 11, 13, 17, 19, 23, 29\},\ n(S) = 10$
f $S = \{12, 14, 15, 16, 18, 20, 21, 22, 24, 25, 26, 27, 28\}$,
$n(S) = 13$
5 a $n(A) = 12$ **b** $n(A) = 21$
6 a $\{\ \}, \{x\}$ **b** $\{\ \}, \{8\}, \{9\}, \{8, 9\}$
c $\{\ \}, \{p\}, \{q\}, \{r\}, \{p, q\}, \{p, r\}, \{q, r\}, \{p, q, r\}$
7 a $A = \{$red, blue, green$\}$,
$B = \{$orange, red, yellow, blue, pink, green$\}$
b yes   **c** **i** $n(A) = 3$ **ii** $n(B) = 6$ **d** yes
8 $x = 3, 6, 7,$ or 11

EXERCISE 2B

1 a $A' = \{1, 3, 5, 7, 8, 9\}$ **b** $B' = \{2, 4, 6, 8\}$
c $C' = \{1, 2, 5, 6, 9\}$ **d** $D' = \{4, 5, 6\}$ **e** $E' = \varnothing$
2 a $P = \{1, 2, 3, 4, 6, 12\}$ **b** $Q = \{11, 13, 17, 19\}$
c $P' = \{5, 7, 8, 9, 10, 11, 13, 14, 15, 16, 17, 18, 19\}$
d $Q' = \{1, 2, 3, 4, 5, 6, 7, 8, 9, 10, 12, 14, 15, 16, 18\}$
3 a $X = \{$I, N, H, A, B, T$\}$ **b** $Y = \{$M, I, L, O, N$\}$
c $X' = \{$C, D, E, F, G, J, K, L, M, O, P, Q, R, S, U, V, W, X, Y, Z$\}$
d $Y' = \{$A, B, C, D, E, F, G, H, J, K, P, Q, R, S, T, U, V, W, X, Y, Z$\}$
4 a **i** $U = \{$Rugby, Athletics, Volleyball, Baseball, Cricket, Archery, Netball, Tennis$\}$
  ii $K = \{$Baseball, Cricket, Tennis$\}$
  iii $K' = \{$Rugby, Athletics, Volleyball, Archery, Netball$\}$
b K' represents sports played at school which do not involve hitting a ball with a bat or racquet.
5 a **i** $n(U) = 7$ **ii** $n(A) = 4$ **iii** $n(A') = 3$
  iv $n(B) = 2$ **v** $n(B') = 5$
b **i** $n(U) = 12$ **ii** $n(A) = 6$ **iii** $n(A') = 6$
  iv $n(B) = 8$ **v** $n(B') = 4$
c $n(S) + n(S') = n(U)$
6 Yes, any positive whole number is either even or odd.
7 No, 1 is neither prime nor composite.

EXERCISE 2C

1 a $A \cap B = \{3, 5\}$ **b** $A \cup B = \{1, 3, 5, 6, 7, 9\}$
2 a $P \cap Q = \{$Dragons, Tigers$\}$
$P \cup Q = \{$Dragons, Tigers, Roosters, Raiders, Storm, Knights$\}$
b $P \cap Q = \{1\},\ \ P \cup Q = \{1, 3, 4, 6, 9, 10, 15, 16\}$
c $P \cap Q = \{g, m\},\ \ P \cup Q = \{d, e, g, h, k, l, m, p\}$
3 B and C
4 a $X = \{2, 3, 5, 7, 11, 13, 17, 19\}$,
$Y = \{1, 2, 4, 5, 10, 20\}$
b **i** $X \cap Y = \{2, 5\}$ **ii** $n(X \cap Y) = 2$
  iii $X \cup Y = \{1, 2, 3, 4, 5, 7, 10, 11, 13, 17, 19, 20\}$
  iv $n(X \cup Y) = 12$
5 a $\varnothing$ **b** A
6 A contains all the elements in $A \cap B$, and may contain elements which are not in B. Also, $A \cup B$ contains all the elements in A, and may contain elements in B which are not in A.
$\therefore\ n(A \cap B) \leqslant n(A) \leqslant n(A \cup B)$

7 **a** $U = \{$January, February, March, April, May, June, July, August, September, October, November, December$\}$

b $G = \{$September, October, November, December$\}$
$R = \{$October, November, December, January, February, March$\}$

c $G \cap R = \{$October, November, December$\}$
This set represents the months when both gardenias and roses are in flower.

d $G \cup R = \{$September, October, November, December, January, February, March$\}$
This set represents the months when either gardenias or roses (or both) are in flower.

e $(G \cup R)' = \{$April, May, June, July, August$\}$
This set represents the months when neither flower is in bloom.

8 **a** $S = \{$flour, butter, milk, jam, sultanas$\}$
$C = \{$margarine, flour, caster sugar, eggs, milk, jam$\}$

b $S \cap C = \{$flour, milk, jam$\}$
This set represents ingredients common to both scones and cake.

c $S \cup C = \{$flour, butter, milk, jam, sultanas, margarine, caster sugar, eggs$\}$

d 8 different ingredients   (which is $n(S \cup C)$)

9 **a** $A = \{4, 8\}$, $B = \{1, 2, 4, 8, 16\}$ **b** yes

c **i** $A \cap B = \{4, 8\}$ **ii** $A \cup B = \{1, 2, 4, 8, 16\}$

d "If $A \subseteq B$ then $A \cap B = A$ and $A \cup B = B$."

EXERCISE 2D

1 **a** 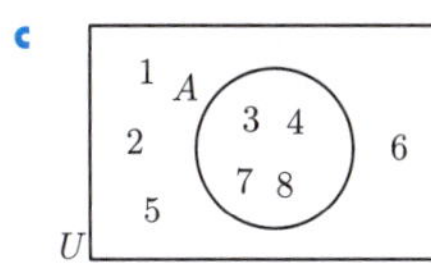   **b**

c

2 **a** $A = \{1, 2, 5, 6, 8\}$ **b** $B = \{2, 4, 6\}$
c $A \cap B = \{2, 6\}$ **d** $A \cup B = \{1, 2, 4, 5, 6, 8\}$
e $A' = \{3, 4, 7, 9\}$ **f** $B' = \{1, 3, 5, 7, 8, 9\}$
g $U = \{1, 2, 3, 4, 5, 6, 7, 8, 9\}$

3 **a** 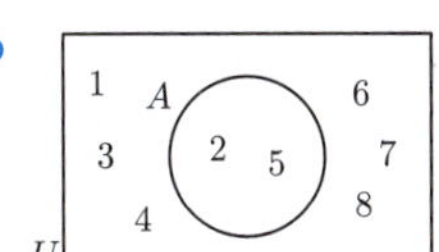   **b**

c 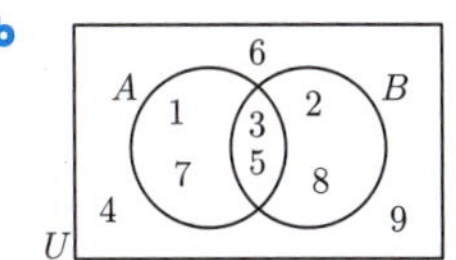   **d**

e 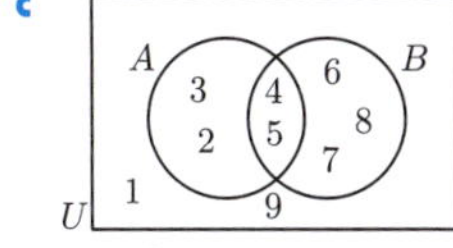   **f** 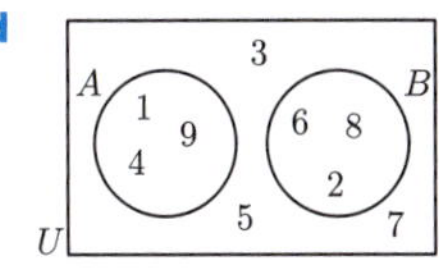

4 **a** 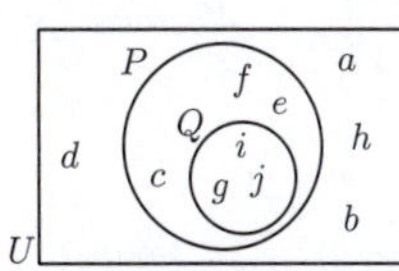   **b**

5 **a** $G = \{$Monday, Thursday, Saturday$\}$
$H = \{$Tuesday, Thursday, Sunday$\}$

b

c $G' = \{$Sunday, Tuesday, Wednesday, Friday$\}$
This set represents the days on which George does not play cricket.

d $G \cap H = \{$Thursday$\}$
This set represents the days on which both George and Hugh play cricket.

6 **a** $A = \{2, 3, 5, 7\}$ **b** 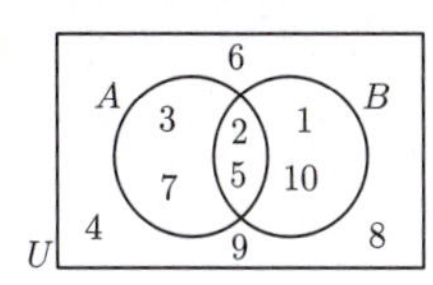
$B = \{1, 2, 5, 10\}$

c **i** $A \cup B = \{1, 2, 3, 5, 7, 10\}$
ii $(A \cup B)' = \{4, 6, 8, 9\}$

7 **a** 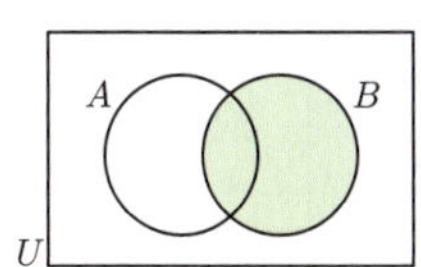   **b**

c 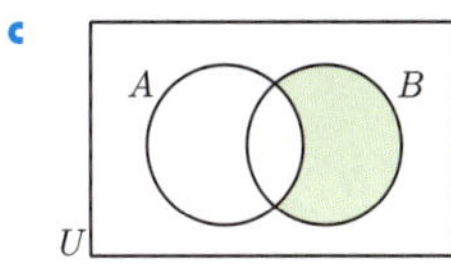   **d** 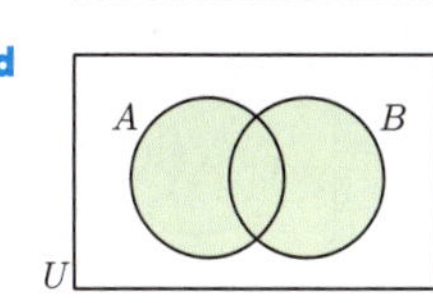

8 **a** in either X or Y (or both)   **b** in Y but not in X
c in X

9 **a** $X = \{2, 4, 6, 8, 10, 12\}$, $Y = \{2, 3, 5, 7, 11\}$
b **i** $X \cap Y = \{2\}$
ii $X \cup Y = \{2, 3, 4, 5, 6, 7, 8, 10, 11, 12\}$

c 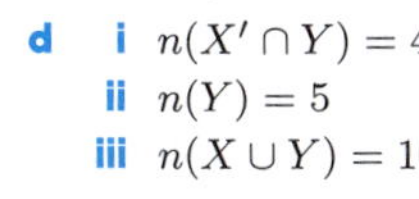   **d** **i** $n(X' \cap Y) = 4$
ii $n(Y) = 5$
iii $n(X \cup Y) = 10$

EXERCISE 2E

1 **a** 9   **b** 5   **c** 3   **d** 11   **e** 6   **f** 18

2 **a** 9   **b** 7   **c** 9   **d** 18   **e** 14

3 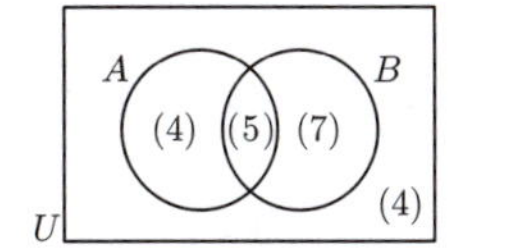   **4** 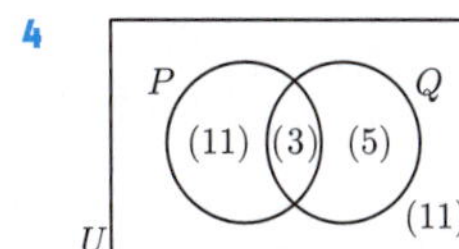

5 a 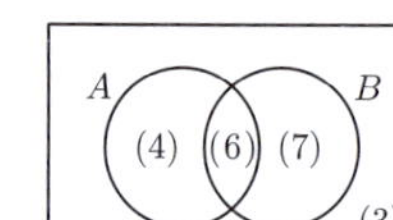   **b i** 7   **ii** 17

6 There are 3 possible Venn diagrams.

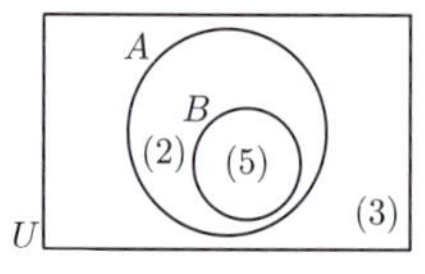   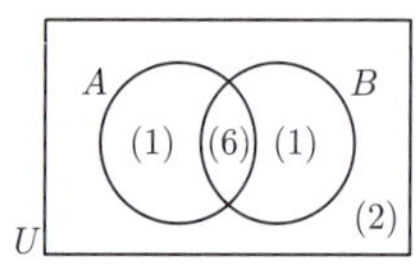

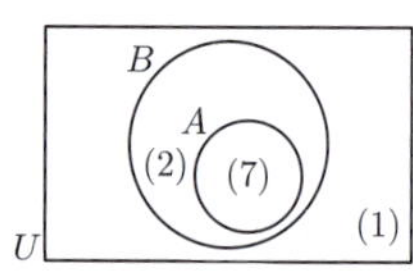

7 a 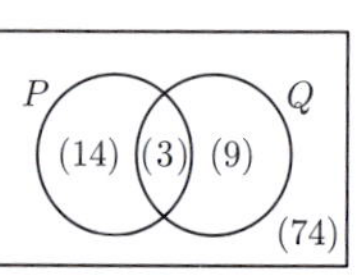   **b** $x = 60$

c i $P \cap Q = \{2, 3, 5\}$
ii $P' \cap Q = \{1, 4, 6, 10, 12, 15, 20, 30, 60\}$

EXERCISE 2F

1 a   **b i** 25 people
ii 7 people
iii 5 people
iv 1 person

2 a   **b i** 8 trees
ii 11 trees
iii 25 trees

3 a   **b i** 22 people
ii 41 people
iii 9 people

4 a   **b i** 3 types
ii 7 types
iii 5 types

5 a   **b** 18 nations
c 4 nations
d 6 nations

6 a   **b i** 8 patients
ii 16 patients
iii 29 patients

REVIEW SET 2A

1 a $A = \{1, 2, 3, 4, 6, 12\}, \quad B = \{2, 4, 6, 8\}$
b i $n(A) = 6$ **ii** $n(B) = 4$
c i true   **ii** false   **iii** true   **iv** false

2 a A' represents the months of the year that do not contain the letter R.
b $A' = \{\text{May, June, July, August}\}$ **c** $n(A') = 4$

3 a $X \not\subseteq Y$ **b** $X \subseteq Y$ **c** $X \subseteq Y$ **d** $X \subseteq Y$

4 $\{\ \}, \{2\}, \{4\}, \{6\}, \{2, 4\}, \{2, 6\}, \{4, 6\}, \{2, 4, 6\}$

5 a $S \cap E = \{4, 16\}$
b $S \cup E = \{1, 2, 4, 6, 8, 9, 10, 12, 14, 16, 18, 25, 36, 49\}$

6 a i $A = \{1, 3, 5, 15\}$ **ii** $B = \{3, 6, 9, 12, 15\}$
iii $A \cap B = \{3, 15\}$
iv $A \cup B = \{1, 3, 5, 6, 9, 12, 15\}$
v $(A \cup B)' = \{2, 4, 7, 8, 10, 11, 13, 14\}$

b 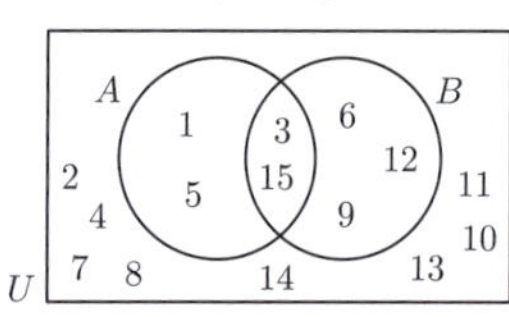

7 a 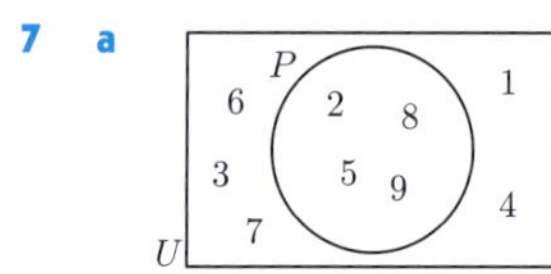   **b** 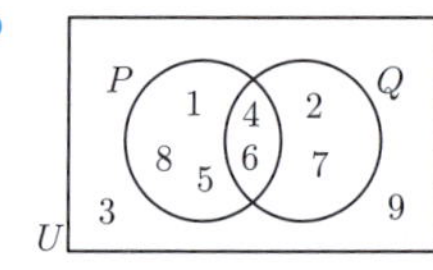

c 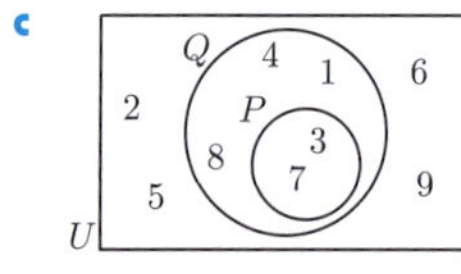   **d** 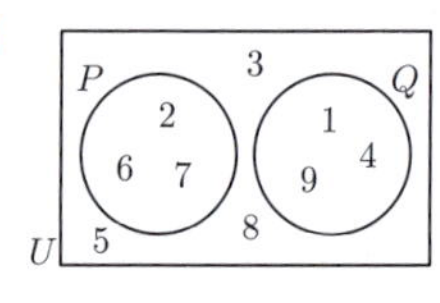

8 a 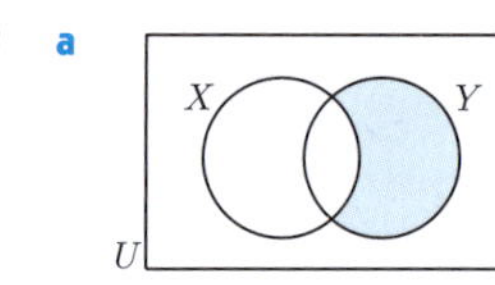   **b** 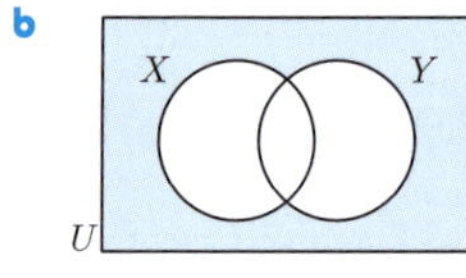

9 a 15   **b** 10   **c** 6   **d** 7   **e** 19   **f** 26
g 9   **h** 4

10 a i $P = \{2, 3, 5, 7, 11, 13, 17, 19\}$
ii $S = \{1, 4, 9, 16\}$
b $P \cap S = \varnothing$
No square number is prime.

c 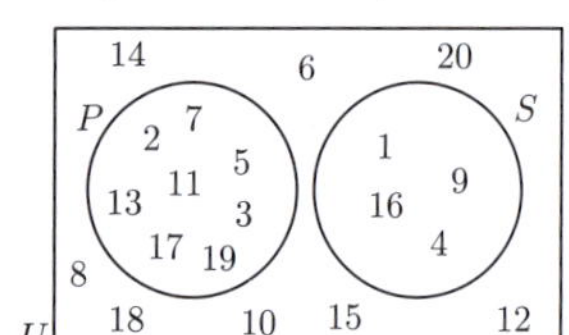   **d i** $n(P) = 8$
ii $n(S) = 4$
iii $n(P \cup S) = 12$

e $n(P \cup S) = n(P) + n(S)$

11 a $M \cap S = \{\text{Craig, Kylie, Nigel}\}$
This set represents the people who both Mark and Stephen want to invite.
b $M \cup S = \{$Michael, Bradley, Craig, Sally, Alistair, Kylie, Emma, Nigel, William, David, Sam, Luke$\}$
c 12 guests

REVIEW SET 2B

1 **a** {6, 12, 18, 24, 30, 36, 42, 48}
   b {1, 2, 4, 5, 8, 10, 20, 40}

2 $n(C) = 11$

3 **a** **i** $A' = \{1, 2, 4, 6, 8, 9\}$ **ii** $B' = \{1, 8, 9\}$
   b **i** true   **ii** true

4 $x = 3, 4, 5, 7,$ or 15 **5** A and D, B and C

6 **a** $S \cup P = \{1, 2, 4, 8, 9, 16, 25, 32, 36\}$
   b $S \cap P = \{1, 4, 16\}$

7 **a** **i** $A = \{A, E, F, H, I, K, L, M, N, T, V, W, X, Y, Z\}$
      ii $V = \{A, E, I, O, U\}$ **iii** $A \cap V = \{A, E, I\}$
   b

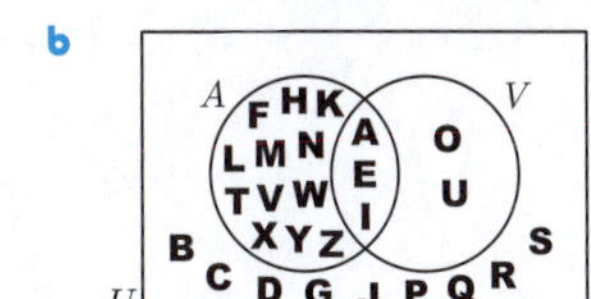

8 **a** in both P and Q **b** in neither P nor Q

9

10 **a**   **b** **i** 12 students
            ii 15 students
            iii 0 students

11 **a**

   b The sets M and T are not disjoint since the two sets have elements in common, that is, $M \cap T \neq \varnothing$.
   c $(M \cup T)' = \{$Drama, Music, Foreign Movies, Cartoons$\}$
   This set represents the channels which neither Tom nor Margot wish to see.
   d History, News, Sci-Fi

EXERCISE 3A

1 **a** proper fraction   **b** improper fraction
   c proper fraction   **d** mixed number
   e improper fraction   **f** mixed number

2 **a** $\frac{7}{4}$ **b** $\frac{7}{2}$ **c** $\frac{21}{5}$ **d** $\frac{19}{8}$ **e** $-\frac{4}{3}$ **f** $-\frac{16}{5}$

3 **a** $12 \div 4 = 3$ **b** $42 \div 6 = 7$ **c** $45 \div 5 = 9$
   d $108 \div 9 = 12$ **e** $132 \div 11 = 12$ **f** $85 \div 5 = 17$

4 **a** $56 \div 7 = 8$ **b** $-56 \div 7 = -8$
   c $56 \div -7 = -8$ **d** $-56 \div -7 = 8$
   e $99 \div 11 = 9$ **f** $-99 \div 11 = -9$
   g $99 \div -11 = -9$ **h** $-99 \div -11 = 9$

5 **a** $1\frac{4}{7}$ **b** $3\frac{1}{5}$ **c** $2\frac{3}{8}$ **d** $-1\frac{1}{9}$ **e** $5\frac{1}{5}$ **f** $-3\frac{1}{4}$

6 **a**

b

7 **a**

   b $-1\frac{4}{5}, -\frac{7}{5}, -\frac{2}{5}, \frac{3}{5}, 1\frac{1}{5}$

8 **a** 9 **b** 3 **c** 4 **d** -4 **e** -4 **f** 2
   g 6 **h** -3

9 **a** -5 **b** 3 **c** -6 **d** 6 **e** -6 **f** 2

EXERCISE 3B

1 **a** $\frac{14}{20}$ **b** $\frac{15}{20}$ **c** $\frac{13}{20}$ **d** $\frac{3}{20}$

2 **a** $\frac{21}{30}, \frac{18}{30}, \frac{20}{30}$ **b** $\frac{7}{10}, \frac{2}{3}, \frac{3}{5}$

3 **a** $\frac{2}{3}$ **b** $\frac{5}{6}$ **c** $\frac{5}{3}$ **d** $\frac{3}{5}$ **e** $-\frac{1}{6}$ **f** $-\frac{1}{3}$
   g $\frac{3}{4}$ **h** $-\frac{2}{5}$ **i** $\frac{11}{16}$ **j** $-\frac{3}{2}$ **k** $\frac{3}{13}$ **l** $-\frac{3}{11}$
   m $\frac{10}{3}$ **n** $-\frac{5}{2}$ **o** $-\frac{8}{3}$

4 **a** $\frac{2}{5}$ **b** $\frac{2}{3}$ **c** $\frac{5}{2}$ **d** $-\frac{1}{4}$ **e** $-\frac{4}{5}$ **f** $\frac{4}{3}$
   g $-\frac{7}{3}$ **h** $\frac{14}{3}$

5 **a** 60 players **b** **i** $\frac{5}{12}$ **ii** $\frac{2}{15}$ **iii** $\frac{11}{20}$

EXERCISE 3C

1 **a** $\frac{7}{9}$ **b** $\frac{5}{11}$ **c** $\frac{7}{10}$ **d** $\frac{1}{12}$ **e** $\frac{11}{8}$ or $1\frac{3}{8}$
   f $\frac{1}{12}$ **g** $\frac{48}{35}$ or $1\frac{13}{35}$ **h** $-\frac{1}{18}$ **i** $\frac{1}{4}$ **j** $-\frac{8}{21}$
   k $-\frac{7}{8}$ **l** $\frac{2}{15}$

2 **a** $\frac{11}{9}$ or $1\frac{2}{9}$ **b** $\frac{11}{8}$ or $1\frac{3}{8}$ **c** $\frac{23}{30}$ **d** $\frac{7}{60}$

3 **a** $\frac{4}{7}$ **b** $\frac{15}{8}$ or $1\frac{7}{8}$ **c** $\frac{7}{5}$ or $1\frac{2}{5}$ **d** $\frac{11}{24}$
   e $\frac{43}{45}$ **f** $\frac{9}{20}$

4 **a** $\frac{23}{40}$ **b** $\frac{17}{40}$ **5** **a** $2\frac{1}{6}$ cups **b** $\frac{5}{6}$ cup more

6 **a** $\frac{107}{210}$ **b** $-\frac{17}{120}$ **c** $\frac{31}{44}$

EXERCISE 3D

1 **a** 20 **b** 12 **c** $\frac{12}{5}$ or $2\frac{2}{5}$ **d** $\frac{5}{4}$ or $1\frac{1}{4}$
   e $\frac{35}{3}$ or $11\frac{2}{3}$ **f** 6 **g** $\frac{9}{2}$ or $4\frac{1}{2}$ **h** 14

2 **a** 20 **b** 9 **c** 25 **d** \$15 **e** 25 m
   f 24 kg **g** 24 kg **h** 21 ha **i** \$120

3 490 km

4 **a** $\frac{8}{15}$ **b** $\frac{15}{28}$ **c** $\frac{1}{2}$ **d** $\frac{1}{2}$ **e** $\frac{8}{21}$ **f** $\frac{3}{7}$
   g $\frac{39}{80}$ **h** $\frac{1}{5}$ **i** $\frac{1}{4}$ **j** $\frac{8}{27}$ **k** $\frac{121}{25}$ or $4\frac{21}{25}$
   l $\frac{2}{3}$ **m** 1 **n** $\frac{1}{3}$ **o** 4 **p** 3

5 **a** $-\frac{8}{15}$ **b** $-\frac{3}{4}$ **c** $\frac{4}{5}$ **d** $\frac{1}{16}$ **e** $-\frac{8}{125}$ **f** $-\frac{1}{6}$

6 $\frac{3}{10}$ **7** **a** $\frac{4}{27}$ **b** $\frac{20}{27}$ **8** $\frac{39}{80}$ of a bottle

9 **a** $\frac{16}{81}$ **b** $\frac{22}{25}$ **c** 1

EXERCISE 3E

1 **a** $\frac{21}{20}$ or $1\frac{1}{20}$ **b** $\frac{8}{3}$ or $2\frac{2}{3}$ **c** $\frac{11}{8}$ or $1\frac{3}{8}$
   d $\frac{5}{18}$ **e** 10 **f** $\frac{16}{9}$ or $1\frac{7}{9}$ **g** $\frac{2}{5}$ **h** $\frac{20}{33}$

2 **a** $-\frac{18}{35}$ **b** $-\frac{5}{6}$ **c** $-\frac{5}{8}$ **d** $-\frac{35}{12}$ or $-2\frac{11}{12}$

3 6 loaves **4** **a** $\frac{23}{3}$ or $7\frac{2}{3}$ **b** $\frac{1}{2}$ **c** $\frac{77}{570}$

EXERCISE 3F

1 **a** $1 + \frac{2}{10}$ **b** $1 + \frac{2}{100}$ **c** $1 + \frac{2}{100} + \frac{3}{1000} + \frac{4}{10\,000}$
d $9 + \frac{9}{100} + \frac{9}{10\,000}$ **e** $\frac{3}{100} + \frac{8}{1000} + \frac{2}{10\,000}$

2 **a** 4.5   **b** 2.69   **c** 0.502   **d** 0.038
e 2.005   **f** 4.0104

3 **a** 7000   **b** 70   **c** $\frac{7}{10}$ **d** $\frac{7}{100}$ **e** $\frac{7}{1\,000\,000}$

4 **a** $\frac{1}{5}$ **b** $\frac{17}{100}$ **c** $\frac{37}{50}$ **d** $\frac{1}{25}$ **e** $\frac{1}{40}$ **f** $\frac{1}{125}$
g $\frac{5}{8}$ **h** $-\frac{4}{5}$

5 **a**

b

c

EXERCISE 3G

1 **a** 6   **b** 8   **c** 19   **d** 200
2 **a** 4.2   **b** 6.4   **c** 11.5   **d** 6.9
3 **a** 7.54   **b** 3.97   **c** 2.16   **d** 9.00
4 **a** 5.24   **b** 37.1   **c** 0.241   **d** 0.0193
5 **a** 27.04   **b** 0.2572   **c** 4.002   **d** 0.010 20
6 **a** $\approx 0.046\,511\,627\,91$
b **i** 0.0465   **ii** 0.047   **iii** 0.046 51

EXERCISE 3H

1 **a** 6.88   **b** 11.2   **c** 22.84   **d** 24.976   **e** 13.1
f 29.98   **g** 7.996   **h** 31.3232   **i** 39.966
2 **a** 4.62   **b** 4.44   **c** 7.88   **d** 12.15   **e** 4.773
f 7.902   **g** 43.138   **h** 1.763   **i** 0.026 21
3 18.35 kg   **4** 3.86 m
5 **a** \$18.70   **b** \$1.30   **c** \$7.10 more

EXERCISE 3I

1 **a** 153.28   **b** 1532.8   **c** 15 328   **d** 153 280
e 1 532 800
2 **a** 93.67   **b** 9.367   **c** 0.9367   **d** 0.093 67
e 0.009 367
3 **a** 57   **b** 4.6   **c** 600   **d** 0.01   **e** 0.07
f 0.0032   **g** 0.0022   **h** 0.000 009 1   **i** 0.000 200 5

EXERCISE 3J

1 **a** 4.8   **b** 6.3   **c** 0.36   **d** 0.84   **e** 0.09
f 0.0088   **g** -6 **h** -0.12 **i** 0.72
2 **a** 5.12   **b** -7.14 **c** 60.8   **d** 2.378   **e** 71.76
f -6.992
3 17.6 litres   **4** 0.2 g   **5** \$12.85

EXERCISE 3K

1 **a** 0.59   **b** 3.49   **c** 0.6374
2 **a** 8   **b** 5   **c** 8   **d** 40   **e** 0.15
f 0.2   **g** 0.016   **h** -17 **i** 6
3 **a** 11.8   **b** 6.7   **c** 6.4
4 4 times heavier   **5** 300 cups

EXERCISE 3L

1 **a** 1   **b** 4   **c** 5   **d** 10   **e** 7   **f** 8
g 11   **h** 9
2 **a** 17   **b** 22   **c** 24   **d** 39   **e** 48
3 **a** 2 and 3   **b** 3 and 4   **c** 5 and 6   **d** 8 and 9
e 10 and 11
4 **a** ≈ 2.6458 **b** ≈ 4.7958 **c** ≈ 10.2470
d ≈ 13.6382 **e** ≈ 17.6068
5 **a** ≈ 1.7321 **b** ≈ 17.3205 **c** ≈ 173.2051
d ≈ 1732.0508

EXERCISE 3M

1 **a** 1   **b** -1 **c** 2   **d** -2 **e** 3   **f** -3
g 5   **h** -5 **i** 0   **j** 10   **k** -10
2 **a** 6   **b** 7   **c** 9   **d** 13   **e** ≈ 1.260
f ≈ 2.080 **g** ≈ 3.530 **h** ≈ 6.300

EXERCISE 3N

1 **a** $\frac{4}{3}$ **b** $\frac{7}{1}$ **c** $\frac{6}{1}$ **d** $\frac{-4}{1}$ **e** $\frac{3}{10}$ **f** $\frac{-11}{4}$
g $\frac{7}{3}$ **h** $\frac{1}{100}$ **i** $\frac{0}{1}$ **j** $\frac{-21}{1}$

2 **a** terminating   **b** recurring   **c** terminating
d terminating   **e** recurring   **f** terminating
g recurring   **h** recurring

3 3 is also a factor in the numerator and will cancel with the factor of 3 in the denominator. Oliver should have written the fraction in lowest terms as $\frac{3}{5}$.

4 **a** 0.9   **b** 0.85   **c** 0.4   **d** 0.12   **e** 0.62
f 0.375   **g** 0.788   **h** 0.544

5 **a** $0.\overline{3}$ **b** $0.\overline{2}$ **c** $0.\overline{36}$ **d** $0.1\overline{6}$ **e** $0.\overline{571\,428}$
f $0.\overline{5}$ **g** $0.\overline{81}$ **h** $0.\overline{857\,142}$

6 **a** $0.\overline{615\,384}$ **b** $0.\overline{162}$ **c** $0.\overline{740}$
d $0.4\overline{3}$ **e** $0.68\overline{1}$ **f** $0.785\,714\,2$

EXERCISE 3O

1 **a** **i** rational   **ii** irrational   **iii** rational   **iv** irrational
v rational   **vi** irrational   **vii** rational   **viii** irrational
b

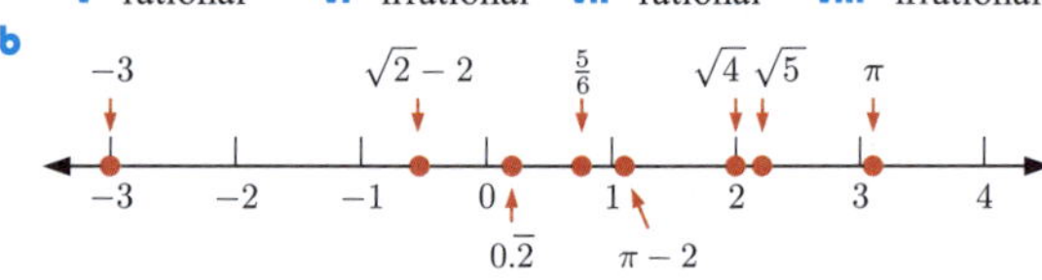

2 **a** true   **b** true   **c** true   **d** false
3 $\sqrt{2} - 1$ (Other answers are possible.)
4 No. A counterexample is $\sqrt{2} + (-\sqrt{2}) = 0$ which is rational.
5 $0 = \frac{0}{1}$ is rational, but its reciprocal $\frac{1}{0}$ is not rational as it is undefined.

REVIEW SET 3A

1

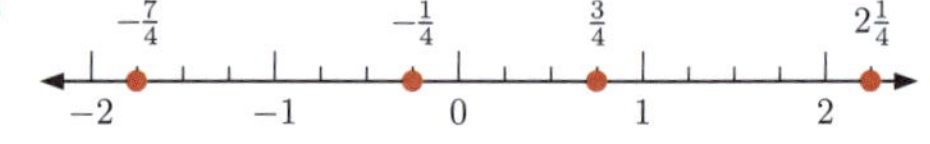

2 **a** $54 \div 9 = 6$ **b** $88 \div 8 = 11$ **c** $120 \div -12 = -10$
3 **a** $\frac{1}{6}$ **b** $\frac{2}{5}$ **c** $\frac{7}{4}$ **d** $-\frac{21}{23}$ **e** $-\frac{2}{9}$
4 **a** $\frac{7}{9}$ **b** $-\frac{1}{12}$ **c** $\frac{14}{15}$ **d** $\frac{25}{36}$ **e** $-\frac{3}{5}$ **f** $\frac{27}{7}$ or $3\frac{6}{7}$
5 **a** 5   **b** $\frac{5}{100}$ **c** $\frac{5}{10\,000}$ **d** $\frac{5}{10}$

6 **a** 2.57 **b** 10.1 **c** 0.0515

7 **a** 8.455 **b** 7.66 **c** 1.73

8 **a** 634.9 **b** 0.052 97 **c** 5.6 **d** −11.55
e 0.08 **f** 18.7

9 \$39.24

10 **a** 72.66 m **b** 3.66 m farther **c** 9.54 m

11 **a** $\frac{3}{4}$ **b** $\frac{1}{4}$ **c** $\frac{1}{4}$ **d** Yes, as everyone gets $\frac{1}{4}$ of the pie.

12 13 and 14 **13** **a** 100 **b** −100

14 **a** 0.8 **b** 0.95 **c** 0.44 **d** 0.54 **e** 0.075

15 **a** **i** recurring **ii** recurring **iii** terminating
iv terminating **v** recurring
b **i** $0.\overline{432}$ **ii** $0.58\overline{3}$ **iii** 0.3125 **iv** 0.476
v $0.329\,\overline{54}$

REVIEW SET 3B

1 **a** $2\frac{1}{5}$ **b** $3\frac{5}{6}$ **c** $-2\frac{3}{8}$

2 **a** 4 **b** $\frac{1}{2}$ **c** $-\frac{3}{5}$

3 **a** $\frac{76}{15}$ or $5\frac{1}{15}$ **b** 2.512 **c** 6
d 15.325 **e** $\frac{3}{2}$ or $1\frac{1}{2}$ **f** 2.78

4 $\frac{7}{32}$

5

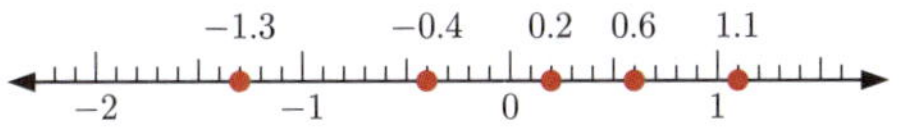

6 **a** $\frac{11}{50}$ **b** $\frac{13}{20}$ **c** $\frac{23}{250}$

7 **a** 0.359 **b** 3.59 **c** 359

8 **a** 0.0084 **b** 5 **c** 28.1

9 **a** $\frac{37}{84}$ **b** pages 64 to 105 **c** $\frac{25}{63}$

10 $0.\overline{5}$ **11** 8 and 9 **12** **a** 11 **b** ≈ 12.164

13 **a** rational, $\frac{8}{1}$ **b** irrational **c** rational, $\frac{9}{1}$
d rational, $\frac{20}{3}$ **e** rational, $\frac{-7}{2}$ **f** irrational

14 **a** $0.\overline{3}$ **b** **i** 0.6 **ii** $0.0\overline{6}$ **iii** $1.\overline{6}$ **iv** $0.0\overline{6}$
c $\frac{1}{6} = 0.1\overline{6}$, $\frac{1}{7} = 0.\overline{142\,857}$
d **i** $\frac{1}{3} \div \frac{1}{7} = 2.\overline{3}$ **ii** $\frac{1}{6} \div \frac{1}{3} = 0.5$

EXERCISE 4A

1 **a** $2b$ **b** $3q$ **c** $2x+4y$ **d** $3c+e$
e $3+2y+z$ **f** $4a+7$ **g** $2+4g$ **h** $3-3d$
i s **j** $s-2t$ **k** $5+3r$ **l** $2+2a+2b$

2 **a** $5x$ **b** $2c$ **c** $7q$ **d** $4fg$ **e** $6pq$ **f** $9rs$
g $6ab$ **h** $4mn$ **i** $5ab$ **j** $2pq$ **k** jkl **l** dhp

3 **a** $pq+r$ **b** $4x+5y$ **c** $2a-b$ **d** $ab-c$
e $b-ac$ **f** $f-7g$ **g** $ac+ad$ **h** $12-6rs$
i $x-3xy$

4 **a** $3(x+y)$ **b** $5(d-1)$ **c** $8(w-x)$ **d** $pq(r-2)$

EXERCISE 4B

1 **a** x^3 **b** n^4 **c** $3k^2$ **d** $4a^3$ **e** $2d^4$
f $4pq^2$ **g** st^3 **h** $3f^2g^2$ **i** w^2xy^3

2 **a** $a \times a \times a \times a$ **b** $4 \times p \times p \times p$
c $3 \times t \times t \times t \times t \times t$ **d** $5 \times x \times x \times y$
e $7 \times f \times f \times g \times g \times g$ **f** $5 \times a \times 5 \times a$
g $5 \times a \times a$ **h** $p \times p + 2 \times q$
i $p \times p \times p - 3 \times q \times q$

3 **a** x^2+3 **b** c^2+2d **c** $m+m^2$ **d** n^3+n
e y^2-z^4 **f** a^2+7a **g** $8b-b^3$ **h** $pq-2p$
i $2pq^2+6r^2s$ **j** $2h^2-hj$ **k** $3x+5x^3$
l $a^2+2b^3-ab^2$

EXERCISE 4C

1 **a** y plus 2 **b** x plus y
c x minus the product of 2 and y
d 3 minus the product of 2 and x
e the square root of m **f** 4 times b squared
g 3 times x times y **h** 3 times y minus 2 times x
i x times y times z **j** the square root of all of x minus 3
k a plus b plus c **l** b divided by a
m 3 divided by the square root of x
n the product of 3 and a, divided by the product of 5 and b
o the sum of 5 and a, divided by a

2 **a** x squared plus y **b** the product of 5 and x, all squared
c 3 minus 2 times x squared
d y squared minus x squared
e x squared, divided by 2
f the quotient of x and 2, all squared
g 2 plus b squared **h** the sum of 2 and b, all squared
i a squared minus 7
j the square root of all of 3 times a minus 7
k the square root of all of x squared plus y squared
l x cubed minus y cubed

3 **a** $p+7$ **b** $yz-x$ **c** $\dfrac{x}{y}$ **d** $3z-15$

4 **a** $a+b^2$ **b** a^2+b **c** $(a+b)^2$ **d** $a+b^2$
e $(a+b)^2$

5 **a** $7x^2$ **b** $\left(\dfrac{7}{x}\right)^2$ **c** a^2-b^2 **d** $(a-b)^2$
e $\sqrt{x}-p$

6 **a** x^2+4 **b** $2(3-x)$ **c** $\sqrt{x+2}$

EXERCISE 4D

1 **a** \$100 **b** \$$20a$ **c** \$$ad$

2 **a** 8 years old **b** $(14-x)$ years old

3 **a** \$55 **b** \$$(100-15h)$ **c** \$$(100-hp)$

4 $(w-6)$ kg **5** $(20+m-n)$ people

6 \$$(0.6a+0.9p)$ **7** **a** $0.8x$ m **b** $(600-0.8x)$ m

8 **a** 45 km **b** st km

9 **a** 12 cupcakes **b** $\dfrac{c}{n}$ cupcakes

10 **a** $(bx+3)$ apples **b** $(by+4)$ apples
c $(bt+7)$ apples **d** $(bm+n)$ apples

11 \$$(bm+nr)$

EXERCISE 4E

1 **a** 3 **b** 12 **c** −3 **d** 15 **e** −2 **f** 7
g 15 **h** 32

2 **a** 2 **b** −10 **c** 2 **d** 14 **e** 1 **f** −2
g 3 **h** 12 **i** −2 **j** 12 **k** 22 **l** −14

3 **a** 2 **b** 3 **c** 1 **d** −4 **e** $\frac{3}{8}$ **f** $-\frac{1}{9}$
g $\frac{6}{11}$ **h** $-\frac{11}{4}$ **i** $\frac{9}{14}$ **j** $\frac{2}{7}$ **k** 3 **l** 4

4 **a** 10 **b** 0 **c** −21 **d** 84

5 a 1   **b** 2   **c** 4   **d** 3

6 a i 11 years old   **ii** 39 years old   **iii** 43 years old

b i $(n-2)$ years old   **ii** $3n$ years old

iii $(3n+4)$ years old

c Yes, by substituting $n=13$ into the expressions in **b**.

7 27 cm^2 **8 a** 5 g/cm^3 **b** 7.8 g/cm^3

9 a i 4.43 m/s   **ii** 9.90 m/s   **b** 15.34 m/s

EXERCISE 4F

1 a expression   **b** equation   **c** equation

d equation   **e** expression   **f** expression

2 a 3 terms   **b** 4 terms   **c** 5 terms

3 a 3   **b** -8 **c** 1   **d** -1 **e** 4   **f** -5

g 3   **h** -2

4 a 5   **b** -5 **c** 14   **d** -7 **e** -1 **f** 6

g 3   **h** -2

5 a 2   **b** -4 **c** 3   **d** 2

6 a expression   **b** 6 terms

c i 2   **ii** 5   **iii** -7 **iv** -2 **d** 1

7 a $3x$ and $5x$ **b** $2y$ and $-3y$ **c** $2x$ and $-5x$

d xy and $-2xy$, $3x^2y$ and x^2y **e** $5x$ and $\dfrac{x}{2}$

f $3x^2y^2$ and x^2y^2, 7 and -2

EXERCISE 4G

1 a $x+6$ **b** $q+11$ **c** $3b+3$ **d** $2a+7$

e $2d$ **f** $2q+5$ **g** $2y$ **h** $3z$

i $2g^2$ **j** cannot be simplified   **k** w^2

l cannot be simplified   **m** $2x$ **n** $9ab$ **o** $4m$

2 a 0   **b** $7p$ **c** cannot be simplified

d $6pq$ **e** $4ab$ **f** cannot be simplified

g $12w$ **h** $13xy$ **i** $3z$ **j** 0

k $9d-9$ **l** $-10n$ **m** $-5j-4$ **n** $a+a^2$

o $7g-7g^2$ **p** $4s+4s^2$ **q** cannot be simplified

r cannot be simplified

3 a $6x$ **b** $-8x$ **c** $-15m$ **d** $7y$

e $9y$ **f** $5y$ **g** $-3a$ **h** x

4 a $-2x-1$ **b** $5t+3$ **c** $-2x-y$ **d** $6pq-4$

e $10cd$ **f** $-a$ **g** $5x^2-2$ **h** $-3n-3$

i $-5v-5w$ **j** $-3x^2$ **k** $5a-7b$ **l** $-5z-7$

m $p-2pq$ **n** $-4n-5$ **o** $-7mn-2m$

EXERCISE 4H

1 a $20k$ **b** x^5 **c** $28x^3$ **d** $30z^2$ **e** $8m^4$

f $10g^4$ **g** $25y^2$ **h** $12x^2$ **i** $24x^4$

2 a $-8x^2$ **b** $5a^2b$ **c** $-6c^2$ **d** $3m^3n$

e $-12a^2b^2$ **f** $20p^2$ **g** $-35x^3$ **h** $16g^3$

i $24x^3y^3$ **j** $-7x^3$ **k** $9x^2$ **l** $-2s^3t$

m $-x^5$ **n** $5r^3$ **o** $6x^3$ **p** $27x^3y^4$

q $8y^3$ **r** $-8z^3$

EXERCISE 4I

1 a $2x$ **b** $\dfrac{3}{x}$ **c** 5   **d** $\dfrac{3}{5}$ **e** x

f $\dfrac{1}{x}$ **g** $2c$ **h** $\dfrac{3}{m^2}$

2 a x^3 **b** $5x^2$ **c** $\dfrac{x^2}{3}$ **d** $\dfrac{6}{x}$ **e** y **f** $\dfrac{1}{b}$

g $\dfrac{2d}{c}$ **h** $\dfrac{4x}{5y}$ **i** 1   **j** $\dfrac{1}{7m}$ **k** $5p^2q$ **l** $\dfrac{x^2}{3y^2}$

m $\dfrac{3t}{2}$ **n** 1   **o** $\dfrac{xy^2}{z}$ **p** $\dfrac{5ac^3}{6b}$

EXERCISE 4J

1 a $\dfrac{ab}{12}$ **b** $\dfrac{x^2}{7y}$ **c** 5   **d** $\dfrac{30}{xy}$ **e** $\dfrac{p}{3}$ **f** $\dfrac{a}{4b}$

g $\dfrac{1}{xy}$ **h** t^2

2 a $\dfrac{x^2}{4y}$ **b** $\dfrac{xy}{2}$ **c** $\dfrac{2x^2}{3y^3}$ **d** $\dfrac{2x^2y^2}{5}$

EXERCISE 4K

1 a $\dfrac{4a}{7b}$ **b** $\dfrac{40}{xy}$ **c** $\dfrac{c^2}{8}$ **d** $\dfrac{6p}{q^2}$ **e** $\dfrac{3}{2}$ **f** $\dfrac{jk}{4}$

g $\dfrac{3s}{t^3}$ **h** 1

2 a $\dfrac{3a^3}{20}$ **b** $\dfrac{11m}{5k^3}$ **c** $\dfrac{4a^3}{3b^3}$ **d** $\dfrac{3a^2}{10}$

EXERCISE 4L

1 a $2a$ **b** $5b$ **c** $4xy$ **d** $4x$ **e** x **f** $-2x$

g $-b$ **h** $4a$ **i** $-3xy$

2 a 2   **b** c **c** 1   **d** k **e** 3   **f** $5x$

g $5x$ **h** $8y$ **i** x **j** $5x$ **k** r **l** 3

3 a 18   **b** q **c** x **d** $12a$ **e** ab **f** abc

g $3pq$ **h** $2ab$ **i** $6xy$ **j** 5   **k** $12wz$ **l** $12pqr$

REVIEW SET 4A

1 a $21a^2b$ **b** b^3+2b **c** $2d^3-3d^2$

d $2a^2-3(2b-a)$

2 a $7\times t\times t\times t\times t$ **b** $3\times b\times 3\times b\times 3\times b$

c $2\times b\times b-3\times c\times c\times c$

3 a m divided by 3   **b** the square of the product of 3 and x

c a plus x squared

4 a $a+b+c$ **b** $\sqrt{x}+5$ **c** $k+h^2$

5 a 12 apples   **b** $(18-3a)$ apples   **c** $(18-ad)$ apples

6 a 9   **b** -72 **c** $\dfrac{1}{12}$

7 a 72 newtons   **b** 0.27 newtons

8 a expression   **b** expression   **c** equation

9 a -1 **b** 7

10 a $13k-6$ **b** $x+2$ **c** $17f-5g$

11 a 5 terms   **b** $12gh$ and $3gh$ **c** -2

d 4   **e** $5g+15gh-2h+4$

12 a $20x^3$ **b** $-2a^4$ **c** $-8x^2y^2$

13 a 2   **b** $\dfrac{x}{2y}$ **c** $\dfrac{7a^2}{4b^2}$

14 a $\dfrac{5}{2m}$ **b** $\dfrac{6}{x^2}$ **c** $\dfrac{10n^3}{7p^2}$

15 a 6   **b** $3ab$

REVIEW SET 4B

1 a $2a + a^2$ **b** $4p^2 - 6p^2q$ **2** $q - p^2$

3 $a \times a \times a - 7 \times a \times a \times b \times b \times b$

4 a \$14   **b** $\$(20 - 3p)$ **c** $\$(20 - cp)$

5 a -9 **b** 49   **6** $(20 - 2n + m)$ muffins

7 a 2   **b** -3

8 a $2 - 5z$ **b** $12u - 5t$ **c** $3x^2 - 3x + 5$

9 a 5 terms   **b** -10 **c** -2 **d** $3s - 2t - 10st - 2$
  e i -8 **ii** -21

10 a $(6n + 3)$ fish   **b i** 15 fish   **ii** 45 fish   **c** 9 fish

11 a $9d^2$ **b** $-5x^5$ **c** $-x^3y^2$

12 a $5x^2$ **b** $\dfrac{7a}{2b}$

13 a $-\dfrac{6}{x^2}$ **b** $\dfrac{a}{3}$ **c** $56c$ **14 a** $2y$ **b** $2x$

EXERCISE 5A

1 a 0.48   **b** 1.1   **c** 1.01   **d** 0.625
  e 0.049   **f** 0.02   **g** 5.52   **h** 0.0001
  i 10   **j** 0.000 025   **k** 2.352   **l** 0.273

2 a 0.065   **b** 0.138

3 a $\dfrac{11}{20}$ **b** $\dfrac{8}{25}$ **c** $\dfrac{21}{10}$ **d** $\dfrac{4}{25}$ **e** $\dfrac{18}{25}$ **f** $\dfrac{3}{8}$
  g $\dfrac{13}{400}$ **h** $\dfrac{36}{25}$ **i** $\dfrac{1}{3}$ **j** $\dfrac{6}{5}$ **k** $\dfrac{1}{10\,000}$ **l** $\dfrac{3}{400}$

4 a $\dfrac{1}{20}$ **b** $\dfrac{19}{20}$ **c** $\dfrac{6}{25}$ **d** $\dfrac{71}{100}$

EXERCISE 5B

1 a 2%   **b** 40%   **c** 37%   **d** 65%
  e 41.6%   **f** 320%   **g** 125%   **h** 1.5%
  i 33.3%   **j** 0.62%   **k** 3.04%   **l** 105.7%

2 a 30%   **b** 18%   **c** 84%   **d** 25%
  e 200%   **f** 65%   **g** 80%   **h** 42.5%
  i 87.5%   **j** 175%   **k** 145%   **l** 112.5%

3 a $66\tfrac{2}{3}\%$ **b** $22\tfrac{2}{9}\%$ **c** $83\tfrac{1}{3}\%$ **d** 6.4%   **e** 6.25%

4 a $\approx 42.9\%$ **b** $\approx 61.5\%$ **c** $\approx 6.7\%$ **d** $\approx 105.9\%$

EXERCISE 5C

1 a 85%   **b** 40%   **c** $\approx 13.3\%$ **d** 25%
  e 16%   **f** $\approx 2.67\%$ **g** 6.25%   **h** $\approx 16.7\%$
  i 1.4%   **j** 4.6%

2 30%   **3** Jemima, as she saved 40%, Henry saved $\approx 38.9\%$

4 a 50%   **b** 10%   **c** $\approx 26.7\%$ **d** $\approx 86.7\%$

5

Name	Steps climbed	% already run	% remaining
Marcel	813	$\approx 73.7\%$	$\approx 26.3\%$
Ariel	672	$\approx 60.9\%$	$\approx 39.1\%$
Shane	901	$\approx 81.7\%$	$\approx 18.3\%$
Emma	866	$\approx 78.5\%$	$\approx 21.5\%$

6 a

Year	Votes	Registered voters	Voter turnout %
2019	9 359 673	18 286 865	$\approx 51.2\%$
2014	9 723 232	18 284 066	$\approx 53.2\%$
2009	9 946 748	18 293 277	$\approx 54.4\%$
2004	10 791 215	18 449 344	$\approx 58.5\%$
2000	11 559 458	17 699 727	$\approx 65.3\%$

  b i 2000   **ii** 2019

EXERCISE 5D

1 a 24   **b** 9   **c** 27.5 kg   **d** 280 m
  e \$87.50   **f** 1.8 L   **g** 0.9 km   **h** 6.8 m
  i £138.60   **j** ¥2 880 000   **k** 17.5 kg   **l** \$74.25

2 \$16 640   **3** 5.28 L   **4** \$437.50   **5** 588 pears

6 \$17 325 000   **7 a** \$15 750   **b** \$24 500

8 a 138 g   **b** Tom

EXERCISE 5E

1 a 400 mL   **b** 500 g   **c** €1200   **d** 70 kg
  e \$300   **f** 40 minutes

2 600 students   **3** $\approx 231\,000$ km^2

4 a 56 mL   **b** 126 g   **c** 640 mL   **d** \$210

5 a 250 kg   **b** 212.5 kg   **6** 35 cookies

7 60 children   **8 a** 1.68 kg   **b** 250 g

9 a 22%
  b i 144 ice creams   **ii** 30 ice creams
    iii 400 ice creams

EXERCISE 5F.1

1 a 176 kg   **b** 9 km   **c** \$29 925   **d** 390 mL
  e 205 L   **f** €369

2 a 13 students   **b** 78 students

3 a \$210.6 million   **b** \$113.4 million

EXERCISE 5F.2

1 a 1.05   **b** 0.94   **c** 1.12   **d** 0.75   **e** 0.51
  f 1.34   **g** 0.925   **h** 1.038

2 a \$840   **b** 133.4 kg   **c** 672 m   **d** 1320 L
  e \$1 122 000   **f** 1.875 L   **g** 217.2 m   **h** \$352.60

3 790.5 km^2 **4** \$265 per week   **5** 173.6 cm

6 a \$5.50   **b** \$6.05
  c The 10% increase was applied to a higher original price in November than in May.

7 Multipliers are 1.1 for 10% increase and 0.9 for 10% decrease. The overall multiplier is $1.1 \times 0.9 = 0.99$. This represents a 1% decrease.

8 a 1080 tonnes   **b i** 2018   **ii** 2019

EXERCISE 5G

1 a 20% increase   **b** 20% decrease   **c** 35% increase
  d 35% decrease   **e** 63% increase   **f** 63% decrease
  g 140% increase   **h** 0.5% decrease

2 a 10% increase   **b** 15% decrease   **c** 40% decrease
  d 50% increase   **e** 15% decrease   **f** $\approx 45.2\%$ increase

3 a 15% increase   **b** 28% decrease   **c** 12.5% increase
  d 7.5% increase   **e** 12.5% decrease

4 $\approx 14.3\%$ decrease

5 Year 8 ($\approx 26.5\%$ compared with $\approx 22.5\%$)

6 a i $\approx 36.2\%$ **ii** $\approx 19.0\%$ **iii** $\approx 13.8\%$
  b $\approx 84.5\%$

7 a $\approx 830\%$ **b** $\approx 71.1\%$ **c** $\approx 1610\%$ **d** $\approx 149\%$

8 a year 2000: $\approx 448\,000$ students,
    year 2020: $\approx 1\,120\,000$ students
  b $\approx 150\%$ increase
  c i $\approx 150\%$ increase   **ii** $\approx 300\%$ increase
    iii $\approx 162.5\%$ increase

9 $\approx 38.6\%$

EXERCISE 5H

1 **a** 20 cm   **b** 60 kg   **c** $125   **d** 4000 L
   e 50 000 people
2 80 nations   **3** £140   **4** 450 songs
5 $420   **6** ≈ 2475 black rhinoceroses

EXERCISE 5I

1

	Profit or loss
a	$30 profit
b	£65 loss
c	$150 profit
d	€6500 profit

2

	Cost price	Selling price	Profit or loss
a	$56	$76	$20 profit
b	$420	$345	$75 loss
c	$385	$580	$195 profit
d	$265	$200	$65 loss

3 **a** **i** profit of $20   **ii** $\approx 44.4\%$ profit
   b **i** loss of €1150   **ii** $\approx 16.4\%$ loss
   c **i** loss of $23   **ii** $\approx 60.5\%$ loss
   d **i** profit of $60   **ii** $\approx 31.6\%$ profit
   e **i** profit of $90   **ii** 37.5% profit
4 $\approx 26.9\%$ **5** $\approx 64.7\%$ **6** **a** RM 4500   **b** $33\frac{1}{3}\%$
7 **a** CNY5100   **b** $\approx 309\%$
8 **a** loss of $1.50   **b** $\approx 0.469\%$ loss
9 **a** $780   **b** €360   **c** $2975   **d** $28 000
10 $2592   **11** $140
12 **a** $7.50   **b** $166\frac{2}{3}\%$
   c Theo might want to know how much profit he is making *relative* to the cost price of each figurine.
   d No, it decreases by 3%.
13 $167.20

EXERCISE 5J

1 **a** $34   **b** $249   **2** $2254   **3** £158.40
4 16%   **5** **a** $171   **b** $296   **c** $801

6

	Marked price	Discount	Selling price	Discount as a % of marked price
a	$160	$40	$120	25%
b	$500	$170	$330	34%
c	$2.40	$0.36	$2.04	15%
d	$4.15	$0.75	$3.40	$\approx 18.1\%$
e	$252	$89	$163	$\approx 35.3\%$
f	€490	€88.20	€401.80	18%
g	$5450	$2071	$3379	38%

EXERCISE 5K

1 **a** 156 krona   **b** 1710 rubles
2 **a** $29.04   **b** 5400 cedi   **3** **a** 7% tax   **b** 19% tax
4 **a** HRK61 000   **b** CHF2000

REVIEW SET 5A

1 **a** 37.5%   **b** 250%   **c** $\approx 41.67\%$ **d** 1.5%
   e 2.5%   **f** $\approx 88.89\%$
2 **a** 0.83   **b** 0.274   **c** 1.52   **d** 0.004

3 **a** $\frac{12}{25}$ **b** $\frac{3}{20}$ **c** $\frac{11}{200}$ **d** $\frac{1}{1000}$
4 72%   **5** **a** 150 kg   **b** 240 kg   **6** 150 people
7 **a** 59%   **b** $\frac{7}{20}$ **c** "Poor"   **d** 500 guests
8 **a** 1.45   **b** 0.25   **c** 3.27   **9** 640 kg
10 **a** $\approx 6.67\%$ increase   **b** $\approx 23.9\%$ decrease
11 $5.40 per kg   **12** 54 900 people
13 **a** $480 loss   **b** 80% loss
14 $78   **15** $\approx 32.9\%$ **16** $60

REVIEW SET 5B

1

	Percentage	Decimal	Fraction (in lowest terms)
a	32%	0.32	$\frac{8}{25}$
b	87.5%	0.875	$\frac{7}{8}$

2 $\frac{43}{50}$ **3** 20%   **4** 506 g   **5** 1800 students
6 **a** 6 students   **b** 32%   **c** 62.5%
7 **a** 41 marks   **b** 50 marks
8 **a** $2067.36   **b** $6546.64   **9** $\approx 10.7\%$ decrease
10 18%

11

Offence	Old fine	New fine
Speeding	$200	$210
Drink driving	$840	$882
Not wearing seatbelt	$260	$273
Illegal parking	$50	$52.50

12 **a** $52.50   **b** $56.70   **13** $60
14 **a** 28 serves   **b** SportsLife $382.50, is cheaper by $1.50
   c 175 km/h
15 **a** $\approx 73.3\%$ **b** **i** 40%   **ii** $\approx 7.14\%$ **c** $2860
16 17% VAT

EXERCISE 6A

1 **a** $3^5 = 243$ **b** $2^4 = 16$ **c** $5^6 = 15\,625$
   d $3^7 = 2187$ **e** a^5 **f** n^3 **g** x^9 **h** y^8
2 **a** $2^2 = 4$ **b** $3^2 = 9$ **c** $7^3 = 343$
   d $10^1 = 10$ **e** x^4 **f** y^4 **g** c^2 **h** b^3
3 **a** $2^6 = 64$ **b** 10^8 **c** $3^6 = 729$
   d $2^{12} = 4096$ **e** x^{10} **f** p^9 **g** t^{12} **h** z^{12}
4 **a** c^6 **b** b^5 **c** y^{15} **d** y^{10} **e** q^7 **f** z^{30}
   g t^3 **h** a^{2+n} **i** g^3 **j** n^{10} **k** k^7 **l** p^6
5 **a** $(7^2)^3 = 7^6$ **b** $2^4 \div 2 = 2^3$ **c** $2^2 \times 2^7 = 2^9$
   d $(x^3)^4 = x^{12}$ **e** $a^5 \times a^5 = a^{10}$ **f** $5^9 \div 5^3 = 5^6$
   g $b^4 \times b^3 \div b^2 = b^5$ **h** $c^8 \div c^2 \div c^3 = c^3$
   i $(x^3)^2 = x^{10} \div x^4$
6 **a** $5a^2$ **b** $15q^3$ **c** $16x^3y^4$ **d** $7t$ **e** $4z$
   f jk^2 **g** x^4 **h** m^5 **i** h^8
7 **a** 2^7 **b** 2^{11} **c** 2^7 **d** 2^3
8 **a** x^{13} **b** x^8 **c** x^6 **d** x^{26} **e** x^{19} **f** x^6
9 **a** The bases should not be multiplied. $2^5 \times 2^3 = 2^8$
   b The exponents should be multiplied. $(3^2)^3 = 3^6$
   c The exponents should be added. $x^5 \times x^3 = x^8$
   d The exponents should be subtracted. $\dfrac{x^{12}}{x^3} = x^9$
   e The bases should not be divided. $\dfrac{3^5}{3^3} = 3^2$

EXERCISE 6B

1 **a** p^2q^2 **b** x^4y^4 **c** a^6b^6 **d** $a^3b^3c^3$
   e $8a^3$ **f** $243d^5$ **g** $32k^5$ **h** $25g^2h^2$

2 **a** $\dfrac{a^2}{b^2}$ **b** $\dfrac{b^3}{8}$ **c** $\dfrac{j^4}{k^4}$ **d** $\dfrac{16}{z^4}$
   e $\dfrac{16}{x^2}$ **f** $\dfrac{32}{b^5}$ **g** $\dfrac{q^4}{16}$ **h** $\dfrac{27}{b^3}$

3 **a** $\dfrac{4}{25}$ **b** $\dfrac{27}{64}$ **c** $\dfrac{16}{81}$ **d** $\dfrac{1}{32}$

4 **a** $4a^4$ **b** $27b^9$ **c** $16c^8$ **d** $32d^{10}$
   e j^2k^6 **f** x^3y^6 **g** $18g^3$ **h** $25r^4s^2$
   i $64a^2b^6$ **j** $27h^6k^3$ **k** $8b^3c^6$ **l** $49e^6f^2$

5 **a** $\dfrac{j^2k^2}{4}$ **b** $\dfrac{4}{c^2d^2}$ **c** $\dfrac{27p^3}{q^3}$ **d** $\dfrac{z^4}{25}$
   e $\dfrac{49d^4}{e^2}$ **f** $\dfrac{w^4}{9v^2}$ **g** $\dfrac{16r^2}{9s^4}$ **h** $\dfrac{125g^3}{8h^9}$

6 $a^2b^2 = (ab)^2$ which is a perfect square.
   For example, $2^2 \times 3^2 = (2 \times 3)^2$.

7 $a^3b^3 = (ab)^3$ which is a cubic number.
   For example, $2^3 \times 3^3 = (2 \times 3)^3$.

EXERCISE 6C

1 **a** 1 **b** 1 **c** 1 $\{x \neq 0\}$ **d** 5 **e** 9 **f** 5
   g 11 **h** p^6 $\{p \neq 0\}$ **i** 1 **j** 1 **k** 7 **l** 1

2 **a** 1 **b** 1 **c** x **d** 3 **e** a^3 **f** n^2 **g** q
   h $\dfrac{y^2}{4}$

EXERCISE 6D

1 **a** $\dfrac{1}{5}$ **b** $\dfrac{1}{4}$ **c** $\dfrac{1}{8}$ **d** $\dfrac{1}{9}$ **e** $\dfrac{1}{4}$
   f $\dfrac{1}{121}$ **g** $\dfrac{1}{49}$ **h** $\dfrac{1}{27}$ **i** $\dfrac{1}{32}$ **j** $\dfrac{1}{128}$

2 **a** $\dfrac{1}{10} = 0.1$ **b** $\dfrac{1}{100} = 0.01$ **c** $\dfrac{1}{1000} = 0.001$
   d $\dfrac{1}{10\,000} = 0.0001$ **e** $\dfrac{1}{100\,000} = 0.000\,01$

3 **a** $\dfrac{8}{3} = 2\frac{2}{3}$ **b** $\dfrac{4}{3} = 1\frac{1}{3}$ **c** $\dfrac{6}{7}$ **d** $\dfrac{29}{5} = 5\frac{4}{5}$

4 **a** $\dfrac{1}{x}$ **b** $\dfrac{1}{k}$ **c** $\dfrac{1}{a^2}$ **d** $\dfrac{1}{t^3}$ **e** $\dfrac{1}{r^5}$

5 **a** 3^{-2} **b** 5^{-1} **c** 2^{-2} **d** 3^{-2} **e** 7^{-1}

6 **a** $\dfrac{2}{x}$ **b** $\dfrac{1}{2x}$ **c** $\dfrac{s}{t}$ **d** $\dfrac{1}{st}$ **e** $\dfrac{7}{a^2}$ **f** $\dfrac{1}{25z^2}$
   g $\dfrac{g}{h^3}$ **h** $\dfrac{1}{g^3h^3}$ **i** $\dfrac{4c}{d^2}$ **j** $\dfrac{1}{16c^2d^2}$ **k** $\dfrac{4}{c^2d^2}$

7 **a** 1 **b** x **c** $\dfrac{1}{x}$ **d** x

EXERCISE 6E

1 **a** $2x + 14$ **b** $3x - 6$ **c** $4a + 12$ **d** $5a + 5c$
   e $6b - 18$ **f** $7m + 28$ **g** $2n - 2p$ **h** $4p - 4q$
   i $15 + 3x$ **j** $5y - 5x$ **k** $8t - 64$ **l** $6d + 6e$
   m $40 - 4j$ **n** $7y + 7n$ **o** $2n - 24$ **p** $88 - 8d$

2 **a** $18x + 9$ **b** $3 - 9x$ **c** $10a + 15$ **d** $11 - 22n$
   e $18x + 6y$ **f** $5x - 10y$ **g** $12b + 4c$ **h** $2a - 4b$
   i $7a - 35b$ **j** $24 + 36d$ **k** $24 - 32y$ **l** $30b + 18a$
   m $22x - 11y$ **n** $7c - 63d$ **o** $6m + 42n$ **p** $64a - 8c$
   q $4x^2 - 4$ **r** $10 + 15a^2$

---

3 **a** $x^2 + 2x$ **b** $5x - x^2$ **c** $2a^2 + 4a$ **d** $5b - 3b^2$
   e $ab + 2ac$ **f** $a^3 + a$ **g** $3x - 4x^2$ **h** $18x - 3x^2$
   i $5x^2 - 20x$ **j** $4a - 4a^2$ **k** $7b^2 + 14b$ **l** $a^2 + ab$
   m $3b - 8b^2$ **n** $m^2 - 3mn$ **o** $c^2 - 4ac$ **p** $24p - 42p^2$
   q $x^3 + 4x^2$ **r** $6x - 2x^3$

4 **a** ac units^2 **b** $a(b - c)$ units^2 **c** ab units^2
   d area of blue region = total area − area of red region

5 **a** $3x + 3y + 12$ **b** $14x + 21z - 7y$
   c $10l - 20m + 35n$ **d** $2a^2 + 2ab - 16a$
   e $3x^2 + 6xy - 15x$ **f** $8p^2 + 12pq + 20p$

6 **a** $-2x - 4$ **b** $-3x - 12$ **c** $-4x + 8$ **d** $-25 + 5x$
   e $-a - 2$ **f** $-x + 3$ **g** $-5 + x$ **h** $-2x - 1$
   i $-12 + 3x$ **j** $-20x + 8$ **k** $-15 + 20c$ **l** $-14 + 10x$

7 **a** $-a^2 - a$ **b** $-b^2 - 4b$ **c** $-5c + c^2$
   d $-2x^2 - 4x$ **e** $-2x + 2x^2$ **f** $-3y^2 - 6y$
   g $-20a + 4a^2$ **h** $-18b + 12b^2$ **i** $-2xy^2 + x^2y$

8 **a** $3x + 11$ **b** $7x + 2$ **c** $25 - 12x$ **d** $10x - 2$
   e $11 - 10x$ **f** $8 + 2x$ **g** $9 + 21x$ **h** $4x + 18$
   i $4x + 7$ **j** $6 + 2x$ **k** $-5x - 5$ **l** $-47 + 15x$
   m $x^2 + 7x$ **n** $x^2 + 7x$ **o** $4x - x^2$ **p** $2x^2 - 2x$
   q $x^2 + 7x$ **r** $x^2 + 4x$

9 **a** $8x + 28$ **b** $8 + 4x + x^2$ **c** $6x + 18$
   d $5x - x^2 + 10$ **e** $7a + 7$ **f** $-4x - 2 + x^2$
   g $2a^2 - a + 15$ **h** $7x^2 + 7x$ **i** $-2y^2 - y - 5$
   j x **k** $4x - x^2$ **l** $-22 - n$

EXERCISE 6F

1 **a** $2x + 4 = 2(x + 2)$ **b** $3a - 12 = 3(a - 4)$
   c $15 - 5p = 5(3 - p)$ **d** $18x + 12 = 6(3x + 2)$
   e $4x^2 - 8x = 4x(x - 2)$ **f** $2m + 8m^2 = 2m(1 + 4m)$
   g $4x + 16 = 4(x + 4)$ **h** $10 + 5d = 5(2 + d)$
   i $5c - 5 = 5(c - 1)$ **j** $cd - de = d(c - e)$
   k $6a + 8ab = 2a(3 + 4b)$ **l** $6x - 2x^2 = 2x(3 - x)$

2 **a** $3(x + 3)$ **b** $4(x + 3)$ **c** $7(x + 5)$ **d** $4(2x + 1)$
   e $5(2x + 3)$ **f** $4(7b + 2)$ **g** $x(1 + a)$ **h** $a(b + c)$
   i $y(x + 1)$

3 **a** $4(x - 2)$ **b** $5(y - 2)$ **c** $3(b - 8)$ **d** $8(a - 4)$
   e $5(2n - 3)$ **f** $6(2a - 3)$ **g** $a(1 - b)$ **h** $x(y - z)$
   i $2b(3 - a)$

4 **a** $3(a + b)$ **b** $8(x - 2)$ **c** $3(p + 6)$ **d** $14(2 - x)$
   e $7(x - 2)$ **f** $6(2 + x)$ **g** $c(a + b)$ **h** $6(2y - a)$
   i $a(5 + b)$ **j** $c(b - 6d)$ **k** $x(7 - y)$ **l** $a(1 + b)$
   m $y(x - z)$ **n** $p(3q + r)$ **o** $c(d - 1)$

5 **a** $x(x + 2)$ **b** $3x(x + 4)$ **c** $2x(x - 4)$
   d $x(9 - x)$ **e** $9x(2 - x)$ **f** $2x(7 - 3x)$
   g $2x^2(x + 3)$ **h** $x^2(3x + 7)$ **i** $2ab(2b - 3a)$

6 **a** $2x(x^2 + 2x + 2)$ **b** $x^2(x^2 + 3x + 6)$
   c $ab(b + 1 + a)$

7 $(3x - x^2)$ still has the common factor x.

REVIEW SET 6A

1 **a** k^9 **b** b^{12} **c** p^8

2 **a** a^4b^4 **b** $27z^3$ **c** $8x^3y^3$

3 **a** $42g^6$ **b** $8ab^2$ **c** m^3

4 **a** 5^5 **b** 5^{14} **c** 5^0

5 **a** $\dfrac{c^2}{d^2}$ **b** $\dfrac{q^3}{64}$ **c** $\dfrac{a^2b^2}{64}$

6 **a** 1 **b** 13 **c** 6 **d** $\dfrac{2}{3}$

7 **a** $\dfrac{1}{9}$ **b** $\dfrac{1}{36}$ **c** $\dfrac{1}{y^3}$ **d** $\dfrac{1}{x^2y^2}$

8 **a** $7x+42$ **b** $-5x-10$ **c** $4y^2-7y$

9 **a** $pq+5p$ **b** $2a+6b+10$ **c** $3x^3+21x^2-15x$

10 **a** $11x-24$ **b** $5x+4$ **c** $9a+10$
    d $-y^2-2y-5$ **e** $3x^2+17x+36$ **f** $5p^2-10p$

11 **a** $4(x+5)$ **b** $6(x-6)$ **c** $x(7-y)$

12 **a** $a(a+7)$ **b** $6x(2x-3)$ **c** $5x^2(4x+3)$
    d $4(x+6y)$ **e** $2x(x-4)$ **f** $3a(1+2b+3a)$

13 **a** **i** a^2 cm^2 **ii** ab cm^2 **b** $a(2a+b)$ cm^2
    c total area of rug = area of 2 brown squares
$$+ \text{ area of white rectangle}$$
$$= a^2 + a^2 + ab$$
$$= 2a^2 + ab$$
    So, $2a^2 + ab = a(2a+b)$ {from **b**}

REVIEW SET 6B

1 **a** x^6 **b** c^5 **c** d^{33}

2 **a** $3^2 \times 3^3 = 3^5$ **b** $(x^4)^2 = x^8$ **c** $a^5 \div a^2 \times a^3 = a^6$

3 **a** $10a^3b$ **b** $5x$ **c** t

4 **a** x^6y^6 **b** $\dfrac{125a^3}{b^3}$ **c** $27x^6y^3$ **d** $100x^2y^4$
    e $\dfrac{m^3}{64n^3}$ **f** $4k^{38}$

5 **a** 1 **b** x^3 **c** b^2

6 **a** $\dfrac{1}{81}$ **b** $\dfrac{1}{16}$ **c** $\dfrac{1}{1\,000\,000}$ **d** $\dfrac{1}{x^3}$

7 **a** $\dfrac{1}{x^2y^2}$ **b** $\dfrac{2}{ab}$ **c** $\dfrac{3x}{y}$ **d** $\dfrac{1}{y}$

8 **a** **i** 36 **ii** 12 **b** **i** 81 **ii** 27
    c When a number is tripled then squared, the result is 3 times larger than when the number is squared then tripled.
    d $(3a)^2 = 9a^2 = 3(3a^2)$

9 **a** $3x-18$ **b** $8x+12y$ **c** $-3x^2+24x$

10 **a** $9x+27$ **b** $7z+18$ **c** $6x^2-19x$

11 **a** $6x-3$ **b** $-y^2-6y+4$ **c** $5x^2+23x+21$
    d $5x+10$ **e** $-2x^2+x-6$

12 **a** $3(x+9)$ **b** $8(x-6)$ **c** $a(b-d)$

13 **a** $x(x-10)$ **b** $a(a+8)$ **c** $5y(y+6)$
    d $8x(x-3)$ **e** $5z(2z-3)$ **f** $2x(4-11x)$

EXERCISE 7A

1 **a** $x=1$ **b** $x=9$ **c** $x=5$ **d** $x=5$
    e $x=-2$ **f** $x=4$ **g** $x=-4$ **h** $x=13$
    i $x=40$ **j** $x=-175$ **k** $x=-5$ **l** $x=-4$

2 **a** $x=2$ **b** $x=3$ **c** $x=-2$ **d** $x=5$
    e $k=-4$ **f** $m=1$

3 **a** C **b** F **c** E **d** A **e** B **f** D

4 **a** $z=0$ or 1 **b** $k=3$ **c** no real values
    d no real values **e** $q=4$ **f** $d=0$
    g all values of t **h** all values of g **i** $w=0$

5 **a** B **b** D **c** B **d** D **e** B **f** C
    g B **h** C **i** D **j** A **k** B **l** A

6 **a** identity **b** not an identity, $k=1$ **c** identity
    d identity **e** not an identity, $t=0$
    f not an identity, $p=0$ **g** identity **h** identity
    i not an identity, $x=0$

EXERCISE 7B

1 **a** $x=5$ **b** $x=9$ **c** $5x=15$ **d** $7x=x+6$

2 **a** $x=4$ **b** $2x=4$ **c** $3x=-3$ **d** $4x=3x+2$

3 **a** $x=16$ **b** $x-1=5$ **c** $3x=14$ **d** $3x-4=-40$

4 **a** $x=-10$ **b** $x=-9$ **c** $2-x=5$ **d** $2x-1=11$

5 **a** $2 \times 12 - 5 = 24 - 5 = 19$ ✓ **b** $2x=24$
    c $x=12$ **d** no

6 **a** $\dfrac{5+1}{2} = \dfrac{6}{2} = 3$ ✓ **b** $x+1=6$ **c** $x=5$ **d** no

EXERCISE 7C

1 **a** -3 **b** $+8$ **c** $\div 2$ **d** $\times 5$ **e** -12 **f** $\times 6$
    g $+5$ **h** $\div 9$ **i** $-\dfrac{2}{3}$ **j** $\times 13$ **k** $\div 15$ **l** $+\dfrac{4}{5}$

2 **a** x **b** x **c** x **d** x **e** x **f** p
    g q **h** $8r$ **i** s **j** x **k** x **l** $3y$

3 **a** $a=-10$ **b** $b=1\dfrac{1}{3}$ **c** $c=-9$ **d** $d=-3$
    e $e=10$ **f** $f=60$ **g** $z=-2$ **h** $w=-72$

4 **a** $x=6$ **b** $x=10$ **c** $x=7$ **d** $x=6$
    e $x=6$ **f** $x=-2$ **g** $x=-35$ **h** $x=-11$
    i $x=-32$ **j** $x=-3$ **k** $x=-7$ **l** $x=-9$
    m $x=-9$ **n** $x=4$ **o** $x=9$ **p** $x=-18$
    q $x=-4$ **r** $x=-27$ **s** $x=-4$ **t** $x=-90$
    u $x=36$ **v** $x=0$ **w** $x=44$ **x** $x=14$

5 $x=3$

EXERCISE 7D

1 **a** BU: $x \xrightarrow{\times 7} 7x \xrightarrow{+3} 7x+3$
    UD: $7x+3 \xrightarrow{-3} 7x \xrightarrow{\div 7} x$

  b BU: $x \xrightarrow{+3} x+3 \xrightarrow{\times 7} 7(x+3)$
    UD: $7(x+3) \xrightarrow{\div 7} x+3 \xrightarrow{-3} x$

  c BU: $x \xrightarrow{-2} x-2 \xrightarrow{\times 5} 5(x-2)$
    UD: $5(x-2) \xrightarrow{\div 5} x-2 \xrightarrow{+2} x$

  d BU: $x \xrightarrow{\times 5} 5x \xrightarrow{-2} 5x-2$
    UD: $5x-2 \xrightarrow{+2} 5x \xrightarrow{\div 5} x$

  e BU: $x \xrightarrow{\div 3} \dfrac{x}{3} \xrightarrow{+1} \dfrac{x}{3}+1$
    UD: $\dfrac{x}{3}+1 \xrightarrow{-1} \dfrac{x}{3} \xrightarrow{\times 3} x$

f BU: $x \xrightarrow{+1} x+1 \xrightarrow{\div 3} \dfrac{x+1}{3}$

UD: $\dfrac{x+1}{3} \xrightarrow{\times 3} x+1 \xrightarrow{-1} x$

g BU: $x \xrightarrow{\div 8} \dfrac{x}{8} \xrightarrow{-5} \dfrac{x}{8}-5$

UD: $\dfrac{x}{8}-5 \xrightarrow{+5} \dfrac{x}{8} \xrightarrow{\times 8} x$

h BU: $x \xrightarrow{-5} x-5 \xrightarrow{\div 8} \dfrac{x-5}{8}$

UD: $\dfrac{x-5}{8} \xrightarrow{\times 8} x-5 \xrightarrow{+5} x$

2 a BU: $x \xrightarrow{\times 2} 2x \xrightarrow{-6} 2x-6$

UD: $2x-6 \xrightarrow{+6} 2x \xrightarrow{\div 2} x$

b BU: $x \xrightarrow{\div -3} \dfrac{x}{-3} \xrightarrow{+10} \dfrac{x}{-3}+10$

UD: $\dfrac{x}{-3}+10 \xrightarrow{-10} \dfrac{x}{-3} \xrightarrow{\times -3} x$

c BU: $x \xrightarrow{-7} x-7 \xrightarrow{\times 8} 8(x-7)$

UD: $8(x-7) \xrightarrow{\div 8} x-7 \xrightarrow{+7} x$

d BU: $x \xrightarrow{-3} x-3 \xrightarrow{\div 4} \dfrac{x-3}{4}$

UD: $\dfrac{x-3}{4} \xrightarrow{\times 4} x-3 \xrightarrow{+3} x$

3 a BU: $x \xrightarrow{\times 3} 3x \xrightarrow{+2} 3x+2 \xrightarrow{\div 5} \dfrac{3x+2}{5}$

UD: $\dfrac{3x+2}{5} \xrightarrow{\times 5} 3x+2 \xrightarrow{-2} 3x \xrightarrow{\div 3} x$

b BU: $x \xrightarrow{\times 3} 3x \xrightarrow{\div 5} \dfrac{3x}{5} \xrightarrow{+2} \dfrac{3x}{5}+2$

UD: $\dfrac{3x}{5}+2 \xrightarrow{-2} \dfrac{3x}{5} \xrightarrow{\times 5} 3x \xrightarrow{\div 3} x$

c BU: $x \xrightarrow{+2} x+2 \xrightarrow{\times 3} 3(x+2) \xrightarrow{\div 5} \dfrac{3(x+2)}{5}$

UD: $\dfrac{3(x+2)}{5} \xrightarrow{\times 5} 3(x+2) \xrightarrow{\div 3} x+2 \xrightarrow{-2} x$

d BU: $x \xrightarrow{\times 7} 7x \xrightarrow{-1} 7x-1 \xrightarrow{\div 6} \dfrac{7x-1}{6}$

UD: $\dfrac{7x-1}{6} \xrightarrow{\times 6} 7x-1 \xrightarrow{+1} 7x \xrightarrow{\div 7} x$

e BU: $x \xrightarrow{\times 7} 7x \xrightarrow{\div 6} \dfrac{7x}{6} \xrightarrow{-1} \dfrac{7x}{6}-1$

UD: $\dfrac{7x}{6}-1 \xrightarrow{+1} \dfrac{7x}{6} \xrightarrow{\times 6} 7x \xrightarrow{\div 7} x$

f BU: $x \xrightarrow{-1} x-1 \xrightarrow{\times 7} 7(x-1) \xrightarrow{\div 6} \dfrac{7(x-1)}{6}$

UD: $\dfrac{7(x-1)}{6} \xrightarrow{\times 6} 7(x-1) \xrightarrow{\div 7} x-1 \xrightarrow{+1} x$

g BU: $x \xrightarrow{\times 5} 5x \xrightarrow{\div 6} \dfrac{5x}{6} \xrightarrow{-3} \dfrac{5x}{6}-3$

UD: $\dfrac{5x}{6}-3 \xrightarrow{+3} \dfrac{5x}{6} \xrightarrow{\times 6} 5x \xrightarrow{\div 5} x$

h BU: $x \xrightarrow{-3} x-3 \xrightarrow{\times 5} 5(x-3) \xrightarrow{\div 6} \dfrac{5(x-3)}{6}$

UD: $\dfrac{5(x-3)}{6} \xrightarrow{\times 6} 5(x-3) \xrightarrow{\div 5} x-3 \xrightarrow{+3} x$

i BU: $x \xrightarrow{\times 5} 5x \xrightarrow{-3} 5x-3 \xrightarrow{\div 6} \dfrac{5x-3}{6}$

UD: $\dfrac{5x-3}{6} \xrightarrow{\times 6} 5x-3 \xrightarrow{+3} 5x \xrightarrow{\div 5} x$

j BU: $x \xrightarrow{\times -2} -2x \xrightarrow{\div 3} \dfrac{-2x}{3} \xrightarrow{+1} 1-\dfrac{2x}{3}$

UD: $1-\dfrac{2x}{3} \xrightarrow{-1} \dfrac{-2x}{3} \xrightarrow{\times 3} -2x \xrightarrow{\div -2} x$

k BU: $x \xrightarrow{\times -2} -2x \xrightarrow{+1} 1-2x \xrightarrow{\div 3} \dfrac{1-2x}{3}$

UD: $\dfrac{1-2x}{3} \xrightarrow{\times 3} 1-2x \xrightarrow{-1} -2x \xrightarrow{\div -2} x$

l BU:

$x \xrightarrow{\times -1} -x \xrightarrow{+1} 1-x \xrightarrow{\times 2} 2(1-x) \xrightarrow{\div 3} \dfrac{2(1-x)}{3}$

UD:

$\dfrac{2(1-x)}{3} \xrightarrow{\times 3} 2(1-x) \xrightarrow{\div 2} 1-x \xrightarrow{-1} -x \xrightarrow{\div -1} x$

EXERCISE 7E

1
a $x=2$ **b** $x=5$ **c** $x=4$ **d** $x=-8$
e $x=4$ **f** $x=\frac{3}{8}$ **g** $x=1\frac{1}{2}$ **h** $x=17$
i $x=-2$ **j** $x=-2$ **k** $x=2$ **l** $x=-4\frac{1}{4}$

2
a $x=-3$ **b** $x=14$ **c** $x=-3$ **d** $x=4$
e $x=-2$ **f** $x=4$ **g** $x=5$ **h** $x=\frac{5}{8}$
i $x=7$ **j** $x=15$ **k** $x=7\frac{1}{4}$ **l** $x=4\frac{1}{5}$

3
a $x=4$ **b** $x=22$ **c** $x=16$ **d** $x=9$
e $x=-21$ **f** $x=-24$ **g** $x=63$ **h** $x=-56$
i $x=22$

4
a $x=9$ **b** $x=-6$ **c** $x=-12$ **d** $x=-4$
e $x=20$ **f** $x=24$

5 **a** $x = 11$ **b** $x = 9$ **c** $x = 3\frac{1}{2}$ **d** $x = -7$
 e $x = -4$ **f** $x = -28$ **g** $x = -5$ **h** $x = 2$
 i $x = -11$

6 **a** $x = -14$ **b** $x = 5$ **c** $x = 2$ **d** $x = 4$
 e $x = 0$ **f** $x = 33$

7 **a** $x = 10$ **b** $x = 2$ **c** $x = 4\frac{1}{2}$ **d** $x = -\frac{2}{3}$
 e $x = -3$ **f** $x = 0$ **g** $x = 1\frac{1}{3}$ **h** $x = \frac{1}{2}$
 i $x = 0$ **j** $x = -1$ **k** $x = 7\frac{1}{2}$ **l** $x = \frac{1}{6}$

8 **a** $a = 3$ **b** $x = 448$ **c** $x = 5$ **d** $x = 10$
 e $x = 1$ **f** $n = 4$ **g** $a = -8$ **h** $x = 3$
 i $x = 2\frac{1}{2}$ **j** $x = 75$ **k** $x = 3$ **l** $n = -3$
 m $k = 7$ **n** $z = -\frac{4}{5}$ **o** $x = 5$

9 **a** $x = 1$ **b** $x = -7$

EXERCISE 7F

1 **a** $x = 2$ **b** $x = -5$ **c** $x = 15$ **d** $y = 0$
 e $x = 3$ **f** $m = 2\frac{1}{2}$ **g** $n = 4$ **h** $x = 2\frac{8}{9}$
 i $b = 4$ **j** $p = 15$ **k** $t = 6$ **l** $d = -10$
 m $x = -6$ **n** $x = 12$

2 **a** $x = 6$ **b** $x = 2$ **c** $x = -4$ **d** $x = 3$
 e $x = 4$ **f** $x = -2\frac{1}{2}$ **g** $x = 1$ **h** $x = 1$
 i $x = \frac{1}{10}$ **j** $x = 4$ **k** $x = 1$ **l** $x = -1\frac{1}{3}$

3 **a** $x = 3$ **b** $t = 5$ **c** $x = 1$ **d** $y = \frac{1}{2}$
 e $a = 3$ **f** $p = -3$

4 **a** $x = 3$ **b** $x = 6$ **c** $x = 5$ **d** $x = 2$
 e $x = -1$ **f** $x = -2\frac{4}{5}$

5 **a** $7(a + 3) = 21 + 7a$ reduces to $0 = 0$.
 b The equation is true for all real values of a. So, infinitely many values of a satisfy this equation.
 c It is an identity.

6 **a** We get $3 = 4$ which is absurd. So, this equation has no solution.
 b no values of a

7 **a** $x = \frac{1}{3}$ **b** $x = 2\frac{2}{9}$ **c** $x = 3$ **d** $x = -\frac{1}{4}$
 e $x = -8$ **f** $x = -2$ **g** $x = 1\frac{3}{5}$ **h** $x = 3$
 i $x = -1$ **j** $x = -13$ **k** $x = -2$ **l** $x = 3$

8 **a** $x = 5$ **b** $x = -1$ **c** $x = 1$ **d** $x = 3$
 e $x = -2$ **f** $x = 0$ **g** $x = 2$ **h** $x = -2$
 i $x = 1$ **j** $x = -3$

EXERCISE 7G

1 **a** $x = \pm 3$ **b** $x = \pm 7$ **c** $x = \pm 6$
 d $x = 0$ **e** $x = \pm 1$ **f** $x = \pm\sqrt{17}$
 g $x = \pm\sqrt{23}$ **h** $x = \pm 10$ **i** no real solutions
 j no real solutions   **k** $x = \pm\sqrt{27}$ **l** no real solutions

2 **a** $x = 2$ **b** $x = 3$ **c** $x = 0$
 d $x = 4$ **e** $x = -1$ **f** $x = -5$
 g $x = \frac{1}{2}$ **h** $x = \sqrt[3]{36}$ (≈ 3.30)

3 **a** $x = \pm 2$ **b** $x = \pm 3$ **c** $x = \pm 5$
 d $x = \pm 4$ **e** $x = \pm 6$ **f** $x = \pm 5$
 g $x = \pm\sqrt{50}$ **h** $x = \pm\sqrt{44}$ **i** $x = \pm\sqrt{12}$

4 **a** $x = 1$ **b** $x = -2$ **c** $x = 6$
 d $x = 100$ **e** $x = \sqrt[3]{16}$ (≈ 2.52)
 f $x = \sqrt[3]{-800}$ (≈ -9.28)

5 **a** $x = \pm 3$ **b** $x = \pm 4$ **c** $x = \pm 2$
 d $x = \pm 4$ **e** $x = \pm\sqrt{30}$ **f** $x = \pm\sqrt{20}$
 g $x = \pm 3$ **h** $x = \pm\sqrt{7}$ **i** $x = \pm\sqrt{\frac{1}{5}}$

REVIEW SET 7A

1 **a** D   **b** A   **c** B   **d** C
2 $x = -6$ **3** **a** $3x = -9$ **b** $4x = -10$
4 **a** $x = 15$ **b** $x = 7$ **c** $a = -13$ **d** $t = -21$

5 **a**

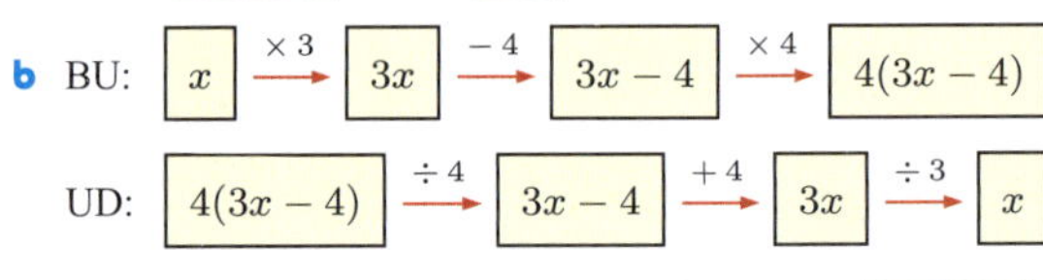

6 **a** $x = 2$ **b** $x = 6$ **c** $x = -2$
7 **a** $x = 7$ **b** $x = 30$ **c** $x = -2\frac{1}{5}$
8 **a** $x = 9$ **b** $x = -4\frac{2}{3}$ **c** $x = 10$
9 **a** $a = 5$ **b** $b = -5$ **c** $c = \frac{3}{5}$
10 **a** $x = -9$ **b** $x = 1$ **c** $x = \frac{6}{7}$
11 **a** $x = 10$ **b** $x = 10$
 c They both give the answer $x = 10$.
   Both methods are correct.
12 **a** $x = \pm 8$ **b** $x = \sqrt[3]{-10}$ (≈ -2.15) **c** $x = \pm\sqrt{35}$

REVIEW SET 7B

1 **a** not an identity, $x = 0$ **b** identity
 c not an identity, $t = 2$
2 $x = -3$ **3** $2x + 3 = -2$
4 **a** dividing by 5   **b** adding 7
5 **a** $a = 7$ **b** $b = -9$ **c** $c = -10$ **d** $d = -96$

6 BU: $x \xrightarrow{\times -3} -3x \xrightarrow{+1} 1 - 3x \xrightarrow{\div 8} \dfrac{1 - 3x}{8}$

 UD: $\dfrac{1 - 3x}{8} \xrightarrow{\times 8} 1 - 3x \xrightarrow{-1} -3x \xrightarrow{\div -3} x$

7 **a** $x = 4$ **b** $x = -48$ **c** $x = -1$
8 **a** $x = -1$ **b** $x = -2$ **c** $x = 20$
9 **a** $x = 7$ **b** $x = -\frac{3}{4}$ **c** $x = 4\frac{1}{4}$
10 **a** $x = 1\frac{2}{3}$ **b** $x = \frac{2}{3}$
11 **a** **i** $\frac{x}{3} = 5$ **ii** $x = 15$
 b **i** $3\left(\frac{x}{3} + 2\right) = 21$ **ii** $x + 6 = 21$ **iii** $x = 15$
12 **a** no real solutions   **b** $x = \sqrt[3]{50}$ (≈ 3.68) **c** $x = \pm\sqrt{5}$

EXERCISE 8A

1 **a** true **b** true **c** false **d** false **e** false **f** true
2 **a** $\hat{MON}$, obtuse **b** $\hat{CAB}$, acute **c** $\hat{ABC}$, right
3 **a** $62°$ **b** $13°$ **c** $(90 - y)°$ **d** $x°$
4 **a** $95°$ **b** $53°$ **c** $(180 - q)°$ **d** $n°$ **e** $(90 - x)°$
5 **a** $\hat{AOC}$ and $\hat{BOC}$ are supplementary angles.
   b

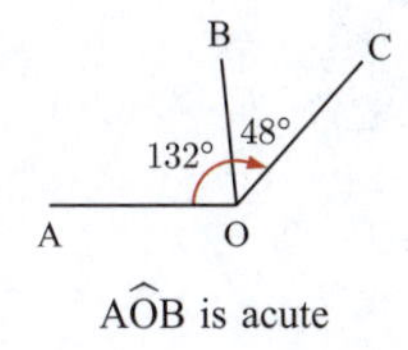

EXERCISE 8B

1 **a** (PQ) ∥ (SR) **b** (SQ) ⊥ (PR) **c** (PX) ⊥ (XR)
   d (XR) ∥ (PS)
2 P, Q, and R are collinear.

EXERCISE 8C

1 **a** $a = 61$ {angles in a right angle}
   b $b = 45$ {angles on a straight line}
   c $x = 100$ {vertically opposite angles}
   d $c = 318$ {angles at a point}
   e $x = 60$ {angles at a point}
   f $n = 65$ {angles at a point}
   g $a = 135$ {angles at a point}
   h $x = 35$ {angles on a straight line}
   i $x = 48$ {vertically opposite angles}
   j $x = 17$ {angles on a straight line}
   k $x = 38$ {angles on a straight line}
   l $x = 78$ {angles at a point}
2 **a** $x + y = 90$ **b** **i** $x = 90$ **ii** $x = 0$
   c $x = 14$ **d** $y = 35$

EXERCISE 8D

1 **a** $x = 115$ {equal corresponding angles}
   b $f = 41$ {supplementary co-interior angles}
   c $h = 58$ {equal alternate angles}
   d $j = 153$ {supplementary co-interior angles}
   e $t = 155$ {equal alternate angles}
   f $d = 50$ {supplementary co-interior angles}
   g $x = 35$ {equal alternate angles}
   h $x = 52$ {equal corresponding angles}
   i $c = 60$ {angles on a straight line, $(180 - c)°$ and $2c°$
        are equal corresponding angles}
2 **a** no {co-interior angles are not supplementary}
   b yes {corresponding angles are equal}
   c yes {alternate angles are equal}

REVIEW SET 8A

1 **a** $a = 142$ {vertically opposite angles}
   b $b = 130$ {angles on a straight line}
   c $c = 45$ {angles in a right angle}
2 **a** $27°$ **b** $154°$
3 **a** (AB) ∥ (DC) **b** (BE) ⊥ (AD)
4 **a** $a = 120$ {equal corresponding angles}
   $b = 60$ {angles on a straight line}

b $x = 100$ {vertically opposite angles}
   $y = 80$ {supplementary co-interior angles}
c $p = 70$ {equal alternate angles}
   $q = 70$ {equal corresponding angles}
5 **a** yes {corresponding angles are equal}
   b yes {co-interior angles are supplementary}
   c no {alternate angles are not equal}
6 **a** $\hat{AOB}$ and $\hat{COD}$ are supplementary.
   b

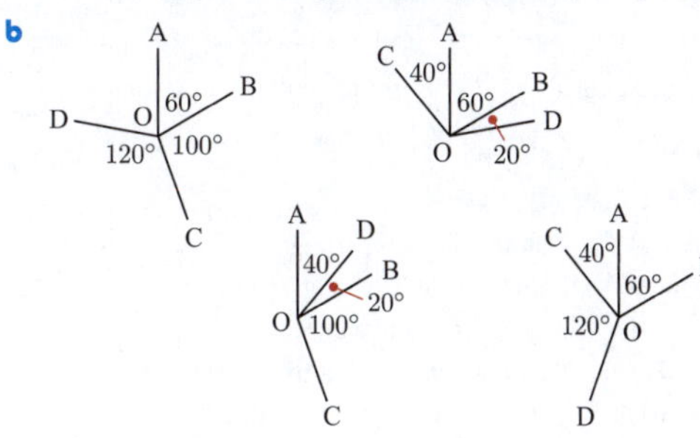

   c $\hat{BOD} = 140°$

REVIEW SET 8B

1 **a** false **b** true **c** false
2 **a** $\hat{AOB}$, straight **b** $\hat{RQP}$, right **c** reflex $\hat{XYZ}$, reflex
3 **a** co-interior angles **b** [AB] and [CD] are parallel.
4 **a** $x = 55$ {vertically opposite angles}
   b $x = 100$ {angles at a point}
5 **a** $x = 50$ {equal alternate angles}
   b $m = 75$ {supplementary co-interior angles}
   c $k = 45$ {vertically opposite angles, $(k + 55)°$ and
                $(2k + 10)°$ are equal corresponding angles}
6 **a** [AB] ⊥ [AD] and [AD] ⊥ [DC]
   b Yes, co-interior angles $\hat{BAD}$ and $\hat{ADC}$ are supplementary.
   c $\hat{ABC}$ and $\hat{BCD}$ are also supplementary co-interior angles.
   $\therefore \ \hat{BAD} + \hat{ADC} + \hat{ABC} + \hat{BCD} = 180° + 180°$
   $= 360°$

EXERCISE 9A

1 **a** E **b** A **c** H **d** F **e** C **f** D
   g G **h** B
2 a diameter
3 **a** A diameter is made up of two radii, one on each side of the
      centre.
   b **i** 8 cm **ii** 6 cm
4 **a** 3 cm **b** 5 cm **c** 6 cm **d** 10 cm **e** 2 cm
   f 8 cm
5 **a, b**

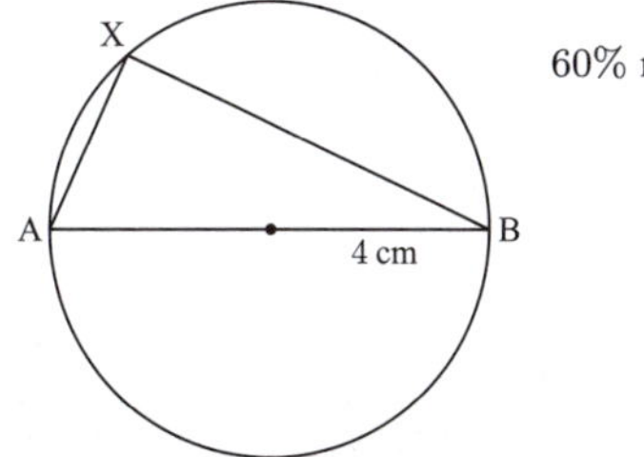

   c $\hat{AXB}$ is a right angle.
   e The angle in a semi-circle is a right angle.

6 **a, b**

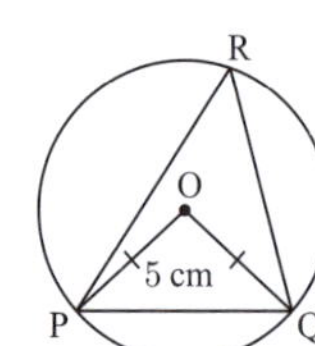

80% reduction

c, e $\widehat{POQ}$ is double the size of $\widehat{PRQ}$.

EXERCISE 9B

1 **a** isosceles **b** scalene **c** equilateral

2 **a** acute angled **b** right angled **c** obtuse angled

3 **a** [AB] **b** [BC] **4** **a** $\widehat{PRQ}$ **b** $\widehat{PQR}$

EXERCISE 9C

1 **a** $a = 130$ {angle sum of a triangle}
    b $b = 90$ {angle sum of a triangle}
    c $c = 43$ {angle sum of a triangle}
    d $d = 129$ {exterior angle of a triangle}
    e $e = 53$ {exterior angle of a triangle}
    f $x = 131$ {exterior angle of a triangle}

2 $50°$ **3** **a** true **b** false **c** false **d** false

4 **a** [AB] **b** [BC] **c** [AC]

5 **a** [AC] **b** [AC] **c** [BC]

6 **a** $a = 55$ **b** $b = 55$ **c** $x = 20$ **d** $a = 55,\ b = 77$
    e $x = 45,\ y = 70$ **f** $a = 45,\ b = 85,\ c = 130,\ d = 50$

EXERCISE 9D

1 **a** $x = 69$ **b** $x = 24$ **c** $x = 45$ **d** $x = 22.5$
    e $x = 103$ **f** $x = 33$

2 **a** $x = 12$ **b** $x = 8$ **c** $x = 90$ **d** $x = 50$
    e $x = 5$ **f** $x = 90$ **g** $x = 140$ **h** $x = 4$
    i $x = 50$

3 $67°$ **4** **a** $x = 66$ **b** isosceles, $BC = AC$

5 **a** $\widehat{ABD} = 48°$ **b** isosceles, $BD = AD$ **c** 3 cm

6 base angles = $58°$, vertical angle = $64°$

7 $x = 56,\ 59,$ or 62

8 **a** $\triangle OPR$ is isosceles {$OP = OR =$ radius}
    $\therefore\ \ \widehat{PRO} = \alpha°$ {base angles of isosceles triangle}
    $\triangle OQR$ is isosceles {$OQ = OR =$ radius}
    $\therefore\ \ \widehat{QRO} = \beta°$ {base angles of isosceles triangle}
    b $2\alpha + 2a = 180$ {angle sum of $\triangle PQR$}
    c From **b**, $2(\alpha + \beta) = 180$
        $\therefore\ \ \alpha + \beta = 90$
    $\therefore\ \ \widehat{PRQ} = (\alpha + \beta)° = 90°$

9 Label a triangle ABC so that $\widehat{BAC}$ is the largest angle.
Suppose we construct an isosceles triangle BAC' with vertical angle $\widehat{ABC}$ and base $[AC']$.

The base angles are the average of $\widehat{BAC}$ and $\widehat{BCA}$. They are **less** than $\widehat{BAC}$ since $\widehat{BAC}$ is the **largest** angle.
So, C' must lie on $[BC]$.
But $AB = BC'$ {**isosceles** triangle}
    $\therefore\ \ AB < BC$
Using the same procedure, we can construct an isosceles triangle $B'AC$ with vertical angle $\widehat{ACB}$ and base $[AB']$.

Following the same argument, $\mathbf{AC} < BC$.
So, BC is the longest side, and it is opposite the **largest angle**.

EXERCISE 9E

1 **a**

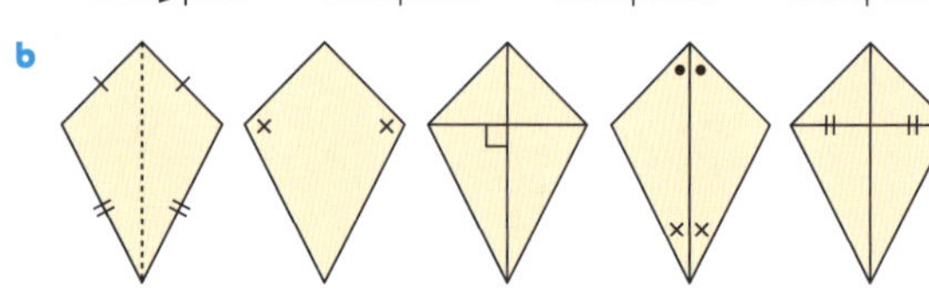

    b

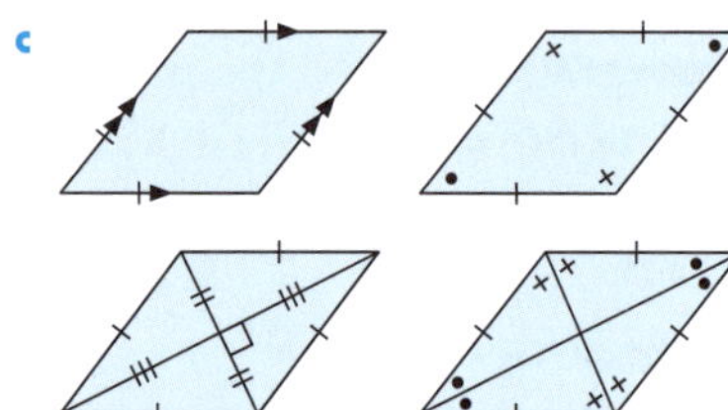

    c

2 **a** $x = 90,\ y = 5$ **b** $x = 50$ **c** $x = 20$
    d $x = 60,\ y = 90$ **e** $a = 6,\ b = 8$ **f** $x = 14$

3 **a** true **b** true **c** false **d** true

4 **a** a rhombus {diagonals bisect each other at right angles, and diagonals are not equal in length}
    b

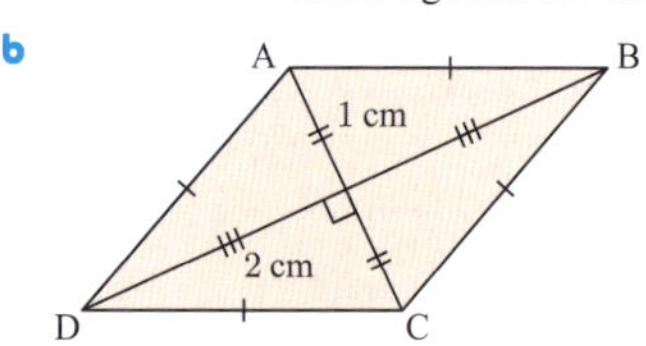

5 **a** $a = 50,\ b = 120$ **b** $a = 42,\ b = 96$
    c $a = 40,\ x = 2$

6 **a** $x = 90$ {opposite angles of a parallelogram are equal}
    The figure is a rectangle.
    b $x = 45$ {angle sum of a triangle}
    The figure is a square.

EXERCISE 9F

1 **a** $x = 65$ **b** $x = 93$ **c** $x = 35$ **d** $x = 40$
    e $x = 50$ **f** $x = 90$

2 **a** $a = 65,\ b = 91$ **b** $m = 60,\ n = 65$
    c $a = 70$ **d** $x = 41,\ y = 59$

EXERCISE 9G

1 **a** $540°$ **b** $900°$ **c** $720°$ **d** $1800°$ **e** $2340°$

2 **a** $x = 108$ **b** $x = 150$ **c** $x = 60$ **d** $x = 120$
    e $x = 125$ **f** $x = 135$

3 $135°$ **4** 13 sides

5 **a**

Regular polygon	Number of sides	Sum of angles	Size of each angle
triangle	3	$180°$	$60°$
quadrilateral	4	$360°$	$90°$
pentagon	5	$540°$	$108°$
hexagon	6	$720°$	$120°$
octagon	8	$1080°$	$135°$
decagon	10	$1440°$	$144°$

b i The sum of the angles of an n-sided polygon is $(n-2) \times 180°$.
  ii The size of each angle of a regular n-sided polygon is $\dfrac{(n-2) \times 180°}{n}$.
  c $150°$

6 a 3 reflex angles   **b**

7 $1260°$

8 a $1080°$ {angle sum of an octagon}
  b $\widehat{A} = 80°$, $\widehat{B} = 100°$, $\widehat{C} = 260°$, $\widehat{D} = 100°$, $\widehat{E} = 80°$, $\widehat{F} = 100°$, $\widehat{G} = 260°$, $\widehat{H} = 100°$ with sum $1080°$ ✓

REVIEW SET 9A

1 a, c   **b** 44 mm

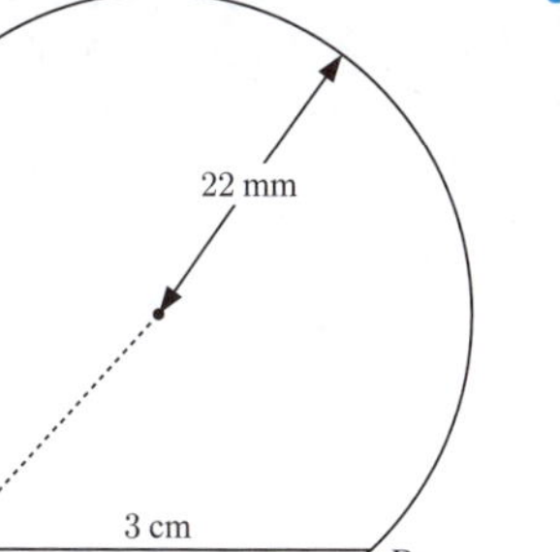

2 a $z = 55$ {angle sum of a triangle}
  b $x = 70$ {equal corresponding angles, angle sum of a triangle}
  c $a = 55$ {vertically opposite angles}
    $b = 80$ {angles on a straight line}
    $c = 45$ {angle sum of a triangle}

3 a [XZ]   **b** [XY]

4 a a square   **b** a rhombus   **c** a rectangle

5 a scalene, right angled   **b** isosceles, obtuse angled
  c equilateral, acute angled

6 a $x = 90$, $y = 42$ {isosceles triangle theorem}
  b $y = 70$ {equal corresponding angles}
    $x = 40$ {angle sum of a triangle}
  c $x = 4$ {isosceles triangle theorem}

7 a $x = 90$, $y = 60$ **b** $x = 55$ **c** $a = 70$

8 $1620°$ **9 a** $x = 70$ **b** $x = 130$ **c** $x = 60$

10 a i $70°$ **ii** $70°$ **iii** $40°$
  b It is isosceles, as it has two equal sides.   **c** $70°$
  d [BC] ∥ [AE]   **e** a trapezium

11 a $\triangle$ABC is right angled isosceles, so $\widehat{BAC} = 45°$.
  b $\widehat{GAF} = 45°$ {diagonals of a square bisect its angles}
  c Diagonals of a square bisect each other at right angles.
  d $\widehat{IGA} = 135°$ {$\triangle$GJI is right angled isosceles, exterior angle of a triangle}

REVIEW SET 9B

1 8 cm   **2 a** isosceles   **b** obtuse angled

3

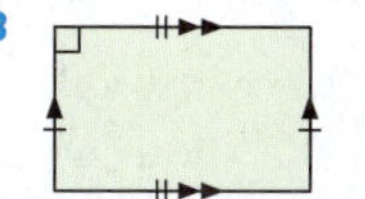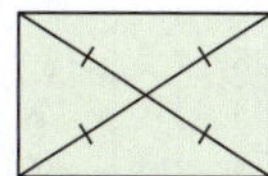

4 a $x = 138$ **b** $x = 20$ **c** $x = 82$, $y = 16$

5 a true   **b** true   **6** $49°$

7 a $a = 90$ {diagonals of a kite cut each other at right angles}
    $b = 50$ {angle sum of a triangle}
    $c = 70$ {one diagonal of a kite bisects one pair of angles at the vertices, angle sum of a triangle}
    $d = 120$ {one pair of opposite angles of a kite are equal}
  b $x = 45$ {angles on a straight line, angle sum of a triangle}
    $y = 4$ {equal sides of isosceles triangles}

8 a $x = 90$ {co-interior angles are supplementary}
    The figure is a rectangle.
  b Diagonals bisect each other at right angles.
    {isosceles triangle theorem}
    The figure is a rhombus.
  c One pair of sides are parallel.   {equal alternate angles}
    The figure is a trapezium.

9 $135°$

10 a 2 reflex angles   **b**

11 a right angled isosceles   **b** right angled isosceles
  c Yes, $\widehat{AXF} = 90°$
    ∴ [AE] ∥ [BD]   {equal corresponding angles}
  d No, FX $= 10.5$ cm, so AB $=$ ED $= 19.2$ cm but BD $= 21$ cm.

12 a, b

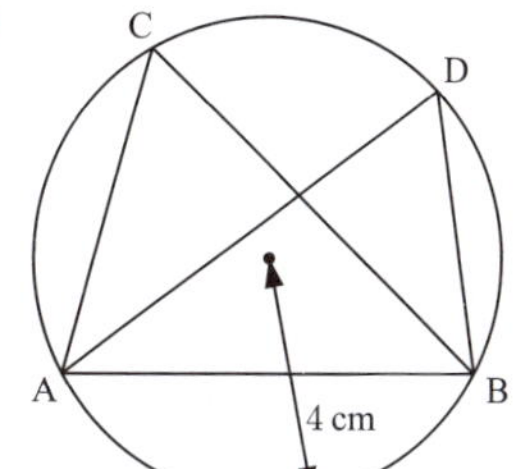

60% reduction

c, e $\widehat{ACB}$ and $\widehat{ADB}$ are equal.

EXERCISE 10A

1 a

Input	Calculation	Output
1	$1 + 6$	7
2	$2 + 6$	8
3	$3 + 6$	9
4	$4 + 6$	10

b

Input	Calculation	Output
2	3×2	6
3	3×3	9
6	3×6	18
9	3×9	27

c

Input	Calculation	Output
1	$1 \times 4 - 3$	1
4	$4 \times 4 - 3$	13
9	$9 \times 4 - 3$	33
12	$12 \times 4 - 3$	45

d

Input	Calculation	Output
0	$(0 + 5) \times 2$	10
3	$(3 + 5) \times 2$	16
7	$(7 + 5) \times 2$	24
10	$(10 + 5) \times 2$	30

e

Input	Calculation	Output
2	2×2	4
5	5×5	25
8	8×8	64
10	10×10	100

f

Input	Calculation	Output
4	$4 \div 2$	2
8	$8 \div 2$	4
12	$12 \div 2$	6
20	$20 \div 2$	10

g

Input	Calculation	Output
5	$(5-4) \times 10$	10
7	$(7-4) \times 10$	30
9	$(9-4) \times 10$	50
11	$(11-4) \times 10$	70

h

Input	Calculation	Output
7	$(7+8) \div 3$	5
16	$(16+8) \div 3$	8
22	$(22+8) \div 3$	10
28	$(28+8) \div 3$	12

2 **a** $M = 5n$ **b** $M = 3n - 4$ **c** $M = \dfrac{1+n}{2}$
d $M = n^2 + 7$

EXERCISE 10B

1 **a** $M = n + 3$ **b** $M = 5n$ **c** $M = 2n + 3$
d $M = 4n + 2$ **e** $M = 13 - n$ **f** $M = 3n - 1$
g $M = 20 - 5n$ **h** $M = 6n - 1$

2 **a** $M = 2n + 4$ **b** $y = 11x$ **c** $P = 7a - 4$
d $N = 27 - 4m$ **e** $C = 3d - 2$ **f** $L = 9h - 1$
g $D = 4t + 1$ **h** $y = 34 - 3x$ **i** $y = 17 - \frac{3}{2}x$
j $P = 6n - 19$

EXERCISE 10C

1 **a** $P = 19$ **b** $t = 9$
2 **a** **i** $C = 8$ **ii** $C = -12$ **iii** $C = 38$
b **i** $d = 3$ **ii** $d = \frac{28}{5}$ or $5\frac{3}{5}$
3 **a** **i** $A = 1$ **ii** $A = 5$ **iii** $A = 14$
b **i** $b = 28$ **ii** $b = 4$ **iii** $b = -24$
4 **a** \$224 **b** \$320 **c** \$620
5 **a** $B = 400$. There are 400 bread rolls in the bakery at the start of the day.
b **i** 300 rolls **ii** 150 rolls **iii** 375 rolls **c** 8 hours
6 **a** **i** \$4.20 **ii** \$13.20 **b** 23 minutes
7 **a** $M = 60.4$ kg **b** $x = 55.7$ kg
8 **a** **i** $M = 1$ **ii** $M = 7$ **iii** $M = 31$
b **i** $n = 0$ **ii** $n = \pm 3$ **iii** $n = \pm 6$
9 **a** D **b** E **c** B **d** A **e** C
10 **a** **i** $P = 14$ **ii** $P = 1$ **b** **i** $x = 7$ **ii** $x = 8$
c **i** $y = 6$ **ii** $y = 9$
11 **a** 56 people **b** $a = 7$ **c** $b = 13$
12 **a** 40 000 newtons **b** 300 Pa
13 **a** ≈ 1.33 g/cm^3 **b** 24 969.6 g or 24.9696 kg
14 **a** n must be a positive integer.
b **i** $L = 27$ **ii** $L = 75$ **iii** $L = 192$
c **i** $n = 6$ **ii** $n = 7$ **iii** $n = 11$
15 **a** **i** 2 regions

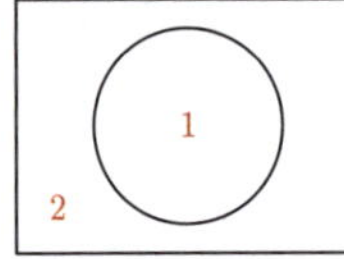

ii 8 regions

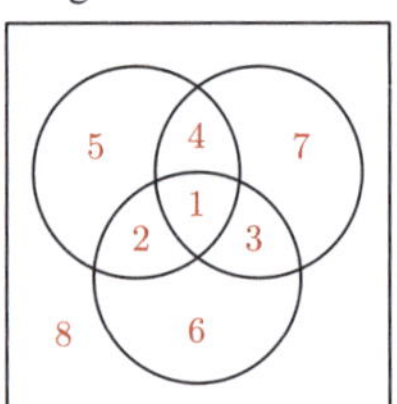

iii 14 regions

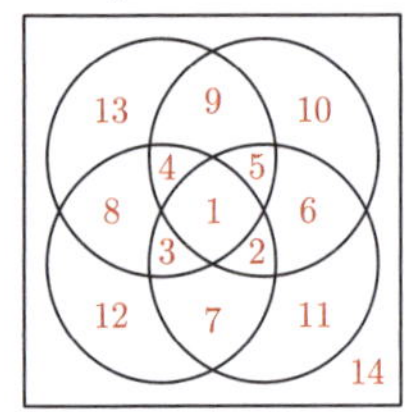

b 92 regions **c** 13 circles

EXERCISE 10D

1 **a**

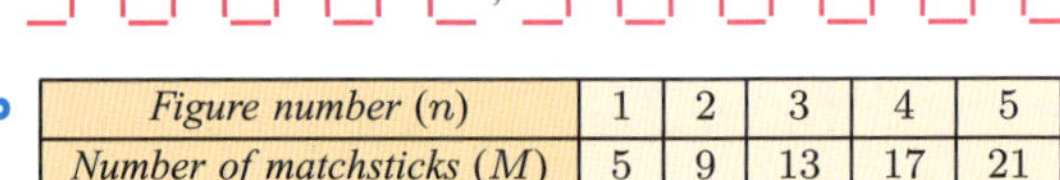

b

Figure number (n)	1	2	3	4	5
Number of matchsticks (M)	5	9	13	17	21

c $M = 4n + 1$ **d** 69 matchsticks

2 **a**

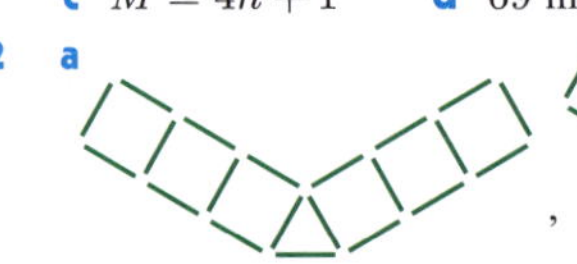

b

Figure number (n)	1	2	3	4	5
Number of matchsticks (M)	3	9	15	21	27

c $M = 6n - 3$
d **i** 57 matchsticks **ii** 147 matchsticks

3 **a**

Number of rungs (n)	1	2	3	4	5
Number of metal pieces (M)	5	8	11	14	17

b $M = 3n + 2$ **c** 38 metal pieces
d A ladder with 16 rungs.

4 **a**

Pigeonholes (p)	1	2	3	4	5
Cardboard pieces (C)	4	7	10	13	16

b $C = 3p + 1$ **c** 82 pieces of cardboard
d 19 complete pigeonholes

5 **a**

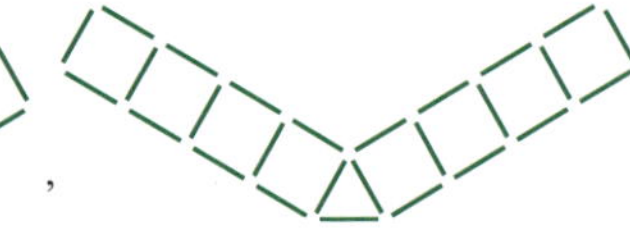

b

Figure number (n)	1	2	3	4	5
Number of dots (D)	5	8	13	20	29

c $D = n^2 + 4$ **d** 125 dots **e** figure 9

6 **a**

b **i** $H = 5n + 2$ **ii** $V = 4n + 2$ **iii** $T = 9n + 4$
c **i** 94 toothpicks **ii** 42 toothpicks
d 122 toothpicks

EXERCISE 10E

1 **a** \$15p **b** $C = 15p + 5$ **c** **i** \$50 **ii** \$125
2 **a** Each of the b boxes has mass 20 kg and the pallet has mass 30 kg.
$\therefore$ the total mass is $W = 20b + 30$ kg.
b 250 kg **c** 38 boxes

3 **a** $S = 7n + 200$
   b **i** 221 shells   **ii** 284 shells   **iii** 389 shells
   c 4 years 2 months
4 **a** $75n   **b** $M = 5000 - 75n$
   c **i** $4400   **ii** $1100   **d** 40 weeks
5 **a** $5t$ people   **b** $P = 300 - 5t$
   c **i** 280 people   **ii** 255 people
6 **a** **i** 20 000 cards   **ii** 28 000 cards   **iii** 36 000 cards
   b $C = 8000n + 4000$ **c** 84 000 cards   **d** 17 years
7 **a** $A = 8400 + (2300 - e)n$ **b** $A = 12\,600$ **c** $e = 1650$
8 **a** $T = 4(n - 2)$ **b** $I = (n - 2)^2$
   c **i** $T = 28$ **ii** $I = 49$
   d **i** $n = 11$ **ii** $I = 81$ **e** $n = 12$

REVIEW SET 10A

1
Input	Output
2	3
5	15
8	27
12	43

2 **a** $M = 4n + 1$
   b $y = 7x - 2$
3 **a** $W = 11$
   b $n = 10$
4 **a** **i** $P = 19$ **ii** $P = 147$
   b **i** $n = \pm 1$ **ii** $n = \pm 5$
5 **a** $188   **b** $75.50   **6** **a** $55   **b** $b = 5$ **c** $a = 9$
7 **a**

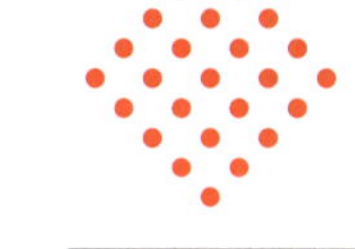

   b
Figure number (n)	1	2	3	4	5
Number of matchsticks (M)	4	7	10	13	16

   c $M = 3n + 1$ **d** 76 matchsticks
8 **a**

   b
Figure number (n)	1	2	3	4	5
Number of dots (D)	1	4	9	16	25

   c $D = n^2$ **d** 144 dots   **e** figure 22
9 **a** $25w   **b** $A = 25w + 450$ **c** **i** $550   **ii** $700
10 **a** There is a fixed credit card fee of $3, and each ticket costs $19, so n tickets will cost $C = 19n + 3$ dollars.
   b **i** $41   **ii** $98   **c** 4 tickets
11 **a** $N = 20 + (12 - b)d$ **b** $N = 11$ **c** $b = 14$
12 **a**

Figure number (n)	1	2	3	4	5
Number of matchsticks (M)	6	10	14	18	22

   c $M = 4n + 2$ **d** 42 matchsticks   **e** figure 17
   f **i** When $n = 1$, $T = 2(1)^2 + 4(1) = 6$ ✓
      When $n = 2$, $T = 2(2)^2 + 4(2) = 16 = 6 + 10$ ✓
      When $n = 3$, $T = 2(3)^2 + 4(3) = 30$
      $$= 6 + 10 + 14 \checkmark$$
   ii 880 matchsticks

REVIEW SET 10B

1 **a** $M = \dfrac{n}{6}$ **b** $M = 4n + 3$
2 **a** $Q = 6p - 2$ **b** $D = 3t + 2$
3 **a** $L = 22$ **b** $n = 8$
4 **a** 300 kg m/s   **b** 20 m/s
5 **a**

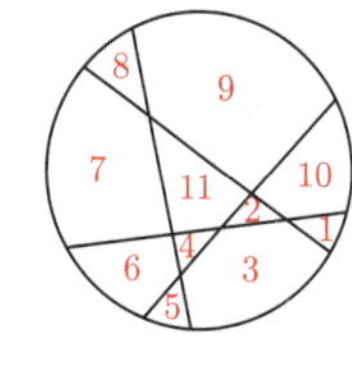

   b
Figure number (n)	1	2	3	4	5
Number of matchsticks (M)	2	6	10	14	18

   c $M = 4n - 2$
   d **i** 58 matchsticks   **ii** 166 matchsticks
6 **a**
Figure number (n)	1	2	3	4	5
Number of matchsticks (M)	4	13	22	31	40

   b $M = 9n - 5$ **c** 67 matchsticks   **d** figure 11
7 **a** **i** $P = 11$ **ii** $P = 123$ **iii** $P = 681$
   b **i** $n = 5$ **ii** $n = 10$ **iii** $n = 8$
8 **a** **i** 450 km   **ii** 225 km   **b** 8 hours
9 **a** 95 grams   **b** 635 grams   **c** 9 plums
10 **a** There is $60 in the money box, and each of the c cupcakes sold adds $1.50.
      ∴ the total amount is $A = 60 + 1.50c$ dollars.
   b **i** $90   **ii** $135
11 **a** $20n$ pages   **b** $(150 + 20n)$ pages
   c 290 pages   **d** 18 days of the holidays
12 **a** **i** 7 regions   **ii** 11 regions

   b 79 regions   **c** $M = n + 1$

EXERCISE 11A

1 **a** 7 cm   **b** 12 000 m   **c** 1.2 m   **d** 382 cm
   e 1.25 km   **f** 35 cm   **g** 434 mm   **h** 3200 mm
   i 1.24 m
2 **a** 0.356 km   **b** 35 600 cm
3 **a** 737.1 cm   **b** 4828.6 m   **4** 24.1 mm
5 **a** Mens: 1829 cm,   Womens: 1550 cm
   b 279 cm farther
6 45 biscuits   **7** **a** 2.97 cm   **b** 600 coins
8 247.5 cm   **9** 38.448 m

EXERCISE 11B

1 **a** 23 cm   **b** 15 cm   **c** 22 cm   **d** 26 cm
   e 56 m   **f** 38 cm
2 **a** 3900 m or 3.9 km   **b** 470 cm or 4.7 m
   c 79.5 mm or 7.95 cm
3 **a** 1500 m   **b** 6000 m   **c** $2292   **4** 77 m
5 5 laps

6 **a** $P = 3x$ cm **b** $P = 4l$ m
  c $P = (2x + 2y)$ m **d** $P = 4x$ cm
  e $P = (3x + 10)$ cm **f** $P = (4x + 14)$ cm
  g $P = (6x + 4)$ cm **h** $P = 16x$ cm
  i $P = (8x + 6)$ cm

7 **a** $P = (5x - 2)$ cm **b** 13 cm **c** $x = 12$

8 **a** $P = (6x + 1)$ m **b** **i** $x = 7$ **ii** 9 m

9 **a** $P = (5x + y)$ cm **b** $P = 50$ **c** $x = 4$

EXERCISE 11C

1 **a** ≈ 12.6 m **b** ≈ 20.7 cm **c** ≈ 28.9 cm
  d ≈ 102 mm

2 **a** ≈ 1030 mm **b** ≈ 87.9 cm

3 **a** ≈ 23.88 m **b** ≈ 106.81 cm

4 **a** ≈ 56.5 m **b** ≈ 50.3 cm **c** ≈ 42.7 km

5 ≈ 12.6 cm **6** ≈ 9.425 m **7** ≈ 4.71 m

8 ≈ 56.5 cm

9 **a** ≈ 220 cm **b** ≈ 44.0 km **c** $\approx 20\,500$ times

10 ≈ 10.2 cm **11** Both paths have equal length, 4π metres.

12 ≈ 63.7 m **13** **a** ≈ 15.9155 cm **b** $\approx 0.0628\%$

14 **a** $P = (3x + 6 + 2\pi x)$ cm **b** ≈ 43.1 cm
  c $x \approx 5.630$

EXERCISE 11D

1 **a** **B** **b** **C** **c** **C**

2 **a** 6 m^2 **b** **i** 200 rows, 300 columns **ii** 60 000 cm^2

3 **a** 500 mm^2 **b** 25 cm^2 **c** 70 000 m^2
  d 36 000 cm^2 **e** 40 ha **f** 8300 mm^2
  g 800 000 m^2 **h** 1.56 m^2 **i** 12 km^2
  j 9 cm^2 **k** 7.6 ha **l** 280 mm^2
  m 25 ha **n** 124 800 cm^2 **o** 9200 mm^2

4 15 000 mm^2 **5** **a** Bruno **b** Carlos

6 **a** 3420 kg **b** \$2719.45 **7** ≈ 4050 m^2

EXERCISE 11E

1 **a** 2.25 m^2 **b** 40 cm^2 **c** 108 m^2 **d** 36 mm^2
  e 9 cm^2 **f** 42.5 cm^2 **g** 120 m^2 **h** 15 cm^2
  i 450 cm^2

2 **a** 36 m^2 **b** 72 cm^2 **c** 576 m^2 **d** 42 cm^2
  e 85 m^2 **f** 52 cm^2

3 **a** 60 mm^2 **b** 9000 cm^2 **c** 9000 cm^2

4 **a** **i** 57 912 cm^2 **ii** 67 130 cm^2 **b** $\approx 15.9\%$ larger

5 **a** 13.5 m^2 **b** \$263.25 **6** 9.9 m^2 **7** 37.5%

8 640 bricks **9** **a** 144 m^2 **b** **i** 24 m^2 **ii** 12 m^2

10 **a** 5400 cm^2 **b** 13.5 m^2

11 **a** $A = 3x^2$ cm^2 **b** $A = \left(\dfrac{x^2}{2} + 5x\right)$ cm^2
  c $A = (6a^2 + 12a)$ cm^2

12 12 cm

EXERCISE 11F

1 **a** ≈ 78.54 cm^2 **b** ≈ 4778.36 cm^2 **c** ≈ 1661.90 mm^2
  d ≈ 21.24 m^2 **e** ≈ 39.27 m^2 **f** ≈ 63.62 mm^2

2 **a** ≈ 15.08 cm **b** ≈ 18.10 cm^2

3 ≈ 531 cm^2 **4** ≈ 40.7 m^2

5 **a** $\dfrac{\theta}{360}$ **b** $A = \left(\dfrac{\theta}{360} \times \pi r^2\right)$ cm^2

EXERCISE 11G

1 **a** 44 m^2 **b** 46 m^2 **c** 38 cm^2 **d** 21 mm^2
  e 126 cm^2 **f** 19.5 cm^2

2 **a** 33.5 m^2 **b** 31 cm^2

3 **a** ≈ 4463.5 m^2 **b** ≈ 37.7 cm^2 **c** ≈ 14.7 m^2
  d ≈ 22.7 cm^2 **e** ≈ 34.1 m^2 **f** ≈ 64.2 mm^2

4 180 tiles **5** ≈ 320 mm^2 **6** ≈ 2.89 m^2

7 ≈ 526 mm^2 **8** ≈ 258 cm^2

9 **a** **i** 2400 cm^2 **ii** ≈ 314 cm^2 **b** $\approx 13.1\%$
  c ≈ 1040 cm^2

10 **a** $A = (10x + 65)$ cm^2 **b** $A = (2x^2 + 80x)$ cm^2
  c $A = (3x^2 + 9x)$ cm^2 **d** $A = \frac{3}{4}\pi r^2$ cm^2

11 **a** $\dfrac{3\pi}{50}$ **b** $\approx 81.2\%$

12 **a** $A = \left(60 - \dfrac{3x}{2}\right)$ m^2 **b** $x = 4$

13 **a** 48 cm^2 **b** 21 m^2

REVIEW SET 11A

1 **a** 9.5 cm **b** 4800 m **c** 2.7 m

2 **a** 32.8 cm **b** 8.32 m **c** ≈ 31.4 m

3 **a** ≈ 35.2 cm **b** ≈ 98.5 cm^2

4 **a** 5.5 m^2 **b** 34 000 cm^2 **c** 1.85 ha

5 **a** 50 cm^2 **b** ≈ 28.3 m^2 **c** 22.5 m^2

6 **a** 184 ha **b** 3680 trees

7 **a** **i** 350 cm^2 **ii** 175 cm^2
  b 19×350 cm$^2 + 2 \times 175$ cm$^2 = 350$ cm $\times 20$ cm
  $= 7000$ cm^2

8 **a** 915 mm^2 **b** ≈ 23.4 m^2 **c** ≈ 31.7 cm^2

9 **a** ≈ 6.68 m **b** ≈ 2.73 m^2

10 **a** ≈ 2.01 m^2 **b** ≈ 0.534 m^2

11 **a** $A = (30 - 3x)$ km^2 **b** $x = 2$

12 **a** **i** ≈ 141 cm **ii** ≈ 1590 cm^2
  b **i** 434 mm **ii** ≈ 1360 mm

13 **a** 1500 cm^2 **b** **i** 495 cm^2 **ii** 66.9 cm^2
  c 376.2 cm^2 **d** 25.08%

REVIEW SET 11B

1 150 days

2 **a** ≈ 18.8 km **b** ≈ 73.7 m **c** 100 mm or 10 cm

3 **a** 34 000 mm^2 **b** 0.327 ha

4 **a** 60 cm^2 **b** 42 cm^2 **c** 120 m^2 **5** 20 cm^2

6 $P = (4x + \pi x)$ m, $A = \left(2x^2 + \dfrac{\pi x^2}{2}\right)$ m^2

7 Eliza's is ≈ 254.5 cm^2, Claire's is 256 cm^2
  $\therefore$ Claire's is larger.

8 124 floorboards

9 **a** 128 cm^2 **b** 8 m^2 **10** 43.73 m^2 **11** 1566 cm^2

12 **a** 135 cm **b** 2 m **c** 2125 cm^2

13 **a** **i** 7140 m^2 **ii** 664.95 m^2 **iii** ≈ 284 m^2
  b $\approx 3.97\%$ **c** 83 m

EXERCISE 12A

1 **a** 54 cm^2 **b** 24 mm^2 **c** 13.5 m^2

2 **a** 432 cm^2 **b** 128 m^2 **c** 970 mm^2 **3** \$960

4 **a** 120 cm^2 **b** 1120 cm^2 **c** 520 m^2

5 **a** 950 cm^2 **b** 480 m^2 **c** 984 cm^2

6 **a** 156 cm^2 **b** 360 mm^2 **c** 564 m^2

7 184 cm^2 **8** $974.12

EXERCISE 12B

1 **a** $\approx$ 188 cm^2 **b** $\approx$ 1730 cm^2 **c** $\approx$ 126 cm^2
    d $\approx$ 1600 mm^2 **e** $\approx$ 100 m^2 **f** $\approx$ 0.754 m^2

2 $\approx$ 1355 cm^2 **3** $\approx$ 35.6 L of paint

4 **a** $\approx$ 25 900 cm^2 **b** $\approx$ 103 000 cm^2

5 **a** 106 jars
    b Assumes that 106 whole labels can be cut (without joins).

EXERCISE 12C

1 **a** $\approx$ 50.3 m^2 **b** $\approx$ 11 300 cm^2 **c** $\approx$ 763 cm^2

2 **a** 36π m^2 **b** 144π cm^2 **c** 75π cm^2

3 $\approx$ 154 cm^2 **4** $\approx$ 515 000 000 km^2

5 $\approx$ 2.82 cm   **6** **a** $\approx$ 27.1 cm^2 **b** $\approx$ 67.9 cm^2

7 $\approx$ 4880 km   **8** 565 balls

EXERCISE 12D

1 **a** cm^3 **b** m^3 **c** cm^3 **d** m^3 **e** cm^3 **f** mm^3

2 **a** 48 000 mm^3 **b** 0.029 m^3 **c** 1 200 000 cm^3
    d 12.485 cm^3 **e** 450 cm^3 **f** 0.0145 m^3
    g 295 mm^3 **h** 0.001 43 cm^3 **i** 5600 cm^3

3 235 000 cm^3 **4** 200 000 pieces   **5** $\approx$ 1.14 m^3

EXERCISE 12E

1 **a** 96 m^3 **b** 343 mm^3 **c** 12 cm^3

2 **a** $\approx$ 226 cm^3 **b** $\approx$ 754 mm^3 **c** $\approx$ 56.5 m^3
    d $\approx$ 942 cm^3 **e** $\approx$ 15.7 m^3 **f** $\approx$ 4.00 m^3

3 $\approx$ 8140 cm^3 **4** $\approx$ 4320 cm^3 **5** 1.4 cm

6 $\approx$ 0.813 cm   **7** $\approx$ 1.50 cm

8 **a** 60 cm^3 **b** 272 m^3 **c** 82.5 m^3
    d 112.5 cm^3 **e** 2.6 m^3 **f** 2.8 cm^3

9 **a** 168 cm^3 **b** 960 cm^3 **c** $\approx$ 0.188 m^3
    d 65 536 cm^3 **e** 15 000 mm^3 **f** $\approx$ 2390 cm^3

10 10.5 m^3 **11** 18 cm   **12** **a** 8352 cm^3 **b** 7938 cm^3

13 **a** $\approx$ 129 mm^3 **b** 62 rings   **c** $\approx$ 2.49 g   **d** $152.34

14 $\approx$ 93.2 cm^3 **15** 30 cm^3

EXERCISE 12F

1 **a** 108 cm^3 **b** $\approx$ 1900 mm^3 **c** 120 cm^3
    d $\approx$ 3390 cm^3 **e** 480 cm^3 **f** 12 cm^3

2 **a** $\approx$ 838 m^3 **b** $\approx$ 1530 t   **c** $\approx$ $10 300 000

3 **a** $\dfrac{x^2(x+5)}{3}$ cm^3 **b** $\dfrac{\pi x^3}{6}$ mm^3 **c** $\dfrac{x^3}{6}$ m^3

EXERCISE 12G

1 **a** $\approx$ 905 cm^3 **b** $\approx$ 14.1 m^3 **c** $\approx$ 268 mm^3
    d $\approx$ 2570 cm^3 **e** $\approx$ 262 m^3 **f** $\approx$ 3620 cm^3

2 972π cm^3 **3** **a** $\approx$ 38 500 mm^3 **b** $\approx$ 38.5 cm^3

4 **a** $\approx$ 382 cm^3 **b** $\approx$ 7.25 cm

EXERCISE 12H

1 **a** mL   **b** mL   **c** ML   **d** kL   **e** L   **f** kL

2 **a** 2.96 L   **b** 3010 kL   **c** 0.0991 kL
    d 1.8 ML   **e** 6600 L   **f** 18 700 mL
    g 0.056 kL   **h** 350 L   **i** 30 000 L

3 111.6 L   **4** 1 h 15 min   **5** 11 900 000 doses

EXERCISE 12I

1 **a** 90 mL   **b** 7.5 L   **c** 33 L   **2** 360 L

3 $\approx$ 217 L   **4** 200 kL   **5** $\approx$ 7.61 L

6 **a** $\approx$ 2710 kL   **b** $\approx$ 2.71 ML   **7** 7 full jars

8 $\approx$ 4.68 cm

REVIEW SET 12A

1 **a** 96 cm^2 **b** 142 mm^2 **c** 528 m^2 **2** $104

3 **a** $\approx$ 170 cm^2 **b** 168 cm^2 **c** $\approx$ 462 cm^2

4 **a** 2 800 000 cm^3 **b** 4.2 m^3 **c** 1100 mm^3

5 **a** $\approx$ 1130 cm^3 **b** 195 cm^3 **c** 168 cm^3

6 **a** $\approx$ 275 m^2 **b** $\approx$ $6 600 000

7 **a** 64π cm^2 **b** $\dfrac{256}{3}\pi$ cm^3

8 **a** 150 cm^3 **b** 230 cm^2 **c** 130 bases

9 750 mL   **10** 480 L   **11** $\approx$ 1.87 kL

12 **a** a circle and a rectangle   **b** $\approx$ 817 cm^2
    c $\approx$ 2510 cm^3 **d** no

REVIEW SET 12B

1 **a** 76 cm^2 **b** 84 m^2 **c** 123.6 cm^2

2 $\approx$ 13.6 cm^2 **3** 19 vases   **4** $\approx$ 21.1 cm

5 1.2 m^3 **6** 40 000 cm^3 or 0.04 m^3

7 **a** 72.6 mm^3 **b** $\approx$ 170 cm^3 **c** $\approx$ 46.0 m^3

8 **a** $\approx$ 37.7 cm^3 **b** 23 873 hourglasses

9 **a** 0.82 L   **b** 3070 kL   **c** 720 000 mL   **10** 0.72 kL

11 $\approx$ 4.61 cm   **12** **Hint:** Let r be the radius of the cylinder.

EXERCISE 13A.1

1 **a** 1020 min   **b** 23 min   **c** 4320 min
    d 268 min   **e** 693 min   **f** 3180 min

2 **a** 42 days   **b** $\approx$ 2922 days   **c** 6 days
    d 48 days   **e** 3 days

3 **a** 420 s   **b** 825 s   **c** 21 600 s
    d 4560 s   **e** 86 400 s   **f** 16 020 s

4 **a** 1 h 47 min   **b** 4 h 42 min   **c** 8 h 12 min
    d 12 h 17 min

5 **a** 3 min 6 s   **b** 3 min 34 s   **c** 5 min 50 s
    d 14 min 11 s

6 **a** 1 day 19 h   **b** 2 days 13 h   **c** 4 days 6 h
    d 1 day 1 h

7 1 h 20 min   **8** 134 min   **9** $\approx$ 11 days

10 2 h 7 min   **11** **a** 7 h 35 min   **b** 1 h 5 min

12 3 min 16.86 s   **13** 1 min 1.86 s

EXERCISE 13A.2

1 **a** 70 years   **b** 1200 years   **c** 3000 years
    d 450 years   **e** 35 years   **f** 2250 years
    g 230 years   **h** 5400 years

2 **a** 2 centuries   **b** 43 centuries

3 **a** 500 centuries   **b** 50 millennia

4 $\approx$ 165 000 millennia

EXERCISE 13B

1 **a** 8:12 pm   **b** 7:12 am   **c** 4:04 pm   **d** 4:19 am
    e 12:15 pm   **f** 11:50 pm the previous day   **g** 7:44 am
    h 4:30 am the next day

2 1:45 pm   **3** 4:45 pm   **4** 2 pm   **5** 4:45 pm

6 4:57 pm

7 **a** 4 h 25 min **b** 7 h 16 min **c** 2 h 48 min
   d 3 h 46 min **e** 9 h 3 min **f** 1 day 2 h 20 min

8 1 h 42 min **9** 4 h 23 min

10 **a** 10 h 55 min **b** 52 min

11 **a** Deborah **b** Terry **c** no **d** $337.50

12 **a** 40 h **b** $840 **13** 34 h 52 min

14 **a** Otter Odyssey **b** 3D Underwater World
   c 2:05 pm **d** Whale Mania and 3D Underwater World
   e 9:45 am **f** 2 h 25 min

   g **i** 2 h 55 min **ii**

Show	Time
Diving Dolphins	1:40 pm
Whale Mania	3:15 pm
Seal of Approval	11:30 am
Otter Odyssey	10:00 am
3D Underwater World	12:30 pm
Marine Park Parade	4:00 pm

EXERCISE 13C

1 **a** 03:47 **b** 10:12 **c** 18:00 **d** 13:35 **e** 08:41
   f 21:25 **g** 12:00 **h** 18:39 **i** 00:07

2 **a** 2:49 am **b** 11:53 am **c** 3:24 pm **d** 8:00 pm
   e 9:00 am **f** 12:34 pm **g** 11:23 pm **h** 12:20 am

3 **a** 3 h 20 min **b** 5 h 5 min **c** 9 h 45 min
   d 4 h 37 min **e** 7 h 47 min **f** 5 h 41 min

4 **a** 10 h 43 min **b** 19 h 20 min **c** 40 h

5

Departure	Travelling time	Arrival
07:25	2 h 15 min	09:40
11:30	4 h 35 min	16:05
10:39	29 min	11:08
08:49	3 h 41 min	12:30
23:52	5 h 3 min	04:55 (next day)

6 **a** 12 h 48 min **b** 13:07

7 **a** **i** 6 flights **ii** 3 flights **b** 1 h 25 min
   c 30 min **d** 37 min **e** 2 h 5 min

EXERCISE 13D

1 **a** 7 am **b** 2 pm **c** 8 pm **d** 8 pm

2 **a** 11 am Wednesday **b** 3 am Wednesday
   c 8 am Wednesday **d** 7 pm Tuesday

3 **a** 11 pm Thursday **b** 8 am Friday
   c 10 am Friday **d** 3 pm Friday

4 5 pm

5 **a** 4 pm **b** 12 am the next day **c** 10 am **d** 12 pm

6 7:55 pm **7** 7 h 45 min **8** 10:45 pm

9 8 am the next day **10** 7 h 30 min **11** 12 h 10 min

12 **a** 7:50 pm Tuesday HK time **b** 1 h 30 min
   c **i** 1 h 55 min **ii** 3 h 55 min **d** 22 h 30 min

REVIEW SET 13A

1 **a** 14 h **b** 436 s **2** 30 crossings

3 **a** 32 decades **b** 210 decades

4 **a** 1:20 pm **b** 3:46 pm **c** 4:49 pm **d** 7:31 am

5 **a** 5 h 10 min **b** 5 h 15 min **c** 3 h 30 min
   d 3 h 37 min

6 3 h 45 min **7** 2:08 pm

8 **a** 09:47 **b** 12:49 **c** 19:17 **d** 00:26

9 **a** 12 h 23 min **b** 13:04 **10** **a** 9 pm **b** 9 am

11 **a** **i** 10 h **ii** 7 h 30 min **b** Thursday **c** 65 h

12 **a** in the evening **b** 10 h 24 min **c** Prague and Brno
   d 17:36 **e** Dresden to Prague

REVIEW SET 13B

1 **a** 7 h 37 min **b** 3 days 10 h **2** 194 s

3 11.7 millennia **4** 17 h 4 min **5** 8:21 am

6 **a** 8:45 am **b** 10:22 am **7** 7:34 am

8 **a** 7:12 am **b** 3:02 pm **c** 9:59 pm **d** 3:48 am

9 2 h 48 min **10** 4:30 am Saturday

11 **a** 2 h 5 min **b** The Italian Job
   c The Sound of Music, 12:24 am
   d Gone With the Wind, The Sound of Music, Singin' in the Rain, The Godfather
   e 2 h 55 min

12 **a** 4 pm Bangkok time **b** 3 h 35 min
   c **i** 2 h 25 min **ii** 10 h 40 min **d** 16 h 40 min

EXERCISE 14A

1 **a** **i** 2 **ii** 0 **iii** -5
   b **i** 0 **ii** 1 **iii** -4 **c** **i** C **ii** E

2 **a** P(4, 2), Q(-1, -3), R(3, -1), S(-2, 5), T(7, -3)
   b 2nd quadrant **c** R and T

3 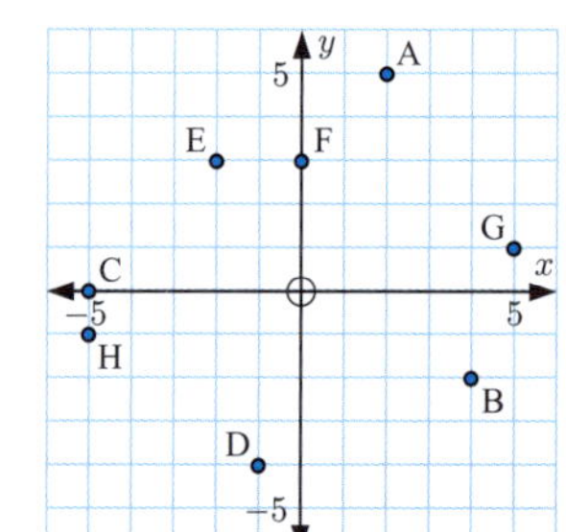

4 **a** 3rd quadrant **b** 1st quadrant **c** 2nd quadrant
   d 4th quadrant

5 **a** $(-1, 3)$ **b** $(3, -5)$

6 **a** 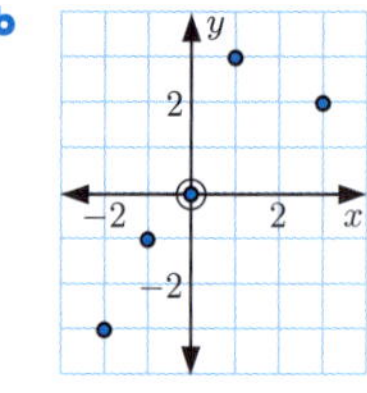 **b**

7 **a** 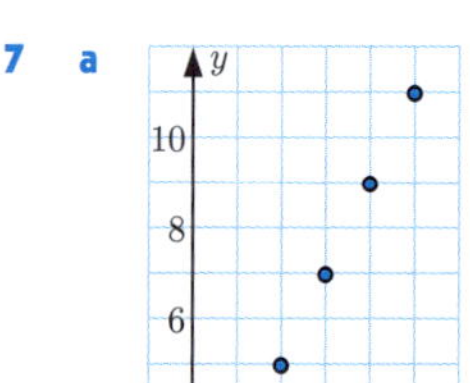 **b** yes
   c **i** 13 cm
   ii 17 cm

8 a
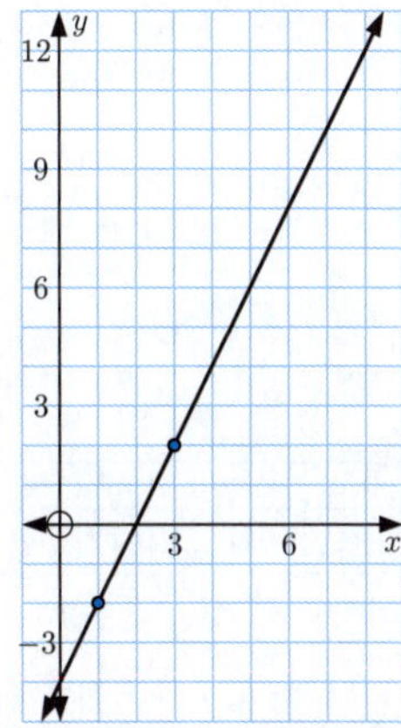

b

x	1	3	5	8
y	-2	2	6	12

c **i** $(2, 0)$ **ii** $(0, -4)$

9 a

x	-2	-1	0	1	2
y	-6	-3	0	3	6

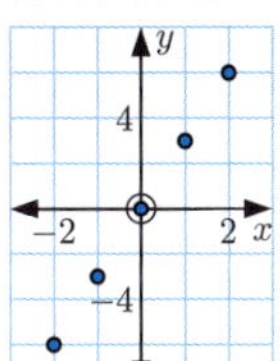

b

x	-2	-1	0	1	2
y	-1	0	1	2	3

c

x	-2	-1	0	1	2
y	5	4	3	2	1

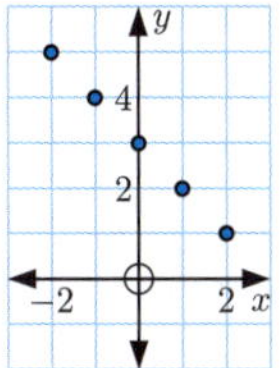

d

x	-2	-1	0	1	2
y	4	1	0	1	4

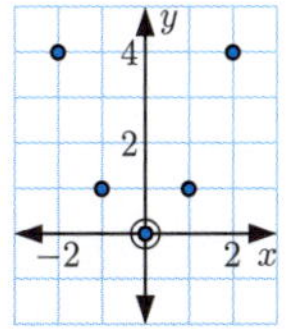

e

x	-2	-1	0	1	2
y	-3	0	1	0	-3

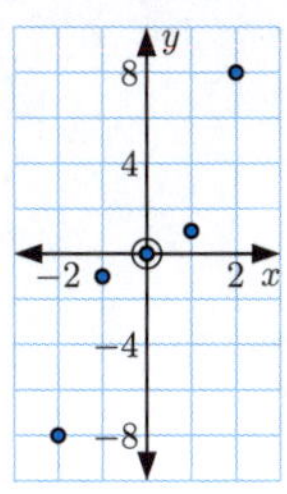

f

x	-2	-1	0	1	2
y	-8	-1	0	1	8

10 a
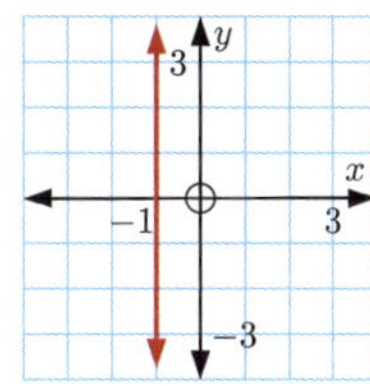

b
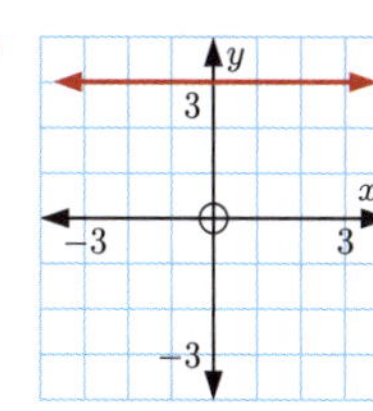

c
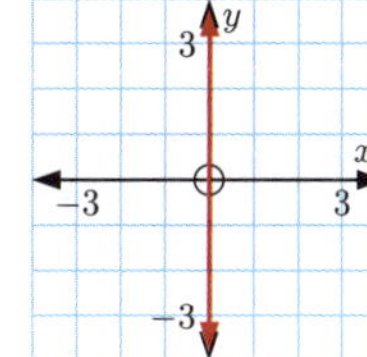

d
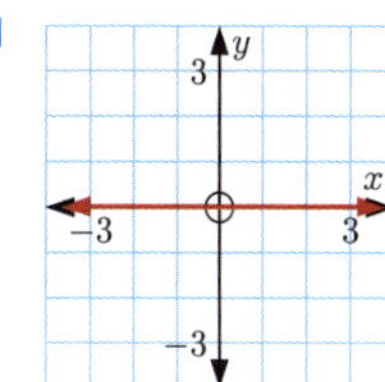

11 a
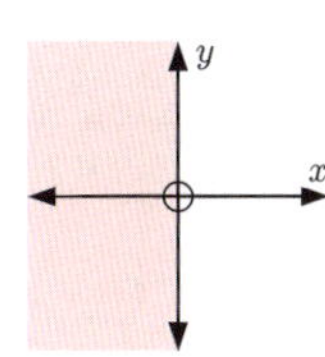

The points on the y-axis are not included.

b
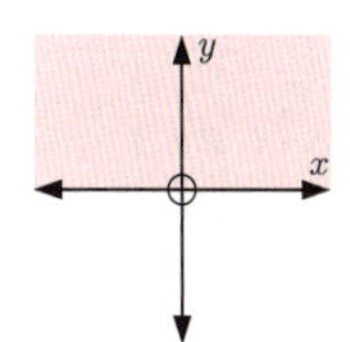

The points on the x-axis are not included.

c
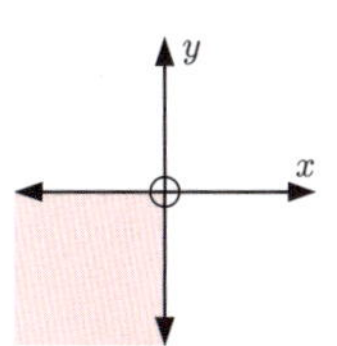

The points on the axes are not included.

d
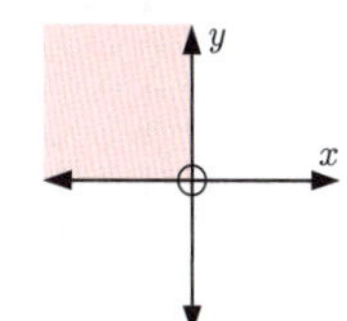

The points on the axes are not included.

12 a quadrants 1 and 3   **b** quadrants 2 and 4

13 a

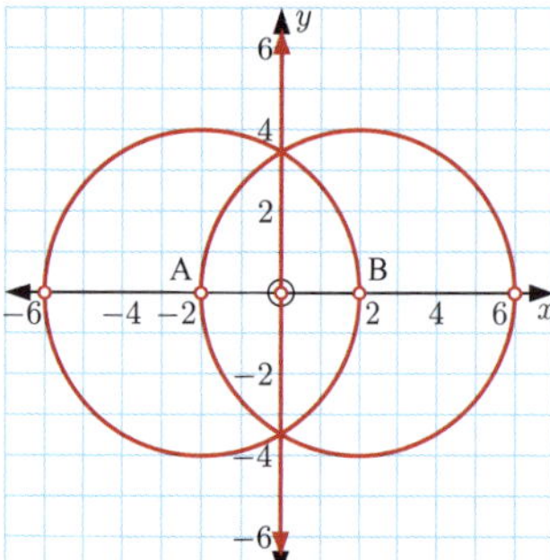

b

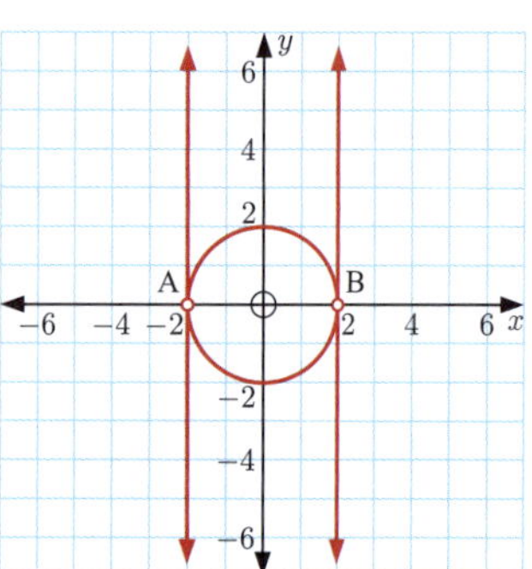

EXERCISE 14B

1 a

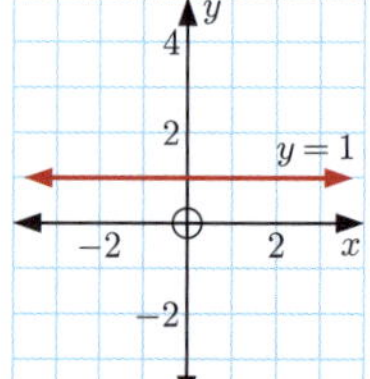

b

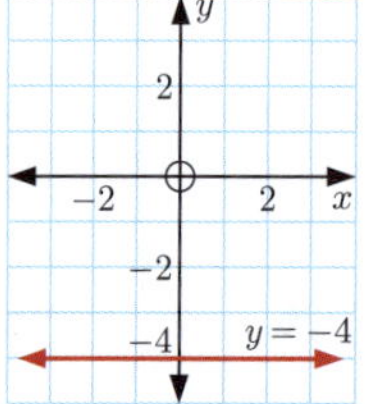

c

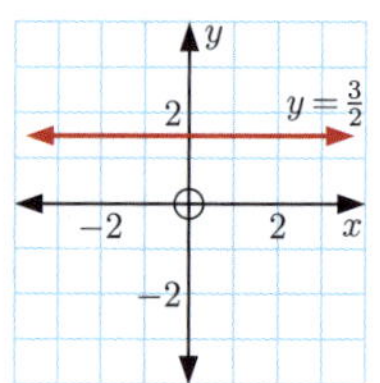

d

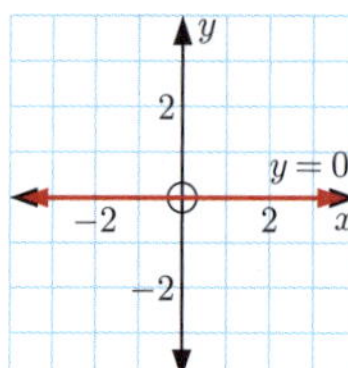

2 a

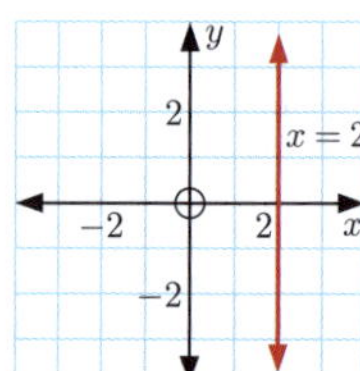

b

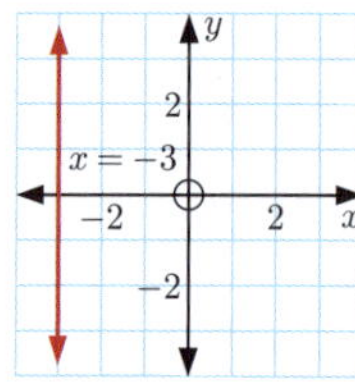

c

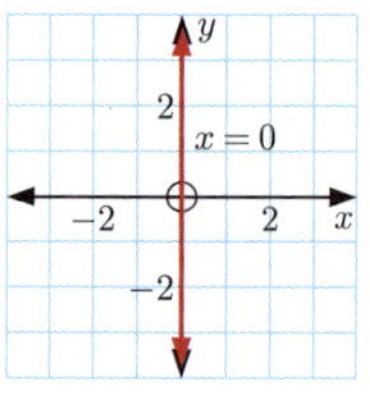

d

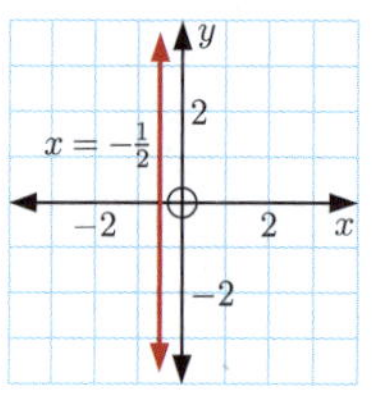

3 a

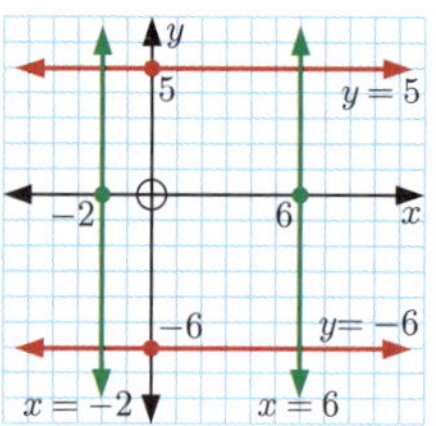

b   i rectangle
ii 88 units^2

4 a i

x	-3	-2	-1	0	1	2	3
y	$-1\frac{1}{2}$	-1	$-\frac{1}{2}$	0	$\frac{1}{2}$	1	$1\frac{1}{2}$

ii

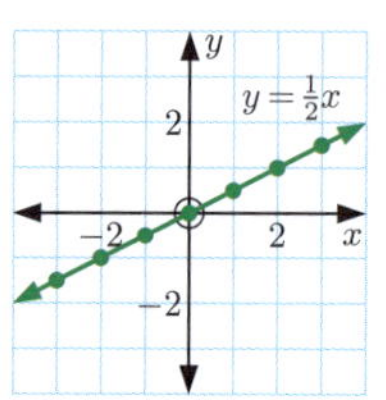

b i

x	-3	-2	-1	0	1	2	3
y	$1\frac{1}{2}$	1	$\frac{1}{2}$	0	$-\frac{1}{2}$	-1	$-1\frac{1}{2}$

ii

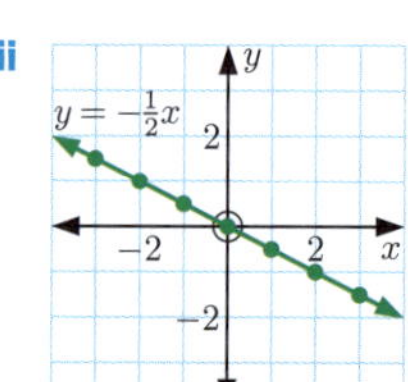

c i

x	-3	-2	-1	0	1	2	3
y	-7	-4	-1	2	5	8	11

ii

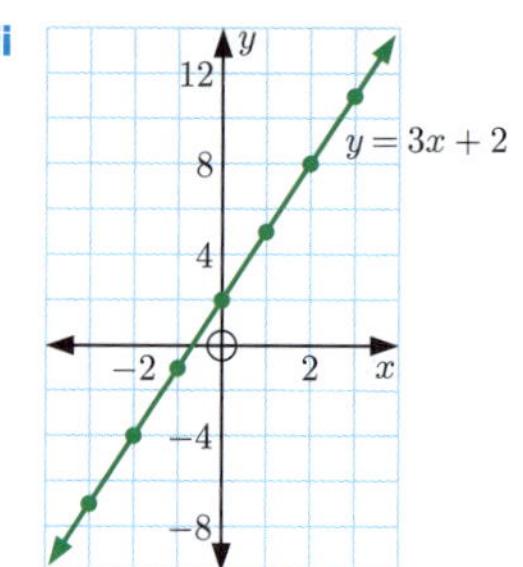

d i

x	-3	-2	-1	0	1	2	3
y	11	8	5	2	-1	-4	-7

ii

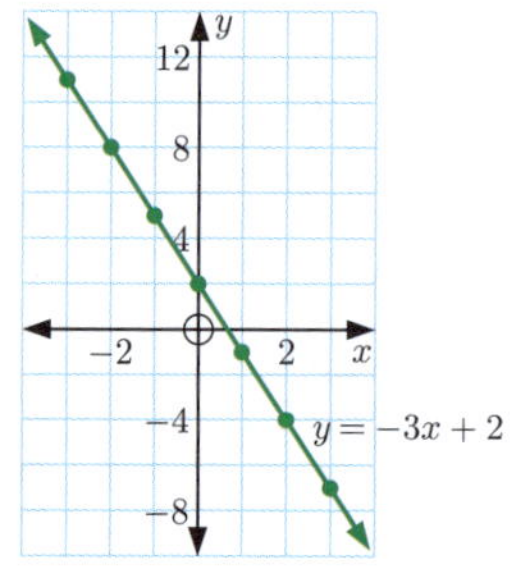

e **i**

x	-3	-2	-1	0	1	2	3
y	-9	-7	-5	-3	-1	1	3

ii
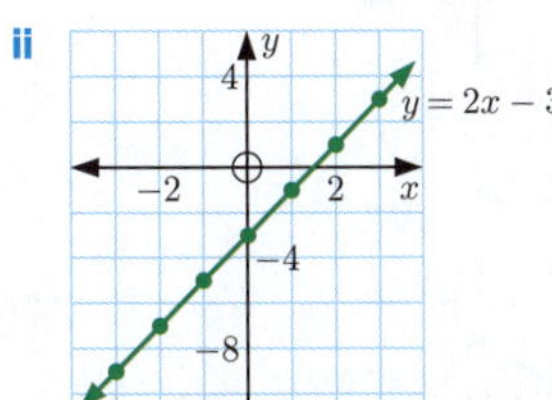

f **i**

x	-3	-2	-1	0	1	2	3
y	3	1	-1	-3	-5	-7	-9

ii
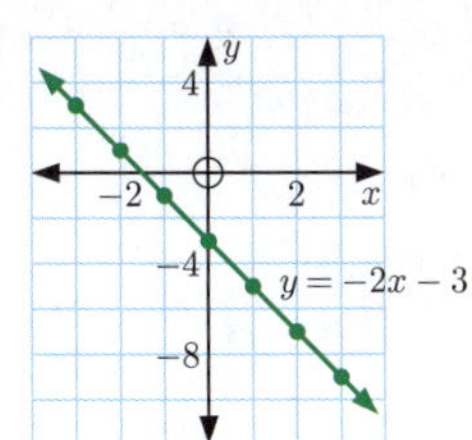

g **i**

x	-3	-2	-1	0	1	2	3
y	$1\frac{1}{4}$	$1\frac{1}{2}$	$1\frac{3}{4}$	2	$2\frac{1}{4}$	$2\frac{1}{2}$	$2\frac{3}{4}$

ii
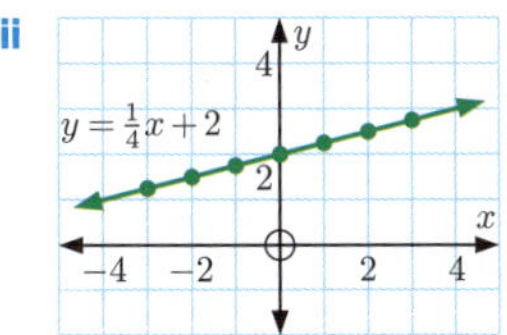

h **i**

x	-3	-2	-1	0	1	2	3
y	$2\frac{3}{4}$	$2\frac{1}{2}$	$2\frac{1}{4}$	2	$1\frac{3}{4}$	$1\frac{1}{2}$	$1\frac{1}{4}$

ii
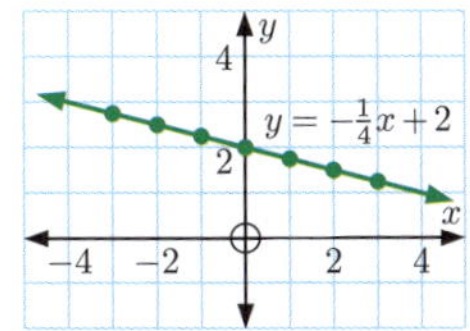

i **i**

x	-3	-2	-1	0	1	2	3
y	$-5\frac{1}{2}$	-4	$-2\frac{1}{2}$	-1	$\frac{1}{2}$	2	$3\frac{1}{2}$

ii
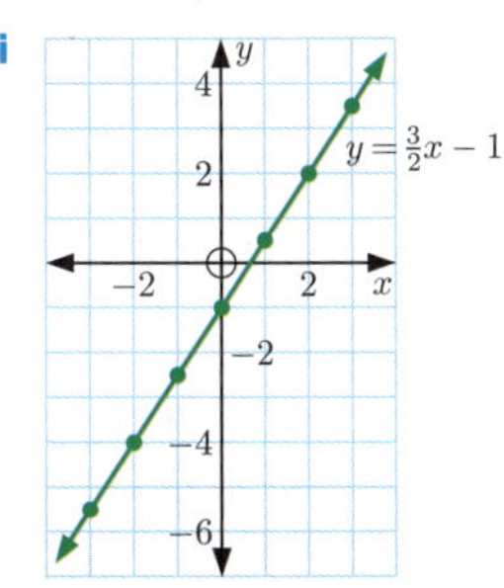

j **i**

x	-3	-2	-1	0	1	2	3
y	$3\frac{1}{2}$	2	$\frac{1}{2}$	-1	$-2\frac{1}{2}$	-4	$-5\frac{1}{2}$

ii
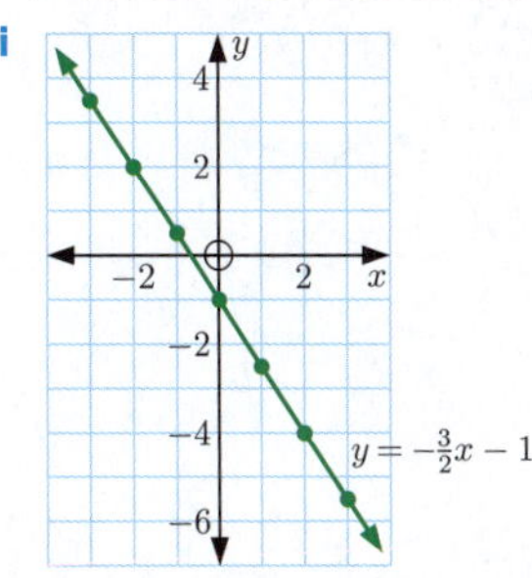

k **i**

x	-3	-2	-1	0	1	2	3
y	-15	-11	-7	-3	1	5	9

ii
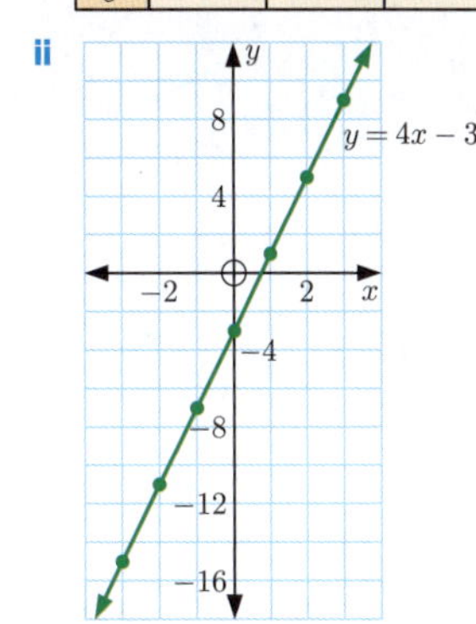

l **i**

x	-3	-2	-1	0	1	2	3
y	-9	-5	-1	3	7	11	15

ii
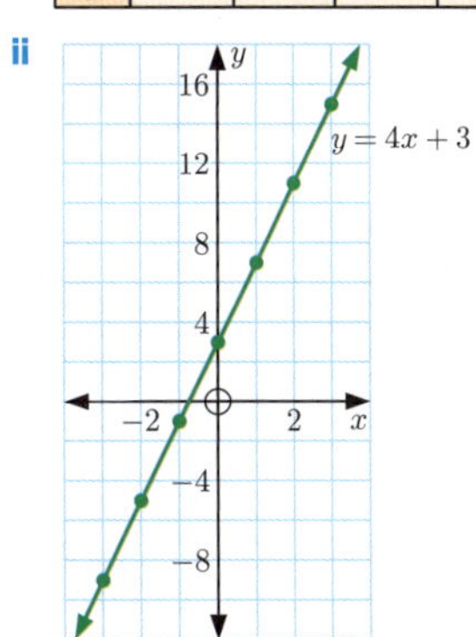

5 **a** $y = 4x$ **b** $y = x + 2$ **c** $y = -2x$
d $x + y = 3$ **e** $y = 2x + 1$ **f** $y = 2x - 1$

6 **a** yes   **b** no   **c** yes   **d** no

7 **a** 5 **b** $-\frac{1}{2}$ **8** **a** 9 **b** 4

EXERCISE 14C

1 **a** $\frac{2}{3}$ **b** -1 **c** $\frac{1}{3}$ **d** $-\frac{4}{3}$ **e** $\frac{1}{6}$
  f undefined   **g** -3 **h** $-\frac{1}{8}$ **i** 0

2 **a**   **b**   **c**   **d**

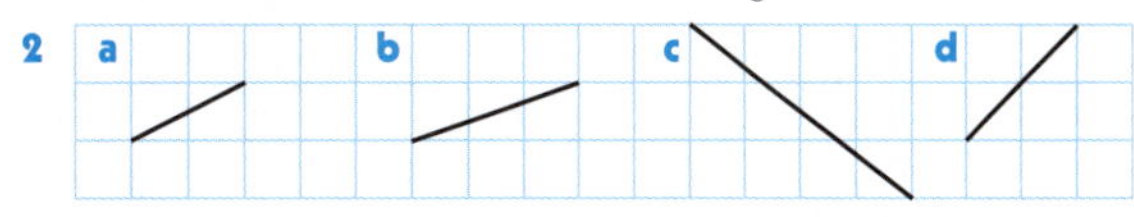

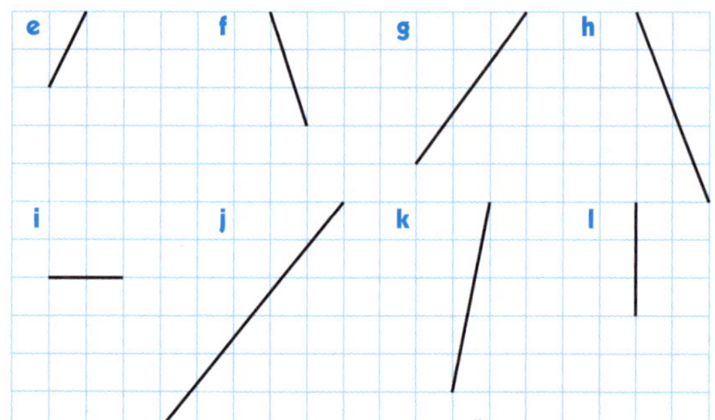

3 **a** 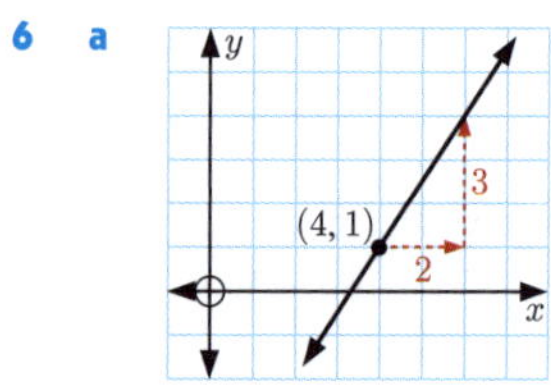 **b** 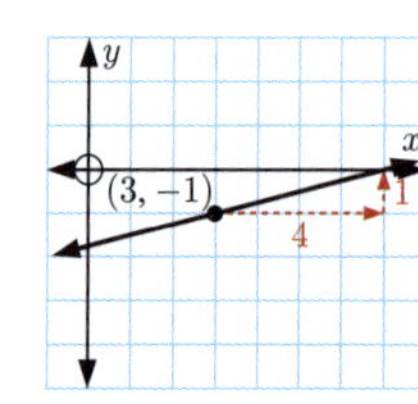

c 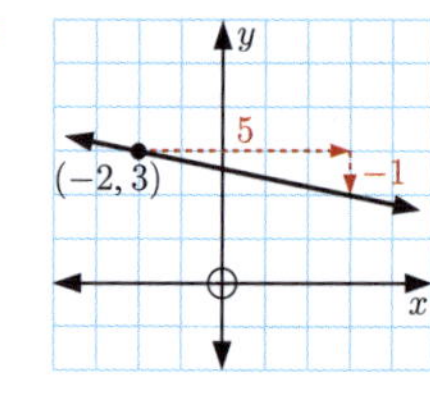 **d** 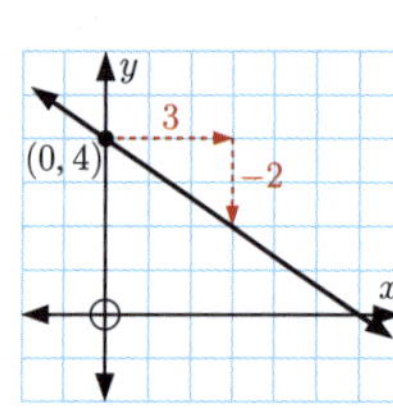

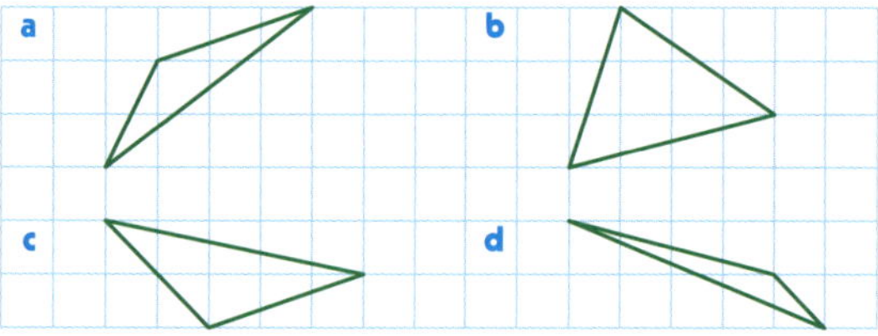

4 **a** $\frac{1}{6}$ **b** $-\frac{3}{4}$ **c** $-\frac{2}{5}$ **d** 14

5 **a** **i** 0   **ii** $\frac{1}{5}$ **iii** $\frac{3}{5}$ **iv** 1   **v** 2   **vi** 4

  vii undefined   **viii** -4 **ix** $-\frac{1}{2}$

  b **i** 0   **ii** undefined

6 **a** 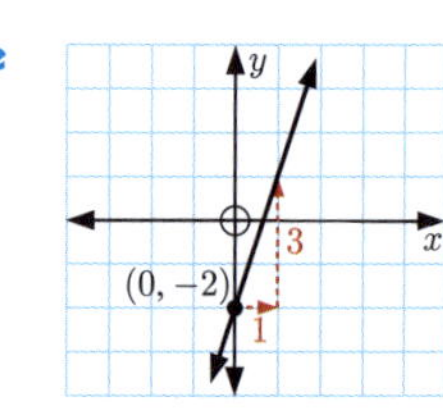 **b** 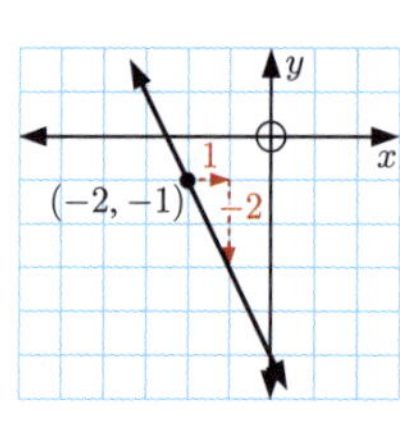

c **d**

e **f**

g 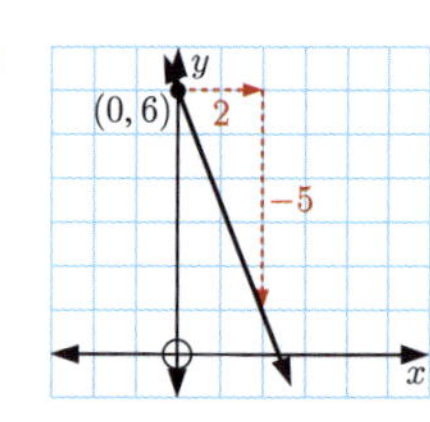 **h** 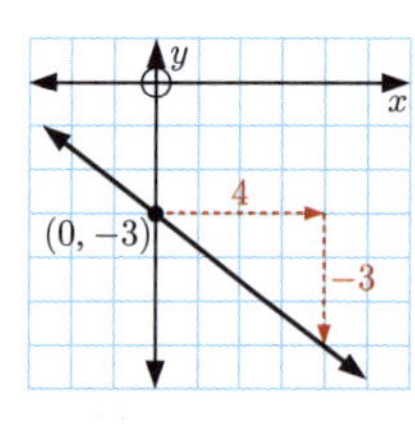

7 **a** 3   **b** $-\frac{1}{2}$ **c** 3   **d** 1   **e** $-\frac{1}{2}$ **f** -1

8 **a** 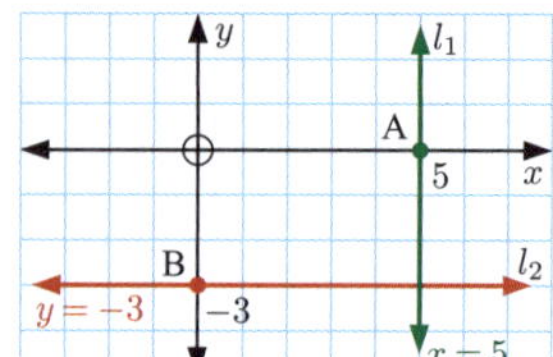   **b** **i** A(5, 0),
  B(0, -3)

  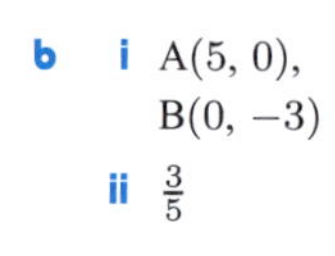 **ii** $\frac{3}{5}$

9 **a** 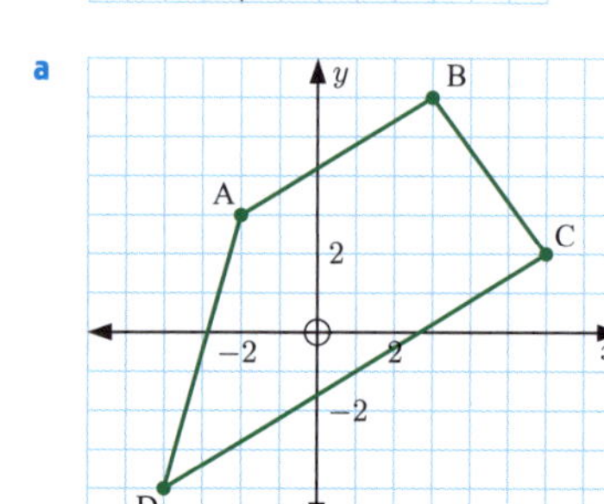   **b** trapezium

  c **i** $\frac{3}{5}$

  ii $-\frac{4}{3}$

  iii $\frac{3}{5}$

  iv $\frac{7}{2}$

  d Parallel lines have the same gradient.

10 **a** $(1, -1)$ and $(3, 1)$ **b** 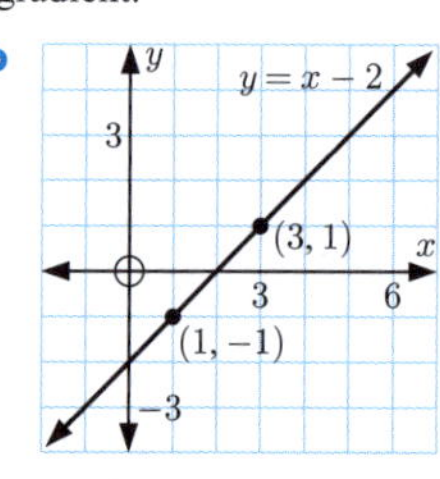

  c 1

11 **a** $(1, 2)$ and $(3, 0)$ **b** 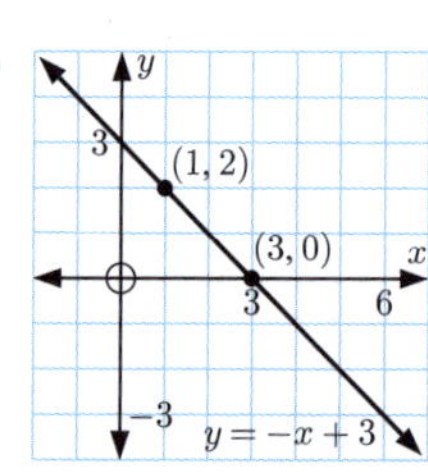

  c -1

12 **a** $(0, 3)$ and $(2, 5)$;   gradient is 1
  b $(0, 1)$ and $(2, -1)$;   gradient is -1
  c $(0, -1)$ and $(2, 3)$;   gradient is 2
  d $(0, 3)$ and $(2, -1)$;   gradient is -2
  e $(0, 0)$ and $(2, 1)$;   gradient is $\frac{1}{2}$
  f $(0, 4)$ and $(2, -2)$;   gradient is -3
  g $(0, 1)$ and $(2, -2)$;   gradient is $-\frac{3}{2}$
  h $(0, -3)$ and $(2, 5)$;   gradient is 4

EXERCISE 14D

1 **a** -2 **b** 1   **c** -4

2 **a** -1 **b** 4   **c** 3   **d** 5   **e** -3 **f** -4

3 **a** gradient is 4, y-int. is 8   **b** gradient is -3, y-int. is 2
  c gradient is -1, y-int. is 6   **d** gradient is -2, y-int. is 3
  e gradient is 0, y-int. is -2
  f gradient is -3, y-int. is 11
  g gradient is $\frac{1}{2}$, y-int. is -5 **h** gradient is $-\frac{3}{2}$, y-int. is 3
  i gradient is $\frac{2}{5}$, y-int. is $\frac{4}{5}$ **j** gradient is $\frac{1}{2}$, y-int. is $\frac{1}{2}$
  k gradient is $\frac{2}{5}$, y-int. is -2
  l gradient is $-\frac{3}{2}$, y-int. is $\frac{11}{2}$

EXERCISE 14E

1 **a** 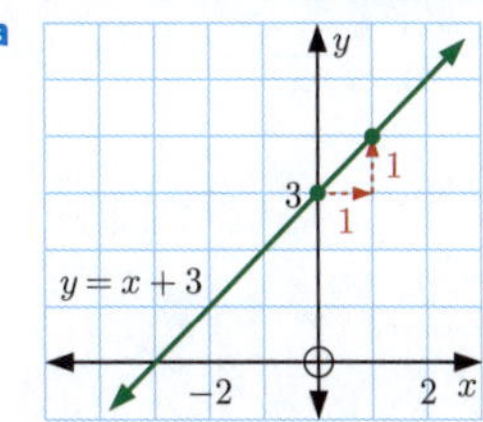
$y = x + 3$

b 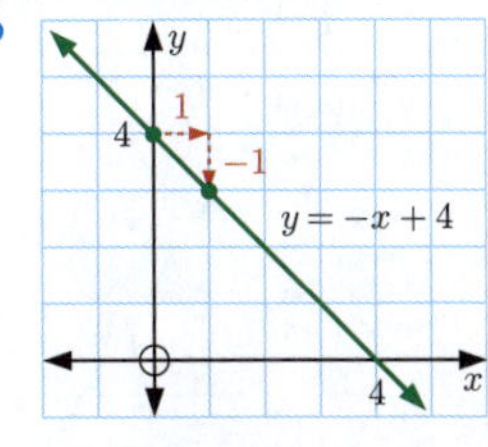
$y = -x + 4$

c 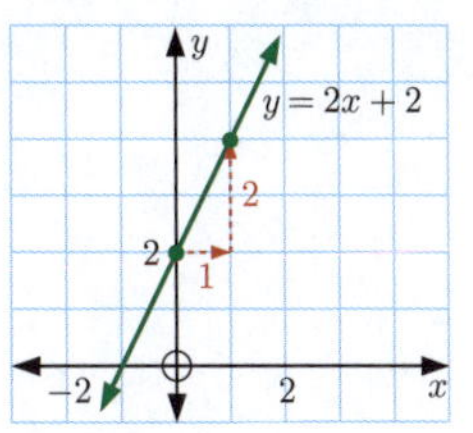
$y = 2x + 2$

d 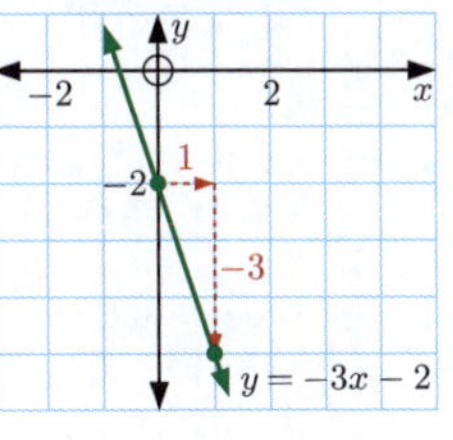
$y = -3x - 2$

e 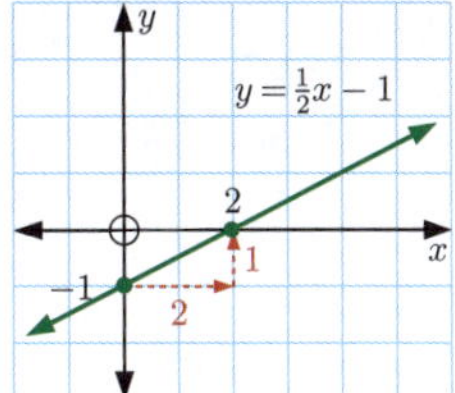
$y = \frac{1}{2}x - 1$

f 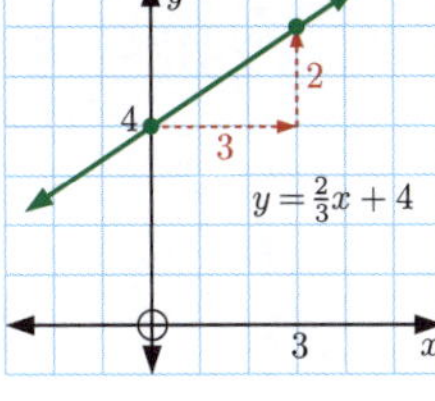
$y = \frac{2}{3}x + 4$

g 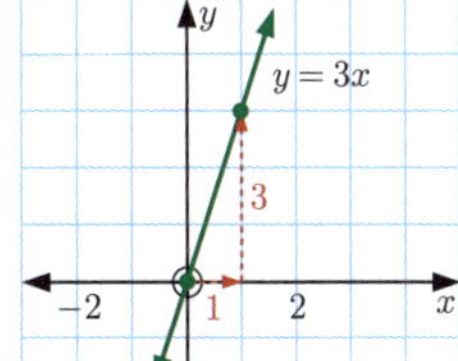
$y = 3x$

h 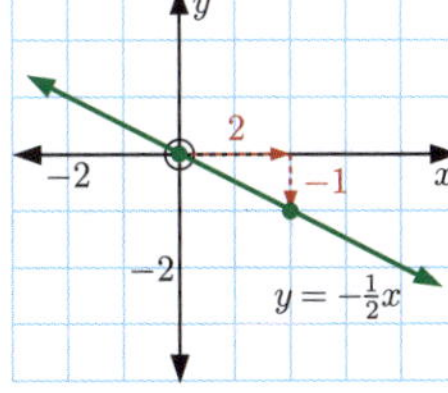
$y = -\frac{1}{2}x$

i 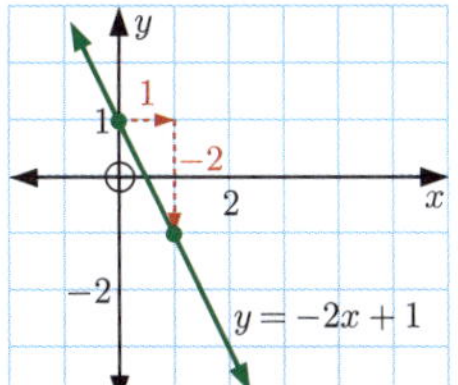
$y = -2x + 1$

j 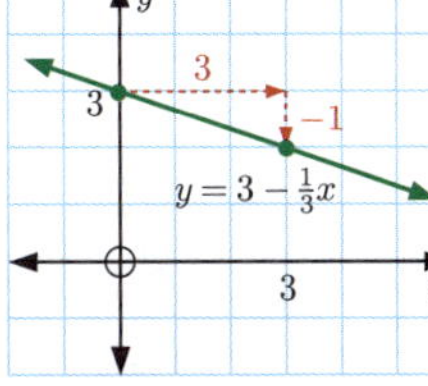
$y = 3 - \frac{1}{3}x$

k 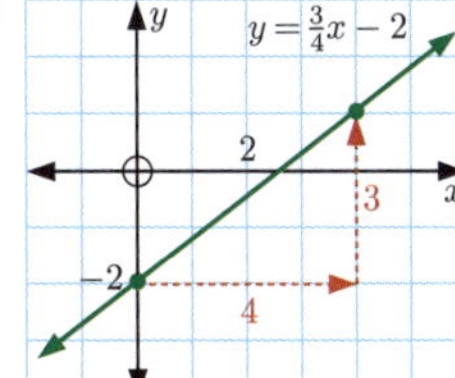
$y = \frac{3}{4}x - 2$

l 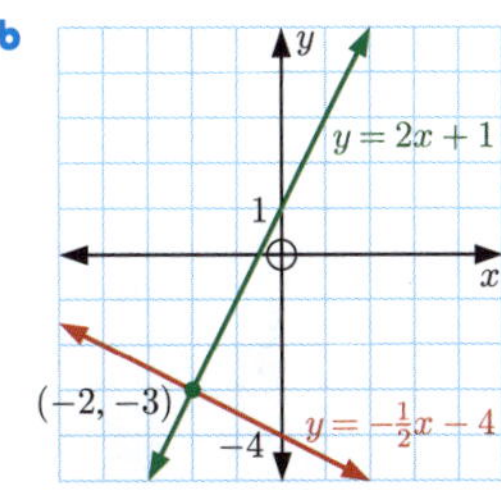
$y = -\frac{1}{4}x - 3$

2 **a** 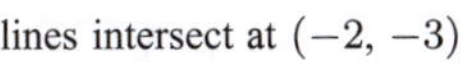
$y = -x + 3$
$y = x + 1$

lines intersect at $(1, 2)$

b
$y = 2x + 1$
$(-2, -3)$
$y = -\frac{1}{2}x - 4$

lines intersect at $(-2, -3)$

c 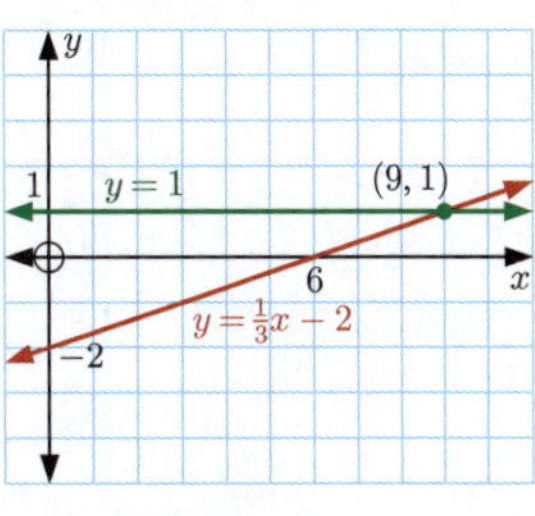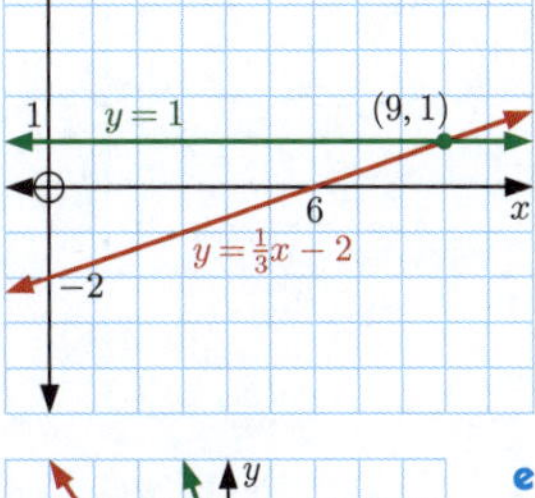
$y = 1$
$(9, 1)$
$y = \frac{1}{3}x - 2$

lines intersect at $(9, 1)$

d 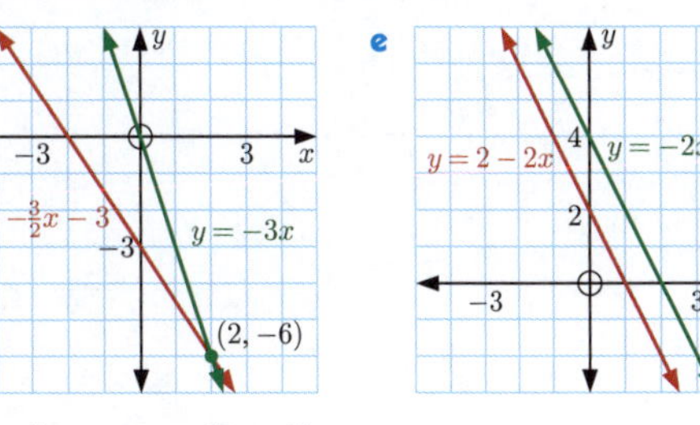
$y = -\frac{3}{2}x - 3$
$y = -3x$
$(2, -6)$

lines intersect at $(2, -6)$

e
$y = 2 - 2x$
$y = -2x + 4$

f 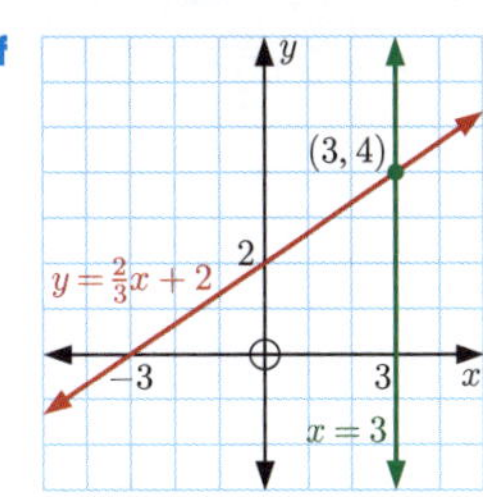
$y = \frac{2}{3}x + 2$
$(3, 4)$
$x = 3$

lines intersect at $(3, 4)$

EXERCISE 14F

1 **a** x-intercept is 1, y-intercept is -2
b x-intercept is -3, y-intercept is -1
c x-intercept is 4, y-intercept is 3
d x-intercept is 3, y-intercept is 2
e x-intercept is 2, y-intercept is -4
f x-intercept is 0, y-intercept is 0

2 **a** **i** x-intercept is -1, y-intercept is 1, gradient is 1
    ii x-intercept is 3, y-intercept is 4, gradient is $-\frac{4}{3}$
    iii x-intercept is -4, y-intercept is 3, gradient is $\frac{3}{4}$

b We cannot determine the gradient of a line from the axes intercepts if the line passes through the origin. In this case the x-intercept and y-intercept are both 0, which only tells us that the line passes through $(0, 0)$. This is not enough information to determine the gradient of the line.

3 **a** 3   **b** -4 **c** $-\frac{1}{4}$ **d** -8 **e** 9   **f** 18

4 **a** $-\dfrac{c}{m}$ **b** $m \neq 0$
   c **i** $m = 0,\ c \neq 0$ **ii** $m = 0,\ c = 0$

EXERCISE 14G

1 **a** 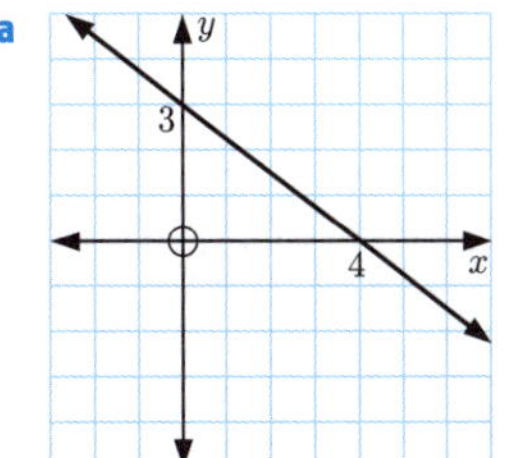
b 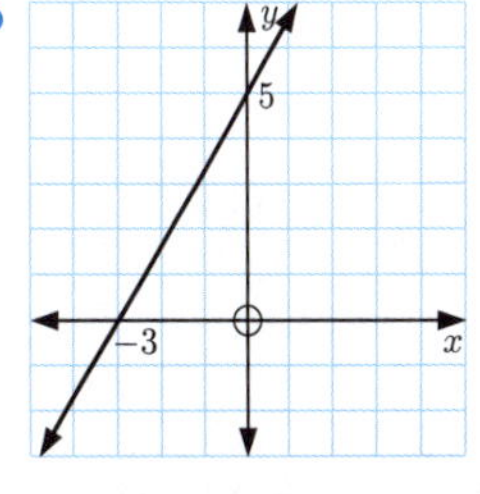

c 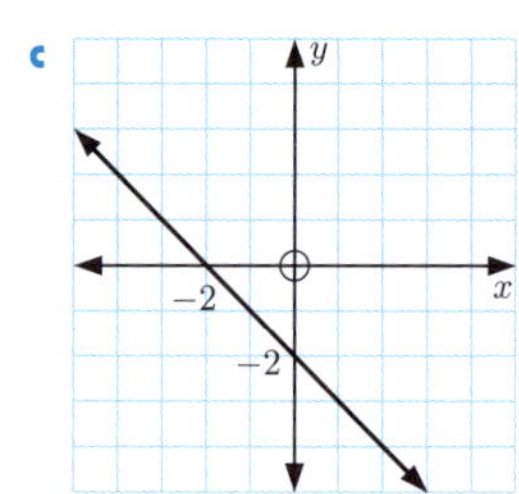   **d** 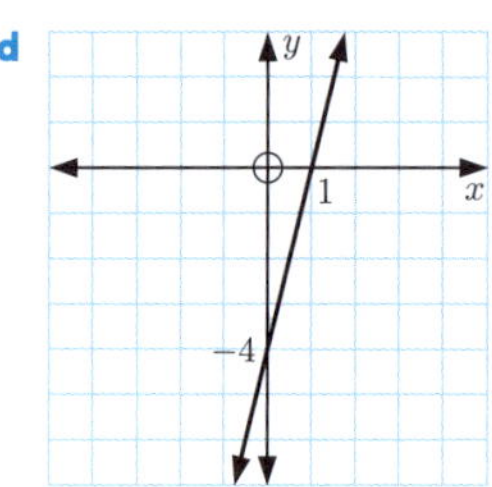

2 a 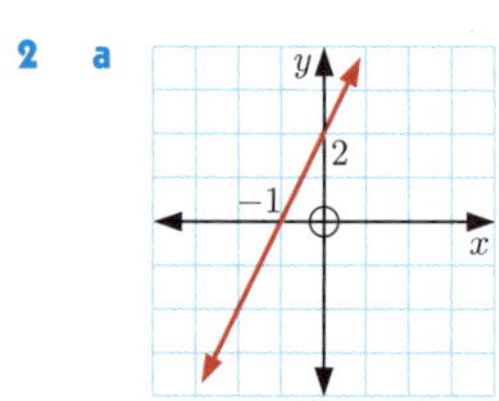

gradient is 2

b 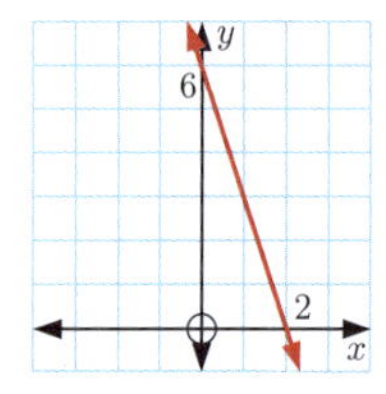

gradient is -3

c 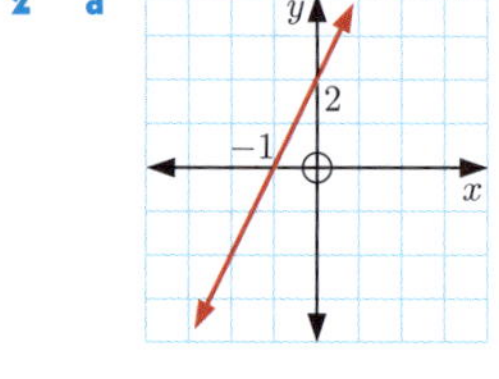

gradient is $\frac{5}{3}$

d 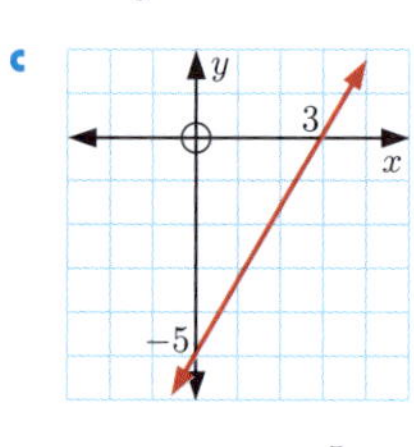

gradient is $-\frac{1}{2}$

EXERCISE 14H

1 a i 2   **ii** $\frac{1}{4}$ **iii** $y = \frac{1}{4}x + 2$

b i 0   **ii** $\frac{1}{2}$ **iii** $y = \frac{1}{2}x$

c i 2   **ii** -1 **iii** $y = -x + 2$

d i 0   **ii** -3 **iii** $y = -3x$

e i -2 **ii** 2   **iii** $y = 2x - 2$

f i -1 **ii** $-\frac{7}{3}$ **iii** $y = -\frac{7}{3}x - 1$

2 a the line is horizontal; $y = 3$

b the line is vertical; $x = -2$

3 a $V = \frac{1}{2}t + 1$ **b** $N = 3 - d$ **c** $C = \frac{4}{5}t + 3$

REVIEW SET 14A

1 a A(2, 3), B(−2, 2), C(3, 0), D(−3, −4), E(1, −3)

b 3rd quadrant   **c** C

2 a 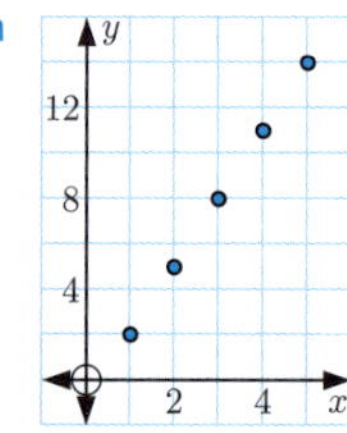

b yes

c i 20 minutes
  ii 23 minutes

3 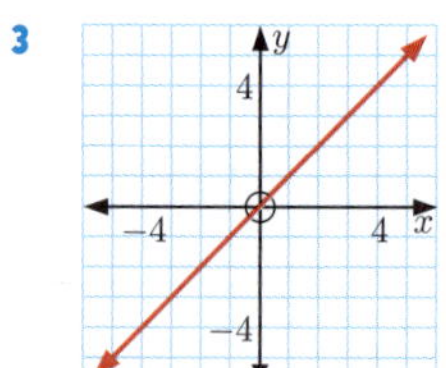

4 a

x	−3	−2	−1	0	1	2	3
y	13	10	7	4	1	−2	−5

b 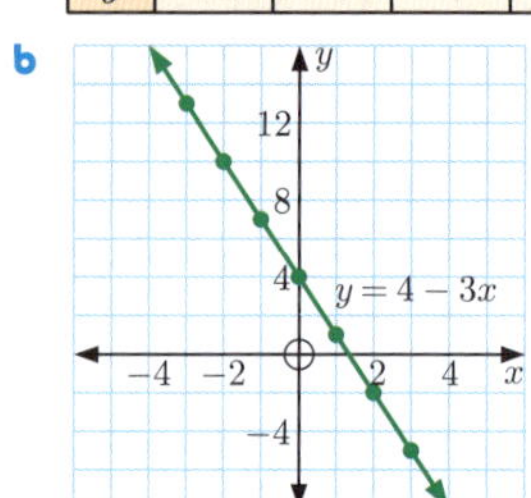

5 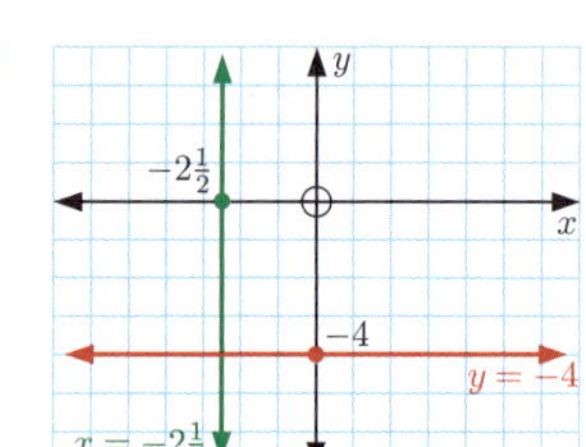

6 a -6 **b** $\frac{1}{2}$ **7 a** 2   **b** $\frac{2}{5}$ **c** $-\frac{4}{3}$

8 a 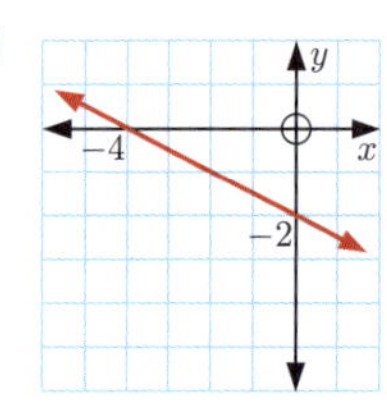   **b**

9 a x-intercept is -3, y-intercept is 6

b x-intercept is 5, y-intercept is -3

c x-intercept is 6, y-intercept is 4

10 a -2 **b** 6   **c** 1

11 a A: 2nd quadrant,   B: 4th quadrant

b 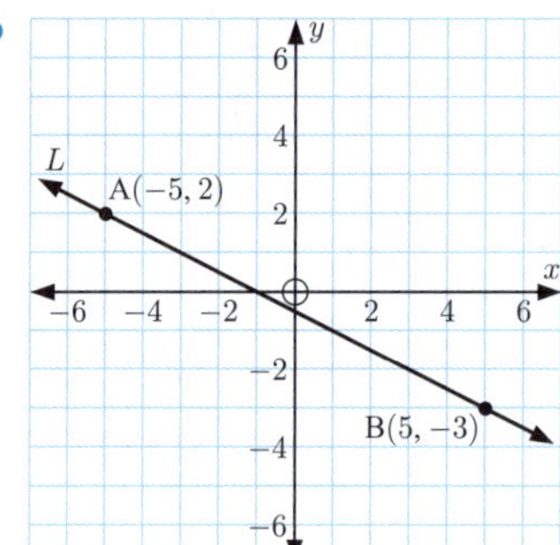   **c** $-\frac{1}{2}$

    d -1

12 a 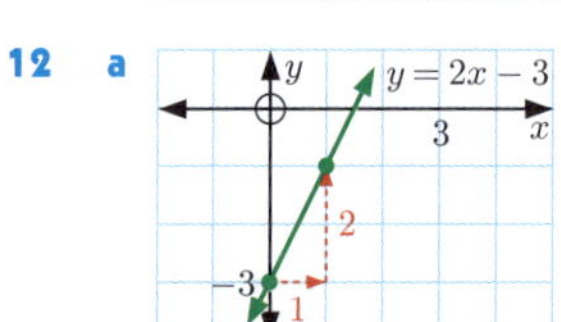   **b** 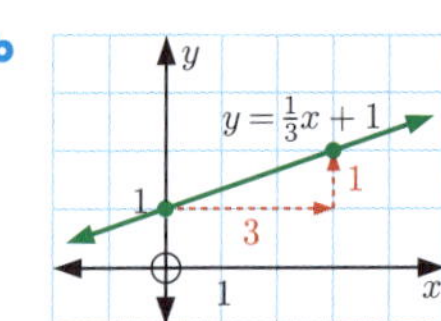

c 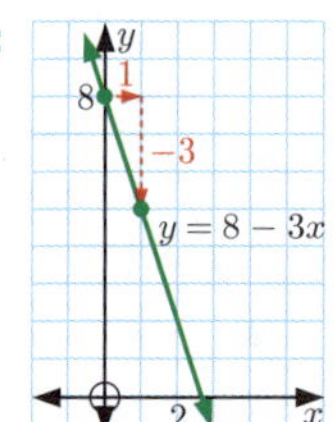

13 **a** 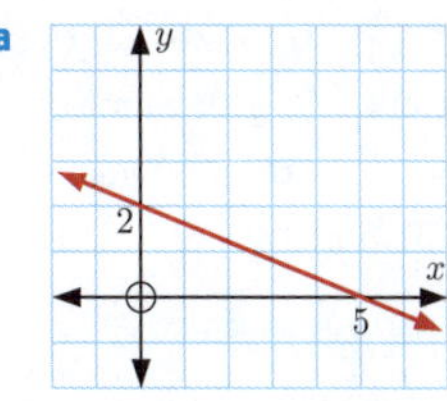 **b** 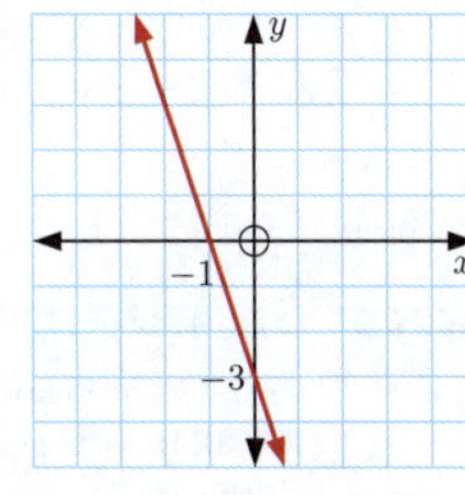

14 **a** **i** -3 **ii** 1 **iii** $y = x - 3$

  b **i** 1 **ii** $-\frac{3}{4}$ **iii** $y = -\frac{3}{4}x + 1$

  c **i** 0 **ii** -2 **iii** $y = -2x$

REVIEW SET 14B

1 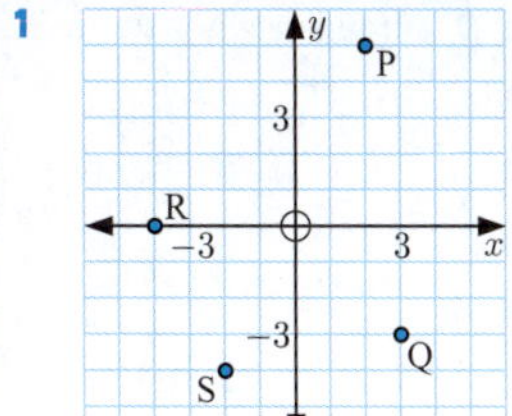

2 $x + y = 5$

3 **a**

x	-3	-2	-1	0	1	2	3
y	$11\frac{1}{2}$	9	$6\frac{1}{2}$	4	$1\frac{1}{2}$	-1	$-3\frac{1}{2}$

  b 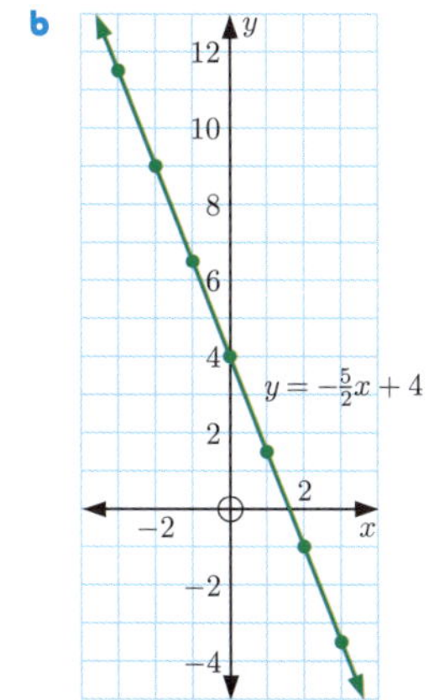

4 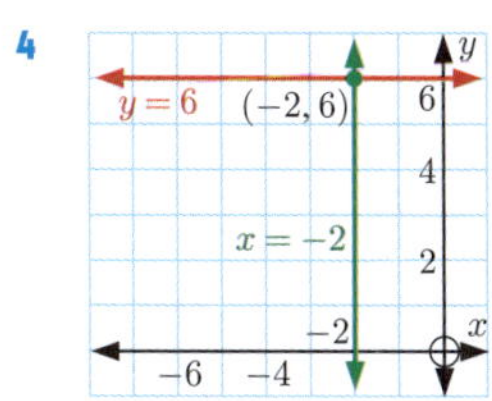

5 yes

6 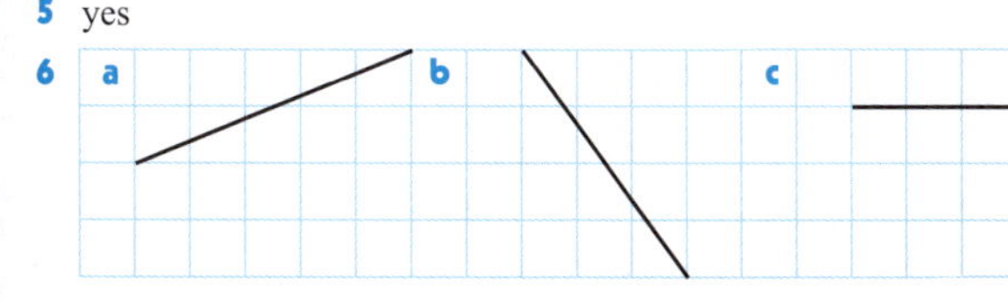

7 **a** 2 **b** $-\frac{5}{3}$

8 **a** 2 **b** -3 **c** $\frac{3}{2}$

9 **a** $(2, 3)$ and $(4, 4)$ **b** 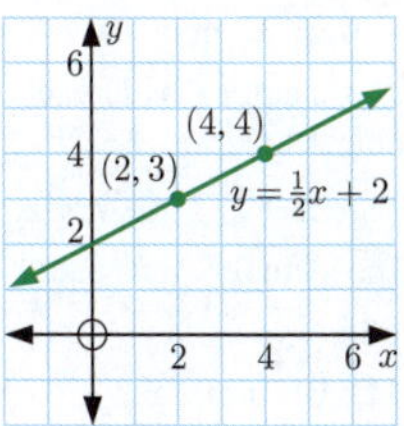

  c $\frac{1}{2}$

10 **a** 5 **b** $-\frac{7}{2}$ **c** 12

11 **a** **i** gradient $= -\frac{1}{3}$, y-intercept $= 2$ **b** **i** gradient $= \frac{3}{4}$, y-intercept $= -3$

  ii 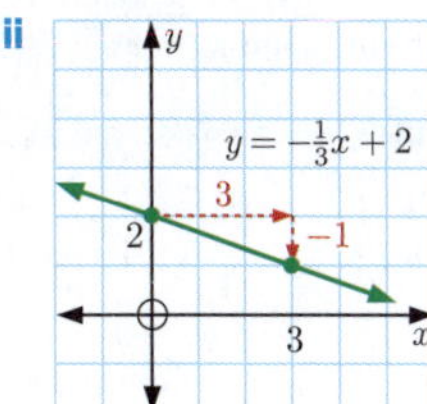 **ii** 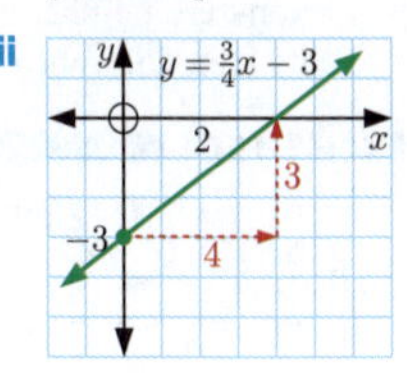

12 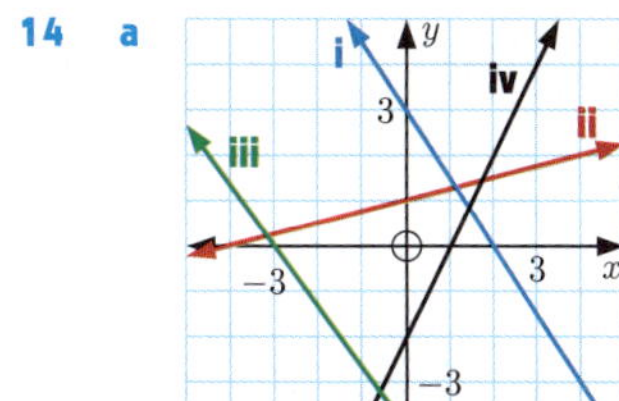

lines intersect at $(3, 7)$

13 **a** $y = -\frac{1}{5}x + 3$ **b** $P = r + 4$

14 **a** 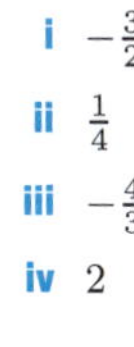

  i $-\frac{3}{2}$

  ii $\frac{1}{4}$

  iii $-\frac{4}{3}$

  iv 2

  b $y = -\dfrac{b}{a}x + b$

EXERCISE 15A

1 **a** $6 : 11$ **b** $5 : 2$ **c** $12 : 25$ **d** $165 : 152$
  e $1 : 3 : 8$ **f** $2 : 5 : 9$

2 **a** $2 : 5$ **b** $4 : 3$ **c** $2 : 3$ **d** $3 : 2 : 2$

3 $5 : 21$

4 **a** $93 : 100$ **b** $36 : 11$ **c** $14 : 31$ **d** $240 : 80$
  e $23 : 120$ **f** $225 : 750$

5 **a** $1600 : 1807 : 2120$ **b** $860 : 1280 : 940$
  c $105 : 120 : 134$

6 **a** $2 : 7$ **b** $3 : 1$ **c** $5 : 4$ **d** $120 : 37$
  e $10 : 7 : 3$ **f** $1000 : 250 : 400$

EXERCISE 15B

1 **a** $12 : 30$ **b** $2 : 5$ **2** **a** $8 : 20$ **b** $0.4 : 1$

3 **a** $14 : 21 : 56$ **b** $0.5 : 0.75 : 2$

4 Each part of $4 : 5$ is multiplied by 3 to obtain $12 : 15$.

5 $12 \div 6 = 2$ and $9 \div 3 = 3$
So, we cannot divide each part of $12 : 9$ by the same non-zero number to obtain $6 : 3$.

6 **a** equal   **b** not equal   **c** not equal   **d** not equal
e not equal   **f** equal   **g** equal   **h** not equal

7 $25 : 35$

8 **a** $4 : 6$ and $8 : 12$ **b** $6 : 9 : 9$ and $24 : 36 : 36$
$4 : 6 = 4 \times 2 : 6 \times 2$ $6 : 9 : 9 = 6 \times 4 : 9 \times 4 : 9 \times 4$
$= 8 : 12$ $= 24 : 36 : 36$

9 **a** $2 : 1$ **b** $4 : 1$
c No, we cannot multiply each part of $2 : 1$ by the same non-zero number to obtain $4 : 1$. When a square is enlarged, its area does not increase in the same ratio as its side length.

EXERCISE 15C

1 **a** $1 : 4$ **b** $2 : 1$ **c** $1 : 6$ **d** $2 : 3$ **e** $7 : 3$
f $3 : 5$ **g** $9 : 7$ **h** $8 : 9$ **i** $3 : 4$ **j** $25 : 12$
k $11 : 7$ **l** $7 : 13$

2 **a** $3 : 4 : 7$ **b** $3 : 8 : 11$ **c** $14 : 4 : 7$

3 **a** $8 : 1$ **b** $4 : 1$ **c** $1 : 1$ **d** $2 : 1$ **e** $1 : 1$

4 **a** $2 : 5$ **b** $5 : 1$ **c** $4 : 1$ **d** $7 : 1$ **e** $1 : 24$
f $6 : 11$ **g** $1 : 2$ **h** $5 : 9$ **i** $7 : 9$

5 **a** $1 : 2$ **b** $4 : 3$ **c** $4 : 9$ **d** $15 : 4$ **e** $4 : 15$
f $3 : 4$

6 **a** $1 : 3$ **b** $4 : 7$ **c** $11 : 9$ **d** $1 : 2$ **e** $4 : 1$
f $3 : 5$ **g** $1 : 9$ **h** $1 : 3$ **i** $27 : 20$

7 **a** $8 : 5$ **b** $5 : 6$ **c** $4 : 3$ **d** $5 : 3$ **e** $3 : 7$
f $1 : 3$ **g** $3 : 16$ **h** $7 : 2$ **i** $4 : 1$ **j** $4 : 7$
k $3 : 10$ **l** $3 : 14$

8 **a** not equal   **b** not equal   **c** equal   **d** equal
e not equal   **f** equal   **g** equal   **h** not equal

EXERCISE 15D

1 **a** $\square = 8$ **b** $\square = 24$ **c** $\square = 10$ **d** $\square = 25$
e $\square = 9$ **f** $\square = 6$ **g** $\square = 3$ **h** $\square = 3$
i $\square = 3$ **j** $\square = 2$ **k** $\square = 5$ **l** $\square = \frac{1}{6}$
m $\square = 15$ **n** $\square = 3$ **o** $\square = 33$ **p** $\square = 12$
q $\square = 32$ **r** $\square = 4\frac{1}{2}$

2 **a** $\square = 12,\ \triangle = 14$ **b** $\square = 6,\ \triangle = 27$
c $\square = 35,\ \triangle = 15$

3 **a** 6 women   **b** 66 men

4 36 chicken kebabs

5 15 hours

6 1.8 L

7 $1\frac{1}{4}$ cups

8

Size	Width	Length
small	6 cm	8 cm
regular	9 cm	12 cm
medium	12 cm	16 cm
large	24 cm	32 cm

9 **a** 210 g   **b** 120 g

10 stocks \$43 200,   shares \$28 800

EXERCISE 15E

1 **a** **i** $\frac{2}{5}$ **ii** $\frac{3}{5}$ **b** **i** 8 balloons   **ii** 12 balloons

2 35 amateurs   **3** \$250 and \$200

4 **a** 14 kg of soil and 6 kg of sand
b 35 kg of soil and 15 kg of sand

5 **a** 20 mL   **b** 30 mL

6 **a** Ben \$1000,   Alice \$1600   **b** \$720   **c** \$975

7 480 g of ricotta,   240 g of mozzarella,   80 g of parmesan

8 6 tonnes   **9** **a** 6000 guests   **b** 4500 guests

10 39 to A and 11 to B   **11** 21 limes

EXERCISE 15F

1 **a** $1 : 100$, scale factor is 100
b $1 : 100\,000$, scale factor is 100 000
c $1 : 3000$, scale factor is 3000
d $1 : 2\,000\,000$, scale factor is 2 000 000
e $1 : 250\,000$, scale factor is 250 000
f $1 : 20\,000\,000$, scale factor is 20 000 000

2 **a** 1 cm represents 2.5 m   **b** 1 cm represents 40 m
c 1 cm represents 5 m   **d** 1 cm represents 250 m
e 1 cm represents 1.5 km   **f** 1 cm represents 220 km

3 **a** $1 : 2000$ **b** $1 : 6\,000\,000$

4 **a** 35 m   **b** 1.6 km   **c** 520 m   **d** 64 m

5 **a** 50 cm   **b** 13 cm   **c** 24 cm   **d** 28 cm

6 **a** 38.4 m by 17.6 m   **b** C

7 **a** 4.88 m   **b** 64 cm   **c** 1.44 m   **d** 1.12 m

8 **a** 38 m   **b** 25 m   **c** 14 m

9 **a**

Scale: 1 cm represents 10 m

b

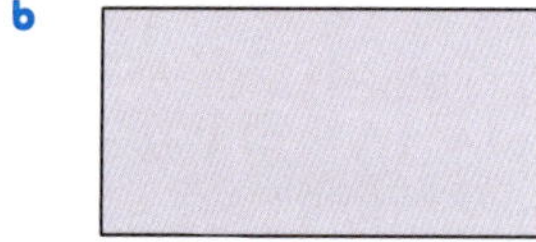

Scale: 1 cm represents 2 km

c

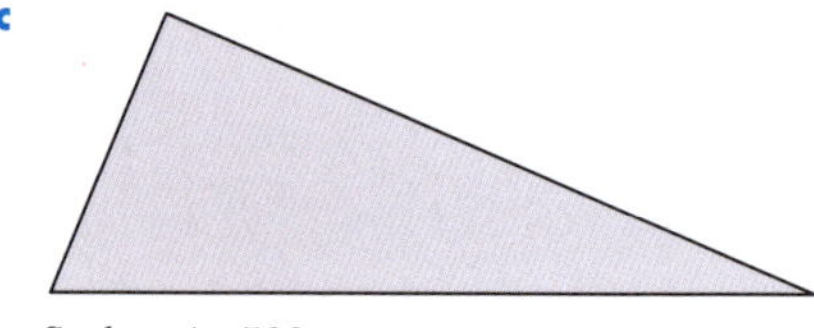

Scale: $1 : 500$

d

Scale: $1 : 250\,000$

10 **Note:**   Other answers are possible.

a

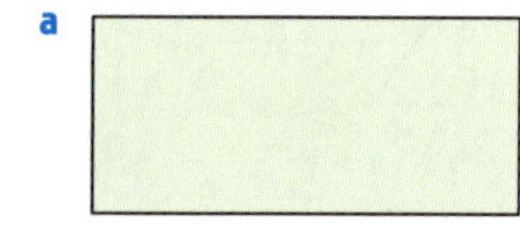

Scale: $1 : 1000$

b

Scale: $1 : 200$

c

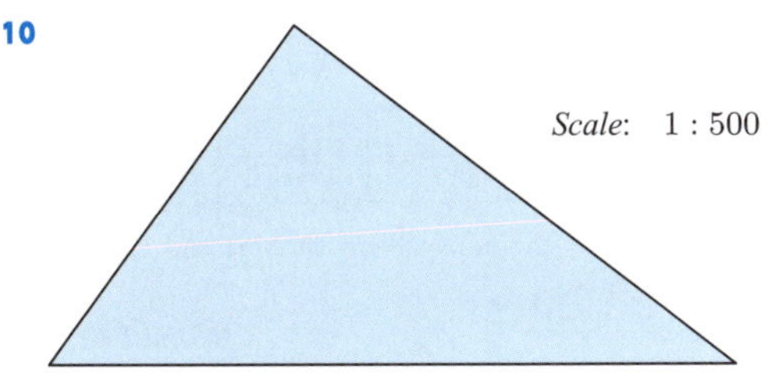

Scale: 1 : 1500

11 a 1 mm represents 1.25 m   **b** 68.75 m   **c** 7.5 m

12 a 120   **b** 10.8 m by 6.6 m   **c** $2735.64

13 a ≈ 1900 km   **b** ≈ 2600 km   **c** ≈ 1600 km

14 a 5.5 mm   **b** 2.25 mm   **c** 1.25 mm

15 a $\frac{1}{8000} = 0.000\,125$

   b   i 0.001 25 mm   **ii** 0.003 mm

REVIEW SET 15A

1 a 5 : 9   **b** 2000 : 839   **c** 141 : 180 : 53

2 a 40 : 120 : 100   **b** 2 : 6 : 5

3 a 4 : 9   **b** 3 : 11   **c** 15 : 11

4 a equal   **b** not equal   **c** equal

5 a □ = 24   **b** □ = 32   **c** □ = $3\frac{1}{2}$

6 a 10 baritone singers   **b** 25 soprano singers

7 $2500 and $1000   **8** 40 kg, 50 kg, 30 kg

9 a 1 cm represents 4 m   **b   i** 12 m   **ii** 16 m

10

Scale: 1 : 500

11 a 20 parts plain flour and 1 part baking powder

   b plain flour   **c** 100 g   **d** 20 g

12 a Izaak $\frac{2}{9}$, Penny $\frac{4}{9}$, Sean $\frac{3}{9} = \frac{1}{3}$ **b   i** $60   **ii** $45

   c Izaak $16, Penny $32, Sean $24   **d** 7 : 11

REVIEW SET 15B

1 a 5 : 3   **b** 60 : 45 : 18   **2 a** equal   **b** not equal

3 a 4 : 11   **b** 14 : 5   **c** 15 : 8

4 a 4 : 7   **b** 7 : 2   **5** □ = 36, △ = 42

6 60 students

7 8 hectares of white grapes and 18 hectares of red grapes

8 $140, $120, $60   **9** 38.3 cm

10 a 40 cm by 22 cm   **b C**

11 a 1 mm represents 20 cm   **b   i** 80 cm   **ii** 3.4 m

12 a   i 3.4 m by 3 m   **ii** 8.6 m by 1 m   **b** 23.2 m²

   c   i 10 mm by 25 mm   **ii** No, it would be too wide.

EXERCISE 16A

1 a $/h   **b** km/h   **c** cents/L   **d** words/min   **e** °C/min

2 75 beats/min   **3** 39 megajoules/day   **4** 1.2 m/year

5 ≈ $6.64/kg   **6 a** ≈ 3.24 g/week   **b** week 2

7 a

City	Population growth	Population growth per year	% population growth
Sao Paulo	2 470 000	247 000	12.7%
Lagos	3 789 000	378 900	37.5%
Cairo	3 946 000	394 600	23.9%
Mexico City	1 714 000	171 400	8.6%
Beijing	4 350 000	435 000	27.7%
Manila	2 077 000	207 700	17.9%

   b Beijing   **c** Lagos

8 a 400 L per day   **b** 8000 L

9 a $21.60 per hour   **b** $410.40   **10** 24 300 L

11 a 322 km   **b** 60 L

12 a $20 per metre   **b** $540   **c** 220 m

13 a   i 18.5 km/L   **ii** ≈ 5.41 L/100 km

   b   i 80 L   **ii** $108

14 $1664.92

EXERCISE 16B

1 a 25 km/h   **b** 30 km/h   **c** 12 km/h   **d** 900 km/h

2 Jason's average speed is 105 km/h which is over the speed limit.
∴ he has broken the law.

3 a 216 km   **b** 45 minutes   **4** ≈ 1.6 m/s

5 a 350 km   **b** 2 hours

6 a Yiren: ≈ 2.67 m/s,   Sean: 2.5 m/s   **b** Yiren by 50 s

7 72 km/h

EXERCISE 16C

1 a 4 g per cm³   **b** 22.6 g per cm³   **c** ≈ 1.06 g per cm³

2 a 11.35 times   **b** ≈ 7.94 times   **c** ≈ 1.47 times

3 die **B**   (6.25 g/cm³ compared with ≈ 5.79 g/cm³)

4 Petrol has a lower density than water so it will float on water. So,
the upper layer is petrol.

5 The density of the doorstop is ≈ 1.33 g/cm³, which is higher
than the density of water. So, the doorstop will sink in water.

6 a 10.5 g/cm³   **b** 3.15 kg   **7** ≈ 11.9 kg

8 ≈ 2.19 g/cm³   **9** ≈ 0.379 g/cm³

EXERCISE 16D

1 10 800 L/hour

2 a 1 beat/s   **b** 3600 beats/h   **c** 86 400 beats/day

3 a 450 L/h   **b** 125 mL/s

4 a 0.3 m/day   **b** 0.0125 m/h   **c** 12.5 mm/h

5 a 1225 g/week   **b** 1.225 kg/week

6 6800 kg/m³   **7 a** 0.179 mg/cm³   **b** less dense

8 a 720 km/h   **b** 162 km/h   **c** 97.2 km/h   **d** 2880 km/h

9 a ≈ 13.9 m/s   **b** ≈ 30.6 m/s   **c** ≈ 5.83 m/s
   d 150 m/s

10 a ≈ 10.42 m/s   **b** ≈ 37.52 km/h

11 a ≈ 37.1 km/h   **b** ≈ 62.1 km/h   **c** ≈ 55.4 km/h
   d 6 km/h   **e** ≈ 106 km/h

12 ≈ 1 h 58 min

EXERCISE 16E

1 a 150 words   **b** 6 minutes   **c** 75 words/min

2 a 100 km   **b** 50 km/h   **c** 100 km   **d** 100 km/h
   e 200 km   **f** ≈ 66.7 km/h

3 a $120   **b** $20/h   **c** $180   **d** $30/h   **e** $22.50/h

4 a 40 km/h
   b The gradient of the line is 40. This corresponds to the speed of the cyclist.
   c The speed of the cyclist is constant.

5 a *Age* is the independent variable and *mass* is the dependent variable.
   b *Time* is the independent variable and *score* is the dependent variable.
   c *Amount of water* is the independent variable and *amount the plant grows* is the dependent variable.

6 a

Weight of chickpeas (kg)	0	1	2	3	4	5
Cost ($)	0	6	12	18	24	30

   b *Weight of chickpeas* is the independent variable and *cost* is the dependent variable.
   c 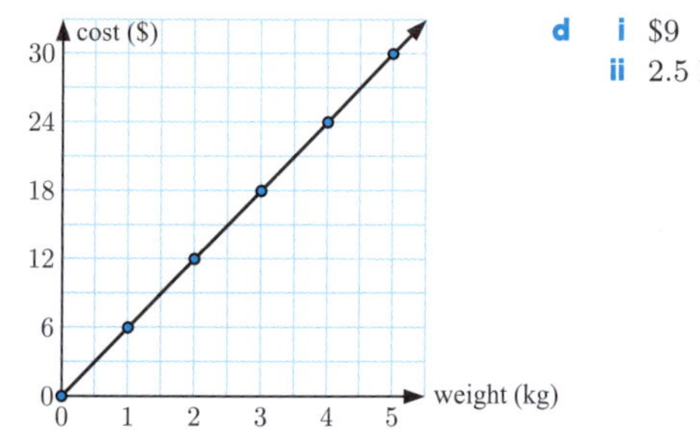
   d i $9
      ii 2.5 kg
   e The gradient of the line is 6. This corresponds to the cost of the chickpeas per kg.

7 a

Time (h)	0	1	2	3	4	5
Distance (km)	0	40	80	120	160	200

   b *Time* is the independent variable and *distance* is the dependent variable.
   c 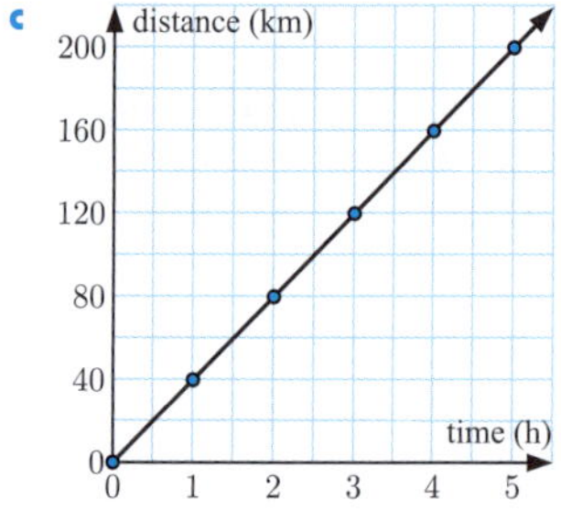
   d 100 km
   e 3.75 hours
   f The gradient of the line is 40. This corresponds to the speed of the cargo ship.

8 a

Fuel used (L)	0	30	60	90
Distance (km)	0	40	80	120

   b *Fuel used* is the independent variable and *distance* is the dependent variable.
   c 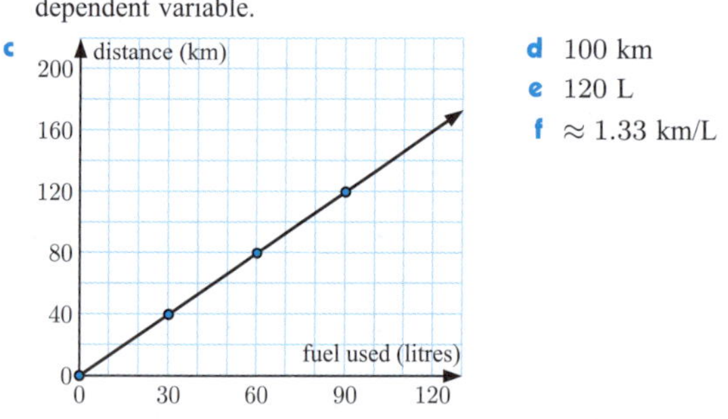
   d 100 km
   e 120 L
   f ≈ 1.33 km/L

9 a

Time (hours)	0	1	2	3	4	5
Charge (%)	80	75	70	65	55	45

   b *Time* is the independent variable and *charge* is the dependent variable.
   c 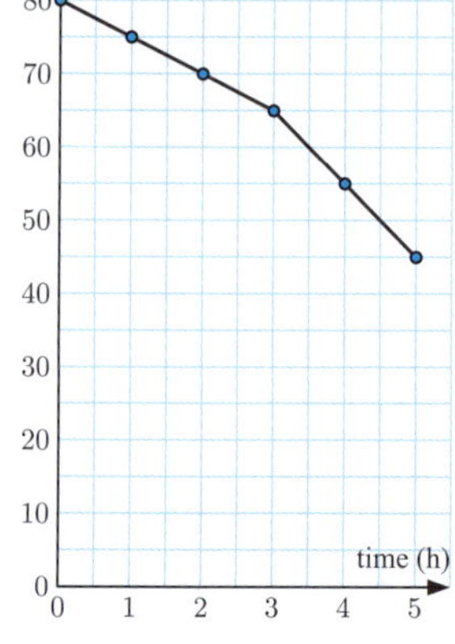
   d 67.5%
   e 4.5 hours
   f 7% per h

REVIEW SET 16A

1 a $8.50/kg   **b** $21.25

2 a 12 L/min   **b** 10 minutes

3 freight train (96.25 km/h compared with ≈ 84.7 km/h)

4 2850 km   **5** 2.8 g/cm^3 **6** ≈ 4.87 kg

7 a 2.5 beats/s   **b** 9000 beats/h

8 a 600 kJ   **b** 5 minutes   **c** 150 kJ/min

9 a 20 km   **b** 30 min   **c** 40 km/h   **d** 52 km/h

10 a

Length of fabric (m)	0	1	2	3	4	5
Cost ($)	0	4	8	12	16	20

   b *Length of fabric* is the independent variable and *cost* is the dependent variable.
   c 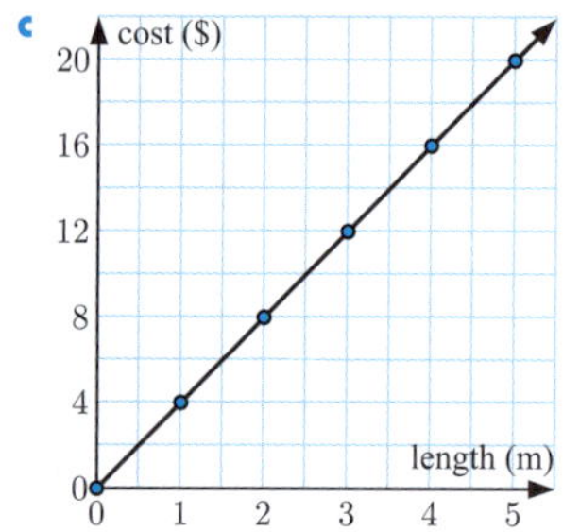
   d i $10
      ii 3.5 m

11 a Speed is a comparison between the distance travelled and the time taken. It is a measure of how fast an object is travelling.
   b train   **c i** ≈ 21.7 m/s   **ii** 114 km/h
   d 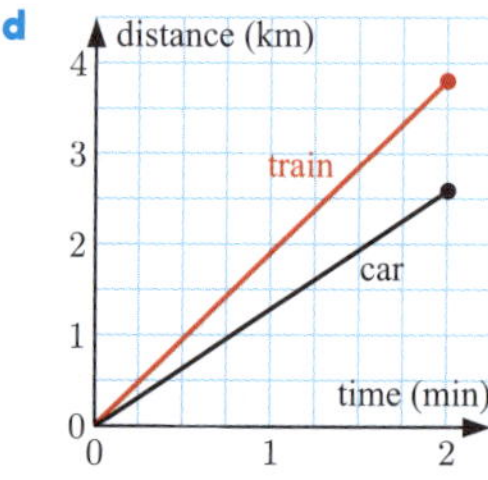

   e The speeds of the car and train correspond to the gradient of the respective line segments.

REVIEW SET 16B

1 a 66 km/h **b** 18 km/L   **2** $12 000/year

3 a 50 km/h **b** 6.5 hours **c** ≈ 13.9 m/s

4 7 km **5** 720 g

6 Density of cylinder ≈ 0.707 g/cm^3, which is lower than the density of water. So, the cylinder will float.

7 12.5 mm/month **8 a** 6000 L/h **b** 1008 kL/week

9 a 150 km **b** 2 hours **c** 50 km/h

10 a

Time (s)	0	5	10	15	20	25
Distance (km)	0	15	30	45	60	75

b 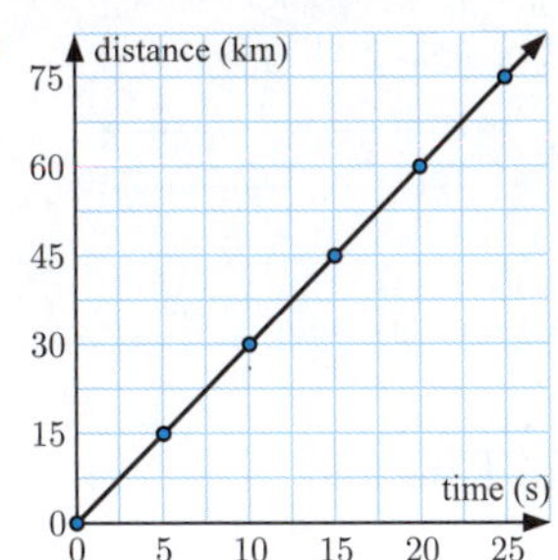 **c** 21 km

11 a **A** ≈ 12.0 g/cm^3, **B** ≈ 7.37 g/cm^3, **C** ≈ 10.5 g/cm^3

b i **A** Palladium, **B** Tin, **C** Silver

ii Measurements made are not exact or have been rounded.

c i 7.722 g **ii** Iridium

EXERCISE 17A

1 a unlikely **b** slightly more than 50-50 chance

c highly unlikely

2 a Lions **b** Strikers **c** 41% **d** false

3 a Patricia **b** Rob

c 100% = 1, which means it is certain that one of the students will win the race.

d 35% = 0.35

e It is highly unlikely that Rob will win the race.

f i P$(E) = 0.22$

ii E' is the event that Edward will *not* win the race.

iii P$(E') = 0.78$

4 a 50-50 chance **b** certain **c** impossible

d highly unlikely

5 a S' is the event that it will *not* snow tomorrow.

b P$(S') = 0.97$

6 No, both R and S could occur at the same time.

EXERCISE 17B

1 a {1, 2, 3, 4, 5, 6}, 6 outcomes

b {A, B, C}, 3 outcomes

c {summer, autumn, winter, spring}, 4 outcomes

d {hearts, spades, clubs, diamonds}, 4 outcomes

e {10, 11, 12, 13, 14, 15, 16, 17, 18, 19}, 10 outcomes

f {1, 2, 4, 5, 8, 10, 20, 25, 40, 50, 100, 200}, 12 outcomes

2 a 6 outcomes **b i** 4 outcomes **ii** 3 outcomes

3 a

6 outcomes

b 18 outcomes

c 20 outcomes

4 a, c, d

b 24 outcomes

5 a, c, d **b** 12 outcomes

EXERCISE 17C

1 a $\frac{1}{6}$ **b** $\frac{3}{6} = \frac{1}{2}$ **c** $\frac{5}{6}$ **d** $\frac{2}{6} = \frac{1}{3}$

2 a $\frac{1}{10}$ **b** $\frac{1}{5}$ **c** $\frac{3}{10}$ **d** $\frac{3}{5}$ **e** $\frac{2}{5}$ **f** $\frac{3}{5}$

3 a $\frac{1}{16}$ **b** $\frac{1}{8}$ **c** $\frac{7}{16}$ **4 a** $\frac{2}{3}$ **b** $\frac{1}{3}$

5 a $\frac{1}{5}$ **b** $\frac{4}{15}$ **c** $\frac{8}{15}$ **d** 0 **e** $\frac{4}{5}$ **f** $\frac{11}{15}$

6 a i $\frac{1}{4}$ **ii** blue is more likely

b i $\frac{3}{8}$ **ii** red is more likely

c i 1 **ii** red is more likely

7 a $\frac{16}{45}$ **b** $\frac{4}{9}$ **c** $\frac{5}{9}$

8 a $\frac{4}{25}$ **b** $\frac{12}{25}$ **c** $\frac{16}{25}$ **d** $\frac{9}{25}$

9 a $\frac{2}{25}$ **b** $\frac{1}{10}$ **c** $\frac{3}{10}$ **d** $\frac{29}{50}$

10 a **b** 4 outcomes

c i $\frac{1}{4}$ **ii** $\frac{1}{2}$

iii $\frac{3}{4}$

11 a **b** 6 outcomes

c i $\frac{1}{6}$ **ii** $\frac{1}{3}$

iii $\frac{1}{3}$ **iv** $\frac{1}{3}$

v $\frac{1}{2}$

12 a 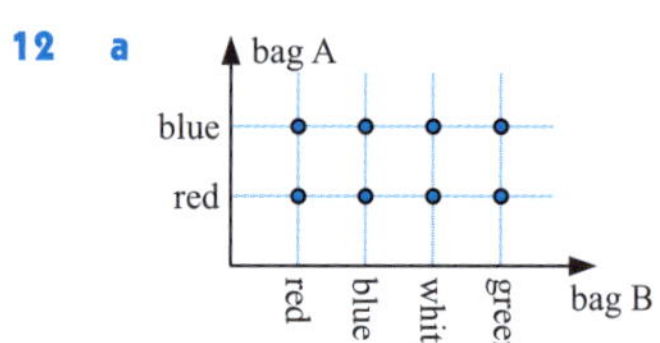   **b** **i** $\frac{1}{8}$ **ii** $\frac{1}{4}$ **iii** $\frac{1}{4}$ **iv** $\frac{3}{4}$ **v** $\frac{3}{4}$

13 a 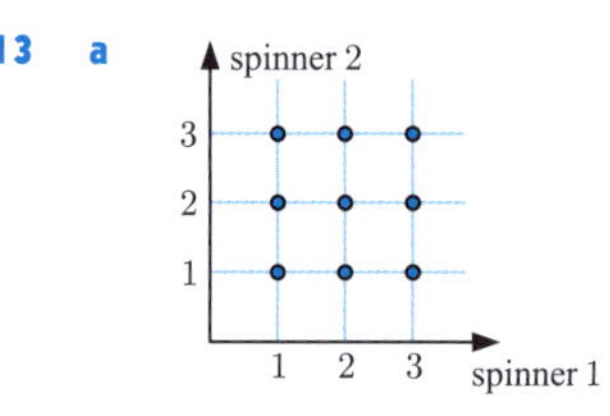   **b** **i** $\frac{1}{9}$ **ii** $\frac{4}{9}$ **iii** $\frac{2}{9}$ **iv** $\frac{2}{9}$ **v** $\frac{1}{3}$ **vi** $\frac{4}{9}$

14 a 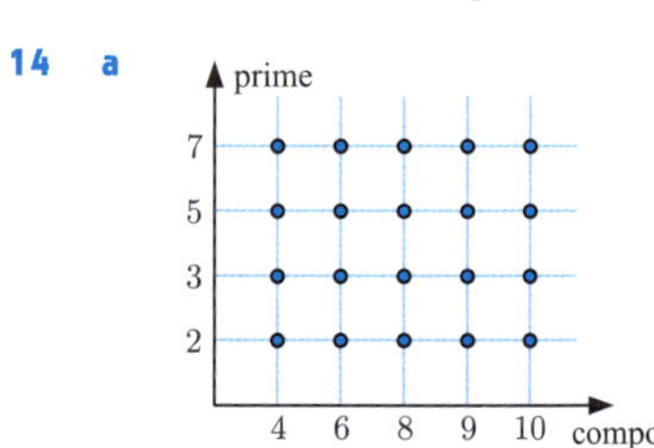   **b** **i** $\frac{1}{20}$ **ii** $\frac{3}{20}$ **iii** $\frac{1}{5}$ **iv** $\frac{3}{20}$ **v** $\frac{1}{4}$ **vi** $\frac{7}{20}$

EXERCISE 17D

1 a $\frac{1}{10}$ **b** $\frac{1}{5}$ **2 a** $\frac{1}{12}$ **b** $\frac{1}{4}$

3 a $\frac{3}{14}$ **b** $\frac{4}{21}$ **c** $\frac{1}{14}$ **d** $\frac{2}{7}$

4 a $\frac{8}{21}$ **b** $\frac{1}{7}$ **c** $\frac{2}{7}$

5 a **i** $\frac{2}{15}$ **ii** $\frac{2}{5}$ **iii** $\frac{1}{5}$ **iv** $\frac{4}{15}$

b They both miss the target.

c $\frac{2}{15} + \frac{2}{5} + \frac{1}{5} + \frac{4}{15} = 1$, exactly one of the events in **a** will occur.

6 a 0.42   **b** 0.58

EXERCISE 17E

1 a ≈ 0.78 **b** ≈ 0.22 **2 a** ≈ 0.325 **b** ≈ 0.675

3 a ≈ 0.0324 **b** ≈ 0.291

4 a ≈ 0.21 **b** **i** ≈ 0.0441 **ii** ≈ 0.6241

5 a **i** ≈ 0.426 **ii** ≈ 0.463 **iii** ≈ 0.446

b We expect **a iii** to be the most accurate because it uses the largest sample size.

EXERCISE 17F

1 a

Type	Frequency	Relative frequency
Shortbread	62	≈ 0.332
Chocolate chip	125	≈ 0.668
Total	187	1

b **i** ≈ 0.332 **ii** ≈ 0.668

2 a

Type	Frequency	Relative frequency
Platinum	419	≈ 0.213
Gold	628	≈ 0.319
Silver	921	≈ 0.468
Total	1968	1

b 1968 members   **c** **i** ≈ 0.468 **ii** ≈ 0.787

3 a

Type	Frequency	Relative frequency
Espresso	49	≈ 0.098
Milk coffee	187	≈ 0.375
Tea	150	≈ 0.301
Hot chocolate	113	≈ 0.226
Total	499	1

b **i** ≈ 0.226 **ii** ≈ 0.399

4 a

City	Frequency	Relative frequency
Helsinki	750	0.3
Tallinn	600	0.24
Klaipėda	400	0.16
Gdańsk	750	0.3
Total	2500	1

b **i** 0.3   **ii** 0.4   **iii** 0.76

EXERCISE 17G

1 a   *Mathematics*   **b** 54 students

English		Yes	No	Total
	Yes	29	6	35
	No	7	12	19
	Total	36	18	54

c 7 students completed their Mathematics homework but not their English homework.

d ≈ 0.667 **e** ≈ 0.171

2 a   *Preference*

Age		Like	Dislike	Undecided	Total
	Under 40	29	49	26	104
	Over 40	48	14	34	96
	Total	77	63	60	200

b 49 people surveyed were under 40 and disliked the new show.

c **i** ≈ 0.315 **ii** ≈ 0.17

3 a   *Gender*

Instrument		Male	Female	Total
	Yes	39	14	53
	No	17	2	19
	Total	56	16	72

b ≈ 0.875 **c** ≈ 0.304

d The estimate in **c** is more likely to be accurate, since it involves a larger sample size.

4 a   *Mathematics*

Science		A	B	C	Total
	A	7	5	3	15
	B	6	8	4	18
	C	4	1	2	7
	Total	17	14	9	40

b **i** ≈ 0.15 **ii** ≈ 0.625 **iii** ≈ 0.425 **c** ≈ 0.2

5 a   *University*   **b** ≈ 0.0875

Children		Yes	No	Total
	Yes	80	21	101
	No	45	14	59
	Total	125	35	160

EXERCISE 17H

1 a 24 students   **b** **i** $\frac{1}{6}$ **ii** $\frac{17}{24}$

2 a 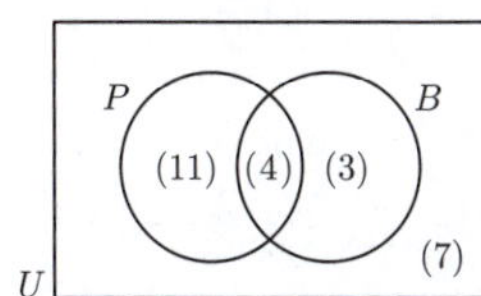 **b** $\frac{18}{25}$

3 a 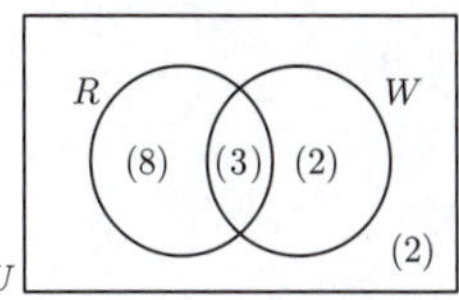 **b** $\frac{2}{15}$

4 a 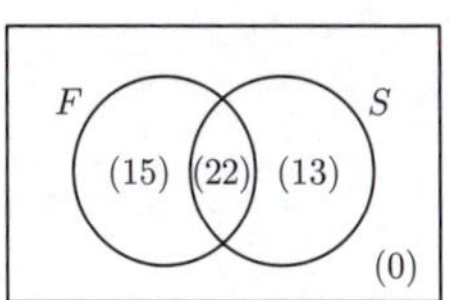 **b** $\frac{3}{10}$ **c** $\frac{13}{35}$

5 a **b** $\frac{49}{75}$ **c** $\frac{14}{17}$

EXERCISE 17I

1 250 times **2** 75 times **3** 5000 hatchlings

4 ≈ 569 books

5 a 80 times **b** 60 times **c** 40 times **d** 20 times

6 a 100 times **b** 300 times **c** 200 times

7 a 25 times **b** 50 times **c** 225 times **d** 225 times
e 75 times **f** 100 times

REVIEW SET 17A

1 a {a, b, c, d, e, f, g, h, i, j, k, l, m, n, o, p, q, r, s, t, u, v, w, x, y, z}, 26 outcomes
b {blue, white, yellow, red, black, green}, 6 outcomes

2 a E' is the event that Ned's bike will *not* get a puncture next week.
b $P(E') = 0.87$

3 a $\frac{1}{4}$ **b** 0 **c** $\frac{3}{20}$ **4 a** $\frac{1}{9}$ **b** $\frac{4}{9}$ **c** $\frac{4}{9}$

5 a 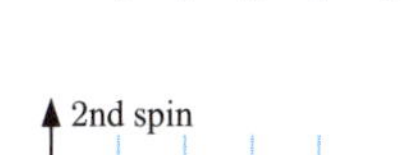 **b i** $\frac{1}{10}$
ii $\frac{3}{5}$

6 a **b i** $\frac{9}{16}$ **ii** $\frac{1}{8}$
iii $\frac{7}{16}$ **iv** $\frac{7}{16}$
c b i and **b iii**

7 a 0.03 **b** 0.68 **c** 0.17

8 a ≈ 0.35 **b** ≈ 0.423

9 a

Reason	Frequency	Relative frequency
Homework	29	≈ 0.186
Social networking	43	≈ 0.276
Playing games	15	≈ 0.096
Downloading music	69	≈ 0.442
Total	156	1

b i ≈ 0.186 **ii** ≈ 0.558

10 a

	Age			
Type	Less than 3 years old	3 - 10 years old	More than 10 years old	*Total*
Automatic	9	17	9	35
Manual	1	7	6	14
Total	10	24	15	49

b i ≈ 0.714 **ii** ≈ 0.265 **c** ≈ 0.257

11 a 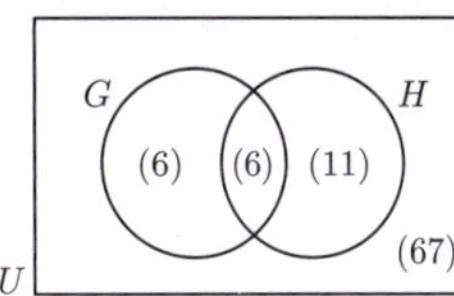 **b** $\frac{23}{90}$
c $\frac{11}{17}$

12 13 times

REVIEW SET 17B

1 a highly likely **b** 0.07

2 a {red, pink, white, brown}
b {10, 15, 20, 25, 30, 35, 40, 45, 50, 55, 60, 65, 70, 75, 80, 85, 90, 95}

3 a $\frac{1}{5}$ **b** $\frac{1}{2}$

4 Choosing a whole number from 1 to 20.

5 a **b i** $\frac{1}{5}$
ii $\frac{1}{5}$

6 a i $\frac{3}{10}$ **ii** $\frac{3}{20}$ **b** No, $\frac{3}{10} + \frac{3}{20} = \frac{9}{20} \neq 1$

7 ≈ 0.417

8 a

Year level	Frequency	Relative frequency
7	38	≈ 0.306
8	29	≈ 0.234
9	57	≈ 0.460
Total	124	1

b i ≈ 0.460 **ii** ≈ 0.766

9 a no **b i** $\frac{2}{5} = 0.4$ **ii** $\frac{9}{25} = 0.36$ **c** $\frac{18}{125} = 0.144$
d Use the relative frequency of students with a food allergy in either or both Class 8A and Class 8B.

10 a

		Radio station			
		Mash	Planet	Gold	*Total*
Age	Under 30	43	14	24	81
	Over 30	12	35	22	69
	Total	55	49	46	150

b i ≈ 0.16 **ii** ≈ 0.367 **iii** ≈ 0.673 **c** ≈ 0.218

11 a **b** $\frac{7}{40}$

12 a $\frac{1}{5}$ **b** 12 times

EXERCISE 18A

1 a a census **b** a sample **c** a sample **d** a census

2 a In general, people at a beach are more likely to be able to swim. People in other places do not get asked.
  b Could be biased since sunlight may not reach the plants because of the barn and so growth could be slower.
  c Workers in Liverpool are likely to take longer to travel to work than workers in the rest of England, as Liverpool is a large city.
  d People who live in the CBD are likely to eat out more than the average person.

3 Maura's sample is very small.

4 a 70% preferred tea
  b The estimate is reliable as the sample is random and sufficiently large.

EXERCISE 18B

1 a Could be: strawberry, lime, chocolate, banana, vanilla, raspberry, orange, and so on.
  b Could be: bicycle, car, bus, train, aeroplane, ship, and so on.
  c Could be: string, wind, brass, percussion, and so on.
  d Could be: talking, phone, email, TV, radio, and so on.

2 a

Response	Tally	Frequency
Beach	$\|\|\|\| \|\|\|$	8
Movies	$\|\|\|\| \ \|\|\|\| \ \|\|\|\|$	14
Park	$\|\|\|\| \ \|$	6
Shopping	$\|\|\|\| \ \|\|\|\| \ \|\|\|$	13
Video games	$\|\|\|\| \ \|\|\|\|$	9
	Total	50

  b 13 students
  c movies
  d $\frac{4}{25}$

3 a 36 guests
  b England. There were more guests who lived in England than in any of the other countries.
  c 4 more guests **d** $\approx 11.1\%$

4 a

Cuisine	Tally	Frequency
Chinese	$\|\|\|$	3
Greek	$\|\|\|\| \ \|$	6
Indian	$\|\|\|$	3
Italian	$\|\|\|\| \ \|\|\|\|$	9
Thai	$\|\|\|\|$	4
	Total	25

  b 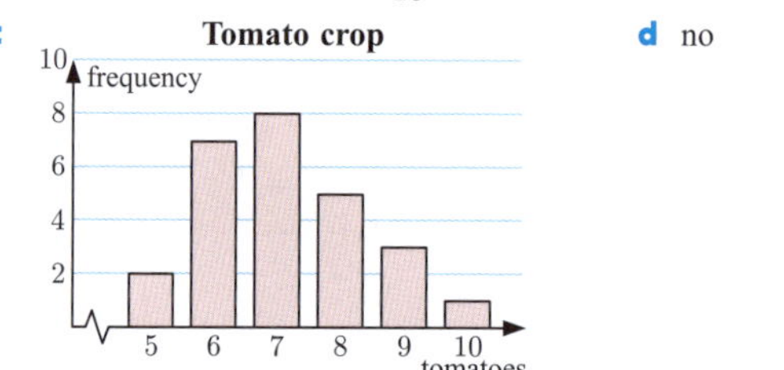

c Italian, Greek, Thai

5 a 50 households
  b C, brand C was the most common brand.
  c 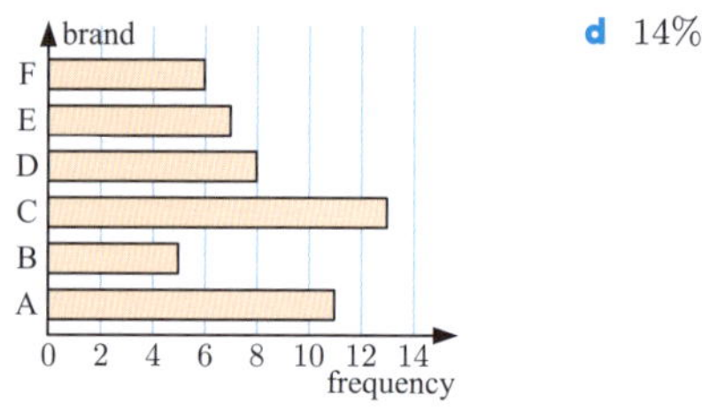

  d 14%

6 a **i** £170 000 **ii** £270 000
  b £80 000 higher **c** $\approx 24.4\%$ lower

7 a Food Galore
  b **i** ≈ 4479 supermarkets **ii** ≈ 617 supermarkets
  iii ≈ 1610 supermarkets

8 a **i** $\approx 38\%$ **ii** $\approx 19\%$
  b **i** ≈ 29 children **ii** ≈ 10 more children

9 a 1500 performances **b** 350 performances
  c $\approx 5.77\%$ decrease **d** 2012

10 a $1307.1 million **b** 6.7% **c** teaching hospitals
  d **i** $\approx \$92.8$ million **ii** $\approx \$48.4$ million

11 a $\approx 18.3\%$ **b** $\frac{3}{20}$ **c** 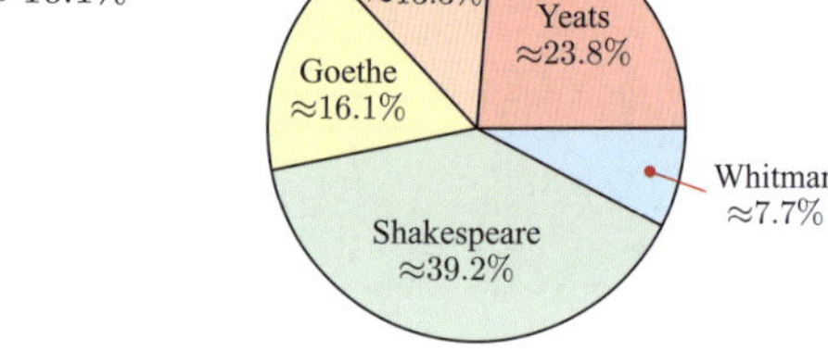

12 a ≈ 2700 students **b** $\approx 25.9\%$ **c** 2017

13 a **i** $\approx 39.2\%$ **ii** $\approx 16.1\%$ **b**

EXERCISE 18C

1 a numerical **b** categorical **c** categorical **d** numerical
  e numerical **f** categorical **g** numerical

2 a 37 teenagers **b** 30 teenagers **c** $\frac{6}{37}$ **d** $\approx 43.2\%$

3 a 3 residents **b** 76% **c** an outlier

4 a 26 tomato plants **b** $\frac{9}{26}$ **c**

  d no

5 **a**

Number sold	Tally	Frequency
8	|	1
9		0
10		0
11	|||	3
12				5
13	|||	3
14	||||	4
15				5
16	|||	3
17	||	2
18	||	2
	Total	28

b

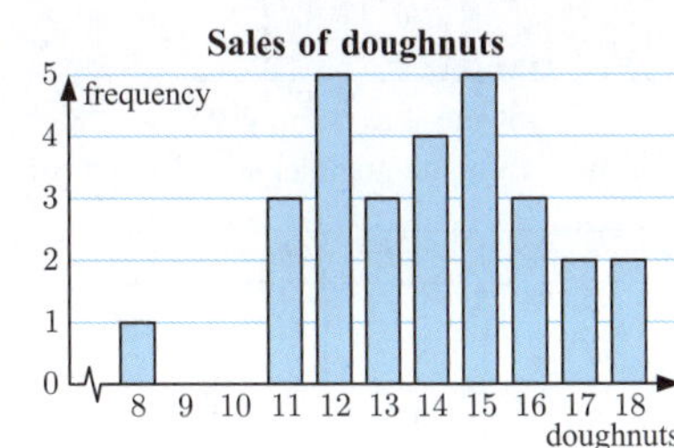

c ≈ 42.9%   **d** "8" is an outlier.

6 **a**

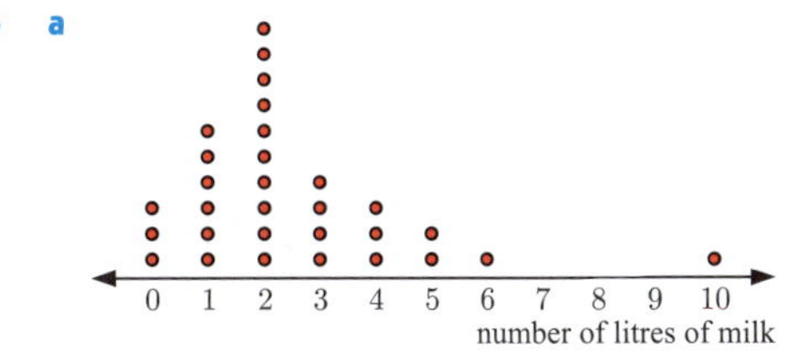

b Yes, "10" is an outlier.   **c** 10%   **d** ≈ 36.7%

EXERCISE 18D

1 **a**

Score	Tally	Frequency
70 to 74	||| 8
75 to 79	|||| 16
80 to 84	||| 12
85 to 89	||||	4
	Total	40

b 4 scores   **c** 20%   **d** ".... 75 to 79 ...."

2 **a** 33 students   **b** 80 to 89 words   **c** 5 students
d 3 students   **e** no

3 **a**

Number of customers	Tally	Frequency
70 to 79	||| 7
80 to 89	||| 8
90 to 99	||| 7
100 to 109	||| 10
110 to 119	|||| 12
120 to 129	||| 6
	Total	50

b

Customers in a delicatessen

frequency — number of customers

c 110 to 119 customers   **d** 56%

EXERCISE 18E

1 **a** 46   **b** 56   **c** 3 data values   **d** 15 data values
2 **a** 73   **b** 122   **c** 12 data values   **d** 4 data values
e 6.25%

3 **a**

1	9 8 9
2	7 9 7 4 1
3	4 6 4 3 0 5
4	2 8 6
5	2 1 6

Scale: 1 | 9 means 19

b

1	8 9 9
2	1 4 7 7 9
3	0 3 4 4 5 6
4	2 6 8
5	1 2 6

c 30 to 39

4 **a**

0	1 7 5 9 6 6 9 4 8
1	3 4 2 4 6 3 9
2	0 3 3 4 7
3	2 0
4	
5	5

Scale: 1 | 3 means 13

b

0	1 4 5 6 6 7 8 9 9
1	2 3 3 4 4 6 9
2	0 3 3 4 7
3	0 2
4	
5	5

c highest: 55 points, lowest: 1 point   **d** 0 to 9 points
e 16 games   **f** Yes, "55" is an outlier.

5 **a**

1	6 8 7 0 9 2 6 7 5 4 4
2	1 3 4 9 8 9 5 3 6 7 4 0 5 3 2
3	5 7 8 3 4 4 3 1 2
4	1 9 3 1
5	2

Scale: 1 | 6 means 16

b

1	0 2 4 4 5 6 6 7 7 8 9
2	0 1 2 3 3 3 4 4 5 5 6 7 8 9 9
3	1 2 3 3 4 4 5 7 8
4	1 1 3 9
5	2

Scale: 1 | 6 means 16

c highest: 52 absences, lowest: 10 absences
d 12.5%   **e** 47.5%

EXERCISE 18F

1

	Mean	Median	Mode	Range
a	6	5	9	10
b	15.4	16	18	9
c	10.6	10	11	14
d	≈ 3.87	3.85	undefined	0.7
e	≈ 8.82	8.9	8.3	4.2
f	128	127.5	115, 127	45

2 **a** 2.3125 bedrooms   **b** 2 bedrooms
c 2 bedrooms   **d** 4 bedrooms

3 **a** The range is 6. It is the difference between the largest and smallest shoe size.
b 8
c 8 and 8.5, these are the most common shoe sizes.
d No, as the owner would need less of the small and large sizes.

4 **a** ≈ \$4.67   **b** \$3.97
    c No, each data value is different.   **d** \$12.04

5

	Median	Mean
a	3.5	3.7
b	2	2.5
c	44	≈ 41.1

6 72   **7** 11 data values   **8** \$1287
9 **a** 19 cm   **b** 26 cm   **c** ≈ 17.7 cm
10 2, 3, 3, 3, 4, 5, 6, 6

EXERCISE 18G

1 **a** 2   **b** 2   **c** ≈ 1.89   **d** 3

2 **a**

Number of occupants	Frequency
1	9
2	15
3	15
4	11
5	5
Total	55

  b 55 cars
  c **i** 2 and 3
        ii 3
        iii ≈ 2.78
        iv 4

3 **a** **i** 3   **ii** 3   **iii** ≈ 2.87   **iv** 8
  b
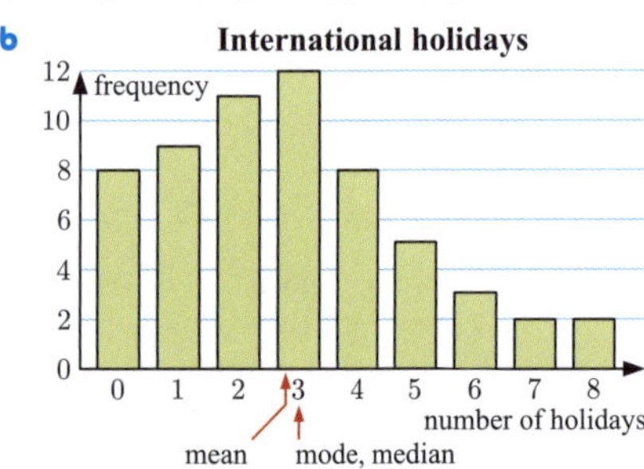

4 **a** 90 students
  b 15, this is the most common age of the students participating in the competition.
  c 15   **d** ≈ 14.8
  e The median age is greater. From the graph, we can see that there is an outlier which is less than the general body of data. This outlier will lower the mean more than the median.

REVIEW SET 18A

1 **a** numerical   **b** categorical   **c** numerical
2 **a** 18 netballers   **b** Yes, "2" is an outlier.   **c** 7 goals
3 **a** adults   **d**
  b 20%
  c $\frac{1}{4}$

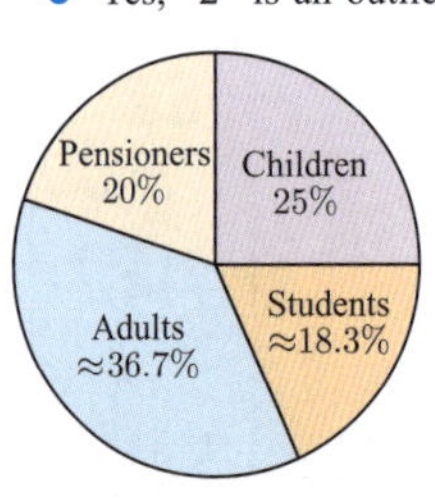

4 **a** The sample does not take into account the views of older students. Also, this sample only takes one location into account so it is not representative of all high school students.
  b Diners on other days and times are not represented in the sample.
5 **a** 1075 billion barrels   **b** 155 billion barrels more
  c ≈ 14.4%

6 **a**

Grade	Tally	Frequency
HD	|	1
D	||||	4
C	ⵌ |||	8
P	ⵌ ⵌ |	11
F	ⵌ	5
	Total	29

  b
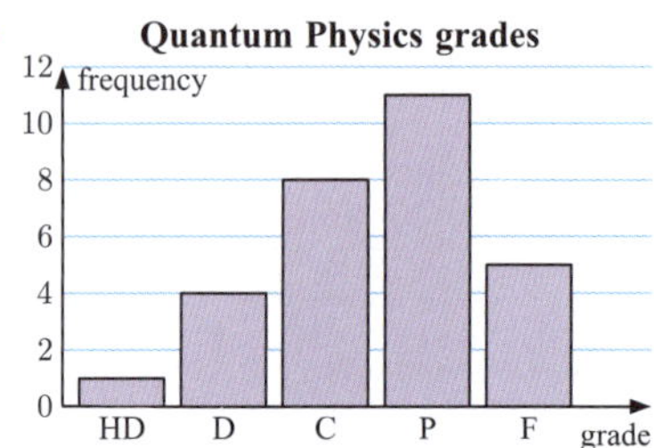

  c P, it was the most common grade obtained   **d** ≈ 17.2%

7 **a**

Satisfaction rating	Tally	Frequency
10 to 19	|	1
20 to 29		0
30 to 39		0
40 to 49		0
50 to 59	||||	4
60 to 69	ⵌ	5
70 to 79	|||	3
80 to 89	ⵌ ||	7
90 to 100	ⵌ ⵌ	10
	Total	30

  b
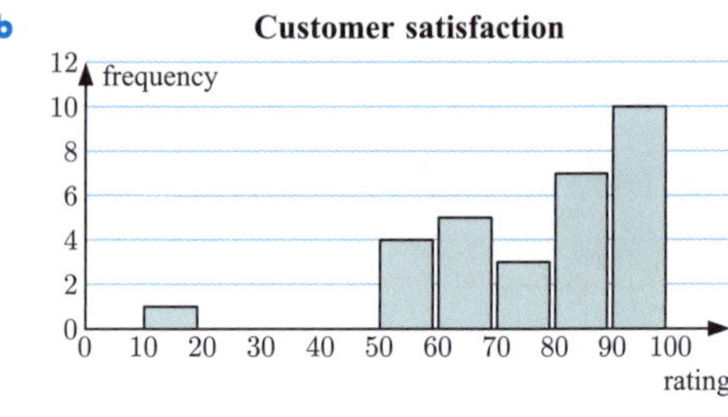

  c 90 to 100   **d** ≈ 56.7%

8 **a**

```
0 | 9
1 | 8 8
2 | 5 8 4 9 2 6 5 0 3 8 7
3 | 5 2 4 9 3 4 9 1 6 5 6 3 5 2
4 | 0 0            Scale:  3 | 5  means 35
```

  b

```
0 | 9
1 | 8 8
2 | 0 2 3 4 5 5 6 7 8 8 9
3 | 1 2 2 3 3 4 4 5 5 5 6 6 9 9
4 | 0 0            Scale:  3 | 5  means 35
```

  c We can see the individual scores in the stem-and-leaf plot.
  d 31 marks   **e** 20%
9 **a** \$24.80   **b** \$6.64
10 **a** 20.6   **b** 10.5
  c Patrons leaving a children's movie are likely to be younger than the average patron.

11 **a** 4.2 and 5.1 **b** 5.1 **c** ≈ 5.23 **d** 3.2

12 **a**

Number of beverages	Tally	Frequency
0	|||	3
1	⊬||| |	6
2	⊬||| ||	7
3	|||	3
4	||||	4
5	|	1
6		0
7		0
8		0
9		0
10	|	1
	Total	25

b **i** 2 beverages **ii** 2 beverages **iii** 2.4 beverages

c
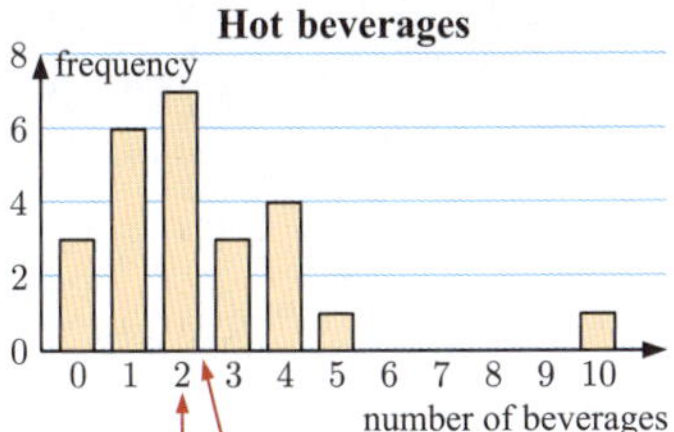

d "10" is an outlier

e The mean is most affected by the outlier because *all* the data values are used to calculate the mean.

f $\frac{16}{25}$

13 **a** Thriller/Crime **b** 26%

c **i** 400 customers **ii** 84 customers

REVIEW SET 18B

1 **a** a census **b** a sample

2 People who visit the library are more likely to want the library upgraded.

3 **a** 28 students

b
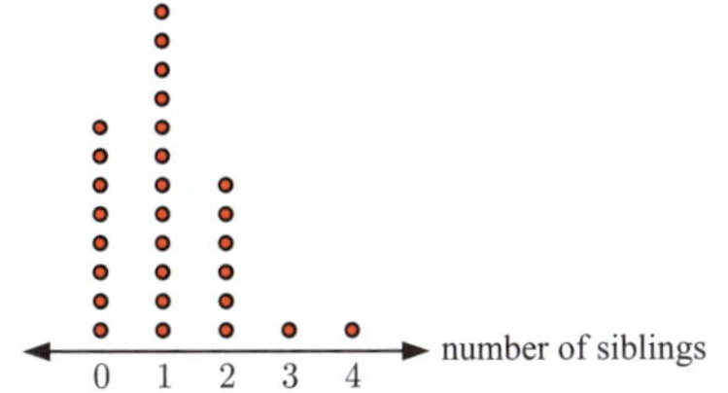

c No, there are no data values that are separate from the main body of data.

d ≈ 28.6%

4 **a** 30 to 39 pages

b

Number of pages read	Frequency
0 to 9	2
10 to 19	5
20 to 29	6
30 to 39	8
40 to 49	3
50 to 59	2
Total	26

c ≈ 19.2%

d 13 students

e No, as we do not know the exact data values.

5 **a**

Position	Tally	Frequency
Striker	⊬||| ⊬|||	10
Midfielder	⊬||| ⊬||| ||||	14
Defender	⊬||| ⊬||| |	11
Goalkeeper	⊬|||	5
	Total	40

b Midfielder, this is the most common position.

c $\frac{7}{20}$

d 40%

e
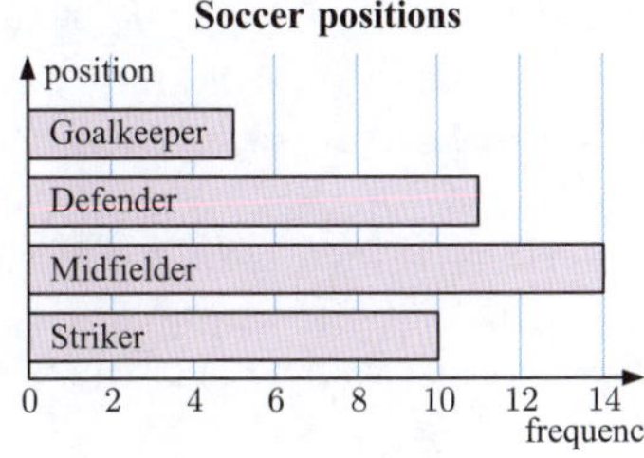

6 **a**

Number of calls	Tally	Frequency
10 to 19	⊬||| |	6
20 to 29	⊬||| ||	7
30 to 39	⊬||| ⊬||| ||	12
40 to 49	⊬|||	5
	Total	30

b
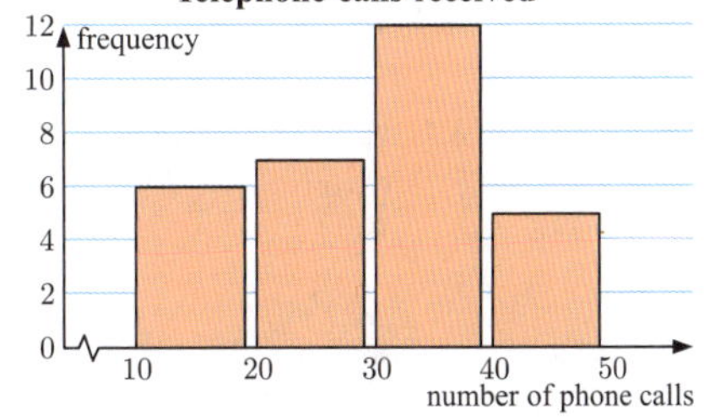

c 30 to 39 phone calls **d** ≈ 56.7%

7 **a** 32 kg **b** 91 kg **c** 55 kg **d** 59 kg

e 5 students **f** 20%

8 **a**

4.1	9
4.2	1 4 7 8
4.3	5 7 8
4.4	3 6 8
4.5	2 4 8 8
4.6	4 8
4.7	5

Scale: 4.1 | 9 means 4.19

b 0.56 m

9 **a** 11 **b** 8.5 **c** 8.65 **d** 7

10 **a** 200 males **b** ≈ 55.1%

c 65 - 74 years and 75 years and over

11 **a** 9 **b** ≈ 8.47 **c** Well above average, much fitter.

12 **a**

Number of assists	Frequency
0	0
1	5
2	10
3	3
4	4
5	6
6	8
Total	36

b 128 assists

c **i** 2 assists

ii 3.5 assists

iii ≈ 3.56 assists

iv 5 assists

EXERCISE 19A

1 **a** yes **b** no **c** no **d** yes **e** yes **f** no

2 **B** and **E**   3 **A**, **C**, and **D**

4 **a** 10 cm **b** 98° **c** 37.8 cm

5 **a** 5 cm **b** 53° **c** 6 cm^2

EXERCISE 19B

1 The equal sides are not in corresponding positions. The angle is included in one triangle, but not the other.

2 **a** yes {SSS} **b** yes {RHS}
   c No, equal sides are not in corresponding positions.
   d yes {AAcorS} **e** yes {AAcorS}
   f No, no equal sides. **g** No, insufficient information.
   h yes {SAS} **i** yes {SAS}
   j No, equal sides are not in corresponding positions.
   k yes {AAcorS} **l** yes {AAcorS}

3 **C**   4 **B** and **D**

5 **a** **i** $\triangle$ABC $\cong$ $\triangle$PQR {AAcorS}
      ii $P\widehat{R}Q = A\widehat{C}B$, BC = QR, and AB = PQ
   b **i** $\triangle$JKL $\cong$ $\triangle$XZY {SSS}
      ii $J\widehat{K}L = X\widehat{Z}Y$, $K\widehat{J}L = Z\widehat{X}Y$, and $K\widehat{L}J = Z\widehat{Y}X$
   c **i** not congruent
   d **i** $\triangle$RST $\cong$ $\triangle$YXZ {SAS}
      ii $R\widehat{S}T = Y\widehat{X}Z$, $T\widehat{R}S = Z\widehat{Y}X$, and RS = YX
   e **i** $\triangle$ABC $\cong$ $\triangle$EDC {AAcorS}
      ii $A\widehat{B}C = E\widehat{D}C$, BC = DC, and AB = ED
   f **i** $\triangle$PQT $\cong$ $\triangle$SQR {AAcorS}
      ii $P\widehat{Q}T = S\widehat{Q}R$, PQ = SQ, and PT = SR
   g **i** $\triangle$ABC $\cong$ $\triangle$DFE {AAcorS}
      ii $B\widehat{A}C = F\widehat{D}E$ and BC = FE
   h **i** not congruent

EXERCISE 19C

1 **a** In triangles ABD and CDB:
   - $A\widehat{D}B = C\widehat{B}D$ {equal alternate angles}
   - $A\widehat{B}D = C\widehat{D}B$ {equal alternate angles}
   - [BD] is common to both triangles
   $\therefore$ $\triangle$ABD $\cong$ $\triangle$CDB {AAcorS}
   Equating corresponding sides, AB = CD and AD = CB.
   b Opposite sides of a parallelogram are equal.

2 **b** In a kite, one diagonal bisects one pair of opposite angles.

3 **c** The diagonals of a square are equal in length.

4 **c** The opposite sides of a rhombus are parallel.

5 **c** The diagonals of a rhombus bisect the angles at each vertex.
   d $\triangle$PSM $\cong$ $\triangle$RSM {SAS}
   f $S\widehat{M}P = S\widehat{M}R = 90°$
   h The diagonals of a rhombus bisect each other at right angles.

7 **Hint:** Draw P on [AB] such that [CP] ∥ [MX] and [NY]. Then show that $\triangle$MAX $\cong$ $\triangle$ACP and $\triangle$CBP $\cong$ $\triangle$BNY.

EXERCISE 19D

1 **a** reduction, scale factor $\frac{1}{2}$ **b** enlargement, scale factor 2
   c enlargement, scale factor 4
   d enlargement, scale factor 3
   e reduction, scale factor $\frac{3}{4}$ **f** reduction, scale factor $\frac{2}{5}$

2 **a**
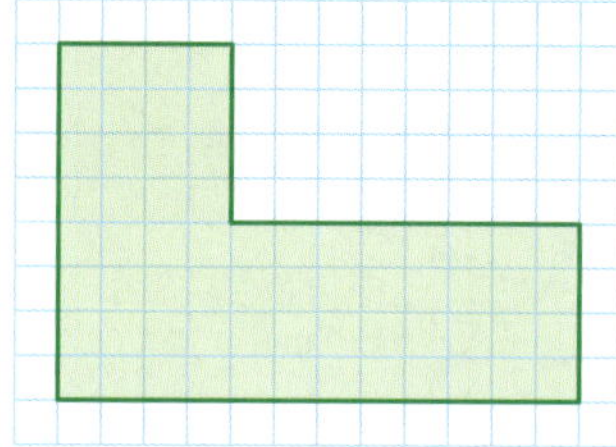
   b 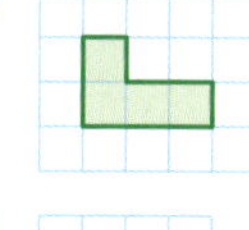 **c** 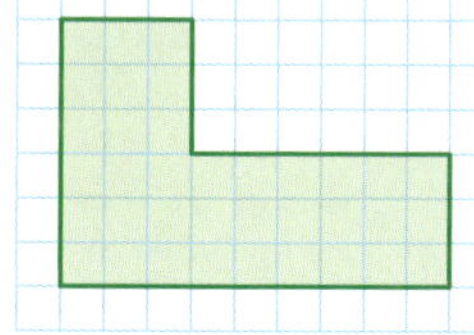
   d

3 **a** $0 < k < 1$
   b Enlarge the image with scale factor $\frac{1}{k}$.

4 **a** $\dfrac{9}{6} = \dfrac{4.5}{3} = \dfrac{7.5}{5} = 1.5$, $\therefore$ $k = 1.5$
   b corresponding angles are the same size; $\approx 94\frac{1}{2}°, 55\frac{1}{2}°, 30°$

EXERCISE 19E

1 **a** similar **b** not similar **c** not similar **d** similar

2 no, $\dfrac{140}{100} \neq \dfrac{100}{60}$

3 **a** true **b** false

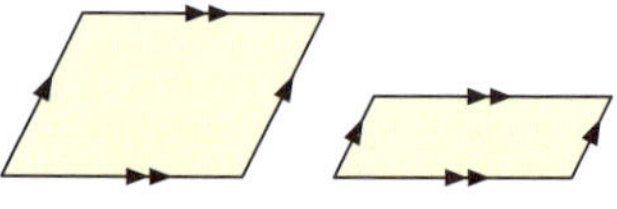

   c true **d** false

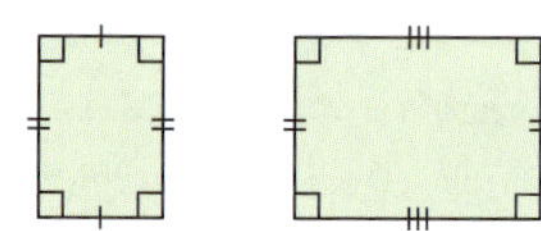

4 **a** $x = 8.75$ **b** $x \approx 3.86$ **c** $x = 96$ **d** $x = 127$

5 6 cm   6 $k = 1.5$ 7 **a** 6 cm **b** $\frac{1}{60}$

8 EF = 2.4 m

9 **a** $x = 4.8$ **b** $x = 9$ **c** $x = 100$ **d** $x = 16$

10 **a**
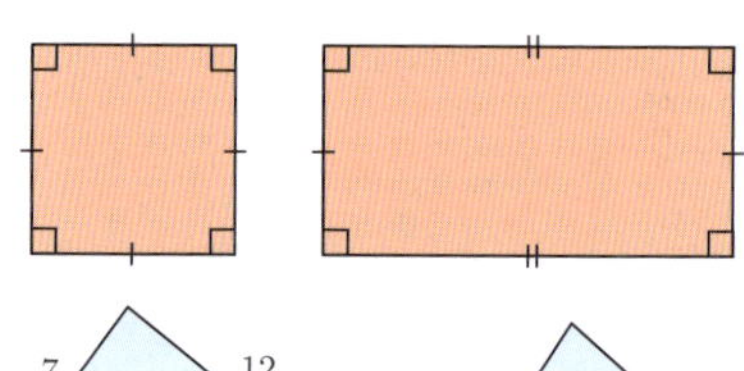
   b
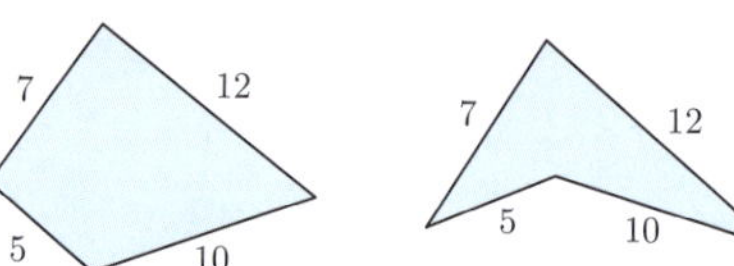

11 no

12 **Hint:** The large rectangle is $(y + 2z)$ cm by $(x + 2z)$ cm.

EXERCISE 19F.1

1 **a** $V\widehat{U}W = X\widehat{Y}W$ {given}
      $V\widehat{W}U = X\widehat{W}Y$ {vertically opposite angles}
   b $F\widehat{G}H = F\widehat{D}E$ {given},   angle F is common

c $\widehat{RST} = \widehat{RQP}$ {given}
   $\widehat{SRT} = \widehat{QRP}$ {vertically opposite angles}

d $\widehat{ABC} = \widehat{EDC}$ {equal alternate angles}
   $\widehat{ACB} = \widehat{ECD}$ {vertically opposite angles}

e $\widehat{LKM} = \widehat{LNJ}$ {given}, angle L is common

2 a $YZ = 2WV$, $XY = 2UW$, $XZ = 2UV$

b $\widehat{CAB} = 70°$ {angle sum of a triangle}
       $= \widehat{QRP}$

3 Hint: Find the unknown angles in terms of α.

EXERCISE 19F.2

1 a $\widehat{ABC} = \widehat{EDC}$ {given}
   $\widehat{ACB} = \widehat{ECD}$ {vertically opposite angles}; $x = 5$

b $\widehat{PQR} = \widehat{TSR}$ {given}
   $\widehat{PRQ} = \widehat{TRS}$ {vertically opposite angles}; $x = 2.8$

c angle Z is common, $\widehat{XYZ} = \widehat{UVZ}$ {given};
   $x = \frac{36}{11} \approx 3.27$

d $\widehat{AEB} = \widehat{DCB}$ {given}
   $\widehat{ABE} = \widehat{DBC}$ {vertically opposite angles};
   $x = \frac{20}{3} \approx 6.67$

e angle X is common,
   $\widehat{XYZ} = \widehat{XUV}$ {equal corresponding angles}; $x = 7$

f $\widehat{PTQ} = \widehat{SRQ}$ {given}
   $\widehat{PQT} = \widehat{SQR}$ {vertically opposite angles}; $x = 7.2$

2 20 m

EXERCISE 19G

1 a 7 m   **b** 7.5 m   **2** 180 cm or 1.8 m

3 a ≈ 2.67 m   **b** ≈ 5.67 m   **4** 39 m   **5** 1.44 m

6 1.52 m   **7** 9.9 s   **8** ≈ 117 m   **9** 2.4 cm

10 a i SU = 5.5 m   **ii** BC = 8.2 m
   b No, the ball's centre is ≈ 11 cm on the D side of C.

11 BN : CM = 2 : 1
   Hint: Draw D on [BC] such that [MD] ∥ [AB].

REVIEW SET 19A

1 **B** and **E**   **2** **C**

3 a i $\triangle ABC \cong \triangle ZYX$ {SSS}
   ii $\widehat{ABC} = \widehat{ZYX}$, $\widehat{BCA} = \widehat{YXZ}$, and $\widehat{BAC} = \widehat{YZX}$

   b i $\triangle DEF \cong \triangle STU$ {AAcorS}
   ii $\widehat{DFE} = \widehat{SUT}$, FE = UT, and DE = ST

4 d In a kite, one diagonal bisects the other.

6 a enlargement, scale factor $= 2$
   b reduction, scale factor $= \frac{2}{3}$
   c enlargement, scale factor $= \frac{4}{3}$

7 a similar   **b** not similar   **8 a** $x = 2.8$ **b** $x = 94$

9 a $\widehat{AXB} = \widehat{DXC}$ {vertically opposite angles}
   $\widehat{XAB} = \widehat{XDC}$ {equal alternate angles}
   $\widehat{XBA} = \widehat{XCD}$ {equal alternate angles}

   b $\widehat{BAC} = \widehat{DEC}$ {given}
   $\widehat{ACB} = \widehat{ECD}$ {vertically opposite angles}

   c angle R is common, $\widehat{RQS} = \widehat{RTP}$ {given}

10 a $\widehat{ABC} = \widehat{DEC}$ {equal alternate angles}
   $\widehat{ACB} = \widehat{DCE}$ {vertically opposite angles}; $x \approx 6.47$

   b $\widehat{PSQ} = \widehat{QSR}$ {given}
   $\widehat{PQS} = \widehat{QRS}$ {angle sum of a triangle}; $x \approx 2.67$

11 625 m

REVIEW SET 19B

1 a no   **b** yes   **c** yes

2 a yes {AAcorS}   **b** yes {RHS}   **c** yes {SAS}

3 b In a kite, one pair of opposite angles are equal.

4 c A and D are 50 cm^2.   B and C are 150 cm^2.

5 a 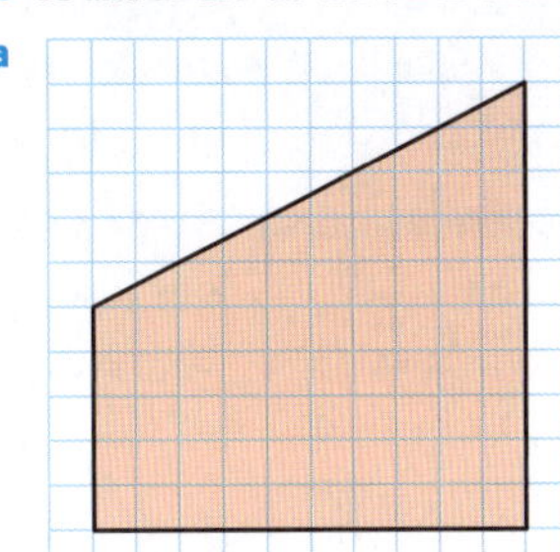 **b** 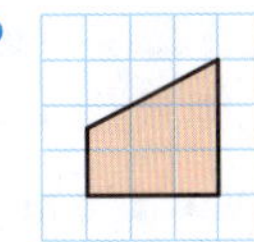

6 No, even though corresponding side lengths of rhombuses are always in the same ratio, they are not always equiangular.

7 Hint: Let the length of the longer side be k times the length of the shorter side in each shaded rectangle.

8 a angle A is common,
       $\widehat{AEB} = \widehat{ADC}$ {equal corresponding angles}

   b $\widehat{RPQ} = \widehat{RTS}$ {given}
       $\widehat{PRQ} = \widehat{TRS}$ {vertically opposite angles}

9 ≈ 66.7 m

10 angle R is common, $\widehat{RSQ} = \widehat{RPT}$ {given}; $x = 12$

11 a Hint: Use $\triangle BCF$.

EXERCISE 20A

1 a 5 cm   **b** 15 m   **c** 8.5 cm

2 a ≈ 3.61 cm   **b** ≈ 8.60 m   **c** ≈ 8.39 cm

3 a 8 cm   **b** 10 cm   **c** 12 m

4 a ≈ 5.66 cm   **b** ≈ 6.71 m   **c** ≈ 8.92 km

5 a ≈ 8.60 mm   **b** ≈ 7.14 cm   **c** ≈ 9.85 cm
   d ≈ 5.71 m   **e** 20 m   **f** ≈ 11.7 m

6 a ≈ 19.06 cm   **b** ≈ 15.57 m   **7** ≈ 12.2 km^2

8 a $x \approx 5.37$ **b** $x \approx 2.90$

EXERCISE 20B

1 ≈ 3.16 m   **2** ≈ 9.49 km   **3** ≈ 35.8 m

4 a ≈ 660 m   **b** ≈ 1580 m or ≈ 1.58 km

5 ≈ 6.51 m   **6** ≈ 20.5 m   **7** ≈ 42.4 cm by 42.4 cm

8 48 mm   **9** 6000 cm^2 **10** ≈ 1.63 m

11 130 cm   **12 a** ≈ 26.9 m   **b** ≈ 25.1 m

EXERCISE 20C

1 a no   **b** yes   **c** no   **d** yes   **e** yes   **f** no

2 a The street lamp and the ground are both straight (no bumps or bends).

b Yes there is sufficient information to conclude that the street lamp is *not* at right angles to the ground.
$(2^2 + 1.5^2 = 6.25 \neq 5.76 = 2.4^2)$

3 $1^2 + 2^2 = 1 + 4 = 5 \neq 4 = 2^2$
$\therefore$ the triangle is not right angled.

4 a $\sqrt{2x^2}$ cm
b In any right angled isosceles triangle, the hypotenuse is always $\sqrt{2}$ times the length of the other sides.

REVIEW SET 20A

1 a 13 m **b** 10 cm **c** 5.3 cm
2 a ≈ 13.27 cm **b** ≈ 11.36 m **c** ≈ 3.05 m
3 a ≈ 24.30 cm **b** ≈ 2.90 m **4** ≈ 3.39 m
5 a yes **b** no **c** yes **d** no **6** ≈ 157 cm

REVIEW SET 20B

1 a ≈ 7.21 cm **b** ≈ 13.1 cm **c** ≈ 8.93 m
2 ≈ 20.03 cm^2 **3 a** $x \approx 2.43$ **b** $x = 1.4$
4 ≈ 2.17 km **5** ≈ 110.7 cm by 62.3 cm
6 No, $15^2 + 28^2 = 1009 \neq 1024 = 32^2$.

EXERCISE 21A

1 a $x - 7 = 11$ **b** $\dfrac{x}{8} = 5$ **c** $3x + 5 = 23$
d $\dfrac{x+1}{4} = 10$ **e** $9(x - 3) = 36$ **f** $\dfrac{x}{4} + x = 20$
g $5x - 8 = x + 4$
2 a $x + 3 = 5$ **b** $x + 15 + x = 37$
c $x + x + 6 = 100$ **d** $x + 100 = 3x$
e $x + 4 = 2\left(\dfrac{x}{3} + 4\right)$

EXERCISE 21B

1 The number is 14. **2** The number is 1.
3 The number is 43. **4** The number is -2.
5 The number is -6. **6** 10 potatoes **7** 35 lollies
8 28 roses **9** 60 goals **10** 6 students **11** 3 apples
12 $87.50 **13** 9 cm **14** 15 cm **15** 36 cm^2
16 12 cm **17** 162 m^2

EXERCISE 21C

1 a If x is greater than 9 and $3x + 4y = 30$, then y will be negative or 0.

b

x	1	2	3	4	5	6	7	8	9
y	$6\frac{3}{4}$	6	$5\frac{1}{4}$	$4\frac{1}{2}$	$3\frac{3}{4}$	3	$2\frac{1}{4}$	$1\frac{1}{2}$	$\frac{3}{4}$

c $x = 2$, $y = 6$ and $x = 6$, $y = 3$

2

x	2	4	6	8
y	12	9	6	3

3 a 23
b The lowest common multiple of 4 and 5 is 20, so the number is $20 + 3 = 23$.

4 12 cm $\times$ 4 cm
5 3, 6, 7, 11, 13, 14, 15, 18, 19, 20, 21, 22, 23, 29, 30, 32, 34, 35, 38, 39
6 64

7 a

Width (cm)	Length (cm)	Area (cm^2)
1	30	30
2	28	56
3	26	78
4	24	96
5	22	110
6	20	120
7	18	126
8	16	128
9	14	126
10	12	120

b length: 16 cm, width: 8 cm
8 28
9 Group value $69.
{$20, $18, $24, $5, $2}, {$10, $1, $8, $35, $15}
Note: Other solutions are possible.
10 4732, when $a = 7$ and $b = 26$

EXERCISE 21D

1 The number is 9. **2** 16 horses
3 Xia: 134 cm, Xuen: 130 cm **4** 5 peppermints
5 Bonnie **6** 72 km
7 Anders: 18 lollies, Bree: 10 lollies, Charlie: 8 lollies
8 X: 300 mL, Y: 100 mL

EXERCISE 21E

1 24 and 29 **2** 33 **3** 15 nails
4 2, 3, and 8 years old **5** 27 biscuits
6 7.2 cm **7** $289 **8** 37 years old
9 13 questions **10** 9, when $x = 4$ **11** $900
12 The number is 32. **13** 800 mL **14** The number is 12.
15 18 **16 a** $35 **b** $75 **17** 6 g
18 $12 = 4 \times (1 + 2)$, $24 = 4 \times (2 + 4)$, $36 = 4 \times (3 + 6)$, $48 = 4 \times (4 + 8)$
19 142857 **20** 1:12 pm

EXERCISE 21F

1 a i 9 paths **ii** 6 paths **iii** 15 paths **b** 30 paths
2 12 positive integers
3 We cut each link of one chain to create 4 open links. We use each open link to join a pair of chains as shown, then weld each open link.

Total cost $= 4 \times \$2 + 4 \times \$5$
$\qquad\qquad = \$28$

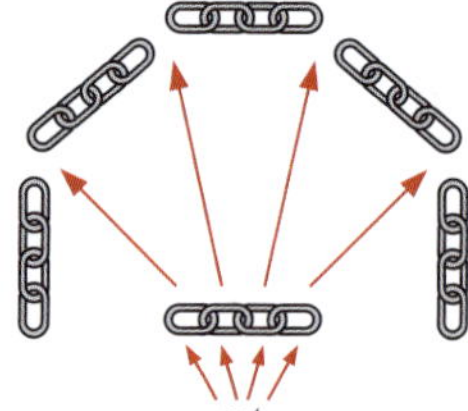

4 A = 9, B = 5, C = 7, D = 0
5

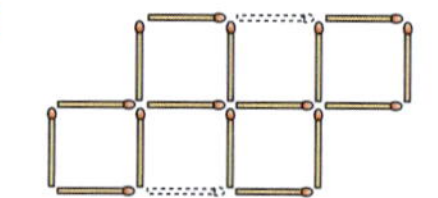

6 yes,

7 The problem is an addition.
The solution is **A** $= 5$, **C** $= 9$, **D** $= 1$, **E** $= 4$, **G** $= 7$, **N** $= 8$, **O** $= 2$, **R** $= 6$, and **S** $= 3$.

8 1005 m   **9** 1 day   **10** 13 triangles   **11** 36 cubes

12 $E = 5$, $H = 2$, $S = 6$

13

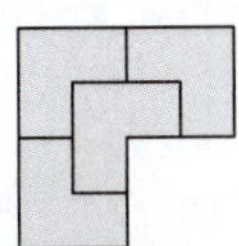

14 155 844 holes   **15**

16 In the table below, the mass to be measured is placed on the *left* side of the scale.

Mass (kg)	Left	Right
1	-	1 kg
2	1 kg	3 kg
3	-	3 kg
4	-	1 kg, 3 kg
5	1 kg, 3 kg	9 kg
6	3 kg	9 kg
7	3 kg	1 kg, 9 kg
8	1 kg	9 kg
9	-	9 kg
10	-	1 kg, 9 kg
11	1 kg	3 kg, 9 kg
12	-	3 kg, 9 kg
13	-	1 kg, 3 kg, 9 kg

17 9 ostriches, 8 giraffes   **18** $2^5 = 32$ ways

19 **a** anticlockwise   **b** 2 revolutions

20 Not possible for 10, 11, 13, 14, 18, and 19.

$1 = \frac{4}{4} \div \frac{4}{4}$, $2 = \frac{4}{4} + \frac{4}{4}$, $3 = \frac{4+4+4}{4}$, $4 = \frac{4-4}{4} + 4$,

$5 = \frac{4 \times 4 + 4}{4}$, $6 = \frac{4+4}{4} + 4$, $7 = 4 + 4 - \frac{4}{4}$, $8 = 4 + 4 + 4 - 4$,

$9 = 4 + 4 + \frac{4}{4}$, $12 = \left(4 - \frac{4}{4}\right) \times 4$, $15 = 4 \times 4 - \frac{4}{4}$,

$16 = \frac{4 \times 4 \times 4}{4}$, $17 = 4 \times 4 + \frac{4}{4}$, $20 = \left(4 + \frac{4}{4}\right) \times 4$

Note: Other answers are possible.

21 $\approx 11.1\%$ **22** $3\frac{1}{2}$ minutes

23 **a** 20 balls   **b** 120 balls   **24** $1\frac{7}{8}$ days

25 **a** 35 km per hour   **b** 60 km   **26** $-\frac{1}{2}$

27

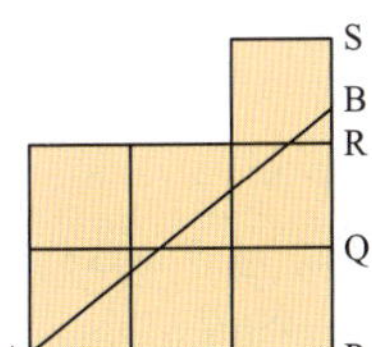

B is located $\frac{1}{3}$ of the distance between R and S.

28 Celine   **29** 14 days   **30** $35^2 = 1225$

31

32 Adam should leave a multiple of 11 marbles on the table after each of his turns.

REVIEW SET 21A

1   **a** The number is 38.   **b** The number is 4.

2 45 minutes   **3** 15 cm

4   **a** If x is greater than 13 and $3x + 4y = 40$, then y will be negative.

   b

x	1	2	3	4	5	6	7	8	9	10	11	12	13
y	$9\frac{1}{4}$	$8\frac{1}{2}$	$7\frac{3}{4}$	7	$6\frac{1}{4}$	$5\frac{1}{2}$	$4\frac{3}{4}$	4	$3\frac{1}{4}$	$2\frac{1}{2}$	$1\frac{3}{4}$	1	$\frac{1}{4}$

   c $x = 4$, $y = 7$; $x = 8$, $y = 4$;   and $x = 12$, $y = 1$

5 15 marbles

6   **a** 16 cm   **b** 10 cm   **7** 13 cupcakes   **8** 8 pm

9   **a** $\$(x - 0.80)$

   b doughnuts: $1.75 each,   lamingtons: 95 cents each

   c 26 lamingtons

10   **a** If one side of the triangle is longer than 10 cm, then the sum of the remaining sides is at most 10 cm, which is impossible. Each side in a triangle must be shorter than the sum of the lengths of the remaining sides.

   b

Side 1 (cm)	1	2	3	3	4	4	5	5	5	6	6	7
Side 2 (cm)	10	9	8	9	7	8	6	7	8	6	7	7
Side 3 (cm)	10	10	10	9	10	9	10	9	8	9	8	7

   c   **i** equal sides: 6 cm,   base: 9 cm

      ii 2 cm, 9 cm, 10 cm

REVIEW SET 21B

1 The number is 12.   **2** bracelet: $11,   necklace: $22

3 18   **4** 12 minutes   **5** 8, when $a = 2$ and $b = 4$

6 10, 14, 15, 21, 22, 25, 26, 33, 34, 35, 38, 39, 46, 49, 51, 55, 57, 58, 62, 65, 69, 74, 77, 82, 85, 86, 87, 91, 93, 94, and 95.

7 $73   **8** 14 years old   **9** $k : x = 1 : 3$

10   **a**   **i** $x + (x - 3)$ coins   **ii** $20x + 50(x - 3)$ cents

   b   **i** 9 20 cent coins,   6 50 cent coins   **ii** $4.80

11   **a** Dean   **b** Ange   **c** Ange, Chad, or Eve

INDEX